ENVIRONMENTAL HYDROLOGY
SECOND EDITION

ENVIRONMENTAL HYDROLOGY
SECOND EDITION

Andy D. Ward
Stanley W. Trimble

FOREWORD BY M.G. WOLMAN

LEWIS PUBLISHERS

A CRC Press Company
Boca Raton   London   New York   Washington, D.C.

Pictured on the front cover: a winter scene of the Portage River in northern Ohio (see Chapter 6 and related topics in Chapters 5, 7, 8, 9, and 12), photo courtesy Andy Ward; inset top right: slab failure of the banks of a river in New Mexico (see Chapters 6 and 9 and related topics in Chapters 9 and 12), photo courtesy John Lyon; middle: students conducting a stream reconnaissance with Professor Ward (see Chapter 6), photo courtesy Andy Ward; bottom: contour strip cropping in Wisconsin (see Chapter 9 and related topics in Chapter 5), photo courtesy Stanley Trimble; back cover: great egret at Reelfoot Lake in Northwest Tennessee (see Chapter 6 and related topics in Chapter 12), photo courtesy Northwest Tennessee Tourism.

Library of Congress Cataloging-in-Publication Data

Ward, Andrew D.
  Environmental hydrology.-- 2nd ed. / Andrew Ward, Stanley Trimble.
    p. cm.
  Rev. ed. of: Environmental hydrology / edited by Andy D. Ward, William J. Elliott. c1995.
  Includes bibliographical references (p.).
  ISBN 1-56670-616-5
    1. Hydrology. 2. Hydrology--Environmental aspects. I. Trimble, Stanley Wayne. II.
Environmental hydrology. III. Title.

GB665.W28 2003
551.48--dc22

2003061133

Visit the CRC Press Web site at www.crcpress.com

© 2004 by CRC Press LLC
Lewis Publishers is an imprint of CRC Press LLC

No claim to original U.S. Government works
International Standard Book Number 1-56670-616-5
Library of Congress Card Number 2003061133
Printed in the United States of America   2 3 4 5 6 7 8 9 0
Printed on acid-free paper

Dedication

To my parents, Alan and Beryl, for their love, wisdom, understanding, high moral standards, compassion for humanity, and ability to make the best of any situation. To my children, Samantha and Trevor, and my wife, Clover, for their love, support, tolerance, understanding, and the sacrifices they made during the many hours, weeks, and months I spent working on this book.

Andy Ward

To my parents, to my family, to my students, and to my teachers, especially my geography graduate mentor, Kirk H. Stone, 1914–1997. To the taxpayers of California and the United States. While most of them will never read this book, they have supported the activities that have given me the knowledge and experience to write my contributions to it.

Stan Trimble

In Loving Memory
Jarrett Mynear
April 16, 1989–October 4, 2002
A kind and strong spirit who inspired everyone he touched.

Jarrett (left) on his 12th birthday with Samantha, Trevor, and Andy Ward.
A portion of Andy's royalties will be donated to Jarrett's Joy Cart.
Be inspired and read about the Joy Cart and Jarrett's story at www.thejoycart.com.

In Appreciation

The Blair School of Music at Vanderbilt University, which has been so important to Stan's daughters, Alicia and Jennie, and, in a broader sense, important to the cultural life of middle Tennessee. A portion of Stan's royalties will be donated to Blair. Read more about Blair at www.vanderbilt.edu/blair

Foreword: For the Student

This is not your ordinary textbook. *Environmental Hydrology* is indeed a textbook, but five elements often found separately combine here in one text to make it different. It is eclectic, practical, in places a handbook, a guide to fieldwork, engagingly personal and occasionally opinionated.

The material covered includes expected chapters on basic aspects of the hydrologic cycle such as rainfall, runoff, and evapotranspiration. There is, at the same time, extensive coverage of stream processes, soil erosion and sedimentation, and human impacts on the hydrologic cycle.

Throughout, the authors have chosen to emphasize the practical rather that the theoretical aspects of hydrology. Many texts do this in hydrology, perhaps too many, but in this book that orientation is enhanced with a content and style that evidence their commitment to what the authors call a "student-centered" view. Many, if not most of us, in the academic world profess this centered vision but quickly lodge our teaching on the periphery. Not so here.

Chapters devoted to the application of remote sensing and geographic systems to hydrology and to conducting and reporting hydrologic studies nicely capture the practical flavor. At the same time, a focus on the practical leads not only to the inclusion of many approaches to solving specific problems but also to the inclusion in text and appendices of the vital statistics of hydrology, thus the attribute of a modest handbook.

I hope any student will enjoy, just not profit from the careful advice given to those involved in their first observations of rivers in the field. For example, the authors "stoop" to telling the student reader how many persons should be engaged in carrying out a task; perfectly laughable unless you have nearly drowned by failing to do the obvious, or finished a day's fieldwork and notes by lantern in the dark.

And, perhaps most engaging to me, in places the authors offer personal views as well as more strongly worded opinions. The former often relate to evaluation of alternative approaches, or formulations, of specific solutions to specific hydrologic problems. One or the other of the authors, not the anonymous royal we, states his choice based on his own experience. More rarely, it is noted that contrary to the notoriety of some hydrologic findings, the findings are grossly in error and the resulting policies foolish and misdirected.

This thick book is a labor of love. No doubt it contains errors of commission and perhaps even omission. One hopes that students, as they struggle with the material in the text, will warm both to what inspires the authors and through that to the subject — hydrology.

Professor M. Gordon Wolman
The Johns Hopkins University

Acknowledgments

Preparation of this book would not have been possible without the outstanding collective contributions of many people. We regret that it is not possible for us to fully express our gratitude for these efforts or to adequately recognize every contribution. Particular thanks are extended to Dawn Farver for her extraordinary efforts in coordinating the final preparation of the book materials, contributing to Chapter 1 (some of the statistics) and Chapter 12 (bioassessment methods), and providing many valuable suggestions. Chase Langford brilliantly produced many of the final illustrations in the book. Diane Yagich also produced many fine illustrations and assisted Dawn Farver with preparing all the illustrations in their final format. Dan Mecklenburg reviewed Chapter 6 and parts of Chapter 12, provided technical assistance with several of the Excel illustrations, provided data for several examples and figures, and provided some photographs of streams. Several illustrations were prepared by Jeff Blatt and Julie Thomas. Bill Elliot, John Lyon, David Montgomery, and the Center for Watershed Protection provided a large number of the photographs. Additional photographs were provided by Yuichi Kayaba, Don McCool, Gary Sands, Dawn Farver, David Derrick, the Ohio Department of Natural Resources, and Salix Applied Earthcare. Northwest Tennessee Tourism provided photographs of Reelfoot Lake and the photograph on the back cover. Chapter 13 was written by John Lyon, who also prepared the information on data sources that are presented in the appendices, Charles Luce wrote parts of Chapter 10, and E. Scott Bair wrote most of Chapter 11. Bill Elliot wrote some of the materials on soil erosion (the first part of Chapter 9) and contributed to Chapters 1 and 2. Portions of Chapters 3, and 4 are based on material from the "Purple Book" that were prepared by Jay Dorsey and Sue Nokes. We would like to thank Terry Logan who contributed to the 1st Edition of the book.

Important contributions were provided by Jan Boll (part of the discussion on cold climate hydrology in Chapter 9), Anne Christy and Julie Weatherington-Rice (fracture flow, Chapters 5 and 11), Kerry Hughes (landfills, Chapter 12), Tim Lawrence (NEMO, Chapter 12), and Lance Williams (stream biota, Chapters 6 and 12). The section on constructed wetlands in Chapter 12 was primarily based on information presented in fact sheets written by Ted Tyson and Ken Simeral. We would also like to thank Dave Rosgen and Wildland Hydrology, for allowing use of various materials on stream processes.

Reviews of portions of the book were provided by Felicia Federico, John D. Hewlett, Anand Jayakaran (several chapters), David Montgomery (classification section of Chapter 6), Frank Parker (part of Chapter 12), Mike Singer (part of Chapter 3), Terry Stewart (several chapters), Peter Whiting (Chapter 6), Lance Williams (Chapter 6), and Jon Witter (several chapters). Other important contributions were provided by Jean Boenish, Jan Boll, Tess Brennan, Larry Brown, Ann Chin, Jeff Harvey, John Hewlett, Barbara Hoag, Linda O'Hirok, Norman Meek, Alex Mendel, Paul Price, Lisa Lindenmann, Ken Schwarz, and Steve Workman. Terry Stewart assisted with the research for several topics, produced a few of the Excel figures, and did all the collating and formatting of the references, glossary of terms, and some of the other materials in the appendices. The many hours that Anand Jayakaran and Jon Witter spent obtaining materials from various libraries is greatly appreciated. Additional help was provided by Erick Powell. Thanks to Joan Wu who provided an errata for materials from the 1st edition and some helpful suggestions.

We are indebted to Professor M. Gordon (Reds) Wolman for writing a preface to the book. We hope that this book will aid society and partially repay the many excellent scientists and engineers who have shared their knowledge and wisdom with us throughout our careers. Every effort has been made to accurately and fairly represent the collective knowledge of the countless people whom we have been fortunate to work with and learn from.

The corrections, suggestions, and patience of the many students who have used parts of the book during the past few years are also greatly appreciated. The authors would like to recognize the valuable support and many sacrifices that were made by students and their families during the development of the book. The permission that was granted by many people and organizations to use a diversity of copyright materials helped us to produce a book of this quality. Acknowledgment of each contribution of work published by others is presented in the text or caption for each table and figure where appropriate. We apologize if we have inadvertently failed to recognize anyone for their contribution and for any omissions in obtaining a necessary copyright permission. In particular, we apologize for any omissions in citing the page and figure or table number from an original sources. Contributions to the 1st and 2nd editions were prepared by many people during a period of almost 10 years. Tracking down all oversights became an overwhelming task.

In some cases it was not possible in the body of the book to adequately identify the contribution, permission or copyright ownership of an individual or organization. We would like to thank the family of Filip Hjulström for permission to include some of the work of this eminent Swedish scientist; Elsevier, for permission to include summaries of some of the descriptions of the Rosgen Stream Classification that were originally published in Catena; the Geological Society of America for permission to include summaries of some of the descriptions of the stream work by Montgomery and Buffington that were originally published in the GSA Bulletin; and the many federal agencies, and the scientists and engineers in these agencies, who have collected, analyzed and published a wealth of data and information that has contributed to our knowledge and this book. We would also like to thank the organizations and individuals who provided information and permissions to use their materials in the "Blue Book" and the 1st edition of Environmental Hydrology. Some of these materials have been retained in this the 2nd edition

Preparation of this book was made possible through the support of the Food, Agricultural and Biological Engineering Department at The Ohio State University, the College of Food, Agricultural and Environmental Sciences at The Ohio State University, the Ohio Agricultural Research and Development Center, the Ohio State University Extension, and the Department of Geography at UCLA.

About The Authors

Andy D. Ward, Ph.D., is a Professor in the Department of Food, Agricultural, and Biological Engineering, The Ohio State University and member of the faculty since 1986. As a child he lived in Zambia and Zimbabwe. He then spent 2 years in Switzerland before completing his final 2 years of high school in Southend-on-Sea, England.

In 1971, he obtained a B.Sc. degree in civil engineering from Imperial College, London, England. He then worked as an engineer on a construction project in London and as a schoolteacher in New York City. Dr. Ward earned M.S. and Ph.D. degrees in agricultural engineering from the University of Kentucky in 1977 and 1981, respectively. He then worked for 3 years with an international consulting company in South Africa before joining Ohio State.

Dr. Ward is a registered professional engineer and a member of the American Society of Agricultural Engineering and the Soil and Water Conservation Society. He has authored more than 100 manuscripts and coauthored (with John Lyon and coworkers) a paper that received the 1994 Autometric Award from the American Society of Photogrammetry and Remote Sensing for the best interpretation of remote sensing data.

Dr. Ward has provided leadership to the development of several hydrologic computer programs, including the WASHMO storm hydrograph model, the DEPOSITS reservoir sedimentation model, and the ADAPT agricultural water quality model. He has 25 years of international experience in the areas of watershed hydrology, stream geomorphology, reservoir sedimentation, modeling hydrologic systems, drainage, soil erosion, water quality, and the development and implementation of techniques to prevent or control adverse impacts of land use changes on water resources, streams, and drainage networks. Together with his graduate students, he has also conducted research on remote sensing applications in agriculture and hydrology.

Dr. Ward is an advocate of student-centered learning; in his courses, he incorporates teamwork, solving real-world problems, applying engineering and scientific judgment, and enhancing communication skills. He is an avid recreational jogger and has completed 22 marathons or ultramarathons, including the 56-mile Comrades Marathon.

Stanley W. Trimble, Ph.D., is a Professor in the Department of Geography at the University of California, Los Angeles, and a member of the faculty since 1975. Among his interests is historical geography of the environment, especially human impacts on hydrology, including soil erosion, stream and valley sedimentation, and stream flow and channel changes. His regional interests are the humid U.S. states and western and central Europe. In 1963, he received a B.S. in chemistry from the University of North Alabama. Taking an Army ROTC commission, he spent 2 years as an intelligence research officer and served with the 101st Airborne Division from 1964 to 1965. After a year teaching in Europe, he earned M.A. and Ph.D. degrees in geography from the University of Georgia in 1970 and 1973, respectively.

Dr. Trimble was a research hydrologist (adjunct appointment) with the U.S. Geological Survey from 1973 to 1984 and a visiting professor at the Universities of Chicago (1978, 1981, 1990), Vienna (1994, 1999), Oxford (1995), London (University College, 1985), and Durham (1998). Currently, he is the joint editor of *Catena*, an Elsevier international journal of soils, hydrology, and geomorphology. He has also taught courses in environmental geology/hydrology for the U.S. Army Corps of Engineers, and he is a hydrologic/geomorphologic consultant for several agencies. His research awards include a Fulbright to the U.K. He also has to his credit more than 100 published articles, ranging from several in *Science* to one in *The Journal of Historical Geography;* some of his research work appears in this book. Recently, he served on the National Research Council Committee on Watershed Management.

Dr. Trimble believes in student-centered learning and giving his students as much hands-on and problem-oriented learning as possible with a lot of written work. His outside interests are music, classical and early American architecture, and English landscape gardens.

M. Gordon (Reds) Wolman has taught at Johns Hopkins University since 1958 and is the B. Howell Griswold, Jr. Professor of Geography and International Affairs in the Department of Geography and Environmental Engineering, with a joint appointment in the Department of Environmental Health Sciences. He received his bachelors' degree in 1949 from Johns Hopkins, and his doctorate in 1953 from Harvard.

His research has focused on human activities and their interactions with the natural processes impacting the Earth's surface, specifically the control of quantity and quality of streamflow and the behavior of rivers. His studies of environmental processes have involved him in the environmental policy work for water, land and energy resources. His publications are legion with many being of classic stature, and several are cited in this book. Students interested in streams will quickly understand the importance and usefulness of the Wolman Pebble Count Method.

Professor Wolman has been recognized through many awards including the Cullum Geography Medal of the American Geographical Society, Rachel Carson Award from the Chesapeake Appreciation, Ian Campbell Medal of the American Geological Institute, Penrose Medal of the Geological Society of America, and the Horton Medal of the American Geophysical Union.

Past president of the Geological Society of America, Professor Wolman was elected to the National Academy of Sciences in 1988, to the American Philosophical Society in 1999, and the National Academy of Engineering in 2002.

Introduction

Interest in water and related problems has grown markedly throughout all segments of society. An understanding of the occurrences, distribution, and movement of water is essential in agriculture, forestry, botany, soil science, geography, ecology, geology, and geomorphology. In short, water is an important element of the physical environment. We should all seek knowledge of hydrology as an aid in understanding the physical environment in which humanity has developed and in which we now live. Furthermore, the population of the world could increase by 50% some time this century. The majority of this global population increase of more than 3 billion people will occur in developing countries that already face shortages of potable water and food.

Questions we are often asked are: Why did we write a hydrology book? and How does this book differ from other books? A decade ago, we wrote the "Purple Book," which was the precursor of this book. It was written by an interdisciplinary team of authors who used as its foundation what we affectionately call the "Blue Book." The Blue Book was a wonderful collection of information and analyses that was put together in the 1970s by a group of prominent scientists and engineers because, at that time, there was no available textbook suitable for use in an introductory hydrology class. When we wrote the Purple Book, there were still very few introductory hydrology books, and the Blue Book was still widely used. However, the scope of the Blue Book was rather narrow, the materials needed updating, and the format and style of the materials were very old fashioned.

Since publishing the Purple Book, many fine hydrology books have been written; we felt that simply updating the book would not be adequate. After much thought, we decided to write a new text that would still include the most useful and interesting parts of both the Blue and the Purple Books. The main motivation for writing the new text was that we felt there were topics that needed to be included in the book and that were not adequately addressed in most hydrology books. In addition, we collectively have about 60 years of experience with a wide range of topics relating to hydrology and believe that society could benefit from the knowledge we acquired from our mistakes and successes. Furthermore, in the last decade, society has begun to recognize better the value of using interdisciplinary teams and knowledge from several branches of science and engineering to solve complex environmental and ecological problems. We have included extensive information on stream processes, sediment budgets, land use and human impacts on the hydrologic cycle, issues of concern to society, and management strategies. When possible, we also tried to foster an appreciation for the biology of hydrologic systems. Unlike many scientific texts, every effort was made to engage the reader in the discussion of each topic. The book contains many examples, illustrations, and accounts of personal experiences.

The rest of this introduction presents information on the organization of the book and then a perspective on hydrology by each author.

BOOK ORGANIZATION

The purpose of this book is to provide a qualitative understanding of hydrologic processes and an introduction to methods for quantifying hydrologic parameters and processes. It has been prepared for use in introductory hydrology courses taught at universities to students of environmental science, natural resources, geology, geography, agricultural engineering, and environmental engineering. A comprehensive understanding of the presented topics and problems should provide sufficient knowledge for students to make an assessment of hydrologic processes associated with environmental systems and to develop initial conceptual evaluations that are part of most assessments. We hope that the book will also serve as a reference resource.

In the problem sets and especially in the practical exercises (Chapter 14), we attempted to give students a "hands-on" feel for the matters at hand. Having actually dealt with data rather than simply having talked about it gives a better and more confident perspective. We also highly recommend attaining skills in surveying and field measurements, and some of the practical exercises require those skills. We believe it is important to be able to visualize a liter, a 1% slope, an acre-foot, or 10 ft^3/sec.

Readers will note that we frequently use English units of measure in this book. Although the rest of the world has converted to SI (International System of Units [*Système international d'unités*], metric) units, most field-level work in the U.S. is still done in English units. Moreover, the wealth of hydrologic data produced and held by the U.S. is mostly in English units. While most scientific journals use SI units, several U.S. scientific journals have switched back to English units so that published papers will have more applied impact. In any case, in the U.S., we need to know how to use and convert quickly between both systems. It is quite analogous to living in a bilingual nation. There, two languages must be spoken and understood; similarly, we must be bimensural or bimetric. Thus, we have made little effort to convert units and sometimes both systems are used in the same sentence. We believe that it will help prepare students for the real world, where they must deal with both systems, often under pressure, and where the ability to make mental conversions can provide a great advantage. All of the foregoing may be an unfortunate imposition, but it is a reality. By incorporating English units, we realize that we diminish this book's marketability in countries outside the U.S., but not to incorporate English units, in our view, would be to abrogate our responsibilities as educators. A table of unit conversion factors is presented in Appendix A, and we urge students to learn the more salient ones.

The topic of hydrology contains many different terms that may not be familiar to the reader. Therefore, we present a glossary of terms in Appendix B.

A PERSPECTIVE BY ANDY WARD

I was a few years old, and the heat and humidity were unbearable as we walked stealthily through the parched scrubland and thorn bushes. A short distance below us was a muddy river. Impala (antelope) were cautiously drinking, and a group of women and children walked along a narrow, well-trodden path. There was a sudden commotion and loud splash as a young impala was dragged into the river by an enormous crocodile. We were in the Luangwa River Valley in Zambia, a small country in the southern part of Africa, a continent where I spent much of my youth, and there are only two seasons — dry or rainy. The women and children were walking to the river to bathe, wash their few clothes, and get water to take back to their village, which was many miles away. On their return, the women balanced huge open containers of water on their heads — a remarkable feat, as one false step and the precious contents would spill. Amazingly, hundreds of thousands of other women throughout Africa repeat a similar journey daily. Many hours of walking and the perils of snakes, lions, elephants, and crocodiles have to be endured to obtain relatively small amounts of muddy, polluted water.

It was the early 1960s; our lifestyle was spartan compared to living conditions for most of the population in the U.S. Electrical power failures occurred often. When water did flow through the taps, it always seemed to be a new shade of brown, yellow, or orange. There were concerns about the population explosion; how the world might run out of food before the year 2000; how a nuclear war or some dreadful disease would wipe out most of humanity; and how, by the end of the century, civil wars and anarchy might occur throughout the world because of unstable economies and limited food and water. Some of these concerns are now reality as there are severe food and water shortages throughout Asia and even in developed parts of the world, such as eastern Europe and the former Soviet Union.

My interest in hydrology was not stirred just by childhood images of drought and food shortages. Following thunderstorms, I often saw huge, raging rivers, such as the mighty Zambezi, running the color of chocolate with sediment. I was amazed that any soil remained on the land. However, erosion rates in Africa are lower than in parts of Asia, South America, Central America, and North America. In the U.S., sediment is the main pollutant in surface water systems, and agricultural activities contribute more than 50% of the nation's sediment.

In the mid-1960s, I spent 2 years in Switzerland, replacing the relatively flat landscape of Africa with huge mountains, lush vegetation, frequent rainfall, and cold snowy winters. In midsummer, the rivers ran full and were icy cold for they were fed by snowmelt from the Alps. Snowmelt is the main source of water for great rivers such as the Rhine, which runs through Germany, France, Belgium, and Holland. I lived near the headwaters of the Rhine where it forms the border between Switzerland and Austria. Farming was a year-round activity in this region, and cows were everywhere. "Honey" wagons seemed to have endless loads of fresh manure to deposit on pastures and crops already darkened from earlier loads. The Rhine was already polluted by high discharges of nitrates from the manure.

Water and hydrologic phenomena have always been part of my life. I was born in Southend-on-Sea, Essex, England, which has the longest pier in the world (about 1.2 miles long) because when the tide goes out mud banks stretch more than a mile from the shore. While studying for a degree in agricultural engineering at the University of Kentucky, I came face to face with the environmental debate on how to maintain the high standard of living many Americans enjoy while preserving our environment. The focus of my research was on developing a method to predict how sediment deposits accumulate in sediment ponds. These ponds are constructed downstream from surface mining operations to trap soil and spoil materials that rainfall washes away from the mining operations. I worked in the Appalachian Mountains of eastern Kentucky and Tennessee, where enormous machinery reshapes mountains to reach the rich underlying coal seams. The benefits of this environmental disturbance are affordable electricity, transportation, food, and manufactured commodities.

Surface mining activities and environmental pollution are global problems. An example of problems associated with mining can be found in Tasmania, a small island the size of West Virginia located about 140 miles southeast of Australia. The interior of Tasmania is very mountainous; to the west of these mountains, there is heavy annual rainfall and impenetrable rain forests, home of the notorious Tasmanian devil, a small carnivorous bear. Much of Tasmania's mineral wealth of silver, zinc, and gold is located in the northwest. Trees have been removed from the mountainsides to provide fuel for deep underground mining operations and the mining families. The removal of trees and acid rain from the processing of ore have denuded the area of all vegetation. High rainfall and steep mountain slopes have resulted in the removal of all soil and thwarted efforts to revegetate the area. This is perhaps the bleakest, most desolate landscape I have ever seen — it is comparable to scenes from the moon.

During visits to the Kingdom of Swaziland, located on the east side of southern Africa, I have seen stark granite mountains and rivers and reservoirs choked with sediment. The need to cut wood to provide fuel for domestic use and overgrazing by cattle have resulted in gradual denuding of many areas and the formation of huge gullies called *dongas*. Throughout the world, deforestation is a serious problem (see Chapter 10).

In 1985, Tom Haan (an author of several prominent hydrology books; a friend, mentor, and advisor for my master's degree) and I found ourselves drifting along a small chain of lakes on the northeast coast of South Africa. We were evaluating the feasibility of growing rice in a nearby wetland area that was virtually untouched by civilization. If constructed, this project would drain much of the area and convert it into cropland. Suddenly there was commotion all around us, and we discovered that we had drifted into a family of hippopotamuses. Fortunately, they had little interest in us, and we continued on our way. As we slowly moved from one lake into the next, we saw a wide variety of birds and an occasional crocodile. We stopped on a small mound of soggy land and watched several natives catching freshwater fish in crude handmade wooden traps. As we continued toward the sea, the water became more brackish and the current stronger. Here, those fishing used poles and lines to catch saltwater fish. It was sad to think that this beautiful chain of lakes, the nearby wetland area, swamps to the south, and the birds and wildlife might also be destroyed due to political uncertainties and the need to feed a nation; fortunately, this has not occurred here, but other parts of the world have not been as fortunate.

In 1996, I had the opportunity to visit China (see Chapter 1). This vast nation faces enormous challenges to provide food and potable water to an expanding population that already exceeds 1.5 billion people. During the past decade, I have focused my professional interests on stream systems. Engineers have constructed some amazing dams, bridges, canals, and water control systems. However, historically we have struggled to manage small stream systems adequately. Many modifications to streams have not been self-sustaining and have had a severe impact on the ecology and biology of these systems. A desire by society to develop and construct more natural systems has produced mixed results and often has been rather costly.

Therefore, much of my recent work has focused on looking at these issues and at strategies that are low cost and result in more self-sustaining stream and watershed systems.

A PERSPECTIVE BY STAN TRIMBLE

Hydrology, as the art and science it is, was a long time coming to me in any formal way. My father, Stanley D. Trimble, was a power service maintenance foreperson for the Tennessee Valley Authority (TVA). As such, he traveled hundreds of miles each week, ranging from the giant Shawnee Steam Plant on the Ohio River to the tiny hydroelectric projects in southwestern Virginia. I grew up hearing long phone conversations about "thrust bearing number 3," "[turbine] rotor bushings," or a "failed surge tank." We often traveled with him in the early days, and one of my earliest memories is living in Cleveland, TN, driving up the Ocoee River, and seeing the Ducktown desert, which was created by sulfuric acid fumes from copper smelting. But, while all this did not make me a hydrologist, it did give me a good intuitive idea of the actions and power of running water, especially when under a hundred feet of head from a reservoir; that is probably what planted the hydrology seed, so to speak.

After an undergraduate program of math and physical science, military service, and a year of travel, I entered graduate school (1967) in geography at the University of Georgia, fully intent on working in historical geography with Louis DeVorsey, Jr. My original masters of art thesis was to be on historical water-powered industry (there's my Dad's influence!) on the Georgia Piedmont. The industrial archeology aspect of this interested me very much, and I set out to visit and analyze the water power technology and arrangement at each of the some 50 or so sites that I had identified from archival work. Imagine my bewilderment when I visited site after site and, in most cases, found only a swampy morass rather than the dams, flumes, buildings, shafting, penstocks, and water wheels I had expected. I positively knew that I was at the right locations. Finally, one day an old farmer, when asked, said, "Son, that mill was buried under the mud years ago." The light suddenly came on; I had been looking at the effects of culturally accelerated erosion and sedimentation from the historical farming of corn and cotton. It was a big switch, but from then on, my interest was on the stream processes rather than the mills and that required some retooling.

The time I spent studying historical geography and learning archival and field research techniques was not wasted; indeed, I was often able to use these same techniques to date the fluvial processes, and I have continued to champion this methodology (as one of many) throughout my career. Moreover, the appreciation of the historical–cultural landscape on which humans have altered the fluvial system has held me in good stead throughout my career. This interest in fluvial geomorphology did not make me a hydrologist, but it eventually convinced me that I had to learn hydrology, which I did, much of it after graduate school. For this learning process, I had three important mentors. One is A.P. Barnett, a research agricultural engineer with the U.S. Department of Agriculture (USDA) who taught me about soil erosion processes and prediction and who made sure that I got to meet erosion scientists on the research frontier, such as George Foster and the late Walt Wischmeier, both at Purdue University.

The second is John D. Hewlett, generally long acknowledged as the world's foremost forest hydrologist because of his revolutionary concepts of runoff and a gentle, thoughtful intellectual. Incredibly, I never had a course with Hewlett at the University of Georgia, but we spent many productive hours in the field and talking about problematic hydrologic processes and fluvial landforms. Our association has continued for 35 years, and he continues to influence me, as evidenced in this book and acknowledgments. Among the many things I learned from him was to always be skeptical and to question conventional wisdom, especially "bandwagons."

The third mentor was the late Stafford C. Happ, an engineering geologist who spent his doctoral time at Columbia University in the company of such people as J. Hoover Mackin, C.F.S. Sharpe, Arthur Strahler, Louis Peltier, and Douglas Johnson. Although Happ's written work is limited, it is outstanding. I discovered it as a graduate student, and he became for me a distant personal role model for his work on accelerated sedimentation. Thanks to A.P. Barnett, I met Happ in 1970 where he worked at the USDA Sedimentation Laboratory in Oxford, MS; he became a colleague and then a collaborator when I began the Coon Creek work in 1973 (see Chapter 9). Happ was simply the best field person (not to mention one of the best writers) I have ever known, and I am sure that I learned only a fraction of what he had to teach. His standards for publication were incredibly high, which is the main reason he published so little; he often lamented the "superfluous clutter" in the literature. Happ had a profound effect on me; even 30 years later, I never close a field survey or send out a manuscript for publication without wondering if Dr. Happ would have approved.

I also owe much to contemporary disciplinewide colleagues and former students, although I would not dare single any out for fear of slighting someone. But, a great influence whom I never met was and is the late Ven Te Chow of the University of Illinois. His edited *Handbook of Hydrology* (McGraw-Hill, 1964) was and remains, in my view, the most important hydrology book ever published. Much of what I know about hydrology came from its more than 1000 pages of small print and beautifully clear diagrams; I would encourage anyone interested in hydrology to access this book.

Of course, every scientist thinks his or her field is the "Queen of Science," but as a reluctant hydrologist, I can say that the past 35 years have convinced me that water is the core of, at least, environmental science. Like any late convert, I am a proselytizer and find it difficult to overemphasize hydrology — I remain convinced that most would-be environmental scientists need a healthy dose. To that end, I hope that this book contributes something.

Contents

1 The Hydrologic Cycle, Water Resources, and Society

1.1 INTRODUCTION

The term *hydrology* can be divided into two terms: *hydro*, relating to water, and *loge*, a Greek word meaning knowledge. Thus, hydrology is the study, or knowledge, of water. Questions we might ask are: Why should we acquire knowledge of water? How might this knowledge help society? A simple starting point in understanding our societal need to study water is that, at many points in time, every place in the world will experience an excess or deficit of water that will have an adverse impact on society or a fragile ecosystem. At each location, the available mean water resources, the magnitude and frequency of high and low values of these resources, and other prevailing factors will have an impact on the communities (human and other) that inhabit a location and strategies that might be implemented to help protect these communities. Of particular concern is providing adequate sustainable water resources and food for an ever-expanding global population. Furthermore, factors like global warming could have a significant impact on global food and water supplies.

Every minute, the population of the world increases by about 200 people. This growth amounts to an annual increase of more than 70 million people or an annual population growth rate of about 1% (**Figure 1.1**). While it is anticipated that this annual growth rate will continue to decline, it is possible that some time this century the population of the world will exceed 10 billion people. Population increases in developed countries such as the U.S. will be offset by declines in other developed countries in Europe. Therefore, the majority of the global population increase of more than 3 billion people in the next 50 to 100 years will occur in developing countries that already face shortages of potable water and food. Both India and China will face huge population increases, and it is probable that India at some point will have the world's largest population. In Africa, it is projected that the population of Nigeria could double and approach that of the U.S.

Many developing countries are not self-sufficient and depend on aid from developed countries. Other countries, like China, are very dependent on declining water resources to provide sufficient food for their expanding population. A few years ago Andy Ward visited China as part of a scientific group that had been asked to look at strategies for making Chinese agricultural systems more sustainable. The visit focused on the Yellow River Plain (also called the Hwang Ho, Huang He, or North China Plain), where groundwater levels are dropping at rates that exceed 1 m annually, and the area is frequently ravaged by floods (**Figure 1.2**). It is now common for the Yellow River to run dry many times annually, and during each irrigation event, water obtained from the Yellow River deposits 1 to 2 m of sediment in irrigation canals (see Chapter 6). Typically, in the Yellow River Plain one crop annually is primarily rain fed, and one or two other crops depend on irrigation. Current water use for irrigation in this region is not sustainable, yet more food needs to be produced.

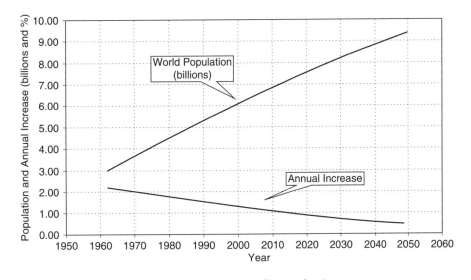

FIGURE 1.1 World population increase. (Based on 2000 U.S. Census Bureau data.)

FIGURE 1.2 Flooding in the Yellow River Plain.

FIGURE 1.3 People and animals plowing a small field in China.

Food production problems in countries like China include labor-intensive and inefficient farming practices (**Figure 1.3**) and postharvest yield losses that often exceed a third of the harvest. A dilemma with using methods that are less labor intensive is that a rapidly expanding labor force would be put out of work. Postharvest yield losses are associated with limited resources and infrastructure to dry, store, and process crops. For example, in China, Ward observed grain threshed and dried by spreading it across roads (**Figure 1.4**). There was much spoilage and contamination of the grain. Often, it was then stored in bags underneath a farmer's house, where some of it would spoil, and some would be eaten by rodents. Typically, resources to refrigerate, process, or transport fruit and vegetables were limited. What was not consumed fairly shortly after harvest in localized regions quickly spoiled. These types of problems are common in Asia and Africa.

In developed countries like the U.S., the food supply, distribution, and use problem is very different from that in developing countries. Generally, production systems are efficient, but depend on machinery and the use of chemicals. In recent years, there has been a focus on making these systems more efficient, particularly on methods to use fewer chemicals and more efficient irrigation practices. Practices such as precision agriculture are seeing increased attention and application. For example, in **Figure 1.5** the dark patches are weeds, and it can be seen that spot applications of herbicides at those locations would only use a small fraction of the herbicides commonly applied across the whole field.

Problems with food supply and use in developing countries are high waste by the consumer, a diet that contains a lot of meat, and high energy costs associated with processing and transportation of food products hundreds and sometimes

FIGURE 1.4 Grain is dried and threshed by spreading it across a busy road in China.

FIGURE 1.5 False color image, from a sensor mounted on a low-altitude aircraft, shows patches of weeds in a soybean field.

thousands of miles. The scientific literature and various Internet Web sites are replete with "facts" on how much land is needed to feed a nonvegetarian and how much less land and water is needed to feed a vegetarian. A review of these various sources suggests that 1 to 3 acres of land are needed to provide each person a diverse diet, perhaps 10 times this amount of land is used to feed each person in the U.S., and the resources needed to feed a vegetarian might be 5 to 20% of those needed to support a nonvegetarian in a developed country. We are not suggesting that people in developed countries should change their diet, but we need to recognize that the majority of the people in the world eat a diet that is high in carbohydrates and low in protein. Much of the diet of many people consists of a staple food such as rice, maize (corn), potatoes, or sweet potatoes. As populations grow and our arable land base shrinks, there will be further need to

develop more efficient food supply systems and greater pressure to switch from high land, water, and energy livestock production systems, such as those needed for beef production, to more efficient animal production systems, such as those associated with poultry. However, poultry and egg production systems are not without their critics and often face intensive focus from environmental groups, animal activist groups, and local citizens. For further information on these issues, refer to the work of P. Foster (1992), L.R. Brown (1995), Smil (2000), and Rosegrant et al. (2002).

The Food and Agriculture Organization of the United Nations (FAO, www.fao.org) reports that one in five developing countries will face water shortages by 2030. During the past 30 years, irrigated land increased at a rate of about 1.6% each year. However, during the next 30 years the FAO projects that the annual increase will only be 0.6%.

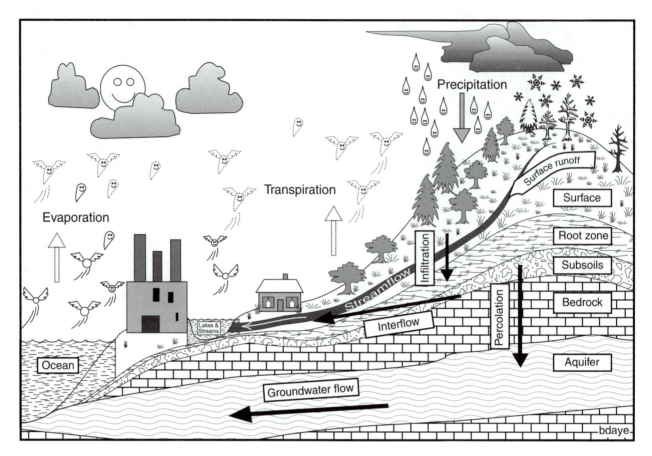

FIGURE 1.6 The hydrologic cycle.

The reasons for the drop in these increases are because in many countries much of the arable land is already irrigated or many of the renewable water resources are already in use. Food production constitutes 65 to 70% of global water use. In North America, water use by industry and agriculture is similar, and each accounts for 40 to 45% of all withdrawals. In Africa, Asia, and the Near/Middle East, withdrawals for agricultural production account for more than 80% of water use. However, in developing countries in Asia and Africa only about 30 to 45% of the arable land is irrigated (FAO, 2002). An interesting and useful study on World Water and Food to 2025 was reported by Rosegrant et al. (2002). This work was performed by the International Food Policy Research Institute (IFPRI) and the International Water Management Institute (IWMI), which developed the IMPACT-WATER model to study global water and food policy and investment issues.

The potential exists that each of us will be impacted by global water resource and food supply issues, but for many of us, our interest in hydrology will have a more local or regional focus. Regardless of the magnitude of the problem we consider, a basic understanding of the complexity of the hydrologic cycle will be helpful. In particular, this book attempts to illustrate how the environment behaves as a connected system of components that inter-

acts at different spatial and temporal scales. Therefore, a piecemeal understanding or attempt to manage a few parts of the system, at a very localized scale, will often result in an unsatisfactory outcome.

1.2 THE HYDROLOGIC CYCLE

The earth holds more than 300 million mi^3 of water beneath the land surface, on the surface, and in the atmosphere. This vast amount of water is in constant motion, known as the *hydrologic cycle*. Hydrology is concerned with the transport of water through the air, over the ground surface, and through the strata of the earth. Knowledge of hydrology is important in practically all problems that involve the use and supply of water. Therefore, hydrology is of value not only in the field of engineering, but also in forestry, agriculture, and other branches of the environmental sciences.

The hydrologic cycle, illustrated in **Figure 1.6**, shows the pathways water travels as it circulates throughout global systems by various processes. The visible components of this cycle are precipitation and runoff. However, other components, such as evaporation, infiltration, transpiration, percolation, groundwater recharge, interflow, and groundwater discharge, are equally important. A summary of the world water balance by continent is presented in

TABLE 1.1
Annual World Water Balance by Continent (Expressed as Inches of Water)

Water Balance Elements[a]	Europe	Asia	Africa	North America	South America	Australia	Weighted Average
Area, millions of square miles	3.8	17.6	11.8	8.1	4.0	3.4	
Precipitation	28.9	28.6	27.0	26.4	64.9	29.0	34.1
Total river runoff	12.6	11.5	5.5	11.3	23.0	8.9	12.1
Groundwater runoff	4.3	3.0	1.9	3.3	8.3	2.1	3.8
Surface water runoff	8.3	8.5	3.6	8.0	14.7	2.8	7.6
Infiltration and soil water	20.6	20.0	23.4	18.4	50.2	22.2	25.8
Evaporation	16.3	17.0	21.5	15.1	41.9	20.1	22.0

[a] Total river runoff is the sum of the groundwater runoff and surface water runoff. The sum of the total river runoff and evaporation equals the precipitation.

Source: Summarized from van der Leeden, F., F.L. Troise, and D.K. Todd, *The Water Encyclopedia*, 2nd ed., Lewis, Chelsea, MI, 1991.

Table 1.1. In the following sections, we discuss each of these components and their relationships.

1.2.1 PRECIPITATION

Water that evaporates from the earth is temporarily stored as water vapor in the atmosphere. While in the atmosphere, this vapor and small water droplets form clouds. As the atmosphere becomes saturated, water is released back to earth as some form of *precipitation* (rain, snow, sleet, or hail). Some of the water might evaporate before it reaches the ground. Precipitation reaching the ground can evaporate from anywhere, including bare soil surfaces, plant surfaces, and the surfaces of ponds, lakes, and streams. Precipitation is a natural phenomenon that humans can do very little to control.

1.2.2 EVAPORATION

Evaporation occurs when water is changed from a liquid to a vapor. Increases in air and water temperatures, wind, and solar radiation all increase evaporation rates, while a high water vapor percentage in the air (high relative humidity) decreases the potential for evaporation. Through the process of evaporation, water moves back to the atmosphere in the form of vapor.

1.2.3 TRANSPIRATION

Water can take several paths after it enters the soil. Some water becomes part of the soil storage. This water is not stationary and moves downward at a rate that depends on various soil properties, such as hydraulic conductivity and porosity. While in storage near the surface, some of this water is used by plants and is eventually returned to the atmosphere as water vapor. The process by which plants release water vapor to the atmosphere is called *transpiration*. This water vapor is a natural by-product of photosynthesis.

1.2.4 EVAPOTRANSPIRATION

Because of the difficulty in separating the processes of evaporation and plant transpiration, we usually view these two processes as one process called *evapotranspiration* (ET). This term includes both the water that evaporates from soil and plant surfaces and the water that moves out of the soil profile by plant transpiration. More than half of the water that enters the soil is returned to the atmosphere through evapotranspiration.

1.2.5 INFILTRATION

Infiltration is the entry of water into the soil. The amount of water that infiltrates into the ground varies widely from place to place. The rate at which water infiltrates depends on soil properties such as soil water content, texture, density, organic matter content, hydraulic conductivity (permeability), and porosity. *Hydraulic conductivity* is a measure of how fast water flows through certain soils or rock layers. Infiltration and hydraulic conductivity are greater in porous materials, such as sands, gravels, or fractured rock than in clay soils or solid rock. *Porosity* is a measure of the amount of open space in soil or rock that may contain water.

Conditions at the soil's surface also influence infiltration. For example, a compacted soil surface restricts the movement of water into the soil profile. Vegetation can play a prominent role in infiltration. The surface soil layer in a forest or a pasture will generally have far greater infiltration rates than a paved parking area or a compacted soil surface. Topography, slope, and the roughness of the surface also affect infiltration, as do human activities in urban and agricultural areas, where alteration of soil properties and surface conditions have occurred.

1.2.6 PERCOLATION AND GROUNDWATER RECHARGE

Another path that water can take after it enters the soil surface is *percolation*, which is water moving downward through the soil profile by gravity after it has entered the soil. Water that moves downward through the soil below the plant root zone toward the underlying geologic formation is called *deep percolation*. For the most part, deep percolation is beyond the reach of plant roots and this water contributes to replenishing the groundwater supply. The process of replenishing or refilling the groundwater supply is called *groundwater recharge*.

1.2.7 RUNOFF AND OVERLAND FLOW

Runoff is the portion of precipitation, snowmelt, or irrigation water that flows over and through the soils, eventually making its way to surface water systems. Contributions to runoff might include overland flow, interflow, and groundwater flows. Once the precipitation rate exceeds the infiltration rate of the soil, depressions on the soil surface begin to fill. The water held in these depressions is called *surface storage*. If surface storage is filled and precipitation continues to exceed infiltration, water begins to move downslope as overland flow or in defined channels. A large percentage of surface runoff reaches streams, where we typically then describe it as *streamflow* or *discharge*. Overland flow can also occur when the soil is saturated (soil storage is filled). In this case, all the voids, cracks, and crevices of the soil profile are filled with water, and the excess begins to flow over the soil surface.

1.2.8 INTERFLOW

As water percolates, some of it may reach a layer of soil or rock material that restricts downward movement. Restrictive layers can be formed naturally (clay pan or solid bedrock) or because of human activities. Once water reaches a restrictive layer, it may move laterally along this layer and eventually discharge to a surface water body such as a stream or lake. The lateral movement of water is called *interflow*.

1.2.9 GROUNDWATER FLOW

Groundwater comprises approximately 4% of the water contained in the hydrologic cycle and can flow to surface water bodies such as oceans, lakes, and rivers. This process creates a baseflow for a surface water body that is an important contribution to groundwater and surface water. More than 50% of the population depends on groundwater as the primary source of drinking water. Approximately 75% of American cities derive their supplies totally or partially from groundwater. In 1980, the U.S. used 88

billion gal/day of groundwater, and 68% of this total was used for irrigation.

1.3 WATER SUPPLY

As a society, we face increasing demands for water and increasing threats to water quality. As water circulates between earth and the atmosphere, chemicals and particles contained and transported by it are modified by natural processes and human activities. Chemicals and particles from dust, smoke, and smog in the atmosphere eventually fall back to the earth with precipitation. Runoff from roadways and parking lots wash grit and metal particles directly into storm sewers and streams. Water moving across the soil surface as overland flow can detach soil particles and transport them to a stream or lake. Some chemicals attach to soil particles and are transported to receiving waters. Runoff from lawns, pastures, and agricultural fields can also carry dissolved nutrients and pesticides. An assessment of the impact of nonpoint pollution on U.S. surface water resources is summarized in **Table 1.2** (see http://www.epa.gov/305b/). The percentage of rivers, lakes, and estuaries that are threatened or already impaired is very high.

Water that flows to a groundwater system can be polluted by the leaching of chemicals, nutrients, or organic wastes from the land surface or from materials buried in landfills. Groundwater close to the surface or in porous sands and gravels is vulnerable to pollution. Deep groundwater is also vulnerable, especially if connected to the surface by fissures or sinkholes in underlying formations, such as in limestone rock areas.

Hillel (1992) provided the following insight on the escalation in water use:

> Historical records show that in 1900, the World's annual water use was about 400 billion cubic meters, or 242 cubic meters per person. By 1940, the total water use had doubled, while the per capita use had grown by some 40% to about 340 cubic meters. … By 1970, it had reached 700 cubic meters per capita. Both agricultural and industrial water use grew twice as much in the 20 years between 1950 and 1970 as they had during the first half of the century.

Worldwide, agriculture is the major user of water, accounting for about 70% of all withdrawals. The amount of irrigated land per capita in various parts of the world is presented in **Table 1.3** (FAO, 1999). Between 1960 and 1980, the majority of the world's increased food production was achieved through additional irrigation. Unfortunately, irrigation can also cause damage to the land through waterlogging and salinization. For example, in India and Pakistan, 13% and 22% of the irrigated land,

TABLE 1.2
Year 2000 Water Quality Assessment with the Percentage of Each Water Resource Impaired by Specific Constituents

Attribute	Lakes, Reservoirs, Ponds	Rivers and Streams	Tidal Estuaries	Great Lakes
Assessed	43.0	6.0	36.0	92.0
Impaired	45.0	14.0	51.0	78.0
Pollutants				
Nutrients	49.9	9.9	0.0	10.0
Metals	41.8	10.6	51.5	0.0
Siltation and suspended solids (SS)	20.6	11.5	0.0	8.9
Total dissolved solids (TDS)	19.4	0.0	15.9	0.0
Oxygen depletion	14.6	23.5	34.0	6.7
Excess algae	12.4	0.0	0.0	0.0
Pesticides	8.2	0.0	38.2	0.0
Pathogens	0.0	88.5	30.4	9.3
Polychlorinated biphenyls (PCBs)	0.0	0.0	16.7	3.9
Flow alterations	0.0	0.0	0.0	0.0
Habitat alterations	0.0	0.0	0.0	0.0
Oil and grease	0.0	11.1	0.0	0.0
Turbidity	0.0	12.2	0.0	0.0
Toxic organics	0.0	12.2	0.0	45.4
Taste and odor	0.0	0.0	0.0	3.9

Note: Percentages add to more than 100% because most water resources are impaired by several pollutants.
Source: http://www.epa.gov/305b /.

TABLE 1.3
Amount of Irrigated Land per Capita by World Region in 1999

Continent	Irrigated Area (Mha)	Population (Million)	Irrigated Land/Capita (ha/capita)
America	40.40	819	0.05
Asia	191.20	3634	0.05
Europe	24.60	729	0.03
Africa	12.50	767	0.02
Oceania	2.70	30	0.09
World	271.40	5979	0.05

Source: Food and Agriculture Organization of the United Nations, *FAO Yearbook: Production–1999*, Food and Agriculture Organization of the United Nations, Rome, 1999. With permission.

respectively, is unusable because of salinity problems (Tarrant, 1991).

The hydrologic cycle for the U.S. is summarized in **Table 1.4** (Federal Council for Science and Technology, 1962). The water available in surface water systems is somewhat misleading as more than 90% of the total is located in the Great Lakes. However, we are fortunate to have a vast network of rivers. A map showing the relative size and location of the largest rivers in the U.S. is presented in **Figure 1.7** (Iseri and Langbein, 1974). Water undergoes repeated cycles of withdrawal, treatment, and partial consumption; the remaining amount is returned to the rivers. Some portion of the flow might be used 15 to 20 times before it discharges into the ocean. Rivers such as the Mississippi and St. Lawrence are also important transportation systems that link inland agricultural and commercial areas with ocean ports.

Average annual water use in selected countries is shown in **Table 1.5** (Rosegrant et al., 2002). It can be seen that the U.S. has the highest per capita use, primarily due to the high level of development, the standard of living

TABLE 1.4
Distribution of Water in the Continental U.S.

Water Source	Volume (×10⁹ m³)	Volume (%)	Annual Circulation (×10⁹ m³/year)	Replacement Period (years)
Groundwater				
Shallow (<800 m deep)	63,000	43.2	310	>200
Deep (>800 m deep)	63,000	43.2	6.2	>10,000
Freshwater lakes	19,000	13.0	190	100
Soil moisture (1-m root zone)	630	0.43	3,100	0.2
Salt lakes	58	0.04	5.7	>10
Average in stream channels	50	0.03	1,900	<0.03
Water vapor in atmosphere	190	0.13	6,200	>0.03
Frozen water, glaciers	67	0.05	1.6	>40

Source: Federal Council for Science and Technology, *Report by an Ad Hoc Panel on Hydrology and Scientific Hydrology*, Federal Council for Science and Technology, Washington, DC, 1962.

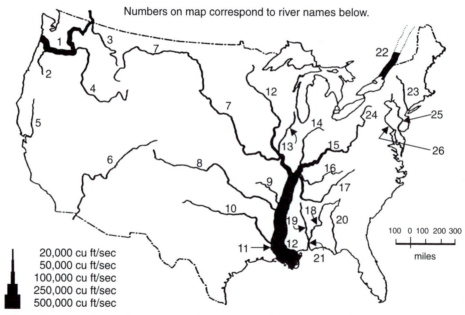

Numbers on map correspond to river names below.

20,000 cu ft/sec
50,000 cu ft/sec
100,000 cu ft/sec
250,000 cu ft/sec
500,000 cu ft/sec

100 0 100 200 300
miles

Rivers shown are those whose average flow at the mouth is 17,000 cu ft/sec or more. Average flow of Yukon River, Alaska (not shown) is 240,000 cu ft/sec.

1 Columbia	8 Arkansas	15 Ohio	22 St. Lawrence
2 Willamette	9 White	16 Cumberland	23 Hudson
3 Pend Oreille	10 Red	17 Tennessee	24 Allegheny
4 Snake	11 Atchafalaya	18 Alabama	25 Delaware
5 Sacramento	12 Mississippi	19 Tombigbee	26 Susquehanna
6 Colorado	13 Illinois	20 Apalachicola	
7 Missouri	14 Wabash	21 Mobile	

FIGURE 1.7 Relative size and location of the largest rivers in the U.S.

enjoyed by most Americans, and the need to use water for electric power cooling and agriculture. Let us look closer at how water is used in the U.S.

Approximately two thirds of the precipitation in the U.S. is used to satisfy evaporation and evapotranspiration requirements (**Figure 1.8**). The remaining third enters surface and groundwater systems. Only a small portion of this water supply (8.1% of the precipitation or about 25% of the surface flow) is withdrawn for use in thermoelectric power generation, industrial and mining purposes, commercial and

TABLE 1.5
Projected Average Annual Water Withdrawal and Water Use under Business-as-Usual Scenarios

Country	Water Withdrawal (km³)		2021–2025 Consumptive Water Use (km³)			
	1995 Baseline	2021–2025 Projections	Irrigation	Domestic	Industrial	Livestock
China	679	844	233	58	32	7
India	674	822	338	40	16	8
Sub-Sahara Africa	128	207	62	23	2	4
West Asia/North Africa	236	294	137	13	9	3
Asia	1953	2420	842	146	79	23
Developed countries	1144	1272	278	68	115	18
Developing countries	2762	3481	1216	215	122	43
World	3906	4752	1493	283	237	61

Source: Based on Rosegrant, M.W., X. Cai, and S.H. Cline, *World Water and Food to 2025: Dealing with Scarcity*, International Food Policy Research Institute, Washington, DC, 2002.

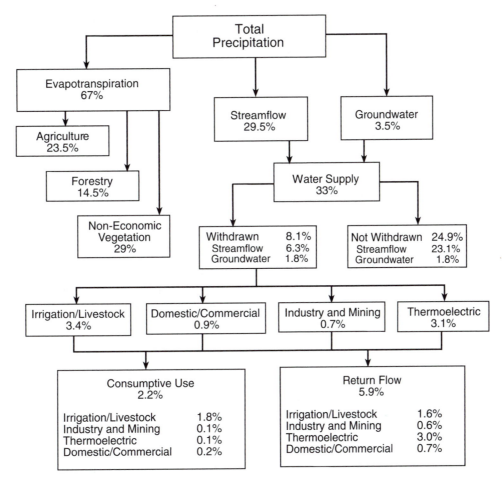

FIGURE 1.8 Water use in the U.S.

domestic water use, and for agricultural purposes. Depending on location, water will cost the consumer less than one dollar per 1000 gallons to more than two dollars per 1000 gallons.

Average urban domestic water use is presented in **Table 1.6** (van der Leeden et al., 1991). Typically, families in many European countries use less than half the amount of water used by families in the U.S., while in developing

TABLE 1.6
Typical Urban Water Use by a Family of Four

Type of Household Use	Daily Family Use Gallons/day	%	Daily per Capita Use (gal/day)
Drinking and water used in kitchen	8	2	2.00
Dishwasher (3 loads per day)	15	4	3.75
Toilet (16 flushes per day)	96	28	24.00
Bathing (4 baths or showers per day)	80	23	20.00
Laundering (6 loads per week)	34	10	8.50
Automobile washing (2 car washes per month)	10	3	2.50
Lawn watering and swimming pools (180 h per year)	100	29	25.00
Garbage disposal unit (1% of all other uses)	3	1	0.75
Total	346	100	86.50

Note: Percentages added.

Source: van der Leeden, F., F.L. Troise, and D.K. Todd, *The Water Encyclopedia*, 2nd ed., Lewis, Chelsea, MI, 1991, based on U.S. Water Resources Council, 1975.

countries, the rate of domestic water use is sometimes less than 10% of the average domestic amount used in U.S. households. It can be seen that a 33% savings could be obtained by simply eliminating lawn watering, swimming pools, automobile washing, and garbage disposal units. In European countries, other savings are obtained by using more efficient toilet systems that reduce water use for this purpose to less than one third of U.S. consumption. In 1994, federal legislation was passed that specified the maximum flow rates of new shower and toilet installations. These maximum rates are similar to current European levels. In developing countries, only small amounts of water are used for bathing and laundering, primarily because of limited access to running water, hand bathing, and bathing and laundering in rivers by part of the population.

A summary of amounts of water used to produce various commodities is presented in **Table 1.7** (Kollar and MacAuley, 1980). Large quantities of water are needed to manufacture most commodities, and water is one of the main limiting factors in furthering worldwide industrial development. It can be seen from **Table 1.7** that although a lot of water is needed to manufacture a commodity, only a small amount of this water is consumed. The rest is discharged or recycled, treated, and then used again for this or another purpose. Improved technologies are required to reduce the amount of water needed and consumed.

1.4 THE IMPORTANCE OF HYDROLOGY TO SOCIETY

Water has been central to the history of humanity. Civilizations have persisted or perished as they experienced situations of too much or too little water. Abandoned irrigation projects the world over, including Native American works in the southwestern U.S. (1100 A.D.), illustrate that there was a woeful lack of knowledge of hydrology on which to base the development of water supply systems. Salinity problems in agriculture of semiarid and arid regions of the world are also evidence of the lack of hydrologic principles of water management. Locating costly developments in floodplains of large river systems indicates a lack of understanding of river hydrology or direct ignorance of the hydrologist's forecast of flooding. In the past, there were signs of deficiencies in the science of hydrology, but they affected only a few people and went almost unnoticed. It has been quite different since the mid-20th century.

Water is still central to our life today, but we have grown more aware of the accelerated growth of industry and population and their use of water as related to the earth's fixed supply. We are coming to understand that our prosperity and prospects for survival vary with the amount and distribution of fresh, unpolluted water, and that each year there are millions more of us, but no more water. Multiple demands for the use of the same gallon or liter of water, and prospects for even greater demands tomorrow, indicate that more and more people of different disciplines need to have knowledge of hydrology. Today's student will be tomorrow's water management decision maker.

As the science of hydrology becomes more generally understood, parts of the hydrologic cycle are examined to determine how and to what extent the cycle can be modified by human activity in practical ways. Attempts to modify the weather to increase rainfall in specific areas, for example, have been made through research, trial, and demonstration. However, success in this effort has been limited. Extensive drainage projects in low-lying swamps open up land for food production, thus lowering groundwater levels, reducing evaporation, changing rainfall–runoff relationships, and affecting wildlife habitats. In other areas, large-scale

TABLE 1.7
Water Use vs. Industrial Units of Production in the U.S.

Industry	Parameters of Water Use	Intake by Unit of Production	Consumption by Unit of Production	Discharge by Unit of Production
Meatpacking	gal/lb carcass weight	2.2	0.1	2.1
Dairy products	gal/lb milk processed	0.52	0.03	0.48
Canned fruits and vegetables	gal/case 24-303 cans e/case	107	10	98
Frozen fruits and vegetables	gal/lb frozen product	7.1	0.2	6.9
Wet corn milling	gal/bu corn grind	223	18	205
Cane sugar	gal/ton cane sugar	18,250	950	17,300
Malt beverages	gal/barrel malt beverage	420	90	330
Textile mills	gal/lb fiber consumption	14	1.4	12.8
Sawmills	gal/bd ft lumber	3.3	0.6	2.7
Pulp and paper mills	gal/ton pulp and paper	38,000	1,800	36,200
Paper converting	gal/ton paper converted	3,900	270	3,600
Industrial gases	gal/1000 ft³ industrial gases	226	31	193
Industrial inorganic chemicals	gal/ton chemicals	4,570	470	4,300
Plastic materials and resins	gal/lb plastic	6.7	0.6	6.1
Synthetic rubber	gal/lb synthetic rubber	6.5	1.4	5.1
Cellulosic man-made fibers	gal/lb fiber	68	4.6	63
Organic fibers, noncellulosic	gal/lb fiber	38	1.1	37
Paints and pigments	gal/gal paint	7.8	0.4	7.4
Industrial organic chemicals	gal/ton chemical building	54,500	2,800	51,700
Nitrogenous fertilizers	gal/ton fertilizer	4,001	701	3,299
Petroleum refining	gal/barrel crude oil input	289	28	261
Tires and inner tubes	gal/tire car and truck tires	153	14	139
Hydraulic cement	gal/ton cement	830	150	680
Steel	gal/ton steel net production	38,200	1,400	36,800
Iron and steel foundries	gal/ton ferrous castings	3,030	260	2,760
Primary copper	gal/lb copper	17	4.1	13
Primary aluminum	gal/lb aluminum	12	0.2	11.8
Automobiles	gal/car domestic automobiles	11,464	649	10,814

Source: Based on Kollar, K.L., and P. MacAuley, *J. Am. Water Works Assoc.*, 72, 1980.

irrigation programs increase soil water content, evaporation, and crop use of water. In watersheds, where the vegetative cover can be considerably modified by afforestation or deforestation and where huge tracts of grassland can be plowed under for grain production, the movement of water into and over the land surface may be drastically altered. Numerous dams and reservoirs provide water, recreation, and flood protection, but modify natural streamflow regimes.

A practical knowledge of the science of hydrology will help the decision maker and public understand the overall effect of humanity's influences on the hydrologic cycle and the side effects of projects on other people, their activities, and the environment. The informed decision maker will be able to weigh the advantages of each proposed change in the hydrologic cycle against the disadvantages.

1.5 DATA ANALYSIS AND STATISTICS

The following section is a general review of basic statistics terminology and application. It is not meant to be a comprehensive guide to statistics, but an aid to understanding how statistics can be used to analyze and interpret data and how results can be communicated to others. In many of the chapters of this book, we present results that are based on a statistical analysis of measured data. Also, we often use empirical models and methods in hydrology because hydrological systems are very complex; even when we can use mathematical equations to describe the processes that occur, it may not be possible or practical to obtain the input information for these equations. More pragmatic, but approximate, methods that are based on statistics are the most commonly used tools in hydrology and are often incorporated in even the most sophisticated and complex computer models. For the study of the application of statistics in hydrology, a good starting point is the text by Haan (1997).

Statistics is, very generally speaking, the science of data. More specifically, it is the science of collecting, classifying, summarizing, organizing, analyzing, exploring, and interpreting data. The word statistics in its plural form can also be used to describe a collection of numerical information.

There are two main branches of statistics, descriptive and inferential. To understand the differences between them, it is important to know the definitions to two terms. A *population* is the data set that is the focus of interest and involves all elements under investigation (e.g., all of the streams in your state). However, a *sample* is a subset of data collected from that population or the collection of observations actually available for analysis. It is a finite part of a population with properties studied to gain information about the "whole" (e.g., all streams in your state accessible from a main road). The size of the sample can be dependent on goals of the study, time, funding, and accessibility of the population. *Descriptive statistics* involves the organization, summarization, and description of data sets. *Inferential statistics,* on the other hand, is the process of drawing conclusions about an entire population based only on the results obtained from a small sample of that population.

1.5.1 Plotting Data

Most people who read this book will tabulate, plot, analyze, and discuss data at some point. Many of us will need to do this numerous times. Deciding how best to analyze and present that data is often a challenge. A common mistake is to tabulate the data, analyze it, and then decide whether to create a plot of the results. Creating a plot of the data before performing the analysis is often helpful in understanding the data and selecting the correct statistical test to use in performing the analysis. We need to be careful that we do not then rush into an analysis. With modern spreadsheet technologies, we can rapidly perform statistical tests with little or no knowledge of the test that is performed or knowledge as to whether it is appropriate. Another problem is deciding what type of plot to create. If we are unsure, there is no harm and very little extra work involved in generating several different plots. Basic rules to keep in mind are as follows:

1. The plot should aid in understanding the data and should provide more information or insight than tabulated data.
2. The plot should be logical.
3. Many different things should not be plotted on the same plot.
4. Three-dimensional plots might look nicer, but are often harder to understand.
5. You should only use statistical methods you understand.

Example 1.1

A study is conducted to measure the monthly soil erosion (see Chapter 9 for a detailed discussion on this topic) from a research plot. The plot is located in the Midwest region of the U.S., no tillage is used, and the plot is planted with corn at the end of April. In addition to measuring actual soil losses from the plot, two predictive methods are used to estimate the monthly soil erosion. The measured and predicted data are presented in **Table 1.8**. Does the measured data seem intuitively correct, and which predictive method is best?

TABLE 1.8
Monthly Soil Erosion Data for Example 1.1

Month	Measured Soil Loss	Method 1		Method 2	
		Predicted Loss	Residual 1	Predicted Loss	Residual 2
January	0.0	0.0	0.0	0.1	−0.1
February	0.1	0.0	0.1	0.0	0.1
March	0.3	0.3	0.0	0.3	0.0
April	0.6	0.9	−0.3	0.5	0.1
May	0.5	0.7	−0.2	0.8	−0.3
June	0.4	0.5	−0.1	0.4	0.0
July	0.3	0.3	0.0	0.2	0.1
August	0.2	0.1	0.1	0.1	0.1
September	0.1	0.0	0.1	0.1	0.0
October	0.2	0.1	0.1	0.1	0.1
November	0.2	0.1	0.1	0.1	0.1
December	0.1	0.0	0.1	0.3	−0.2
Mean	0.25	0.25	0.0	0.25	0.0

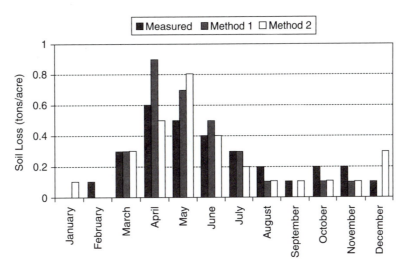

FIGURE 1.9 Bar chart of measured and predicted monthly soil loss (tons/acre) for Example 1.1.

Solution

From the tabulated data, we can see that the most erosion occurs in April, May, and June. This seems reasonable as the Midwest region has high-intensity and sometimes-severe storms during those months (see Chapter 2), and ground cover will be limited in April and May. Both of the predictive methods overpredict erosion in some months and underpredict erosion losses in other months. The predicted mean monthly erosion losses for both methods are the same as the measured mean (calculating means is discussed later in this chapter). We can better see the magnitude of the over- and underpredictions by calculating the differences in the measured and predicted monthly losses. We have called these differences *residuals*. Let us look at the creation of some plots.

Figure 1.9 shows a bar chart of the observed and predicted values for each month. Method 1 overpredicts the peak values, which all occur in the spring. Method 2 overpredicts the soil losses in December and January. It also has a tendency to underpredict losses in the late summer and fall. March is the only month when both predicted values are the same as the measured value.

Figure 1.10 shows a plot of the residual values. It can be seen that the magnitude of the overprediction by Method 1 increases as the erosion loss increases. We call this a *biased* or *skewed* model. Visually, Method 2 does not appear to be biased as both small and large erosion losses are over- or underpredicted.

Answer

If the reason for the bias of Method 1 could be determined, it might be possible to modify the method so that it provides improved results. Otherwise, Method 2 is preferred as there is no bias.

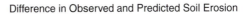

Difference in Observed and Predicted Soil Erosion

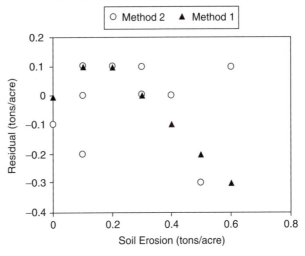

FIGURE 1.10 Residual soil losses for Example 1.1.

1.5.2 DESCRIPTIVE STATISTICS

The normal curve that we discuss more in depth in this chapter illustrates two important types of descriptive statistics: (1) statistics of location/central tendency and (2) statistics of dispersion and variability.

The statistics of location/central tendency include three of the most commonly used values in statistics. The *sample mean* $\bar{y}$ is the arithmetic average of all samples collected. The *median* M_d is the midpoint or middle observation in an ordered distribution. If there is an even number of observations, the median is the average of the middle two observations. The *mode* M_o is the value that occurs most frequently in a distribution. Of these three central tendency measurements, the sample mean is the most commonly used and is represented by the equation

TABLE 1.9
Average Daily Rainfall Values per Month for a Location in the Mojave Desert, U.S.

Month	Average Daily Rainfall (inches)
January	0.029
February	0.020
March	0.000
April	0.008
May	0.000
June	0.000
July	0.005
August	0.045
September	0.015
October	0.002
November	0.002
December	0.027

TABLE 1.10
Statistical Results for Example 1.1

Rank	Values	Unique Values	Occurrence of Each Unique Value
1	0.000		
2	0.000		
3	0.000	0.000	3
4	0.002		
5	0.002	0.002	2
6	**0.005**	0.005	1
7	**0.008**	0.008	1
8	0.015	0.015	1
9	0.020	0.020	1
10	0.027	0.027	1
11	0.029	0.029	1
12	0.045	0.045	1

Note: The two middle values are in bold.

$$\bar{y} = \frac{\Sigma y_i}{n} \qquad (1.1)$$

where $\bar{y}$ is the sample mean, Σy_i is the sum of all measurements (i denotes any one measurement in a series), and n is the number of measurements made.

The normal curve or normal distribution illustrates two types of descriptive statistics. Common measures of dispersion include range, sample variance s^2, and standard deviation s. The *range* is equal to the difference between the smallest and the largest measurement in a data set. The *sample variance* is described by the equation

$$s^2 = \frac{\Sigma(y_i - \bar{y})^2}{n-1} \qquad (1.2)$$

where n is the size of sample, and $\bar{y}$ is the sample average.

Sample standard deviation is the square root of the sample variance and is described by the equation

$$s = \sqrt{\frac{\Sigma(y_i - \bar{y})^2}{n-1}} \qquad (1.3)$$

where $n - 1$ represents the degrees of freedom for a sample or number of independent observations n in a data set.

EXAMPLE 1.2

Find (a) the sample mean, (b) the median, (c) the mode, (d) the sample variance, and (e) the standard deviation for the data set of average daily rainfall values for a location in the Mojave Desert in the U.S. (**Table 1.9**).

Solution

(a) To find the sample mean $\bar{y}$ of the data set, we need to use Equation 1.1:

$$\bar{y} = \frac{\Sigma y_i}{n} \qquad (1.1)$$

where n represents the number of samples, so for this example, n equals 12.

$$\bar{y} = \frac{\begin{array}{c} 0.029 + 0.020 + 0 + 0.008 + 0 + 0 + 0.005 \\ +0.045 + 0.015 + 0.002 + 0.002 + 0.027 \end{array}}{12}$$

$$\bar{y} = \frac{0.153}{12} = 0.01275 \text{ (in.)} = 0.013 \text{ (in.)}$$

(b) To find the median of the data set, we need to rank all of the values in the data set numerically as shown in **Table 1.10**.

The median is the middle value. In this example, there are two middle values because there is an even number of values; therefore, we have to take the average of the two middle values:

$$M_d = \frac{0.005 + 0.008}{2} = .0065 \text{ (in.)}$$

If there were an odd number of values, the median would be represented by the middle value.

(c) To find the mode M_o of the data set, we need to record the number of times each value appears in the data set. So, from **Table 1.10**, the mode is 0.000 (in.) because it appears the most frequently in the data set, with three occurrences.

(d) The sample variance is calculated using Equation 1.2:

$$s^2 = \frac{\Sigma(y_i - \bar{y})^2}{n-1} \tag{1.2}$$

After entering the data points and using the result from Part (a), we obtain

$$s^2 = \frac{\begin{array}{c}(0.029 - 0.01275)^2 + (0.020 - 0.01275)^2 + ... \\ + (0.002 - 0.01275)^2 + (0.027 - 0.01275)^2\end{array}}{12-1}$$

$$s^2 = \frac{0.00225}{11} = 0.000205$$

(e) Finally, we can calculate the sample standard deviation from Equation 1.3 in the test, and it is simply the square root of the sample variance:

$$s = \sqrt{\frac{\Sigma(y_i - \bar{y})^2}{n-1}} \tag{1.3}$$

Therefore, s equals $\sqrt{s^2}$, which we calculated in Part (d):

$$s = \sqrt{0.000205} = 0.0143$$

Answer

(a) The sample mean is 0.013 in.; (b) the median is 0.0065 in.; (c) the mode is 0.0000 in.; (d) the sample variance is 0.000205 in.; and (e) the standard deviation is 0.0143 in. for the average daily rainfall per month for a location in the Mojave Desert in the U.S.

1.5.3 EXPERIMENTAL ERROR

Whenever samples are collected or measurements are taken, there is some experimental error present. *Experimental error* is the fluctuation or discrepancy in replicate observations from one experiment to another. In this context, error does not necessarily mean a mistake was made. The measuring equipment or the measurer can cause these errors even following the correct procedure. When describing the quality of data collected, different terms are used to relate which types of error can be associated with the data. Many of these terms are used interchangeably, but it is important to understand the differences and

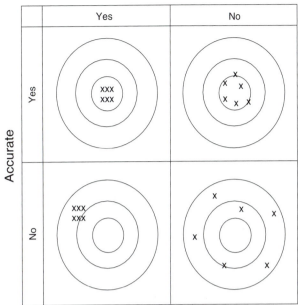

FIGURE 1.11 Illustration of the difference between accuracy and precision.

relationship among accuracy, precision, and bias. A measurement can be accurate and precise, precise but not accurate, or accurate but not precise. The differences between accuracy and precision can be seen in **Figure 1.11**. *Precision* is the scatter between repeated measurements. The smaller the difference between repeated measurements, the more precise the measurements. *Accuracy*, however, is the proximity of the measured value to the actual value. Accuracy is very dependent on the measurer and the measuring device. Taking measurements with an uncalibrated or incorrectly calibrated piece of equipment can cause *bias*, also known as systematic error. Simply repairing or calibrating the equipment can correct this bias in the measurements.

1.5.4 STANDARD ERROR

Standard error is the estimate of the variation of a statistic. The estimate of the *standard error of the mean* (SEM) is described by the following equation:

$$s_y = \frac{s}{\sqrt{n}} \tag{1.4}$$

where s is the sample standard deviation, and n is the number of samples.

Another measure of statistical variation is the *coefficient of variation* (CV), also known as the relative standard deviation. The coefficient of variation is the standard deviation expressed as a percentage mean.

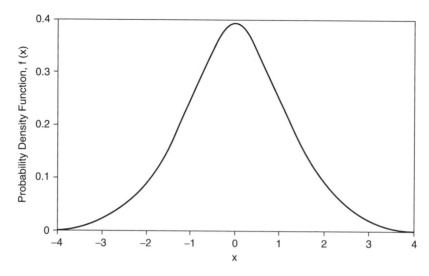

FIGURE 1.12 A normal probability distribution that is called a density function (PDF).

$$CV = \frac{s}{\bar{y}} \cdot 100 \qquad (1.5)$$

1.5.5 CONFIDENCE INTERVALS

The sample mean $\bar{y}$ is an estimate of the true mean of some parameter of a population μ. A way to include the true value of the mean by relating it to an expression of the sample mean is by expressing the estimated value $\bar{y}$ as a range (e.g., $\bar{y} \pm 10$). Therefore, the *confidence interval* (*CI*) of the true population mean estimated by the value of the sample mean is represented by the equation

$$CI = \bar{y} \pm t_{n-1}\left(\frac{s}{\sqrt{n}}\right) \qquad (1.6)$$

where t_{n-1} is the chosen t for $n - 1$ degrees of freedom.

When calculating the *CI*, the t values are chosen for the desired level of significance. The following steps are used to calculate the *CI*:

1. Calculate the sample mean $\bar{y}$ using Equation 1.1.
2. Select the desire desired t value from a table for the degree of freedom ($n - 1$) and desired level of significance.
3. Calculate the sample standard deviation s using Equation 1.3.
4. Obtain the sample size n.
5. Calculate the *CI* using Equation 1.4.

Note that the more confidence we wish to have in our final answer, the larger the interval between values.

1.5.6 PROBABILITIES AND DISTRIBUTIONS

Probabilities and distributions are ways to view data collected in graphic form and to predict the likelihood of a measured value occurring in a population. A *probability distribution* is a function with arguments that are values of a random variable and with functional values that are probabilities. In defining a probability distribution (**Figure 1.12**), the following are always true:

1. Each value of $f(x)$ must lie between 0 and 1, inclusive.
2. The sum of all values of $f(x)$ must equal 1.

A *probability histogram* (**Figure 1.13**) is a pictorial representation of a probability distribution for which the area of each bar represents the probability. Histograms are also often plotted to show the frequency of occurrence, as illustrated in Example 1.3.

EXAMPLE 1.3

In **Figure 1.13**, a histogram has been plotted of annual peak discharges for 50 years of flow data for a stream. Discharge intervals of 1000 ft³/sec were used to group the data. What information does the histogram provide? The data set for this example is presented in Problem 1.7 at the end of the chapter.

Answer

The histogram has a bell shape that is similar to **Figure 1.12**. Therefore, the data are normally distributed (see the next section). The most frequent annual peak discharges are in the range of 6000 to 7000 ft³/sec. More than 60% of the peak discharges are in the range 4000 to 9000 ft³/sec.

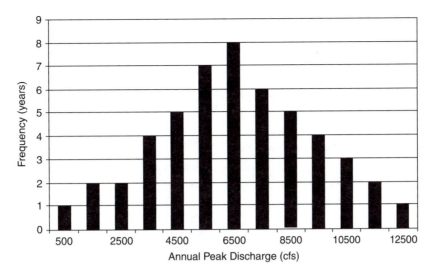

FIGURE 1.13 Histogram of annual peak discharges for Example 1.3. Discharges are grouped by 1000-ft³/sec increments.

1.5.7 Normal Distribution

Probably the most well known and most often referenced type of distribution is the *normal distribution* (**Figure 1.12**), which is characterized completely by its mean and variance. Notation for defining a normal curve appears as $N(\eta, \sigma^2)$ and is read as "a normal distribution with mean η and variance σ^2." The standard deviation σ is equal to the distance from the mean to the point of inflection on the normal distribution curve, and as with a probability distribution, the area under the curve is equal to 1.0. Additional definitions of a normal distribution/curve include:

1. Data are as likely to fall above or below the mean.
2. 68% of data are found within 1 standard deviation.
3. 95% of data are found within 2 standard deviations.
4. 99.7% of data are found within 3 standard deviations.

1.5.8 *t* Distribution

The *t distribution* is also known as the *student t distribution* and is a way of creating a distribution for samples collected from a population when actual population values are unknown. Standardizing a normal random variable requires that the mean η and the standard deviation σ are known. When these values are known, a normal distribution can be used; the data are scaled in terms of standard deviation as defined above and are represented by the following equation:

$$z = \frac{y - \eta}{\sigma} \qquad (1.7)$$

where z is the standardized normal random variable, y is a data value from the sample, η is the mean of the population, and σ is the standard deviation of the population.

However, in practice, σ is unknown, and the sample standard deviation s is substituted for σ, resulting in an equation to calculate the *t statistic*. Using this method, t has a known distribution under the following conditions:

1. y is normally distributed about η with variance σ^2.
2. s is distributed independently of the mean (the variance of sample does not increase or decrease as the mean increases or decreases).
3. The quantity s^2, which has v degrees of freedom, is calculated from normally and independently distributed observations having variance σ^2.

1.5.9 Properties of Statistical Procedures

The following terms are defined to aid in the understanding of the requirements for different statistical procedures. *Normality* describes the fact that the error term in a measurement e_i is assumed to come from a normal probability distribution, although many statistical procedures tend to yield correct conclusions even when applied to data that are not normally distributed. Another important term to understand when it comes to sampling procedure is *random*. Random sampling means that when samples are drawn from a population, they are taken in a manner that ensures that every member or element of a population has an equal chance to be drawn. In some cases, this is lacking in environmental data for a few reasons. One is that it is often inconvenient or impossible to randomize the data, and another is that there are times and studies when it is actually undesirable to randomize the sampling because the dynamic behavior of the population needs to be studied. For example, if runoff from storms is studied, it does not make sense to sample randomly (e.g., every other day); it makes sense to sample when there is a storm event and perhaps more intensively during different parts of a storm. The property of *independence* states that simple multiplicative laws of probability are true. This means that the product of the probabilities of each individual occurrence gives the probability of the joint occurrence of two events.

1.5.10 INFERENTIAL STATISTICS AND HYPOTHESIS TESTING

Inferential statistics is the process of drawing conclusions about an entire population based only on the results obtained from a small sample of that population. *Statistical inference* is defined as making an assessment about an unknown population parameter from experimental data. There are two basic methods of applying inferential statistics: (1) significance tests and (2) examination of confidence intervals.

A traditional decision-making approach of inferential statistics is hypothesis testing. This process provides an objective framework for making decisions and provides uniform criteria for decision making that are consistent for all investigators, present and future. The first step in the process is to develop a *statistical hypothesis*. A statistical hypothesis is a statement that something is true, for example, "an increase in urbanization in a watershed results in an increase in runoff." In general, the hypothesis testing process is made up of five components: (1) null hypothesis H_o, (2) alternative (research) hypothesis H_A, (3) test statistic, (4) criterion for rejecting hypotheses, and (5) conclusion. This overview is based on the concept of "proof by contradiction" (Ott, 1984).

The *null hypothesis* is the hypothesis to be tested and is generally a statement that a population parameter has a specified value or that parameters from two or more populations are similar, as in

$$H_o : \mu_1 = \mu_2 \qquad (1.8)$$

For this example, the alternative hypothesis would be

$$H_A : \mu_1 \neq \mu_2 \qquad (1.9)$$

Another way to state this alternative hypothesis is as a two-sided hypothesis, in which case the alternative hypothesis is true if

$$\mu_1 > \mu_2 \text{ or } \mu_1 < \mu_2 \qquad (1.10)$$

If, for example, the null hypothesis was

$$H_o : \mu_1 \leq \mu_2 \qquad (1.11)$$

then the alternative hypothesis is represented by a *one-sided hypothesis* and will be defined as

$$H_A : \mu_1 > \mu_2 \qquad (1.12)$$

1.5.11 PROBABILITY VALUES

Probability values, often called *p values,* refer to the odds that we are wrong when we state that the null hypothesis is correct. The lower the *p* value, the more likely that the null hypothesis is incorrect. The *threshold value* or *level of significance* is the risk of falsely rejecting the null hypothesis. Generally, the threshold value is set equal to 0.05 or 0.01, and the null hypothesis is rejected when the *p* value is equal to or less than the threshold value. For example, if the threshold value is equal to 0.05 and the *p* value is 0.05 or less, we can reject the null hypothesis with a 1 in 20 chance to be wrong.

1.5.12 STATISTICAL ERRORS

There are two types of statistical errors associated with hypothesis testing, Type I and Type II. First, we have to define the results. If, through statistical testing, we determine that H_o is true, we can more correctly state that we "failed to reject" H_o. If we instead reject H_o, as a result we then must accept that H_A is true. Following the statistical analysis of our data and our hypothesis, there are four potential statistical outcomes (**Table 1.11**).

The probability of a Type I error occurring is equal to α, which is equal to the significance level of a test. The probability of a Type II error occurring is represented by β. The ideal situation is that we minimize both α and β; however, Type I and Type II errors are inversely related, so as α increases, β decreases and vice versa. The way we control for these errors is instead by assigning them small probabilities (e.g., 0.01, 0.05).

1.5.13 LINEAR REGRESSION

Applying a *linear regression* to a data set is a very common way to determine the presence or absence of a correlation between the data points. Linear regression applies

---

TABLE 1.11
Hypothesis Testing and Statistical Errors

Outcome	Statistical Error
H_o is accepted and is true	None
H_o is accepted, but H_A is true	Type II error
H_o is rejected, but H_o is true	Type I error
H_o is rejected, and H_A is true	None

to *bivariate data*, for which two variables are related systematically such that one is a fairly constant multiple of the other. Linear regression, also known as the least-squares best-fit line, can be calculated for bivariate data that have a linear appearance when plotted in scatter plots. The best-fit line through these data is defined by the *straight-line equation*:

$$y = a + bx \qquad (1.13)$$

This equation represents the relationship between an independent variable x and dependent variable y. Given a value for the independent variable, use of this equation allows prediction of the dependent variable. The statistical parameters of this equation are the intercept of the line a and the slope of the line b. To calculate the slope of the line, the following equation is used:

$$b = \frac{n(\Sigma xy) - (\Sigma x)(\Sigma y)}{n(\Sigma x^2) - (\Sigma x)^2} \qquad (1.14)$$

where n is the number of measurements.

Once the slope of the line is calculated, the intercept can be calculated using the equation

$$a = \bar{y} - b\bar{x} \qquad (1.15)$$

where $\bar{y}$ is the sample mean of the y values sampled, and $\bar{x}$ is the sample mean of the x values sampled.

The *correlation* between the two variables is defined as the intensity or level of association between the two variables. For two variables, the calculation is named Pearson's product or the moment correlation coefficient r. This method can be applied to both interval and ratio data. The correlation coefficient values range from -1.0 to $+1.0$, and the following statements hold true:

1. A value of r near $+1.0$ indicates that, as one variable increases, the other variable increases in a similar trend, and a strong correlation exists.
2. A value of r near -1.0 indicates that, as one variable increases, the other decreases, and a strong correlation between the two exists.
3. If r is near 0, there is little correlation between the two variables.

The correlation coefficient r can be calculated using the following equation:

$$r = \frac{n(\Sigma xy) - (\Sigma x)(\Sigma y)}{\sqrt{n(\Sigma x^2) - (\Sigma x)^2} \cdot \sqrt{n(\Sigma y^2) - (\Sigma y)^2}} \qquad (1.16)$$

Plotting the x and y values in a graphing program like Microsoft Excel can yield both the straight-line equation and the r value for the data set. The coefficient of determination r^2 (commonly reported as R^2) is the fraction or percentage of the total variation in the data that is accounted for by the regression equation. For example, a regression equation with an R^2 of 0.8 accounts for 0.8 or 80% of the variability in the data. The 20% that is not accounted for could be due to measurement error or factors not included in the equation. Example 1.4 relates to this issue. In the section that follows Example 1.4, how additional factors can be added to a regression equation is shown. When regression equations are used as predictive tools, we often refer to them as *regression* or *empirical* models.

EXAMPLE 1.4

A study is conducted to determine if the size of the trout in a mountain stream is related to the depth of the pools. The mean trout size and pool depth for each pool are summarized in **Table 1.12**. Is there a linear relationship between mean fish size and mean pool depth? Is this equation adequate for use in predicting the size of trout as a function of pool depth?

Solution

No information is reported on the standard deviation or other statistics, such as the number of fish, maximum and minimum sizes in each pool, and the volume or length and maximum and minimum depths in each pool. Therefore, we treat each mean trout size and pool depth as a discrete pair of values and plot them on a graph as shown in **Figure 1.14**.

The data were plotted using a spreadsheet (Microsoft Excel) and a linear trend line was fitted to the data. This is the same as performing a linear regression analysis. The $R^2 = 0.946$ which means virtually all the variability in the data is accounted for by the regression equation.

The adequacy of the equation is a difficult question to answer. Certainly, the high R^2 suggests that this might be a useful equation. It is helpful to obtain additional statistics. In Excel, this can be done using the Analysis Pak Add In and then using the regression analysis option and requesting output on the residuals and a plot of the residuals. The residuals are shown in **Table 1.12** and **Figure 1.15**. The residuals could have been calculated by inserting the measured mean pool depths into the regression equation and then subtracting the calculated trout sizes from the measured sizes. The results plotted in **Figure 1.15** show that, across the whole range of pool depths, there is no bias in the results. However, we can also see that there is about a 6 to 22% difference in the

TABLE 1.12
Mean Trout Size vs. Mean Pool Depth Data for Example 1.4

Mean Pool Depth (meters)	Mean Trout Size (g)	Calculated Size (g)	Residual (measured minus estimated grams)
0.4	121	107.0	14.0
0.5	110	134.4	−24.4
0.6	152	161.8	−9.8
0.7	213	189.1	23.9
0.9	219	243.8	−24.8
1	302	271.2	30.8
1.1	281	298.6	−17.6
1.2	347	325.9	21.1
1.4	367	380.6	−13.6

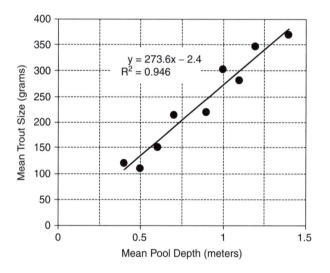

FIGURE 1.14 Regression equation and plot of mean trout size vs. mean pool depth.

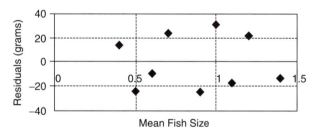

FIGURE 1.15 Plot of the residual difference between regression estimates and measured mean trout sizes.

calculated and measured trout sizes. If errors in the estimate as large as 22% are not a concern, then the equation is adequate. Another question might be whether the equation is logical (rational). When the pool size is zero, the equation predicts there will be trout with a negative weight. Clearly, this is not logical. However, the extra statistics we can obtain

using the Analysis Pak or using more sophisticated statistical software will show us that the intercept term of −2.4 grams is insignificant. By inspection of the values in **Table 1.12**, we see that by ignoring this term the largest error we might produce is just over 2%.

Answer

There is a linear relationship between the mean trout size, and the coefficient of determination R^2 is 0.946.

Care needs to be taken in using regression equations to make scientific and engineering decisions. For example, the results in this example suggest that if we construct some really deep pools, we might have a haven for large trout. Certainly, this is a possibility, but this type of approach is extending the equation beyond the range for which it is valid and ignores certain factors that will control the size, presence, and absence of trout. The equation suggests that a pool 2 m deep will have trout with a mean size of about 545 g — almost 50% bigger than those measured in the deepest pool of 1.4 m. However, the possibility exists that we would find no trout in this deep pool because the oxygen, temperature, light, bed material, or food sources might not be suitable for this trout species in pools that deep. Another possibility is that there is a much slower increase in size as the depth increases, and the relationship becomes nonlinear. Inspection of the measured results for pools with mean depths of 1.2 m and 1.4 m show only a small increase in the size of the trout.

1.5.14 THE USE AND MISUSE OF REGRESSION EQUATIONS

In the previous section, we began to look at how we might use a regression equation to help make scientific and engineering decisions. This is a very common practice, but we have already seen that it is easy to misuse statistical models. If we consider Example 1.4, we notice that from the outset that we started to "smooth" the data. Each pair of values is actually a set of measurements. We do not know how the mean values were obtained or how many measurements were taken to acquire the mean values. Probably a seining or electroshocking method was used to collect fish from each pool. The fish were then sorted into species, and all or maybe only a sample of the fish in each species were weighed. Some measure was then made of the number of fish in each pool. This could have been a small number in some pools and a very large number in others. Clearly, many more statistics could be obtained if we had the "raw" field data. Similar issues could be raised regarding the mean pool depths. Were the number of measurements in each pool a function of the pool's size, or was one or more transects (lines of measurements) made from one bank to the other? In fact, we do not even know if the mean value is an arithmetic or geometric mean. If we smooth the data enough, we can eliminate all the variability, but end up with a result that is not very useful.

Perhaps the most common mistake is to use a regression model to predict a result and then to compare this result with some independent measurements. If we get a "good" fit, we assume the original model is valid and we can use it to make a decision. What we fail to realize is that we have actually now created a second model that relates the observed and predicted values to each other. This is probably easier to understand if we consider an example.

EXAMPLE 1.5

We plan to use the regression equation obtained in Example 1.4 to help size some pools on a mountain stream that has been recently impacted by mining operations. The stream is located in the same region, but has different land uses (because of the mining) and slightly different topography and bed slopes. We are unsure if it is valid to use the equation, so we test its usefulness by making some measurements of trout sizes in a nearby stream that has recovered from some earlier mining and logging operations. We then plan to compare statistically the measured trout sizes with the sizes predicted by the regression equation. Can we use the regression equation to size pools in the impacted stream?

Solution

First, we need to make it clear that there are three streams. One stream from Example 1.4 that was used to develop the regression equation, a second stream that will be used to "test" or "calibrate" this equation, and a third stream for which we want to use our calibrated equation to size pools that will be constructed. The measured and predicted data are presented in **Table 1.12** and **Figure 1.17**. Measurements were made in six pools to evaluate the regression equation (**Figure 1.16**). When we fit a trend line through the measured and calibrated data, we see that the R^2 equals 0.952, and we now have an equation with a slope of 0.52 and an intercept term of 57.8. This is a new model that says that we must multiply the values that were calculated by the original regression equation by 0.52 and then add 57.8 g. If the original regression equation had given perfect results for the second stream, the slope would have been 1, the intercept term would have been 0, and the R^2 would be 1. Although, we have a higher R^2 value than the original equation, the results actually show that the original equation worked poorly. One way to evaluate the usefulness of the original equation statistically is to see if the intercept term, in the calculated vs. measured equations, is significantly different from 0 and to see if the slope term is significantly different from 1. If either of these results is significant, it means that statistically the original equation cannot be used "as is." We can see that a new equation is needed by looking at the results. Both the slope and intercept terms obtained by developing a new equation are very different from the original equation. For deep pools, we have much smaller fish than the original equation predicted; for shallow pools, we have larger fish than the original equation estimated. We now have two regression models that relate fish size to pool depth. Unfortunately, we do not know whether either is suitable for use in designing pools for the stream that has been recently impacted by mining. What we have done is evaluate the performance of the first equation, but we have not calibrated it so that it can be applied to the stream of interest.

Answer

It is not recommended that the original equation or the new equation be used without further testing.

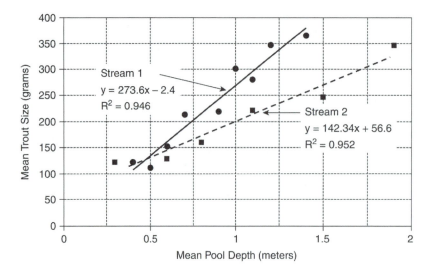

FIGURE 1.16 Measured-and-calculated trout sizes vs. pool depth.

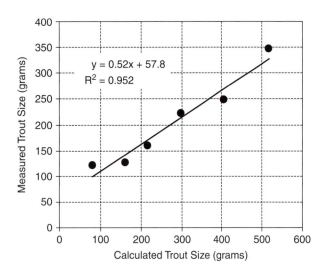

FIGURE 1.17 Regression analysis of measured vs. calculated trout sizes.

In Example 1.5, a regression analysis was performed to evaluate the performance of a regression model. Often, engineers and scientists mistakenly think that if this test yields a high R^2, the predictive model is performing well. There are better tests that can be used, and reference should be made to statistics books written to solve hydrology, environmental, and ecological problems, such as the work of I. Miller and Miller (1995), Mendenhall and Sincich (1995), Haan (1997), or Berthouex and Brown (2002).

1.5.15 DATA TRANSFORMATIONS

Many relationships in hydrology and other branches of science are nonlinear. However, it is often practical and valid to transform these data mathematically. There is almost an unlimited number of mathematical transformations that might be considered. One common method is to make power and exponential relationships linear by performing a logarithmic transformation. For example, consider the equation

$$y = ax^b \tag{1.17}$$

If we take logarithms (to the base 10 in this case) of both sides of the equation, we obtain:

$$\log_{10} y = \log_{10} a + b \log_{10} x \tag{1.18}$$

we now have a linear equation. When we use a spreadsheet program such as Microsoft Excel, we sometimes transform the x or y axis to produce a semilog or log–log plot. When we fit a linear trend line through data plotted with a log–log scale, this will provide the same result as transforming the data first and then using a normal x and y scale that is not transformed. Example 1.6 in the next section evaluates how transforming data can help us perform a regression analysis.

1.5.16 MULTIPLE REGRESSION ANALYSIS

Multiple regression analysis is similar to simple linear regression models except multiple regressions contain more terms and can be used to represent relationships that are more complex. An example is the quadratic model, also known as a second-order model:

$$y = ax^2 + bx + c \tag{1.19}$$

Often, in watershed, hydrology, and ecological studies the data are related by power functions, such as

TABLE 1.13
Data for Example 1.6

Measurement	Velocity (ft/sec)	Bed slope (ft/ft)	Depth (ft)
1	0.4	0.002	0.2
2	0.9	0.003	0.4
3	1.6	0.005	0.6
4	1.8	0.008	0.5
5	1.9	0.01	0.5
6	1.2	0.007	0.3
7	1.4	0.015	0.2
8	2.6	0.009	0.8
9	2	0.007	0.7
10	2.1	0.025	0.3

$$y = ax_1^b x_2^c x_3^d \ldots\ldots \qquad (1.20)$$

A logarithmic transformation of Equation 1.20 results in the linear equation:

$$\log_{10} y = \log_{10} a + b \log_{10} x_1$$
$$+ c \log_{10} x_2 + d \log_{10} x_3 + \ldots \qquad (1.21)$$

An application of this type of equation is illustrated in Example 1.6.

EXAMPLE 1.6

The mean velocity of the flow is measured in several streams that have sand and gravel beds. The bed slope and depth of flow are also measured. A multiple regression analysis is performed to see if the mean velocity is related to the depth of flow and the bed slope. The data are presented in **Table 1.13**.

Solution

A review of Chapter 7 indicates that the data might be related by an equation similar to Equation 1.20, so we make a logarithmic transformation of the data. The Analysis Pak in Microsoft Excel was issued to conduct a multiple regression analysis. The results of the statistical analysis are reported in **Table 1.14**. Most computer software, including Excel, will report the results to more decimal places than reported in **Table 1.14**. Reference should be made to a statistics book to understand all the information in the table. The information presented is called an analysis of variance (ANOVA). The coefficient

of determination R^2 is 0.987. The F, t stat, and p value are all based on statistical tests to see if the regression equation and various coefficients in the equation are significant. The very low significance value for F of 1.05×10^{-07} means that the F value of 341.8 is highly significant. The very low p values for each of the variables in the regression equation also mean that each variable is very significant. The logarithmic regression equation is

$$\log_{10} y = 1.6 + 0.70 \log_{10} x_1 + 0.55 \log_{10} x_2$$

where y is the velocity v in ft/sec, x_1 is the depth d in ft, and x_2 is the bed slope s as a fraction. If we take the antilog of this equation, we obtain the following equation:

$$v = 40 d^{0.7} s^{0.55}$$

If we refer to Chapter 7, we can see that this equation is similar to Manning's equation, and the high R^2 indicates that all of the streams have a similar bed roughness that might correspond to a Manning's n value of 0.025.

Answer

The multiple regression equation that relates velocity to depth and bed slope is $v = 40 d^{0.7} s^{0.55}$, and the coefficient of determination for this equation is 0.987.

1.6 MODELING THE HYDROLOGIC CYCLE

Traditional methods for estimating hydrologic responses using tables, charts, and graphs are currently being replaced with computer models in research, in the classroom, and in the field. Since the mid-1960s, engineers and scientists have been developing hydrologic applications for computers. Government and university scientists with access to mainframe computers began to develop models to describe climate, infiltration, runoff, erosion, soil water movement, drainage, groundwater flow, and water quality. With the advent of the personal computer in the mid-1980s, the use of models spread from laboratories to agencies concerned with hydrologic systems. As portable computers develop, scientists and technicians in regulatory agencies and consulting companies will be taking models directly to the field to assist managers of farms, forests, mines, land development projects, and other hydrologic units in analyzing hydrologic problems. Computer applications feature strongly in current research and future development in all areas of hydrologic science. An understanding of the importance and applications of computer modeling of hydrologic processes is important for all interested in hydrology.

TABLE 1.14
Statistical Analysis Results for Example 1.6

Summary Output

Regression Statistics

Multiple R	0.995
R^2	0.990
Adjusted R^2	0.987
Standard error	0.027
Observations	10

ANOVA

	df	SS	MS	F	Significant F
Regression	2	0.491	0.246	341.8	1.05×10^{-7}
Residual	7	0.005	0.0007		
Total	9	0.497			

Variable	Coefficients	Standard Error	t Stat	p value
Intercept	1.60	0.063	25.61	3.54×10^{-8}
X variable 1	0.70	0.042	16.84	6.38×10^{-7}
X variable 2	0.55	0.028	19.39	2.42×10^{-7}

Residual Output

Observation	Predicted Y	Residuals
1	–0.366	–0.032
2	–0.058	0.012
3	0.187	0.0167
4	0.243	0.012
5	0.296	–0.018
6	0.055	0.024
7	0.112	0.034
8	0.415	-2.13×10^{-5}
9	0.314	–0.013
10	0.358	–0.035

1.6.1 TYPES OF MODELS

Models may be empirical, deterministic, or stochastic. Empirically based models are developed by analyzing a large set of data and developing statistical relationships between the inputs and the outputs (Woolhiser, 1982). An example of an empirical model is the universal soil loss equation (USLE; Wischmeier and Smith, 1978; see Chapter 9) and its computer counterpart, the revised universal soil loss equation (RUSLE; Renard et al., 1991). Empirical models are not easily transferable between geographic regions.

Deterministic models, sometimes described as theoretical (Woolhiser, 1982), deterministic, or process-based models, mathematically describe the processes modeled, such as runoff. As the processes are independent of geographic variations, deterministic models can be applied to a wider range of conditions than empirical models can. In some instances, however, it may not be possible to describe a process adequately. An example is the gain or loss of dissolved chemicals as water percolates through a soil. In some cases, the amount of data required to describe a process may restrict the use of the model. Excessive input data may be necessary to provide all the information necessary to describe a process fully. This restriction was particularly apparent in earlier modeling attempts when computer memory was limited.

Stochastic models seek to identify statistical probabilities of hydrologic events (Woolhiser, 1982), like rainfall or flood flows, and to predict the probability of a given outcome. They also consider the natural variability that might occur in some model input parameters. As users become more acquainted with the statistical nature of hydrology, stochastic modeling will increase. Recent developments in fractal theory and spatial distribution statistics will lead to further developments in stochastic modeling.

1.6.2 ADVANTAGES AND DISADVANTAGES

Computers allow managers to consider many more options than would be possible with hand calculations. For example, a recreational park planner can compare the effects of subsurface drainage depth and spacing on the number of days that a playing field may be available for use. Many combinations of depth and spacing can be considered, and the benefits of improved drainage can be compared to the cost of each system. Computer models are particularly helpful in determining which erosion management practices may best suit a given farm.

Research scientists can identify critical areas for further research and study and the sensitivity of hydrologic systems to the various parameters needed for a given model. Researchers are aware of which parameters they require for a model to operate, but the importance of each of the parameters to the model predictions is not known. If a model shows that the rate of plant canopy growth is not very important in the operation of a computer model, then research resources can be directed to other areas in which a model may be highly sensitive to changes in inputs. The development of computer models requires a sound understanding of the system modeled and the dominant processes in the system. The need to understand such processes is increasingly dictating the direction of much research (e.g., Laflen et al., 1991).

Scientists and engineers can better understand key parameters in a hydrologic system by using a computer model of that system. By developing the necessary input files to run a model, users gain a much greater appreciation of the importance of considering the entire system and not simply concentrating on one or two aspects of that system.

Computer models can lead to wrong conclusions if not properly applied. Numerous models have not been developed to provide absolute answers, but rather to give relative results to allow the user to compare the effect of different management systems on some hydrologic response. The water quality model GLEAMS (Groundwater Loading Effects of Management Systems) is such a model (Leonard et al., 1987), although numerous users have found that the absolute predictions by the model have been good. A user may also input unrealistic combinations of values into a model, which can lead to misleading results. Altering the clay content in an input file without altering a property closely related to clay, like the cation exchange capacity, can lead to misleading results with an erosion model.

Considerable time may be required to build the input files necessary to run a computer model. Sometimes, the necessary data are not available for the conditions under which the model is applied. Many models have been developed for specific geographic areas and, when applied outside that area, have not been successful. In other cases, users may not understand fully the importance of some of the inputs, and poor estimates of some parameters may lead to poor model performance.

1.6.3 TYPICAL MODEL ARCHITECTURE

Most hydrologic computer models consist of three main components: (1) the input file; (2) the output file; and (3) the computer model (Woolhiser, 1982). The user will either develop the input file with a word processor or text editor, or the model will include a file builder to assist in assembling the input file. More sophisticated models have developed user-friendly interfaces to assist with building the input files and interpreting the output files.

In some cases, the output file from one part of the model may serve as the input file for another. This practice is more common with complex models like CREAMS (Chemicals, Runoff, and Erosion from Agricultural Management Systems; Knisel, 1980). Users may wish to access these intermediate files to get a better grasp on the hydrologic system or modify the files to study the performance of a particular component of the model.

Typical input files for most hydrologic models include climate, soil, topography or structure, and management files. With some models, some of the files may be combined. Some models may require additional files describing chemical properties for water quality.

The climate file may describe a single storm or runoff hydrograph. Some models have the option of daily or hourly rainfall distributions. Models with crop components will require information to estimate evapotranspiration or plant growth rate, like daily temperatures, wind speeds, or solar radiation.

The soil files generally include the necessary information to describe infiltration and water movement within the soil profile. Specific infiltration and permeability values or properties that are more easily measured and will predict those values may be required. In programs for which water-holding characteristics are important, like drainage and irrigation programs, a method of describing water content is required.

The topographic or structure file is necessary to describe slope lengths and steepnesses, spacing of drains, or the relationship between model elements. In watershed models, the structure file can become quite complex, and users need to be extremely methodical in their approach. A structure file for the WEPP (Water Erosion Prediction Project) water erosion model (Nearing et al., 1989) describing a 40 ha- (100 acre-) watershed was found to require 35 elements to describe fully the combination of soils, crops, and drainageways.

The management files contain the necessary information to describe surface roughness as it affects runoff and water surface storage. It may also contain information describing plant growth rates and residue conditions as affected by tillage and decomposition. Plant water use is frequently required for models that predict soil water conditions for drainage or erosion prediction.

TABLE 1.15
Annual Peak Discharge Data for Problem 1.7

Annual Peak Discharges (ft³/sec)				
990	4,610	5,740	7,170	9,010
1,310	4,710	6,210	7,380	9,180
1,720	4,780	6,360	7,460	9,500
2,120	4,780	6,540	7,760	9,790
2,580	5,050	6,660	7,930	10,300
3,160	5,150	6,700	8,250	10,700
3,180	5,380	6,700	8,350	10,810
3,270	5,500	6,930	8,670	11,350
3,760	5,530	6,930	8,730	11,900
4,550	5,580	7,020	8,860	12,900

TABLE 1.16
Channel and Drainage Area Data for Problem 1.8

Drainage Area (mi²)	Width (ft)	Depth (ft)
0.5	4	1
2	7	1.5
5	10	2
12	16	3.5
28	32	5
40	40	5.5
90	65	7.5
110	68	8.5
200	85	9
280	95	10

Generally, the better the quality of the data in the input files, the better the model will perform. Most model users soon develop sets of input files to describe the range of conditions they are modeling. From these typical data sets, only minor file modifications are then necessary to describe each new application.

PROBLEMS

1.1. Select an environmental system of interest to you (such as a forest, crop production system, urban area, surface mine, etc.) and then list the components of the hydrologic cycle most likely to cause water contamination problems for this system. Briefly discuss how the land use you have selected will influence the components of the hydrologic cycle you listed. As you learn more about hydrology in the next few chapters, you might wish to refer to your answer to this question.

1.2. In the future, it is anticipated that, in the U.S., the amount of water used for irrigation and livestock will be four times higher than the levels reported in **Figure 1.8**. Also, the amount of water used for industry and mining will double. For this future use, determine the consumptive use, return flow, and amount of water not withdrawn from streams and groundwater systems. Assume other water uses do not change.

1.3. **Table 1.1** provides information on the world water balance by continent. Determine the evaporation rates in North America and Africa as percentages of the precipitation in each of these continents. Why do you think the rate is higher in Africa?

1.4. The chapter contains a quotation from Hillel (1992) on water use changes. Convert the quantities expressed in this quotation in cubic meters to cubic feet and to gallons. (Note that 1 m³ is

35.27 ft³, and there are 7.48 gal in a cubic foot; see Appendix A.)

1.5. Based on the information in **Table 1.6**, determine the urban water use by a family of four if a dishwasher is used twice daily, toilets use 50% less water, one car is washed monthly, and water use for lawns and swimming pools is only 30% of reported levels. What percentage reduction in the reported total daily water use has been achieved?

1.6. Select from **Table 1.7** one industry that is of interest to you, then visit a library and determine the annual number of units of production for this industry in the U.S. Determine the total annual water use and consumptive use for this industry. Convert this volume into cubic feet and then, acre-feet. Suggest how the needed consumed amounts of water might be reduced. (An acre is 43,560 ft². Acre-feet are often used to express large volumes of water in the environment. For example, 5 acre-ft can be visualized as a depth of 5 ft of water covering an area of 1 acre. The volume contained in 5 acre-ft would be 5 times 43,560 ft² or 217,800 ft³.)

1.7. Calculate the mean, median, and standard deviation of the annual peak discharge data presented in **Table 1.15**. Group the data by dividing it into 500-ft³/sec increments. What is the magnitude of the discharge range that occurs most often, and how often does it occur?

1.8. A study is conducted on a watershed located in the Midwest region of the U.S. At different locations in the watershed, measurements are made of the width and mean depth of the main channel. A GPS (global positioning system) is used to obtain the coordinates of each location. The locations are marked on a topographic map, and the area of the watershed draining to each location is then calculated. The data are reported in **Table 1.16**. Develop a relationship between the dimensions (width, depth, and cross-sectional area of the main channel) and the drainage area.

1.9. Calculate the mean, median, mode, and standard deviation of the mean monthly temperature at a location in the Mojave Desert or a desert near your location.

1.10. Conduct a literature review and identify common computer models used to determine (a) soil erosion; (b) surface runoff from single storm events; (c) daily information on runoff, evaporation, and infiltration; (d) steady-state groundwater flow; (e) the hydrology of watersheds; and (f) water quality information for either agricultural or surface mining land uses.

2 Precipitation

2.1 INTRODUCTION

Precipitation is any form of solid or liquid water that falls from the atmosphere to the earth's surface. Rain, hail, sleet, and snow are examples. Precipitation is the single strongest variable driving hydrologic processes and events, but it must be considered in the context of other variables. For example, areas with more than 20 in. of annual precipitation are often considered humid, while those areas with less are considered arid. However, 20 in. of precipitation in South Dakota is usually adequate to grow wheat, while in Texas it would be quite inadequate. The difference lies in differences of evapotranspiration losses between the two regions (see Chapter 3). Likewise, a 50-year return period storm on dry soil might produce a 10-year runoff event because much of the water would be absorbed, while a 10-year storm falling on an already-saturated soil might produce a 50-year runoff event.

Precipitation is formed from water vapor in the atmosphere. As air in the atmosphere cools, its capacity to hold water decreases. When the capacity of air to hold water is reached, it is said to be *saturated*. For example, at 70°F, a given volume of saturated air contains about 16 times the water vapor that it contains at 0°F. **Figure 2.1** shows the relationship between air temperature and the amount of water contained at saturation. Moist, unsaturated air becomes saturated as it cools. Cooling beyond saturation will cause some water vapor in the air to condense and change into a liquid or solid water (ice). The condensation process can be seen when droplets of water form on the outside of a glass containing a cold beverage because the moist air next to the glass has cooled below the temperature of saturation, or *dewpoint*.

If all the atmospheric water were precipitated at one time, it would result in an average depth of only 1 in. (2.5 cm) of water on the earth's surface, but there would be tremendous variance from that average. Some arid regions, or extremely cold regions, have little atmospheric moisture; other areas, especially the wet tropics, have several inches. Very intense storms over long periods (e.g., 24 h) require large, wet air masses. Even with only moderate atmospheric moisture, however, a 1-h storm can result in precipitation of several inches over land areas of a few square miles because of the lateral flow of moist air from surrounding areas to the storm center. Some of these *frequency–magnitude* relationships and their geography are discussed in this chapter.

A *meteorologist* is generally concerned with the day-to-day movement of moist air masses from their sources (lakes, oceans, transpiration from land areas) and their associated effects on precipitation and temperature. When longer time periods, such as years and decades, are considered, we speak of *climatology*. A *hydrologist* considers such phenomena in relation to water supplies and movement to, on, and beneath the earth's surface. Hydrologists usually become involved in climatology in the planning, study, and evaluation of the interaction among the climate, surface water, and subsurface water.

Knowledge of precipitation amounts, patterns, and frequency are needed at many different temporal and spatial scales. For example, the design of a stormwater system in an urban subdivision might only require knowledge of

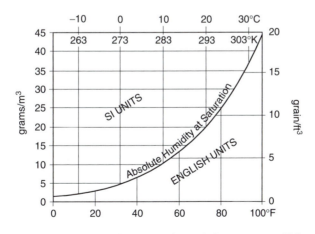

FIGURE 2.1 The relationship between water content of air at saturation and air temperature. Values are in absolute humidity (weight of water/volume of air).

TABLE 2.1
Typical Precipitation and Temperature Conditions for Koeppen Climatic Types[a]

Climatic Type	Latitude (°)	Precipitation	Temperature
A. Tropical moist	15 to 25	More than 1500 mm; either uniform year around or a wet/dry season	Above 18°C in all months
B. Dry	20 to 35	Evapotranspiration (ET) exceeds precipitation; often less than 1000 mm annually and less than 100 mm in deserts	Diurnal and annual temperatures can vary greatly in tropical deserts, with lows of 10 to 15°C and highs of 28 to 37°C
C. Moist subtropical midlatitude	30 to 50	Winter midlatitude cyclones; Summer convective storms; annual precipitation depends on latitude and proximity to coast — often <1500 mm	Mild winters; warm, humid summers
D. Moist continental midlatitude		Three types: wet all seasons, dry summers, or dry winters; precipitation often 200–1200 mm depending on latitude, elevation, and proximity to coast	Cold winters; cool-to-warm summers
E. Polar climates	Antarctica, Greenland, and northern coasts of Asia, Europe, and North America	Annual precipitation is less than 250 mm, primarily in the summer; polar desert climates occur at high altitudes in these regions and have year-round blizzard conditions with several hundred millimeters of annual precipitation	Warmest month below 10°C and coldest months colder than −35°C

[a] Climatic subtypes: M = rainforest; W = desert; S = steppe; a = hot summer; b = warm summer; c = short, cool summer; d = very cold winter; f = moist; h = dry-hot; k = dry-cold; s = dry summer; w = dry winter .

the area draining to the system and the characteristics of a single storm event that will be used to develop the design. In contrast, the selection of appropriate farming practices for a region will require knowledge of climatic information for a period that will range from several weeks to several months, depending on the crop and the probable annual variability in these data. Developing strategies to reduce hypoxic conditions in the Gulf of Mexico requires knowledge of many factors, including climatic conditions across the Mississippi River Basin during a period of several decades. Evaluating global warming requires knowledge of air currents, temperature changes, and water vapor in the air as it circulates around the world. To aid in understanding, describing, and using climatic data, a number of climatic classification systems have been developed. Geographers, meteorologists, and climatologists commonly use the Koeppen Climate Classification System developed by a German scientist in 1918 (A.H. Strahler and Strahler, 2002). This classification system has five major climatic types (A to E) that are a combination of latitude, mean annual temperature, and mean annual precipitation (**Table 2.1**). Since the classification system was developed, a highland climatic type (H) has been added. Subtypes have been developed that differentiate more finely and may show degree, extremes, or seasonality of temperature or moisture (**Table 2.1**). Thus, Csa is the dry–summer subtropical or Mediterranean climate, typical of southern California, while Daf is the humid continental, long-summer phase, located across the northern U.S. from

Iowa to New England (A.H. Strahler and Strahler, 2002, pp. 222–225).

In many branches of science, people, such as ecologists and botanists, are interested in the vegetation that might naturally occur at a location in the world. Vegetation can be related to seasonal and annual precipitation and temperature. Whittaker (1975) reported 21 unique terrestrial biological communities, called *biome types*, that are related to climatic zones and vegetation structure. A simplified interpretation of the typical climatic zones associated with Whittaker's biome types is summarized in **Table 2.2**. Arms (1990) presented the useful relationship between biomes and climate shown in **Figure 2.2**. Her book contains an informative presentation on the vegetation and biological communities that inhabit these biomes. The vegetation that will grow in a region will also be a function of seasonal differences in the climate, soils, topography, hydrogeology, land use practices, and the use of irrigation.

The Food and Agricultural Organization of the United Nations (FAO), in collaboration with the International Institute for Applied Systems Analysis (IIASA), has developed global agro-ecological zones (IIASA, 2002). Climatic parameters, topography, soil and landform, and land cover data have been combined to map the biophysical potentials and limitations for crop production. Around the world, many countries have mapped ecological zones and regions for use in the development of planning, management, policy, protection, and economic strategies.

TABLE 2.2
Biome Types for Different Climatic Zones[a]

Mean Annual Temperature (°C)	Mean Annual Precipitation (mm)	Biome Type
Arctic–Alpine		
−10 to −15	0 to 700	Arctic–alpine deserts
−12 to −8	0 to 300	Arctic–alpine semideserts
−10 to −6	300 to 600	Alpine grasslands and tundra (shrubland)
Cold Temperate		
−7 to −3	600 to 1200	Alpine shrublands
−5 to 3	700 to 1400	Taiga (forest)
−2 to 3	1600 to 2400	Elfin woods
Warm Temperate		
3 to 18	0 to 300	Cool semideserts
3 to 18	300 to 800	Temperate shrublands
3 to 18	700 to 1100	Temperate woodlands and grasslands
3 to 18	1100 to 2100	Temperate deciduous and evergreen forests
3 to 18	2000 to 3000	Temperate rain forests
Tropical		
18 to 29	0 to 400	Warm semidesert scrublands
18 to 29	400 to 800	Thornwoods
18 to 23	800 to 1300	Savanna
22 to 29	800 to 1300	Tropical broadleaf woodlands
18 to 29	1300 to 2400	Tropical seasonal forests
18 to 28	2400 to 4500	Tropical rain forests
28+	<300	True deserts

[a] Temperature and precipitation ranges are approximate.

Source: Modified from Whittaker, R.H., *Communities and Ecosystems*, 2nd ed., Macmillan, New York, 1975.

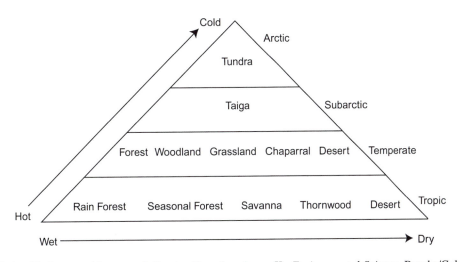

FIGURE 2.2 Relationship between biomes and climate. (Based on Arms, K., *Environmental Science*, Brooks/Cole, 1990.)

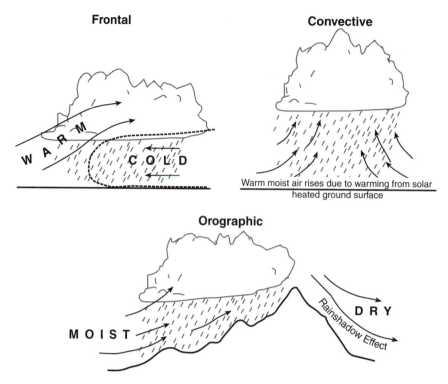

FIGURE 2.3 Main mechanisms causing air to rise and cool, resulting in frontal, convective, or orographic precipitation (precipitation mechanisms). For more complete and technical (but accessible) explanations, see A.H. Strahler and Strahler (2002).

2.1.1 PRECIPITATION DATA

Precipitation data are necessary for most land use plans and for hydrologic planning. The supply of water for human, agricultural, or industrial use; the disposal of wastewater; and the control of excess rainfall are key elements in most planning processes. In addition, recreational areas may include lakes or ponds that require precipitation data for both supply and overflow considerations.

In agriculture, precipitation data can indicate when and where a lack or a surplus of water for crops may be expected. Provision for irrigation or drainage systems may be necessary. Soil erosion from fields or urban development may be worse where large and intense rainstorms can be expected. In forestry, precipitation predictions assist planners in determining forest growth rates and downstream water yields.

State and federal governments have collected and published precipitation data for many years. These publications are available in some libraries. Data sets can also be purchased on electronic tapes or CD-ROM formats from government and private sources. Many are now available on the Web (see Exercise 14.1). These data aid researchers in studying characteristics and distributions of precipitation events. Precipitation records may report amounts of precipitation by years, months, days, and sometimes shorter time periods and indicate the form (i.e., rain or snow). Special reports are occasionally published to provide precipitation data on specific topics like flood events of major importance, rainfall rate–duration–frequency, or droughts.

2.2 CAUSES OF PRECIPITATION

There is always some water vapor in the air, and some condensation always occurs in the atmosphere, even on the fairest day. Clouds may be composed of water vapor, water droplets, ice crystals, or a combination. The precipitation process begins by the concentration of water molecules on condensation nuclei such as smoke, dust, or sea salt particles. As air cools, the amount of condensed water increases. Precipitation takes place when the air is cooled, cloud formation increases, and the condensed water droplets or ice crystals reach a size that causes them to fall toward the earth's surface. Some of these droplets may collect additional condensed water vapor as they fall. Other droplets may evaporate and return to the atmosphere.

In the atmosphere, the cooling of the air is mainly caused by the lifting of the air mass. At higher elevations, lower atmospheric pressures and temperatures cause cooling of the rising air mass. Air may be lifted by three processes, or *precipitation mechanisms* (**Figure 2.3**). Extensive information on climatic processes is provided by the University of Illinois Department of Atmospheric Sciences (2000).

2.2.1 FRONTAL PRECIPITATION

When a warm, moist air mass meets a colder, heavier air mass, the warm moist air will ride up over the heavier air. The zone where the two unlike air masses meet is commonly called a *front*, and the resulting precipitation is *frontal*. Frontal systems are typically described as warm or cold and are associated with low- or high-pressure systems. The warmer the air mass, the lower the pressure, and the lighter it will be. The terms warm and cold or low and high pressure are relative terms. Sometimes, the difference in temperature or pressure between the two air masses might be small.

Atmospheric pressure is the force per unit area exerted on a surface by the weight of the air above that surface. It is measured with an instrument called a *barometer* and is sometimes called the *barometric pressure*. Atmospheric pressure is expressed in inches of mercury (inHg or ″Hg), atmospheres (atm) or bars, and kilopascals (kPa) or millibars (mbar). At sea level, the average pressure is 1 atm, 29.92 inHg, 101.325 kPa, or 1013.25 mbar. This is also the same as a pressure of about 14.7 lb/in.^{2}. The density of mercury is 13.6 times that of water, so 29.92 inHg is equivalent to 406.6 in. (33.9 ft or 1034 cm) of water. Therefore, a pressure of 1 atm is the same as that exerted by a 1 ft^2 column of water that is 33.9 ft high. When pressure is referred to as a height of water, this is called a *pressure head*, a head of water, or just a head.

Frontal precipitation is usually the dominant type of precipitation in the northern U.S. and other midlatitude areas of the world. These fronts are associated with midlatitude cyclones (low pressure) that move from west to east. In the Northern Hemisphere, air circulates counterclockwise around low-pressure areas, so warm air is most often found to the south and east of low-pressure areas. Low- and high-pressure systems that move into a region are steered by upper-level winds. The layer of the atmosphere closest to the earth is the *troposphere,* and it varies in thickness from 11 mi (17.6 km) at the equator to 4 mi (6.4 km) at the poles. This layer produces our precipitation. The *stratosphere* lies above the troposphere and contains very little moisture, but is the location of some of the fast-moving, upper-level winds, called *jet streams*, that circulate around the world and are one of the steering mechanisms for low- and high-pressure systems.

2.2.2 CONVECTION

Air expands when heated by solar energy and becomes lighter than the air around it. The lighter air rises by convection, potentially causing *convective* precipitation. Convection increases and can even become violent with *atmospheric instability,* for which dewpoints and lapse rates (vertical temperature gradients) are high and the added heat energy from the latent heat of condensation

helps drive the convection. In humid climates, convective precipitation frequently occurs on hot midsummer days in the form of late-afternoon thunderstorms. Convective precipitation also results from the movement of air into a low-pressure atmospheric system. Air flowing to the center of the low-pressure system causes the air in the system to rise and cool until it reaches saturation. Severe thunderstorms that produce rain and hail may follow. Convective storms are common during the summer in the central U.S. and other continental climates with moist springs and summers.

Hurricanes are severe convectional storms that form in the wet tropics during late summer and fall and that produce very heavy rainfall. These often move onto the Gulf and Atlantic Coasts and can produce prodigious amounts of rainfall over large areas, even at places distantly inland.

2.2.3 OROGRAPHIC

Prevailing winds force moist air masses to rise over hills or mountains, thus allowing them to cool. Precipitation that results from this process is called *orographic* precipitation. This often is the dominant source of precipitation in the mountains in the western U.S and contributes to precipitation in most mountainous areas.

2.3 PRECIPITATION EVENTS

A storm is described by several key parameters. The most common of these are the *type* (hail, rain, snow, sleet, etc.) and the total amount of precipitation, or *depth*, usually in inches, centimeters, or millimeters. The time from the beginning of the storm until the end of the storm is the *duration*. The *average rate* of precipitation, or *intensity*, is found by dividing the amount of precipitation that occurs during a given period by the length of that period. Common units of intensity are inches per hour, millimeters per hour, or centimeters per hour. Less common, but of importance to some studies, are the area covered by a given event, usually given in acres or hectares or square miles or square kilometers; the changes in intensity during the event; and the likelihood of the event occurring, commonly called the *return period* or the *recurrence interval* (RI).

2.3.1 GEOGRAPHICAL AND SEASONAL VARIATIONS

Annual precipitation in the contiguous U.S. averages 30 in. (75 cm), but there is great spatial variation of amounts and seasonality across the country due to availability of moisture, temporal variances, and differences of precipitation mechanisms (**Figure 2.4** and **Figure 2.5**). In the mountains of the extreme Pacific Northwest, mean annual precipitation of up to 140 in. is a function of eastward-moving wet and cool air masses, midlatitude cyclones,

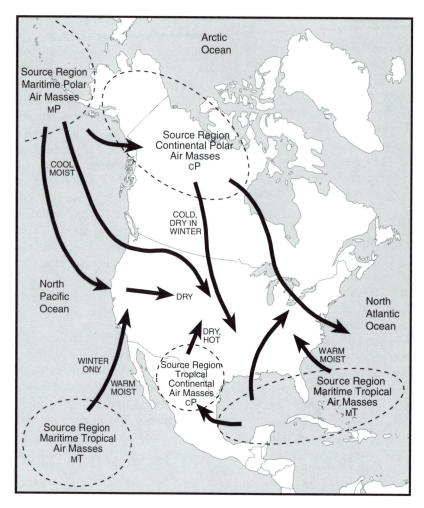

FIGURE 2.4 Major wet and dry air masses influencing the precipitation pattern in the U.S.

and orographic lifting over mountains (**Figure 2.4** and **Figure 2.6**). These mountains, along with the high pressure further south, block most Pacific moisture from the continental interior, with the effect noticeable as far east as the midcontinent. The humid east is supplied with moisture by warm, moist air (Maritime Tropical or MP air masses) from the Gulf of Mexico and the South Atlantic, with both midlatitude and tropical cyclones (hurricanes), along with convection, serving as precipitation mechanisms. Average precipitation amounts range from less than 20 in. along the eastern foothills of the Rocky Mountains to nearly 80 in. in the southeastern mountains.

The map of mean annual precipitation (**Figure 2.6**) is only a rough guide to natural water supply because there is much temporal variance in the annual regime (**Figure 2.6**). The West Coast exhibits a summer minimum of precipitation, with the summer dry season increasing southward, where it is largely controlled by the large and permanent subtropical high-pressure cell of the northern Pacific Ocean. Further north, cool, wet air (Maritime Polar [MP] or Maritime Tropical [MT]) air masses are brought in by the Westerlies, especially in winter, and their incursions

increase northward. In great contrast, the northern, continental part of the country from the Rockies eastward has a summer maximum of precipitation driven by the monsoonal incursion of warm, moist (MT) air masses from the Gulf and South Atlantic and set off by both the frontal and convectional mechanisms. Conversely, the winter is relatively dry, with the region controlled by cold, dry (Continental Polar, cP) air masses from northern Canada. The southeast, especially the Gulf and Atlantic Coasts, is usually controlled by Maritime Tropical air and exhibits fairly well-distributed rainfall, with the autumn rainfall often supplemented by hurricanes moving inland.

2.3.2 Historic Time Trends

Climate tends to fluctuate in cycles. Even now, there is concern about long-term global warming with unknown regional effects on precipitation. Over the past few hundred years, precipitation has tended to fluctuate in cycles of roughly 3 to 7, 15 to 20, and 100 or so years. Thus, the hydrologist must often do time-trend studies to see if there has been a tendency for precipitation to increase or

FIGURE 2.5 Monthly precipitation bar chart (inches) and mean temperature solid line (°F) for selected states in the U.S. (From Teigen, L.D., and F. Singer, Weather in U.S. Agriculture: Monthly Temperature and Precipitation by State and Farm Region, 1950–1990, USDA Statistical Bulletin 834, 1992.)

decrease over some period of interest. This is done to isolate hydrologic variability over time for a study area. For example, are increasing floods for some region due to changes in land use, or are the floods due to increasing precipitation (along with storm size)?

An actual research example is from the upper midwestern U.S. (**Figure 2.7**). In this region, soil erosion increased rapidly in the early 20th century, but declined sharply by midcentury. Were these fluctuations of erosion rates due to changes of land use or to changes of precipitation over the same period? Since the area was opened to European agriculture in the mid-19th century, it was necessary to analyze the precipitation as far back as possible. Although a few precipitation stations existed in the U.S. before 1850 (Trimble and Cooke, 1991), most were installed after 1850, and there was a general increase up to about 1950, when the number stabilized. This is reflected in **Figure 2.7A**, which shows that the number of weather stations in the region increased from 2 to 40 during the period 1867 to 1974. The trend of the annual average of all stations for each year shows that annual precipitation decreased during the 1920s and 1930s and then increased (**Figure 2.7B**).

To show time trends more clearly, a *trend line* (simply a regression line; see Exercise 14.1, Chapter 14) can be fitted. In this example, two trend lines were statistically calculated to show the initial decrease and the ensuing increase (**Figure 2.7B**). Of course, the rainfall characteristics that drive soil erosion are not annual averages, but rather the size and intensity of storms (see Section 2.9). However, there is a positive correlation between annual precipitation and size of storms (**Figure 2.7C**). Note that all large storms of more than 4.5 in. (18 cm) in 24 h were recorded at stations having an above-average rainfall year. Thus, big, erosive storms tend to occur in wetter years.

Yet another way of displaying time trends, especially for water supply questions, is to smooth the time trend by a 3- or 5-year moving average. For example, the rainfall values for 1867 to 1869 were averaged and plotted in the center year of 1868 (**Figure 2.7D**). With the 3-year moving average, note that 1 year will be "lost" at either end of the record or series. Annual precipitation for the period before rainfall records can sometimes be estimated from tree ring growth records, a science known as *dendrochronology*. Rings are thin for dry years and thick for wet years, and the record can be extended back for centuries (Egan and Howell, 2001).

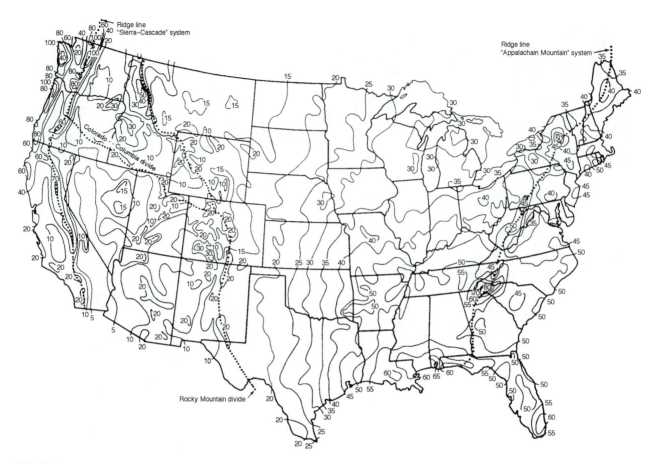

FIGURE 2.6 Mean annual precipitation in inches, 1889 to 1938. (From U.S. Department of Agriculture Soil Conservation Service, *SCS National Engineering Handbook on Hydrology*, Washington, DC, 1972.)

2.3.3 Storm Area Patterns

Rainfall amounts, durations, and intensities vary spatially within the area covered by a given storm. Large-area storms, such as large frontal systems, tend to be more uniform in distribution and have longer durations. Such storms are known by farmers as "general rains," referring to their widespread and prolonged nature. Small-area storms tend to be convectional and to have much higher rainfall amounts near the center of the storm compared to the edges and tend to be shorter. **Figure 2.8** shows the relationship among the area covered by the storm, the duration, and the average amount of precipitation as a percentage of the maximum precipitation at the center. Hydrologists generally select a maximum storm for a point, but based on **Figure 2.8** will reduce the size of storm for larger watersheds. For example, for an area of 50 mi^2, the average rainfall depth for a 30-min event is about 68% of the maximum point rainfall. The average depth of a 24-h event is about 96% of that at the storm center.

2.4 MEASUREMENT OF PRECIPITATION

Measurement of rainfall at a particular site, *point rainfall*, is sufficient for studies for which spatial distribution is of no concern. Point observations may be used to define rain or snowfall for a specific area, such as a watershed, or to identify the characteristics of storms, irrespective of defined land areas.

Rain gages must be established in such a way that their sample is a measure of true precipitation at a point and is essentially unaffected by the surroundings. The Environmental Science Service Administration (ESSA; formerly U.S. Weather Bureau, USWB) checks its sites for possible wind eddy currents from trees, buildings, or other objects or from sharp topographic ridges or valleys. As a rule, no object should be closer to the gage than two times the height of the object above the receiver's surface. If the top of the nearest object is 30 ft above the gage receiver's surface, the gage should be located at least 60 ft from the object. As strong wind currents affect the accuracy of most common gages in the U.S., it might be desirable to locate gages in sheltered spots, like a clearing of sufficient size in an orchard or grove of trees.

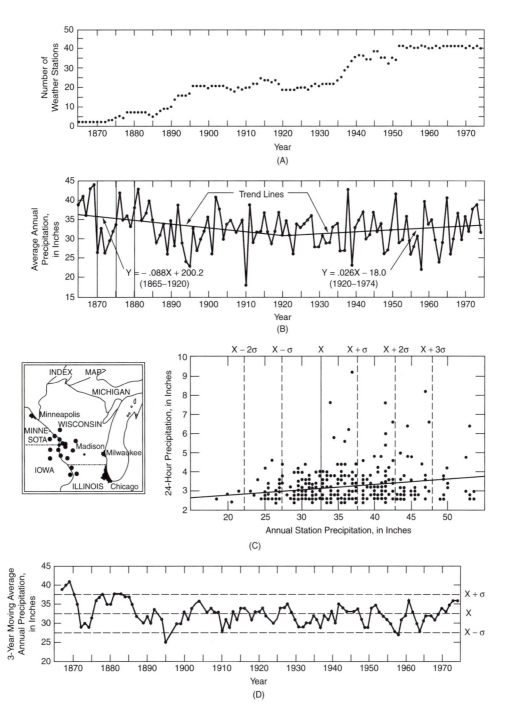

FIGURE 2.7 Precipitation time trends in the Driftless Area of the upper Midwest, 1867–1974. The sigma notation denotes 1 standard deviation from the mean. (A) Number of weather stations; (B) Average annual precipitation and time trends; (C) Relation of annual station precipitation and storms exceeding 2.5 in (10 cm) in 24 h; (D) 3-year moving average annual precipitation. (From Trimble, S.W., and S.W. Lund, Soil Conservation and the Reduction of Erosion and Sedimentation in the Coon Creek Basin, Wisconsin, USGS Professional Paper 1234, 1982. With permission.)

Hydrologic studies of precipitation involving areas or watersheds of many square miles require data from a number of gages. If there are no gages in the study area, then the hydrologist may follow the less than satisfactory approach of using precipitation data from nearby gages. If the hydrologist is initiating a watershed study, a network of gages can be established to meet the needs. However, the establishment of the ideal network is likely to be limited by economics and accessibility. As the area per gage increases, so does the standard error associated with each gage reading. To limit measurement error, smaller watersheds also require a greater density of gages than larger basins.

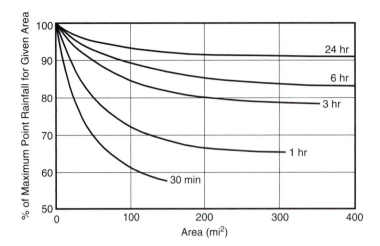

FIGURE 2.8 Relationship among storm area, storm duration, precipitation at the center of a storm, and average precipitation. (From Myers, V.A., and R.M. Zehr, *A Methodology for Point-to-Area Rainfall Frequency Ratios*, NOAA Technical Report NWS 24, National Oceanic and Atmospheric Administration, U.S. Department of Commerce, Washington, DC, 1980.)

TABLE 2.3
Guide for Network Gage Numbers

Size of Watershed	Number of Gage Sites
Acres	
40	2
100	3
600	4
Square Miles	
5	10
10	15
100	50
300	100

Source: From Brakensiek, D.L., H.B. Osborn, and W.J. Rawls, *Field Manual for Research in Agricultural Hydrology*, USDA Agriculture Handbook 224, 1979. With permission.

There are numerous recommendations for gage spacings; these depend on the reason for gaging, the topography, and the climate. Generally, more rugged areas require more gages, as do climates that tend to experience a significant amount of precipitation from small-area convective thunderstorms. Flatter areas receiving the majority of significant rainfall from frontal storms generally tolerate wider gage spacing. Brakensiek et al. (1979) developed a guide for planning precipitation gage networks for watershed studies (**Table 2.3**). The distribution of gages over the study area should be fairly uniform, with each gage representing about the same area as any other. However, local conditions, such as accessibility, may make it impractical to distribute the gages evenly. If **Table 2.3** is used to determine the number of gages needed for a 300-acre watershed, it can be seen that 3 are required for a

100-acre area and 4 are required for a 600-acre area. Therefore, at least 3 gages are needed for a 300-acre watershed.

Often, hydrologists will be working with larger basins than those shown in **Table 2.3**, and the gage net will be preexisting. Hence, it will be important to have an idea of the statistical probability of error, especially for hydrologic modeling. Developed for southern Ohio, **Figure 2.9** shows the relationship among rain gage spacing, basin area, and percentage standard error.

EXAMPLE 2.1

You are working with a watershed of 1050 mi² with 7 rain gages. According to the gages, the area-weighted average rainfall for a large 24-h storm is 4.0 in. How much is the standard error, and what does that mean?

Solution

Divide 1050 mi² by 7 gages to obtain 150 mi²/gage. Enter the graph (**Figure 2.9**) from the abscissa and go almost to the 1000-mi² diagonal line. Move left to the ordinate and obtain the reading of 14+% (call it 15%) standard error.

Answer

The standard error is 4.0 in. times 0.15, which equals 0.6 in. standard error. Thus, using standard bell-curve statistical characteristics, there is a 68% probability that the actual value lies between 3.4 and 4.6 in., 95% probability that the actual value lies between 2.8 and 5.2 in., 2.5% probability that the actual value is above 5.2 in., and a 2.5% probability that the actual value lies below 2.8 in.

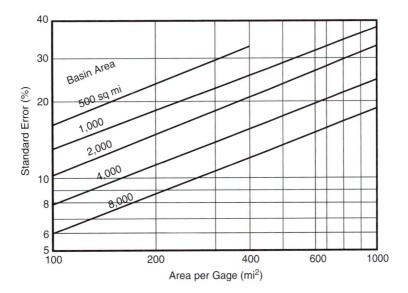

FIGURE 2.9 Relationship among rain gage spacing, basin area, and percentage standard error of rain gages.

Because this gage density, or a lower density, is not uncommon in the U.S., this gives an idea of how imprecise, and even inaccurate, reported precipitation depths from a storm can be. This uncertainty can be a significant problem in hydrologic modeling, for which the output can be no more accurate than the input. Fortunately, Doppler radar is becoming available; it can measure precipitation intensities in real time, and these measures can be integrated to give the spatial distribution of precipitation for a basin.

2.4.1 MEASUREMENT OF SNOWFALL

The measurement of snow on the ground, termed *snowpack*, is important in hydrology for determining water supply and flood runoff potential. In some mountainous regions, snow may be 80 to 90% of the annual precipitation. Foresters are frequently involved in snowpack measurements because, in the U.S., much of the high-elevation snow country is forested, and about 75% of the West's water supply comes from forest land.

Hydrologists are concerned with snowpack depth and its water equivalent (water depth after melting). Freshly fallen snow has a water equivalent ranging from 5 to 20% or 0.5 to 2.0 in. of water per 10 in. of snow, averaging 10%. Old snow may reach a density of 60%.

Inventories of snowpack depth and density are made at periodic intervals throughout the snow season to estimate total water supply in the watershed snowpack. Several snow courses (sampling lines), usually about 50 ft long, are laid out over a watershed to represent the expected snowpacks. Depth of snow is measured, and a known amount of snow in a sampling tube is weighed to determine its water equivalent.

EXAMPLE 2.2

The snow depth at one site was recorded as 5.0 ft, and its measured density was 20%. What is the equivalent depth of water?

Solution

The equivalent depth of water would be Depth = 5 ft × 12 in./ft × 20% = 12 in.

Answer

The equivalent depth of water is 12 in.

Snow pillows of white neoprene placed on the ground before the snow season are useful in snowpack inventories. Pillows may vary from 6 to 12 ft in diameter and be 2 to 4 in. deep. The pillows are filled with antifreeze solution (alcohol–water mix) and contain electrical sensors that allow the remote monitoring of the pressure on the pillows because of the snowpack. The pressure is proportional to the water equivalent of the snowpack on top of the pillow. Snowpack depth and water equivalent data may be compiled into estimates of total watershed water storage. The expected snowmelt seasonal water supply can be estimated from these data.

2.4.2 RAIN GAGES

The purpose of rain gages is to measure the depth of rain falling on a horizontal surface. They may also be used, with some difficulty, to measure snowfall. There are several common types of nonrecording and recording rain

FIGURE 2.10 Weather station at the Priest River Experimental Forest in northern Idaho. The instrumentation from left to right is as follows: tipping bucket rain gage; sunshine recorder; solar radiation pyranometer; standard 8-in. rain gage; screened box containing thermometers for maximum, minimum, and wet-bulb temperatures; and strip-charge recording rain gage. The striped post with a diamond-shape sign in the background is one of the points on the snow survey course around this site. (Photograph courtesy W. Elliot.)

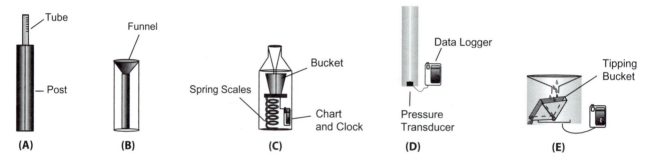

FIGURE 2.11 Types of rain gages: (A) test tube on post; (B) standard; (C) weighing gage with strip recorder; (D) 8-in. pipe with pressure sensor; and (E) tipping bucket recording rain gage with electronic data logger.

gages (**Figure 2.10** to **Figure 2.12**). Standard nonrecording rain gages are low cost and usually maintenance free. The simplest are tapered plastic and test-tube-type rain gages. These may be attached to fence posts or set vertically in the ground. They are fairly reliable, and their data (to 0.1-in. accuracy) are of some value to hydrologists in areas where no full-size standard gages are located. Many of these tube gages are made of glass and should not be used in freezing weather because water that freezes in the gage may break the glass.

The most common rain gage used in the U.S. is the standard, 8-in. diameter, sharp-edged horizontal catchment receiver placed about 36 in. above the ground. In areas where deep snow is expected, the receiver surface is raised so that it will always be exposed. During seasons of no snow, a funnel inside the receiver passes rainwater into a small measuring tube or storage bucket. The small hole in the funnel outlet reduces evaporation of rainwater from the tube or bucket. The funnel is removed in the snowy season. These gages are not good for measuring snowfall because wind tends to blow the snow across the receiver, allowing only part of the snow to fall into it. Wind shields

are advisable where measurements of snowfall are important (**Figure 2.12**).

In standard rain gages, rainfall amounts are obtained by measuring the depth of water in the collecting tube with a scale provided with the gage or with a ruler graduated in inches and dividing the number by 10 (since the area of the tube is 1/10th the area of the gage catch surface). Readings are accurate to 0.01 in. Some rain gages may use a graduated cylinder calibrated to measure the rainfall amount directly. Others measure the volume in the gage, which is then converted to a depth by dividing by the collecting area of the gage. Observations are usually made at the beginning of the workday and at the same time on weekends and holidays.

Recording rain gages may be of several types. Traditionally, weighing recording gages were preferred in small-area research, and many are still in use today (**Figure 2.10** and **Figure 2.11**). This gage has a spring scale beneath the collecting bucket platform that is calibrated to show the rainfall depth on a paper chart. The chart is rotated by a spring-driven clock at speeds of 1 revolution in 6, 9, 12, 24, or 192 h. The record on the

FIGURE 2.12 Weather station at the Manitou Experimental Forest near Denver, CO. The instrumentation from left to right is as follows: screened box with thermometers; strip-chart recording rain gage; tower with tipping bucket rain gage; electronic temperature sensor solar panel to recharge system batteries; antenna to relay recordings to central site via satellite; d box with data logger instrumentation; and 8-in. standard rain gage with windshields. In the background is Pike's Peak. (Photograph courtesy W. Elliot.)

charge shows the accumulation of rainfall depth over time. A variation on the weighing bucket is the use of a single, large-diameter pipe, about 5 ft high, with an electronic pressure sensor in the base that regularly records the pressure with an electronic data logger. The record of changes in pressure with time allows the hydrologist to determine the cumulative rainfall depth over extended periods. Careful interpretation of the results is necessary to account for the evaporation of water from the pipe that will occur between storms. A concentrated antifreeze solution can be added to these gages during the winter so any snow that falls in the gage will melt, allowing direct reading of water equivalent.

The tipping bucket gage connected to a data logger is the most common type of recording rain gage presently installed. Two small calibrated buckets of equal size and weight are balanced on opposite sides of a fulcrum. Rain falls through a collector funnel into one bucket, causing it to tip and empty. The rainwater is then delivered to the other bucket until it is full, tips back, and empties, and so on. At each tip, a signal is sent to an electronic data logger. The data logger calculates the amount of rainfall that occurs for the time increment desired and provides the user with the resulting rainfall distribution.

2.5 STORM TIME TRENDS

Some recording rain gages provide information to determine rainfall intensity distribution within a storm and a storm duration. For some hydrologic projects, hourly rainfall

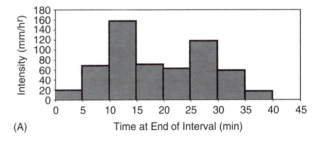

(A)

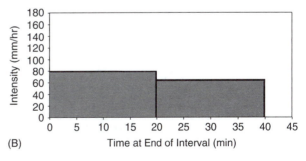

(B)

FIGURE 2.13 Hyetographs of the storm presented in **Table 2.4** showing (A) 5-min and (B) 20-min breakpoint data.

data are sufficient. For others, 30-min or even 5-min rainfall amounts may be desired. Rainfall rates may be plotted with time to produce a histogram of rainfall. **Figure 2.13** shows histograms for 5-min and 20-min increments from the data given in **Table 2.4**.

EXAMPLE 2.3

For the rainfall amounts shown in **Table 2.4**, calculate (a) the total rainfall amount; (b) the duration; (c) peak intensities for 5 min, 10 min, 20 min, and 30 min; and (d) the average intensity.

Solution

(a) The total rainfall amount is the sum of the individual amounts in **Table 2.4**.

$$\text{Amount} = 1.5 + 5.6 + 13.0 + 5.8 + 5.1$$
$$+ 9.7 + 4.8 + 1.3 = 46.8 \text{ mm}$$

$$46.8 \text{ mm} \div 25.4 \text{ mm/in.} = 1.84 \text{ in.}$$

(b) The duration is the length of time from the beginning of the storm until the end of the storm. In **Table 2.4**, the time is given in minutes from the beginning of the storm, which gives a total duration of 40 min.

(c) The 5-min peak intensity occurs between 10 and 15 min, when 13 mm of precipitation is measured:

$$\text{Intensity} = (13 \text{ mm})\left(\frac{60 \text{ min/hr}}{5 \text{ min}}\right) = 156 \text{ mm/hr}$$

or

$$\frac{156 \text{ mm/hr}}{25.4 \text{ mm/hr}} = 6.14 \text{ inches/hr}$$

The 10-min intensity can be found by comparing each 10-min period to find the period that receives the most rainfall. The period is between 10 and 20 min, with a total rainfall of 13.0 + 5.8 mm = 18.8 mm. The intensity is

$$\left(\frac{18.8 \text{ mm}}{25.4 \text{ mm/inch}}\right)\left(\frac{60 \text{ min/hr}}{10 \text{ min}}\right) = 4.44 \text{ inches/hr}$$

By the same procedure, the maximum 20-min and 30-min rainfall intensities are determined. The maximum 20-min rainfall is between 10 and 30 min = 13.0 + 5.8 + 5.1 + 9.7 = 33.6 mm; the maximum 30-min rainfall is between 5 and 35 min = 5.6 + 13.0 + 5.8 + 5.1 + 9.7 + 4.8 = 44.0 mm.

(d) The average intensity is the total rainfall amount divided by the storm duration:

$$\text{Average Intensity} = (1.84 \text{ inch})\left(\frac{60 \text{ min/hr}}{40 \text{ min}}\right)$$

$$= 2.76 \text{ inches/hr}$$

TABLE 2.4
Distribution of Rainfall during a Storm on the University of Georgia Whitehall Forest in June 1990

	Time (min)							
	5	10	15	20	25	30	35	40
Depth (mm)	1.5	5.6	13.0	5.8	5.1	9.7	4.8	1.3

Source: Brown, R.E., Runoff and Sediment Production from Forested Hillslope Segments in the Georgia Piedmont, master's thesis, University of Georgia, Athens, 1993.

Answer

The total rainfall is 1.84 in. in 40 min. The peak intensities for 5, 10, 20, and 30 min are 6.14 in./h, 4.44 in./h, 3.96 in./h, and 3.46 in/h. The average intensity is 2.76 in./h.

2.5.1 TIME SEQUENCE PATTERNS

The time distribution of rainfall rates in a storm may appear random. On closer examination, thunderstorms often begin with relatively high intensities, followed by periods of decreasing intensity. Winter rainfalls often have relatively low intensity, with little variation in intensity throughout the storm. Other storms may have the maximum intensity at an intermediate time or a delayed time. The delayed-intensity distribution is more common on the West and Southeast Coasts of the U.S. **Figure 2.14** shows four typical histograms of rainfall patterns. The distribution of runoff rates from a given storm can be influenced by the rainfall pattern, with delayed patterns tending to result in greater runoff rates.

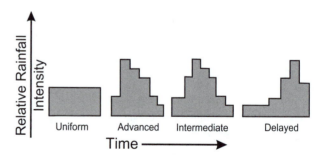

FIGURE 2.14 Histograms of typical storm patterns. (From Schwab, G.O., D.D. Fangmeier, W.J. Elliot, and R.K. Frevert, *Soil and Water Conservation Engineering*, 4th ed., John Wiley & Sons, New York, 1993. With permission.)

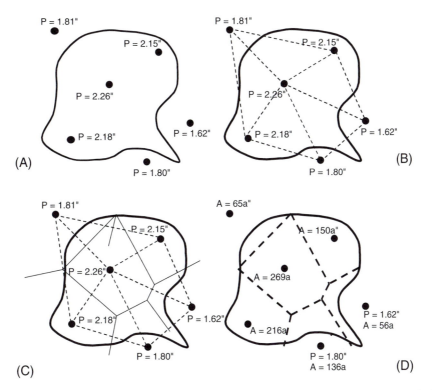

FIGURE 2.15 Use of Thiessen method to find average rainfall: (A) distribution of rain gages in a watershed located on a map; (B) connection lines drawn between rain gage positions; (C) lines perpendicular to connection lines drawn until they intersect to form polygons; and (D) areas measured for each polygon.

2.6 AVERAGE PRECIPITATION OVER AN AREA

Hydrologists frequently need to determine with multiple rain gages the rainfall distribution over an area. The distribution may be for specific storms, days, months, or even years. It may be necessary to determine an average depth or a spatial distribution of depths. There are three methods commonly used to determine the average rainfall amount from a series of rain gages. If the rain gages are evenly distributed, a simple arithmetic average will be adequate. If the gages are not evenly distributed, the most common method of determining an average depth is the Thiessen method.

The use of the Thiessen method is illustrated in **Figure 2.15**. In **Figure 2.15A**, the locations of each rain gage in and immediately adjacent to the area of interest are located on a map. The rainfall depth at each gage is noted. The straight dashed lines of **Figure 2.15B** are drawn between each adjacent gage site; in **Figure 2.15C**, solid perpendicular bisectors to these lines are constructed so that the area around each gage is enclosed by the bisectors or the area boundary. The enclosed areas around each gage are known as *Thiessen polygons*. The entire area within each polygon is closer to the rain gage in that polygon than to any other rain gage. The rainfall depth

measured by the gage within each polygon is assumed to be representative of the rainfall for that polygon. As shown in **Figure 2.15D**, the areas of each polygon are calculated by geometry, with a planimeter, with graph paper, or with a computer-aided drawing package. The average rainfall for the entire area is then assumed to be a weighted average of the observed rainfalls, calculated by Equation 2.1:

$$P = \frac{\sum_{i=1}^{n} A_i P_i}{\sum_{i=1}^{n} A_i} \tag{2.1}$$

where P represents the average depth of rainfall in the watershed with a total area of $\sum_{i=1}^{n} A_i$, and A_i is the area of the ith polygon with precipitation of P_i in that polygon.

EXAMPLE 2.4

Use the Thiessen method and an arithmetic average to determine the average rainfall for the watershed shown in **Figure 2.15** for the depths and areas shown.

TABLE 2.5
World's Greatest Point Rainfall Depths and Corresponding Average Intensities

Duration	(in.)	(mm)	Location	Date
1 min	1.5	38	Barot, Guadeloupe	November 26, 1970
15 min	7.8	198	Plumb Point, Jamaica	May 12, 1916
20 min	8.1	206	Curtea-de-Arges, Roumania	July 7, 1889
30 min	11.0	280	Sikeshugou, Hebei, China	July 3, 1974
60 min	15.8	401	Shangdi, Inner Mongolia, China	July 3, 1975
2 h	19.3	489	Yujiawanzi, Inner Mongolia, China	July 19, 1975
3 h	23.6	600	Duan Jiazhuang, Hebei, China	June 28, 1973
6 h	33.1	840	Muduocaidang, Inner Mongolia, China	August 1, 1977
10 h	55.1	1,400	Muduocaidang, Inner Mongolia, China	August 1, 1977
24 h	71.8	1,825	Foc Foc, Reunion	January 7–8, 1966
2 days	97.2	2,467	Aurere, Reunion	April 8–10, 1958
3 days	127.6	3,240	Grand Ilet, Reunion	January 24–27, 1980
4 days	146.5	3,721	Cherrapunji, India	September 12–15, 1974
7 days	183.2	4,653	Commerson, Reunion	January 21–27, 1980
14 days	239.5	6,082	Commerson, Reunion	January 15–28, 1980
31 days	366.1	9,300	Cherrapunji, India	July 1–31, 1861
12 months	1,041.8	26,461	Cherrapunji, India	August 1860–July 1861
2 years	1,605.1	40,768	Cherrapunji, India	1860–1861

Source: Summarized from the NOAA Web site based on World Meteorological Organization, *Manual for Estimation of Probable Maximum Precipitation*, 2nd ed., Operational Hydrology Report 1, WMO Number 332, Geneva, Switzerland, 1986; Updated by the National Weather Service, Office of Hydrology, Hydrometeorological Branch, 1992.

Solution

By the Thiessen method, the areas represented by the various rain gages are determined and used in Equation 2.1 as follows:

$$P = \frac{\begin{array}{c}65 \times 1.81 + 150 \times 2.15 + 269 \times 2.26 \\ +216 \times 2.18 + 56 \times 1.62 + 136 \times 1.8\end{array}}{65 + 150 + 269 + 216 + 56 + 136} = 2.08 \text{ inches}$$

The arithmetic average is the mean of the observations:

$$P = \frac{1.81 + 2.15 + 2.26 + 2.18 + 1.62 + 1.8}{6} = 1.97 \text{ inches}$$

Answer

The arithmetic average rainfall is 1.97 in. The average rainfall determined with the Thiessen method is 2.08 in.

Another method of determining average rainfall is the isohyetal method. With this method, lines of equal rainfall (isohyets) are drawn similar to contour lines on the watershed map (see **Figure 2.6** and Appendix C). From the resulting map (**Figure 2.6**), a weighted average based on the area within each of the contour lines can be calculated.

This method may have some benefit in mountainous regions where rainfall variation with elevation is fairly consistent, making isohyets a better representation of the rainfall distribution than Thiessen polygons.

The rain gage distribution in Example 2.4 is fairly uniform with relatively small areas for each rain gage. The average rainfall calculated should be relatively reliable. For watersheds where there are no rain gages within their boundaries, averages by any method is less reliable. Generally, hydrologists have preferred the Thiessen method, but computer mapping and measurement make the isohyetal method easier to use, assuming adequate rain gages are present.

2.6.1 RAINFALL DISTRIBUTIONS FOR SHORT TIME PERIODS

Storms with longer durations tend to have a greater total amount of precipitation because there is more time for atmospheric water to be advected to the storm. Shorter duration storms, however, tend to have greater intensities, assuming the local air is high in water vapor. The maximum amounts of rainfall for periods of 1 min to 1 year recorded over the world prior to 1992 are listed in **Table 2.5**.

Most hydrologic studies with smaller watersheds require rainfall amount–frequency (termed *frequency–magnitude*) data for time periods or durations shorter than a year. Rainfall amounts for these shorter periods can be found by analyzing weather records in a manner similar to that demonstrated above, but it is usually much easier and more accurate to use the extensive published data. In the next sections, we begin our look at these data with the simplest and most available and then move toward the more complex.

2.7 RAINFALL FREQUENCY DISTRIBUTIONS

Hydrologists need to estimate the probability that a given rainfall event will occur to assist planners in determining the likelihood of the success or failure of a given project. The main parameters needed to describe rainfall frequency distributions are the duration, intensity, and return period. The *return period* is the average period of time in years expected either between high-intensity storms or between very dry periods. Generally, plans for agricultural drainage ditches and grassed waterways consider an event with a return period of 5 to 10 years sufficient. Larger, less-frequent events will overtop the banks and might result in damage to the crop, but this risk is offset by having smaller waterways that occupy less land or ditches that cost less to construct. However, hydrologic plans for urban areas are more commonly concerned with adverse impacts of flooding on buildings and the potential for injuries or loss of life associated with the flooding. Therefore, extreme events associated with a 100-year return period, or probable maximum precipitation (PMP), might be used in developing land use management strategies and designs. Culverts under roads are usually sized based on the risk of flooding of the road, the type of road, and the anticipated volume of traffic on the road.

For projects concerned with storm runoff from small- to medium-size watersheds (under 1 to 100 mi²), plans generally require a storm event with a duration varying from 5 min to 24 h. Larger watersheds (over 100 mi²) may require storm durations or cumulative rainfall amounts for time periods of up to 1 month. For river basin projects with large-scale irrigation, drainage, river flooding, or river drought flows, rainfall amounts from time periods of 1 month to 1 year may be needed.

If long-term records are available, such as 100 or more years, it may be sufficient to determine the highest or lowest rainfall from that record. Because data from recording rain gages have not generally been available for more than 60 years, it is frequently necessary to estimate a 50- or 100-year event from 30 or fewer years of rainfall records. There are numerous methods for estimating the return period for a given rainfall duration. The Hazen method, discussed here, is commonly used by the U.S. Department of Agriculture (USDA) Natural Resources Conservation Service (NRCS) and is similar to other

TABLE 2.6
Annual Precipitation for Los Angeles, CA, 1934–1953

Year	Depth (in.)	Year	Depth (in.)
1934	14.6	1944	19.2
1935	21.7	1945	11.6
1936	12.1	1946	11.6
1937	22.4	1947	12.7
1938	23.4	1948	7.2
1939	13.1	1949	8.0
1940	19.2	1950	10.6
1941	32.8	1951	8.2
1942	11.2	1952	26.2
1943	18.2	1953	9.5

TABLE 2.7
Numerical Ranking of Annual Precipitation, Probabilities or Occurrence, and Return Periods for Los Angeles, CA, 1934–1953

Rank	Precipitation (in.)	Probability F_a (%)	Return Period (years)
1	32.8	2.5	40.0
2	26.2	7.5	13.3
3	23.4	12.5	8.0
4	22.4	17.5	5.7
5	21.7	22.5	4.4
6	19.2	27.5	3.6
7	19.2	32.5	3.1
8	18.2	37.5	2.7
9	14.6	42.5	2.4
10	13.1	47.5	2.1

methods. The method consists of determining the statistical distribution of rainfall amounts for the duration of interest, plotting that distribution on probability paper, and interpolating or extrapolating from the graph to determine the storm associated with the return period of interest. The relationship developed is called an *intensity–duration–frequency* (IDF) *curve*.

The first step in the Hazen method is to assemble rainfall records for as many years as possible for the duration of interest. An example set of data for annual rainfall records is for Los Angeles, 1934 to 1953 (**Table 2.6**). Analyses of annual data are useful for planning water supply and waste disposal projects.

The annual values are first listed in order from the highest to the lowest, as shown in **Table 2.7**. A ranking number is then given each rainfall amount, with 1 for the

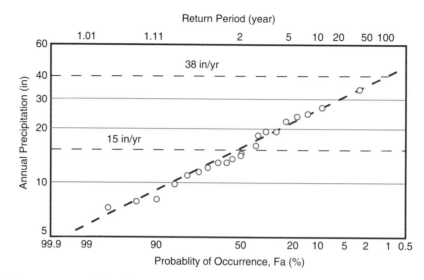

FIGURE 2.16 Annual frequency–magnitude of precipitation for Los Angeles, CA, 1934–1953. Log-normal plot of data is shown in **Table 2.7** for Example 2.5, plotting annual rainfall vs. probability of occurrence F_a. Note the relationship between return period and probability. A 5-year event may be expected to occur or be exceeded on average every 5 years, or better expressed, 20 times per century. Hence, there is a 20% probability for it to occur or be exceeded in any 1 year. *Note:* Return period = 100/probability.

highest, 2 for the second highest, and so on. From the ranking, a plotting position is determined from

$$F_a = \frac{100(2n-1)}{2y} = \frac{100}{\text{Return Period, T}} \quad (2.2)$$

where F_a is the plotting position or probability of occurrence (%) for each event, y is the total number of events, and n is the rank of each event. The precipitation amounts are plotted against the probability of recurrence on probability paper (**Figure 2.16**). A straight line is drawn through the plotted points. The line can be extended to estimate larger return periods. The size of event for a given return period can be estimated from the graph.

EXAMPLE 2.5

For the annual rainfall given in **Table 2.7**, calculate the 2-year and 100-year annual rainfall amounts.

Solution

Rank the values as in **Table 2.7**. For each ranking, calculate the probability of occurrence F_a using Equation 2.2. The first value for F_a would be

$$F_a = \frac{100\big((2\times1)-1\big)}{2\times20} = 2.5$$

The return period is found from the second part of Equation 2.2:

$$\text{Return Period} = \frac{100}{F_a} = \frac{100}{2.5} = 40 \text{ years}$$

The values for F_a and return period are found for each of the other years by the same process. The precipitation is then plotted against F_a on log-normal probability paper, as shown in **Figure 2.16**.

Figure 2.16 shows that the 2-year amount would be about 15 in. Extending the prediction line in **Figure 2.16** until it intersects the 100-year return period shows that the wettest year in 100 years would experience about 38 in. of rainfall, or in a given year, based on these data there is a 1% chance that there will be more than 38 in. of rainfall.

Answer

The annual rainfall for a 2-year return period is 15 in. The annual rainfall for a 100-year return period is 38 in.

In Example 2.5, the probabilities of the occurrence of larger events were predicted. The extreme wet-year probabilities are useful in planning reservoir spillway capacities and river flood-routing projects. On the lower end of the probability scale, the probabilities of dry years are used to assist in planning volumes of storage reservoirs or the capacity of irrigation systems that may be needed to meet demands during dry years. Low river flow estimates are also needed to assist in determining the quality of sewage discharge that a river can accept without adversely affecting the river ecosystem.

An alternative plotting method that has seen widespread application is the Weibull method (Chin, 1999):

$$T = \frac{y+1}{n} \qquad (2.3)$$

The Weibull and Hazen methods will give similar IDF curves if the period or record for a particular application is long (at least 20 years). Commonly used spreadsheet programs often do not provide a simple option to use a probability scale. For many applications, useful results can be obtained by plotting the IDF on a log–log or semilog scale and then fitting a trend line (based on a least-squares analysis) through the data. Care should be taken to ensure that there is a good fit ($r^2 > 0.9$) to the data with the range that the IDF will be used.

Sometimes, hydrologists need to estimate the probability that a given return period storm will occur at least once within a given number of years, such as, estimation of the probability that a 100-year storm will occur at least once during the next 10 years. The relationship to determine that probability is

$$P(T,n) = 1 - \left[1 - \frac{1}{T}\right]^n \qquad (2.4)$$

where $P(T,n)$ is the probability that a T-year return period storm will occur at least once during n years (Barfield et al., 1981).

EXAMPLE 2.6

What is the probability that a 50-year storm will occur during the first 5 years following the construction of a drainage ditch?

Solution

Apply Equation 2.4 with T = 50 years and n = 5 years:

$$P(50,5) = 1 - \left[1 - \frac{1}{50}\right]^5 = 0.096 \ (9.6\%)$$

Answer

The probability that a 50-year storm will occur during the first 5 years is 9.6%.

Example 2.6 is an example of *calculated risk*. We encounter this concept again in flooding, flood design, and reservoir design (Chapters 7 and 8), but will use a graphical solution.

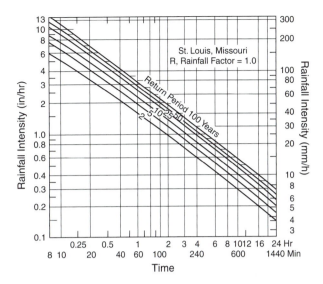

FIGURE 2.17 Rainfall rate–duration–frequency distribution for St. Louis, MO. (From Hershfield, D.N., Rainfall Frequency Atlas of the United States, U.S. Weather Bureau Technical Paper 40, 1961, and Weiss, L.L., *Monthly Weather Rev.*, 90:87–88, 1962.)

2.7.1 ST. LOUIS FREQUENCY RELATIONSHIP TRANSFERRED TO THE CONTIGUOUS U.S.

Figure 2.17 shows the relationship between storm duration, return period, and rainfall intensity for St. Louis, MO. Rainfall amounts can then be found by multiplying the duration and intensity. The value obtained from **Figure 2.17** for St. Louis is then multiplied by a factor from **Figure 2.18** to estimate a storm of the same duration and return period for a different locality in the continental U.S. For example, Chicago, IL, would have a storm intensity approximately 0.75 the size of that predicted for St. Louis, MO, for the same duration and return period.

EXAMPLE 2.7

Estimate the amount of rainfall expected once in 25 years for a watershed near Chicago, IL, for a storm with a duration of 30 min.

Solution

From **Figure 2.17**, determine that the 25-year, 30-min duration storm in St. Louis has an intensity of 4.1 in./h or a total rainfall of

$$P = (4.1 \text{ inches/hr}) \left(\frac{30 \text{ min}}{60 \text{ min/hr}} \right) = 2.05 \text{ inches}$$

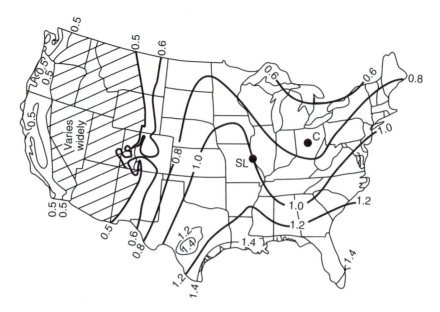

FIGURE 2.18 Factor to adjust rainfall amount determined for St. Louis, MO. (from **Figure 2.17**) for other sites in the U.S. (From Hamilton, C.L., and H.G. Jepson, Stock–Water Developments; Wells, Springs, and Ponds, USDA Farmers Bulletin 1859, 1940.)

For Chicago, **Figure 2.18** shows that both the rainfall intensity and the amount will be 0.75 of that for St. Louis; $i = 4.1$ in./h × 0.75 = 3.1 in./h, and $P = 2.05$ in. × 0.75 = 1.54 in.

Answer

The 25-year, 30-min rainfall is 1.54 in.

2.7.2 REGIONAL FREQUENCY–MAGNITUDE RELATIONSHIPS: SEATTLE, SANTA FE, CHICAGO, AND NEW ORLEANS

Four graphs similar to **Figure 2.17** for Seattle, WA; Chicago, IL; Santa Fe, NM; and New Orleans, LA (**Figure 2.19**) demonstrate the highly disparate frequency–magnitude conditions that exist in different climatic regions of the country. Note that these are the climatic conditions imposed by nature on the regional landscapes, and these are the conditions with which hydrologists and planners must contend. While it is important to be able to read such graphs, it is more useful if there is at least some understanding of the climatology involved (see the section on regional climatology and Koeppen climatic classification).

First, consider the spatial disparities in the short-duration rainfall. For the 100-year return period curve, find the 60-min (1-h) rates. For Seattle, it is only 0.8 in./h, a value that reflects the cooler, more stable air found there (Koeppen Cfb), which is incapable of delivering copious short-term rain. Santa Fe (Koeppen BSk) is normally controlled

by dry, warm-to-hot, Continental Tropical air masses that have their source in the Southwest. Warm, moist air must travel long distances from the Pacific or the Gulf and, while often unstable, cannot supply copious rain for periods of even an hour. The graph (**Figure 2.19**) indicates about 1.5 in. for the 100-year, 60-min event at Santa Fe. This is about twice as high as for Seattle, but still not high.

Moving to the east, Chicago (Koeppen Dfa) and New Orleans (Koeppen Cfa) are both supplied with warm, moist Maritime Tropical air from the Gulf; the difference is that New Orleans is on the Gulf, while Chicago is almost 800 miles inland. Thus, the 100-year, 60-min rainfall for New Orleans is truly impressive, 4.5 in./h. Despite the distance, the masses of Gulf air reaching Chicago are still large, warm, wet, and unstable enough to give a 60-min rate of 3.0 in./hr, understandably less than New Orleans, but still twice as high as Santa Fe and almost four times greater than Seattle. The implications are obvious for the planner/hydrologist, who must in some way shunt the runoff from these storms without flooding and also avoid environmental damage such as soil erosion from new urban developments.

A look now at the 24-h precipitation will show perhaps unexpected disparities between short- and long-duration events that will not show up in simpler techniques, like the transfer from St. Louis to U.S. shown in **Figure 2.17** and **Figure 2.18**. Looking at the 100-year, 24-h storm for Seattle, we see a rate of about 0.15 in./h or about 3.6 in. total in 24 h. While the low moisture-holding capacity of the cool air there cannot produce high intensities, the humid westerly winds coming directly off the ocean can provide continuous lower rates that, given enough time,

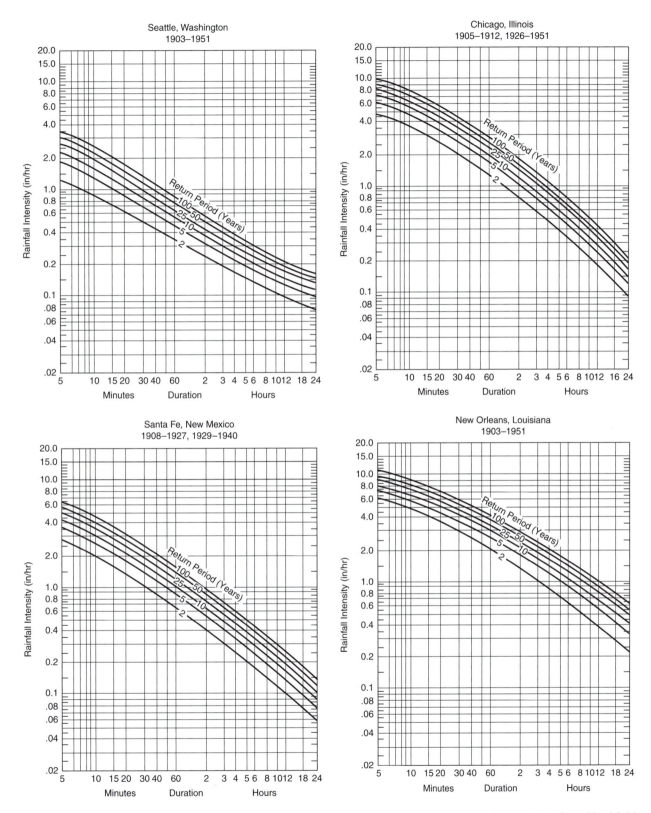

FIGURE 2.19 Frequency–magnitude graphs for Seattle, WA; Santa Fe, NM; Chicago, IL; and New Orleans, LA. (From Hershfield, 1961. Rainfall Frequency Atlas of the United States, U.S. Weather Bureau Technical Paper 40.)

can provide a lot of rain. Seattle is the turtle of the turtle–hare precipitation race. In Santa Fe, there is simply not enough atmospheric moisture, either locally or regionally, to deliver large, longer duration events. There, the 100-year, 24-h rate is about 0.13 in./h or a 24-h total of about 3.1 in.. So, while semiarid Santa Fe can deliver 1-h storms almost twice as large as Seattle, the 24-h storm is only about 85% of the value for Seattle.

Returning to the humid east, New Orleans, on the Gulf, often has an unlimited supply of warm, moist, unstable air. Hence, the 100-year, 24-h rate is over 0.6 in./h, or close to 15 in. in 24-h, about five times that of Santa Fe. Chicago, 800 miles inland, cannot obtain the unlimited supply of warm, moist, unstable air often available to New Orleans, so its 100-year, 24-h rate is about 0.22 in./h or about 5.3 in., only about half that of New Orleans. Chicago and the continental interior of the U.S. are the rainfall hare. That is, they have high short-duration rates, but relatively lower long-duration rates. The turtle, Seattle, will catch up with Chicago and pass it after about 10 days of rain. The long- and short-term rates have important implications for regional hydrology. High-intensity rates for even short periods can cause severe soil erosion and flooding in small-stream basins. Low intensities over long periods, such as found in Seattle or Western Europe (Koeppen Cfb), rarely cause severe soil erosion or flooding in small basins, but they can cause severe flooding in larger rivers, as demonstrated in the huge Central European floods of 2000.

Frequency–magnitude graphs like those of **Figure 2.19** are desirable for use at any permanent location of the hydrologist/planner, but there is a treasury of easily obtained and accessible series of precipitation maps from which the professional can quickly and easily find detailed frequency–magnitude relationships for any location in the contiguous U.S. for frequencies of 1 year to 100 years and durations from 30 min to 10 days (Hershfield, 1961; J.F. Miller, 1964; see Exercise 14.1, Chapter 14 for some examples). Internet access to data is available at many sites, such as that of the Western Regional Climate Center (www.wrcc.dri.edu/pcpnfreq.html). More modern compilations and analyses have given more detail and changed some values. The professional will need to keep easy access to these data, whether in hard copy or by Web because they are indispensable.

2.8 PROBABILITY OF AN EVENT OCCURRING

2.8.1 SEASONALITY OF STORM PROBABILITY

Probability is normally stated for a year with the implicit assumption that the event has an equal chance of occurring at any time of the year. But, discussion in Section 2.5 showed how the seasonal presence of air masses controls the occurrence of storms. Thus, it would helpful to know the time of year a storm of any magnitude might be likely to occur. Hershfield (1961) again masterfully worked this out for different duration storms for various climatic regions of the U.S. Such a seasonality graph for a 1-h event in the Great Lakes region is shown in **Figure 2.20**.

2.8.2 PROBABLE MAXIMUM PRECIPITATION

In some situations, the cost of failure of a structure, such as a large dam, is so great that planners desire to ensure that the structure can store or pass the runoff from the PMP likely to occur. PMP amounts are determined from long-term records using principles described in Section 2.7, as well as considering the extreme fundamental meteorological events likely to lead to such an event. Values from **Figure 2.21** are generally adjusted based on the size of a watershed and other site-specific conditions. **Figure 2.21** is an example of a map that is available to assist in determining the PMP for a given duration and area of coverage. Specialized references (e.g., Viessman et al. 1977) should be consulted for more detailed discussion of this topic.

2.9 RAINFALL EROSIVITY

The ability of rainfall to detach and transport soil particles is the rainfall *erosivity* (*EI*). Soil erosion researchers have found that the erosivity of a given storm can be estimated from the product of the maximum 30-min intensity and the amount of kinetic energy delivered by the storm. The total storm energy is found by calculating the amount of energy within each time increment and multiplying that energy by the rainfall that fell during that time increment. **Figure 2.22** gives the relationship between rainfall energy and the rainfall intensity. The total energy is the sum of each of the time increments. This total is multiplied by the maximum 30-min intensity and, to keep the numbers smaller, the product is divided by 100.

EXAMPLE 2.8

Calculate the *EI* value of the storm summarized in **Table 2.8**.

Solution

Table 2.8 is set up to determine the energy for each of the time increments. The table was shortened for this example by considering two 20-min increments instead of each 5-min increment separately. Generally, the storm should be divided into increments with approximately similar intensities.

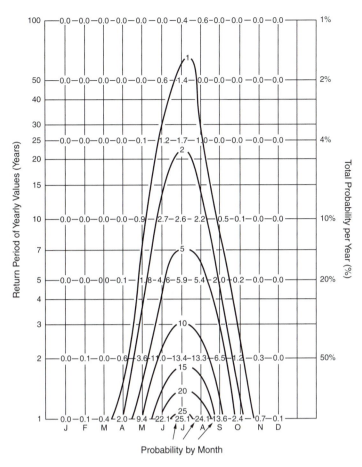

FIGURE 2.20 Seasonal probability of intense rainfall of 1-h duration, Great Lakes region. (From Hershfield, D.N., Rainfall Frequency Atlas of the United States, U.S. Weather Bureau Technical Paper 40, 1961.) The total probability for the year is shown on the right ordinate. In the graph matrix, the probability is allocated or distributed by month. Note that adding the monthly probabilities will give the total probability. The graph is best understood by first ignoring the curved lines. After the graph is understood, it is seen that the curves are merely isolines of probability. Similar graphs for longer durations have a greater annual distribution.

The total energy is multiplied by the maximum 30-min intensity of 3.46 in./h, and the result is divided by 100:

$$EI = 1948 \times 3.46/100 = 67.4 \ 100\text{-ft-tons/acre}$$

Answer

The *EI* value is 67.4 100-ft-tons/acre.

At each location throughout the U.S., such values are added to give an annual cumulative value. The annual long-term average (termed R) is indicative of the role of precipitation alone in predicting soil erosion (Chapter 8). These values indicate, for example, that areas along the Gulf Coast would have about three times as much soil erosion as Iowa in an average year, everything else being equal. Such is the erosive power of rainfall, and areas that have high R values must have more care taken in agriculture and urban development.

PROBLEMS

2.1. The following cities are located on a similar latitude: Akita, Argan, Baku, Beijing, Columbus, Denver, Hungnam, Istanbul, Lincoln, Madrid, Naples, New York, Rome, Sacramento, and Salt Lake City. Select four cities from the list that are in different countries, then find their latitude, mean annual temperature, and mean annual precipitation. What climatic zone and biome type is associated with each of the four cities? *Note:* Refer to A.H. Strahler and Strahler, 2002; MacDonald, 2002; or an Internet search might be helpful.

2.2. Select from the list in Problem 2.1 four cities that are located in the U.S., then find their latitude, mean annual temperature, and mean annual precipitation. What climatic zone and biome type are associated with each of the four cities?

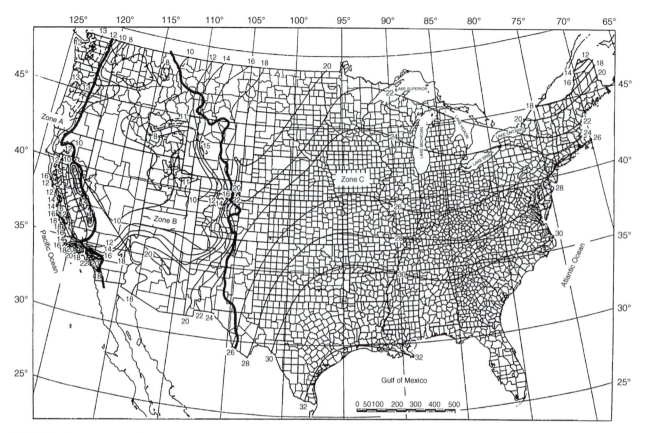

FIGURE 2.21 The 10 mi² or less probable maximum precipitation (PMP) for 6-h duration (inches). (From U.S. Weather Bureau, 1947.)

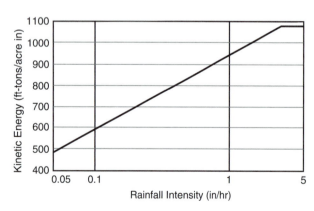

FIGURE 2.22 Kinetic energy vs. rainfall intensity, $E = 916 + 331 \log_{10}$ (intensity); if $i > 3$, $E = 1074$ ft-ton/acre-in. (Based on Wischmeier, W.H., and D.D. Smith, *Predicting Rainfall Erosion Losses — A Guide to Conservation Planning*, USDA Handbook 537, USDA–Science and Education Administration, Washington, DC, 1978.)

2.3. What agricultural crops are grown in rural areas near the four cities you considered in Problem 2.1 or 2.2? Are the crops irrigated, and if so, why?

2.4. In the late 1800s, you decided to leave New England and explore the interior of the U.S.

Your journey went through the areas where the following cities are located: Buffalo, Columbus, St. Louis, New Orleans, Oklahoma City, Phoenix, Los Angeles, and Seattle. What vegetation (biome types) would you have found in each of these locations, and why? How long do you think this journey would have taken? *Note:* Refer to A.H. Strahler and Strahler, 2002, or MacDonald, 2002.

2.5. After reaching Seattle, WA, in the late 1800s, you felt that this location was too wet and decided to follow the Columbia River inland. You reached the Snake River, which is the main tributary to the Columbia River, and followed it into Idaho. How did the annual distribution and amount of precipitation change as you moved from the coast eastward across Washington State and into Idaho? How did these changes influence the vegetation and the selection of crops now grown in this region?

2.6. The maximum rainfall for a 3-h storm was estimated as 0.85 in. What would be the average rainfall amount and average intensity if the watershed area is 200 mi²? *Hint:* Look at **Figure 2.8**.

TABLE 2.8
Energy Analysis Summary for Example 2.8

Time (min)	Depth (mm)	Amount (in.)	Duration (min)	Intensity (in./h)	Energy (per acre-in.)	Total Energy (ft-ton/acre-in.)
0 to 20	25.9	1.02	20	3.06	1074	1095
20 to 40	20.9	0.82	20	2.47	1040	853
Total	46.8	1.84	40			1948

TABLE 2.9
Data for Problem 2.11

Time (min)	Amount (mm)	Time (min)	Amount (mm)	Time (min)	Amount (mm)
10	3.0	110	2.8	210	2.0
20	0.8	120	10.9	220	0.3
30	0.5	130	3.0	230	0.8
40	3.3	140	2.0	240	0.3
50	0	150	1.8	250	0.5
60	0	160	0.3	260	0.3
70	0	170	0.8	270	0.3
80	0.3	180	0.8	280	0.3
90	0.3	190	9.9	290	0.5
100	0.5	200	12.2	300	0.8

TABLE 2.10
Data for Problem 2.13

Rain Gage	Acres per Gage	Rainfall (in.)
A	400	2.1
B	300	2.5
C	200	2.3
D	100	2.7

TABLE 2.11
Rainfall Data for Problem 2.13

Year	Rainfall (mm)	Year	Rainfall (mm)
1981	32.73	1991	31.53
1982	31.16	1992	38.72
1983	34.43	1993	39.45
1984	39.04	1994	32.54
1985	32.26	1995	21.55
1986	23.61	1996	27.21
1987	26.25	1997	26.33
1988	28.01	1998	27.97
1989	27.31	1999	28.09
1990	35.08	2000	34.63

2.7. A forest researcher wishes to locate a series of rain gages around the edge of a newly harvested area. The height of the surrounding unharvested trees is 100 ft. What is the minimum distance that should be allowed between the unharvested trees and the rain gages?

2.8. A hydrologic study is planned for a watershed of 6 mi². How many rain gages are recommended for this study? *Hint:* Look at **Table 2.3**.

2.9. You are studying a basin of 2000 mi² that has 20 rain gages. What percentage standard error would you expect for a given rain event? If a rainstorm there was measured as 3.0 in., what is the probability that the storm was actually greater than 3.6 in.? What is the probability that it was less than 2.7 in.? *Hint:* Look at **Figure 2.9** and Example 2.1.

2.10. On a snow course, the depth is measured as 7.5 ft. The density is measured as 30%. What is the water equivalent? *Hint:* Look at Example 2.2.

2.11. From the data provided in **Table 2.9**, calculate the storm duration, total rainfall amount (maximum intensities for 10, 30, and 60 min), and average intensity.

2.12. Compute the average rainfall for a 1000-acre watershed by the arithmetic average method and by the Thiessen method with data from **Table 2.10**. *Hint:* Read Section 2.6.

2.13. Use the Hazen method and probability paper to determine the expected annual rainfall for return periods of 2, 20, and 100 years for the annual rainfall records in **Table 2.11**. *Hint:* Look at Equation 2.2.

2.14. Use the Weibull method and a spreadsheet (use a log–log or semilog scale) to solve Problem 2.13. Do the two methods give similar answers? *Hint:* Look at Equation 2.3.

TABLE 2.12
Rainfall Amounts Equaled or Exceeded for Periods of 5 min to 24 h
for Expected Return Periods of 2 to 100 years for Columbus, OH

Duration	2 years	5 years	10 years	25 years	50 years	100 years
5 min	0.35	0.45	0.51	0.59	0.61	0.71
10 min	0.55	0.72	0.83	0.98	1.08	1.18
15 min	0.65	0.88	1.05	1.22	1.38	1.50
30 min	0.90	1.20	1.40	1.70	1.87	2.07
1 h	1.10	1.50	1.75	2.10	2.32	2.60
2 h	1.26	1.72	2.00	2.40	2.65	3.00
4 h	1.42	1.93	2.26	2.68	2.96	3.30
8 h	1.61	2.16	2.52	2.96	3.30	3.64
12 h	1.80	2.36	2.74	3.20	3.53	3.88
24 h	2.14	2.76	3.18	3.75	4.08	4.50

2.15. An irrigation reservoir was designed to withstand a 50-year storm event. During the first 3 years, the reservoir withstood the design storm runoff once. What was the probability that this reservoir would experience this 50-year storm at least once in those 3 years? *Hint:* Look at Equation 2.4.

2.16. The maximum recorded point precipitation in southern Pennsylvania is about 31 in. in a period of 6 h. How does this value compare to the information in **Figure 2.21**. (Remember that **Figure 2.8** can be used to convert point data to an area estimate.)

2.17. Use a spreadsheet to make a plot of precipitation depth vs. duration on a log–log scale for the data in **Table 2.7**. Plot the extreme precipitation value from Problem 2.16 on your graph and draw a line through the point that is parallel to your plot of the world's extreme point data. If we assume that this new line gives approximations of PMPs for Pennsylvania, estimate the 24-h PMP at a point and across a 200-mi^2 watershed.

2.18. Rainfall rate–duration frequency data for Columbus, OH, are presented in **Table 2.12**. How well does the 100-year, 6-h duration value compare to estimates obtained using **Figure 2.17** and **Figure 2.18**? What percentage of the PMP for Columbus, from **Figure 2.21**, is your value? If you live below a dam designed based on 100-year values, should you be concerned?

3 Infiltration and Soil Water Processes

3.1 INTRODUCTION

Infiltration plays an important role in nature and human activities. *Infiltration* is defined as the passage of water through the surface of the soil, via pores or small openings, into the soil profile. Water infiltrating into the soil profile is a necessity for vegetative growth, contributes to underground water supplies that sustain dry-weather stream-flow, and decreases surface runoff, soil erosion, and the movement of sediment and pollutants into surface water systems. Infiltration directly affects deep percolation, groundwater flow, and surface runoff contributions to the hydrologic balance in a watershed. Accounting for infiltration is fundamental to understanding and evaluating the hydrologic cycle.

Infiltration processes are complex and difficult to quantify. Infiltration methods may be classified by theoretical equations based on the physics of soil water movement, empirical equations based on parameters with physical significance, empirical equations that contain parameters with no physical significance, or *in situ* measurement methods. Much research has been devoted to measuring infiltration and developing and verifying empirical methods. In recent years, advances in computer technologies have made the solution and application of complex equations that describe infiltration processes more feasible.

This chapter presents a description of soil water relationships, infiltration processes, factors affecting water movement through soils, methods for estimating infiltration rates, and methods for measuring infiltration rates and soil physical properties that influence soil water movement.

3.2 SOIL WATER RELATIONSHIPS

A soil profile consists of a mixture of solid, liquid, and gaseous materials. It is also the home of billions of microorganisms, such as bacteria, fungi, and protozoa. Solid materials in a soil profile include particles of different sizes, shapes, and mineral composition. The soil particles consist primarily of disintegrated and decomposed rock fragments. Simply crushing rock into fine fragments will not produce soil, but might accelerate its formation. Rock decomposition into soil occurs mainly due to oxidation, carbonation, hydration, chemical corrosion, and biological processes associated with microorganisms, the roots of plants, and other life forms that inhabit the soil environment. Other solid materials in the soil profile are organic matter from plants, animals, and microorganisms.

Description of the solid soil phase is commonly based on the size of the soil particles. Soils usually contain a mixture of clay, silt, and sand particles. Sand has large particles, while clay has very small particles. Commonly used soil texture classification systems are presented in **Figure 3.1**. The USDA has established a classification

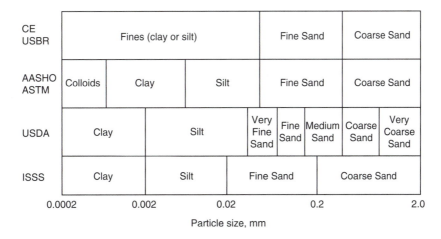

FIGURE 3.1 Soil classification based on particle size (AASHO = American Association of State Highway Officials; ASTM = American Society for Testing and Materials; CE = Corps of Engineers; ISSS = International Society of Soil Science; USBR = U.S. Bureau of Reclamation; USDA = U.S. Department of Agriculture).

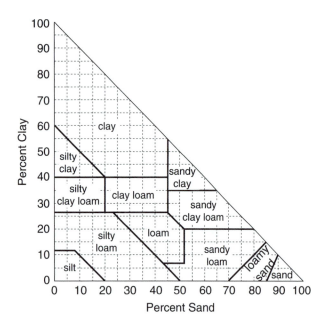

FIGURE 3.2 USDA triangle for determining textural classes.

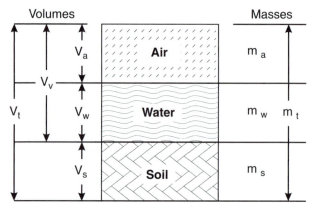

FIGURE 3.3 Soil matrix constituents.

scheme based on different particle size mixtures (**Figure 3.2**). For example, a soil that is 30% clay, 10% silt, and 60% sand is classified as a sandy clay loam. This classification system is based on less than a few percent of the soil mass having particles larger than coarse sand.

Between the soil particles and organic matter are open spaces called *voids* or *pores*. Water that fills part or all of the pores is commonly called *soil water*. When the pores are completely filled, the soil is described as *saturated*. Usually a small percentage of the water, called *hygroscopic* soil water, is held so tightly by molecular attraction to the surface of soil particles that it is not removed under normal climatic conditions. Pores that are empty or only partially filled with water contain oxygen, water vapor, and other gases. Chemicals and solid materials associated with climatic processes and surface activities might also be found in a soil profile. These materials and chemicals might be attached to soil particles, trapped in pores, or in the soil water.

If we approximate this complex system as three phases consisting of air, water, and solids (**Figure 3.3**), then several useful relationships can be established. The total volume V_t is the sum of the volumes for the three phases:

$$V_t = V_a + V_w + V_s \qquad (3.1)$$

where the subscripts *a*, *w*, and *s* refer to the air, water, and solid phases, respectively. Volume can be expressed in cubic feet, cubic centimeters (cc), or as a depth (inches, feet, or centimeters) per unit cross-sectional area. For example, it is not uncommon in agriculture to talk about

the change in the soil water content of a soil profile as a matter of some measured number of inches. To convert this amount into a volume, it is necessary to multiply it by the surface area of the land under consideration.

The volume of the voids or pores is

$$V_v = V_a + V_w \qquad (3.2)$$

If we substitute Equation 3.2 into Equation 3.1, rearrange the equation, and divide the whole equation by V_t, a relationship is established for the *porosity*:

$$n = \frac{V_v}{V_t} = 1 - \frac{V_s}{V_t} \qquad (3.3)$$

Porosity is an important property in problems involving water volumes or water movement. It is commonly used in calculations made by hydrologists, soil scientists, and agricultural drainage engineers.

It is often necessary to know the actual soil water content in a soil profile. Soil water contents can be expressed either by volume or by mass. The soil water content by volume θ_v is the volume of water in a soil sample divided by the total volume of the sample:

$$\theta_v = \frac{V_w}{V_t} \qquad (3.4)$$

Normally, the volumetric soil water content is expressed as a percentage, and the answer from Equation 3.4 will need to be multiplied by 100.

The fraction or percentage of the pores that are filled with water is called the *degree of saturation*. The volumetric soil water content, degree of saturation, and porosity are related as follows:

$$\theta_v = Sn \qquad (3.5)$$

where S is the degree of saturation, and n is the porosity. Use of the relationships presented in Equation 3.1 to Equation 3.5 is illustrated in Example 3.1.

EXAMPLE **3.1**

Determine the maximum depth of water that can be stored in the top 3 ft of a soil profile with a porosity of 0.5. Also, determine the volumetric soil water content if the degree of saturation is 60%.

Solution

From Equation 3.3, the volume of voids $V_v = nV_t$. Therefore,

$V_v = 0.5 \times 3$ ft $\times 12$ in. $= 18$ in./unit surface area

The volumetric soil water content when the degree of saturation is 60% can be determined from Equation 3.5:

$\theta_v = Sn = 0.6 \times 0.5 = 0.3$ (in./in. or in.^{3}/in.^{3})

Answer

The top 3 ft of the soil profile can store 18 in. of water. The volumetric soil water content is 0.30 (in.^{3}/in.^{3}) when the degree of saturation is 60%.

The total mass m_t of a soil profile is

$$m_t = m_a + m_w + m_s \qquad (3.6)$$

where m_a is the mass of air, m_w is the mass of water, and m_s is the mass of solids. The mass of the air is negligible and the total mass is approximated as the sum of the water and soil masses.

In Equation 3.4, the soil water content is expressed by volume. However, the most common way of determining the soil water content is to obtain a sample of the soil, weigh the sample, dry the sample, and then reweigh the dry sample. This procedure is discussed in more detail at the end of the chapter. When the soil water content is determined as a function of mass or weight, it is called the *gravimetric soil water content* θ_g and is equal to

$$\theta_g = \frac{m_w}{m_s} \qquad (3.7)$$

Use of Equation 3.7 is illustrated in Example 3.2.

EXAMPLE **3.2**

A sample of soil is placed in a soil tin with a mass of 25 g. The combined mass of the soil and tin is 140 g. After oven drying, the soil and tin had a mass of 120 g. Calculate the soil water content by mass.

Solution

By subtracting the mass of the tin, the mass of wet sample is calculated as 115 g (140 − 25 g), and the mass of the dry sample is 95 g (120 − 25 g). Therefore, the mass of water in the sample is 20 g (115 − 95 g). Using Equation 3.7, the soil water content by mass is

$$100 \times \frac{20}{95} = 21\%$$

Answer

The soil water content by mass (or weight) is 21%.

The behavior of a soil is very dependent on the *bulk density* of the soil profile ρ_b, which is the mass of soil per unit volume and is related to other soil physical properties as follows:

$$\rho_b = \frac{m_t}{V_t} \qquad (3.8)$$

The dry bulk density is the mass of dry soil divided by the total volume:

$$\rho_{\text{dry}} = \frac{m_s}{V_t} \qquad (3.9)$$

The dry bulk density of the profile can be related to the density of soil particles and porosity as follows:

$$\rho_{\text{dry}} = \rho_p (1 - n) \qquad (3.10)$$

The density of the soil particles ρ_p is defined as the mass of dry soil divided by the volume of the soil (m_s/V_s). The density of most soil particles is 160 to 170 lb/ft^3 (2.6 to 2.75 g/cm^3). If the soil particle density is not measured, a value of 165 lb/ft^3 (2.65 g/cm^3) is often assumed.

The ratio of the density of the soil particles to the density of water is called the *specific gravity*:

$$G_s = \frac{\rho_p}{\rho_w} \qquad (3.11)$$

where ρ_w, the density of water, is usually assumed to be 1.0 g/cm^3 or 62.4 lb/ft^3. Use of several of these equations is illustrated in Example 3.3.

EXAMPLE 3.3

The dry bulk density of a soil sample is 84.24 lb/ft^3, and the density of the soil particles is 168.48 lb/ft^3. Calculate the porosity of the soil profile and specific gravity of the soil particles. Also, determine the bulk density of the sample if it is saturated.

Solution

In Equation 3.10, substitute 84.24 lb/ft^3 for the dry bulk density ρ_{dry} and 168.48 lb/ft^3 for the density of the soil particles ρ_p; then,

$$\rho_{dry} = \rho_p (1 - n)$$

and

$$84.24 = 168.48 \, (1 - n)$$

Dividing both sides by 168.48 lb/ft^3 and rearranging gives $n = 0.5$.

The specific gravity of the soil particles is found by substituting into Equation 3.11 the particle density of 168.48 lb/ft^3 and the density of water, 62.4 lb/ft^3:

$$G_s = \frac{\rho_p}{\rho_w} = \frac{168.48}{62.4} \left(\frac{lb/ft^3}{lb/ft^3} \right)$$

Therefore, the specific gravity of the soil particles is 2.7.

If the sample is saturated, all the pores are filled with water, and the volume of water will be equal to the volume of pores. From Equation 3.3, the volume of water will be equal to the porosity times the total volume:

$$n = \frac{V_v}{V_t} \quad \text{and} \quad V_w = V_v$$

Therefore,

$$V_w = 0.5 \, V_t$$

The mass of water is equal to the density of water (62.4 lb/ft^3) times the volume of water. The mass of soil is the density of the soil particles (168.489 lb/ft^3) times the volume of soil or the dry bulk density (84.24 lb/ft^3) times the total volume, Equation 3.9.

Therefore, using Equation 3.8 and Equation 3.6, the saturated bulk density is

$$\rho_b = \frac{m_t}{V_t} = \frac{m_w + m_s + m_a}{V_t}$$

and

$$\rho_b = \frac{0.5 \times 62.4 \times V_t + 84.24 \times V_t + 0}{V_t} = 115.44 \; (lb/ft^3)$$

Answer

The porosity n is equal to 0.5, and the specific gravity G_s is 2.7. V_t in the numerator and denominator cancels and $\rho_b = 115.44$ lb/ft^3.

3.3 INFILTRATION AND SOIL WATER RETENTION

The downward movement of water through a soil profile is caused by *tension* and *gravitational* forces in the soil matrix. The soil matrix is generally heterogeneous and consists of a labyrinth of pores of varying shape and size connected by porous fissures and channels. Water is held in the soil matrix by tension forces at the air–water interfaces in pores. These tension forces are also frequently called suction, matric, or capillary forces. The term *suction* is often used because water is sucked or pulled into the pores. A common example of capillary flow (flow due to tension forces) occurs when a person's finger is pricked, and a blood sample is taken. The blood sample is drawn into the thin sample tube due to capillary forces. To illustrate water movement due to tension forces, dip a dry blotter or paper towel into water and note how quickly water is sucked into the dry material against the pull of gravity. Also, note that the rise of water is slower as the height of the wetting front above the water surface increases.

Soil water tension varies from less than 1 in. of water head for a soil near saturation to as much as 10 million in. of head for a very dry condition. The effect of surface tension in a soil matrix during drainage can be described by considering water held in a single pore within the soil profile connected to the groundwater table (**Figure 3.4**). The figure describes an equilibrium state at the meniscus with radius r_1. If it is assumed that the meniscus with radius r_1 supports a column of water h, then the gravitational forces will equal the surface tension forces and

$$\left(\pi r_1^2 \right) h \rho_w \, g = \left(2\pi \, r_1 \right) \tau \, \cos \alpha \qquad (3.12)$$

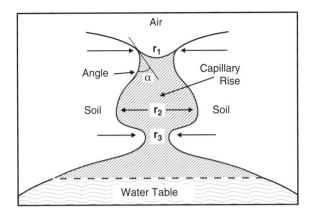

FIGURE 3.4 Water retention in a heterogeneous porous media.

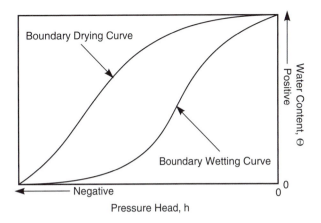

FIGURE 3.5 Hysteresis during drying and wetting.

where ρ_w is the density of water, g is the gravitational constant, τ is the surface tension force, and α is the angle of contact of the meniscus with the soil.

Surface tension forces are due to adhesion and cohesion. The attraction of water to the sides of the pore is called *adhesion,* and the attraction of water to water is called *cohesion.* Water is at a higher level on the sides of a pore because of adhesion and creates a curved surface that is called the *meniscus.* The drop in the surface of the water at the center of the pore pulls water into the pore until the weight of the water in the pore balances the difference in the force of the atmosphere inside and outside the pore. The height (head) that water rises in the pore is inversely proportional to the radius of the pore (or area of the meniscus). Therefore, the smaller the radius, the longer the column of water that can be supported. The smallest pores will fill first as they exert the largest surface tension forces. If the water table falls, h will increase, and Equation 3.12 will no longer hold true. The largest pores containing water will begin to drain until a smaller controlling meniscus is reached and equilibrium is attained between the suction and gravitational forces. For the soil profile illustrated in **Figure 3.4**, a slight drop in the water table might cause no drainage from the pore with radius r_2 because the water in this pore is controlled by the two smaller openings with radii r_1 and r_3. Drainage would only occur when the water table falls far enough that the small opening r_3 begins to drain because it cannot maintain the water column to the water table. It is evident that the soil water content under drainage conditions is different from that under adsorption (wetting) for a given soil water tension.

During wetting, the small pores fill first, while during drainage and drying, the large pores empty first. This causes hysteresis in the system (**Figure 3.5**). Soil water hysteresis, a different relationship between soil water and soil suction during wetting and drying, will vary depending on the wetting and drying history of the soil. Other hysteresis relationships are discussed separately. Retention

and movement of water during wetting or drainage is a function of the shape and size of the pores. They are also a function of how well the pores are connected. Drainage is also a function of the moisture or tension gradient. When water is present, soil air is displaced from the pores, the soil water content increases, and the soil tension decreases. This results in decreased infiltration rates. Provided enough water is present, this process will continue until the soil is saturated (all the pores are filled). At saturation, the soil suction will be zero. However, in most natural environments, a small amount of air will be trapped in the pores and prevent complete saturation. The final degree of saturation will probably be 90 to 95%. When water moves down into a pore the gravitational forces and tension forces work together; gravity pushes water into the pore, and tension forces suck or pull water into the pore. However, when a pore empties, the two forces work against each other, and there is a tug-of-war between the gravitational and suction forces. A similar tug-of-war occurs when the root of a plant tries to extract water from a pore. The plant has to apply sufficient pulling force to overcome the opposite pulling force due to tension in the pore.

Following wetting, there will be a redistribution of soil water, and pores will drain due to capillary and gravity flow. When gravity flow becomes negligible, the soil water content of the profile will be at *field capacity* (FC). Field capacity typically occurs at soil suctions of 0.1 to 0.33 bar (**Figure 3.6**). It will vary depending on the wetting and drying history of the soil, soil texture, porosity, and subsurface characteristics of the soil profile. The elusive nature of field capacity is illustrated in **Figure 3.7**.

Without realizing it, you have probably experienced a situation when gravity drainage occurred following wetting of a soil. Consider watering a plant in a pot and adding the water until you see water ponded at the top of the container. If you come back a short time later, you discover that the water has drained through the potting soil, filled the saucer below the pot, and perhaps overflowed onto

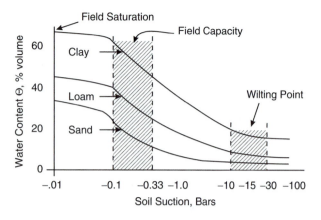

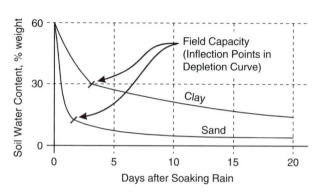

FIGURE 3.6 Soil water and soil suction relationships to field capacity wilting point and plant-available water.

FIGURE 3.7 The elusive nature of field capacity.

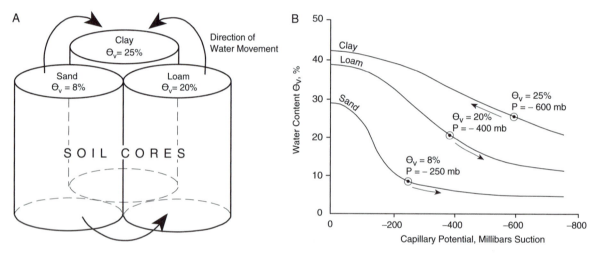

FIGURE 3.8 Direction of water movement when three dissimilar soils are placed together. Equilibrium between the three soils will occur only when the capillary potentials are equal. (Redrawn from Hewlett, J.D., *Principles of Forest Hydrology*, University of Georgia Press, Athens, 1982.)

your table or desk (note that some of this water may have simply flowed around the soil mass if there is any space between the soil mass and the container).

Redistribution of water due to capillary flow will continue after gravity flow ceases. Such redistribution of soil water is driven by soil water tension and not by soil water content θ, measured either by volume or by weight. Thus, a seemingly "drier" soil may actually give up water to an ostensibly "wetter" soil. This is demonstrated in **Figure 3.8A**, which shows three soil cores of varying texture and wetness placed in mutual contact. Note that soil water content by volume is highest in the clay (25%), next highest in the loam (20%), and driest in the sand (8%). A question that might be asked is, "In what direction will the soil water move?" Intuitively, one would expect the clay and loam to give up water to the sand, but it does not happen that way. The reason is because the relation

of θ to suction is largely a function of soil texture (**Figure 3.6**). An example of that is given in **Figure 3.8B**, which shows that, for a given value of soil water content (say 20%), the clay will have the highest suction, sand the lowest, and loam somewhere in between. Thus, the clay is dynamically or functionally driest, while the sand is wettest. For a given soil suction, say 200 mb, the clay, conversely, will have the highest proportion of water, while the soil will have the least. Thus, the answer to the question regarding the soil cores becomes apparent. Since soil water is drawn toward the relatively higher suctions, water will move from the sand to both the loam and the clay, but the loam will give up water only to the clay. Given perfect capillary action within the three cores, their capillary potential or suction will come to a common equilibrium, as suggested in **Figure 3.8B**.

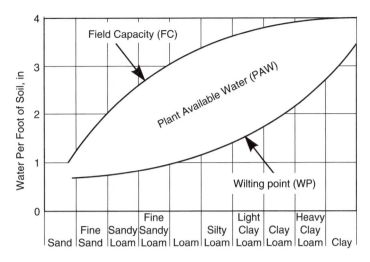

FIGURE 3.9 Water-holding characteristics of soils with different textures.

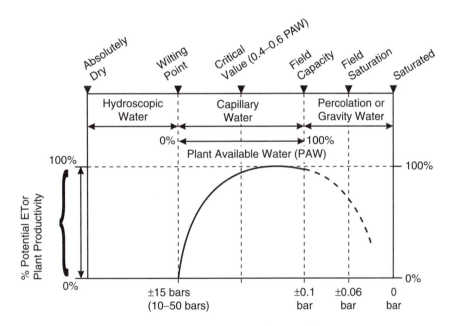

FIGURE 3.10 Yield responses to soil water content for typical mesophytic crop plants.

Usually, upward movement of soil water will occur due to drying of upper soil levels from evapotranspiration, with a resulting higher capillary potential or head. The depth that plants can remove water from the soil will depend on the root distribution and mass, profile characteristics, and climatic conditions. The soil water content in the root zone will eventually reach the wilting point (WP) unless there is further wetting or the soil water content within 1 to 3 ft of the root zone is near saturation. The soil suction at wilting point is typically about 15 bars (15 Mpa) and is the point at which plants begin to wilt and do not regain turgor if the soil is rewetted. For sand, the soil water content at WP is very low. For clay, it is much higher (**Figure 3.6** and **Figure 3.9**). Soil water in excess of the amount at the wilting point is available for evapotranspiration (*ET*). The maximum amount of water that can be extracted from the soil by plants is termed the plant available soil water (PAW) (**Figure 3.9** and **Figure 3.10**). It is related to field capacity and wilting point as follows:

$$\theta_{paw} = \theta_{fc} - \theta_{wp} \qquad (3.13)$$

where θ_{paw}, θ_{fc}, and θ_{wp} are the volumetric or gravimetric plant available soil water content, the soil water contents at field capacity, and at the wilting point, respectively.

Note that field capacity is primarily a function of the soil profile and soil properties; the wilting point also depends on the type, growth stage, and health of the plants. Some vegetable plants will wilt at soil suctions much less than 15 bars, while some trees can extract water at suctions that are several times larger than 15 bars.

Knowledge of the wilting point, field capacity, and plant available soil water content is particularly important to agriculture. In arid areas, this type of information is needed to design irrigation systems and determine irrigation schedules. For most plants, there is a reduction in yield when the soil water content falls below a critical value, which is 40 to 70% of the plant available soil water content. A typical relationship is shown in **Figure 3.10**, and use of this type of information is illustrated in Example 3.4. Note that the use of "plants" in this text generally refers to mesophytic plants of humid to semihumid mid-latitude regions (**Figure 3.10**). These include most field crops. For the plant physiology and water use characteristics of plants from other, more exotic regions, refer to a text such as that by Nobel (1983).

EXAMPLE 3.4

Determine the amount of water needed to increase the soil water content in the root zone of a clay loam soil from the critical value for plant growth to field capacity. The critical value occurs at 50% of the plant available water, the root zone depth is 2 ft, and the soil has the soil water properties presented in **Figure 3.9** at the interface between a clay loam and heavy clay loam.

Solution

From **Figure 3.9**, the soil water content at wilting point is 2 in. of water in each foot of the soil profile. The soil water content at field capacity is 3.8 in. of water in each foot of the soil profile. Therefore, using Equation 3.13, the plant available water content is

$$\theta_{paw} = (3.8 - 2.0) \text{ in. water per foot of soil profile}$$

The root zone is 2 ft deep, and PAW = $1.8 \times 2 = 3.6$ in.

Answer

If the critical value is 50% of the PAW, then 1.8 in. of water (0.5×3.6 in.) are needed to increase the soil water content in the root zone from the critical value to field capacity.

Bodman and Coleman (1943) showed that soil water movement into a uniform dry soil under conditions of surface ponding could be divided into four zones as shown

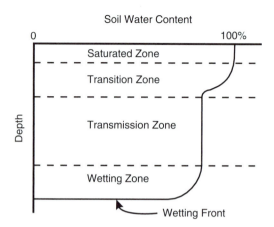

FIGURE 3.11 The infiltration zones of Bodman and Coleman (Based on Bodman, G.B., and E.A. Coleman, *Soil Sci. Soc. Am. Proc.*, 7:116–122, 1943).

in **Figure 3.11**. The *saturated zone* usually only extends from the surface to a depth of less than 1 in. Below the saturated zone is the *transition zone*, which represents a zone of rapid decrease in soil water content. Underlying the transition zone is a zone of nearly constant soil water called the *transmission zone*. This zone increases in length as the infiltration process continues. Below the transmission zone is the *wetting zone*. The wetting zone maintains a nearly constant shape and moves downward as the infiltration process proceeds. The wetting zone culminates at the *wetting front,* which is the boundary between the advancing water and the relatively dry soil. The wetting front represents a plane of discontinuity across which a high suction gradient occurs (for more detail, see Figures 6-4 and 6-6 in Singer and Munns, 2002).

3.4 FACTORS AFFECTING WATER MOVEMENT INTO AND THROUGH SOILS

Infiltration may involve soil water movement in one, two, or three dimensions even though it is often approximated as one-dimensional vertical flow. Water movement into and through a soil profile is dependent on many interrelated factors, both natural and human induced.

As mentioned, soil texture, bulk density, heterogeneity, cracks, and surface conditions will all influence water movement. Also, *hydraulic conductivity* is an important property of soils and is defined as the ability of a soil to transmit water under a unit hydraulic gradient. Hydraulic conductivity is often called *permeability* and is a function of soil suction and soil water content. Fine-grained soils tend to have lower hydraulic conductivity values than coarse-grained soils. However, fine-grained soils, such as clays, have smaller pores than coarse-grained soils. Remember, the smaller the pores, the greater the suction forces.

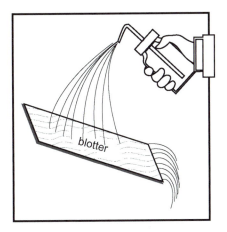

FIGURE 3.12 Infiltration capacity of blotter is low, and there is much runoff.

FIGURE 3.13 Infiltration capacity of sponge is high, and there is little runoff.

Therefore, water might move readily into a dry, fine-grained soil even though the hydraulic conductivity is low.

If there is no spatial variability in soil properties, the soil is described as *homogeneous*. Most soils are *heterogeneous* because they exhibit a considerable amount of variability in properties, both laterally and vertically. Soils are also described as isotropic or anisotropic. Isotropic soils exhibit the same hydraulic conductivity in all directions, while anisotropic soils have different values in the vertical and lateral directions.

Land surfaces can be likened to a blotter (**Figure 3.12**) or a sponge (**Figure 3.13**). Infiltration rates of the blotter are low. To visualize this situation, apply water to a blotter that is tilted; a plastic squeeze bottle can be used. Note how little water goes into the blotter and how much runs off. The pores of the blotter are very fine and transmit water slowly. Now, apply water from the squeeze bottle onto a sponge. Note how little water runs off, even though the water application is high. Most of the added water infiltrates quickly into the sponge. Infiltration rates are high and might exceed 2 in./h, as found in sandy areas or in soil with many macropores and stable structure, such as found in mature woodlands on well-drained soil. Scientific literature is filled with a confusing number of different descriptions and definitions of macropores. We suggest you think of *macropores* as places in the soil profile where the dominant flow process is gravity flow, and soil tension forces are negligible. You might hear people refer to *preferential flow*, which does not have to occur in macropores and simply means that, rather than moving uniformly through the whole soil profile, there is a preferred path the flow will take. This might be the path that offers the least resistance to flow; certainly, macropore flow is one common and important type of preferential flow. A good discussion on preferential flow mechanisms is presented by Nieber (pages 1–10) and the work of other authors in a volume by the American Society of Agricultural Engineers (ASAE, 2001).

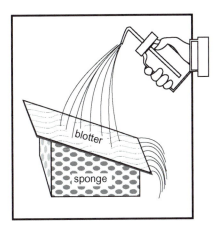

FIGURE 3.14 Infiltration capacity of the sponge is limited by the overlying layer with low permeability.

The blotter might represent a slowly permeable clay with very few large pores or a well-drained silt loam in which the surface had been compacted by excessive tillage and puddled and compacted by rainfall. Its infiltration capacity might be less than 0.2 in./h. Now, place the blotter on top of the sponge (**Figure 3.14**) and apply water. Note the effect of the simulated thin, compacted layer of low infiltration capacity overlying a soil with high infiltration rates. Water will not enter the coarse-textured subsurface layer until the suction forces at the interface between the two layers is equal. This condition, called *soil crusting* (dry) or *sealing* (wet) affects many soils, especially those with low organic matter and unstable structure.

The sponge on top of the blotter (**Figure 3.15**) exemplifies a field situation in which a grass-covered, well-drained soil overlays a layer of soil of low permeability. This situation affects infiltration more than the sponge alone or the blotter on top of the sponge. In this case, at the start of the storm, infiltration rates into the sponge are high. However, the blotter restricts the percolation rate,

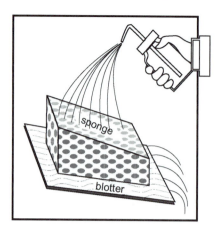

FIGURE 3.15 Infiltration capacity of the sponge is limited by the underlying layer.

causing infiltrated water to accumulate in the pores of the sponge, depleting its capacity for storing more water, and slowing the infiltration rate until it corresponds with that of the blotter.

Indeed, the soil and surface infiltration process is the interface between the atmosphere and the earth's hydrologic processes, both surface and subsurface. While *infiltration capacity* of soils is due to many natural factors, it is the critical zone at which the human impact has one of the greatest roles in affecting hydrologic processes as well as geomorphological change (Chapters 5 and 6). Any precipitation not infiltrated becomes overland flow, which can do all sorts of mischief. Factors of importance for infiltration include surface conditions, subsurface conditions, factors that influence surface and subsurface conditions, flow characteristics of the water or fluid, and hydrophobicity (**Figure 3.16**).

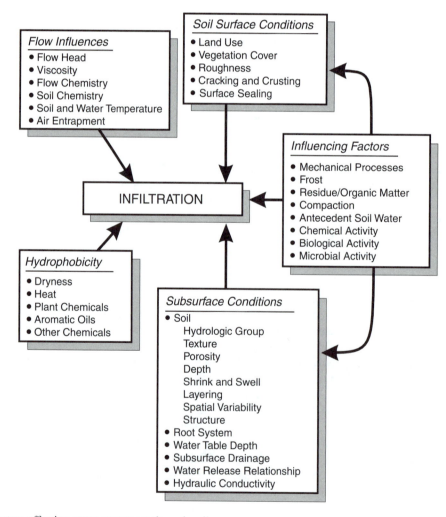

FIGURE 3.16 Factors affecting water movement through soils.

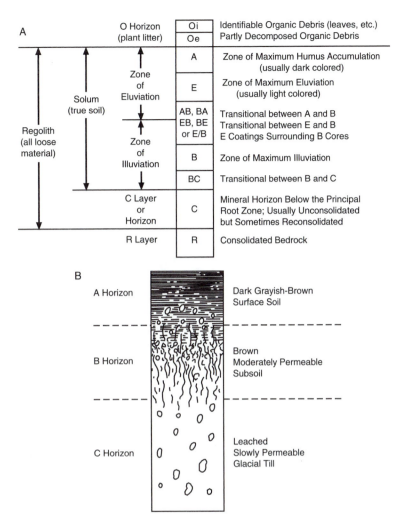

FIGURE 3.17 Typical soil horizons. A. Soil horizon nomenclature. B. Typical soil profile formed in glacial till with grass vegetation. (Based on Troeh and Thompson, 1993. *Soils and Fertility.* Oxford University Press, New York.)

Note that many of the factors discussed below are also significant to other characteristics, such as percolation, water storage and transmission, and plant growth.

3.4.1 SOIL PROPERTIES

Soil is the most important natural variable that influences infiltration. Among the several properties of soils affecting infiltration capacity are texture (coarse texture normally permits more infiltration), category of clay (swelling or nonswelling; see Section 3.4.2.7, Antecedent Soil Moisture), structure, and organic material. We normally look at the surface horizon or topsoil, but any horizon or layer may be limiting. Soil horizons or layers are typically described and represented by the capital letters A, B, C, E, H, O, and R. An example of soil layers that might occur in many parts of the U.S. is presented in **Figure 3.17**. The A horizons are primarily mineral horizons that form at or near the ground surface. They are characterized by humidified organic matter mixed with the mineral fraction. They have properties associated with a land management practice

such as tillage, crop or vegetation type and rotation, and a morphology that is different from the underlying layers. Details on soil horizons and soil taxonomy can be found in most soil texts. A particularly useful book is the *Soil Survey Manual* (USDA, 1993), which describes how field measurements are made, soils are mapped, and soils are described.

In the U.S., the NRCS has conducted detailed soil surveys that provide a wealth of information on the general properties of soils within each county (see Exercise 14.9, Chapter 14). There are numerous ways and levels of detail that can be used to group and describe soils. If you are not a soil scientist, an understanding at least of the terms *taxonomic class*, *soil series,* and *soil map unit* is useful. A taxonomic class has a set of soil characteristics with precisely defined limits. In the U.S., these classes are mainly based on the kind and character of soil properties and the arrangement of the soil horizons in a profile. In the U.S., the classification is based on the following six categories: order, suborder, great group, subgroup, family, and series. For example, a Gilpin Soil is a fine-loamy, mixed, mesic, Typic Hapludults for which fine-loamy mixed is the family, Typic identifies the subgroup, Haplu identifies the great group, Udults represents the suborder and is from Ultisol, which is the order (for more details, see USDA, 1975). The soil series is a group of soils that have similar characteristics and a similar arrangement of soil layers (horizons). A soil map unit is a delineation on a map of an area dominated by one major kind of soil or several soils with similar properties.

3.4.2 SOIL CONDITION

In addition to general knowledge of the soil type (texture, taxonomy, or series), specific knowledge of the physical, chemical, and biological condition of the soil is often needed. These conditions are discussed separately below, but there are many interconnections.

Compaction.   Compaction is a process that increases density and reduces the pore size and pore space in the soil so that less water can move through in a given time period. Mechanical compaction can come from operating heavy equipment over land, especially when the soil is wet. But, on a worldwide scale, overgrazing is the biggest culprit. A mature cow, for example, can weigh as much a ton, but have all the weight on the relatively small basal area of four hooves. Quite often, under conditions of momentum (like climbing a hill), they have most of their weight on one hoof, thus exerting tremendous pressure for soil compaction. In more extreme cases, the net effect can increase bulk density of the soil by 50% and reduce infiltration by 80% over field-size areas (Trimble and Mendel, 1995). Surveyors know the effects of soil compaction: In noncompacted soil, "chaining pins" (steel stakes 4 mm in diameter, acting here like a soil penetrometer) can be easily pushed in the ground with the pressure from one finger, while the body weight and strength of a large man can hardly drive them into a highly compacted soil. Often, the surface of two contiguous fields may lie at different elevations, with the grazed one several inches lower, with the difference due to compaction and perhaps also to erosion.

Soil Structure.   The textural particles of soils tend to be formed into larger units (such as crumbs, plates, and prisms) that allow more pore space and more infiltration. These structural particles are held together by electrochemical bonds and chemical glues (see Section 3.4.2, "Organic Material"). The particles may be broken apart by mechanical compaction. Also, a reduction of organic material, accumulation of sodium (see Section 3.4.2, "Water Quality"), or soil nutrient depletion, especially of divalent cations such as calcium and magnesium, can reduce the structural strength and allow the soil to compact.

Organic Material.   The role of organic material in infiltration is at least threefold. First, the organic material can absorb moisture and form porous spaces through which water can move. This is especially true of roots, particularly dead ones. Second, the organic material can be digested by bacteria to become organic glues (polysaccharides) that help hold the structural particles together (think of organic gels used for hair styling). Third, organic material helps create a healthy habitat for soil flora and fauna, which in turn can increase infiltration capacity. Organic material in soil is reduced by conventional cultivation with the removal of crops and crop residues and with conventional plowing, which exposes the organic material to oxidation (Singer and Munns, 1999).

Soil Fauna (Endopedofauna).   Soil fauna (endopedofauna), microscopic to large in size, generally have positive effects on the hydraulic conductivity of soil by (1) increasing the porosity and permeability, (2) improving soil structure, and (3) increasing fertility, which increases other biological activity (see later sections). Soil fauna range from insects, to moles, to woodchucks, but the most common and significant is the earthworm (*Lumbricina*). It has been observed that a good pasture may have a biomass of earthworms beneath the surface almost equal to the biomass of cattle grazing on the surface. Not only does the earthworm have a strong role in creating holes through the soil and reducing bulk density, but it also helps support higher vertebrates such as moles and voles that help keep soil permeable (Wallworth, 1970; Hole, 1981). Soil fauna populations can be vastly reduced by plowing, overgrazing, and poor use of pesticides.

Soil Profile Truncation.   The topsoil (A horizon, **Figure 3.17**) tends to be coarser, have more organic material, and have better structure than the subsoil (B horizon). Thus, it usually has a higher infiltration capacity than the subsoil and acts to relay water into the subsoil. When part or all of the topsoil has been eroded, the soil takes on the characteristics of the subsoil. Thus, soil erosion often provides a positive-feedback loop so that erosion proceeds more rapidly as more topsoil is eroded.

Hydrophobicity (Water Repellancy).   Water falling on the surface of a hydrophobic soil will bead up rather than enter the soil, even when suction forces are high. The phenomenon occurs due to waxy organic materials (aliphatic hydrocarbons, among several others identified so far) on the soil surface or at some depth, which create a negative contact angle between the waxes and any applied water. Fungi and other microorganisms may also affect hydrophobicity. This effect, usually at the surface, can come from leaching of organic material on the ground. This phenomenon was originally identified in the Sclerophyllous brushland (chapparral) of Mediterranean climates such as southern California (DeBano, 1981; Imeson et al., 1992) More recent work shows that it occurs significantly in the coniferous forests of the western U.S. and even appears to some degree in grasslands and with certain crops (DeBano, 2000; Doerr et al., 2000). While hydrophobicity is temporary, it may last from hours to years. Thus, hydrophobicity is a widespread, if not ubiquitous, phenomena, and its significance may be far greater than suspected in the past.

Fire can exacerbate the effects of hydrophobicity, especially under brushland and forest. Some hydrophobic substances may be vaporized and driven to depths of up to 20 cm, where they condense on soil particles. The result can be a layer at some depth through which little water can move. Water "piles up" in the permeable layer above, and the result is greatly enhanced saturated overland flow and erosion, sometimes with the removal of a soil layer several inches deep (Shakesby et al., 2000). Soil texture is also an important variable, with coarse texture less affected because of lower specific surface area of particles. That is, a finite amount of hydrophobic substance may cover the relatively small surface area of sand and make it hydrophobic, while the infinitely larger surface area of clays might permit only partial coverage with reduced effects. A third variable is soil moisture, with dry soils most affected by hydrophobicity. As a dry soil gradually "wets up" in the presence of water, however, it tends to lose its hydrophobicity.

A similar, but often different, process to hydrophobicity is a phenomenon recognized by many veteran farmers as "dry soil hydrophobicity." As water infiltrates, not all air in the pores is freely displaced by the flow, and air bubbles are trapped in the pores. Trapped air will block some pores and will retard infiltration as air bubbles try to move upward against the direction of water flow. If you observe a puddle on bare soil immediately following a storm event, it is possible that you will see bubbles of air coming to the surface of the puddle. Entrapment of air is also dependent on the physical properties of the soil, particularly soil pore sizes. The percentage of pore space filled by trapped air will increase as the pore size decreases. This is because the high suction forces produced by small pores oppose upward movement of air and, when combined with the small pore sizes, result in air trapped or pushed downward. The likelihood that air gets pushed downward into the soil profile ahead of the wetting front increases, and this causes a buildup of air pressure that opposes the downward movement of water.

Antecedent Soil Moisture. Apart from the conditions of hydrophobicity discussed in the previous section, initial infiltration rates are higher for dry soil than for wet soil. This is illustrated in **Figure 3.18**. Infiltration rates decrease as the soil water content increases. This is because soil suction decreases with increases in soil water

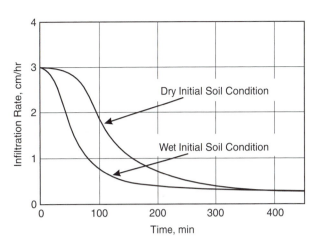

FIGURE 3.18 Infiltration in top soil and spoil materials for dry initial conditions and wet initial conditions.

FIGURE 3.19 Surface crusting and cracking can cause macropore flow.

content. Some clay soils will swell during wetting and will then shrink during drying. Swelling will inhibit infiltration; shrinking will create cracks and increase macropore flow. As a soil becomes wetter during a rainstorm, the infiltration capacity tends to decrease. Thus, previous storms or irrigation can increase the runoff from an area. The soil water content following a rainfall event will depend on the rainfall depth, duration, and intensities. Intermittent light rain will produce higher soil water content than if the same amount of rain occurred in a short period. This is due, in part, to the fact that during light intermittent rain, it is easier for air bubbles to escape upward throughout the soil profile.

Surface Crusting and Surface Sealing. A crust will form at the surface of some soils as they dry (**Figure 3.19**). Crusting occurs when clay and silt-size particles dominate the upper soil materials. The crust will inhibit water movement into the soil matrix. In other settings, especially where clay-size soil particles dominate, infiltration is dominated by macroporosity and surface fracturing of finer-grained materials, and a lack of cover may actually enhance infiltration rates. If flow through a pore occurs primarily due to gravity flow, the pore is commonly called a *macropore*. Gravity flow will occur down the cracks provided a positive head of water is established at the top of the crack. This might be due to water ponding on the surface or surface runoff that flows across a crusted area and into a crack. Where coarser grains of soil in the fine-to-coarse sand sizes wash into the fracture structures, they can create a bridging condition that holds the fractures open even when the soils are rehydrated during times of higher soil moisture. These settings may also be more susceptible to the formation of ice wedging and freeze–thaw cycles, which increases the presence of fractures, than a setting where a well-developed root mass stabilizes the surface soils.

Cracks might also form in the absence of crusting, and gravity flow might occur in the absence of cracks. Worm holes, coarse sands and gravels, and dry organic matter (residue) extending to the soil surface might all result in gravity flow. Water movement through macropores will result in more rapid wetting at deeper depths. In some cases, wetting of the soil matrix might be due primarily to lateral movement of water that has ponded in cracks or upward movement of water from an impeding layer because of water that reached the impeding layer by macropore flow.

Infiltration can also be impeded by surface sealing due to the breakdown of surface structure. A breakdown can be caused by the dispersion of clays or the physical destruction of aggregates by the beating action of rain or irrigation water. Infiltration through loosely packed, air-dried, sandstone materials from Kentucky surface mines was very rapid. However, after wetting and drying, it was noted that infiltration rates were extremely slow during any future applications of water. A very fine seal formed at the soil surface, but water would flow readily through any breaks in the seal (A.D. Ward et al., 1983). During high-intensity rainfall or surface applications of water, soil might be detached and then moved across the soil surface. Some of the detached soil might then be redeposited in cracks and large pores. This deposition of soil in cracks and pores will tend to seal the surface, and infiltration or percolation will be reduced.

Frozen Soil. The effect of frost on infiltration will depend on the soil condition and soil water content at the time of freezing. A very wet, frozen soil can be practically impervious, causing a condition commonly called *concrete frost*. This is common on cropland, especially when the soil is degraded and compacted. On the other hand, a dry forest or grassland frozen soil is likely to be porous, providing nearly normal infiltration rates and a condition called *lattice frost*.

Vegetation and Residue. Vegetation can play a role in infiltration commensurate with that of natural soil characteristics. First, of course, is the fact that vegetation protects the ground from direct raindrop impacts. Excellent, ungrazed deciduous forests of the eastern U.S., with their profuse and deep rooting systems, organic material, and encouragement of soil biological activity, can develop infiltration and transmission rates higher than even the very rainfall intensities characteristic of that region (Hewlett, 1969). The result is that most runoff can be transmitted in the subsurface as interflow (discussed in Chapter 5).

Excellent grassland, such as the original prairie of the U.S. Midwest, also has very high infiltration capacities. In general, grass has an all-around beneficial effect on soils. It provides a dense cover, imparts a large amount of neutral to slightly basic organic material, and concentrates basic, divalent cations (such as calcium and magnesium) near the surface. Perhaps most hydrologically important, it provides a welcoming habitat for soil fauna, which keep the soil open and porous, as indicated by low bulk densities. If the ability of the soil profile to transmit infiltrating water is not limiting, soil surface conditions usually govern infiltration. This is illustrated by curves of cumulative infiltration for vegetation ranging from bare ground to old, well-established pasture during the first 60 min of a rainstorm (**Figure 3.20**). Vegetation provides a ground cover that absorbs the energy of falling raindrops and prevents soil puddling or packing. In more natural environments, the type of vegetation, season of the year, and land management practices (such as tillage) that temporarily modify the near-surface soil conditions will greatly influence infiltration processes. Some practices, such as no-till or conservation tillage, are used to increase water movement into the soil (**Figure 3.21**). In other situations, such as sites used for land disposal of toxic wastes, efforts are made to compact underlying soil to minimize water movement into the soil profile.

Surface Detention (Roughness). It is important to note that overland flow occurs as soon as the infiltration rate is less than the water application rate, and surface depressions have been filled with water. Once water is lost due to runoff, it is obviously not available for infiltration. Tillage practices that leave the soil surface rough with many pockets for water storage are likely to have more infiltration than where the surface has been worked down and smoothed by tillage (**Figure 3.22**). In the latter case, infiltration occurs primarily while it is raining. In the first case with the rough soil surface, infiltration continues after the rain has stopped until all stored water is absorbed. This prolonged time for infiltration opportunity may be many

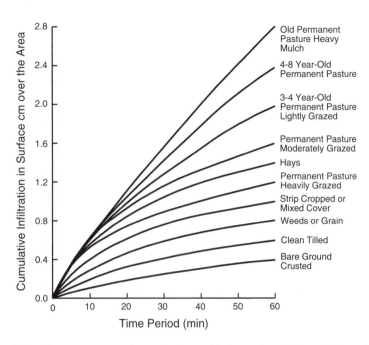

FIGURE 3.20 Typical mass infiltration curves according to land use. (Redrawn from Holtan, H.N., and M. Kirkpatick, *Trans. Am. Geophys. Union*, 31:771–779, 1950.)

FIGURE 3.21 Crop residue left on the ground surface after harvesting will reduce runoff and detachment of soil due to raindrop impact.

FIGURE 3.22 High surface roughness due to tillage will result in more infiltration and hence less runoff.

minutes, and in some cases, significant amounts of infiltration occur. Some specific agricultural practices to create a storage opportunity for infiltration are contour plowing, deep plowing, and terracing.

Water Viscosity. The viscosity of water can affect infiltration. The colder the temperature, the higher the viscosity is, and the slower is the infiltration. In watershed studies, the viscosity of water flow is often neglected by the practicing hydrologist. However, in most laboratory research, viscosity of water is a significant factor.

Water Quality. Under conditions of irrigation, sediment concentration becomes very important because suspended sediment particles can block the soil pores and severely restrict infiltration. A primary role of an irrigation engineer is to remove the sediment from water before it is applied to a field. Indeed, several prominent sedimentation specialists in the past came from the irrigation field. Another problem the irrigation engineer usually encounters is salt, primarily NaCl. Under certain conditions, the monovalent cation sodium can cause collapse of soil structure or *puddling*, which can vastly reduce infiltration. Just

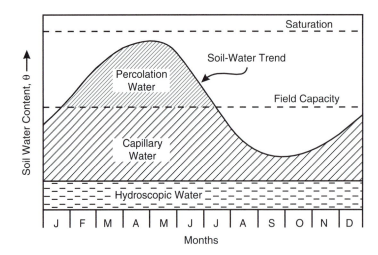

FIGURE 3.23 Generalized seasonal soil water content trends and occurrence of percolation water.

to underline this point, salt or soda ash is often applied to pond bottoms to keep them from leaking.

Hydraulic Depth.   In addition to all the qualitative factors listed above, there is also the depth of soil or overburden through which water can move both vertically and laterally. Most commonly, the restriction is depth to bedrock. Less common is depth to some impermeable strata, such as a soil horizon.

Urbanization.   The hallmark of urbanized land is inadvertent waterproofing of the surface. Roofs, streets, sidewalks, and parking lots are impermeable. Indeed, so much uninfiltrated water can accumulate in an urbanized area during an intense storm that civil engineers must design extensive storm sewer networks to convey the overland flow away. An unanswered challenge is to develop a "porous pavement" that will allow significant infiltration. The hydrologic effects of urbanization can be highly significant and receive considerable coverage in Chapter 5, Chapter 6, and Chapter 12.

Time.   Long-term soil management has effects on the biophysical qualities that affect infiltration. For example, continuous row crops such as corn or soybeans with poor soil conservation management will cause a reduction in organic material, faunal activity, and soil structure with increases of bulk density. The result can be a significant reduction in infiltration capacity on a decadal scale. Conversely, continuous grassland or even well-managed cropland can enjoy increases of organic material, faunal activity, and soil structure with decreases of bulk density and increases of infiltration on a decadal scale (Trimble, 1988, 1990). There are obvious implications for soil conservation, streamflow, and stream channel stability, which are covered in other chapters, especially Chapter 9.

3.5 SOIL WATER BALANCE

Changes in soil water storage occur throughout the year. Increases occur due to precipitation, irrigation, and subsurface inflows. Depletion is caused by percolation, gravitational drainage, and evapotranspiration. A typical cycle for Ohio is illustrated in **Figure 3.23**.

A water balance equation describing these changes for any period of time is expressed as

$$\Delta SM = P + IR - Q - G - ET \qquad (3.14)$$

where ΔSM is the change in soil water storage in the soil profile, P is precipitation, IR is irrigation, G is percolation water, ET is evapotranspiration, and Q is surface runoff. All quantities are expressed as a depth (inches or millimeters) of water over a study area for a specific period of time. Monthly water budget data in **Table 3.1** illustrates changes in soil water as related to factors given in Equation 3.14. An increase in soil water of 2.78 in. in January resulted from precipitation P of 3.48 in., runoff Q of 0.31 in., evapotranspiration ET of 0.31 in., and percolation to groundwater of 0.08 in. These data were obtained on an 8-ft deep natural soil profile lysimeter for which the entire soil block was weighed automatically at 10-min intervals.

For Ohio, soil water storage (**Figure 3.23** and **Table 3.1**) typically increases through January and February, stays at high levels in March, depletes throughout the warm growing season, and then increases in the fall (September to December). Excessive rainfall in August started the trend of increasing soil water for the remainder of the year. The soil water content curve in **Figure 3.23** is only a trend, and actual changes in soil water fluctuate markedly from day to day above and below the trend. Three rainy periods in June caused soil water increases (accretion) that interrupted the general depletion trend for the month. In January,

TABLE 3.1
Monthly Accretion, Depletion, and Storage of Soil Water (in.) as Determined by Monolith Lysimeters, Coshocton, OH, 1962

Month	Precipitation P (in.)	Runoff Q (in.)	Evapotranspiration ET (in.)	Percolation G (in.)	Total Depletion (in.)	Profile Storage Change ΔS (in.)
January	3.48	0.31	0.31	0.08	0.70	2.78
February	4.10	0.57	0.71	0.83	2.17	1.93
March	3.52	0.02	1.12	2.41	3.55	−0.03
April	1.73	0.00	2.44	1.49	3.93	−2.20
May	2.86	0.01	5.71	0.83	6.55	−3.69
June	1.84	0.00	4.25	0.30	4.55	−2.71
July	2.72	0.00	4.56	0.02	4.58	−1.86
August	2.12	0.00	3.59	0.01	3.60	−1.48
September	5.36	0.00	2.88	0.00	2.88	2.48
October	2.39	0.00	2.25	0.00	2.25	0.14
November	3.19	0.00	0.74	0.00	0.74	2.45
December	3.42	0.00	−0.34[a]	0.00	−0.34	3.76
Year	36.73	0.91	28.22	5.97	35.16	1.57

[a] Negative ET resulted from blowing snow.

there was a change in soil water content of 2.78 in. This represented the net amount of water that was added to the soil profile during the month of January.

3.6 ESTIMATING INFILTRATION RATES

3.6.1 HORTON EQUATION

One of the most widely used infiltration models is the three-parameter equation developed by R.E. Horton (1939):

$$f = f_c + (f_o - f_c)\, e^{-\beta t} \qquad (3.15)$$

where f is the infiltration rate at time t, f_o is the infiltration rate at time zero, f_c is the final constant infiltration capacity, and β is a best-fit empirical parameter.

Advantages of the method are that the equation is simple and usually gives a good fit to measured data because it is dependent on three parameters. Main disadvantages are that the method has no physical significance, and field data are required to calibrate the equation. The equation does not describe infiltration prior to ponding.

Horton's equation has seen widespread application in storm watershed models. The most commonly used model that uses Horton's method is the Environmental Protection Agency (EPA) Storm Water Management Model (SWMM; Huber et al., 1981).

3.6.2 GREEN–AMPT EQUATION

In 1911, Green and Ampt developed an analytical solution of the flow equation for infiltration under constant rainfall.

The method was developed directly from Darcy's law and assumes a capillary tube analogy for flow in a porous soil. The equation can be written as

$$f = K(H_o + S_w + L)/L \qquad (3.16)$$

where K is the hydraulic conductivity of the transmission zone, H_o is the depth of flow ponded at the surface, S_w is the effective suction at the wetting front, and L is the depth from the surface to the wetting front. The method assumes *piston flow* (water moving down as a front with no mixing) and a distinct wetting front between the infiltration zone and soil at the initial water content (refer to **Figure 3.11**).

The Green–Ampt method is often approximated by the equation

$$f = \frac{A}{F} + B \qquad (3.17)$$

where f is the infiltration rate, F is the accumulative infiltration, and A and B are fitted parameters that depend on the initial soil water content, surface conditions, and soil properties. This form of the method is used in the DRAINMOD water table management simulation model (NCCI, 1986).

EXAMPLE 3.5

The results of an infiltration test are presented in **Table 3.2**. Determine the empirical parameters A and B in the Green–Ampt equation. Determine the infiltration rate and accumulated infiltration after applying water for 6 h.

Solution

The empirical parameters *A* and *B* can be estimated by substituting values of the infiltration rate *f* and the accumulative infiltration *F* into Equation 3.17. As there are two unknown parameters, values need to be substituted into the equation for two different times to form a pair of equations that are then solved simultaneously.

Substitute values of *f* and *F* at times 0.5 h and 4.0 h. At 0.5 h,

$$1.41 = \frac{A}{0.85} + B \quad \text{(inches/hr)}$$

and at 4.0 h,

$$0.53 = \frac{A}{3.58} + B \quad \text{(inches/hr)}$$

Subtracting the two equations from each other will eliminate *B* and gives

$$(1.41 - 0.53) = \frac{A}{0.85} - \frac{A}{3.58}$$

Multiplying both sides by (0.85×3.58) gives

$$0.88 \times 0.85 \times 3.58 = 3.58A - 0.85A$$

Therefore, $2.73A = 2.68$, and $A = 0.98$ in²/h. Substituting this value in Equation 3.17 gives

$$0.53 = \frac{0.98}{3.58} + B \quad \text{(inches/hr)}$$

and *B* is equal to 0.26 in./h.

Therefore, the Green–Ampt equation can be written

$$f = \frac{0.98}{F} + 0.26 \quad \text{(inches/hr)}$$

After 6 h, the infiltration rate and accumulative infiltration are unknown. Therefore, a trial-and-error solution needs to be developed. We know that the infiltration rate will be less than the value of 0.53 in./h at 4 h. Assume a value of 0.4 in./h.

From 4 to 6 h, the average infiltration rate will be (0.53 in./h + 0.40 in./h)/2, which is 0.465 in./h. Therefore, the infiltration volume during these 2 h will be 2 h × 0.465 = 0.93 in. The accumulated infiltration at 6 h will be the accumulated infiltration at 4 h (3.58 in.) plus 0.93 in., which gives 4.51 in.

TABLE 3.2
Infiltration Test Data for Example 3.5

Time (h)	Infiltration Rate (in./h)	Accumulative Infiltration (in.)
0.0	2.0	0.0
0.5	1.41	0.85
1.0	1.09	1.47
2.0	0.69	2.36
4.0	0.53	3.58

Substitute $F = 4.51$ in. in the Green–Ampt equation:

$$f = \frac{0.98}{4.51} + 0.26 \quad \text{(inches/hr)}$$

and $f = 0.48$ in./h. This value is larger than the assumed value of 0.4 in./h. Repeat the calculation with an infiltration rate at 6 h that is between 0.4 and 0.48 in./h. An estimate of 0.47 in./h gives an accumulative infiltration of 4.58 in. that, when substituted into the Green–Ampt equation, gives a similar infiltration rate of 0.474 in./h.

Answer

The Green–Ampt parameter $A = 0.98$ in.²/h, and $B = 0.267$ in./h. After 6 h, the infiltration rate is 0.47 in./h, and the accumulative infiltration is 4.58 in.

The Green–Ampt equation was derived for application when ponded infiltration occurred through a homogeneous soil with a uniform initial soil water content. If the equation parameters are estimated based on fitting infiltration results from field infiltrometer tests, good estimates of infiltration can also be obtained in heterogeneous soils. Applications of this nature result in good estimates because the equation is treated as a fitted parameter procedure rather than because it accounts for physical processes.

The methods described in this section are only applicable to infiltration during ponded conditions. In many situations, application of these models is limited. Mein and Larson (1973) also developed a two-stage form of the Green–Ampt equation to account for conditions prior to ponding.

3.6.3 PHYSICALLY BASED METHODS

In 1856, Henry Darcy developed the basic relationship for describing the flow of water through a homogeneous soil. Darcy used sand as the porous media and concluded that the rate of flow is proportional to the hydraulic gradient. Darcy's law may be written as

$$q = -K \frac{\partial \phi}{\partial z} \qquad (3.18)$$

where q is the *flux* or volume of water moving through the soil in the z direction (vertically or laterally) per unit area per unit time, and $\partial \phi / \partial z$ is the hydraulic gradient in the z direction (vertically or laterally). Darcy's experiments were limited to one-dimensional flow, but his results have been applicable to three-dimensional flow. Further discussion is presented in Chapter 11.

The hydraulic conductivity is also a function of the soil water content θ and the pressure head, ϕ. Generally, the hydraulic conductivity is dependent on the direction of flow, and the soil is defined as anisotropic. For anisotropic soils, a subscript is used to denote the direction of flow. The hydraulic conductivity is usually expressed as $K(\phi)$, a function of the pressure head, or as $K(\theta)$, a function of the soil water content.

A theoretical differential equation for unsaturated flow is obtained by combining Darcy's equation with the continuity equation. This equation is referred to as the diffusion equation or Richards equation (Jury et al., 1991). The equation is expressed as

$$\frac{\partial \theta}{\partial t} = \frac{\partial}{\partial z}\left[D_w(\theta) \frac{\partial \theta}{\partial z} \right] + \frac{\partial K(\theta)}{\partial z} \qquad (3.19)$$

where $D_w(\theta)$, the soil water diffusivity, is defined as $K(\theta)\partial h / \partial \theta$, and h is the matric potential or the negative pressure head. The difficulty with solving the equation is the nonlinear relationship among hydraulic conductivity, pressure head, and soil water content.

Several solutions of the equation can be developed based on the boundary conditions for the system under evaluation. For a saturated soil, $\partial \theta / \partial t$ is zero, $K(\theta)$ reaches a constant, and the pressure head becomes positive. Equation 3.19 reduces to the Laplace equation for one-dimensional saturated flow (Jury et al., 1991). In recent years, many researchers have examined solutions of the Richards equation. Most of these solutions are only defined for a particular aspect of the infiltration process or are approximate solutions obtained using finite-difference or finite-element methods.

3.7 A PERSPECTIVE ON INFILTRATION METHODS

Based on the authors' research, review of the literature, and consulting experiences, it appears that while numerous equations have been developed to model the infiltration process, no single equation works well for all situations. Most methods are inadequate because they require

knowledge of equation parameters that are difficult to determine on a site-specific basis.

The Richards equation and Green–Ampt methods best describe infiltration processes in profiles in which a piston type of infiltration process occurs and air effects may be modeled by determining the model parameters at field saturation. Thomas and Phillips (1979) and Quisenberry and Phillips (1976) have shown that, in many agricultural soils, a piston type of flow is not the primary mechanism of infiltration. Their results, which were based on field measurements, indicated that, initially, macropore flow through fissures in the profile dominate. Surface-sealing effects have also been found very important, especially in clayey soils. In the last two decades, the scientific community has placed considerable emphasis on better understanding flow processes through soil profiles. The results of much of this research are presented in scientific journals and specialty conferences such as those of the ASAE (2001).

3.8 MEASUREMENT OF SOIL PROPERTIES

Understanding the hydrology of environmental or agricultural systems (e.g., subdivision layout, landfill siting, drainage system design) often requires the measurement of one or more soil properties. Although you may not perform the measurement work yourself, it is important to understand measurement procedures, their application, and their limitations. Many agencies and industries have standard methods of measurement outlined in individualized procedures (or available in a book of standard methods; for example, see ASTM, 1993, or Klute and Dirksen, 1986). This overview should help you understand why a test is used for a certain situation or why some other test may be more appropriate. Such tests should be judged (in order of importance) on (1) relevance to the system studied (i.e., purpose for which the measurements are made and the nature of the physical system); (2) reliability/repeatability/accuracy; (3) skills and knowledge of the person/people conducting the test; and (4) time and equipment considerations (availability/cost).

3.8.1 Particle Size Analysis

The size distribution of individual soil particles can tell much about a soil. Many of the soil characteristics (hydrologic and otherwise) reported in soil surveys and other reports or studies have been attributed to a soil based on its particle size distribution. This characteristic of a soil can give a general idea about the suitability of a given soil for agricultural drainage, reservoir or landfill siting, or other purposes or the susceptibility to chemical leaching or soil erosion.

The measurement of particle size is based on the dispersion/destruction of soil aggregates into discrete units and the separation (and measurement) of these particles into size groups by sieving and sedimentation. The size distribution of larger particles (>0.05 mm diameter) is determined by passing the sample through a series of nested sieves of decreasing opening size.

The size distribution of smaller particles is found from the relationship between settling velocity and particle diameter known as Stokes's law:

$$v = \frac{g\,(\rho_s - \rho_l)X^2}{18\,\eta} \qquad (3.20)$$

where v is velocity of the fall; η is fluid shear viscosity; ρ_s and ρ_l are particle and liquid densities, respectively; and X is the "equivalent" particle diameter. Gee and Bauder (1986) described the theory and process in some detail. The two primary methods of particle size analysis are the hydrometer method and the pipet method.

In the *hydrometer method*, the soil sample (±50 g in 250 ml of water) is dispersed using a combination of chemical (Na-HMP) and mechanical (mixer or shaker) or ultrasonic means. The sample is then poured into a 1-l graduated cylinder topped off with water to the 1-l mark. The sample is shaken to disperse the sample completely throughout the cylinder. A hydrometer (a device that measures specific gravity of a liquid) is introduced to the cylinder, and readings are taken at times such as 30 sec; 1, 3, 10, 30 min; and 1, 1.5, 2, and 24 h (Gee and Bauder, 1986). Measuring the specific gravity (or density of the soil solution) allows calculation of the percentage of the soil sample that has settled out at a given time.

In the *pipet method*, the soil sample is dispersed in a fashion similar to the hydrometer method and again is poured into a 1-l cylinder. A small subsample is taken at time t and depth h using a pipet. At that point, all particles coarser than X have settled below the sampling depth, and the percentage of the original soil left in suspension can be determined.

3.8.2 PARTICLE DENSITY

The particle density is measured by either the pycnometer method or the submersion method (Blake and Hartge, 1986b). Both methods are based on the difference in weight between a volume of water and that same volume with some of the water displaced by a known weight of soil.

The particle density can be calculated from the relationship

$$\rho_p = \frac{\rho_w(W_s - W_a)}{(W_s - W_a) - (W_{sw} - W_w)} \qquad (3.21)$$

where ρ_p is the particle density (lb/ft^3); ρ_w is the density of water (lb/ft^3); W_s is the weight of pycnometer (or weighing dish) plus oven-dry soil; W_a is the weight of pycnometer (or weighing dish); W_{sw} is the weight of pycnometer (or weighing dish) filled with soil and water; and W_w is the weight of pycnometer (or weighing dish) filled with water.

3.8.3 BULK DENSITY

Four general methods are available for measuring bulk density (Blake and Hartge, 1986a): (1) core methods, (2) excavation methods, (3) clod method, and (4) radiation method. The first three methods are based on direct measures of weight and volume, whereas the last is based on the empirical relationship between density and the amount of γ-radiation that is transmitted through or reflected (backscattered) by the soil.

In *core methods*, a volume of soil is collected using a specially designed cylinder (called a *coring tube*) and drop hammer or a hydraulic probe (**Figure 3.24**). Knowing the diameter of the coring tube and the length of the sample allows calculation of sample volume V. The sample is weighed, and the bulk density is calculated using Equation 3.8. The dry bulk density can be determined from the same sample by drying it at 105°C for 24 h and reweighing. The oven-dry weight is taken as the weight of the soil solids only, and the dry bulk density is calculated using Equation 3.9. Extreme care must be taken such that the sample volume is not compacted by the sampling process.

In *excavation methods*, a hole is dug with all of the removed soil retained. The volume of the hole is then determined by placing a plastic bag or balloon in the hole and filling it with water (or air) until the fluid reaches the surface of the surrounding soil. The displaced soil is either weighed on a portable scale in the field or taken (in a sealed container to prevent water loss) to the laboratory to be weighed. Dry bulk density can be determined by drying the soil in the laboratory to determine oven-dry weight. The advantage of this method is that sampling errors are minimal if the test is performed properly. A major disadvantage is the amount of work involved, especially if the bulk densities of layers below the surface are desired.

In the *clod method*, naturally forming soil units (*peds* or *clods*) are coated with a substance, such as paraffin, that prevents water from entering the clod. The clods are coated by tying a string to the clod and dipping it into a heated vat of the liquefied waterproofing agent. On cooling, the waterproofing agent will solidify and form a waterproof seal. Each clod is weighed and is then lowered (submersed) into a beaker or graduated cylinder filled with water. The amount of water displaced (as measured by the rise in water level) is equal to the volume of the clod. Equation 3.8 is then used to determine the bulk density of the individual clod. The dry bulk density can also be found by this method by oven-drying the clods before coating and using Equation 3.9. The advantage of this method is that it gives very specific density information

A

B

FIGURE 3.24 A. Uhland drop hammer bulk density sampler. B. Hydraulic soil coring device.

about soil structural units (clods) of various sizes. However, the method has several drawbacks. This method will not work for soils, such as organic soils, with bulk density less than the density of water because the clods will float instead of submerging. Many soils, particularly cultivated or sandy soils, do not have stable structural units for performing the test. The layer of waterproofing agent introduces a small error that will vary, depending on its thickness and the size of the clod. This method is also very time consuming.

In the *radiation method*, an instrument known as a γ-density gage uses a radioactive source (e.g., cesium[137]) to estimate soil density indirectly. The radioactive source emits γ-radiation at a known rate. The rate that this radiation is transmitted through, or reflected by, the soil is related to the density of the surrounding soil, as well as the texture, chemical composition, and water content of the soil. The γ-density gage uses either two probes (one containing an emitter, the other a receiver) that are placed at a fixed spacing to measure transmitted radiation or a single probe (containing both emitter and receiver) that measures reflected radiation. Although calibration curves are typically provided with commercial γ-density gages, the actual relationship will depend somewhat on local conditions. The γ-density gage may be used in conjunction with the neutron probe (see section on measurement of soil water content) to determine soil density and soil water properties *in situ*.

3.8.4 Soil Hydraulic Properties

The properties that describe the movement of water into and through the soil are among the most important in designing agricultural and environmental management systems. They are also among the most difficult to quantify. Measurement and prediction techniques for these hydraulic properties are discussed below. The property or combination of properties measured should depend on the system to be analyzed or designed. **Table 3.3** offers a quick reference on the properties that should be used for a particular application.

3.8.5 Soil Water Content

Methods for measuring soil water content can be divided into direct measures and indirect measures (Gardner, 1986; Jury et al., 1991). Direct methods are based on determination of weight before and after oven-drying. To measure the soil water content directly, a wet sample is weighed, dried in an oven, and reweighed. The sample can be dried in a conventional oven (at 105°C for 24 h) or in a microwave oven (see Gardner, 1986). The gravimetric water content θ_g is then calculated using Equation 3.7 (Jury et al., 1991).

Indirect measures depend on the change in physical and chemical properties of the soil relative to the change in soil water content. Indirect methods include (1) measures of electrical conductivity and capacitance, (2) radiation

TABLE 3.3
Application of Soil Hydraulic Properties

Application	Measurement Techniques
Above Water Table **(Unsaturated Hydraulic Conductivity)**	
Vadose zone models evaporation rate	For soil water/soil suction-type techniques; see Green et al. (1986); Klute and Dirksen (1986); or Bouwer and Jackson (1974)
Below Water Table **(Saturated Hydraulic Conductivity)**	
Agricultural drainage system design groundwater exploration investigation	Auger hole (Amoozegar and Warrick, 1986), pumping test/slug test (Freeze and Cherry, 1979)
Surface Applications **(Infiltration)**	
Irrigation system design	
Sprinkler	Sprinkler infiltrometer (Peterson and Bubenzer, 1986)
Surface	Furrow infiltration test (Kincaid, 1986)
Urban/suburban drainage	Sprinkler or ring infiltrometer (Bouwer, 1986)
Infiltration models	Sprinkler, ring, furrow infiltrometer
Erosion models	Sprinkler infiltrometer
Combination **(Saturated Hydraulic Conductivity and Infiltration)**	
Water table management models	Sprinkler or ring infiltrometer plus auger hole
Pesticide/nutrient movement to ground and surface water	Sprinkler or ring infiltrometers plus shallow well pump-in method

methods, and (3) reflectometry. Gardner (1986) described the process of determining soil water content by means of *electrical conductivity methods*. Electrodes are embedded in blocks of some porous material (typically gypsum). The blocks are then placed in the soil at the desired depth. The porous blocks reach a degree of wetness that is in equilibrium with the surrounding soil. The conductance of the block is then determined by connecting specially designed electrical equipment (commercially available) that contains a power source and a circuit (called a *Wheatstone bridge*) that measures resistance. The resistance (the inverse of conductivity) is empirically related to the water content of the block and thus to the soil. The resistance of the blocks is calibrated against a soil matric potential. To find the soil water content of a given soil, the relationship between soil water content and matric potential must be established. The method is useful for distinguishing between wet and dry conditions, but is not well suited for measuring small changes in soil water content.

In *radiation methods*, soil water is determined by the attenuation of either γ- or neutron radiation. In γ-ray attenuation, a radioactive source emits γ-radiation at a known rate. The rate that this radiation is transmitted through the soil is related to the water content and density of the soil. By measuring the soil density at a known soil water content,

that component can be removed from the calculations, and a relationship between radiation transmitted and water content is established (for further details, see Jury et al., 1991). The γ-radiation methods are normally used for measuring changes in soil water content with time.

In neutron attenuation (Gardner, 1986; Jury et al., 1991), a probe (containing both emitter and detector) that measures "thermalized" radiation is lowered into an access tube in the ground (**Figure 3.25**). The radiation source emits high-energy neutrons that collide with the nuclei of atoms in the surrounding soil. The energy of the neutrons is substantially changed only by hydrogen nuclei (present mostly in the form of water). The percentage of neutrons that have had their energy changed (thermalized) can be recorded by the detector. Calibration curves are then developed to relate the output to water content (removing background sources of hydrogen nuclei such as organic matter and kaolinite).

In *time domain reflectometry* (Jury et al., 1991), the dielectric constant of the soil is related to the soil water content. A special apparatus is used that sends a step pulse of electromagnetic radiation along dual probes inserted into the soil. The pulse is "reflected" and returned to the source with a velocity characteristic of a specific dielectric

FIGURE 3.25 Indirect method for measuring bulk density or soil moisture content of a soil.

constant. This instrument can be calibrated to give soil water content based on the measured dielectric constant.

3.8.6 SOIL SUCTION AND SOIL WATER RELEASE/RETENTION CHARACTERISTICS

As explained in Section 3.4, the movement of water into and out of the soil matrix is controlled by suction (tension) forces. Unfortunately, an accurate measurement of soil suction is difficult and time consuming. Soil suction relationships are a function of the pore size distribution, soil structure, soil water content, organic matter content, and soil solution properties. The soil texture and the porosity of the soil profile will have the most influence on soil suction properties of the profile.

The most common approach to obtain a relationship between soil suction and soil water content is to obtain equilibrium conditions, for different soil water content conditions, that allow the measurement of both soil suction and soil water content. This is most easily accomplished by drying a wet soil because the changes in soil water content and soil suction are slow. However, it should be recalled that there is hysteresis during wetting and drying (see **Figure 3.5**). Therefore, information obtained during drying needs to be adjusted if used to describe wetting. Details in field and laboratory procedures to determine relationships between soil water content and soil suction were presented by Klute and Dirksen (1986).

FIGURE 3.26 Pressure chamber for making laboratory measurements of soil water release relationships.

The general approach that is used in many laboratory procedures is first to obtain disturbed or undisturbed soil samples from the field and then place the soil samples on a porous plate in a pressure cell or chamber. The samples are then wet to saturation, and the pressure in the cell is raised to a selected value above atmospheric pressure. The samples are then left until equilibrium has been reached and water flow from the samples has ceased (normally at least 24 h). The samples are then weighed to obtain the gravimetric soil water content; this procedure is repeated at several different pressures (**Figure 3.26**). If knowledge of wilting point and field capacity are required, the test will be conducted at pressures of 0.1 to 0.33 bar and 15 bars, respectively.

Field measurements of soil suction are made with tensiometers or pressure transducers. Common types of tensiometers are shown in **Figure 3.27**. They consist of a porous cup that is connected to a top and a device to measure pressure changes as water moves into or out of the porous cup. As water moves out of the cup into the soil profile, a partial vacuum is created in the tube. The movement of water into the soil will depend on the size of soil pores near the porous cup and the water content of the pores. As the soil dries, more water will be "sucked" from the porous cup, and the partial vacuum will increase. If the soil water content increases, perhaps due to a rainfall event, water might flow back into the porous cup. Tensiometers will normally function only if the soil suction is less than about 0.8 bar.

FIGURE 3.27 Tensiometer methods for measuring *in situ* soil suction.

FIGURE 3.28 Ring infiltrometer with a Mariotte hydraulic head device.

3.8.7 INFILTRATION

Infiltration into soil is commonly measured by the cylinder (single-ring or double-ring) infiltrometer or by a sprinkler infiltrometer. The cylinder infiltrometer is the most common method used (and misused) because it is relatively inexpensive and simple. This method, when used properly, can provide useful information when the infiltrating water is ponded on the surface. When the system to be studied involves erosion, runoff, and infiltration of rainfall, the sprinkler infiltrometer should be used. To determine infiltration information for surface irrigation systems, the border or furrow irrigation methods described by Kincaid (1986) should be employed.

When using the cylinder infiltrometer method (Bouwer, 1986), a large ring (>1 m diameter) is installed in the soil surface. The infiltration rate is measured using a Mariotte device, which maintains a shallow head on the exposed surface (**Figure 3.28**). The final infiltration rate is taken as the rate of infiltration when soil suction at the bottom of the ring equals zero (saturated condition). Water flow is assumed to be one dimensional (in the vertical direction), although flow in reality is three dimensional. Users often attempt to "buffer" the effect of horizontal infiltration by using a double-ring infiltrometer. A larger ring is placed around the inner ring (**Figure 3.29**), and the same water level is maintained in the inside and outside rings. However, many users of this methodology assume that, by using the second ring, they can use smaller-diameter rings with the same accuracy. Bouwer (1986) stated that: "When lateral capillary gradients below the cylinder infiltrometer cause the flow to diverge, it does not help to put a smaller cylinder concentrically inside the big cylinder and measure the infiltration rate in it, in hopes of

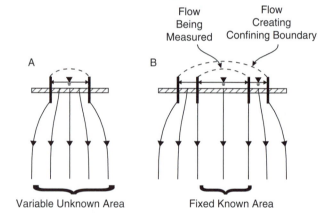

FIGURE 3.29 Divergence of streamlines during infiltration buffered by using a double-ring infiltrometer.

getting a measure of the true vertical infiltration rate." Bouwer (1986) also cautioned that "while cylinder infiltrometers in principle are simple devices ... the results can be grossly in error if the conditions of the system... [i.e., water quality, temperature, soil conditions, surface conditions, etc.] are not exactly duplicated in the infiltrometer test."

The sprinkler infiltrometer as described by Peterson and Bubenzer (1986) provides the most detailed information on the infiltration process. In this method, a rainfall simulator is used to simulate natural rainfall characteristics (rate of application, drop size). The amount and rate of infiltration is determined by monitoring the rate of runoff from a plot of known size and subtracting runoff from rainfall. This method also allows the user to study soil erodibility characteristics. However, this method requires specialized equipment that can be costly to build and operate.

3.8.8 Unsaturated Hydraulic Conductivity

Unsaturated hydraulic conductivity is used to model water movement and water content distributions above the water table. Unsaturated hydraulic conductivity is a function of soil water content and soil suction. Estimation of unsaturated conductivity requires soil water and soil suction values over a range of soil conditions that may require monitoring soil water conditions over several weeks. Techniques for estimating unsaturated hydraulic conductivity were described by Klute and Dirksen (1986) for laboratory techniques; Green et al. (1986) for field techniques; or Mualem (1986) for predictive techniques.

3.8.9 Saturated Hydraulic Conductivity

Saturated hydraulic conductivity can be found directly by measuring water movement through a soil sample or indirectly by estimation from associated soil properties. Methods that measure saturated hydraulic conductivity directly can be divided into laboratory techniques or field techniques. Field techniques can be further divided into methods that require a shallow water table and those that do not. Klute and Dirksen (1986) stated that choosing a method for determining hydraulic conductivity depends on such factors as available equipment, nature of the soil, kind of samples available, the skills and knowledge of the experimenter, the soil water suction range to be covered, and the purpose for which the measurements will be made.

The method for measuring hydraulic conductivity should be selected so that the soil region and flow direction used in the hydraulic conductivity measurement adequately represent the soil and flow direction in the actual system to be characterized (Bouwer and Jackson, 1974). Hydraulic conductivity is difficult to measure, and estimates are often based on other soil properties more readily available or simpler to determine. Methods have been developed to estimate hydraulic conductivity from related soil properties, such as soil texture, porosity, and bulk density. Studies that address estimation of hydraulic conductivity from related properties include those of Rawls et al. (1982), Puckett et al. (1985), Wang et al. (1985), Bouma (1986), and Baumer et al. (1987).

The *core sample method* determines saturated hydraulic conductivity on samples by either a constant head or falling head test. Core samples are taken in the field and kept from drying. In the laboratory, the samples are saturated from the bottom to prevent air entrapment.

For the constant head test, water is supplied to the bottom of the core samples at constant hydraulic head. The volume of outflow is measured with time. The hydraulic conductivity is found by rewriting Darcy's law as

$$K = \frac{qL}{Ah} \quad (3.22)$$

where q is the rate of outflow, L is the length of the core sample, A is the core cross-sectional area, and h is the depth of the constant head applied.

In the falling head test, used with low-permeability soils, the hydraulic conductivity is determined from the rate of change of velocity of a falling water column. A manometer (cross-sectional area **a**) is placed on top of the sample core (cross-sectional area A) and filled with water. The hydraulic conductivity is determined as

$$K = \frac{aL}{A(t_2 - t_1)} \ln \frac{h_1}{h_2} \quad (3.23)$$

where L is the length of the soil core, h_1 is the height of the water column at time t_1, and h_2 is the height of the water column at time t_2.

The laboratory core technique is relatively cheap, does not require a water table, and enables sampling for layers or anisotropy. However, this technique is limited by soil disturbances during handling, small sample size, and loss of head at the water–soil interface. Hydraulic conductivity values resulting from this method are typically lower than those calculated from field measurements, and there is often large variability between samples.

When using field techniques for determining saturated hydraulic conductivity above the water table, the hydraulic conductivity calculations are based on infiltration into unsaturated soil. Even under the best of circumstances, some air is entrapped during infiltration, and the soil does not become fully saturated. These methods give a value for the field-saturated hydraulic conductivity that may be significantly different between different methods, sites, and initial water conditions.

The percolation test is the simplest test to conduct. A hole is dug to the desired depth, water is ponded in the bottom of the hole to saturate the soil, and the vertical velocity of water entering the soil is measured. This technique relates well to subsurface seepage.

The *shallow-well pump-in method*, otherwise known as the *dry auger hole method* or *well permeameter method*, measures the rate of flow of water from a (cased or uncased) auger hole when a constant height of water is maintained in the hole. A float valve is usually used to maintain the water level, with a large water tank providing the water supply. Hydraulic conductivity values are calculated using the steady-state outflow rate and a shape factor determined from nomographs or equations. The position of the water table or impermeable layer below the bottom of the well must be known. This technique is easy to use, but is limited by the time requirements needed to reach steady state and to replicate measurements. A self-contained version of the shallow-well pump-in method has been developed by researchers at Guelph University (Reynolds and Elrick, 1985). The Guelph permeameter uses a built-in Mariotte bottle to control the depth of water in the augured hole. The saturated hydraulic conductivity is determined from the difference in outflow rates for two different water depths.

In the air-entry permeameter (Bouwer, 1986), a small covered cylinder is driven into the ground. Water is applied to the cylinder until all air is driven out. A large constant head is kept in a reservoir at the top of the cylinder until saturation is reached at the bottom of the cylinder. The water level is then allowed to fall, and the conductivity is calculated from falling head equations. The air-entry permeameter measures field-saturated hydraulic conductivity in the vertical direction. An improved (for quickness and ease of use) version of the air-entry permeameter, called the *velocity head permeameter*, has been developed to measure hydraulic conductivity in either the vertical or the horizontal direction. A description of this method, now commercially available, can be found in the work of Merva (1979).

In the presence of a shallow water table, the auger hole method is the most widely used method for determining hydraulic conductivity of soils in the field. A hole is augured to the desired depth below the water table, and water is allowed to rise until equilibrium is reached. The hole is then pumped or bailed, and the rate of rise of the water level in the hole is measured. The saturated hydraulic conductivity is calculated as

$$K = \frac{\pi R^2}{Sh} \frac{dh}{dt} \tag{3.24}$$

where R is the radius of the auger hole, S is a function of hole geometry found from nomographs, h is the depth of water in the auger hole, and dh/dt is the rise in the water level over time increment dt.

The auger hole method is simple in conception and practice, is relatively quick and cheap, and measures a large sample area compared to other techniques. It is also one of the most reliable methods. However, extensive variability in repeated measurements can result from soil heterogeneity, depth of auger hole, water table depth, and depth of water removed from the auger hole. This technique is not suitable for layered soils and is unreliable in some cases due to macropores or side-wall failure.

Other methods can be found in the work of Bouwer and Jackson (1974) and Klute and Dirksen (1986). Several tests are commonly used for exploring the saturated hydraulic conductivity of groundwater systems. These are the *slug test* and the *pumping test*. Laboratory core methods (see above) may also be used to describe individual layers either in or above the aquifer. Special equipment has been developed by some researchers to obtain large-diameter soil cores that can then be transported to a laboratory to conduct infiltration and solute transport studies

(**Figure 3.30**). For more information on measuring saturated hydraulic conductivity for groundwater systems, see Freeze and Cherry (1979).

PROBLEMS

3.1. Based on the USDA classification scheme, determine the textural class of a soil that is 20% clay and 40% sand.

3.2. How much clay would need to be added to the soil in Problem 3.1 to make it a clay loam?

3.3. Determine the soil water content in percentage by weight for a 120-g sample in a moist condition that, when oven-dried, weighed 90 g.

3.4. The combined weight of a soil sample and a tin is 50 g. The tin weighs 20 g. After oven-drying, the soil sample and the tin have a weight of 45 g. Determine the soil water content (by weight) of the soil.

3.5. Determine the porosity of a soil profile if it has a dry bulk density of 1.8 g/cc, and the density of the soil particles is 2.7 g/cc. What would the bulk density of the soil be if it was saturated?

3.6. Based on field tests, the bulk density of a soil profile was 99.8 lb/ft³. The specific gravity of the soil particles is 2.8, and the dry bulk density of the profile is 87.4 lb/ft³. Determine the porosity and soil water content (by weight) of the profile.

3.7. The porosity of a soil is 0.4 (cc/cc). Determine the volumetric soil water content as a fraction if the degree of saturation is 60%. Also, determine the depth of water contained in the top 50 cm of a soil with a degree of saturation of 60% and porosity of 0.4.

3.8. Determine the soil water content at field capacity and plant available soil water content in inches for a 30-in. soil column. The field capacity by volume is 25%, and the wilting point is 10%.

3.9. The field capacity of a soil is 35%, and the wilting point is 15% by weight. The dry bulk density of the soil is 78 lb/ft³. Determine the available water content in percentage by weight and by volume. Also determine the plant available soil water content in the top 2 ft of soil. Express the answer as a depth of water in inches.

3.10. Determine the particle density of a soil based on the following information from a laboratory pycnometer test: $W_s = 32$ g, $W_a = 20$ g, $W_{sw} = 37.5$ g, and $W_w = 30$ g; use Equation 3.21.

3.11. Horton's infiltration is fitted to the infiltration curve, for the dry initial conditions, presented in **Figure 3.18**. Determine the initial infiltration rate,

A

B

FIGURE 3.30 (A) A device that was designed by Professor Larry Brown at Ohio State University to obtain large-diameter undisturbed soil cores. (B) Excavating the core is not simple.

final infiltration rate, and the best-fit empirical parameter β.

3.12. Determine the daily changes in soil water content for the hydrologic conditions presented in **Table 3.4**. What depth of irrigation water should be added after 10 days to raise the soil water content of the root zone to the initial value of Day 1? The initial soil water content in the root zone was 12 in. of water. Assume that all percolation passes through the root zone and is not stored.

3.13. The root zone (1 m deep) of a silt clay loam soil has an average dry bulk density of 1.35 g/cc, and the specific gravity of the soil particles is 2.7. Based on laboratory measurements, it has been established that the degree of saturation at field capacity and wilting point are 60% and 30%, respectively. Irrigation is initiated by an automatic irrometer system when 50% of the plant available water has been depleted. The profile is

irrigated at a constant rate of 10 mm/h for 5 h. Using the Green–Ampt equation, determine (a) the infiltration rate at the end of the irrigation application period; (b) the volume of runoff; and (c) the soil water content of the root zone when gravity flow has ceased. Based on a double-ring infiltration test, the Green–Ampt parameters have been determined as $A = 30$ mm/h and $B = 3$ mm/h. *Hint:* The problem is best solved on a spreadsheet and will require several iterations. Divide the irrigation application time into 30-min time blocks. Assume the initial infiltration rate is the irrigation application rate. Obtain a first estimate of the accumulative infiltration in the first 30 min by assuming a constant infiltration rate. You will then need to estimate infiltration rates at the end of each time block and compare them with values determined from Equation 3.15.

TABLE 3.4
Daily Accretion and Depletion of Soil Water for Use in Problem 3.12

	Accretion		Depletion	
Day	Precipitation P (in.)	Runoff Q (in.)	Evapotranspiration ET (in.)	Percolation G (in.)
1	0.0	0.0	0.15	0.0
2	1.5	1.0	0.05	0.2
3	0.0	0.0	0.10	0.0
4	0.0	0.0	0.15	0.0
5	1.0	0.3	0.05	0.1
6	0.0	0.0	0.10	0.0
7	0.0	0.0	0.15	0.0
8	0.0	0.0	0.10	0.0
9	0.0	0.0	0.15	0.0
10	0.5	0.2	0.05	0.1
11	0.0	0.0	0.10	0.0
12	0.0	0.0	0.15	0.0

3.14. Compute the infiltration rate (inches/hour and millimeters/hour) if 0.1 gal of water moves downward from the central ring of 10-in. diameter double-ring infiltrometer in 30 min (1 gal = 231 in.³).

3.15. Determine the average infiltration rate (inches/hour) into an irrigation furrow 200 ft long with water covering a width of 1 ft. The measured rate of flow into the 200-ft test section is 0.030 ft³/sec, and the outflow rate is 0.028 ft³/sec.

3.16. Determine the depth (inches and millimeters) and rate of infiltration (inches/hour and millimeters/hour) for each time period for a catchment with the rainfall and runoff depth data presented in **Table 3.5**.

TABLE 3.5
Data for Problem 3.16

Period	Time (min)	Rainfall (in.)	Runoff (in.)
A	10	0.8	0.4
B	15	1.0	0.75
C	12	0.6	0.5
D	20	0.4	0.3

4 Evapotranspiration

4.1 INTRODUCTION

Evapotranspiration (ET) is the process that returns water to the atmosphere and therefore completes the hydrologic cycle. Students seem to have an intuitive feel for the hydrologic cycle, and the idea that evapotranspiration completes the hydrologic cycle is easily believed. This was not always the case, however. As late as the mid-1500s there was still a dispute over the source of water in rivers and springs. The age-old theory was that streams originated directly from seawater or from air converted into water. In the mid-1500s, Bernard Palissy, a French potter, stated that rivers and springs could not have any source other than rainfall. Edmond Halley (1656–1742) proved the other half of the concept of the hydrologic cycle by showing that enough water evaporated from the earth to produce sufficient rainfall to replenish the rivers (Biswas, 1969).

Because evapotranspiration is a major component of the hydrologic cycle, this chapter explores the processes by which ET occurs. In addition, we discuss techniques for estimating the amount (depth) and rate (depth/time) of evapotranspiration. You may be asking yourself why you would ever need to calculate evapotranspiration. If you work with water or plants, chances are you will need to estimate ET. The rate and amount of ET is the core information needed to design irrigation projects, and it is also essential for managing water quality and other environmental concerns. Policymakers need to know how estimates of ET are determined because these methods are used in litigation and in negotiations of contracts and treaties involving water. To predict meltwater yields from mountain watersheds or to plan for forest fire prevention, estimates of ET are needed. In urban development, ET calculations are used to determine safe yields from aquifers and to plan for flood control. Anyone involved with resource management will likely need to understand the methods available for estimating evapotranspiration.

Evapotranspiration can be divided into two subprocesses: evaporation and transpiration (**Figure 4.1**). Evaporation essentially occurs on the surfaces of open water, such as lakes, reservoirs, or puddles or from vegetation and ground surfaces. Transpiration involves the removal of water from the soil by plant roots, transport of the water through the plant into the leaf, and evaporation of the water from the leaf's stomata into the atmosphere.

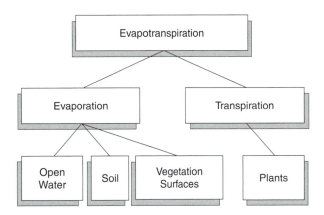

FIGURE 4.1 Evapotranspiration divided into subprocesses.

4.2 EVAPORATION PROCESS

Evaporation occurs when water is changed in state from a liquid to a gas. Water molecules, both in liquid and in gas, are in constant motion. Some water molecules possess sufficient energy to break through the water surface into the atmosphere, and some water vapor molecules may cross back into the liquid. When more molecules are leaving the liquid than returning, net evaporation occurs (**Figure 4.2**). The energy a molecule needs to penetrate the water surface is called the *latent heat of vaporization* (approximately 540 cal/g of water evaporated at 100°C). Evaporation therefore requires a supply of energy, typically provided by received solar radiation or *insolation*, to provide the latent heat of vaporization. In 1802, the English physicist John Dalton (1766–1844) showed that the main driving force of evaporation is *vapor pressure deficit*. That is, the rate of evaporation is largely the function of the difference between (1) the ability of air to hold water based on air temperature and relative humidity and (2) the energy in the water (largely based on temperature) to make it give up water to the air. This is Dalton's law; we discuss it in this chapter in modified form as the Meyer equation.

Consider a newly closed container half full of water. With a sufficient supply of energy, water will evaporate; initially at a high rate, but eventually the container will equilibrate, and no net rate of evaporation will occur. When the container was first closed, the concentration of water molecules in the airspace above the water was very small. The newly evaporated water molecules could move (diffuse) easily into the airspace. The rate of diffusion of water molecules is proportional to the difference in concentration between water molecules at the water surface

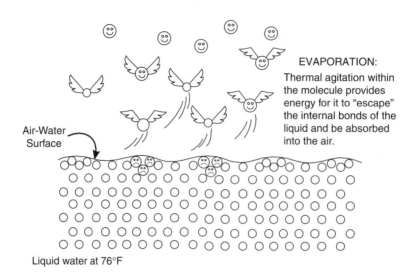

FIGURE 4.2 Schematic of the evaporation process.

and in the air space. As more molecules evaporate and move into the air space, the concentration of water molecules in the airspace increases. Because the concentration of water molecules at the water surface is essentially constant and the concentration in the air space is increasing, the difference between the two concentrations and the rate of evaporation decrease. The net evaporation will eventually cease when the container equilibrates.

When net evaporation ceases, the air is said to have reached the *saturation vapor pressure* e_s. If evaporation is to continue after the saturation vapor pressure is reached, some mechanism to remove water vapor from the air above the evaporating surface is needed. In an open system, such as a lake or crop field, a layer of water vapor can build up adjacent to the water or leaf surface and reduce evaporation. For evaporation to continue, air movement (typically provided by wind) is needed to remove the vapor.

Therefore, two important factors in the evaporation process are (1) a source of energy to supply the latent heat of vaporization and (2) a concentration gradient in the water vapor, typically provided by air movement, that removes the water vapor adjacent to the evaporating surface.

4.2.1 FICK'S FIRST LAW OF DIFFUSION

Fick was one of the first to quantify the movement of molecules from a region of higher concentration to a region of lower concentration (Nobel, 1983), such as water molecules moving from a water surface into air. He developed Fick's first law of diffusion:

$$J_j = -D_j \frac{\partial c_j}{\partial x} \qquad (4.1)$$

where J_j (the flux density) is the amount of species j crossing a certain area per unit time and is typically expressed in units such as moles of particles per square meter per second. D_j is the diffusion coefficient of species j (analogous to resistance in electrical circuits). The term $\partial c_j / \partial x$ represents the concentration gradient of species j and is the driving force that leads to molecular movement (Nobel, 1983). The negative sign indicates that the direction of flow is from high to low concentration.

4.2.2 POTENTIAL EVAPOTRANSPIRATION VS. ACTUAL EVAPOTRANSPIRATION

A sometimes-difficult concept for students is the difference between potential evapotranspiration (PET) and actual evapotranspiration (AET). While agricultural hydrologists may have very precise definitions and distinguish carefully between evaporation and transpiration (discussed later in the chapter), we now look at the PET and AET in simpler terms. PET is the amount of ET that would occur when there is unlimited water available. AET is the amount of ET that actually occurs when water is limited. An example might be for Lake Havasu on the Colorado River between California and Arizona. There, the annual average PET is about 7 ft. Since evaporation from a lake approximates PET, we would expect Lake Havasu, in an average year, to lose by evaporation about 7 ft of water over the entire surface of the lake. That is, for the lake the actual evapotranspiration equals the potential evapotranspiration (AET = PET). However, when we step up on the dry bank, we are in the Sonoran Desert, where the PET is still 7 ft, but the average annual precipitation is only about 8 in. (**Figure 2.6**). So, without irrigation, the maximum ET that could occur on the bank would be the 8 or so in. of water supplied by rain each year. Thus, while the PET everywhere in the locale may be 7 ft, the AET might equal the PET over the water, but the AET on the nonirrigated land could average only 8 in. As we discussed below, AET can be modified by other variables.

4.2.3 EVAPORATION FROM OPEN WATER

Evaporation from open or "free" water has been well studied. Open water in this context refers to lakes, ponds, reservoirs, or evaporation pans (see Section 4.4). While it is true that evaporation from lakes and reservoirs may be of great local interest (for instance, to predict summer water levels), the justification for studying open water in such great detail is that open water provides a reproducible surface. The state of crops and soils is dynamic, which complicates evaporation from these surfaces. The surface of open water is relatively unchanging, and the dependence of evaporation on weather conditions alone is easier to study. Crops and soil evaporation losses can be estimated from open water evaporation predictions (Penman, 1948), but when available, measurements of lake evaporation rates give a far better estimate of PET rates.

There are at least four minor factors that might affect open or free water evaporation rates:

1. **Barometric Pressure**. Free water evaporation increases slightly with decreasing pressure (Veimeyer, 1964), but the effect is usually offset by the cooler temperatures in cyclones and at higher altitudes, where lower pressures exist.
2. **Dissolved Matter**. Dissolved matter lowers the vapor pressure of water, thus reducing evaporation. For salt (NaCl), the evaporation rate is lowered about 1% for each 1% increase of specific gravity (Veimeyer, 1964).
3. **Shape, Site, and Situation of Evaporating Body**. Ponds, lakes, and the like surrounded by hills might decrease wind. Size and especially length along the major axis of the wind (*fetch*) give more near-surface velocity and surface roughness (*waves*).
4. **Relative Depth of Evaporating Body**. Shallower lakes retain more energy near the surface and thus have higher evaporation than a deeper lake. Deep lakes move energy away from the surface due to *rollover*. This factor is considered in the Meyer equation, discussed in Section 4.6.

4.2.4 EVAPORATION FROM BARE SOIL

Evaporation from bare soil is similar to evaporation from open water if the soil is saturated. If the bare soil is not saturated, the process is more complex because water evaporates deeper in the soil, and the vapor must diffuse into the atmosphere. The rate of evaporation from bare soil is typically divided into two distinct stages. During

the first stage, the soil surface is at or near saturation. The rate of evaporation is controlled by heat energy input and is approximately 90% of the maximum possible evaporation based on weather conditions. The duration of the first stage is influenced by the rate of evaporation, the soil depth, and hydraulic properties of the soil. This stage typically lasts 1 to 3 days in midsummer (M.E. Jensen et al., 1990).

The second stage, sometimes called the *falling stage*, begins once the soil surface has started to dry. At this stage, evaporation occurs below the soil surface. The water vapor formed in the soil reaches the soil surface by diffusion or mass flow caused by fluctuating air pressures. The evaporation rate during this stage is no longer controlled by climatic conditions, but rather by soil conditions such as hydraulic conductivity (M.E. Jensen et al., 1990). An important control of hydraulic conductivity is texture, so that for a given period of intensive drying, a given level of soil drying might be found significantly deeper in sand-textured soils than in clay-textured soils. Evaporation rates from bare soil are highest when the soil is wet, so it stands to reason that evaporative losses are greatest from a bare soil that is frequently wetted; under natural conditions, this might come from a climate where rainfall is frequent. In western Europe, a West Coast marine climate (Chapter 2, **Table 2.1**, Koeppen Cfb), many areas have rain about half the days of the year, so that bare soils may have relatively high evaporative losses, similar to vegetated plots. In more continental locations like the central U.S., rainfall is generally much less frequent, so that evaporative losses would be relatively less.

The principles above apply to irrigation, especially spray irrigation. Less-frequent, but more adequate applications of water on bare soil will minimize evaporative water losses from the bare soil surface. The optimum procedure is to bring the soil up to field capacity (FC) without leaching soil nutrients and other chemicals into groundwater.

Most of the concepts in the above paragraphs were known to early settlers in the semiarid regions of the U.S. Before the days of widespread irrigation, farmers often used a technique called "dry farming," by which the surface of the soil was broken up by shallow plowing to reduce upward capillary action and diffusion and thus evaporative loss of the deeper soil moisture. Crops were planted only every other year, with the fallow year used to accrue moisture from rainfall in the hope that the soil would be "filled" to near field capacity by the time the crops started to grow. Note that the considerable water-holding capacity of the soil was used to store the water in a "savings account," a principle we discuss regarding soil water budgets. Dry farming continued in some places into the 1950s, but eventually disappeared because (1) crops were possible only every other year, which limited output and income; (2) the loose surface soil layer was extremely

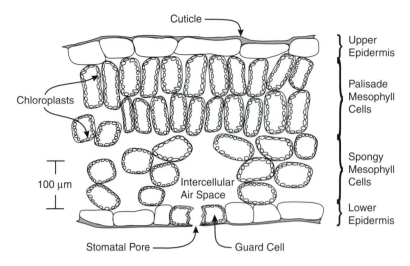

FIGURE 4.3 Schematic transverse section through a leaf.

susceptible to erosion by both wind and water; and (3) irrigation water was increasingly made available, especially by the federal government.

4.3 EVAPOTRANSPIRATION FROM SOIL AND PLANTS

When plants are introduced into the system, the complexity of measuring or predicting ET increases because plants are transpiring, in addition to water evaporating from the soil surface or from the canopy surfaces.

4.3.1 INTERCEPTION

Before precipitation gets to the ground, it normally must pass through vegetation: canopies, stems, grass blades, and vegetative bed litter (duff) on the ground. A small, but significant, amount is intercepted and retained on these surfaces. Evaporation of this retained intercepted water starts even during the storm and continues afterward until all the water is evaporated. Note that about the same amount of energy is required to drive evaporation of free water from the leaf as to drive transpiration from the leaf. Thus, plant transpiration is reduced commensurate with the amount of intercepted water to be evaporated from the leaves. Typical total amounts are 0.1 to 0.2 in. per storm, but greater values can occur. This means that smaller storms may not even wet the soil beneath the vegetation. The proportion of annual precipitation intercepted typically ranges from 10 to 25% (Lull, 1964). As discussed for bare ground, a general rule is the more rain events, the greater the proportion of annual rain intercepted.

Interception is an important consideration in spray irrigation. As with bare ground, applications usually should be of larger amounts and be applied less frequently. To bring a soil up to field capacity, the abstraction or loss

of water due to interception must be accounted for in calculating the depth of water to be applied.

4.3.2 TRANSPIRATION

Transpiration was defined by Kramer (1983) as the loss of water in the form of vapor from plants. We recognize from this definition that transpiration is basically an evaporation process. The transpiration of water that moves out of the leaves and into the atmosphere is responsible for the ascent of water from the roots and the rate at which water is taken in through the roots. The rate at which water moves through a plant is critical to the plant's functioning because water is the vehicle that carries nutrients and minerals into the plant. Water that falls on leaves due to precipitation or irrigation and then evaporates is not associated with transpiration. So, how does evaporation/transpiration of water from within the leaf influence what happens in the roots of the plant?

Figure 4.3 shows the structure of a typical leaf. An epidermis is present on both the upper and the lower sides of a leaf. The epidermal cells usually have a waterproof cuticle on the atmospheric side. Between the two epidermal layers is the leaf mesophyll, which is usually differentiated into two types of cells — the elongated, orderly palisade cells and the more loosely packed, spongy mesophyll cells. The unstructured arrangement of the spongy mesophyll cells allows the intercellular formation of airspaces, which are the primary sites for transpiration. The pore that connects the intercellular spaces with the atmosphere is called a *stoma* or *stomata*. The plural forms are stomata or stomates, respectively. Each stoma can open or close by changing the size of the guard cells at the edges of the pore.

The leaf mesophyll cells are saturated and are typically in contact with the intercellular spaces. Water from a saturated surface in contact with the atmosphere will

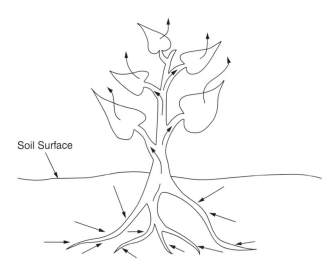

Soil Surface

FIGURE 4.4 Schematic of the path of water through the plant.

evaporate if there is a source of energy and a difference in concentration of vapor in the air (or "room" for new water molecules to enter the air). The needed energy comes primarily from direct solar radiation. Evaporation will increase the vapor pressure in the intercellular space, resulting in a vapor pressure gradient between the leaf and the atmosphere under most atmospheric conditions. This gradient will cause vapor to diffuse out of the stomates into the atmosphere. The larger the vapor pressure gradient, the higher the rate of diffusion, and the faster evaporation will occur (Fick's first law of diffusion). As a practical example of this, consider a hot summer day when the temperature is around 90°F. Is it more comfortable when the relative humidity is 60 or 95%? Why? Most people are more comfortable at 60% because there is more room in the air for water vapor. The sweat on our skin evaporates faster and helps cools us. The latent heat of vaporization is absorbed during the change of state of water, and this removes heat from the air–skin system.

Once some water is evaporated from the interior cells in a leaf, more liquid water moves in to replace that droplet (**Figure 4.4**). The process is similar to using a straw to drink. The plant has many connected cellular pathways that effectively act as the straws. The evaporation at the intercellular space initiates the pull. The divisions and subdivisions of the xylem (the main upwardly conducting tissue) in the leaves allow for mesophyll cells to be no more than three or four cells away from the xylem. Transpiration effectively pulls water into the plant through the roots, up the xylem in the stem, and out the leaves.

Transpiration is considered beneficial to the plant because essential nutrients are brought in solution from the soil water to the plant. Also, the process of transpiring helps the plant absorb minerals and helps cool the leaves. Kramer (1983) countered the second point, however, because leaves in full sun are rarely injured by high temperatures, and transpiration merely increases the amount

of water moved. Kramer also noted that many plants thrive in shaded, humid areas, where the rate of transpiration is very low. Numerous harmful effects of excessive transpiration resulting in water stress have been recorded. So, why do plants transpire? Kramer hypothesized that transpiration is an unavoidable evil because evolution favored high rates of photosynthesis over low rates of transpiration, and the leaf structure favorable for the entrance of carbon dioxide (essential for photosynthesis) also allows water vapor loss.

Many plants do have mechanisms to reduce transpiration if necessary. Plants can reduce the leaf area (rolling leaves or wilting) or change leaf orientation to reduce the amount of tissue exposed to sunlight (reduce the energy received). Plants can close the stomates to block the diffusion of water vapor out of the leaf (this also blocks the entrance of CO_2). The ability of plants to adjust their rate of transpiration constantly is one of the major difficulties encountered when trying to estimate the actual transpiration rate.

4.3.3 FACTORS THAT AFFECT TRANSPIRATION RATES

Following is a brief discussion of some of the factors that affect transpiration rates. The term *plants*, as used in this book, normally refers to indigenous mesophytic plants of humid, temperate, midlatitude regions. While some of the crops mentioned may be exotic, their moisture regimes are mostly mesophytic. For plants with more exotic moisture regimes, refer to the work of Nobel (1983) and Mac-Donald (2002).

Type of Plant. The type of plant has always been controversial. Some have maintained that, as long as the total planar surface of the earth is covered by leaves (thus maximizing the receipt of direct insolation), then the potential transpiration is at a maximum insofar as the plants are concerned. Later work indicated that indirect solar energy also plays a role, meaning that shaded leaves or needles may be transpiring, although not at the higher rate driven by direct solar energy. This would suggest that taller plants would have higher potential transpiration rates. Depth of roots, while often, but not always, correlated with plant height, can give access to water at greater depth. Indeed, often exotic *pheatophytes* (plants with root systems that can reach into groundwater, or the phreatic zone) continue to transpire huge amounts of water in the Southwest with little or no economic return.

The idea that trees consume more water than lower-growing vegetation is no longer in question (Bosch and Hewlett, 1982). For a large part of the southeastern U.S., forested areas, over a period of decades, have continued to transpire about 330 mm (area–depth) more than other land covers (Trimble et al., 1987).

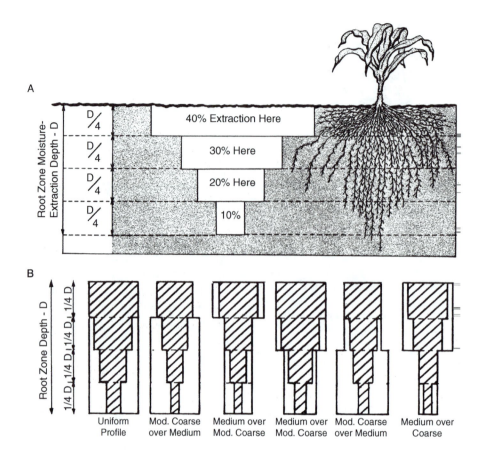

FIGURE 4.5 A. Average moisture extraction pattern of plants growing in a soil without restrictive layers and with an adequate supply of plant-available water (PAW) throughout the root zone. B. Moisture extraction patterns as determined by PAW in various parts of soil profile. Width of each profile represents PAW; gross area of each profile represents total PAW in profile; shaded area shows moisture extraction pattern for each profile. (From USDA-SCS, *National Engineering Handbook, Irrigation*.)

The ability to extract water from the soil is not directly correlated with depth of roots. While more water may be available to lower roots, the energy required to pull that water up to the plant is great, so only a smaller amount may be extracted. This principle is demonstrated in **Figure 4.5**, which also shows that the proportional water drawn from depths may be partially a function of soil texture at those depths. Some small crop plants can have extensive root systems, such as alfalfa, which has been known to extend roots as far as 30 m. Some typical crop root systems are shown in **Figure 4.6**.

Insolation, temperature, and relative humidity are the three factors that drive the energy gradient between the plant leaves and the atmosphere and largely control the potential transpiration from the leaf stomata. As noted here, direct insolation is most effective, but indirect insolation is also important. Temperature and relative humidity also show up in the vapor pressure deficit mentioned in conjunction with Dalton's law. The importance of these three factors is demonstrated under irrigation in southern California, where water consumption rates for several

crops (lemons, oranges, walnuts, and alfalfa) can more than double between April and July (Veimeyer, 1964).

Wind. By breaking up the saturated layer of air closest to the leaf, wind can increase transpiration rates. To preclude excessive rates, however, leaf stomata have guard cells that will curtail transpiration as the wind speed increases, with 8 mph (12 ft or about 4 m/sec) a common threshold. Even so, Santa Ana wind conditions in Los Angeles can rapidly desiccate delicate plants like azaleas.

Plant Available Water. We again need to make the distinction between PET, the amount of ET that could possibly occur under given atmospheric conditions, and AET, with the limitation here available soil water. We know that plants can transpire at maximum rates at soil field capacity and stop transpiring at the wilting point. The main question here is how plant transpiration rates vary in the moisture range between those two values, which is plant available water (PAW; also termed by some as plant available moisture [PAM] or available moisture content [AMC]). Harrold et al. (1986) gave three primary models for mesophytic plant

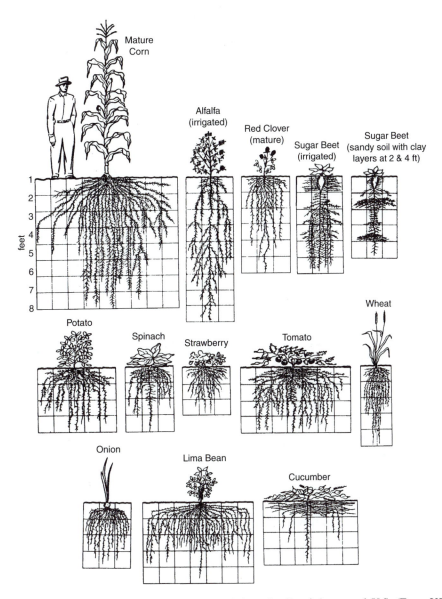

FIGURE 4.6 Root systems of field and vegetable crops in deep irrigated soils of the central U.S. (From USDA-SCS, *National Engineering Handbook, Irrigation.*)

water uptake in the PAW range (**Figure 4.7**). The most commonly accepted of these is the Pierce curve, which shows plants transpiring at maximum rates (AET = PET) from field capacity down to about 50% PAW, at which point the transpiration begins to slow, falling off sharply as the wilting point is approached. With the Veimeyer model, transpiration remains high (AET = PET) through the range of PAW until the WP is approached, when transpiration drops sharply. The Thornthwaite model, rarely used, shows the ratio AET/PET as a linear function of percentage PAW. Because plant or crop growth and productivity are roughly related to transpiration rates, such models are more than just academic exercises. Under irrigation, farmers will apply water to maximize

production, but they need to know the relationship. That is often not clear.

For mesophytic plants, maintaining a soil water content above FC by overirrigating is problematic. Roots are deprived of oxygen, and plant processes, including transpiration, can be curtailed. When soil water content above FC is prolonged, plant diseases may ensue. High levels of gravity water may also be harmful in other ways. If the water table is deep, percolating water may transport nutrients into groundwater. When the water table is near the surface, excess water may raise the water table high enough so that capillary action can "wick" groundwater solutes, such as salt, to the surface, where ET can concentrate them.

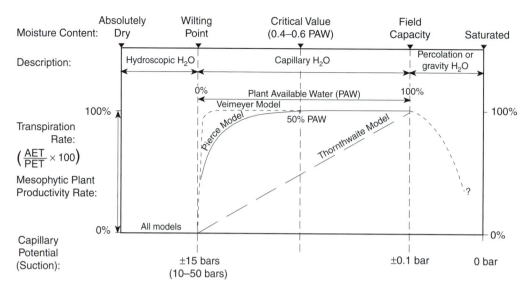

FIGURE 4.7 Relationship of soil moisture to transpiration rate, plant productivity, and capillary potential.

4.3.4 Transpiration Ratio and Consumptive Use

Two terms frequently used relating to transpiration are transpiration ratio and consumptive use. *Transpiration ratio* is the ratio of the weight of water transpired to the weight of dry matter produced by the plant. This ratio is a measure of how efficiently crops use water. For example, the transpiration ratios are approximately 900 for alfalfa, 640 for potatoes, 500 for wheat, 450 for red clover, 350 for corn, and 250 for sorghum. The least-efficient crop, in terms of water use, would be alfalfa because it uses 900 kg of water for every kilogram of dry alfalfa it produces. Sorghum is the most efficient crop listed because it uses only 250 kg of water for every kilogram of dry matter produced.

Consumptive use is the total amount of water needed to grow a crop (the sum of the water used in ET plus the water stored in the plant's tissues). The term consumptive use is generally used interchangeably with ET because the amount of water retained in plant tissue is negligible compared to the amount of ET.

4.3.5 Potential Evapotranspiration and Actual Evapotranspiration Concepts and Practices in Agriculture

Evaporation and, especially, evapotranspiration are complex processes because the rate of water vapor loss depends on the amount of solar radiation reaching the surface, the amount of wind, the aperture of the stomates, the soil water content, the soil type, and the type of plant. To simplify the situation, researchers have attempted to remove all the unknowns, such as aperture of the stomates and soil water content, and focus on climatic conditions. The simplified calculations are termed *potential evaporation* and

TABLE 4.1

Conventional Abbreviations for Terms Relating to Actual and Potential Evaporation and Evapotranspiration

E_p	Potential evaporation
E_{tp}	Potential evapotranspiration
E	Actual evaporation
E_t	Actual evapotranspiration
E_{tr}	Reference crop evapotranspiration

potential evapotranspiration. The definition given by M.E. Jensen et al. (1990) for potential evaporation E_p is the "evaporation from a surface when all surface–atmosphere interfaces are wet so there is no restriction on the rate of evaporation from the surface. The magnitude of E_p depends primarily on atmospheric conditions and surface albedo but will vary with the surface geometry characteristics, such as aerodynamic roughness." *Surface albedo* is the proportion of solar radiation reflected from a soil and crop surface. Conventional abbreviations for the terms used in this section and the subsequent two sections are summarized in **Table 4.1**.

Penman (1956) originally defined potential evapotranspiration E_{tp} as "the amount of water transpired in unit time by a short green crop, completely shading the ground, of uniform height and never short of water." Looking at Penman's definition, we can see the attempt to simplify the situation. The crop is assumed to be short and uniform and completely shading the ground so that no soil is exposed. The crop is never short of water, so soil water content is no longer a variable, and presumably the stomates would always be fully open. These conditions theoretically provide

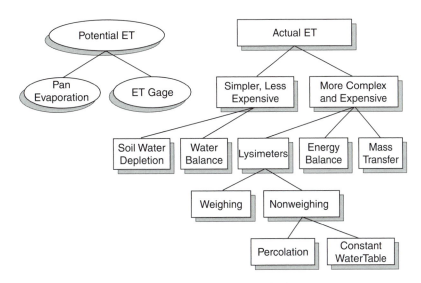

FIGURE 4.8 Options available for measuring potential or actual evapotranspiration.

the maximum evapotranspiration rate based on the given climatic conditions.

Many authors treat potential evapotranspiration and potential evaporation as synonymous, but the original intent was that actual evapotranspiration involved an actively growing crop, and potential evaporation did not.

Potential evapotranspiration and evaporation may be easier to estimate, but do not necessarily represent reality in agriculture. In general, watersheds are not entirely covered by well-watered short green crops. Actual ET or actual evaporation (E) is the amount or rate of ET occurring from a place of interest, and it is the value we want to estimate. In practice, actual ET is obtained by first calculating potential evapotranspiration and then multiplying by suitable crop coefficients to estimate the actual crop evapotranspiration. Crop coefficients are usually residual terms from a statistical analysis of field data, so it is essential that the methods for estimating potential evapotranspiration be consistent with the crop coefficients. Because potential evapotranspiration has been defined vaguely (short green crop), some studies have used alfalfa and some have used grasses to measure potential evapotranspiration. Other methods correlate evaporation from free water surfaces to actual ET. The result is that the methods for determining actual ET are variable and confusing (M.E. Jensen et al., 1990). Scientists have attempted to remedy this problem by introducing reference crop evapotranspiration.

4.3.6 Reference Crop Evapotranspiration

Reference crop evapotranspiration E_{tr} is defined as "the rate at which water, if available, would be removed from the soil and plant surface of a specific crop, arbitrarily called a reference crop" (M.E. Jensen et al., 1990). Typical reference crops are grasses and alfalfa. The crop is assumed to be well watered with a full canopy cover (foliage completely shading the ground). The major advantage of relating E_{tr} to a specific crop is that it is easier to select consistent crop coefficients and to calibrate reference equations in new areas.

Many methods are available for quantifying evaporation and evapotranspiration. Methods are selected based on the accuracy needed, the timescale required, and the resources available. The first decision that needs to be made is whether ET will be measured or estimated from weather data. If we want to determine ET for a time period that has already elapsed, obviously our only choice is to estimate the ET from weather data records. If we want to predict ET for future time periods and money is available to purchase instrumentation, then measuring ET would be an option.

4.4 MEASURING EVAPORATION OR EVAPOTRANSPIRATION

There are several methods available for measuring evaporation or evapotranspiration. Since vapor flux is difficult to measure directly, most methods measure the change of water in the system. **Figure 4.8** shows schematically the options available for measuring potential evaporation, potential evapotranspiration, or actual evapotranspiration. An evaporation pan or ET gage can be used to measure potential evapotranspiration. Actual evapotranspiration can be measured in several ways. If we need a simpler, less-expensive technique, measuring soil water depletion or using a water balance would be possibilities. More precise, but also more complex, methods include lysimeters, either weighing or nonweighing, or using an energy balance or mass transfer technique. These methods are discussed in more detail in the following section.

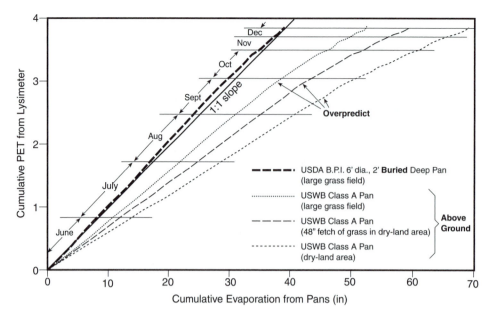

FIGURE 4.9 Relation of cumulative ET for perennial ryegrass to evaporation from two types of pans at Davis, CA (USWB = U.S. Weather Bureau). (From USDA-SCS, *National Engineering Handbook, Irrigation.*)

4.4.1 EVAPORATION PAN

One of the oldest and simplest ways to measure evaporation is with a pan. This involves placing a pan of water outside and recording how much water evaporates during a specific time. The type (steel, plastic, glass) and the size of the pan affect how fast water will evaporate. To standardize pans, the U.S. Weather Bureau has chosen a standard pan, called a Class A pan. This is a cylindrical container made of galvanized steel or Monel™ metal, 10 in. deep and 48 in. in diameter. The pan is placed on an open wooden platform with the top of the pan 16 in. above the soil surface. The platform site should be nearly flat, well sodded, and free from obstructions. The pan should be leveled and filled with water to a depth of 8 in. Periodic measurements are made of the changes in water level with a hook gage set in a still well. When the water level drops 7 in., the pan should be refilled.

The problem with the Class A pan is that it overestimates PET because energy enters the pan through the sides and bottom. Thus, it overestimates lake evaporation by 20 to 40%, depending on where location in the U.S. (see later discussion). As discussed, lake evaporation is a good surrogate for PET, and the closest surrogate for a lake, and thus for PET, is the buried pan. The buried pan is much like the Class A pan, except that it is buried almost up the rim so that the energy exchanges resemble those of a lake. Many buried pans were in use at one time, but no longer. A problem is that leaves and trash can easily blow onto the water surface and affect the readings, so it requires more maintenance than the Class A pan. However, the superiority of the buried pan is clearly shown in **Figure 4.9**, which shows how well a USDA buried pan approximates PET as measured by a lysimeter at Davis, CA. Conversely, the three Class A pans overpredict from 15 to more than 30%, depending on their surrounding vegetation.

So what does a farmer using irrigation who hopes to keep the crops at optimum ET and thus at optimum production do? Many such farmers use a very simple device; they cut a 55-gal barrel in half and bury one half, leaving only 2 to 3 inches of circular rim above ground. They keep the water level at ground level, replacing the loss each day, carefully noting the daily water losses. Knowing the moisture characteristics of their crops and soils, they maintain a running soil water budget and thus know when to water. Such a soil water budget is shown later in this chapter.

Pan evaporation data can be used to estimate actual evapotranspiration of a reference crop using the following equation (M.E. Jensen et al., 1990):

$$E_{tr} = k_p E_{pan} \qquad (4.2)$$

where k_p is a crop or pan coefficient. Many k_p values have been determined in previous studies, but it is important that the study have a similar climate (humid vs. arid) and use the same pan (i.e., Class A pan) with similar nearby surfaces and placement in relation to wind barriers at the site of interest. The coefficients in **Table 4.2** were developed for an alfalfa reference crop. The coefficients can be used to convert pan measurements to an alfalfa reference crop ET estimate or to convert estimates made with the other methods discussed in Section 4.6. **Table 4.2** should be used with caution because the coefficients represent

TABLE 4.2
Ratios of E_{tr}/E_{pan} and $E_{tr}/E_{t(method)}$ Developed from Kimberly, ID, data

Method	E_{tr}/E_{pan} and E_{tr}/E_t (method)						
	April	May	June	July	August	September	October
E_{pan}	0.75	0.86	0.92	0.94	0.92	0.92	0.91
E_{tr}	1.00	1.00	1.00	1.00	1.00	1.00	1.00
$E_{tp}(P)$	0.98	1.14	1.20	1.20	1.15	1.12	1.12
E_{tr}(JH)	1.33	1.25	1.18	1.08	1.08	1.25	1.37

E_{pan} = measured Class A pan evaporation; E_{tr} = reference ET, alfalfa, by combination equation using the 1982 Kimberly wind function (Wright, 1982); $E_{tp}(P)$ = potential ET, grass, modified Penman method using $W_f = (1.0 + 0.0062\, u_2)$ with u_2 in km d^{-1} and vapor pressure deficit Method 3 (M.E. Jensen et al., 1971); E_{tr}(JH) = reference ET, alfalfa, Jensen–Haise method.

Source: Jensen, M.E., R.D. Burman, and R.G. Allen, *Evapotranspiration and Irrigation Water Requirements*, American Society of Civil Engineers, New York, 1990. Copyright 1990 by the American Society of Civil Engineers. Reproduced with permission of the American Society of Civil Engineers.

monthly averages, and the interrelationships of methods may not be the same at all locations. To use **Table 4.2**, select the appropriate month from the top of the table and the appropriate method from the left-hand column, then determine the adjustment coefficient.

EXAMPLE 4.1

If the measured pan evaporation were 4.7 in. for June, what would the reference crop equivalent be for alfalfa?

Solution

From **Table 4.2**, choose the method (E_{pan}), move to June (Column 4), and select the E_{tr}/E_{pan} coefficient (= 0.92). Then, E_{pan} (E_{tr}/E_{pan}) = 4.7 × 0.92 = 4.3 in. of ET from alfalfa.

Answer

The equivalent alfalfa reference crop ET is 4.3 in. of water in June.

4.4.2 E_{tp} Gages

Gages that estimate localized, real-time (as it is actually occurring) potential evapotranspiration are commercially available. A typical gage has a surface that simulates a well-watered leaf. The surface is exposed to the atmosphere, where it will experience the same energy source and wind as the crop. Commercial literature claims excellent agreement between the potential evapotranspiration measured by the gage and potential evapotranspiration calculated by the Penman–Monteith equation for daily ET.

4.4.3 SOIL WATER DEPLETION

AET from a crop can be estimated by observing the change in soil water over a period of time. The average rate of ET in millimeters/day between sampling dates (denoted Δt) can be calculated using the following equation (M.E. Jensen et al., 1990):

$$E_t = \frac{\Delta SM}{\Delta t} = \frac{\sum_{i=1}^{n_{tz}} (\theta_1 - \theta_2)_i \Delta S_i + I - D}{\Delta t} \quad (4.3)$$

where E_t is the actual evapotranspiration (mm/d); ΔSM is the change in soil water content; Δt is the time between sampling dates; n_{rz} is the number of soil layers in the effective root zone; ΔS_i is the thickness of each soil layer (mm); θ_1 is the volumetric water content of soil layer i on the first sampling date (m^3/m^3); θ_2 is the volumetric water content of soil layer i on the second sampling date (m^3/m^3); I is the infiltration (rainfall–runoff) during Δt (mm); and D is the drainage below the root zone during Δt (mm).

Estimating AET by soil sampling may be possible if the following precautions are observed: (1) take multiple soil moisture measurements in the field to obtain an average soil water content representative of the entire field; (2) only use this technique for sites where the depth to the water table is much greater than the depth of the root zone; and (3) use this technique for sampling periods when runoff and drainage out of the root zone are zero. This method can be problematic because (1) measurement of soil moisture is problematic by whatever method and (2) subsurface drainage from a soil may continue even after field capacity has been reached (**Figure 3.7**).

A

B

FIGURE 4.10 Large Chinese weighing lysimeters. A. The underside of the lysimeters. B. Research plots at the surface.

4.4.4 WATER BALANCE

The water balance approach is generally used on large areas such as watersheds. The inflows and outflows are determined from streamflow and precipitation measurements, and the difference between inflow and outflow over a relatively long period of time, such as a season, is a measure of evapotranspiration. The area in question must be confined so that other significant sources of inflow or outflow do not exist. The results are applicable only to the climate, cropping, and irrigation conditions similar to those in the study area (see Exercise 14.2, Part III, Chapter 14).

The equation is

$$AET = P - Q \pm \Delta G \pm \Delta \theta \qquad (4.4)$$

where P is the precipitation depth, AET is actual ET, Q is runoff depth, ΔG is groundwater inflow or outflow, and $\Delta \theta$ is soil water change. For the short term, ΔG and $\Delta \theta$ should be measured, but over a period of years, they become insignificant and can be dropped, so the equation becomes

$$AET = P - Q \qquad (4.5)$$

4.4.5 LYSIMETERS

Lysimeters are devices that allow an area of a field to be isolated from the rest of the field, yet experience similar conditions to that of the growing crop. Typically, the lysimeter is a cylinder inserted into the soil or a tank filled with soil and placed in a field. Crops are grown on the surface of the lysimeter to approximate the conditions of the field. Since the lysimeter is isolated from the remainder of the field, measurement of the individual components of the water balance is possible, so an estimate of actual

evapotranspiration can be calculated. Reference crop ET is typically determined using measurements made with lysimeters planted with alfalfa or grass. Similarly, crop coefficients can be determined by planting the crop of interest in the lysimeter and measuring ET, then relating this to the reference crop ET.

Lysimeters can be grouped into weighing and non-weighing types. The weighing lysimeters allow changes in soil water to be determined either by weighing the entire unit with a mechanical scale or counterbalanced scale and load cell or by supporting the lysimeter hydraulically (M.E. Jensen et al., 1990). A world-renowned example of weighing lysimeters is the large lysimeters located in Coshocton, OH, and the Chinese lysimeters shown in **Figure 4.10**.

Nonweighing lysimeters are either of the percolation type or of the constant water table type. In the percolation type, the changes in soil water are determined either by sampling or by an indirect method such as a neutron probe. The drainage or "percolation" out the bottom of the root zone is also measured. With knowledge of the rainfall, the actual transpiration for the lysimeter can be determined. The constant water table lysimeters are useful in locations where a high water table exists. With this type of lysimeter, the water table is maintained at a constant level inside the lysimeter, and the water added to maintain water level is a measure of actual ET.

4.4.6 ENERGY BALANCE AND MASS TRANSFER

The energy balance and mass transfer methods of determining actual ET require measuring the average gradient of water vapor above the canopy. The instrumentation needed and the technical procedures involved typically limit these methods to research applications. Additional details can be found in the work of M.E. Jensen et al. (1990).

TABLE 4.3
Minimum Climatic Information Needs
of ET Estimation Methods

Method	T^a	RH^b or $e_d{}^c$	Latitude	Elevation	$R_s{}^d$	u^e
Penman	x	x		x	x	x
Jensen–Haise	x			x	x	
SCS Blaney–Criddle	x		x			
Thornthwaite	x					

[a] Air temperature.
[b] Relative humidity.
[c] Actual vapor pressure of the air.
[d] Solar radiation.
[e] Wind speed.

4.5 WEATHER DATA SOURCES AND PREPARATION

Many empirical and physically based methods have been developed for estimating evaporation and evapotranspiration from measured climatic data. For example, evaporation from a lake or reservoir can be estimated if wind speed, relative humidity, and temperature are known. **Table 4.3** summarizes the weather data requirements for the ET estimation methods discussed in the next section.

It is often difficult to find weather data for the area of interest. Many irrigation projects have weather stations from which data may be obtained, and some states have weather station networks that may be a source of data. If interested in data for an area in the U.S., monthly summaries of each state's weather are available from the National Climatic Data Center in Asheville, NC, which is associated with the National Oceanic and Atmospheric Administration (NOAA). The summaries include daily precipitation and daily maximum and minimum temperatures from many observation points in each state. Daily pan evaporation and wind speed are recorded for some stations. These summaries also provide latitude, longitude, and elevation for each recording station. Many libraries subscribe to these climatological data summaries. More detailed sources for climatic data acquisition are given in Chapter 14 in conjunction with the exercises.

Once the data are located, they need to be prepared for calculations using methods identical to those used by the developers of the ET estimation methods. The most obvious preparatory task would be to ensure that data are in the proper units for the equations. In addition, there are multiple ways some parameters could be calculated; the method used must be matched to the ET estimation method. Some of these methods are discussed below.

4.5.1 SATURATION VAPOR PRESSURE

Saturation vapor pressure e_s is a property of air and is a function only of temperature. It is known from everyday experience that warmer air can hold more water vapor than cooler air (see **Figure 2.1**). That is why water condenses out of air as it cools. An example of this is condensation on the side of an iced tea glass in the summer. The glass is cooler than the surrounding air. When the warm, humid air comes in contact with the glass, the glass cools the air. Cooler air holds less water than warm air, so water condenses out of the air onto the side of the glass.

The relationship between saturation vapor pressure and temperature is shown in **Figure 4.11** (note the similarity to **Figure 2.1**). Temperature is shown along the horizontal axis, and saturation vapor pressure in millimeters of mercury, inches of mercury, and kilopascal (units of pressure) is on the vertical axis. Find the temperature of interest, move vertically up to the curved line, and then move horizontally to reach the axis, which shows the saturation vapor pressure.

Alternatively, Equation 4.6 can be used to compute saturation vapor pressure e_s in kilopascal if temperature T is in degrees Celsius:

$$e_s = \exp\left[\frac{16.78\,T - 116.9}{T + 237.3}\right] \qquad (4.6)$$

This equation is useful for writing a computer program to compute some of these values. The equation is valid for temperatures ranging from 0 to 50°C.

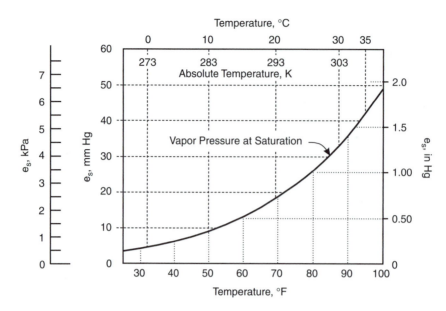

FIGURE 4.11 Saturated vapor pressure as a function of temperature. (From Schwab et al., 1993. With permission.)

EXAMPLE 4.2

Find e_s in kilopascal (using both **Figure 4.11** and Equation 4.6) if the air temperature is 70°F.

Solution

From **Figure 4.11**, at 70°F, the $e_s = 2.5$ kPa. Using Equation 4.6 and with the temperature in Celsius (70°F = 21.11°C), $e_s = 2.5$ kPa.

Answer

The saturated vapor pressure is 2.5 kPa.

4.5.2 ACTUAL VAPOR PRESSURE

Actual vapor pressure e_d is the vapor pressure of the air. Unlike saturation vapor pressure, actual vapor pressure cannot be determined simply by knowing the temperature of the air. To determine e_d, we need to know the air temperature and either the relative humidity or the dewpoint temperature of the air. The actual vapor pressure is dependent on the amount of water currently in the air, whereas the saturation vapor pressure is based on the maximum amount of water that could be absorbed by the air. We need to know relative humidity or dewpoint temperature because they are measures of the amount of water actually in the air. The following equation can be used to find actual vapor pressure:

$$e_d = e_s \times \frac{RH}{100} \qquad (4.7)$$

where e_d is the actual vapor pressure, e_s is the saturation vapor pressure, and RH is the relative humidity in percent.

4.5.3 VAPOR PRESSURE DEFICIT

Many methods exist for calculating the vapor pressure deficit $(e_s - e_d)$ (see M.E. Jensen et al., 1990). Three methods for calculating vapor pressure deficit are shown here.

Method 1. Method 1 is the saturation vapor pressure at mean temperature minus saturation vapor pressure at dewpoint temperature; this can be written as

$$(e_s - e_d) = e_{s(T_{avg})} - e_{s(T_d)} \qquad (4.8)$$

where e_d is the actual vapor pressure, e_s is the saturation vapor pressure, T_{avg} is the mean temperature for time period of interest, and T_d is the mean dewpoint temperature for the time period of interest.

EXAMPLE 4.3

Compute the vapor pressure deficit for June 17, 1993, at Piketon, OH, using Method 1. The data are shown in **Table 4.4**.

Solution

From **Table 4.4**, $T_{avg} = 22.05°C$, and e_s at T_{avg} is computed as follows (Equation 4.6):

$$e_s = \exp\left[\frac{16.78(22.05) - 116.9}{22.05 + 237.3}\right]$$

$$= \exp\left[\frac{253.1}{259.35}\right] = 2.65 \text{ kPa}$$

TABLE 4.4
Piketon, OH, Weather Data for June 17, 1993

Hour	R_s (kJ/m²/h)	T (°C)	e_s (kPa)	e_d (kPa)	$(e_s - e_d)$ (kPa)	RH (%)
1:00 a.m.	0.087	16.08	1.827	1.765	0.062	96.6
2:00 a.m.	0.234	15.28	1.736	1.708	0.028	98.4
3:00 a.m.	0.362	14.68	1.670	1.652	0.018	99.0
4:00 a.m.	0.255	14.25	1.624	1.613	0.011	99.3
5:00 a.m.	0.281	14.01	1.598	1.593	0.005	99.7
6:00 a.m.	1.213	13.13	1.510	1.510	0.000	100
7:00 a.m.	64.25	13.16	1.513	1.513	0.000	100
8:00 a.m.	368.7	14.38	1.638	1.638	0.000	100
9:00 a.m.	720.0	16.81	1.916	1.810	0.106	94.7
10:00 a.m.	1154.0	20.68	2.442	1.988	0.454	81.6
11:00 a.m.	1665.0	23.68	2.931	2.099	0.832	71.8
12:00 noon	2164.0	26.70	3.505	2.199	1.306	62.9
1:00 p.m.	2463.0	28.26	3.837	1.925	1.912	50.2
2:00 p.m.	2568.0	29.19	4.051	1.996	2.055	49.3
3:00 p.m.	2439.0	30.26	4.306	1.995	2.311	46.3
4:00 p.m.	2231.0	30.78	4.437	2.026	2.411	45.7
5:00 p.m.	1776.0	30.96	4.483	2.103	2.380	46.9
6:00 p.m.	1269.0	30.92	4.472	2.139	2.333	47.8
7:00 p.m.	869.0	30.43	4.349	2.193	2.156	50.4
8:00 p.m.	408.3	29.56	4.139	2.424	1.715	58.7
9:00 p.m.	92.70	27.08	3.588	2.496	1.092	69.8
10:00 p.m.	2.035	23.60	2.917	2.42	0.497	83.2
11:00 p.m.	0.020	21.76	2.604	2.327	0.277	89.4
12:00 midnight	0.013	20.78	2.453	2.292	0.161	93.5

$T_{max} = 31.4°C$; $T_{min} = 12.7°C$; $T_{avg} = 22.05°C$.

Next, we need the saturation vapor pressure at the dew-point temperature. If we knew the dewpoint temperature, we would substitute it into Equation 4.8 and obtain the saturated vapor pressure at T_d. However, we do not know T_d, but $e_{s(T_d)}$ was recorded because $e_{s(T_d)} = e_d$. From **Table 4.4**, if we average e_d, we obtain

$$e_{d_{avg}} = \frac{\sum_{i=1}^{24} e_d}{24} = 1.98 \text{ kPa}$$

then,

$$e_s - e_d = 2.65 - 1.98 = 0.67 \text{ kPa}$$

Answer

The vapor pressure deficit is 0.67 kPa.

Method 2. The vapor pressure deficit can be estimated from the saturation vapor pressure at the mean temperature times the quantity 1, minus the relative humidity expressed as a proportion or

$$(e_s - e_d) = e_{s(T_{avg})}\left(1 - \frac{RH}{100}\right) \qquad (4.9)$$

This equation is obtained by writing Equation 4.7 as

$$e_{s(T_s)} = e_{s(T_{avg})}\frac{RH}{100} \qquad (4.10)$$

and then substituting Equation 4.10 into Equation 4.8.

EXAMPLE 4.4

Compute the vapor pressure deficit for June 17, 1993, at Piketon, OH, using Method 2. The data are shown in **Table 4.4**.

Solution:

$$RH_{avg} = \frac{\sum_{i=1}^{24} RH_i}{24} = 76.5\%$$

$$T_{avg} = \frac{T_{min} + T_{max}}{2} = 22.05°C$$

e_s at T_{avg} (from Figure 4.11) = 2.65 kPa

Using Equation 4.9,

$$e_s - e_d = 2.65 \left(1 - \frac{76.5}{100}\right) = 0.62 \quad \text{kPa}$$

Answer

The vapor pressure deficit equals 0.62 kPa.

Method 3. For Method 3, compute the mean of saturation vapor pressure at the maximum and minimum temperatures minus the saturation vapor pressure at the dewpoint temperature determined early in the day, typically at 8 a.m.

$$\left(e_s - e_d\right) = \frac{e_{s_{(T_{max})}} + e_{s_{(T_{min})}}}{2} - e_{s_{(T_{d_{8am}})}} \quad (4.11)$$

EXAMPLE 4.5

Compute the vapor pressure deficit for June 17, 1993, at Piketon, OH, using Method 3. The data are shown in **Table 4.4**.

Solution

From **Table 4.4**, T_{max} = 31.4°C, and T_{min} = 12.7°C. Calculating $e_{s(T_{max})}$ using Equation 4.6,

$$e_{s_{(T_{max})}} = \exp\left[\frac{16.78(31.4) - 116.9}{31.4 + 237.3}\right]$$

$$= \exp\left[\frac{409.82}{268.69}\right] = 4.6 \text{ kPa}$$

Calculating e_s at T_{min} using the same equation,

$$e_{s_{(T_{min})}} = \exp\left[\frac{16.78(12.7) - 116.9}{12.7 + 237.3}\right]$$

$$= \exp\left[\frac{96.21}{250.0}\right] = 1.47 \text{ kPa}$$

Then, averaging these vapor pressures

$$\frac{e_{(T_{max})} + e_{(T_{min})}}{2} = \frac{4.60 + 1.47}{2} = 3.04 \text{ kPa}$$

From **Table 4.4**, we can find e_s at dewpoint temperature at 8 a.m. because it is the same quantity as e_d at 8 a.m. (1.638 kPa). Subtracting the two vapor pressures (3.03 and 1.638 kPa), we find the vapor pressure deficit calculated by Method 3 is 1.39 kPa.

Answer

The vapor pressure deficit is 1.39 kPa.

The average of the 24 hourly vapor pressure deficits is 0.922 kPa and is the most correct estimate of daily vapor pressure deficit. Note that Method 1 underpredicted the correct value by 32.4%, Method 2 underpredicted by 42.6%, and Method 3 overpredicted by 50.8%. The best method would be to average hourly values of vapor pressure deficit to obtain an average for the day. However, it is unlikely that the necessary weather data will be available to do this. The ET estimation methods generally indicate which method is preferred for calculating $(e_s - e_d)$.

4.5.4 MEAN TEMPERATURE

Several methods are available for computing mean or average temperature for a given time period. Two of these are discussed here.

Method 1. Average the individual mean temperatures from the next smallest timescale. For example, if we wanted a daily average of T, we could average the hourly temperatures as follows: From **Table 4.4**, the sum of Column 3 (T) is 536.42°C. Then, divide by 24 h to obtain an average; the average daily temperature is 22.35°C.

Method 2. Average the maximum and minimum temperatures for the period of interest, which in this example is 1 day. The daily maximum temperature was 31.4°C, and the daily minimum temperature was 12.7°C. Averaging these values equals 22.05°C. Similarly, if we were interested in a monthly mean temperature, we could average the mean daily temperatures or average the maximum and minimum temperatures for the month. Notice in the example that the two methods did not result in the same number; they differed by 1.4%. The mean temperatures needed for Method 1 are generally not available, whereas the maximum and minimum temperatures needed for Method 2 are typically available. Method 2 is the method generally used to calculate mean temperature.

4.5.5 SOLAR RADIATION

Some ET estimation methods require either R_s (solar radiation received on a horizontal plane at the earth's surface) or R_n (net radiation). R_s would be measured at a weather station and is more likely to be recorded than R_n, which requires measuring both incoming and outgoing solar radiation.

If R_s is measured, such as in the data shown in **Table 4.4**, we need to determine whether the data are presented as instantaneous readings or as cumulative solar radiation. The units associated with the values will help determine this. If the units are per hour, per second, or the like, then they are the average solar radiation for that time period and are not cumulative.

If neither R_s nor R_n have been measured for a site and these data are unavailable for a location near the area of interest, we can estimate R_s (and subsequently R_n) if we know the ratio of actual to possible sunshine and the latitude of the site. Refer to the work of M.E. Jensen et al. (1990) for a detailed description of this procedure.

The most likely scenario is that R_s has been measured, and we need R_n to use Penman's method or a similar method. It is possible to estimate R_n from R_s since R_n is the net short-wave minus the long-wave components of the radiation.

$$R_n = (1-\alpha)R_s \downarrow - R_b \uparrow \qquad (4.12)$$

where R_b is the net outgoing thermal radiation in megajoules per square meter per day (MJ/m²/d) and α is the albedo or short-wave reflectance, which is dimensionless. The arrows in Equation 4.12 serve as reminders that R_s is incoming and R_b is outgoing. The short-wave reflectance or albedo α is typically set equal to 0.23 for most green field crops with a full cover (M.E. Jensen et al., 1990). Since we know R_s and α, R_b is all that is needed. Equation 4.13 can be used to calculate this value (MJ/m²/d)

$$R_b = \left[a\frac{R_s}{R_{so}} + b\right]R_{bo} \qquad (4.13)$$

The coefficients a and b are determined for the climate of the area of interest. For humid areas, $a = 1.0$, and $b = 0$; for arid areas, $a = 1.2$, and $b = -0.2$; for semihumid areas, $a = 1.1$, and $b = -0.1$. R_{so} is the solar radiation on a cloudless day and can be obtained from **Table 4.5** (in MJ/m²/d) based on the site's latitude. R_{bo} can be computed from Equation 4.14.

$$R_{bo} = \epsilon \sigma T^4 \qquad (4.14)$$

where the Stefan–Boltzmann constant σ is 4.903×10^{-9} MJ/m²/d/K^4, and T is the mean temperature for the period of interest in Kelvin ($273 + T$ in °C). The term ϵ is the net emissivity and is calculated using the Idso–Jackson equation (Equation 4.15) with T in Kelvin.

$$\epsilon = -0.02 + 0.261 \ \exp[-7.77 \times 10^{-4}(273-T)^2] \qquad (4.15)$$

EXAMPLE 4.6

Calculate R_n from the R_s data given in **Table 4.4**.

Solution

Step 1. The first step is to calculate ϵ using Equation 4.15.

$$\epsilon = -0.02 + 0.261 \ \exp[-7.77 \times 10^{-4}(273-295.05)^2]$$
$$= 0.159$$

Step 2. Calculate R_{bo} using Equation 4.14.

$$R_{bo} = \epsilon \sigma T^4 = 0.159(4.903 \times 10^{-9})(295.0)^4 = 5.90 \text{ MJ/m}^2\text{/d}$$

Step 3. Calculate R_b from Equation 4.13 with $a = 1.0$ and $b = 0$ (Piketon, OH, is a humid area). For use in Equation 4.15, R_s (**Table 4.4**) should be in units of megajoules per square meter per day. Totaling the second column gives $R_s = 20,256.45$ kJ/m²/d $= 20.26$ MJ/m²/d. R_{so} is obtained from **Table 4.5** using the latitude of the place of interest (Piketon, OH $= 38°$N) and the month of interest (in our example, June); $R_{so} = 33.49$ MJ/m²/d.

$$R_b = \left[1.0\frac{20.26}{33.49} + 0\right]5.91 = 3.58 \text{ MJ/m}^2\text{/d}$$

Step 4. Then, R_n is calculated using Equation 4.12, assuming the generally used 0.23 value for albedo.

$$R_n = (1 - 0.23)20.26 - 3.58 = 12.0 \text{ MJ/m}^2\text{/d}$$

Answer

The net radiation on June 17, 1993, at Piketon, OH, was 12.0 MJ/m²/d.

4.5.6 EXTRAPOLATING WIND SPEED

Because of friction, wind is typically slower at the ground surface, and the speed increases with height. Most methods for estimating ET that require wind speed specify the height at which the wind speed should be recorded. However, in practice, the data have sometimes been recorded at other heights. To estimate the wind speed u_2 at height z_2, knowing the wind speed u_1 at height z_1, Equation 4.16 can be used (Allen et al., 1989).

$$\frac{u_1}{u_2} = \frac{\ln[z_1 - 0.67h_c] - \ln[0.123h_c]}{\ln[z_2 - 0.67h_c] - \ln[0.123h_c]} \qquad (4.16)$$

TABLE 4.5
Mean Solar Radiation R_{so} for Cloudless Skies

Latitude	Mean Solar Radiation per Month for Cloudless Skies (MJ/m²/d)											
	Jan	Feb	Mar	April	May	June	July	Aug	Sept	Oct	Nov	Dec
60° N	2.51	5.99	13.82	22.32	29.01	31.95	29.85	23.32	15.78	8.50	3.64	1.55
55° N	4.31	8.67	16.33	24.16	29.85	32.66	30.56	25.00	18.00	10.89	5.57	3.22
50° N	6.70	11.43	18.55	25.83	30.98	33.08	31.53	26.67	20.10	13.52	8.08	5.44
45° N	9.34	14.36	20.64	27.21	31.53	33.37	32.36	28.05	22.06	16.04	10.89	8.25
40° N	12.27	17.04	22.90	28.34	32.11	33.49	32.66	29.18	23.74	18.42	13.52	10.76
35° N	14.95	19.55	24.58	29.31	32.11	33.49	32.95	30.14	25.25	20.52	15.91	13.52
30° N	17.46	21.65	25.96	29.85	32.11	33.20	32.66	30.44	26.67	22.48	18.30	16.04
25° N	19.68	23.45	27.21	30.14	32.11	32.66	32.24	30.44	27.63	24.28	20.39	18.30
20° N	21.65	25.00	28.18	30.14	31.40	31.82	31.53	30.14	28.47	25.83	22.48	20.52
15° N	23.57	26.50	29.01	29.85	30.56	30.69	30.56	29.60	29.18	26.92	24.28	22.48
10° N	25.25	27.63	29.43	29.60	29.60	29.31	29.43	28.76	29.60	28.05	25.83	24.41
5° N	26.92	28.47	29.85	29.31	28.18	27.76	27.93	27.93	29.73	28.76	27.21	26.25
0° E	28.18	29.18	30.02	28.47	26.92	26.25	26.67	27.76	29.60	29.60	28.47	26.80
5° S	28.05	29.85	29.85	27.76	25.54	24.58	25.00	26.67	28.09	29.85	30.44	29.31
10° S	30.69	30.44	29.43	26.80	24.70	22.73	22.73	25.00	27.55	29.85	30.44	30.69
15° S	31.53	30.69	28.76	25.54	22.32	20.81	21.48	23.86	26.59	29.73	31.28	31.95
20° S	32.36	30.69	27.93	23.99	20.52	18.71	19.26	22.06	25.54	29.31	31.53	32.95
25° S	32.95	30.69	27.09	22.32	18.13	16.75	17.58	20.64	24.33	28.76	32.11	33.62
30° S	33.37	30.44	25.96	20.81	16.62	14.78	15.49	18.84	22.94	28.05	32.11	34.33
35° S	33.49	29.73	24.58	18.97	14.53	12.56	13.40	16.87	21.48	27.21	32.11	34.88
40° S	33.49	28.76	22.90	17.04	12.41	10.17	11.30	14.78	19.30	26.08	31.82	35.17
45° S	33.49	27.76	21.23	14.95	9.92	7.66	8.79	12.69	18.09	24.70	31.28	35.17
50° S	32.95	26.38	19.26	12.85	7.54	5.32	6.41	10.59	16.20	23.15	30.44	34.88
55° S	32.36	24.83	17.17	10.47	5.32	3.22	4.19	8.25	13.90	21.48	29.60	34.33
60° S	31.53	23.15	15.07	7.83	3.35	1.38	2.22	6.15	11.47	19.68	29.31	34.04

Note: August values in the Southern Hemisphere were corrected to obtain a smooth transition in monthly values. Also, 30-day months were assumed because when actual days per month were used, a smooth transition between January, February, and March did not occur.

Source: Jensen, M.E., R.D. Burman, and R.G. Allen, *Evapotranspiration and Irrigation Water Requirements*, American Society of Civil Engineers, New York, 1990. Copyright 1990 by the American Society of Civil Engineers. Reproduced with permission of the American Society of Civil Engineers.

where h_c is the height of the vegetation, $0.67h_c$ is the height at which the wind velocity approaches zero (known as the roughness height), and $0.123h_c$ is the surface roughness. The variables h_c, z_1, and z_2 are expected to have the same units, then u_2 will have units identical to u_1.

$$u_2 = 3\frac{\ln[25-(0.67\times 6)]-\ln[0.123\times 6]}{\ln[6\,(-0.67\times 6)]-\ln[0.123\times 6]}$$

$$= 3\times\frac{3.35}{0.987} = 10.2\,\frac{\text{mi}}{\text{hr}}$$

Answer

The wind speed at 25 ft was estimated as 10.2 mi/h.

Example 4.7

Find the wind speed at 25 ft if the wind speed at 6 ft is 3 mi/h. The vegetation is corn that is 6 ft tall.

Solution

Using Equation 4.16, with $h_c = 6$ ft,

4.6 ESTIMATING EVAPORATION AND EVAPOTRANSPIRATION

4.6.1 Evaporation from Open Water

Monthly evaporation from lakes or reservoirs can be computed using the empirical formula developed by Meyer

(1915), but based on Dalton's law (1802) (Harrold et al., 1986).

$$E = C(e_s - e_d)\left(1 + \frac{u_{25}}{10}\right) \qquad (4.17)$$

where E is the evaporation in inches/month; e_s is the saturation vapor pressure (inches of mercury) of air at the water temperature at 1 ft deep; e_d is the actual vapor pressure (inches of mercury) of air equal to $e_{s(\text{air T})} \times \text{RH}$; u_{25} is the average wind velocity (miles/hour) at a height of 25 ft above the lake or surrounding land areas; and C is the coefficient that equals 11 for small lakes and reservoirs and 15 for shallow ponds.

EXAMPLE 4.8

Compute the monthly evaporation from a reservoir with a monthly average water temperature of 60°F, measured 1 ft below the surface of the water, and a monthly average wind speed of 3 mi/h, measured at 25 ft above the reservoir. The monthly average air temperature was 70°F, and the monthly average relative humidity was 40%.

Solution

Determine the saturated vapor pressure of air at the water temperature 1 ft deep (60°F) = 0.51 inHg (**Figure 4.11**).

Next determine e_d for the air temperature (70°F) by first finding the saturated vapor pressure of air at 70°F (**Figure 4.11** shows e_s at 70°F = 0.73 inHg). Then, e_d = 0.4(0.73) = 0.292 inHg from Equation 4.7.

Since we are interested in a reservoir, $C = 11$. Using Equation 4.17,

$$E = 11\,(0.51 - 0.292)\left(1 + \frac{3}{10}\right) = 3.1\,\frac{\text{inches}}{\text{month}}$$

Answer

The evaporation was 3.1 in./month.

To estimate daily evaporation from Class A pans, there is a handy nomograph available (**Figure 4.12**). The data requirements are moderate and, for the average day at a particular location in the U.S., are easily available (see Chapter 14, Exercise 14.2, Part II). Consider a location for which the following data are available: (1) mean daily temperature is 80°F; (2) solar radiation is 600 Langleys

(Ly) per day; (3) mean daily dewpoint temperature is 50°F; and (4) wind movement is 100 mi/d 2 ft above the ground. As indicated in **Figure 4.12**, enter the nomograph from the left ordinate at 80°F and proceed to the 600-Ly isoline. From here, move down toward pan evaporation and simultaneously to the right to the dewpoint isoline of 50°F, then down to the 100-mi/d isoline. Then, move left and meet the vertical line intersecting at the value of 0.42 in. (about 1 cm) of Class A pan evaporation per day.

Depending on the source, the wind data may require adjustment before use, but this is easily done using techniques given in this chapter; see Chapter 14, Exercise 14.2, Part II, for details. Conversion factors for Class A pan-to-lake evaporation are given later in the chapter.

4.6.2 ESTIMATING EVAPOTRANSPIRATION

Figure 4.13 summarizes the decision process for selecting a method for estimating ET. It is important to know what weather data are available for the site of interest before selecting an estimation method because each method has different climatic information requirements (**Table 4.3**).

The majority of the ET estimating methods were developed to predict evapotranspiration from a well-watered short green crop (typically alfalfa or grass). Of the methods described in **Figure 4.13**, only the Soil Conservation Service (SCS) Blaney–Criddle method was specifically developed for estimating seasonal actual ET. The SCS Blaney–Criddle method can also be used to estimate actual monthly ET, provided the method has been locally calibrated.

An alternative approach to estimating actual evapotranspiration is to measure or estimate reference evapotranspiration, then adjust that value based on empirical coefficients for converting to actual ET. The reference ET values represent climatic demand, but environmental conditions, such as soil water conditions and the crop canopy status, need to be considered to obtain actual ET. Crop coefficients that relate actual ET to potential or reference crop ET have been derived from experimental data for particular crops, growth stages, and soil water conditions (M.E. Jensen et al., 1990). First, we explore available methods for estimating reference crop ET, then we discuss modifying these estimates with appropriate crop coefficients to obtain estimates of actual evapotranspiration.

Many methods are available to estimate reference crop evapotranspiration. There are two critical questions: What timescale (hourly, daily, 5 day, monthly, or seasonally) needs to be estimated? How much weather data are available? If we need to estimate hourly ET, then the Penman–Monteith method would be most applicable, but this method requires a large amount of weather data. Daily estimations of ET also require large amounts of weather

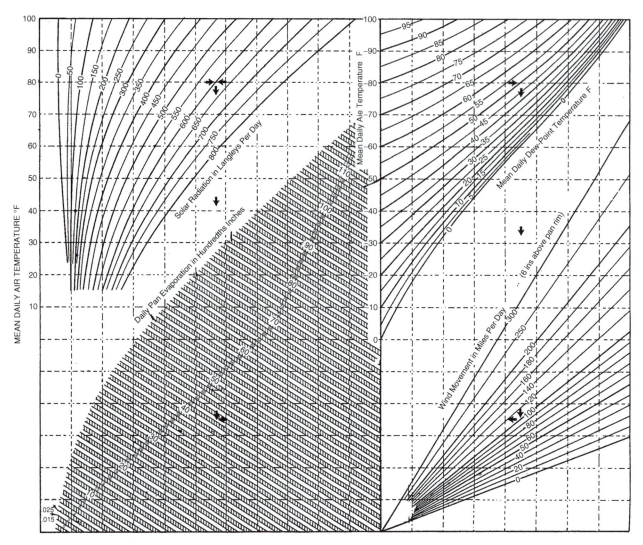

FIGURE 4.12 Nomograph for estimating daily Class A pan evaporation. (From Kohler, M.A., T. Nordenson, and D. Baker, Evaporation Maps for the United States, U.S. Weather Bureau Technical Paper 37, 1959.)

data. Two methods are shown in **Figure 4.13** for daily estimates: the Penman–Monteith (estimating hourly values, then summing over 24 h) or the Penman method. Penman's formula is one of the classics in micrometeorology and hydrology because it incorporated, for the first time, a number of fundamental principles relating the evaporation from a wet surface to its equilibrium temperature (Monteith, 1981).

The next timescale, shown in **Figure 4.13**, convenient for irrigation scheduling is 5 days. Again, if weather data are available, we could compute daily ET with the Penman method or hourly ET with the Penman–Monteith method and sum the values over a 5-day period. Typically, a 5-day time period is chosen because hourly or daily data are not available. Five-day averages can be used in the Penman and Penman–Monteith methods; however, accuracy

is lost compared to using these methods with daily or hourly weather data. The Jensen–Haise method was specifically developed for a 5-day time period and requires elevation and average solar radiation (**Table 4.3**).

Monthly estimates of ET could be obtained by all the methods mentioned, either using shorter time periods and summing over a month or using monthly averages in the individual methods. The SCS Blaney–Criddle method (if calibrated locally) and the Thornthwaite method will predict evapotranspiration on a monthly basis. Fewer weather data are required for the Thornthwaite method, which makes the method popular, but the estimate is less accurate because of the simplified inputs (Amatya et al., 1992). Seasonal ET can be computed by all the methods mentioned, either summing the estimates over the season or using seasonal values for the variables in the equations.

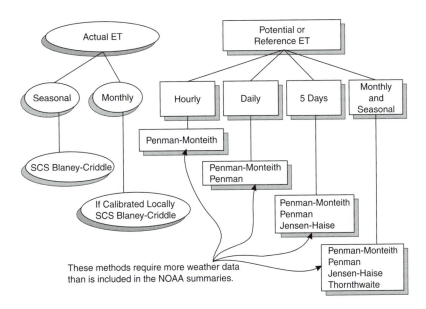

FIGURE 4.13 Methods available for estimating ET using climatic data.

4.6.3 SCS Blaney–Criddle Method

The SCS Blaney–Criddle method estimates seasonal (either growing season or irrigation season) actual evapotranspiration. This method is the standard method recommended by the USDA-SCS, is well known in the western U.S., and is used extensively throughout the world (M.E. Jensen et al., 1990). The original relationships were developed around 1945 and were intended for seasonal estimates. This method may be used to obtain monthly estimates if monthly crop coefficients are locally available.

We know that the amount of evapotranspiration is related to how much energy is available for vaporizing water. The energy is provided by solar radiation, but measuring solar radiation requires instrumentation not available at most field sites. Blaney and Criddle assumed that mean monthly air temperature and monthly percentage of annual daytime hours could be used instead of solar radiation to provide an estimate of the energy received by the crop. They defined a monthly consumptive use factor f as

$$f = \frac{tp}{100} \qquad (4.18)$$

where t is the mean monthly air temperature in °F (average of daily maximum and minimum), and p is the mean monthly percentage of annual daytime hours. The 100 in the divisor converts p from a percentage to a fraction. **Table 4.6** summarizes p for latitudes 0 to 64° N of the equator. Once f is computed for each month, then the actual ET for the season is computed by Equation 4.19:

$$U = K \sum_{i=1}^{n} f_i \qquad (4.19)$$

where K is the seasonal consumptive use coefficient for a crop with a normal growing season (**Table 4.7**), n is the number of months in the season, and U is the seasonal consumptive use in inches/season.

Example 4.9

Estimate the evapotranspiration from alfalfa grown in Hoytville, OH, for the 1992 growing period (June, July, August, and September) using the SCS Blaney–Criddle method.

Solution

Step 1. Obtain p (in percent) for Hoytville, OH, from **Table 4.6**. To use this table, we need to know the latitude for Hoytville (41°13′, obtained from NOAA Climatological Data, Ohio, 1992). From **Table 4.6**,

Latitude	June %	July %	August %	September %
42°	10.24	10.35	9.62	8.40
40°	10.09	10.22	9.55	8.50

We need to interpolate to obtain the required values of p. Interpolating for June,

$$\frac{42 - 40}{42 - 41\frac{13}{60}} = \frac{10.24 - 10.09}{10.24 - x} \quad \text{solving for } x = 10.18$$

TABLE 4.6

Monthly Percentage of Daytime Hours p of the Year for Latitudes 0° to 64° N

Latitude (° N)	Jan	Feb	Mar	Apr	May	Jun	Jul	Aug	Sep	Oct	Nov	Dec
64	3.81	5.27	8.00	9.92	12.50	13.63	13.26	11.08	8.56	6.63	4.32	3.02
62	4.31	5.49	8.07	9.80	12.11	12.92	12.73	10.87	8.55	6.80	4.70	3.65
60	4.70	5.67	8.11	9.69	11.78	12.41	12.31	10.68	8.54	6.95	5.02	4.14
58	5.02	5.84	8.14	9.59	11.50	12.00	11.96	10.52	8.53	7.06	5.30	4.54
56	5.31	5.98	8.17	9.48	11.26	11.68	11.67	10.36	8.52	7.18	5.52	4.87
54	5.56	6.10	8.19	9.40	11.04	11.39	11.42	10.22	8.50	7.28	5.74	5.16
52	5.79	6.22	8.21	9.32	10.85	11.14	11.19	10.10	8.48	7.36	5.92	5.42
50	5.99	6.32	8.24	9.24	10.68	10.92	10.99	9.99	8.46	7.44	6.08	5.65
48	6.17	6.41	8.26	9.17	10.52	10.72	10.81	9.89	8.45	7.51	6.24	5.85
46	6.33	6.50	8.28	9.11	10.38	10.53	10.65	9.79	8.43	7.58	6.37	6.05
44	6.48	6.57	8.29	9.05	10.25	10.39	10.49	9.71	8.41	7.64	6.50	6.22
42	6.61	6.65	8.30	8.99	10.13	10.24	10.35	9.62	8.40	7.70	6.62	6.39
40	6.75	6.72	8.32	8.93	10.01	10.09	10.22	9.55	8.39	7.75	6.73	6.54
38	6.87	6.79	8.33	8.89	9.90	9.96	10.11	9.47	8.37	7.80	6.83	6.68
36	6.98	6.85	8.35	8.85	9.80	9.82	9.99	9.41	8.36	7.85	6.93	6.81
34	7.10	6.91	8.35	8.80	9.71	9.71	9.88	9.34	8.35	7.90	7.02	6.93
32	7.20	6.97	8.36	8.75	9.62	9.60	9.77	9.28	8.34	7.95	7.11	7.05
30	7.31	7.02	8.37	8.71	9.54	9.49	9.67	9.21	8.33	7.99	7.20	7.16
28	7.40	7.07	8.37	8.67	9.46	9.39	9.58	9.17	8.32	8.02	7.28	7.27
26	7.49	7.12	8.38	8.64	9.37	9.29	9.49	9.11	8.32	8.06	7.36	7.37
24	7.58	7.16	8.39	8.60	9.30	9.19	9.40	9.06	8.31	8.10	7.44	7.47
22	7.67	7.21	8.40	8.56	9.22	9.11	9.32	9.01	8.30	8.13	7.51	7.56
20	7.75	7.26	8.41	8.53	9.15	9.02	9.24	8.95	8.29	8.17	7.58	7.65
18	7.83	7.31	8.41	8.50	9.08	8.93	9.16	8.90	8.29	8.20	7.65	7.74
16	7.91	7.35	8.42	8.47	9.01	8.85	9.08	8.85	8.28	8.23	7.72	7.83
14	7.98	7.39	8.43	8.43	8.94	8.77	9.00	8.80	8.27	8.27	7.76	7.93
12	8.06	7.43	8.44	8.40	8.87	8.69	8.92	8.76	8.26	8.31	7.85	8.01
10	8.14	7.47	8.45	8.37	8.81	8.61	8.85	8.71	8.25	8.34	7.91	8.09
8	8.21	7.51	8.45	8.34	8.74	8.53	8.78	8.66	8.25	8.37	7.98	8.18
6	8.28	7.55	8.46	8.31	8.68	8.45	8.71	8.62	8.24	8.40	8.04	8.26
4	8.36	7.59	8.47	8.28	8.62	8.37	8.64	8.57	8.23	8.43	8.10	8.34
2	8.43	7.63	8.49	8.25	8.55	8.29	8.57	8.53	8.22	8.46	8.16	8.42
0	8.50	7.67	8.49	8.22	8.49	8.22	8.50	8.49	8.21	8.49	8.22	8.50

Source: Jensen, M.E., R.D. Burman, and R.G. Allen, *Evapotranspiration and Irrigation Water Requirements*, American Society of Civil Engineers, New York, 1990. Copyright 1990 by the American Society of Civil Engineers. Reproduced with permission of the American Society of Civil Engineers.

The other three values were calculated by similar interpolations.

Latitude	June %	July %	August %	September %
42°	10.24	10.35	9.62	8.40
41°13′	10.18	10.30	9.59	8.396
40°	10.09	10.22	9.55	8.39

Step 2. Next, we need average monthly temperatures for the 4 months. These were also obtained from the NOAA summaries.

June	July	August	September
64.2°F	70.3°F	65.7°F	61.6°F

Step 3. Compute f and U from Equation 4.18 and Equation 4.19:

$$\sum_{i=1}^{4} \frac{tp}{100} = \frac{64.2(10.18)}{100} + \frac{70.3(10.30)}{100}$$
$$+ \frac{65.7(9.59)}{100} + \frac{61.6(8.396)}{100}$$
$$= 25.25$$

TABLE 4.7
Seasonal Consumptive Use Coefficients K for Irrigated Crops in the Western U.S.[a]

Crop	Length of Normal Growing Season[b]	Coefficient K[c]
Alfalfa	Between frosts	0.80 to 0.90
Bananas	Full year	0.80 to 1.00
Beans	3 months	0.60 to 0.70
Cocoa	Full year	0.70 to 0.80
Coffee	Full year	0.70 to 0.80
Corn (maize)	4 months	0.75 to 0.85
Cotton	7 months	0.60 to 0.70
Dates	Full year	0.65 to 0.85
Flax	7 to 8 months	0.70 to 0.80
Grains, small	3 months	0.75 to 0.85
Grains, sorghum	4 to 5 months	0.70 to 0.80
Oil seeds	3 to 5 months	0.65 to 0.75
Orchard crops		
Avocado	Full year	0.50 to 0.55
Grapefruit	Full year	0.55 to 0.65
Orange and lemon	Full year	0.45 to 0.55
Walnuts	Between frosts	0.60 to 0.70
Deciduous	Between frosts	0.60 to 0.70
Pasture crops		
Grass	Between frosts	0.75 to 0.85
Ladino white clover	Between frosts	0.80 to 0.85
Potatoes	3 to 5 months	0.65 to 0.75
Rice	3 to 5 months	1.00 to 1.10
Soybeans	140 days	0.65 to 0.70
Sugar beets	6 months	0.65 to 0.75
Sugarcane	Full year	0.80 to 0.90
Tobacco	4 months	0.70 to 0.80
Tomatoes	4 months	0.65 to 0.70
Vineyard	5 to 7 months	0.50 to 0.60

[a] From USDA (1970).

[b] Length of season depends largely on variety and time of year when the crop is grown. Annual crops grown during the winter period may take much longer than if grown in the summer.

[c] The lower values of K for use in the Blaney–Criddle formula are for the more humid areas; the higher values are for the more arid climates.

Source: Jensen, M.E. et al., *Evapotranspiration and Irrigation Water Requirements,* ASCE, New York, 1990. ©1990 by ASCE. Reproduced with permission.

Step 4. Find K from **Table 4.7** to adjust for crop. K = 0.80 (choose the lower value since Hoytville is a humid area; see table footnote).

Step 5. The estimated ET, $U = K_f$ (Equation 4.19) = 0.8 (25.25 in.) = 20.20 in./season.

Answer

The predicted ET equals 20.2 in./season.

Out of curiosity, let us compare the estimated value with the measured pan evaporation for the 4 months. The pan evaporation recorded at Hoytville was 24.22 in./season, but we need to adjust the pan evaporation by a coefficient to estimate the ET from alfalfa. From **Table 4.2**, to adjust E_{pan} to E_{tr} (alfalfa), multiply by the coefficient 0.92. The estimated ET from alfalfa using the pan method would be 22.28 in., or a 9.3% difference between estimation methods.

If we have monthly consumptive use coefficients available for the specific crop and location, such as those coefficients given in **Table 4.8**, then monthly consumptive use u can be computed as follows:

$$u = k \frac{tp}{100} \qquad (4.20)$$

TABLE 4.8
Blaney–Criddle Monthly Consumptive Use Factors *k*

Crop and Location	Jan	Feb	Mar	Apr	May	Jun	Jul	Aug	Sep	Oct	Nov	Dec
Alfalfa												
Mesa, AZ	0.35	0.55	0.75	0.90	1.05	1.15	1.15	1.10	1.00	0.85	0.65	0.45
Los Angeles, CA	0.35	0.45	0.60	0.70	0.85	0.95	1.00	1.00	0.95	0.85	0.55	0.30
Davis, CA				0.70	0.80	0.90	1.10	1.00	0.80	0.70		
Logan, UT				0.55	0.80	0.95	1.00	0.95	0.80	0.50		
Corn												
Mandon, ND					0.50	0.65	0.75	0.80	0.70			
Cotton												
Phoenix, AZ				0.20	0.40	0.60	0.90	1.00	0.95	0.75		
Bakersfield, CA					0.30	0.45	0.90	1.00	1.00	0.75		
Weslaco, TX			0.20	0.45	0.70	0.85	0.85	0.80	0.55			
Grapefruit												
Phoenix, AZ	0.40	0.50	0.60	0.65	0.70	0.75	0.75	0.75	0.75	0.70	0.60	0.50
Oranges												
Los Angeles, CA	0.30	0.35	0.40	0.45	0.50	0.55	0.55	0.55	0.50	0.50	0.45	0.30
Potatoes												
Davis, CA				0.45	0.80	0.95	0.90					
Logan, UT						0.40	0.65	0.85	0.80			
North Dakota					0.45	0.75	0.90	0.80	0.40			
Grain, small (wheat)												
Phoenix, AZ	0.20	0.40	0.80	1.10	0.60							
Grain, small (oats)												
Scottsbluff, NE					0.50	0.90	0.85					
Sorghum												
Phoenix, AZ						0.40	1.00	0.85	0.70			
Great Plains Field Station, TX						0.30	0.75	1.10	0.85	0.50		

where *k* is an empirical coefficient (**Table 4.8**), and *u* is the monthly consumptive use in inches/month.

4.6.4 JENSEN–HAISE ALFALFA-REFERENCE RADIATION METHOD

The Jensen–Haise method is termed a radiation method because solar radiation is needed in the equation to incorporate the recognized link between a source of energy and evapotranspiration. Jensen and Haise used over 3000 observations of actual evapotranspiration determined by soil sampling and statistically related R_s to E_{tr} as shown in Equation 4.21 (M.E. Jensen and Haise, 1963).

$$E_{tr} = \frac{C_T(T - T_x)R_s}{\lambda} \qquad (4.21)$$

where E_{tr} is the reference evapotranspiration in millimeters/day; C_T is the temperature coefficient (Equation 4.22); λ is the latent heat of vaporization (MJ/kg) (Equation 4.26); R_s is the solar radiation received at the earth's surface on a horizontal surface (MJ/m²/d); T is the mean

temperature for a 5-day period (°C); and T_x is the intercept of the temperature axis (Equation 4.25) (°C).

The temperature coefficient can be calculated as follows:

$$C_T = \frac{1}{C_1 + 7.3\,C_H} \qquad (4.22)$$

and C_1, which is needed to calculate C_T, can be calculated from

$$C_1 = 38 - \frac{2(H)}{305} \qquad (4.23)$$

where H is the elevation above sea level in meters. C_H, which is also needed for Equation 4.22, is calculated as follows:

$$C_H = \frac{5.0 \text{ kPa}}{(e_2 - e_1)} \qquad (4.24)$$

where e_2 and e_1 are the saturation vapor pressures in kilopascals at the mean maximum and mean minimum temperatures, respectively, for the warmest month of the year in an area.

$$T_x = -2.5 - 1.4 \, (e_2 - e_1) - \frac{H}{550} \qquad (4.25)$$

$$\lambda = 2.501 - 2.361 \times 10^{-3} \, T \qquad (4.26)$$

where λ is the latent heat of vaporization (MJ/kg), and T is temperature (°C) (Harrison, 1963).

EXAMPLE 4.10

Estimate the evapotranspiration from alfalfa grown in Hoytville, OH, for August 7 to 11, 1992, using the Jensen–Haise estimating method. Hoytville's elevation is 700 ft above sea level. The climatic data for the period of interest are shown below.

August	7	8	9	10	11	5-Day Average
Max T (°F)	83	85	86	90	93	87.4
Min T (°F)	52	57	61	62	65	59.4
RS (Ly/d)	485	544	609	543	631	562.4
Pan (in.)	0.23	0.16	0.29	0.13	0.34	

Solution

Step 1. Estimate λ from Equation 4.26 using an average temperature of 73.4°F for the 5-day period [(87.2 + 59.4)/2], which is 23.0°C. $\lambda = 2.447$ MJ/kg.

Step 2. Estimate T_x from Equation 4.25. To use this equation, we need the elevation in meters (700 ft = 213.36 m) and the saturation vapor pressure at the mean maximum and mean minimum temperatures for the warmest month of the year in an area. The NOAA summaries for the summer months in Hoytville, OH, show July is the warmest month of the year. The mean maximum temperature for July in Hoytville is 80.3°F, and the mean minimum temperature for July was 60.3°F. Computing e_s for both these temperatures (Equation 4.4) results in $e_2 = 3.532$ kPa, and $e_1 = 1.575$ kPa. Substituting into Equation 4.25, $T_x = -5.63$.

Step 3. Calculate C_H from Equation 4.24 using e_2 and e_1 calculated in Step 2. $C_H = 2.55$.

Step 4. Calculate C_1 knowing the elevation of Hoytville and using Equation 4.23. $C_1 = 36.6$.

Step 5. Calculate C_T from Equation 4.22 using C_1 from Step 4 and C_H from Step 3. $C_T = 0.0181$.

Step 6. Calculate E_{tr} from Equation 4.21. To use R_s, we must first convert Langleys to megajoules/square meter/day (MJ/m²/d) (1 Ly/d = 0.0419 MJ/m²/d). $E_{tr} = 5.0$ mm/d.

Answer

The estimated ET from alfalfa = 5 mm/d.

It is interesting to compare this estimate to the pan measurements for the same period. The pan evaporation was 1.15 in. for the 5 days or 5.84 mm/d. The Jensen–Haise method predicts evapotranspiration from alfalfa, so we need to multiply E_{pan} by a coefficient (**Table 4.2**) of 0.92 to compare the pan measurement with the Jensen–Haise estimate. The pan method estimated E_{tr} as 5.37 mm/d, or a 7.3% difference between the pan and Jensen–Haise estimation methods.

4.6.5 THORNTHWAITE METHOD

Thornthwaite (1948) developed an equation to predict monthly evapotranspiration from mean monthly temperature and latitude data (Equation 4.27). The small amount of data needed is attractive because often ET needs to be predicted for sites where few weather data are available. Based on what we know about ET, we should be skeptical about the general applicability of such a simple equation. Thornthwaite (1948) was not satisfied with the proposed approach: "The mathematical development is far from satisfactory. It is empirical. ... The chief obstacle at present to the development of a rational equation is the lack of understanding of why potential ET corresponding to a given temperature is not the same everywhere."

Taylor and Ashcroft (1972), as cited in Skaggs (1980), provided insight into the answer to Thornthwaite's question. They said:

This equation, being based entirely upon a temperature relationship, has the disadvantage of a rather flimsy physical basis and has only weak theoretical justification. Since temperature and vapor pressure gradients are modified by the movement of air and by the heating of the soil and surroundings, the formula is not generally valid, but must be tested empirically whenever the climate is appreciably different from areas in which it has been tested. ... In spite of these shortcomings, the method has been widely used. Because it is based entirely on temperature data that are available in a large number of localities, it can be applied in situations where the basic data of the Penman method are not available.

M.E. Jensen et al. (1990) warn that Thornthwaite's method is generally only applicable to areas that have climates similar to that of the east central U.S., and it is not applicable to arid and semiarid regions.

Thornthwaite (1948) found that evapotranspiration could be predicted from an equation of the form

$$E_{tp} = 16 \left[\frac{10T}{I} \right]^a \qquad (4.27)$$

where E_{tp} is the monthly ET (mm), T is the mean monthly temperature (°C), a is the location-dependent coefficient described by Equation 4.29, and I is the heat index described by Equation 4.28.

To determine a and monthly ET, a heat index I must first be computed:

$$I = \sum_{j=1}^{12} \left[\frac{T_j}{5} \right]^{1.514} \qquad (4.28)$$

where T_j is the mean monthly temperature during month j (°C) for the location of interest.

Then, the coefficient a can be computed as follows:

$$a = 6.75 \times 10^{-7} I^3 - 7.71 \times 10^{-5} I^2$$
$$+ 1.792 \times 10^{-2} I + 0.49239 \qquad (4.29)$$

EXAMPLE 4.11

Compute the monthly potential ET for August 1992 in Hoytville, OH, using Thornthwaite's method. The average temperature for the month of August 1992 in Hoytville was 18.72°C. Information from the NOAA summaries is given below.

Solution

Step 1. Determine I using Equation 4.28. To use this equation, we first need to know the mean monthly temperature for all the months of the year in Hoytville, OH. These were obtained from the NOAA summaries and are shown below.

1992	Jan	Feb	Mar	Apr	May	Jun
T (°C)	−2.7	−0.1	2.0	8.0	14.1	17.9

1992	Jul	Aug	Sep	Oct	Nov	Dec
T (°C)	21.3	18.7	16.4	9.4	4.7	−0.2

Calculate I using Equation 4.28; $I = 39.9$.

Step 2. Use Equation 4.31 to calculate a, knowing I from Step 1. This results in $a = 1.13$.

Step 3. Substituting into Equation 4.29, $E_{tp} = 9.17$ cm/month.

Answer

The potential ET for August 1992 at Holtville, OH, is 9.17 cm.

The pan evaporation measured in Hoytville in August 1992 was 5.83 in. or 14.8 cm/month. Adjusting by a pan coefficient to obtain E_{tr} (coefficients for E_{tp} were not available), the ET estimated from the pan was 13.6 cm/month, or 32.6% more than the Thornthwaite method predicted.

4.6.6 PENMAN'S METHOD

Penman (1948) first combined factors to account for a supply of energy and a mechanism to remove the water vapor near the evaporating surface. We should recognize these two factors as the essential ingredients for evaporation. Penman derived an equation for a well-watered grass reference crop:

$$E_{tp} = \frac{\dfrac{\Delta}{\Delta + \gamma}(R_n - G) + \dfrac{\gamma}{\Delta + \gamma} 6.43 \, (1.0 + 0.53 \, u_2)(e_s - e_d)}{\lambda} \qquad (4.30)$$

where E_{tp} is the potential evapotranspiration (mm/day); R_n is the net radiation (MJ/m/d); G is the heat flux density to the ground (MJ/m/d); λ is the latent heat of vaporization computed by Equation 4.26 (MJ/kg); u_2 is the wind speed (m/sec) measured 2 m above the ground; Δ is the slope of the saturation vapor pressure–temperature curve (kPa/°C); γ is the psychrometric constant (kPa/°C); and $e_s - e_d$ is the vapor pressure deficit determined by Method 3 (kPa).

The slope of the saturation vapor pressure–temperature curve Δ can be computed if the mean temperature is known:

$$\Delta = 0.200 \, [0.00738 \, T + 0.8072]^7 - 0.000116 \qquad (4.31)$$

where Δ is in kilopascals per degree centigrade, and T is the mean temperature in degrees centigrade.

To calculate the psychrometric constant, we must first calculate P, the atmospheric pressure that Doorenbos and Pruitt (1977) suggested could be calculated by Equation 4.32:

$$P = 101.3 - 0.01055 \, H \qquad (4.32)$$

where P is in kilopascals, and H is the elevation above sea level in meters. Using P, λ calculated from Equation 4.25, and c_p, the specific heat of water at constant pressure (0.001013 kJ/kg/°C), the psychrometric constant (in kPa/°C) can be calculated from Equation 4.33:

$$\gamma = \frac{c_p P}{0.622 \, \lambda} \qquad (4.33)$$

The remaining value to calculate is G, the heat flux density to the ground (MJ/m/d), and this can be determined from Equation 4.34 when the mean air temperature for the time period before and after the period of interest is known:

$$G = 4.2 \frac{(T_{i+1} - T_{i-1})}{\Delta t} \qquad (4.34)$$

where T is the mean air temperature (°C) for time period $i + 1$ and $i - 1$, and Δt is the time in days between the midpoints of time periods $i + 1$ and $i - 1$.

Penman (1963) developed Equation 4.30 using Method 1 for computing the vapor pressure deficit. M.E. Jensen et al. (1971) and Wright (1982) recommended using Method 3 because they found more accurate predictions when Method 3 was used with Penman's equation.

EXAMPLE 4.12

Estimate the E_{tr} for August 2, 1992 from a field close to Hoytville, OH, using Penman's method. Radiation data, wind speed and pan evaporation were measured at the site. The other weather data are taken from the Hoytville daily summaries published by NOAA: T_{avg} (August 2) = 18.6°C; T_{max} (August 2) = 24.4°C; T_{min} (August 2) = 12.8°C; T_{avg} (August 1) = 15.0°C; T_{avg} (August 3) = 20.6°C; e_d (8 a.m.) = 1.1759 kPa; u_2 = 1.83 m/sec; R_s = 19.33 MJ/m²/d; ELEV = 213.4 m; and lat = 41° N.

Solution

Step 1. Calculate G from Equation 4.34.

$$G = \frac{4.2(20.6 - 15.0)}{2} = 11.76 \text{ MJ/m}^2\text{/d}$$

Step 2. Calculate λ from Equation 4.26.

$$\lambda = (2.501 - 2.361 \times 10^{-3} \times (18.6)) = 2.4571 \text{ MJ/Kg}$$

Step 3. Calculate P from Equation 4.32.

$$P = 101.3 - 0.01055 (213.4) = 99.05 \text{ kPa}$$

Step 4. Calculate γ from Equation 4.33, remembering c_p = 0.001013 kJ/(kg°C). γ = 0.0657.

Step 5. Calculate Δ from Equation 4.31. Δ = 0.1340.

Step 6. Determine R_n from R_s as shown in the section on solar radiation. From these calculations, ϵ = 0.1795, R_{bo} = 6.363 MJ/(m²d), R_{so} = 28.95 from **Table 4.5**, R_b = 4.25 MJ/m²/d, and R_n = 10.63 MJ/m²/d.

Step 7. Determine $(e_s - e_d)$ by Method 3 in the vapor pressure deficit section. The calculated vapor pressure deficit is 1.0927 kPa.

Step 8. Substitute into Penman's Equation 4.30:

$$E_{tp} = \frac{\dfrac{0.134}{0.134 + 0.0657}(10.63 - 11.76) + \dfrac{0.0657}{0.134 + 0.0657}(6.43)(1 + 0.53\,(1.83))(1.0927)}{2.4571}$$

$$= 1.54 \text{ mm/d}$$

Answer

The potential ET for a site near Hoytville for August 2, 1992, was 1.54 mm/d. The measured pan evaporation was 1.12 mm/d, and multiplying by the E_{tr}/E_{pan} coefficient of 0.92 results in a predicted potential ET of 1.03 mm/d.

4.6.7 PENMAN–MONTEITH COMBINATION METHOD

Penman (1948) developed an equation of great significance (Equation 4.30) because it combined both aspects of evaporation, namely, the required energy source and a mechanism to move vapor away from the evaporating surface. Penman developed an empirical equation for wind that in practice accounts for the wind removing the water vapor and allowing ET to continue. Penman did not, however, have much theoretical basis for this equation. It did not include an aerodynamic resistance function (to quantify boundary layer resistance), and it did not include surface resistance to vapor transfer (to account for stomatal resistance). Several investigators have proposed equations to remedy these omissions, but the one most often cited is that of Monteith (1981). The modified equation is called the Penman–Monteith equation (M.E. Jensen et al., 1990). Refer to the work of M.E. Jensen et al. (1990, pp. 92–97) for suggestions on applying the Penman–Monteith equation in practice.

4.7 CONVERTING POTENTIAL OR REFERENCE CROP EVAPOTRANSPIRATION USING CROP COEFFICIENTS

The methods described above enable the estimation of potential or reference crop ET. The equations account for meteorological conditions, but do not account for crop or soil water status. A more useful quantity is actual ET, which estimates the amount of water the crop used during a given time period.

TABLE 4.9

Summary of the Type of Estimate Produced by Each of the Methods in the Chapter for Estimating ET

Method	Type of Estimate	Remarks
Penman	Potential evapotranspiration	Short green crop completely shading the ground and never short of water
Penman–Monteith	Reference crop evapotranspiration, either alfalfa or grass	Reference type is dependent on surface roughness and canopy or bulk stomatal resistances used
Jensen–Haise	Alfalfa reference	Reference crop is alfalfa (lucerne) well watered with 30 to 50 cm of growth; coefficients derived from U.S. data
NRCS Blaney–Criddle	Evapotranspiration from a specific crop	Evaluated for alfalfa and grass references using SCS crop coefficients and adjustment to negate the effects of cuttings
Thornthwaite	Potential evapotranspiration for a standard surface	Grass solid cover with no water deficiency for conditions similar to those in eastern U.S.

Source: Jensen, M.E., R.D. Burman, and R.G. Allen, *Evapotranspiration and Irrigation Water Requirements*, American Society of Civil Engineers, New York, 1990. Copyright 1990 by the American Society of Civil Engineers. Reproduced with permission of the American Society of Civil Engineers.

Crop coefficients can be used to estimate actual ET if care and attention are used to ensure that the same procedures and methods are used in applying the crop coefficients as were used in developing the methods. The coefficients were obtained by relating potential or reference crop ET to values of actual ET, typically measured with lysimeters. With careful adherence to recommended procedures, estimates of actual ET within ±10% can be obtained (M.E. Jensen et al., 1990). When procedures and coefficients are applied to a different climate (i.e., arid vs. humid), testing is advisable because the coefficients are likely to be invalid.

To estimate crop actual ET (E_t):

$$E_t = k_c E_{tr} \text{ or } E_t = k_c E_{tp} \qquad (4.35)$$

where E_{tr} is reference crop ET, E_{tp} is potential ET, E_t is actual evapotranspiration, and k_c is the experimentally derived crop coefficient. Typical reference crops used to develop the coefficients are alfalfa or grass. **Table 4.9** shows which reference crop was used for each method discussed in this chapter.

Table 4.10 presents crop coefficients k_c for normal irrigation and precipitation conditions for use with the alfalfa reference E_{tr} (M.E. Jensen et al., 1990). A word of caution is in order because these values were based on data from Kimberly, ID, which is an arid environment. The coefficients should be usable in other areas if the procedure has been verified locally. To use **Table 4.10**, select the appropriate crop from the first column, then select the applicable percentage canopy cover. If the crop

is less than completely closed over, use the first set of rows in the table, but if 100% cover has been established, the second group of rows is appropriate.

EXAMPLE 4.13

The alfalfa reference E_{tr} was calculated for June as 5 mm/day. The grower is interested in corn, however, not alfalfa. Predict the ET if 30% of the time from planting to effective canopy cover has elapsed at the time of ET estimation.

Solution

From **Table 4.10**, select corn from the first column and look in the 30% PTC column. The crop coefficient is 0.20. Multiplying E_{tr} by k_c, 5 mm/d (0.2) = 1.0 mm/d.

Answer

The estimated actual ET expected from corn in June with a 30% canopy cover is 1.0 mm/d.

If the method used to predict ET did not compute alfalfa reference E_{tr} by the combination method, **Table 4.2** can be used to convert the method's prediction to alfalfa E_{tr}. To use **Table 4.2**, select the appropriate month from the top of the table, and the appropriate method from the left-hand column, then determine the adjustment coefficient. Multiply the estimated ET by this coefficient to obtain E_{tr}. Now **Table 4.10** can be used to find kc, and

TABLE 4.10
Mean E_t Crop Coefficients k_c, for Normal Irrigation and Precipitation Conditions, for Use with Alfalfa Reference E_{tr}

Crop	0	10	20	30	40	50	60	70	80	90	100
			PTC, Time from Planting to Effective Cover (%)								
Spring grain[a]	0.20	0.20	0.21	0.26	0.39	0.55	0.66	0.78	0.92	1.00	1.00
Peas	0.20	0.20	0.21	0.26	0.36	0.43	0.51	0.62	0.73	0.85	0.93
Sugar beets	0.20	0.20	0.21	0.22	0.24	0.27	0.33	0.45	0.60	0.80	1.00
Potatoes	0.20	0.20	0.20	0.22	0.31	0.41	0.51	0.62	0.70	0.76	0.78
Corn	0.20	0.20	0.20	0.20	0.23	0.32	0.42	0.55	0.70	0.85	0.95
Beans	0.20	0.20	0.20	0.26	0.35	0.45	0.55	0.66	0.80	0.90	0.95
Winter wheat	0.30	0.30	0.30	0.50	0.75	0.90	0.98	1.00	1.00	1.00	1.00
			DT, Time Since Effective Cover Was Established (in elapsed days)								
Spring grain[a]	1.00	1.00	1.00	1.00	0.90	0.50	0.30	0.15	0.10	—	—
Peas	0.93	0.93	0.70	0.53	0.35	0.20	0.12	0.10	—	—	—
Sugar beets	1.00	1.00	1.00	1.00	0.98	0.94	0.89	0.85	0.80	0.75	0.71
Potatoes	0.78	0.78	0.76	0.74	0.71	0.67	0.63	0.59	0.36	0.25	0.20
Corn	0.95	0.96	0.95	0.94	0.90	0.85	0.79	0.74	0.35	0.25	—
Sweet corn	0.95	0.94	0.93	0.90	0.85	0.75	0.58	0.40	0.20	0.10	—
Beans	0.95	0.95	0.90	0.67	0.33	0.15	0.10	0.05	—	—	—
Winter wheat	1.00	1.00	1.00	1.00	0.95	0.55	0.25	0.15	0.10	—	—

[a] Spring grain includes wheat and barley.

Source: Jensen, M.E., R.D. Burman, and R.G. Allen, *Evapotranspiration and Irrigation Water Requirements*, American Society of Civil Engineers, New York, 1990. Copyright 1990 by the American Society of Civil Engineers. Reproduced with permission of the American Society of Civil Engineers.

Equation 4.37 can be used to calculate actual evapotranspiration. If the method used to estimate E_{tr} or E_{tp} was not developed for alfalfa, we can use the coefficients in **Table 4.2** to convert to an alfalfa reference crop. Multiply the estimated ET by this coefficient to obtain E_{tr}. Now **Table 4.10** can be used to find k_c, and Equation 4.37 can be used to calculate ET.

EXAMPLE 4.14

The Penman method was used to estimate E_{tp} for July as 12 cm/month. Estimate the actual ET that could be expected from corn if it has been 20 days since effective cover was established.

Solution

From **Table 4.2**, the coefficient $[E_{tr}/E_{tp}(P)]$ needed to adjust E_{tp} for the month of July is 1.20. Multiplying 12 cm/month by $1.2 = 14.4$ cm/month $= E_{tr}$. From **Table 4.10**, $k_c = 0.95$. Multiplying E_{tr} by k_c (Equation 4.35) gives 13.7 cm/month for actual ET.

Answer

The actual ET expected from field corn in July is 13.7 cm/month.

4.7.1 Evapotranspiration and Soil Water Budgets in Space and Time

At this point, the factors governing evaporation and transpiration and methods for predicting ET and measuring ET, primarily for short periods and small areas, have been discussed. As the factors governing ET vary greatly in space and time across the U.S., it could be expected that ET would vary greatly across the U.S. in space and time. These broader scales have been addressed by hydrologists, and this section discusses some of this work, so is useful for planning and management as well as regional studies.

Class A Pan. We looked at the Class A evaporation pan as a way of measuring evaporation on a continuing basis for the country. Its utility lies in the reasonably dense network and the monthly reporting service. The average annual values have a huge spatial variance, ranging from as little as 25 in. per year in Maine and Washington to over 140 in. per year in the Sonoran Desert of the Southwest (**Figure 4.14**). Of course, we know that these values greatly overestimate lake evaporation, which is the best surrogate for PET, but this exaggeration also has great spatial variance, as shown by mitigating pan coefficients (**Figure 4.15**). The pan coefficients are multiplied by the

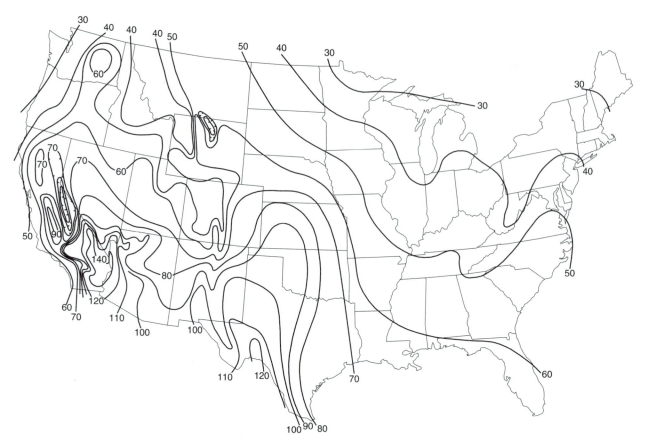

FIGURE 4.14 Mean annual Class A pan evaporation (in inches) 1946–1955. (From Kohler, M.A., T. Nordenson, and D. Baker, Evaporation Maps for the United States, U.S. Weather Bureau Technical Paper 37, 1959.)

pan evaporation to obtain the lake evaporation. Note that they vary from 81% in Maine to less than 60% in the Sonoran Desert, with a mean of about 70%. The coefficients are controlled by climate in relation to water temperature: If solar radiation, wind, dewpoint, and air temperature are such that water in an exposed pan is warmer than the air, the coefficient is greater than 0.7 and vice versa (Kohler et al., 1959).

Lake Evaporation.   Now, we arrive at the most useful map, that of annual average lake evaporation (**Figure 4.16**). This map, based on both measurements and estimates, gives us the best available idea of average annual PET for the U.S. Note that the values range from less than 20 in. in Maine and Washington to over 7 ft in the Sonoran Desert. Indeed, the value is greater than 5 ft for most of the Southwest. This map is a key instrument in water resource planning because it shows the PET that might be expected for any hydrologic venture, such as reservoirs or irrigation, for which water is exposed. Its utility is increased by comparison with **Figure 2.6**, average annual precipitation. For example, these two figures were compared for Lake Havasu in this chapter. We saw there that the average annual PET was over 7 ft, but the AET was limited to a maximum of about 8 in. in the absence of

stored water. However, water is stored there, covering large areas of former desert, and in many other locations in the Southwest where the ET losses become staggering. Using instruments like this to plan for increased water storage in reservoirs along the Colorado River, Langbein (1959) stated that the "gain in regulation [of streamflow] to be achieved by increasing the present 29 million acre-feet to nearly 50 million acre-feet of capacity appears to be largely offset by a corresponding increase in evaporation."

Much has been written about these water losses from reservoirs in the Southwest. Not only is there loss of water, but also the concentration of salt and other solutes is increased. The flow of the Colorado has been so reduced by evaporation and extraction that only a saline trickle now flows across the border into Mexico. While this is so, some people then make the jump in logic that all reservoirs, by evaporation, reduce the quantity and quality of downstream flow. Let us go now to Lake Hartwell on the Savannah River between Georgia and South Carolina. There, PET is about 40 in. (**Figure 4.16**). So, does evaporation from Lake Hartwell reduce streamflow? Even without making measurements, we can say that the reduction is probably not much for most years. Why? Well, the region gets annual average precipitation of about 56 in.

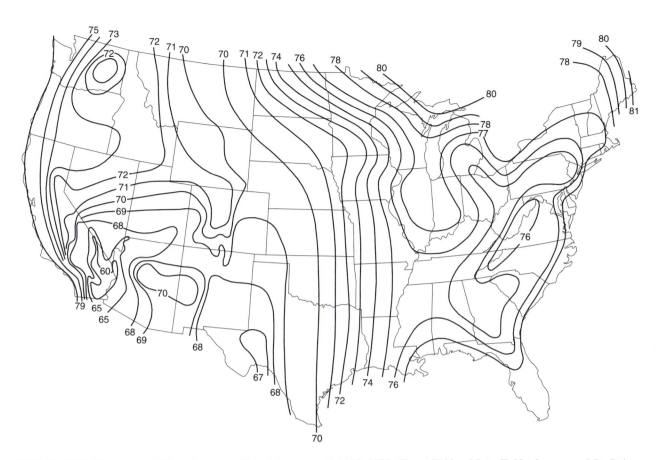

FIGURE 4.15 Mean annual Class A pan coefficient (in percent) 1946–1955. (From Kohler, M.A., T. Nordenson, and D. Baker, Evaporation Maps for the United States, U.S. Weather Bureau Technical Paper 37, 1959.)

(**Figure 2.6**). Comparing the free water reservoir surface with the natural landscape of deep soils with deep-rooted forests from which AET would be considerable (assuming soils remain at optimal PAW), it would appear that such a reservoir would have a moderate or minimal effect on streamflow most years. The exception might be years with low rainfall with higher than normal PET. At such time, the reservoir, still evaporating at potential rates while the PET from forest is suppressed by low PAW, might well have a significant effect.

Time Patterns and Variance. In the preceding discussion, variance over time was mentioned. If the mean value is our predictor for the future, what might future variance about that mean be? That is, what is the standard error of the estimate? We do not have this for lake evaporation, but we do have it for the Class A pan values (**Figure 4.17**). Thus, we can use **Figure 4.14** and **Figure 4.18** to calculate a percentage standard error for pan evaporation. Thus, for Lake Hartwell, the value would (3 in.)/(75 in.) or about 4%. With the reasonable assumption that this proportion applied to lake evaporation, the variance would be low. In central Kansas, however, the standard error is about 20% (**Figure 4.13** and **Figure 4.16**), indicating that PET is liable to much greater variance.

A time variance shown spatially is the growing season (May to October) proportion of Class A pan ET (**Figure 4.18**). This figure gives quantification to the idea that, in areas with cold winters, most of the annual PET occurs during the short period of the growing season, the period when water is critical for crops. Values range from more than 84% in the north central U.S. to less than 60% in Florida.

Soil Water Budgets. We discussed how critical soil moisture is to mesophytic vegetation. The soil water budget, using the annual march of precipitation as "income" and the annual march of PET as potential "expenditures," shows the availability of soil water at all times of the year (**Figure 4.19**). Note that, for most areas, soil water recharge occurs during the water surplus of winter, when precipitation is greater than PET. By late spring, PET increases greatly and usually exceeds precipitation, giving a deficit period. This is not a problem as long as the soil can store enough water in a "savings account" to carry plants through the deficit period, shown in the budgets as "soil water utilization." But, if the deficit is too great for the savings account, soil water is reduced to the wilting point, and vegetational stress begins. For most areas with crops and other mesophytic vegetation, this means irrigation

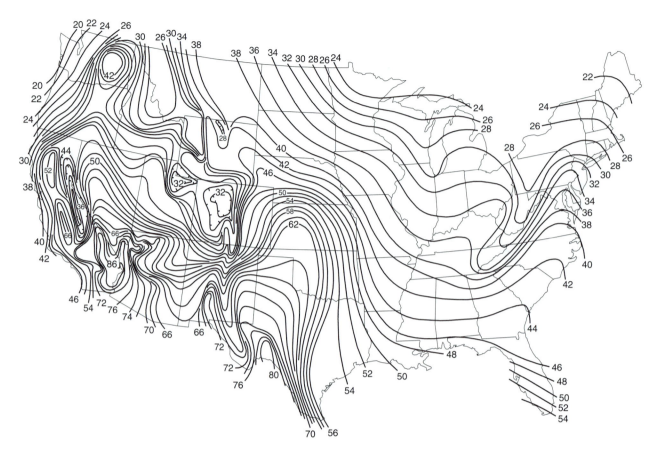

FIGURE 4.16 Mean annual lake evaporation, 1946–1955. (From Kohler, M.A., T. Nordenson, and D. Baker, Evaporation Maps for the United States, U.S. Weather Bureau Technical Paper 37, 1959.)

before the WP is reached. Because the graph ordinates are rates (amount of water per time) and the abscissa is time (days of the year), areas under the graphs indicate volume. Thus, a rough idea of required supplemental water (irrigation) is indicated on each water budget graph by the area of the stress deficit shown. As expected by discussion in this section, the Southwest has the greatest deficits and, noteworthy here, the largest stress deficits. Since much of U.S. agriculture takes place in this region, huge water requirements are manifest.

Conservationists often point out the water costs of extensive irrigation in the Southwest; it is true that the costs to the environment are high. Not only is water diverted from rivers, lakes, estuaries, and groundwater supplies with many associated environmental costs, but also the irrigation process itself is sometimes environmentally problematic, with great ET loss of water and quality problems, like salinization of runoff (return flow) from fields and often groundwater. Improper management can also flush nutrients such as nitrates and contaminants such as selenium into groundwater, streams, and estuaries. On the other hand, the extremely high yields from these irrigated areas release marginal farmland from use in the humid East, where potential problems might be soil erosion from steep slopes under intense rainfall (Chapter 2 and Chapter 9).

Such erosion would bring not only sediment, but also nutrients and agricultural chemicals into eastern streams. Most of these potential farmlands in the humid East are presently in forest. But, as discussed in Chapter 10, even forests can be hydrologically problematic from a water supply standpoint.

4.7.2 EVAPOTRANSPIRATION MANAGEMENT

Perhaps the greatest opportunity for saving water from ET losses in the U.S. is to reduce waste in irrigation. Of all water available annually in the U.S., about 3.4% (1962) was withdrawn for irrigation; of this, about 60% was consumed by ET. Many technical fixes for this are known, such as genetically improved crops, point (drip) irrigation, covering soil with plastic sheeting, applying water at such optimal times as cool periods or night, sealing feeder ditches, and using careful soil water budgets to ensure that, while PAW remains optimal, gravity or percolation water is minimal. But, the main problems are political and economic: Farmers who use irrigation have often been given the rights to government-supplied and subsidized water with the proviso that unused allocation may be taken away from them. The cost of this water to farmers has often been as low as $15 per acre-foot, while open market

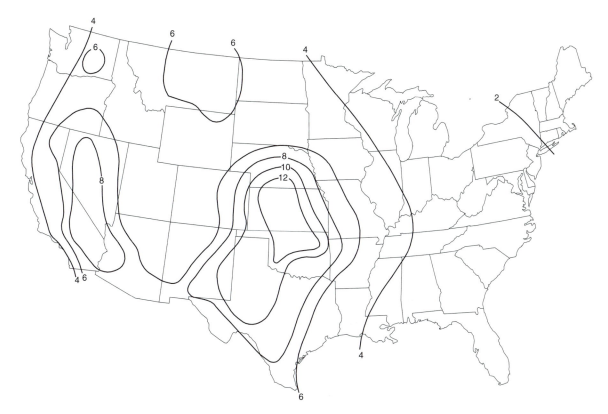

FIGURE 4.17 Standard deviation of annual Class A pan evaporation (in inches). (From Kohler, M.A., T. Nordenson, and D. Baker, Evaporation Maps for the United States, U.S. Weather Bureau Technical Paper 37, 1959.)

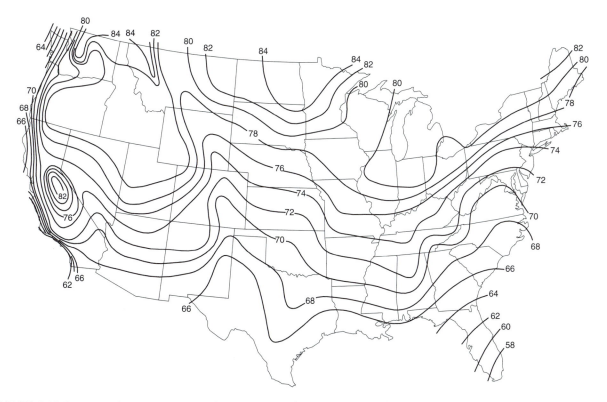

FIGURE 4.18 Mean growing season (May to October) evaporation in percentage of annual. (From Kohler, M.A., T. Nordenson, and D. Baker, Evaporation Maps for the United States, U.S. Weather Bureau Technical Paper 37, 1959.)

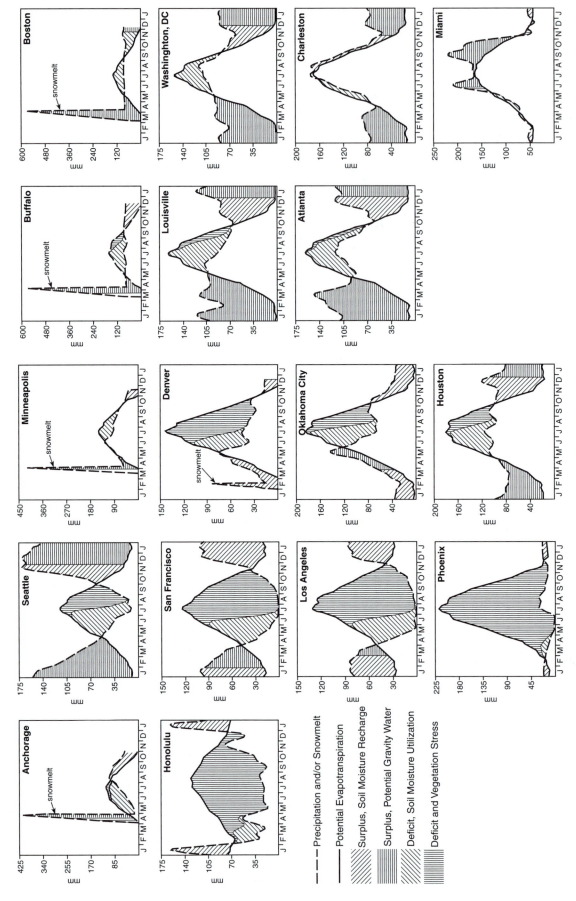

FIGURE 4.19 Soil water budgets from across the U.S. The sharp spikes in spring at northerly locations are snowmelt. Note the large soil water deficits causing vegetation stress in the Southwest. (Data from Mather, J.R., *The Climatic Water Budget in Environmental Analysis*, Lexington Books, Lexington, MA, 1978. With permission.)

value was $300 to $400 per acre-foot. Thus, the incentives often have been to waste water rather than save it (Gottlieb, 1988; Harvey, 1994; Hundley, 1992; National Research Council, 1992). One partial solution has been to allow farmers to sell their unused allocations on the open market, making it available for other uses and ensuring that it would not be applied on fields simply to keep the allocation.

Control of ET on nonirrigated land is rarely used. While wood mulch is often used on gardens, its use on large fields is not practical. Besides its expense and difficulty to spread, fresh wood mulch can tie up much of the nitrogen in soil.

ET can be reduced from some nonirrigated areas by vegetative control. One example considered in Chapter 9 is replacing forests with lower-growing vegetation to reduce ET and increase streamflow. This can work well, but there can be serious resource loss, such as soil erosion and the loss of aesthetic values.

Evaporation from open tanks of water can sometimes be controlled by a 1-molecule thick layer of a fatty alcohol like hexadecanol, but this does not work for larger areas like ponds and lakes because wind will push the alcohol to one side, and the protection is lost. Very little can be done about the ET losses from reservoirs.

PROBLEMS

4.1. If the measured pan evaporation was 3.2 inches for May, what would the alfalfa reference crop equivalent evaporation be if computed by the combination equation?

4.2. If the reference ET for grass was computed as 4.5 in. for July, what would the reference ET be for alfalfa computed by the combination equation?

4.3. (a) Find the saturation vapor pressure of air at 90°F. (b) Find the actual vapor pressure of air at 90°F and 80% relative humidity. (c) Find the vapor pressure deficit. (d) The book lists three methods for finding vapor pressure deficit. Which method did you use in Part c?

4.4. If the average temperature on June 3, 1994, was 25°C, assuming a humid area (latitude 40° N) with $R_s = 23$ MJ/m²/d, calculate the net radiation (assume albedo of 0.23).

4.5. Find the wind speed at 25 ft if the wind speed at 6.6 ft is 2.3 mi/h. The vegetation is soybeans, which are 2.5 ft tall.

4.6. Compute the evaporation for the month of August from a pond that had an average water temperature of 65°F measured 1 ft below the water surface and an average wind speed of 2 mi/h measured 25 ft above the pond. The average air temperature was 78°F, and the relative humidity was 52%.

4.7. The table below gives the average maximum and minimum temperatures for a location close to Columbus, OH. Using the SCS Blaney–Criddle method of Equation 4.20, calculate the monthly ET from corn grown there (latitude 40° N) for the 1992 growing season.

Month	June	July	August	September
Maximum T (°F)	77.6	82.7	79.4	75.3
Minimum T (°F)	56.9	64.2	59.3	54.1

4.8. Using the information presented in the table for Problem 4.7 and the SCS Blaney-Criddle method of Equation 4.20, calculate the seasonal ET (June to September) for corn grown near Columbus, OH.

4.9. Using the Jensen–Haise method, calculate the reference ET in millimeters/day for the data given below if the data were collected at an elevation 219.5 m above sea level. July was the warmest month of the year, with a mean maximum T of 27.8°C and a mean minimum T of 15.5°C.

June	1	2	3	4	5	Average
Maximum T (°C)	27	30	32	31	32	30.4
Minimum T (°C)	13	15	20	18	16	16.4
R_s (MJ/m²/d)	24	20	27	25	26	24.4

4.10. Using Thornthwaite's method, determine the ET for the month of July if the average monthly temperature was 21°C and if the heat index for the location of interest was previously determined as 40.

4.11. Using Penman's method, estimate the E_{tp} for July 15, 1994, from a field at an elevation of 200 m. The average temperature for July 15 was 16°C, with a maximum T of 22°C and a minimum of 10°C. The previous day's average temperature was 14°C, and the following day's average temperature was 21°C. The early morning actual vapor pressure on July 15 was 1.3 kPa; the wind speed was 2 m/sec, and R_n was 15 MJ/m²/d.

4.12. The alfalfa reference E_{tr} was calculated (by the combination equation) as 3.2 mm/day. Predict the E_t from beans that have had full canopy cover for 30 days.

4.13. The Jensen–Haise method was used to estimate E_{tr} for June as 10 cm/month. Estimate the actual ET that could be expected from soybeans when 50% of the time from planting to full cover has elapsed.

5 Runoff and Subsurface Drainage

5.1 INTRODUCTION

Rivers, lakes, reservoirs, and dams provide more than 75% of the water used in the U.S. Surface water systems are also used for recreational purposes and in many cases are important transportation conduits. The importance of surface water in the development of the U.S. is illustrated by looking at a map of the nation. Virtually all cities with populations exceeding 150,000 are located on rivers, and many smaller communities are located on rivers or lakes. There are about 2 million streams and rivers in the U.S., including the mighty Mississippi River, which is the fourth longest river in the world at 3710 mi long. Of the world's 11 freshwater lakes with the largest surface area, 4 are located in the Great Lakes system (Superior is 1st, Huron is 5th, Michigan is 6th, and Erie is 11th). In addition to the Great Lakes, we have many thousands of smaller lakes, dams, reservoirs, and ponds. These include 25 lakes with surface areas greater than 100 mi^2 and more than 200 with surface areas larger than 10 mi^2. All of these water bodies depend on the phenomena of runoff.

Runoff is all water transported out of the watershed by streams. Some of this water may have had its origins as overland flow, while much may have originally infiltrated and traveled through the soil mantle to streams as interflow. In addition, water that infiltrates and goes down into groundwater may later emerge far downstream through seeps and springs to add to streamflow. At that point, it also becomes runoff.

While runoff is usually benevolent, under extreme conditions of rainfall or land use it can cause extensive damage by eroding soils and stream banks; carrying off valuable agricultural nutrients and pollutants; destroying bridges, utilities, and urban developments; and causing flooding and sediment deposits in recreational, industrial, and residential areas along stream systems. The hydrologic effects of deforestation, agriculture, and surface mining have been manifested since European settlement. Since World War II, we have seen a dynamic reshaping of our landscape by rapid urbanization and more intensive agriculture. These activities have changed hydrologic responses, soil erosion, and sedimentation in watersheds. For example, urbanization of farmland and forests causes more rapid runoff, higher peak discharges, and larger runoff volumes.

Chapter 3 made the point that infiltration was the key to maintaining good hydrologic conditions (**Figure 5.1, Ia**). In **Figure 5.1, Ib**, we see a landscape in which most of the rainfall infiltrates, recharges soil water and brings it up to field capacity, and moves down through the soil into groundwater, where it is available for use or eventual flow into streams. Some of it may move down through the soil as interflow, also furnishing water to the stream, but both routes shown are subsurface and generally involve a longer transit time to the stream. Stored ground and even soil water is available to maintain streamflow during the sometimes-lengthy period between storms (*baseflow*). Since the soil and groundwater reservoirs are depleted between storms, baseflow will decrease between storms, a declining rate called the *baseflow depletion curve*. A significant decrease in infiltration (**Figure 5.1, Ib**) changes matters radically. Rather than infiltrating, water is forced to move across the surface as overland flow; moving rapidly, it can erode hillsides (Chapter 9) and quickly flow into streams, creating flooding. Conversely, streams in this situation carry less baseflow because of reduced interflow and groundwater contributions.

Some of the concepts discussed in Chapter 3 are graphed in **Figure 5.1, IIa**, which shows infiltration capacity as a function of land use and time. Overland flow (**Figure 5.1, IIb**) is the complement or mirror image of infiltration rates. That is, any part of rainfall not infiltrating or stored in surface depressions becomes overland flow, bringing with it the problems mentioned here. The compartmentalization of runoff during a storm event and under different watershed conditions is shown in **Figure 5.2**. All three scenarios assume an even rate of rainfall for a few hours and demonstrate how that water is allocated. In the scenario in **Figure 5.2A**, a small constant proportion falls directly into the stream as channel precipitation, and most of the initial rainfall is intercepted. As the intercepting tree and grass foliage become wet, more water gets to the soil and recharges soil water, bringing it up to field capacity. At about this time, gravity or percolation water begins to move downward, where it can recharge groundwater or create interflow, depending on the soil and regolith beneath it. As the soil becomes increasingly wet, infiltration capacities are further reduced so that increasing proportions of the rainfall move off across the surface as overland flow. In the scenario of **Figure 5.2B**, interception, soil water, and channel precipitation are as in the first scenario, but the deep, very porous soil allows the gravity water to move right down into groundwater; from there, it can move into the stream as runoff. In the scenario of **Figure 5.2C**, interception and channel precipitation are the same, but the soil is frozen solid (this is called *concrete frost*), so most water hitting the ground cannot infiltrate and must leave as overland flow.

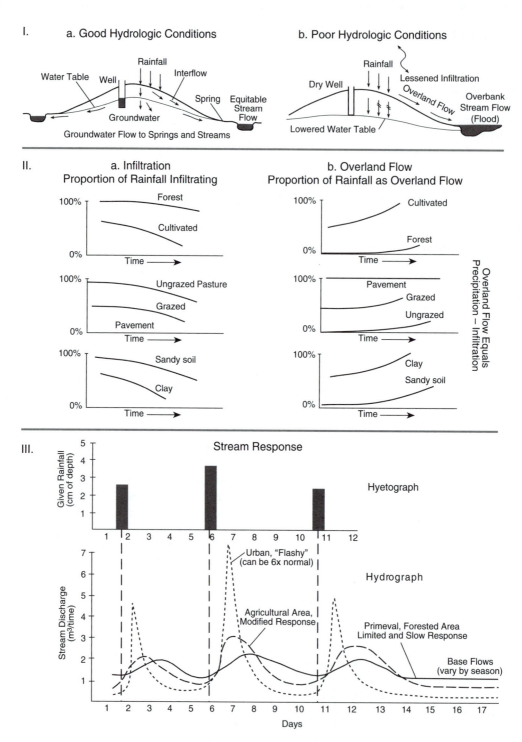

FIGURE 5.1 Good vs. poor hydrologic conditions and their effects on runoff.

The conditions described above are reflected in the stream response, that is, the amount of stormflow (stream-flow during a storm) per unit of rainfall (**Figure 5.1, III**). The scenario shown is how one stream might react to the same-size storms under conditions of different land use and thus different infiltration capacities. The top part of **Figure 5.1, III**, shows the input (precipitation) in a form called a *hyetograph*. It shows a storm of 3 cm on Day 2,

4 cm on Day 6, and another 3 cm on Day 11. The output (stormflow) is the response to the rainfall and is shown as discharge of water in cubic meters per second. The idea here is to show how the same stream might respond to identical rainfall events under three different types of land use. Indeed, one might think of it as the same stream as the landscape changes over time. In the primeval state with excellent hydrologic conditions, the stream responds

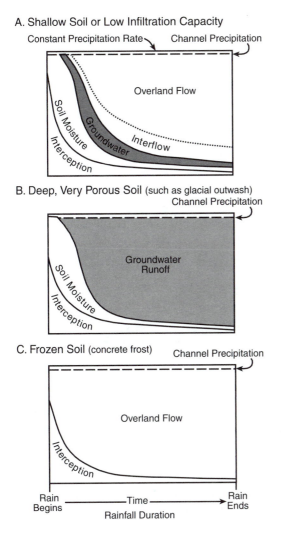

A. Shallow Soil or Low Infiltration Capacity

Constant Precipitation Rate   Channel Precipitation

Overland Flow

Soil Moisture

Groundwater

Interception

Interflow

B. Deep, Very Porous Soil (such as glacial outwash)

Channel Precipitation

Groundwater Runoff

Soil Moisture

Interception

C. Frozen Soil (concrete frost)

Channel Precipitation

Overland Flow

Interception

Rain Begins — Time — Rain Ends

Rainfall Duration

FIGURE 5.2 Schematic of the disposal of storm rainfall in three scenarios. (Redrawn from Ward, R.C., *Principles of Hydrology*, McGraw-Hill, Maidenhead, U.K., 1967. With permission.)

slowly and moderately to rainfall events. Between events, baseflow remains relatively high, although it will decline as the soil and groundwater are depleted. As the watershed is cleared and put into agriculture, however, soil infiltration gradually decreases, stream response increases, while baseflow decreases. Note that a stream with a quick and large response is said to be "flashy." Eventually, the watershed is urbanized, and much of the area is "waterproofed" with roofs, streets, parking lots, and sidewalks. The result, as can be inferred by the graphs, is that most slope runoff is overland flow, and stream response is much faster and greater, truly a condition of "flashiness." The first effect is urban flooding, but the eventual effect is stream channel erosion, enlargement, and destabilization. At the same time, baseflow is radically decreased, sometimes converting permanent or perennial streams to intermittent ones. The combined effects of flooding, channel erosion, and decreased baseflow create ugly streams with little biotic

life. An instructive history, similar to that above, is that of Cherry Creek, a stream that flows through Denver, CO (Costa, 1978).

Growing concern for the adverse effects of land use changes has resulted in the establishment of many federal, state, and local hydraulic and hydrologic design regulations. Extensive research has resulted in the development of numerous methods and computer models to estimate the hydrologic response of watersheds to single storm events or on a more continuous basis, such as a daily time step. Yet, little information is available on how to select an appropriate method and the likely accuracy associated with using the method.

The main sources of water entering surface water systems will vary from one location to another, but might include precipitation falling directly on the water body; surface runoff from storm events; snowmelt; flows from groundwater aquifers; interflow; subsurface drainage; and return flows from water used for domestic, commercial, industrial, mining, irrigation, and thermoelectric power generation purposes. The need to have detailed knowledge of each of these potential contributions will depend on the geographic location, the size of the area that contributes to flows at the location of interest, and the type of study conducted. For example, overland flow will dominate if flooding associated with severe storm events on small watersheds is evaluated. In contrast, overland flow because of storm events might be of less importance for a study of baseline flow during severe drought.

This chapter focuses on flows associated with single storm events and presents details on the factors that influence runoff; descriptions and examples of the use of several common surface runoff techniques; a perspective on the usefulness of these techniques; and information on subsurface drainage that discharges back into surface water systems shortly after a rainfall event.

5.2 FACTORS AFFECTING RUNOFF PROCESSES

There are two basic models of runoff from slopes. The first, the Horton Overland Flow Model, was developed in the 1930s by an engineering hydrologist, Robert Horton, and assumes that infiltration rates are less than rainfall rates. As we shall see, it applies to areas devoid of vegetation, such as drylands and areas impacted by human activity. The second model, the variable source–area or interflow model, assumes that infiltration rates are usually greater than most rainfall rates so that rainfall infiltrates, recharges soil water, and moves through the soil mantle toward the stream as interflow. This model was developed by many researchers in the 1960s and 1970s, but the first person to describe the concept was John Hewlett, a forest hydrologist, so the concept is sometimes known as the

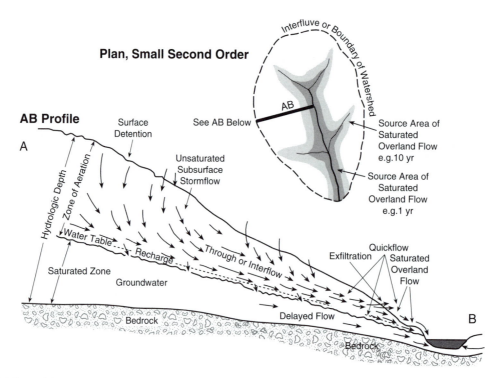

FIGURE 5.3 The variable source–area concept (Hewlett model).

Hewlett Runoff Model (Ward and Robinson, 2000). It applies to forests in excellent condition and on vegetated, deep, well-developed soils where human impacts have been minimal. As discussed here, these models are at two ends of a continuum, and most land surfaces will give runoff processes that fall somewhere between them.

5.2.1 THE HORTON OVERLAND FLOW MODEL

The potential for overland flow exists whenever the rate of water application to the ground surface exceeds the rate of infiltration into the soil. Initially, when water is applied to a dry soil (except with dry soil hydrophobicity), the application rate and infiltration rate might be similar. However, as the soil becomes wetter, the rate of infiltration will decrease, and surface depressions will begin to fill (*surface detention*). If all the surface depressions have been filled and the surface water application rate exceeds the infiltration rate, then overland flow will be initiated. The soil profile near the soil surface need not be saturated for overland flow to occur, and such unsaturated overland flow is often termed *Hortonian overland flow*. By this model, overland flow produces most stormflow in streams.

5.2.2 THE VARIABLE SOURCE AREA CONCEPT OR HEWLETT MODEL

Anyone who has walked in an old forest during a rainstorm will have noticed that no overland flow may occur, but the local stream still carries stormflow. Working in the

humid Southeast with its extremely intense rainstorms (Chapter 2), Hewlett and his U.S. Forest Service colleagues (Hewlett, 1961; Hewlett and Hibbert, 1967) explained this puzzle by showing that water from a rainstorm could often move through soil to furnish stormflow in a stream, even when the soil was not saturated. Variously termed *interflow* or *throughflow* (**Figure 5.3**), this movement of water to streams has been identified by many researchers, but the complete physical processes involved have not yet been fully explained (Ward and Robinson, 2000). Interflow is considered to occur if the lateral flow returns to the land or a stream without first reaching groundwater (Kirkby, 1978). In steep forested areas with shallow soils, interflow might be the primary runoff process.

Many researchers think that the explanation for interflow may lie in infiltrating water reaching discontinuous, less-permeable strata so that some of the water soil is forced to move laterally downslope (much like in a thatch roof), but much more complex explanations have been offered (Ward and Robinson, 2000). Some water will move down into groundwater as explained in the Introduction. As interflow accumulates downslope along streams or ephemeral draws or where the soil may be shallow, the soil will become saturated and may exfiltrate and move across the surface, giving saturated overland flow. This flow, of course, would be augmented by the water from rainfall. Such areas of potential saturation, the *source area* for saturated overland flow, can be mapped over time (e.g., Dunne and Black, 1970; Kirkby and Chorley, 1967; Kirkby, 1978). Their proportional size relative to the entire watershed is a function of rainstorm size and antecedent soil moisture as well as factors already mentioned. Thus, it is the varying size of the source area of saturated overland flow that gives this model its name (**Figure 5.3**). The runoff from this saturated area is known as *quickflow.* as opposed to water coming from groundwater, which is known as *delayed flow.* However, the actual distinction is much more complex. We discuss it further in Section 5.4.

5.2.3 PRECIPITATION

Extensive discussion on precipitation is presented in Chapters 2 and 3, but let us briefly review how vegetation influences precipitation reaching the ground surface and the factors that will most influence runoff. When precipitation falls on plants, some of the precipitation is intercepted by the plant canopy and ultimately evaporates back to the atmosphere. The amount of interception will depend on the season of the year, the wind velocity, and the vegetation type and growth stage. Dense mature forests can intercept significant amounts of precipitation. Perhaps you have walked through a forest during a rainstorm and stayed dry. The season of the year will influence the denseness of the plant canopy. In the autumn, plants that have shed their leaves have little capacity for intercepting precipitation. It is important to distinguish between precipitation that is retained on the canopy and that is only temporarily delayed before reaching the ground. Water that trickles down a plant stem or drips from the canopy is not considered intercepted as it is still available for infiltration and surface runoff. Similarly, water that falls from the canopy when wind shakes the leaves should be added back into that portion of the precipitation that directly reaches the ground. However, precipitation temporarily detained on the canopy sometimes reaches the ground surface long after the precipitation event has ended and might contribute little to runoff. Unfortunately, interception is difficult to measure or estimate.

Precipitation attributes that have the most influence on runoff processes are the precipitation type (rain, hail, sleet, etc.); the duration of the precipitation event; the amount of precipitation; how the precipitation intensity varied during the event; and spatial changes in the precipitation. Dams, runoff conveyance systems, flood control systems, earthworks, and waste disposal systems are designed to control a specified design storm event. Commonly used design storm events are the probable maximum flood (PMF) and events with return periods of 1, 2, 10, 50, and 100 years. On the average, an event with a 10-year return period might be expected to occur or be exceeded once every 10 years. However, in the short term, there might be several events with a 10-year return period in a 10-year period and even a chance of more than one of these events in a single year. It is also necessary to

assign a rainfall duration to the return period of a rainfall event. For example, the 100-year, l-h rainfall depth might be 2 in., while the 100-year, 24-h rainfall depth could be 5 in. (see Chapter 2).

The PMF is defined as the flood that would result from the most severe combination of critical meteorologic and hydrologic conditions that might occur in a region. The rainfall associated with the PMF is known as the *probable maximum precipitation* (PMP) and is defined as the theoretically greatest depth of precipitation for a given duration that is physically possible over a watershed at a particular time of year. The PMF design event is the most stringent hydrologic design criteria that can be practically applied and is only used if the consequences of failure might be catastrophic or if the design life of the structure is long (at least 100 years).

The term *design storm event* is applied to both storm rainfall and storm runoff. It should not be expected that the 10-year rainfall event will produce the 10-year flood event. Storm runoff is very dependent on antecedent soil water conditions. A severe event following very dry conditions may produce only a small amount of runoff, while a small event following a wet period or falling on concrete frost could produce a large volume of runoff. Although not strictly correct, it is generally assumed that flood flows for a given return period will be produced by storm events with the same return period — an assumption that is reasonable for extreme events. Therefore, the design hydrologist will use the rainfall depth associated with a 100-year rainfall event to determine the 100-year peak flood flow. The PMF is determined based on the PMP, and if high runoff conditions are assumed (i.e., wet antecedent soil water conditions or frozen ground), errors in associating the PMF to the PMP might be small. It should be noted that, in regions where peak flows are caused by snowmelt, it is not appropriate to make flood estimations based on storm rainfall information.

The probability of exceeding a design storm event at least once can be determined from Equation 2.3 in Chapter 2. If a channel with a design life of 100 years is designed based on the 1:100 year event, the probability is about 0.63 (63%) that the design capacity will be exceeded (fail) at least once during the design life. A design return period in excess of 1:2000 years would need to be used to provide a 95% confidence that the design condition is not exceeded. For the most common combinations of design life, return period, and probability of *not* failing (or being exceeded), we offer a graph (**Figure 5.4**) to give you those values quickly (to obtain probability of failing, subtract probability of not failing from 100%). The graph may be used in two ways. The first is to predict probable failure or exceedence over some lifetime as discussed. The second way is to show how big a structure must be built (and thus how much money is to be spent). Remember that Ben Franklin pointed out that the only things certain (100%

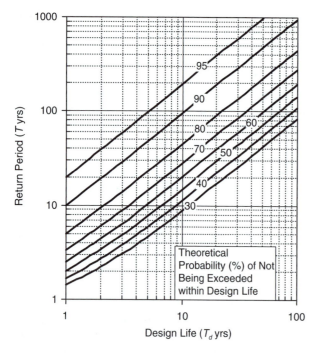

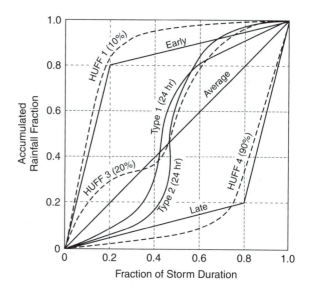

FIGURE 5.5 Commonly used rainfall distributions.

FIGURE 5.4 A calculated risk diagram. The relationship between design return period T years, design period T_d, and probability of not being exceeded (or failing) in T_d years. (From Hershfield, D.N., Rainfall Frequency Atlas of the U.S., U.S. Weather Bureau Technical Paper 40, 1961.)

probability) are death and taxes. Since we cannot guarantee that a channel, reservoir, or any hydrologic structure will never be exceeded or fail, we must deal with some probability that it will, commensurate with how much we and the client, or the taxpayers, are willing to spend to keep it from failing.

Let us suppose the client wants to build a structure and wants to ensure that there is a 5% or less probability that it will fail or be exceeded in the next 50 years. Enter the abscissa at 50 years, go to 95% probability of not being exceeded, and it is clear that you must build a structure to contain the 1000-year event. Now, we are not even sure, in most cases, how big the 1000-year flood is, but let us assume you do know, and the cost is $500 million. However, the client will spend only $250 million; for that, you can build a channel that will contain only the 500-year flood. Again, enter the graph, and you can see that there is a 10% probability that it will be exceeded in the next 50 years, but that is all our client is able, or willing, to finance. This provides better understanding of why most things built or manufactured fail more often than we would like.

Rainfall depth, duration, time distribution, and areal distribution influence the rate and amount of runoff. If runoff information is determined from an actual storm event, then most of these rainfall characteristics will be known. However, in making design or environmental

impact assessments, the rainfall characteristics need to be specified. In the U.S., the most commonly used synthetic rainfall distributions are the NRCS Type 1 and Type 2 curves (**Figure 5.5**). These curves should be used as follows:

Type 1: Hawaii, coastal side of Sierra Nevada in southern California, and the interior regions of Alaska.

Type 1A: Storm distribution represents the coastal side of the Sierra Nevada and the Cascade Mountains in Oregon, Washington, and northern California and the coastal regions of Alaska.

Type 2: Remaining U.S., Puerto Rico, and Virgin Islands.

The rainfall distribution used will have a major impact on the analysis. A steady rainfall throughout the storm duration might result in no runoff if the rainfall rate never exceeded the infiltration rate. However, a storm with the same duration and amount of rainfall might generate some runoff if there is a period of intense rainfall at some point during the event. The point in an event when the intense rainfall occurs is very important. At the start of an event, soils are the driest and infiltration rates are the highest, while near the end of an event, soils are wetter and infiltration rates are lower. Therefore, the advanced-type rainstorm in which the highest rainfall rates occur at the beginning is likely to result in less runoff than the delayed-type storm, in which the highest rainfall rates occur later in the storm.

The Huff 1 distribution is an example of an advanced-type distribution, while the Huff 4 distribution is a delayed-type storm (Huff, 1967). For storms in central Illinois, Huff divided the storms into four groups based on the time quartile most of the rain occurred. The frequency of occurrence of each group is as follows: Huff 1 events occur 30% of the time; Huff 2 events occur 36% of the time; Huff 3 events occur 19% of the time; Huff 4 events occur 15% of the time. Within each group, there are families of curves with different probabilities of occurrence (Barfield et al., 1981). For example, in **Figure 5.5** the 90% Huff 4 storm is illustrated. This signifies that, if a Huff 4 event occurs, then 90% of the time it will have this distribution. Note that, of all storm events, there is only a 15% probability that it will be in the Huff 4 group. Therefore, of all storm events, the probability of the 90% Huff 4 storm is only 13.5% (0.90 × 0.15).

The areal (spatial) distribution of rainfall in a storm is often large. Methods for determining areal distribution are discussed in Chapter 2. On small areas (less than 1 mi^2) such as a farm field or the subdivision of an urban area, it is usually necessary to assume that the rain falls uniformly across the area of interest. In making environmental assessments or developing design details, this assumption will have little influence on the analysis. Due to a lack of information or scattered climatic stations, the assumption of uniform rainfall coverage is often made for very large drainage basins. In this case, the assumption of uniform coverage can result in significant errors. In many parts of the U.S., the most severe events are associated with thunderstorms. These types of events result in different rainfall amounts and time distributions throughout a watershed. Fortunately, rainfall characteristics for regulatory assessments are usually specified. However, there are many applications for which specifications for the rainfall characteristics are not adequate. For example, a small community might require that stormwater drains be designed to convey flow from a 1:10-year event. No information is provided on the duration, time distribution, or assumptions on the areal distribution of this event.

5.3 WATERSHED FACTORS THAT AFFECT RUNOFF

The watershed factors affecting runoff are drainage area size, topography, shape, orientation, geology, interflow, and, perhaps most important, the soil and land use. The depth from the ground surface to an impeding layer and the interaction between surface and groundwater flow regimes in some locations will have a major influence on flows in streams and the response to precipitation of these connected systems. Extensive discussion on land use is presented in Chapter 3, and it is not discussed in this section.

5.3.1 SIZE

The terms river basin, catchment, watershed, or subwatershed are used to delineate areas of different sizes that contribute runoff at their outlets. The terms are used interchangeably, and no clear guidelines are available to identify when each term should be used. We suggest that the term *river basin* be used for large rivers that have several tributaries, while the term *watershed* be used for tributaries and small streams or creeks. Often, it is desirable to subdivide a watershed into subwatersheds to reflect changes in land use, soil type, or topography. For convenience in this book, the term watershed is used to reflect drainage areas of any size.

Determining the boundary or *interfluve* (between the streams) of a watershed can be a challenging task and is normally done using topographic maps and field surveys. In the U.S., a typical starting point might be to obtain a 1:24,000-scale (1 in. = 2,000 ft) USGS (U.S. Geological Survey) topographic map with 10 to 20-ft contour intervals. A practical exercise on this topic is presented in Chapter 14, Exercise 14.8.

A watershed boundary might consist of natural topographic features such as ridges or artificial features such as ditches along roads and stormwater drains. No surface flow will occur across a watershed boundary. All flow within the boundary will drain to the watershed outlet, while flow outside the boundary will be associated with a different watershed. The boundary of a watershed is often called the *watershed divide* or interfluve because it divides the direction of flow. At a watershed divide, which is a natural feature (such as a ridge), flow is perpendicular to a tangent drawn anywhere along the watershed divide (**Figure 5.6**).

In this chapter, empirical equations are presented that relate discharge to climatic and watershed attributes (see Section 5.6.2). The most significant factor in all these relationships is the drainage area. Also, river or "fluvial" geomorphologists often develop or use relationships among drainage area, discharge, and the dimensions of a channel (see Chapter 6).

5.3.2 TOPOGRAPHY

Common topographic features are illustrated in **Figure 5.6**. The only time that a natural watershed divide will cross a contour line will be when the divide runs along a ridge. Surface runoff will occur in the direction of the land slope, and the flow direction will normally be perpendicular to the contours. Topographic maps of a watershed show areas of steep and gentle slope, ridges, valleys, and stream systems. They also show the location of depressions with no surface outlet that contribute to runoff from the watershed. Overland flow slopes can be determined by measuring the lateral change in distance between contours (**Figure 5.7**). The land slope has little effect on infiltration rates or the depth of runoff. However, it has a significant influence on the velocity of water flow on land surfaces and in channels.

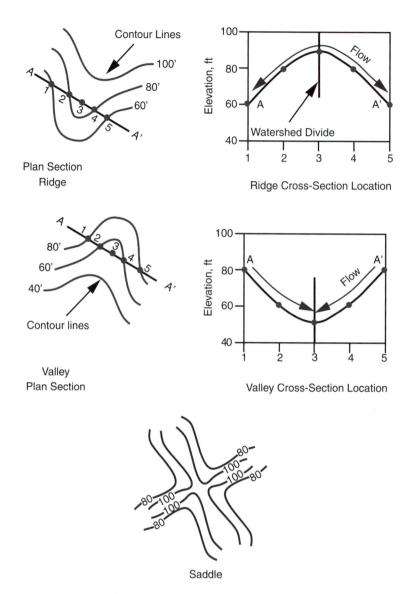

FIGURE 5.6 Common features on a topographic map.

5.3.3 SHAPE

Circular or fan-shape watersheds (**Figure 5.8A**) have high rates of runoff when compared to other shapes because runoff from different points in the watershed are more likely to reach the outlet at similar times. High rates of runoff on small catchments of this shape are short-lived. For long, narrow (elongated) watersheds (**Figure 5.8B**), tributaries join the main stream at intervals along its length. High flow rates from the downstream tributaries pass the gage before high flows from the upper tributaries arrive. Thus, peak flow at the gage is less than for a fan-shape catchments, but persists for a longer time. The influence of watershed shape was found to be significant in several studies by the USGS in which they statistically related climatic and watershed properties to stream discharge data. For example, in southern Arkansas, the peak

discharge associated with recurrence intervals of 2 to 500 years were significantly related to drainage area, elevation, and a shape factor (Hodge and Tasker, 1995). The 100-year recurrence interval discharge Q_{100} (m³/sec) is related to the elevation E (m) and the basin shape factor SH, defined as the drainage area divided by the square of the main channel length, as follows:

$$Q_{100} = 0.471 A^{0.715} E^{0.827} SH^{0.472} \qquad (5.1)$$

The narrower the watershed, the longer the main channel and the smaller the shape factor becomes. This results in lower peak discharge. For example, a square watershed with the main channel running parallel to a boundary would have a shape factor of 1, while a long rectangular watershed with the main channel 10 times the mean width

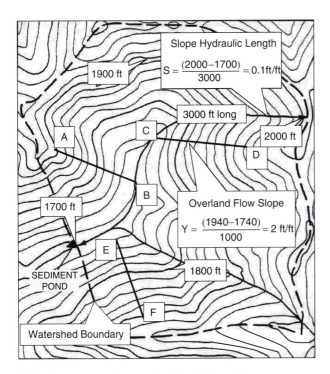

FIGURE 5.7 Measuring overland and channel slopes.

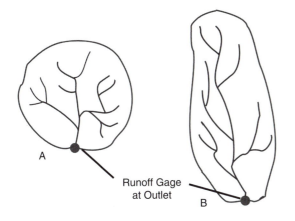

FIGURE 5.8 Catchment shapes: (A) fan; and (B) elongated.

would have a shape factor of 0.1. From Equation 5.1, this would result in the long watershed having a 100-year recurrence interval peak discharge that is only 34% of the discharge from the square watershed.

5.3.4 ORIENTATION OR ASPECT

The compass direction that the land surface faces is termed the *aspect* or *orientation*. The aspect can be important when there is only one major slope face. On large areas, there is a multitude of aspects, and the average tends to balance out to no specific direction. The aspect affects soil water content, vegetation, and soil development. In the Northern Hemisphere, north-facing slopes receive less

solar energy, soil freezes to greater depth, and the period of frost is longer.

Stream basins themselves also have orientation, and east–west orientation can be important. Because weather systems in midlatitudes move from west to east, west-facing slopes and basins may receive more rain. Another consideration, largely uninvestigated, is the movement of a storm upstream or downstream. In an east-facing basin, storms generally move downstream, so that the storm moves with (but not necessarily at the same speed as) the crest of runoff. That is, runoff from the head of the basin as well as the foot, or outlet, of the basin gets to the outlet at more nearly the same time. In a west-facing basin, however, the storm and runoff are moving in opposite directions. Thus, runoff from the foot of the basin may have already left the basin before runoff from the head arrives. Theoretically, the east-facing basin should have a more peaked hydrograph than the west-facing one, but few data are available for demonstration.

5.3.5 GEOLOGY

Geology, the science of the earth's crust and rocks, is a significant factor in the historic formation of soil and the physical characteristics of watersheds and the establishment of surface and subsurface flow systems of the hydrologic cycle. Geological features (type of rock formations, fractures, faults, etc.) and processes (such as the movement of glaciers more than 10,000 years ago) have helped define the surface divide or ridge between watersheds, establish and control stream channel gradients, and formed the subsurface boundaries that control the movement of groundwater to surface streams. Groundwater flow boundaries do not always coincide with watershed divides and are more difficult to define.

The geology of a site may be described by identifying and characterizing the bedrock and any deposits overlying that bedrock. These unconsolidated deposits may include alluvial sediments left behind by surface water processes, windblown loess, and glacial deposits. During the last 2 million years, most of the northern half of the North American continent was repeatedly covered with glaciers. When the glacial ice melted, it left deposits (a few inches to hundreds of feet thick) of loose sediments on top of the buried landscapes. Glacial tills are the most common and heterogeneous of the deposits. Most tills are fine grained, consisting of clay, silt, and sand with some gravel and boulders. After the glaciers melted, the land dried out, and vertical cracks began to form or expand in the deposits. In addition to desiccation, fractures were also caused by freeze–thaw cycling, shear stress from advancing ice, isostatic rebound due to stress release when the glacier melted, and propagation from fractures in the bedrock below (Brockman and Szabo, 2000).

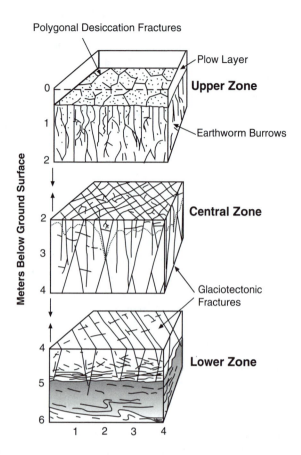

Polygonal Desiccation Fractures

Plow Layer
Upper Zone

Earthworm Burrows

Central Zone

Glaciotectonic Fractures

Lower Zone

Meters Below Ground Surface

FIGURE 5.9 Typical distribution of fractures in clay till. (Modified from Klint, K.E.S., and C.D. Tsakiroglou, A new method of fracture aperture characterisation, in *Protection and Restoration of the Environment V. Proceedings of an International Conference, Thassos, Greece*, Vol. 1, V.A. Tsihrintzis, G.P. Korfiatis, K.L. Katsifarakis, and A.C. Demetracopoulos, Eds., 2000, pp. 127–136. With permission.)

Figure 5.9 shows examples of these fractures. Fractures may be vertical, horizontal, or angled, depending on the forces that formed them. These features can extend from the soil into the subsurface geologic strata, acting as conduits for water and contaminant flow. In **Figure 5.9**, the upper zone is weathered with polygonal desiccation fractures and vertical *biopores* (root holes and earthworm burrows). The central zone, above the groundwater table, is characterized by *glaciotectonic fractures* (shearing, isostatic rebounding, or bedrock initiated), as well as desiccation fractures. The lowermost zone is saturated, and only glaciotectonic fractures are present. In these cracked deposits, the soils formed.

In areas of soluble limestone, Karstic drainage may form where subsurface flow may develop along fractures and bedding planes. In such terrain, surface water may enter bedrock through sinkholes so that there are few small streams. Even stormflow may be accommodated in extreme cases so that it moves directly through the limestone and into larger streams. Such a landscape is found near Mammoth Cave, KY.

5.3.6 SOIL

Soil properties that influence runoff are the same as those that influence infiltration and are discussed in detail in Chapter 3. Traditional soil parameters such as particle size, bulk density, carbonate content, and clay mineralogy may not be predictive of the effective hydraulic conductivity expected in a fractured setting. Tornes et al. (2000) determined that fractures could be maintained in finer-grained materials with as little as 10% clay-size materials included. Tornes et al. provided a readable primer of basic soil characteristics as they relate to macropores and preferential flow in soils and underlying geologic materials.

The NRCS has conducted detailed soil surveys for each county in the U.S.; paper copies of these studies are available from the NRCS, county extension agents, or state departments of natural resources. Digital maps of this information are in preparation and can be acquired through their Web site. **Figure 5.10A** shows a map of soil types overlain onto a 1:24,000 ortho aerial photo mosaic that is exactly the same scale and projection as the USGS 1:24,000 topographic map. The letters correspond to particular soil series, which can then be mapped according to their hydrologic soil groups (see discussion in Section 5.5), as shown in **Figure 5.10B**. See also Exercise 14.9 on soils surveys in Chapter 14.

Earthworm populations can greatly affect soil structure and porosity and may contribute to a decrease in runoff. In particular, the size and number of *Lumbricus terrestris* burrows found in some agricultural fields suggest that they may have a major impact on hydrology. Field research indicates that the amount of rainfall transmitted by *L. terrestris* burrows increases with storm intensity and can amount to as much as 10% of the total rainfall (Shipitalo et al., 2002). In the case of subsurface drained fields, direct connection between earthworm burrows and drainage tile may substantially increase the risk of surface water contamination by surface-applied agricultural chemicals and injected animal wastes. Likewise, soil macropores may connect to subsoil fractures and contribute to rapid water and chemical movement to groundwater.

5.3.7 INTERFLOW AND BASEFLOW

Interflow and baseflow are two important flow processes that contribute to streamflows. Interflow was discussed in Section 5.2. Measuring or estimating interflow is difficult, and

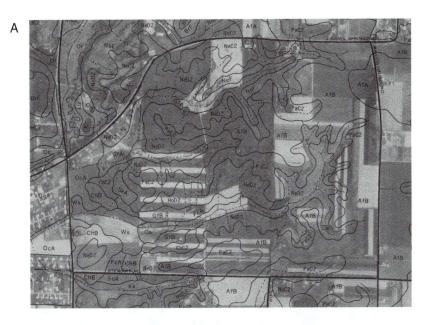

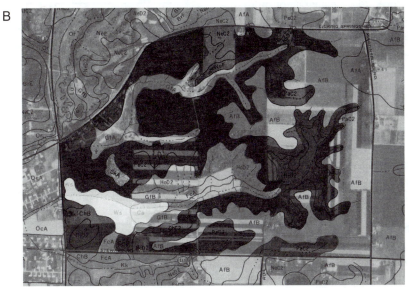

Soil Type	Hydrologic Soil Group	Soil Type	Hydrologic Soil Group
AfB - Alford	B	NeD2, NeF - Negley	B
Ca - Carslisle	A/D	OcA - Ockley	B
ChB, ChC2 - Chili	B	PaC2 - Parke	B
FcA - Fitchville	C	SkA - Sleeth	C
GfA , GfB - Glenford	C	Ws - Westland	B/D
HoD2 - Homewood	C		

FIGURE 5.10 Soil survey map for Licking County, Ohio. A. Soil type; the differences in background color are due to changes in landcover and vegetation. B. Hydrologic soil groups. Original color: soil group C. Dark shading: soil group B. Light shading: other soil groups.

because the flow has the potential to mix with surface runoff, it is often quantified as part of runoff. The volume and rate (or velocity) of the water moving as baseflow can be significantly influenced or even controlled by macropore and other fractured conditions (both vertical and horizontal) in the soil and underlying shallow geologic materials. This topic is discussed further in Section 5.7.

5.4 RUNOFF CHARACTERISTICS — THE HYDROGRAPH

When evaluating or applying knowledge of runoff responses to precipitation, we might be interested in the depth or volume of runoff, the peak runoff rate, or the relationship between runoff rate and time. Numerous

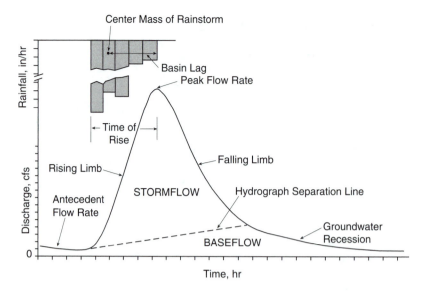

FIGURE 5.11 Storm hydrograph relationships.

methods have been developed to estimate one or more of these characteristics. Before looking at specific estimation techniques, it is important to understand the hydrologic response of complex climatic, watershed, and groundwater interactions. Understanding these responses is facilitated by considering the processes presented in Chapter 2 and Chapter 3 and in the beginning of this chapter and then studying a hydrograph, which is a plot of runoff rates against time (**Figure 5.11**). A rainfall excess (runoff depth) hyetograph for a single block of time is plotted in the upper left corner of the diagram. The runoff hydrograph is developed by routing blocks of rainfall excess to the watershed outlet. The volume of flow under the storm hydrograph will be equal to the sum of the rainfall excess in all the blocks of time associated with the storm event, multiplied by the watershed area. In many cases, there will be flow prior to the start of the storm event. The flow associated with the storm event is added to this baseflow. Baseflow separation lines are drawn from the rising limb of the hydrograph to a point on the recession curve. If there is significant interflow, groundwater flow, subsurface drainage, or delayed runoff (such as from a forest), the receding limb will decay slowly, and it is difficult to determine the point in time that flow from the rainfall event ceased. A practical exercise on this topic is presented in Chapter 14, Exercise 14.4.

Before getting into the clinical details of hydrographs, it is instructive to review schematically some ways that streamflow, as represented by the hydrograph, might respond to a rainfall event under different conditions of rainfall amount and intensity, soil infiltration capacities, and antecedent soil moisture conditions (**Figure 5.12**).

Type 0 shows no response. Shown on the graph is only the normal baseflow recession curve (A–B). In this case, some combination of low antecedent

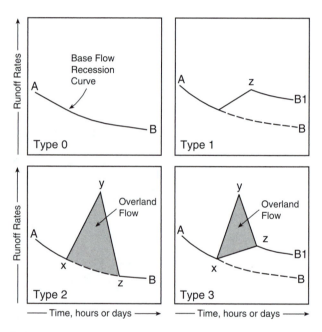

FIGURE 5.12 Four basic hydrograph types. (From Ward, R.C., *Principles of Hydrology*, McGraw-Hill, Maidenhead, U.K., 1967. With permission.)

soil moisture and low rainfall intensity allows the soil to absorb the entire amount of rainfall, so the hydrograph does not respond at all (of course, channel precipitation is ignored).

In **Type 1**, infiltration rates are high or intensities are not excessive, so that all the water infiltrates and moves as interflow or groundwater flow into the stream to create the hydrograph peak. Since soil is wetted above field capacity and groundwater storage has increased, baseflow also increases, and a new, higher, baseflow recession curve (Z–B1) ensues.

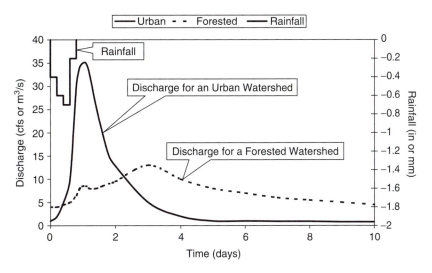

FIGURE 5.13 Streamflow from forested and urban watershed.

In **Type 2**, soil infiltration capacity is low or rainfall is short and intense. The result is that most of the precipitation leaves as overland flow (XYZ). Soil and groundwater are not recharged enough to increase baseflow (soil water was not increased above field capacity).

Type 3 is a composite of Types 1 and 2. It shows both overland flow (XYZ) and recharge of soil and groundwater with increase of baseflow (Z–B1).

The hydrograph tells more about the hydrology of a small catchment than any other measurement. For example, **Figure 5.13** shows that urban and forested watersheds had runoff for a few days after rain. However, the forested watershed continued to have streamflow above baseflow levels more than 5 days after the rainfall, indicating the presence of soil or groundwater storage. The forested watershed's soil reservoir was recharged by stormwater infiltration. The outflow into streams built up slowly and continued at a high rate after rainfall. Storm runoff rates from the urban watershed increased sharply during each rain period, then fell back rapidly after rain stopped. These sharp rises and falls indicate that much of the storm runoff came as surface or quick-return flow with little or no storm recharge to groundwater. As stated in Section 5.1, the hydrograph of the urban watershed is termed flashy, which indicates peak runoff rate shortly after the most intense rainfall occurs. Conditions that can cause this type of rapid response are a small watershed area, impervious areas located adjacent to the stream channel, an impervious land use (such as a parking lot), or steep land slopes.

A stream channel acts as a temporary storage reservoir, subduing (attenuating) the flashiness of runoff from small headwater watersheds (**Figure 5.14**). The shape of the headwater hydrograph reflects the individual rainstorm characteristics. But, a point 50 mi or more downstream

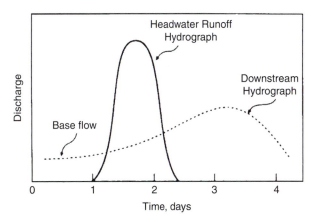

FIGURE 5.14 Hydrograph attenuation in a stream.

shows mostly the effect of temporary storage on runoff rates and practically none of the rainfall variations. Simple assumptions of this sort are usually adequate to help identify, on the hydrograph, sources of water flow from small headwater watersheds.

Runoff assessments often need to be made for watersheds where there is little or no streamflow information or in areas that will undergo a land use modification, such as an urban development or surface mining. Evaluation of observed runoff information is of limited value if the land features are going to be changed. Many surface runoff estimation techniques have been developed for these situations and used on ungaged watersheds. The most commonly used approaches are empirical equations, unit hydrograph techniques, the time–area method, and kinematic methods. When assessing surface runoff, it is necessary to determine one or more of the following attributes: (1) the depth or volume of runoff, (2) the peak runoff rate, and (3) a storm hydrograph. Common approaches to determining these attributes are presented in the next three sections.

5.5 PREDICTING VOLUME OF STORMFLOW AND TOTAL RUNOFF

Rainfall excess (volume of runoff) could be determined using one of the infiltration equations described in Chapter 3, but this approach has only been incorporated in a few hydrologic computer models. The method most commonly used in the U.S. is the NRCS curve number procedure (NRCS, 1972).

In this approach, infiltration losses are combined with surface storage by the relationship

$$Q = \frac{(P - I_a)^2}{(P - I_a + S)} \qquad (5.2)$$

where Q is the accumulated runoff or rainfall excess in inches, P is the rainfall depth in inches, and S is a parameter given by

$$S = \frac{1000}{CN} - 10 \qquad (5.3)$$

where CN is known as the curve number. The term I_a is the initial abstractions in inches and includes surface storage, interception and infiltration prior to runoff. The initial abstractions term I_a is commonly approximated as $0.2S$, and Equation 5.2 becomes

$$Q = \frac{(P - 0.2 S)^2}{(P + 0.8 S)} \qquad (5.4)$$

A graphical solution of Equation 5.4 is presented in **Figure 5.15**.

The NRCS curve number is a function of the ability of soils to infiltrate water, land use, and the soil water conditions at the start of a rainfall event (antecedent soil water condition). To account for the infiltration characteristics of soils, the NRCS has divided soils into four hydrologic soil groups, which are defined as follows (NRCS, 1984):

Group A (low runoff potential): Soils with high infiltration rates even when thoroughly wetted. These consist chiefly of deep, well-drained sands and gravels. These soils have a high rate of water transmission (final infiltration rate greater than 0.3 in./h).

Group B: Soils with moderate infiltration rates when thoroughly wetted. These consist chiefly of soils that are moderately deep to deep, moderately well

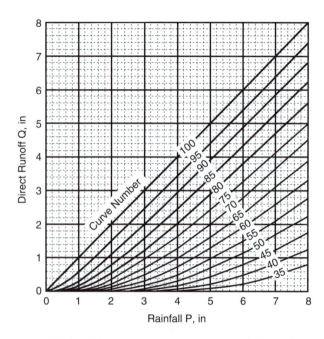

FIGURE 5.15 Relationships of runoff to rainfall based on NRCS curve number method.

drained to well drained with moderately fine to moderately coarse textures. These soils have a moderate rate of water transmission (final infiltration rate 0.15 to 0.30 in./h).

Group C: Soils with slow infiltration rates when thoroughly wetted. These consist chiefly of soils with a layer that impedes downward movement of water or soils with moderately fine to fine texture. These soils have a slow rate of water transmission (final infiltration rate 0.05 to 0.15 in./h).

Group D (high runoff potential): Soils with very slow infiltration rates when thoroughly wetted. These consist chiefly of clay soils with a high swelling potential, soils with a permanent high water table, soils with a claypan or clay layer at or near the surface, and shallow soils over nearly impervious materials. These soils have a very slow rate of water transmission (final infiltration rate less than 0.05 in./h).

Typical curve numbers for Antecedent Soil Moisture Condition II (AMC II) are shown in **Table 5.1** (NRCS, 1984). A summary of common U.S. soils and their hydrologic soil groups is presented in Appendix D. Prior to estimating rainfall excess for a storm event, the curve numbers should be adjusted based on the season and 5-day antecedent precipitation. Antecedent soil moisture conditions are defined as follows:

TABLE 5.1
Curve Numbers for Antecedent Soil Moisture Condition II

Land Use Description	Hydrologic Soil Group			
	A	B	C	D
Commercial, row houses and townhouses	80	85	90	95
Fallow, poor condition	77	86	91	94
Cultivated with conventional tillage	72	81	88	91
Cultivated with conservation tillage	62	71	78	81
Lawns, poor condition	58	74	82	86
Lawns, good condition	39	61	74	80
Pasture or range, poor condition	68	79	86	89
Pasture or range, good condition	39	61	74	80
Meadow	30	58	71	78
Pavement and roofs	100	100	100	100
Woods or forest thin stand, poor cover	45	66	77	83
Woods or forest, good cover	25	55	70	77
Farmsteads	59	74	82	86
Residential quarter-acre lot, poor condition	73	83	88	91
Residential quarter-acre lot, good condition	61	75	83	87
Residential half-acre lot, poor condition	67	80	86	89
Residential half-acre lot, good condition	53	70	80	85
Residential 2-acre lot, poor condition	63	77	84	87
Residential 2-acre lot, good condition	47	66	77	81
Roads	74	84	90	92

Source: From NRCS, 1984.

AMC I: Dormant season antecedent soil moisture less than 0.5 in. Growing season antecedent soil moisture less than 1.4 in.

AMC II: Dormant season antecedent soil moisture between 0.5 and 1.1 in. Growing season antecedent soil moisture between 1.4 and 2.1 in.

AMC III: Dormant season antecedent soil moisture greater than 1.1 in. Growing season antecedent soil moisture greater than 2.1 in.

Curve number adjustments for antecedent soil moisture conditions can be made using the information presented in **Table 5.2**. If Equation 5.4 is used to determine the runoff depth from an observed rainfall event, then the 5-day antecedent rainfall can be determined from rainfall records. However, when the procedure is used to determine runoff associated with a design storm event, an appropriate antecedent soil water condition will need to be specified. A conservative design approach is to assume wet conditions (AMC III) as they produce the most runoff. It should be recognized that, in arid and semiarid areas, the likelihood of these conditions occurring would be much lower than in semihumid and humid areas.

Determining the depth or volume of runoff from watersheds with several land uses can be obtained by determining the volume of runoff from each land use and then summing these amounts or calculating an area-weighted

TABLE 5.2
Adjustments to Runoff Curve Number (CN) for Dry or Wet Antecedent Soil Moisture Conditions

Curve Number (AMC II)	Factors to Convert Curve Number for AMC II to AMC I or AMC III	
	AMC I (dry)	AMC III (wet)
10	0.40	2.22
20	0.45	1.85
30	0.50	1.67
40	0.55	1.50
50	0.62	1.40
60	0.67	1.30
70	0.73	1.21
80	0.79	1.14
90	0.87	1.07
100	1.00	1.00

curve number for the watershed and then using this single value with Equation 5.3 and Equation 5.4. The first approach is preferred as it retains the runoff characteristics of the watershed, and small areas with high runoff potential will not be dwarfed by large areas with low runoff potential. However, for convenience, the second approach is often included in hydrologic computer models.

EXAMPLE 5.1

Estimate the depth and volume of runoff from 2 in. of rainfall on an 8-acre, grassed area in an urban park in California. The grass is grown in San Joaquin soils with good hydrologic conditions. Determine the runoff depth for AMC I, AMC II, and AMC III conditions.

Solution

From Appendix D, it is determined that a San Joaquin soil is in Hydrologic Soil Group D. From **Table 5.1**, the NRCS curve number for grass (lawns) on soils in Hydrologic Soil Group D with good hydrologic conditions is 80.

For AMC II, substitute a curve number CN of 80 in Equation 5.3:

$$S = \frac{1000}{80} - 10 = 2.5 \text{ in.}$$

Substitute $S = 2.5$ in. and $P = 2.0$ in. in Equation 5.4:

$$Q = \frac{[2.0 - (0.2 \times 2.5)]^2}{[2.0 + (0.8 \times 2.5)]} = 0.5625 \text{ in.}$$

Therefore, $Q = 0.5625$ in., or 28% of the rainfall for AMC II. (Note that an estimate of about 0.55 in. would be obtained using **Figure 5.15**.) Generally, Q would not be reported to more than two decimal places, and curve numbers should not be determined to more than one decimal place.

The volume of runoff is determined by multiplying the runoff depth by the area contributing to runoff. Therefore, $Q = 0.5625$ in. $\times$ 8 acres = 4.5 acre-in. Runoff volumes are normally expressed in acre-feet, so divide by 12 to convert inches to feet, and the runoff volume is 0.375 acre-ft. For AMC I, the curve number becomes $0.79 \times 80 = 63.2$ (factor of 0.79 from **Table 5.2**). Substitute 63.2 in Equation 5.3:

$$S = \frac{1000}{63.2} - 10 = 5.8 \text{ in.}$$

Substitute $S = 5.8$ inches and $P = 2.0$ inches in Equation 5.4:

$$Q = \frac{[2.0 - (0.2 \times 5.8)]^2}{[2.0 + (0.8 \times 5.8)]} = 0.11 \text{ in.}$$

Therefore, $Q = 0.11$ inches or 5.5% of the rainfall for AMC I. Multiplying by 8 acres and dividing by 12 to convert inches to feet gives a runoff volume of 0.073 acre-ft.

For AMC III, the curve number becomes $1.14 \times 80 = 91.2$ (factor of 1.14 from **Table 5.2**). Using the same approach, the runoff depth for AMC III is 1.18 in., or 59% of the rainfall for AMC III. The runoff volume is 0.787 acre-ft.

Answer

AMC I, $Q = 0.11$ in. or 0.073 acre-ft. For AMC II, $Q = 0.56$ in. or 0.375 acre-ft, and for AMC III, $Q = 1.18$ in. or 0.787 acre-ft.

EXAMPLE 5.2

Determine the depth and volume of runoff for a 1 in 25-year, 24-h storm on an agricultural watershed in north central Ohio. The watershed has a 100-acre area of Hydrologic Group D soils that are cultivated with conventional tillage and a 50-acre wooded area, with good cover, on soils in Hydrologic Group C.

Solution

From **Table 5.1**, the cultivated area with conventional tillage has a curve number of 91, while the woods have a curve number of 70. No information is provided on the AMC. However, in northern Ohio, severe storms often occur in the spring when there are wet antecedent soil water conditions (AMC III).

Solving for the cultivated area, from **Table 5.2** the correction factor is 1.07 for a curve number of 90 and 1.0 for a curve number of 100. Use linear interpolation to determine the correction factor for a curve number of 91:

$$\frac{(X - 1.07)}{(1.00 - 1.07)} = \frac{(91 - 90)}{(100 - 90)}$$

and $X = 1.0063$. Therefore, the AMC III curve number is $1.063 \times 91 = 96.7$. Substitute 96.7 in Equation 5.3:

$$S = \frac{1000}{96.7} - 10 = 0.34 \text{ in.}$$

From Appendix C, the 1 in 25-year, 24-h rainfall is 4.0 in. Substitute $S = 0.34$ in. and $P = 4.0$ in. in Equation 5.3:

$$Q = \frac{[4.0 - (0.2 \times 0.34)]^2}{[4.0 + (0.8 \times 0.34)]} = 3.62 \text{ in.}$$

The runoff depth Q is 3.62 in., and the runoff volume is

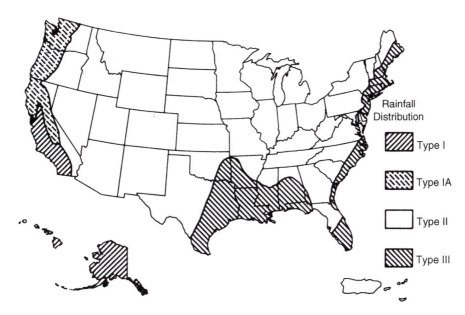

FIGURE 5.16 Geographic regions for NRCS Types I, IA, II, and III unit peak discharge responses q_u for the graphical peak discharge method.

$$Q = 3.62 \text{ inches} \times 100 \text{ acres} \times \frac{1 \text{ ft}}{12 \text{ inches}}$$

Therefore, the runoff volume is 30.17 acre-ft. Solving for the woods, from **Table 5.2**, the correction factor is 1.21 for a curve number of 70. Therefore, the AMC III curve number is $1.21 \times 70 = 84.7$. Substitute 84.7 in Equation 5.3:

$$S = \frac{1000}{84.7} - 10 = 1.81 \text{ in.}$$

Substitute $S = 1.81$ in. and $P = 4.0$ in. in Equation 5.4:

$$Q = \frac{[4.0 - (0.2 \times 1.81)]^2}{[4.0 + (0.8 \times 1.81)]} = 2.43 \text{ inches}$$

The runoff depth Q is 2.43 in., and the runoff volume is

$$Q = 2.43 \text{ inches} \times 50 \text{ acres} \times \frac{1}{12} \text{ ft} = 10.12 \text{ acre-ft}$$

The total runoff volume is 40.29 acre-ft (30.17 + 10.12). The average runoff depth is

$$Q = 3.62 \text{ inches} \times \frac{100 \text{ acres}}{150 \text{ acres}} + 2.43 \text{ inches}$$
$$\times \frac{50 \text{ acres}}{150 \text{ acres}} = 3.22 \text{ in.}$$

Answer

The average runoff depth is 3.22 in., and the total runoff volume is 40.29 acre-ft.

5.6 PREDICTION OF PEAK RUNOFF RATE

Information on peak runoff rates is often used to design ditches, channels, and stormwater control systems. A practical exercise on this topic is presented in Chapter 14, Exercise 14.6. For presentation purposes, commonly used approaches have been grouped into the following three categories: (1) empirical runoff methods, which are best suited to watershed areas smaller than a few square miles; (2) empirical regression models based on stream discharge data; and (3) computer methods.

5.6.1 GRAPHICAL PEAK DISCHARGE METHOD

The graphical peak discharge method (NRCS, 1986) was developed for application in small rural and urban watersheds. It was developed from hydrograph analyses with TR-20 *Computer Program for Project Formulation - Hydrology* (NRCS, 1973b) and has seen widespread application. The peak discharge equation is

$$q = q_u A Q F \tag{5.5}$$

where q is the peak discharge (runoff rate, ft³/sec), q_u is the unit peak discharge (ft³/sec per square mile per inch of runoff, csm/in.), A is the drainage area (mi²), Q is the runoff depth (in.) based on 24 h, and F is an adjustment factor for ponds and swamps. The method depends on the NRCS curve number method (Equation 5.2 to Equation 5.4) to obtain Q and the necessary information to determine q_u. The unit peak discharge q_u is determined from **Figure 5.16** and **Figure 5.17** and requires knowledge of the time of concentration and the initial abstraction I_a from

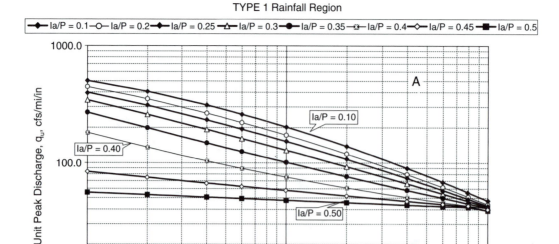

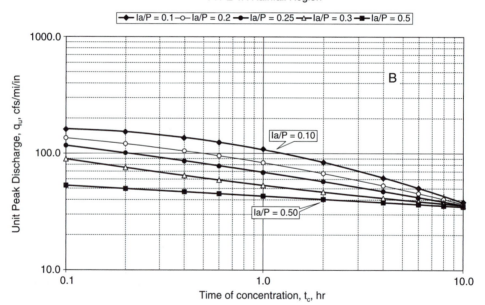

FIGURE 5.17 Unit peak discharge q_u for rainfall distribution: A, NRCS Type I; B, NRCS Type IA; C, NRCS Type II; D, NRCS Type III.

Equation 5.2 and Equation 5.4. **Figure 5.16** shows the geographic location of the regions for the different runoff types presented in **Figure 5.17A,B**.

The method is intended for use in watersheds that are hydrologically homogeneous and have only one main stream or stream branches that have equal times of concentration. The F factor can only be applied to swamps and ponds that are not along the main flow path used to determine the time of concentration. Values of the adjustment factor F are presented in **Table 5.3**.

Intuitively, we would expect to calculate a runoff depth Q by selecting an appropriate antecedent soil moisture condition as presented in Section 5.5. However, this poses a problem when using **Figure 5.17** for AMC III conditions. Often, for AMC III the I_a/P ratio will be less than 0.1. An interpolation approach could be used to develop a solution, or additional curves could be added to the plot. However, it will be found that use of AMC III conditions with the graphical peak discharge method will usually result in overestimations of runoff. Commonly

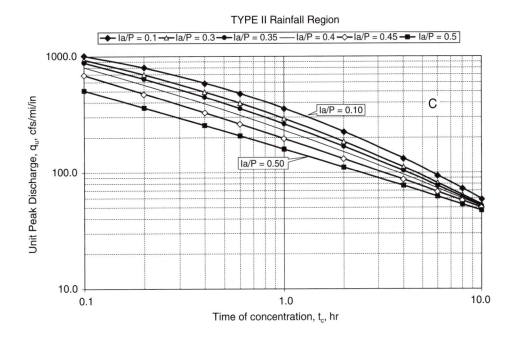

TYPE II Rainfall Region

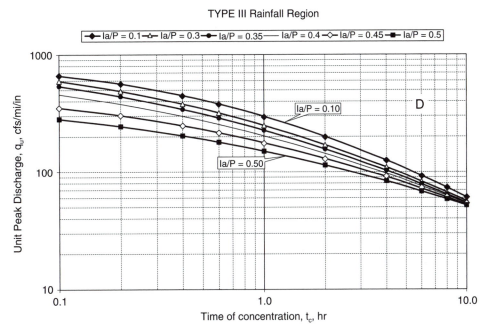

TYPE III Rainfall Region

FIGURE 5.17 (continued)

used computer methods that include the graphical peak discharge method use AMC II to develop a solution (NRCS, 2002).

Numerous equations have been developed for determining the time of concentration and lag time. The *time of concentration* is the time it takes water to travel along the hydraulic length; the *lag time* is the average of the flow times from all locations in the watershed. These methods may give results that are one or two orders of magnitude different from each other. Commonly used methods include those of the NRCS (1972), Kirpich

(1940), and A.D. Ward et al. (1980). Several different methods developed by the NRCS are commonly used with the graphical peak discharge method. One recommended approach is to use the NRCS lag equation:

$$t_L = \frac{L^{0.8}(S+1)^{0.7}}{1900Y^{0.5}} \quad (50 < CN < 95) \qquad (5.6)$$

where t_L is the lag time in hours, L is the hydraulic length of the watershed in feet, S is a function of the NRCS curve

TABLE 5.3
Adjustment Factor *F* for Ponds and Swamps Spread Throughout the Watershed

Swamp and Pond Areas (%)	*F*
0.0	1.00
0.2	0.97
1.0	0.87
3.0	0.75
5.0	0.72

TABLE 5.4
Flow Characteristics for Sheet, Overland, and Channel Flow, Equation 5.9 to Equation 5.11 and Figure 5.17

Type	Surface Description	Manning's n[a]	Flow Depth[b] (ft)	K_2
6	Smooth surfaces	0.01	0.05	2
4	Fallow (no residue)	0.05	0.2	1
	Cultivated Soils			
4	Residue cover <20%	0.06	0.2	1
2	Residue cover >20%	0.17	0.5	0.5
	Grass			
5	Grass waterway	0.025–0.035	0.2–0.4	1.5
3	Short prairie grass	0.15	0.5	0.65
	Dense grasses	0.24		
	Bermudagrass	0.41		
3	Range (natural)	0.13	0.4	0.65
	Woods and Forest			
1	Light underbrush	0.40	0.5	0.25
	Dense underbrush	0.80		
	Streams and Rivers			
7	Cobble bed or weedy	0.04–0.07	1–3	3
8	Sand and gravel bed	0.03–0.05	1–3	5
9	Deep, fast flowing	0.03–0.04	4–5	10
10	Concrete lined and large river	0.016 and 0.035	2 and 7	15

[a] The n values are a composite of information compiled by Engman (1986). Values were rounded to two decimal places.
[b] Flow depths used to estimate K_2.

number (see Equation 5.2), and Y is the average land slope in percent. The lag time is related to the time of concentration as follows:

$$t_L = 0.6t_c \qquad (5.7)$$

Lag time is an estimate of the average flow time for all locations on a watershed. The coefficient of 0.6 in Equation 5.7 accounts for the fact that the average flow time will be 0.45 to 0.65 of the maximum flow time, depending on the watershed shape. Often, the average land slope is approximated as the slope along the hydraulic length.

The TR-55 computer model (NRCS, 1986) calculates the time of concentration as the sum of the travel times due to sheet flow, shallow concentrated flow, and open channel flow along the hydraulic length. For sheet flow, the travel time t_t in hours is estimated as

$$t_t = \frac{0.007(nL)^{0.8}}{(P_2)^{0.5}(S)^{0.4}} \qquad (5.8)$$

where n is Manning's roughness factor, L is the flow length in feet, P_2 is the 24-h 2-year return period rainfall in inches, and S is the land slope or slope of the hydraulic grade line in feet/feet. Manning's n values for sheet flow are presented in **Table 5.4**. For shallow concentrated flow,

$$t_t = \frac{L}{K_1(S)^{0.5}} \qquad (5.9)$$

where K_1 is 58,084 for unpaved surfaces and 73,182 for paved surfaces.

For open channel flow,

$$t_t = \frac{nL}{5364(a/p_w)^{0.67}(S)^{0.5}} \qquad (5.10)$$

where a is the cross-sectional area of flow in square feet, L is the channel length in feet, and p_w is the wetted perimeter in feet. Equation 5.8 to Equation 5.10 were developed based on Manning's equation (see Chapter 7). An approximate simplification of Equation 5.9 and Equation 5.10 is obtained by determining the flow velocity from the equation

$$v = K_2(S)^{0.5} \qquad (5.11)$$

where v is the flow velocity (ft/sec), K_2 is obtained from Table 5.4, and S is the slope in percent.

EXAMPLE 5.3

Use the graphical peak discharge method to determine the 1:25-year peak storm runoff rate (ft³/sec) for a 300-acre watershed located near Minneapolis, MN. The watershed has an overland slope of 0.5%, is Hydrologic Soil Group B, and is cropped in corn-conservation practice. It has a 4000-ft hydraulic length with a grassed waterway that has an average bed slope of 0.2%.

Solution

The graphical peak discharge method is described by Equation 5.5 as follows:

$$q = q_u A Q F$$

The time of concentration can be determined using the NRCS lag time method described by Equation 5.6:

$$t_L = \frac{L^{0.8}(S+1)^{0.7}}{1900\ Y^{0.5}} \quad (50 < CN < 95)$$

From **Table 5.1**, the NRCS curve number for cultivated land with conservation tillage and Hydrologic Soil Group B is 71. Therefore, S is equal to

$$S = \frac{1000}{CN} - 10 = \frac{1000}{71} - 10 = 4.08 \text{ inches}$$

The average land slope is 0.5%, so by substituting for S and Y,

$$t_L = \frac{4000^{0.8}(4.08+1)^{0.7}}{1900 \times 0.5^{0.5}}$$

and $t_L = 1.74$ h. Using Equation 5.7,

$$t_L = 0.6\ t_c; \text{ therefore, } 1.74 = 0.6\ t_c$$

and t_c is equal to 2.9 h.

Using Appendix C, the rainfall amount P for a 25-year, 24-h storm event is about 4.5 in. (note that, for this method, the time of concentration is not used to calculate the rainfall amount P). The initial abstraction I_a is 0.2S or 0.82 in. (0.2 × 4.08 in.). Therefore, the ratio of I_a/P is 0.82/4.5 or 0.18. From **Figure 5.17C**, the unit peak discharge q_u is about 170 csm/in.

The runoff volume Q can be estimated from **Figure 5.15** or calculated from Equation 5.3:

$$Q = \frac{(P-0.2S)^2}{(P+0.8S)} \text{ and } Q = \frac{(4.5-0.82)^2}{(4.5+3.26)} = 1.75 \text{ inches}$$

Therefore, the peak discharge is

$$q = q_u A Q F_p = 170 \times \frac{300}{640} \times 1.75 \times 1.0 = 139 \text{ cfs}$$

Answer

The peak discharge based on the graphical peak discharge method is about 139 ft³/sec.

EXAMPLE 5.4

Redo Example 5.3, but use the velocity approach described by Equation 5.11, **Table 5.4**, and **Figure 5.18**.

Solution

From **Table 5.1**, the NRCS curve number for cultivated land with conservation tillage and Hydrologic Soil Group B is 71. Therefore, S is equal to:

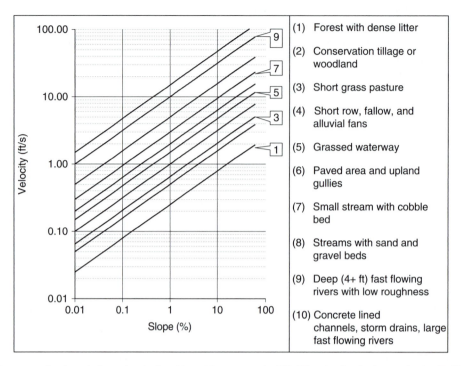

FIGURE 5.18 Sheet, overland, and channel velocity. (Based in part on the NRCS upland velocity method; NRCS, 1986.)

$$S = \frac{1000}{CN} - 10 = \frac{1000}{71} - 10 = 4.08 \text{ inches}$$

The time of concentration can be estimated from **Figure 5.18** based on a bed slope of 0.2% and Curve 5 for a grassed waterway. It also can be calculated using Equation 5.11; K_2 is obtained from **Table 5.4** and is 1.5 for a grassed waterway. Therefore, the flow velocity is 0.67 ft/sec. The time of concentration is then calculated by dividing the hydraulic length by the velocity and converting to hours:

$$t_c = \frac{4000}{(3600)(0.67)} = 1.66 \text{ hours}$$

Using Appendix C, the rainfall amount P for a 25-year, 24-h storm event is about 4.5 in. (note that, for this method, the time of concentration is not used to calculate the rainfall amount P). The initial abstraction I_a is 0.2S or 0.82 in. (0.2 × 4.08 in.). Therefore, the ratio of I_a/P is 0.82/4.5 or 0.18. From **Figure 5.16C**, the unit peak discharge q_u is about 250 csm/in.

The runoff volume Q can be estimated from **Figure 5.15** or calculated from Equation 5.3:

$$Q = \frac{(P - 0.2S)^2}{(P + 0.8S)} \text{ and } Q = \frac{(4.5 - 0.82)^2}{(4.5 + 3.26)} = 1.75 \text{ inches}$$

Therefore, the peak discharge is

$$q = q_u A Q F_p = 250 \times \frac{300}{640} \times 1.75 \times 1.0 = 205 \text{ cfs}$$

Answer

The peak discharge based on the graphical peak discharge method and the velocity method for calculating the time of concentration is 205 ft³/sec.

5.6.2 THE RATIONAL EQUATION

In the U.S., the rational equation is the most widely used empirical method:

$$q = 1.008 \, CiA \qquad (5.12)$$

where q is the peak flow (ft³/sec), C is an empirical coefficient, i is the average rainfall intensity (in./h) during the time of concentration, A is the drainage area (acres), and 1.008 is a unit conversion factor (1 in./h × 1 ft/12 in. × 1 h/3600 sec × 1 acre × 43,560 ft²/1 acre). The unit conversion factor of 1.008 is usually approximated as 1.0 because of the uncertainties associated with determining each of the other equation parameters. The time of concentration is the time it takes flow to move from the most remote

TABLE 5.5
Runoff Coefficients *C* for Use in the Rational Equation

Land Use	Hydrologic Soil Group and Slope Range											
	A			B			C			D		
	0–2%	2–6%	6%+	0–2%	2–6%	6%+	0–2%	2–6%	6%+	0–2%	2–6%	6%+
Industrial	0.67[a]	0.68	0.68	0.68	0.68	0.69	0.68	0.68	0.69	0.69	0.69	0.70
	0.85[b]	0.85	0.86	0.85	0.86	0.86	0.86	0.86	0.87	0.86	0.86	0.88
Commercial	0.71	0.71	0.72	0.71	0.72	0.72	0.72	0.72	0.72	0.72	0.72	0.72
	0.88	0.89	0.89	0.89	0.89	0.89	0.89	0.89	0.90	0.89	0.89	0.90
High-density[c] residential	0.47	0.49	0.50	0.48	0.50	0.52	0.49	0.51	0.54	0.51	0.53	0.56
	0.58	0.60	0.61	0.59	0.61	0.64	0.60	0.62	0.66	0.62	0.64	0.69
Medium-density[d] residential	0.25	0.28	0.31	0.27	0.30	0.35	0.30	0.33	0.38	0.33	0.36	0.42
	0.33	0.37	0.40	0.35	0.39	0.44	0.38	0.42	0.49	0.41	0.45	0.54
Low-density[e] residential	0.14	0.19	0.22	0.17	0.21	0.26	0.20	0.25	0.31	0.24	0.28	0.35
	0.22	0.26	0.29	0.24	0.28	0.34	0.28	0.32	0.40	0.31	0.35	0.46
Agricultural	0.08	0.13	0.16	0.11	0.15	0.21	0.14	0.19	0.26	0.18	0.23	0.31
	0.14	0.18	0.22	0.16	0.21	0.28	0.20	0.25	0.34	0.24	0.29	0.41
Open space[f] (grass/forest)	0.05	0.10	0.14	0.08	0.13	0.19	0.12	0.17	0.24	0.16	0.21	0.28
	0.11	0.16	0.20	0.14	0.19	0.26	0.18	0.23	0.32	0.22	0.27	0.39
Freeways and expressways	0.57	0.59	0.60	0.58	0.60	0.61	0.59	0.61	0.63	0.60	0.62	0.64
	0.70	0.71	0.72	0.71	0.72	0.74	0.72	0.73	0.76	0.73	0.75	0.78

[a] Lower runoff coefficients for use with storm recurrence intervals less than 25 years.

[b] Higher runoff coefficients for use with storm recurrence intervals of 25 years or more.

[c] High-density residential areas have more than 15 dwelling units per acre.

[d] Medium-density residential areas have 4 to 15 dwelling units per acre.

[e] Low-density residential areas have 1 to 4 dwelling units per acre.

[f] For pastures and forests, we recommend using the lower runoff coefficients listed for open spaces (our addition to original source).

Source: Erie and Niagara Counties Regional Planning Board, *Storm Drainage Design Manual*, Erie and Niagara Counties Regional Planning Board, Grand Island, NY, 1981.

point on a watershed to the outlet of the watershed. This longest flow path is called the *hydraulic length*.

The rational equation has been in use since 1851, is the simplest of the available methods, and has spawned numerous ways of computing the coefficient *C* and the time of concentration. The method has many limitations and is based on the following assumptions: (1) rainfall occurs uniformly over the drainage area; (2) peak rate of runoff can be reflected by the rainfall intensity, averaged over a time period equal to the time of concentration of the drainage area; (3) time of concentration is the time required for flow to reach the point in question from the hydraulically most remote point in the drainage area; and (4) frequency of runoff is the same as the frequency of the rainfall used in the equation. The method is best suited for use on watershed areas smaller than 1 mi^2.

Statistical fitting of data to Equation 5.12 results in dependency of the input parameters on each other. For example, if a particular approach is used to determine the time of concentration, then the fitted *C* values will be a function of the time of concentration. This can result in erroneous predictions and inconsistencies in tabulated *C* values. The empirical coefficient *C* can be determined from the information presented in **Table 5.5**. If the rational method is applied on a watershed with several different land uses, a single-area weighted *C* value should be calculated for use with Equation 5.13.

A commonly used time-of-concentration method that is used in conjunction with the rational method is the following equation, which was developed by Kirpich (1940):

$$t_c = 0.00778 \, L^{0.77} S^{-0.385} \qquad (5.13)$$

where *L* is the hydraulic length (maximum length) in feet, *S* is the mean slope along the hydraulic length expressed as a fraction, and t_c is the time of concentration in minutes. Solutions to Equation 5.13 are presented in **Table 5.6**.

TABLE 5.6
Small Watershed Time of Concentration t_c Determined with the Kirpich Method, Equation 5.13

Mean Slope S along the Hydraulic Length		Time of Concentration t_c (min)					
		Hydraulic Length L (ft)					
%	ft/ft	500	1000	2000	3000	4000	5000
0.05	0.0005	18	30	52	69	86	102
0.1	0.001	13	23	39	53	66	78
0.2	0.002	10	17	30	40	50	60
0.5	0.005	7	11	20	28	33	42
1	0.01	6	9	16	22	27	32
2	0.02	4	7	12	17	21	25
5	0.05	3	5	9	12	15	17

EXAMPLE 5.5

Use the rational method to solve the question in Example 5.3.

Solution

Determine the time of concentration using the Kirpich equation, Equation 5.13:

$$t_c = 0.00778 \, L^{0.77} S^{-0.385}$$

where L is the hydraulic length (maximum length) in feet, S is the mean slope along the hydraulic length expressed as a fraction, and t_c is the time of concentration in minutes.

$$t_c = 0.00778(4000)^{0.77}(0.002)^{-0.385} = 50 \text{ minutes}$$

Using **Figure 2.17** and **Figure 2.18**, the rainfall intensity is 3.0 in./h × 0.8 = 2.4 in./h. Using **Table 5.5** to determine C, then $C = 0.21$ for an agricultural land use, slopes of 5%, and a 1:25-year storm event. Therefore, from Equation 5.12:

$$q = CiA = 0.21 \times 2.4 \times 300 = 150 \text{ cfs}$$

Answer

The peak runoff rate based on use of the Kirpich equation and the rational method is 150 ft³/sec.

A difference in prediction of 139 ft³/sec vs. 150 ft³/sec is not meaningful given the various approximations used. Also, for Example 5.3, knowledge that the flow was collected and concentrated in a grassed waterway resulted in the highest discharge rate. If a grassed waterway had not been used, the discharge based on the velocity of flow across the agricultural field would have given similar answers to Example 5.2 and Example 5.4. Therefore, all three results are surprisingly close. However, the reader might wish to evaluate what happens if the bed slope is 0.4%.

In Example 5.2 to Example 5.4, the graphical peak discharge method was solved using different time-of-concentration methods, and a solution obtained with the rational method gave very different answers for the same set of hydrologic conditions. The main reasons for the differences are the methods used to compute time of concentration. Many studies have shown that the Kirpich equation estimates time-of-concentration values that are too rapid. For example, on 48 watersheds located in 16 states, McCuen et al. (1983) compared 14 different time-of-concentration methods with the NRCS upland velocity method (NRCS, 1986; see **Figure 5.18**). It was found that the Kirpich method (Equation 5.13), NRCS lag method (Equation 5.6), and the NRCS velocity method (**Figure 5.18**) had mean time-of-concentration values of 0.81, 1.81, and 3.09 h, respectively. Also, of all methods evaluated, the NRCS lag method had one of the lowest biases, while the Kirpich method had one of the largest biases.

5.6.3 USGS EMPIRICAL REGRESSION MODELS

USGS rural regression equations are available for all 50 states, Puerto Rico, and American Samoa. Urban regression equations have been developed for the nation and the following states: Alabama, Georgia, Montana, North Carolina, Ohio, Oregon, Tennessee, Texas, and Wisconsin. The rural and urban regression equations are based on observed stream discharges, are simple to use, and are accepted by the Federal Emergency Management Agency (FEMA) and many state and federal agencies. Due to the short record length of the stream gage data used to develop many of the equations, there are 15 to over 100% estimation errors in peak discharges estimated using the USGS equations. The period of record of stream gage data used to develop equations for urban watersheds is shorter than for rural watersheds. This leads to standard errors associated with urban regression equations that are larger than those for rural regression equations — typically in the range of 30 to over 100%. Most of the urban equations are only suitable for use on small watersheds (less than a few square miles). The application of the rural equations varies from watersheds that are smaller than a few square miles to more than a thousand square miles. Therefore, it is recommended that to check with the USGS for the most current equations available for the region in question and for the applicable size of watershed.

Most of the USGS regression equations have the following common form:

$$Q_{RI} = a(DA)^b (S)^c (P - X)^d (13 - BDF)^e$$
$$(I)^f (U)^g (STR + 1)^h (E)^i (Q_{rt}.)^j ...... \tag{5.14}$$

where Q_{RI} is the peak discharge for a particular recurrence interval (ft³/sec), DA is the drainage area, S is the channel bed slope, P is an index of precipitation (often the 2-year 24-h precipitation depth or the mean annual depth in inches), BDF is the basin development factor for urban areas, I is the percentage impervious area, U is an index (or indices) of land use (percentage forest, etc.), STR is the percentage of the contributing drainage area occupied by lakes, ponds, and wetlands, E is a function of the elevation (thousands of feet for central coastal region of California), where Q_{rt} is the peak discharge for a return period T years and an equivalent rural watershed, and X, a, b, c, d, and so on are empirical coefficients that are a function of the recurrence interval and region. All equations contain the drainage area DA and then just a few (sometimes none) of the other variables. Examples of equations for a few locations are presented later in this section.

The USGS urban and rural equations have been incorporated in the National Flood Frequency (NFF) computer program; the WMS, which was developed at Brigham Young University; and various commercially available software applications. The NFF program is a Windows program for estimating the magnitude and frequency of peak discharges for unregulated rural and urban watersheds. NFF includes about 2065 regression equations for 289 flood regions nationwide. It can be used to estimate peak discharges with recurrence intervals between 2 and 500 years and contains procedures to calibrate the regression equations based on measured stream data. At the time this book was prepared, the USGS hydrologic analysis

software was available at no charge through the World Wide Web at http://water.usgs.gov/software/. Examples of rural and urban equations for several locations in the U.S. are presented in **Table 5.7** and **Table 5.8**.

EXAMPLE 5.6

Calculate the 2-year recurrence interval discharge for a 3-mi² watershed in a rural area outside Los Angeles, CA. The mean elevation of the watershed is 330 ft, and the mean annual precipitation is 14.8 in.

Solution

Use Equation 5.14 with coefficients for the central coastal region of California from **Table 5.7**, which gives

$$Q_2 = 0.0061(DA)^{0.92} (P - 0)^{2.54} (E)^{-1.1}$$

and $DA = 3$ mi², $P = 14.8$ in., $E = 330$ ft or 0.33 thousand ft; therefore,

$$Q_2 = 0.0061(3)^{0.92} (14.8 - 0)^{2.54} (330/1000)^{-1.1} = 53.3 \text{ cfs}$$

Answer

The 2-year recurrence interval discharge is 53.3 ft³/sec.

Sauer et al. (1983) developed national empirical peak discharge relationships for urban areas based on an analysis of discharge for 269 gages in 56 cities and 31 states. A similar approach for urban areas in Ohio with drainage areas less than 6.5 mi² was developed by Sherwood (1993). The BDF varies from 0 to 12 and is a measure of the urban development within a watershed. The empirical coefficients for the nation, Ohio, and several other locations are summarized in **Table 5.8**. For Sauer's national method, the rural peak discharge Q_{rt} can be computed from regional flood frequency data or regional empirical methods for rural watersheds.

The BDF is based on the occurrence of the following features in each third of the watershed: (1) channel improvements, (2) channel lining, (3) storm drains, and (4) curb and gutter streets. The BDF is a measure of urban development within the watershed. The drainage area is divided into thirds (lower, middle, and upper) by drawing two lines across the watershed that are approximately perpendicular to the main channel and principal tributaries. Flood-peak travel times for streams in each third should be about equal.

More than 50% of a feature within each third results in an incremental BDF of 1, while less than 50% of a feature within each third gives an incremental BDF of 0.

TABLE 5.7
USGS Equations for Estimating Peak Discharges on Ungaged Rural Watershed

Regression Coefficient	Recurrence Interval (years)					
	2	5	10	25	50	100
Central Coast California						
a	0.0061	0.118	0.583	2.91	8.20	19.7
b	0.92	0.91	0.90	0.89	0.89	0.89
d	2.54	1.95	1.61	1.26	1.03	0.84
i	−1.10	−0.79	−0.64	−0.50	−0.41	−0.33
X	0	0	0	0	0	0
Central Minnesota						
a	7.15	14.1	19.8	28.2	35.1	42.5
b	0.796	0.796	0.794	0.792	0.791	0.790
c	0.449	0.475	0.488	0.503	0.513	0.522
h	−0.401	−0.411	−0.414	−0.416	−0.416	−0.416
Ohio						
a (Region A, central Ohio)	56.1	84.5	104	129	148	167
a (Region B, northwest)	40.2	58.4	69.3	82.2	91.2	99.7
a (Region C, southeast)	93.5	133	159	191	214	236
b	0.782	0.769	0.764	0.76	0.757	0.756
c	0.172	0.221	0.244	0.264	0.276	0.285
h	−0.297	−0.322	−0.335	−0.347	−0.355	−0.363
Eastern Tennessee						
a	411	556	648	757	833	905
b	0.523	0.550	0.563	0.577	0.586	0.595
Northwest Nevada						
a	13.1	22.4	4.86	4.11	3.31	2.62
b	0.713	0.723	0.727	0.737	0.746	0.752
i	0	0	−0.353	−0.438	−0.511	−0.584
New York (Northwest, Region 6)						
a	8.8	13.3	16.2	19.7	22.1	24.1
b	0.870	0.869	0.869	0.869	0.869	0.870
c	0.233	0.302	0.334	0.360	0.374	0.385
d	0.481	0.408	0.379	0.360	0.356	0.359
h	−0.217	−0.216	−0.217	−0.220	−0.224	−0.228
X	20	20	20	20	20	20
North Carolina Coastal Plain						
a	64.7	129	188	281	367	468
b	0.673	0.635	0.615	0.593	0.579	0.566
Northwest Washington State (Region 2)						
a	0.09	—	0.129	0.148	0.161	0.174
b	0.877	—	0.868	0.864	0.862	0.861
d	1.51	—	1.57	1.59	1.61	1.62
X	0	0	0	0	0	0
Eastern Washington State (Region 9)						
a	0.803	—	15.4	41.1	74.7	126
b	0.672	—	0.597	0.570	0.553	0.536
d	1.16	—	0.662	0.508	0.420	0.344

Note: Coefficient applies to Equation 5.14; unless stated, all other coefficients are zero.

TABLE 5.8
USGS Equations for Estimating Peak Discharges on Ungaged Urban Watershed in Ohio

Regression Coefficient	Recurrence Interval (years)					
	2	5	10	25	50	100
Georgia Region 2: Columbus/Atlanta/Augusta/Macon						
a	145	258	351	452	548	644
b	0.70	0.69	0.70	0.70	0.70	0.70
f	0.30	0.26	0.21	0.20	0.18	0.17
National and New York Urban						
a	13.2	10.6	9.51	8.68	8.04	7.70
b	0.21	0.17	0.16	0.15	0.15	0.15
e	−0.43	−0.39	−0.36	−0.34	−0.32	−0.32
j	0.73	0.78	0.79	0.80	0.81	0.82
North Carolina Coastal Plain						
a	26.9	68.2	109.0	209	280	363
b	0.722	0.655	0.625	0.570	0.558	0.547
f	0.686	0.572	0.515	0.436	0.396	0.358
Tampa Region						
a	3.72	7.94	12.9	214	245	282
b	1.07	1.03	1.04	1.13	1.14	1.16
c	0.77	0.81	0.83	0.73	0.74	0.76
e	1.04[a]	0.87[a]	0.75[a]	−0.59	−0.55	−0.51
h	−0.11	−0.10	−0.10	0	0	0
Ohio Urban						
a	155	200	228	265	293	321
b	0.68	0.71	0.74	0.76	0.78	0.79
d	0.50	0.63	0.68	0.72	0.74	0.76
e	−0.50	−0.44	−0.41	−0.37	−0.35	−0.33
X	30	30	30	30	30	30

Note: Coefficient applies to Equation 5.14; unless stated, all other coefficients are zero.

[a] $(BDF)^e$ not $(13 − BDF)^e$ for the Tampa Region 2-, 5-, and 10-year recurrence interval equations and (STR) not $(STR + 1)$.

Therefore, the minimum and maximum BDF for each third are 0 and 4, respectively. BDF scores for each third must be totaled to give a score for the whole watershed. It should be noted that a BDF of 0 for the whole watershed does not equal a pristine natural environment or even a typical rural watershed; it simply means that, in each third of the watershed, none of the four features has exceeded the 50% threshold. There can still be some channel improvements, channel lining, storm drains, and curbs and gutter. Also, a BDF of 12 does not imply that the watershed is completely developed. Suburban areas with large lots, parks, playing fields, and storm drains and ditches connected to streams often have a BDF in the 6 to 10 range.

EXAMPLE 5.7

Calculate the 2-year recurrence interval discharge for a 1.5-mi² watershed on the southeastern side of Atlanta, GA.

Of the watershed, 2% is impervious. Calculate the percentage increase in the discharge if the watershed is developed and the impervious area increases to 10%.

Solution

Use Equation 5.14 with coefficients for Atlanta from **Table 5.8**, which gives

$$Q_2 = 145(DA)^{0.7}(I)^{0.3}$$

with $DA = 1.5$ mi², and $I = 2\%$ for the predevelopment condition. Therefore,

$$Q_2 = 145(1.5)^{0.7}(2)^{0.3} = 237 \text{ cfs}$$

Following development, I increases to 10%; therefore,

$$Q_2 = 145(1.5)^{0.7}(10)^{0.3} = 384 \text{ cfs}$$

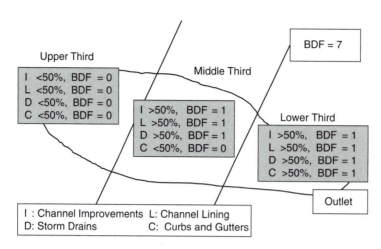

FIGURE 5.19 Illustration of basin development factor (BDF) calculation for Example 5.9.

The percentage increase is

$$\frac{100(384 - 237)}{237} = 62\%$$

Answer

The 2-year recurrence interval pre- and postdevelopment discharges are 237 and 384 ft³/sec, respectively, and the percentage increase is 62%.

EXAMPLE 5.8

Calculate the 2-year recurrence interval discharge for a 0.5-mi² watershed in Tampa, FL. The BDF is 8, 1% of the watershed is a lake, and the main channel slope is 0.6%.

Solution

Use Equation 5.14 with coefficients for Tampa from **Table 5.8**, which gives

$$Q_2 = 3.72(DA)^{1.07}(S)^{0.77}(BDF)^{1.04}(STR)^{-0.11}$$

where $DA = 0.5$ mi², $BDF = 8$, S is 0.6% of 5280 = 31.7 ft/mi, and $STR = 1\%$. Therefore,

$$Q_2 = 3.72(0.5)^{1.07}(31.7)^{0.77}(8)^{1.04}(1)^{-0.11} = 220.5 \text{ ft}^3/\text{sec}$$

Answer

The 2-year recurrence interval discharge is 220.5 ft³/sec.

EXAMPLE 5.9

Calculate the 100-year recurrence interval discharge for the pre- and postdevelopment conditions on a 0.9-mi² watershed just outside Cincinnati, OH. The mean annual precipitation is 40 in. Initially the *BDF* is 0. Following development, the upper third of the watershed will remain unchanged. In the middle third, more than 50% of the channels will be improved and lined, and most of the secondary tributaries will be storm drains. In the lower third, most of the channels will be improved and lined with concrete, the majority of the secondary tributaries will be storm drains, and most of the streets will have curbs and gutters.

Solution

Use Equation 5.14 with coefficients for Ohio from **Table 5.8**, which gives

$$Q_{100} = 321(DA)^{0.79}(P - 30)^{0.76}(13 - BDF)^{-0.33}$$

and $DA = 0.9$ mi², $P = 40$ in., and the $BDF = 0$ for the predevelopment condition. Therefore,

$$Q_{100} = 321(0.9)^{0.68}(40 - 30)^{0.5}(13 - 0)^{-0.5} = 82.9 \text{ cfs}$$

Calculate the postdevelopment *BDF* from **Figure 5.19**. The *BDF* for the upper third is 0; for the middle third, the *BDF* is 3; and for the lower third, it is 4. Therefore, the *BDF* for the whole watershed is 7, and

$$Q_{100} = 321(0.9)^{0.68}(40 - 30)^{0.5}(13 - 7)^{-0.5} = 385.8 \text{ cfs}$$

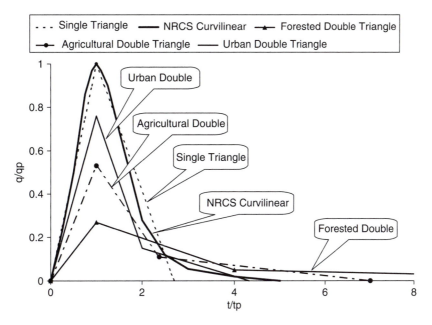

FIGURE 5.20 Curvilinear, single-, and double-triangle unit hydrographs.

Answer

The 100-year recurrence interval discharge is 82.9 ft³/sec before development and 385.8 ft³/sec following development.

5.7 STORMWATER HYDROGRAPHS

Stormwater hydrographs provide information on the change in runoff rates with time, the peak runoff rate, and the volume of runoff. Storm hydrograph methods should be used whenever possible as they provide the most comprehensive information and the best estimates of runoff rates. However, these methods are also more complex, require more information to use, and are best solved using a computerized procedure.

5.7.1 UNIT HYDROGRAPH METHODS

One approach to developing a storm runoff hydrograph is the unit hydrograph method (Sherman, 1932). The unit hydrograph method is about 60 years old and was so named because the area under the hydrograph is equal to 1 in. or 1 mm of runoff (Q). Hydrologists around the world have studied, developed, reported on, and applied unit hydrographs to practical problems for over 50 years. The approach is empirical and is based on the following assumptions: (1) uniform distribution of rainfall excess over the watershed, (2) uniform rainfall excess rate, and (3) the runoff rate is proportional to the runoff volume for a rainfall excess of a given duration.

The basic factors that need to be determined to develop a unit hydrograph are the time to peak, the peak flow, and the shape of the hydrograph. The time to peak is defined as

$$t_p = t_L + D/2 \qquad (5.15)$$

where t_L is the lag time, and D is the time increment of the rainfall excess. A plot of the rainfall excess vs. time for each time increment D is called a *hyetograph*.

The peak flow of the unit hydrograph can be estimated by the equation (NRCS, 1972)

$$q_p = \frac{484\,A}{t_p} \qquad (5.16)$$

where q_p is the peak flow in ft³/sec (per inch of runoff), A is the watershed area in square miles, and t_p is the time to peak in hours.

The shape of the unit hydrograph is usually modeled as curvilinear or triangular. The NRCS has developed single-triangle and curvilinear unit hydrograph procedures (**Figure 5.20**). For a practical exercise on NRCS methods, see Chapter 14, Exercise 14.5. Curvilinear procedures have also been developed by Haan (1970) and DeCoursey (1966). A single triangle can be developed by approximating the base time as 2.67 times the time to peak t_p; using Equation 5.16 and a lag time equation Equation 5.6); and using Equation 5.16 to determine the peak flow. Remember, the depth of rainfall excess represented by the unit hydrograph is 1 in. or 1 mm. A simple application of a single-triangle method is illustrated in Example 5.10.

In a steep, forested watershed, the peak initial runoff response might be due to runoff from rock outcrops, bare rock areas along the streams, precipitation directly on the streams, or rapid interflow. A delayed response is conceptualized as due to streamflow, overland flow delayed by bed litter (duff), or delayed interflow. To account for the delayed response due to interflow, several scientists and engineers have proposed using a double-triangle concept to represent the runoff hydrograph (Tennessee Valley Authority [TVA], 1973; Overton and Troxler, 1978; A.D. Ward et al., 1980; Wilson et al., 1983). Examples of double-triangle unit hydrographs for different land uses are presented in **Figure 5.20**.

EXAMPLE 5.10

Use a single-triangle unit hydrograph procedure to develop a storm hydrograph for a 1-h storm event on a proposed 500-acre commercial and business watershed with soils in Hydrologic Soil Group D. The watershed has an average overland slope of 1% and a hydraulic length of 6000 ft. The rainfall depth during the 1-h storm event was 2.5 in.

Solution

Use the NRCS curve number method (Equation 5.3 and Equation 5.4) to develop a rainfall excess (runoff) hyetograph.

$$S = \frac{1000}{CN} - 10$$

From **Table 5.1**, CN = 95 (Hydrologic Soil Group D, commercial land use). Assume Antecedent Moisture Condition II, then

$$S = \frac{1000}{95} - 10 = 0.53 \text{ inches}$$

and

$$Q = \frac{[2.5 - (0.2 \times 0.53)]^2}{[2.5 + (0.8 \times 0.53)]} = 1.96 \text{ inches}$$

Assume a triangular unit hydrograph and determine the lag time and time to peak using Equation 5.6 and Equation 5.9:

$$t_L = \frac{L^{0.8}(S+1)^{0.7}}{1900 \times Y^{0.5}}$$

L = 6000 ft, Y = 1%, and S = 0.53 in.:

$$t_L = \frac{6000^{0.8}(0.53+1)^{0.7}}{1900 \times 1^{0.5}}$$

therefore,

$$t_L = 0.75 \text{ hrs} \quad (45 \text{ minutes})$$

and

$$t_p = t_L + \frac{D}{2} = 45 + \frac{60}{2} = 75 \text{ minutes}$$

The peak flow of the unit hydrograph is determined from Equation 5.16:

$$q_p = \frac{484 \, A}{t_p}$$

and

$$q_p = 484 \times \frac{500 \text{ acres}}{640} \times \frac{60}{75 \text{ minutes}}$$

$$= 302.5 \text{ cfs (per inch of runoff)}$$

Note that the 500 acres are divided by 640 to convert them to square miles, and the 75 min are divided by 60 to convert them to hours.

The peak flow is simply 302.5 ft³/sec times 1.96 in. of runoff, which gives 592.9 ft³/sec. The base time of the unit hydrograph is 2.67 t_p = 2.67 × 75 = 200 min.

Answer

The hydrograph contains a runoff depth of 1.96 in., has a peak flow of 592.9 ft³/sec at a time of 75 min, and has a total runoff time of 200 min. The hydrograph is illustrated in **Figure 5.21**.

In Example 5.10, the storm duration was longer than the lag time (60 min vs. 45 min). This causes an error in the solution as the shape of the hydrograph is primarily a function of the storm duration rather than the watershed characteristics. This problem can be prevented if knowledge is available of the rainfall time distribution. The design storm rainfall depth is generally associated with a synthetic rainfall time distribution (refer to **Figure 5.5**).

Rainfall excess can be determined using an infiltration equation or a procedure such as the NRCS curve number method (Equation 5.4). The rainfall event should be divided into blocks of time that have a duration D that is not longer than one third the time to peak of the unit hydrograph. Incremental storm hydrographs are then developed for each block of rainfall excess (**Figure 5.22**). The incremental runoff hydrograph ordinates are equal to the volume of runoff for each block of rainfall excess times

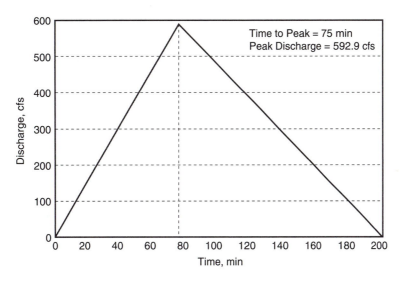

FIGURE 5.21 Storm hydrograph developed as a solution to Example 5.10.

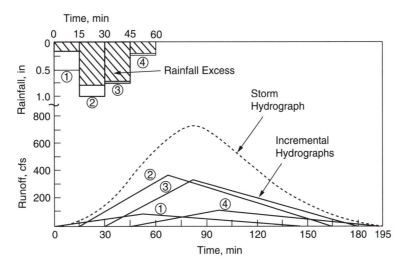

FIGURE 5.22 Development of a storm hydrograph based on knowledge of the rainfall distribution.

the unit hydrograph ordinates. The incremental storm hydrographs each start at the beginning of the block of rainfall excess with which they are associated.

When all the incremental storm hydrographs have been established, the storm runoff hydrograph is obtained by adding the ordinates of the incremental hydrographs at each point in time. This is a time-consuming activity subject to mathematical errors and is best performed by a computer program. The approach presented in Example 5.10 is commonly used and consists of approximating the storm hydrograph using a single time block for the whole event; determining a time to peak and peak runoff rate for the unit hydrograph based on the single time block; and simply multiplying the ordinates of the unit hydrograph by the rainfall excess for the single block. This approach

will underestimate the peak flow rate and will result in a base time for the storm hydrograph that is longer than that obtained if the hydrograph is developed from several incremental time blocks. The best approach is to subdivide the storm event into as many time blocks as practical rather than using as few as possible.

EXAMPLE 5.11

For the same watershed and storm evaluated in Example 5.10, use a unit hydrograph procedure to develop a storm hydrograph. In this case, it is known that the rainfall during each 15 min of the event was 0.5, 1.0, 0.75, and 0.25 in.

Solution

From Example 5.10, a 500-acre watershed has soils in Hydrologic Soil Group D, an average overland slope of 1%, a hydraulic length of 6000 feet, and proposed business and commercial land use. The lag time was calculated as 45 min; therefore. substituting in Equation 5.15,

$$t_p = t_L + \frac{D}{2} = 45 + \frac{15}{2} = 52.5 \text{ minutes}$$

The 15-min time blocks provided can be used as they are less than one third of the time to peak. The peak flow of the unit hydrograph is determined from Equation 5.16:

$$q_p = \frac{484\,A}{t_p}$$

and

$$q_p = 484 \times \frac{500 \text{ acre}}{640} \times \frac{60}{52.5 \text{ minutes}}$$

$$= 432 \text{ cfs (per inch of runoff)}$$

Use the NRCS curve number method (Equation 5.2 and Equation 5.3) to develop a rainfall excess (runoff) hyetograph.

$$Q = \frac{(P - 0.2\,S)^2}{(P + 0.8\,S)}$$

To solve for the first time block,

$$Q = \frac{(0.5 - 0.2 \times 0.53)^2}{(0.5 + 0.8 \times 0.53)} = 0.17 \text{ inches}$$

To solve for the second time block,

$$Q_1 + Q_2 = \frac{(1.5 - 0.2 \times 0.53)^2}{(1.5 + 0.8 \times 0.53)} = 1.01 \text{ inches}$$

and

$$Q_2 = 1.01 - Q_1 = 1.01 - 0.17 = 0.84 \text{ inches} \quad .$$

Note that it is necessary to determine the rainfall excess for the total time from the start of rainfall to the time at the end of the block considered. The accumulated rainfall excess up to the start of the block considered is then subtracted from the total rainfall excess for the time period up to the end of this block. This approach is used so that the curve number does not need to be adjusted for changing antecedent conditions during the event.

To solve for the third time block,

$$Q_1 + Q_2 + Q_3 = \frac{(2.25 - 0.2 \times 0.53)^2}{(2.25 + 0.8 \times 0.53)} = 1.72 \text{ inches}$$

and

$$Q_3 = 1.72 - Q_1 - Q_2 = 1.72 - 0.17 - 0.84 = 0.71 \text{ inches}$$

To solve for the fourth time block,

$$Q_1 + Q_2 + Q_3 + Q_4 = \frac{(2.5 - 0.2 \times 0.53)^2}{(2.5 + 0.8 \times 0.53)} = 1.96 \text{ inches}$$

and

$$Q_4 = 1.96 - Q_1 - Q_2 - Q_3 = 1.96 - 0.17 - 0.84 - 0.71$$

$$= 0.24 \text{ inches}$$

Answer

The solution is presented in **Table 5.9** and **Figure 5.22**. The peak runoff rate is 729.7 ft³/sec and occurs 82.5 min after the start of rainfall. Note that this answer is 136.8 ft³/sec higher than the solution obtained in Example 5.10.

5.7.2 TIME–AREA METHOD

The time–area method is a process-based procedure to determine the volume of runoff and peak runoff rate from small watersheds (generally less than 500 acres). The approach does not consider interflow and assumes that surface and channel flow velocities do not change with time. If used on large watersheds, peak flows may be overestimated because the simple routing procedures do not consider attenuation. The most commonly used computer model based on time–area methods is the Illinois Urban Drainage Simulator (ILLUDAS; Terstriep and Stall, 1974).

The approach is illustrated in **Figure 5.23**. The time–area diagram is obtained by drawing on the watershed isochromes of equal flow time *t* to the outlet. The value of time on each isochrome represents the travel time of water from the isochrome to the watershed outlet. Storage effects are ignored, and the runoff hydrograph is due to a translation of rainfall excess to the outlet. Like the unit hydrograph procedure, incremental runoff hydrographs can then be developed for each increment of rainfall excess. Also, a procedure such as the NRCS curve number procedure can be used to develop a rainfall excess hyetograph. The storm hydrograph is then obtained by combining all the incremental hydrographs.

TABLE 5.9
Development of a Storm Hydrograph for Example 5.11
(discharge, ft³/sec)

Time (min)	Incremental Hydrograph 1	Incremental Hydrograph 2	Incremental Hydrograph 3	Incremental Hydrograph 4	Stormwater Hydrograph [a]
0	0.0	0.0	0.0	0.0	0.0
15	21.0	0.0	0.0	0.0	21.0
30	42.0	103.7	0.0	0.0	145.7
45	62.9	207.4	87.6	0.0	357.9
52.5	73.4	259.3	131.5	14.8	479.0
60	67.2	311.1	175.3	29.6	583.2
67.5	60.9	362.9	219.1	44.5	687.4
75	54.6	331.8	262.8	59.3	708.5
82.5	48.3	300.7	306.7	74.0	729.7
90	42.0	269.6	280.5	88.8	680.9
97.5	35.8	238.5	254.1	103.7	632.1
105	29.6	207.4	227.8	94.8	559.6
120	16.8	145.2	175.3	77.0	414.3
135	4.2	82.9	122.7	59.3	269.1
150	0.0	20.7	70.1	41.3	132.1
165	0.0	0.0	17.5	23.7	41.2
180	0.0	0.0	0.0	5.9	5.9

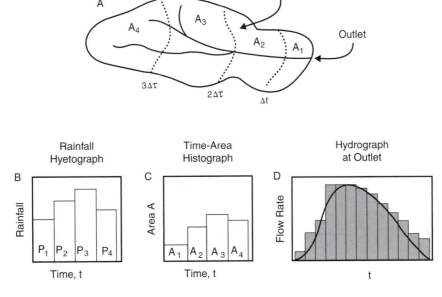

FIGURE 5.23 Illustration of the time–area method.

5.7.3 THE KINEMATIC APPROACH

The physically based kinematic method is theoretically the most complete. Lighthill and Whitham (1955) imagined flow propagation in the form of kinematic waves for describing flood movement in rivers. They also developed kinematic equations for determining overland flow.

However, the first application of kinematic equations to develop runoff hydrographs for steady rainfall excess conditions was by Henderson and Wooding (1964).

Complete details of the method may be found in hydrology texts such as those by Haan et al. (1982) and Bedient and Huber (1988). The method is based on a solution of the continuity and momentum equations, as well as relationships between the geometric characteristics of the watershed and drainage systems (such as slope, roughness, channel shape, etc.). The rainfall excess hyetograph is routed as overland flow to established channels and as streamflow to the watershed outlet.

Unfortunately, the method is data intensive and difficult to use. The approach has mainly been applied in urban areas where the watersheds are small and highly channelized. For most applications, computer models are needed to develop a solution, and it is necessary to make several assumptions and approximations (particularly in the geometry). The most commonly used computer model that incorporates kinematic equations is the Stormwater Management Model (SWMM; Huber et al. 1981).

5.8 ASSESSMENT OF FLOOD ESTIMATION TECHNIQUES

Hydrologists and engineers face two major problems on small catchment hydrology: (1) the availability of small catchment hydrological data and (2) guidelines on the selection and expected accuracy of the different flood estimation techniques.

In general, little information is available on the suitability of the various flood estimation techniques as design tools. Hydrological control structures are frequently designed based on limited data and pragmatic modeling approaches. The success of the design depends largely on experience and judgment. In 1983, a study was conducted in South Africa to evaluate the suitability of commonly used small catchment surface runoff techniques as design methods (Ward et al. 1989). It is probable that the conclusions drawn from this study would be applicable to most countries in the world, including the U.S. The questionnaire survey indicated that a wide variety of flood estimation methods were used in South Africa. The results indicated that the rational method was the most commonly used, but use of such methods as the time–area method, NRCS methods, and kinematic procedures was significant. In general, hydrologists found all the techniques fairly easy to use, but only 50% of the respondents felt the techniques were sufficiently accurate. In addition, flood estimation techniques were primarily used to determine peak flows and volumes of runoff. Storm hydrograph generation was required less frequently, but there was a significant need for this type of data. Despite that it is a peak flow technique, the rational method was used to generate hydrographs by 12% of the respondents. Flood estimation techniques were primarily applied in urban, agricultural, rural, and afforested areas smaller than 10 km². Respondents perceived that the main problem associated with the use of flood estimation techniques is the availability of adequate data. The survey also identified a lack of familiarity with the different techniques and no perception of accuracy.

As part of the same study, an evaluation of seven hydrological methods was made for 102 storm events on 26 diverse catchments. Catchment parameters were obtained from maps and design manuals. The analysis was conducted with both the measured rainfall distribution and with synthetic distributions. This was done to separate the effects of method limitations and parameter inaccuracies from the effects of simplified rainfall. The analysis of the results was based on two measures of method performance: peak flow and volume of runoff.

The results indicated that, while urban catchments could be adequately modeled by most of the methods, flood predictions for rural catchments were far less reliable. In general, the ability of the methods to predict volume of runoff was found to be better than their ability to determine peak flow rates. The rational method was found to be very conservative and gave excessively high peak flow estimates for rural catchments. The NRCS methods gave reasonable peak flow and runoff volume estimates for most of the watersheds and performed slightly better on rural catchments. The time–area and kinematic models gave reasonable estimates for urban watersheds, but performed poorly on rural catchments.

5.9 AGRICULTURAL LAND DRAINAGE MODIFICATIONS

In the U.S., drainage improvements are required on more than 26 % of our cropland (110 of 421 million acres). Egyptian and Greek use of surface ditches to drain land for agriculture dates to about 400 BC (USDA, 1987). In the U.S., extensive drainage of swamps was initiated in the mid-1700s. The Swamp Land Acts of 1849 and 1850 (USDA, 1987) were the first important federal legislation on land drainage. Early subsurface drainage materials were wood, concrete, clay, and bituminized fiber pipe. In 1967, corrugated plastic tubing was commercially produced in the U.S., and by 1973 had largely replaced concrete and clay tile for small-size drains (Schwab et al., 1993). Today, subsurface drains are still commonly called tile drains, even though nearly all the drains installed in the last two decades have been made from perforated corrugated plastic tubing (**Figure 5.24**). In the U.S., much of the tubing used in agriculture has a diameter of 4 in. and is made from high-density polyethylene (HDPE). In Europe, field drains often have diameters of only 2 or 3 in. and are made from polyvinylchloride (PVC).

Agricultural drainage is the removal of excess water from the soil surface or the soil profile of cropland by either gravity or artificial means. The two main reasons for improving the drainage on agricultural land are for soil conservation and for enhancing crop production. Research

FIGURE 5.24 High-speed, and laser-controlled, installation of corrugated plastic drainage pipe.

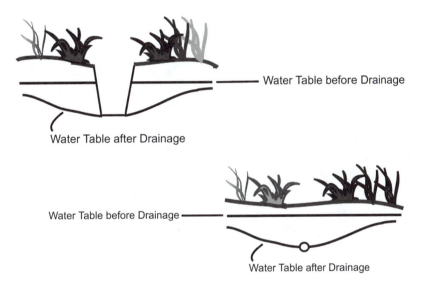

FIGURE 5.25 Types of agricultural drainage improvements.

conducted throughout the midwestern U.S. has documented many benefits of agricultural drainage improvements. Typically, most agricultural producers improve the drainage on their land to help create a healthier environment for plant growth and to provide drier field conditions, allowing farm equipment to access the farm field throughout the crop production season. Healthy, productive plants have the potential to produce greater yields and more food. Also, research in Ohio has shown that agricultural drainage improvements can help reduce the year-to-year variability in crop yields. This helps reduce the risks associated with the production of abundant, high-quality, affordable food. Improved access of farm equipment to the field provides more time for field activities, can help

extend the crop production season, and helps reduce crop damage at harvest.

The two primary types of agricultural drainage improvements are surface and subsurface systems (**Figure 5.25**), used individually or in combination.

5.9.1 SURFACE IMPROVEMENTS

Surface drainage improvements are designed to minimize crop damage resulting from water ponding on the soil surface (**Figure 5.26A**) following a rainfall event and to control surface water runoff without causing erosion. Surface drainage can affect the water table by reducing the volume of water entering the soil profile. This type of improvement includes land leveling and smoothing; the

A

B

FIGURE 5.26 A. Crop damage due to water ponding. B. A grass waterway.

construction of surface water inlets to subsurface drainage improvements; and the construction of shallow ditches and grass waterways (**Figure 5.26B**), which empty into open ditches, streams, and rivers.

Land smoothing or leveling is a water management practice designed to remove soil from high spots in a field or fill low spots and depressions where water may pond. Shallow ditches may be constructed to divert excess water to grass waterways and open ditches, which often empty into existing surface water bodies.

We should also be aware of the disadvantages of surface drainage improvements. First, these improvements require annual maintenance and must be carefully designed to ensure that erosion is decreased. Second, extensive earth-moving activities are expensive, and land grading might

expose less fertile and less productive subsoils. Further, open ditches may interfere with moving farming equipment across a field.

5.9.2 SUBSURFACE IMPROVEMENTS

The objective of subsurface drainage is to drain excess water from the plant root zone of the soil profile by artificially lowering the water table level. Subsurface drainage improvements are designed to control the water table level through a series of drainage pipes installed below the soil surface, usually just below the root zone. In the Midwest region of the U.S., the subsurface drains typically are installed at a depth of 30 to 40 in. and at a spacing of 20 to 80 ft. The subsurface drainage network generally outlets

FIGURE 5.27 The cornfield in the foreground does not have subsurface drainage while the corn field, with large more robust plants, in the background does have subsurface drainage.

to an open ditch, stream, or river. Subsurface drainage improvements require some minor maintenance of the outlets and outlet ditches. For the same amount of treated acreage, subsurface drainage improvements are generally more expensive to construct than surface drainage improvements.

Whether the drainage improvement is surface, subsurface, or a combination of both, the main objective is to remove excess water quickly and to reduce the potential for crop damage safely (**Figure 5.26**). When water is ponded on the soil surface immediately following a rainfall event, a general rule of thumb for most crops is to lower the water table 10 to 12 in. below the soil surface within a 24-h period and 12 to 18 in. below the soil surface within a 48-h period. Properly draining excess water from the soil profile where plant roots grow helps aerate the soil and reduces the potential for damage to the crop roots. Further, it will produce soil conditions more favorable for crop production (**Figure 5.27**). In states that depend heavily on irrigation, subsurface drainage is often used to prevent harmful buildups of salt in the soil.

5.9.3 PERCEPTIONS

In recent years, public concern has increased about the nature of agricultural drainage and the impact of agricultural drainage improvements on the quality of water resources and environment. Few individuals not directly connected with agriculture realize that improved drainage is necessary for sustained agricultural production. In fact, controlled subsurface drainage, an agricultural drainage practice, is considered a sound water quality management and enhancement practice in North Carolina.

Land drainage activities have impacted much of the Midwest region's environment and water resources. For example, early settlers began draining Ohio's swamps in the 1850s; today, approximately 90% of Ohio's wetlands have been converted to other uses. This loss of wetlands is attributed to public health considerations; rural, urban, and industrial development; and agriculture. While the loss of these wetlands has provided many benefits, wetlands also provide many benefits for the environment, including wildlife habitat and enhanced water quality. An important water quality function of wetlands is the trapping and filtering of sediment, nutrients, and other pollutants from agricultural, construction, and other rural and urban sources. Interestingly, subsurface drainage improvements, in a more limited capacity, provide some of these same water quality benefits while providing a necessary element for sustained agricultural production on many soils.

Present agricultural trends are toward more intensive use of existing cropland, with much of the emphasis on management. Maintaining existing agricultural drainage improvements and improving the drainage on wet agricultural soils presently in agricultural production help minimize the need for landowners to convert additional land to agricultural production. In many cases, restoration of previously converted wetlands would be impossible because of large-scale channel improvements, urbanization, and shoreline modification in the Great Lakes region. Focus should be placed on protecting existing wetlands and establishing new wetland areas while maintaining our highly productive agricultural areas.

The use of surface and subsurface drainage improvements is not limited to agricultural lands. Many residential homes use subsurface drainage systems, similar to those used in agriculture, to prevent water damage to foundations

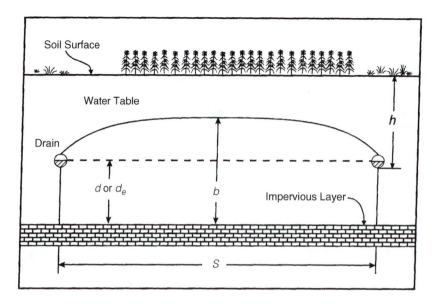

FIGURE 5.28 Subsurface drainage geometry for Equation 5.11 and Equation 5.12.

and basements. Golf courses make extensive use of both surface and subsurface drains. Houses, streets, and buildings in urban areas depend heavily on surface and subsurface drainage systems for protection. These are generally a combination of plastic or metal gutters and concrete pipes or channels.

5.10 DETERMINING SUBSURFACE DRAINAGE FLOWS

Details on the design and installation of subsurface drainage systems are presented in several texts and reports (USDA, 1987; Schwab et al. 1993; NRCS, 1984). A common problem in irrigated agriculture is a buildup of salts in the root zone. A practice to alleviate this problem is to apply additional irrigation water, which will "flush" some of the salt from the root zone. Subsurface drains are then used to intercept leachate from the bottom of the root zone. In nonirrigated agriculture, subsurface drains are used to ensure that elevated water tables are lowered below the root zone within 1 to 4 days. Typically, the drains are located at least 3 ft below the ground surface. The depth and spacing of the drains will depend on the soil characteristics, the crop, agricultural management practices, topography and surface drainage systems, and the depth to an impeding layer. For steady-state conditions in which the water table and rainfall or irrigation rate do not change with time, the ellipse equation can be used to determine drain spacing and flow rate. The geometry of the system is presented in **Figure 5.28**, and the ellipse equation is

$$q = \frac{4K(b^2 - d^2)}{S} \qquad (5.17)$$

where K is the saturated hydraulic conductivity (feet/day), q is the flow rate (square feet/day) into the drain per foot length of drain, S is the drain spacing in feet, b is the height (feet) at the midpoint between the drains of the water table above an impeding layer, and d is the height (feet) of the drains above the impeding layer.

If the flow rate q is set equal to Si, where i is the drainage rate in feet/day, then Equation 5.17 can be written as

$$S = \left[\frac{4K(b^2 - d^2)}{i}\right]^{1/2} \qquad (5.18)$$

An assumption in the development of Equation 5.17 and Equation 5.18 was that resistance to flow resulting from flowlines converging near the drains could be ignored. However, reasonable results are only obtained if the drain spacing is large compared to the depth to the impeding layer. Van Schilfgaarde (1963) and Moody (1966) developed modifications to the ellipse equations for conditions with the drain spacing S small relative to the depth of the impeding layer. We would suggest using an effective depth d_e that is 0.2 to 0.5 of the height of the drains above the impeding layer in place of the depth, d, in Equation 5.17 or Equation 5.18 in the development of provisional assessments, or if detailed knowledge of the subsurface soil profile is not available. If an effective depth is used, then the height, b, of the water table at the midpoint between the drain must be reduced by the difference between d and d_e. Information on recommended drainage rates for different crops and soils was provided by Schwab et al. (1993). Depending on the soil and crop, these rates will usually range between about 0.4 and 2.0 in. per day.

EXAMPLE 5.12

Determine the drain spacing if drains are located 4 ft below the ground surface, the depth of the impeding layer is 12 ft below the drains (16 ft below the ground surface), and the saturated hydraulic conductivity is 0.75 ft/day. For good plant growth, it is required that the water table be at least 2 ft below the ground surface (14 ft above the impeding layer), and it will be necessary to remove irrigation water at a rate of 0.60 in./day.

Solution

The drainage rate needs to be changed to feet/day; therefore,

$$i = \frac{0.6 \text{ inches/day}}{12} = 0.05 \text{ ft/day}$$

then, using Equation 5.18,

$$S = \left[\frac{4K(b^2 - d^2)}{i} \right]^{1/2}$$

and substituting for K, b, d, and i,

$$S = \left[\frac{4 \times 0.75 \, (14^2 - 12^2)}{0.05} \right]^{0.5}$$

and $S = 55.9$ ft.

The diameter of the drains and the flow depth are unknown and were not considered when determining b and d.

Answer

The maximum drainage spacing should be 55.9 ft. If an effective depth d_e of 0.25d had been used instead of d, the height of the water table at the midpoint b would reduce by 0.75d, and the drainage spacing would be 31.0 ft.

5.11 FLOW DURATION AND WATER YIELD

One of the most important characteristics of a stream is flow duration. Does the stream have equitable flow or does it flow at high and low extremes? And, how much of the time (proportion of time) does it flow at various discharges? If the stream is to furnish power, provide water, provide deep water for transportation, transport sewage effluent away, dilute pollution, or just remain "pretty," how much of the time does this happen? Two flow duration curves are shown for the Savannah River at Augusta, GA (**Figure 5.29**). The first flow duration curve is for the

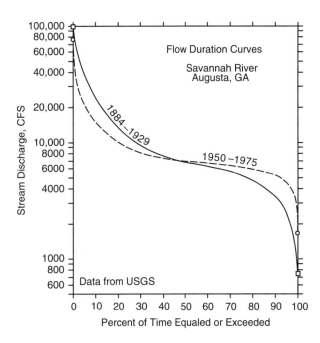

FIGURE 5.29 Savannah River, Georgia. (From USGS.)

period 1884 to 1929. This was the period when the river was "wild" and subject to great variations. Note first that the maximum discharge was 100,000 ft³/sec, while the minimum flow was about 500 ft³/sec. Between those extremes, there is little "median flow." Several large dams were then built upstream, regulating the flow in part. Thus, the second line for 1950 to 1975 is much "flatter," indicating a more equitable flow. That is, flow was within narrower bounds for a greater proportion of the time. For example, it can be seen that the flow remains between about 7,000 and 10,000 ft³/sec about 70% of the time (90% less 20%). In the earlier period, that range of flow occupied only about 40% of the time (70% less 30%). Note also that the extremes have been ameliorated with a minimum of about 2,000 ft³/sec (as opposed to about 500 ft³/sec before) and a maximum of only 80,000 ft³/sec (as opposed to the prior 100,000 ft³/sec).

More detailed information on minimum water yield (streamflow) for different return periods and catchments is useful and important in water management programs to tell the planner that, in very dry periods, it must be planned to manage with only so much flow or to develop storage systems to supply extra water. Those data will tell how often to expect these low flows (see the Ohio example in **Table 5.10**).

The mass curve (accumulation) of monthly runoff given in **Figure 5.30** shows low flow for the period June 1953 through February 1954 and again for June 1954 through February 1955. This type of information is necessary for determining the amount of reservoir storage needed to provide water for a uniform daily demand of a municipality or industry.

TABLE 5.10
Minimum 12-Month Runoff for Catchments of 29 to 17,400 Acres
for Return Period of 2 to 50 years, Coshocton, OH

Drainage Area (acres)	Minimum 12-Month Water Yield (in.) for Return Period of 2 to 50 Years				
	2 years	5 years	10 years	25 years	50 years
29	8.7	6.4	5.2	4.2	3.4
76	9.6	7.4	6.2	5.1	4.3
122	11.2	8.8	7.2	5.9	4.9
349	13.0	10.1	8.7	7.0	6.0
2,570	14.0	10.8	9.2	7.3	6.3
4,580	14.5	11.5	9.9	8.0	6.8
17,400	15.5	12.0	10.5	8.5	7.2

Source: From Harrold, L.L., *Trans. Am. Geophys. Union*, 38:201–208, 1957. With permission.

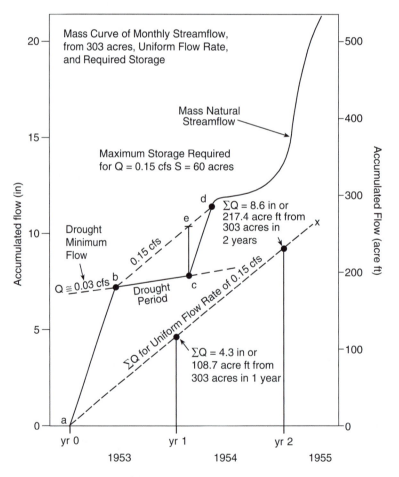

FIGURE 5.30 Mass curve of monthly streamflow at Coshocton, OH. (From Harrold, L.L., *Trans. Am. Geophys. Union*, 38:201–208, 1957. With permission.)

EXAMPLE 5.13

Determine the amount of reservoir storage needed to provide an urban development with an assured supply for a daily water use rate of 97,200 gal/day given the mass curve of monthly streamflow for the continuous 28-month period January 1953 through April 1955 (**Figure 5.30**).

Solution

Calculate the uniform streamflow (cubic feet/second) needed to meet the uniform water use demand rate of 97,200 gal/day (7.48 gal = 1 ft³).

$$\frac{97{,}200 \text{ gal}}{\text{day}} \times \frac{1 \text{ day}}{24 \text{ hr}} \times \frac{1 \text{ hr}}{60 \text{ min}} \times \frac{1 \text{ min}}{60 \text{ sec}}$$

$$\times \frac{1 \text{ ft}^3}{7.48 \text{ gal}} = 0.15 \text{ ft}^3/\text{sec}$$

Total volume of runoff for a year at a uniform streamflow of 0.15 ft³/sec for this catchment of 303 acres is

$$\left(\frac{0.15 \text{ ft}^3}{\text{sec}}\right)\left(\frac{60 \text{ sec}}{1 \text{ min}}\right)\left(\frac{60 \text{ min}}{1 \text{ hr}}\right)\left(\frac{24 \text{ hr}}{1 \text{ day}}\right)\left(\frac{365 \text{ day}}{1 \text{ yr}}\right)$$

$$\left(\frac{1}{303 \text{ acre}}\right)\left(\frac{1 \text{ acre}}{43{,}560 \text{ ft}^2}\right)\left(\frac{12 \text{ in}}{1 \text{ ft}}\right) = 4.3 \text{ in}$$

Start the uniform flowline at the zero point, Point *a* (**Figure 5.30**), and at the end of the year the accumulation is 4.3 in. The slope of this straight line is 0.15 ft³/sec or 97,200 gal/day. Prior to Point *b* on the mass curve (**Figure 5.30**), streamflow accumulates faster than the required uniform use rate of 97,200 gal/day, and no help from storage is needed. From Points *b* to *c*, the flow rate is less than the use rate, and water from storage is needed to augment the natural streamflow rate. The dotted vertical line from Point *c* to the dashed line *b–d* indicates that 60 acre-ft of water from storage is required. After Point *c*, the mass curve of streamflow shows the flow rate exceeds the use rate. If the reservoir had been full at Point *b*, its content would have decreased to a minimum at Time *c* and then would be full again by Time *d*.

Answer

The amount of reservoir storage needed is 60 acre-ft.

PROBLEMS

5.1. Convert 1 in./h of runoff from an area of 10 acres to cubic feet per second (ft³/sec) and to gallons per minute (gal/min) (7.48 gal = 1 ft³).

5.2. Determine the volume of water (acre-feet) that could be added to a reservoir in 1 day from a stream with an average flow rate of 10 ft³/sec.

5.3. Determine the time (hours) required to fill a 20 acre-ft reservoir if the average streamflow was 2 ft³/sec.

5.4. Determine the runoff volume in inches from a 50-year return period storm at your present location assuming Antecedent Moisture Condition III, NRCS Curve Number 75 (Condition II) and storm duration of 4 h.

5.5. Determine the volume of runoff in acre feet for a flood from a 500-acre watershed at your present location during the growing season. Critical duration of storm causing the flood was 3 h, and the return period of the event was 10 years. The antecedent rainfall 5 days prior to the storm was estimated as 1.5 in. Soil groups, land use, and hydrologic conditions for each subarea are presented in **Table 5.11**.

5.6. Determine the runoff depth and runoff volume for Example 5.2 by calculating an area weighted curve number and then using it with Equation 5.2 and Equation 5.3. Determine the percentage difference between your answers and the answers presented in Example 5.2.

5.7. Compute the peak runoff rate (cubic feet/second) that would occur once in 10 years for a 50-acre catchment using the graphical peak discharge method. It is situated near your location and has an average overland slope of 0.5%, a maximum

TABLE 5.11
Problem 5.5: Soil Groups, Land Use, and Hydrologic Conditions for Each Subarea

Subarea Acres	Soil Group	Land Use Treatment and Hydrologic Condition
100	C	Residential 1-acre lots, good condition
300	D	Cultivated conservation tillage
100	B	Forest, good cover

flow length of 2000 ft, an average slope of 0.2% along the hydraulic length, cultivated fields with a 30% residue cover, and soils in Hydrologic Soil Group C.

5.8. Compute the 25-year peak runoff rate from a 30-acre catchment in your area using the graphical peak discharge method. The watershed is a low-density residential area with Hydrologic Soil Group D, an average slope of 1.0%, and a flow length of 1140 feet. Compare answers obtained with AMC II and AMC III.

5.9. Use the Kirpich equation to determine the time of concentration for a watershed with a hydraulic length of 1000 ft and mean slope of 0.05%.

5.10. If the watershed in Problem 5.9 is located near Kansas City, MO, determine the rainfall rate associated with a storm with a 100-year return period and duration equal to the time of concentration (use **Figure 2.17** and **Figure 2.18** in Chapter 2).

5.11. Solve Problem 5.8 using the rational method.

5.12. Compute the peak runoff rate from a 100-acre catchment using the rational method. The estimated 100-year rainfall rate for the catchment is 7.0 in./hour. Corn is grown on 20 acres of Soil Group B and remainder of catchment is in permanent pasture, Soil Group C. (Use a weighted average for the runoff coefficient.)

5.13. Calculate the 2-year recurrence interval discharge for a 1.2-mi^2 watershed in central Ohio. The mean annual precipitation is 37 in., and the basin development factor is 6.

5.14. Use USGS rural regression equations to compare the 2-year recurrence interval discharges on 0.5 mi^2 watersheds located in eastern (Region 9) and northwestern (Region 2) parts of Washington State.

5.15. Use a single-triangle unit hydrograph procedure to develop a storm hydrograph for a 1-h, 50-year storm event on a proposed 100-acre commercial and business watershed with soils in Hydrologic Soil Group C and proposed residential land use (½-acre lots). The watershed has average overland slopes of 1% and a hydraulic length of 6000 feet. The rainfall depth during the 1-h storm event was 2.0 in.

5.16. Write a computer program or spreadsheet solution to develop a storm hydrograph using unit hydrograph procedures. Use your procedure to solve Problem 5.13 if there were 0.25, 1.00, 0.50, and 0.25 in. of rainfall during each 15 min of the rainfall event.

5.17. Determine the drain spacing if drains are located 4 ft below the ground surface, the depth of the impeding layer is 15 ft below the ground surface, and the saturated hydraulic conductivity is 0.25 ft^3/day. For good plant growth, it is required that the water table be at least 3 ft below the ground surface, and it will be necessary to remove excess rainfall at a rate of 0.72 in./day.

5.18. Compute the depth of water in inches from a 303-acre catchment to provide a park with a water usage of 0.1 ft^3/sec. From the mass curve in **Figure 5.30**, estimate graphically the required reservoir storage in acre-feet to ensure an adequate water supply for this 1953 to 1954 drought period.

5.19. From **Table 5.10**, determine the minimum 12-month water yield, in acre-feet, for a 30-acre catchment expected once in 5 years and once in 50 years.

5.20. From **Table 5.10**, determine the minimum 12-month water yield in inches expected once in 10 years from a 30-acre catchment and from a 4600-acre catchment. Explain the greater flow from the larger catchment.

6 Stream Processes

6.1 INTRODUCTION

Seldom does flow move far before entering some kind of channel. Consider, for example, runoff from the roofs of houses in urban areas. The water soon enters gutters, flows into a downspout, and goes through a buried pipe to the street curb. The water then flows along the side of a road and in many parts of the country will then enter a drain leading to a subsurface stormwater system. Even in rural areas, runoff from agricultural fields frequently flows into grassed waterways or open ditches. Does this mean that the procedures described in Chapter 5 have little or no usefulness? Certainly not, for they are used in the design of gutters, pipes, stormwater systems, waterways, and ditches. Also, flow in small conveyance systems is often approximated as similar to overland flow because it is not practical to divide every watershed into subwatersheds representing individual houses and fields.

Of interest is how flow in channels and rivers differs from overland flow and at what point we should start considering the influence of these conveyance systems on flow processes. Also of interest is how these systems function, what makes them stable or unstable, how they interact with the landscape, the ecology of these systems, and their importance to society and to the food chain.

In this text, flow processes are related to watershed areas of less than about 1000 mi². A flood in small streams, rivers, and their tributaries (often called *headwaters*) resulting from small-area storms is termed a *small-area discharge* or *flood.* It is distinguished from a *large-area flood,* which is found in downstream channels where flood control programs are wholly of an engineering nature and are beyond the scope of this text. Headwater floods are typically flash floods of short duration. In smaller headwater areas, the average rainfall intensity can be much higher because of the smaller area. Variations in climatic and physiographic factors that affect runoff are often more extreme and more high gradient than in larger watersheds. The effects of crops, soil, tillage practices, and conservation measures are more important in headwater areas because the surface condition of the entire area might change completely from season to season and from year to year.

The effectiveness of headwater flood control measures decreases rapidly with distance downstream. Most assuredly, the great Mississippi River flood of 1993 was not caused by human use of the land or by lack of flood control techniques on upstream areas. Rather, it was caused by prolonged rain over a large area and would have happened even before European settlement. Indeed, the Mississippi River was in flood and several miles wide when DeSoto first saw it in 1541. Through the Army Corps of Engineers, the federal government develops and supervises most flood control programs on large river systems, while the Natural Resources Conservation Service (NRCS) and U.S. Forest Service (USFS) conduct programs in headwater areas. While the need for flood control is important, we need to recognize that measures used to reduce flood impacts often have a detrimental impact on the geomorphology and ecology of a stream. In fact, in many parts of the U.S., small dams are removed to improve stream health. More discussion on stream structures is presented in Chapters 7, 8, and 12.

Stream processes are to some extent influenced by all the factors that have an impact on the hydrologic cycle throughout a watershed. Failure to consider upland, overland, subsurface, groundwater, upstream, and downstream influences adequately often causes problems in a stream or results in inadequate designs. Understanding some of the discussion presented in this chapter requires an understanding of the processes presented in Chapters 3 and 5 in particular and to a lesser extent Chapters 1 and 2. In this chapter, we first discuss in a general sense how streams interact with the landscape. This is followed by discussion of how both *spatial* (area or space) and *temporal* (time) scale might be considered and why scale is important in terms of stream biota. We then will be in a position to start considering the characteristics of streams and how these characteristics are influenced by *fluvial* processes. Information then is presented on how scientists and engineers have tried to organize or "classify" streams according to their characteristics and degree of stability. We then consider channel evolution, methods for measuring streamflow, and flood management. In Chapter 7, we present methods for calculating discharge rates and applying this knowledge to assessments of open channel flow conditions. In Chapter 8, we consider hydraulic control structures and methods for evaluating their influence on stream systems.

6.2 INTERACTION OF STREAMS WITH THE LANDSCAPE

6.2.1 THE INFLUENCE OF TEMPORAL AND SPATIAL SCALES ON GEOMORPHOLOGY

The landscape is shaped by erosional and depositional processes that occur across a wide range of spatial and temporal scales. The general characteristics of streams usually are associated with complex processes that might have occurred during several decades or hundreds or even thousands of years. For most of us, our view of a stream and how it changes with time is based on anecdotal information or our own observations and experiences. We are therefore considering only a small span of time, which ranges from a few years, decades, or perhaps a century or more. We probably only consider a small portion of the stream system that flows in our neighborhood, county, region, or state. Perhaps when there is a rare event, such as a major flood, we take note of what happens upstream or downstream of the portion of the stream that interests us. Have you ever wondered why "your" stream is located where it is, is bigger or smaller than another stream, or has better or worse stream life than other streams? Perhaps not, but it is helpful to understand the origin and evolution of a stream system to develop strategies to protect, enhance, or sustain these complex and fragile ecosystems. What we discover about the system might come as a surprise.

In Chapter 1, we learned that the Mississippi River and the Great Lakes are two of the largest freshwater resources in the world. Many of the states in the Midwest region of the U.S. have streams that flow northward into one or more of the Great Lakes and other streams that flow southward into tributaries of the Mississippi River. These southern-flowing streams are the headwaters of the mighty Mississippi, which is 3710 miles long, has a drainage area of more than 1.2 million mi^2, and a mean discharge that exceeds 600,000 ft^3/sec (Iseri and Langbein, 1974). Daily, the Mississippi River pours more than a million tons of sediment into the Gulf of Mexico (Hamblin and Christiansen, 1995). The largest tributary is the Ohio River, which provides about 40% of the flow and has a drainage area of 0.2 million mi^2. A major tributary of the Ohio River is the Scioto River, which obtains its name from the Indian word *Scionto*, meaning deer.

Real-time flow data for the Scioto River are available at Higby, OH (USGS Gage 03234500), where the drainage area exceeds 5000 mi^2, and the mean discharge is 4725 ft^3/sec (USGS, NWIS Web site at http://waterdata.usgs.gov/nwis). Upstream of Higby, there are many interesting and wonderful tributaries, including Big and Little Darby Creek, which are national scenic rivers. Flowing through glacial debris, within a 560-mi^2 watershed, these streams are one of the top five warm freshwater habitats in the

Midwest and contain 86 species of fish and 38 species of mussels (Hambrook et al., 1997; Nature Conservancy, 1990). For a decade, Ward had the opportunity to conduct research on identifying agricultural practices that would reduce adverse impacts on the groundwater resource associated with the buried valley aquifer located below the Scioto River near Higby (Nortz et al., 1994). This work was performed as part of the major multiagency Management Systems Evaluation Areas Project (A.D. Ward et al. 1994). Oddities of the Scioto River research site were that there was an Indian mound located on the farm, that wells located at the site provided large volumes of water to the Piketon uranium enrichment facility, that the project had significance to the Midwest, and that the focus was on groundwater, not the river.

In 1650, the Iroquois Indians pushed south from Canada into Ohio and drove out the Algonquian Indians. However, we need to go back much further in time to learn how the river was formed. The buried valley aquifer is associated with the ancient Teays River (**Figure 6.1**). This ancient river was comparable to the size of the modern Ohio River and was destroyed 2 million years ago by the glaciers of the Pleistocene Ice Age.

The Teays River had its headwaters in western North Carolina and flowed northward and westward across Virginia, West Virginia, and Ohio before flowing west across Indiana and Illinois and then south before discharging into the Gulf of Mexico (Ver Steeg, 1946). Near Higby, the elevation of the Teays River was about 640 ft. An early glacier disrupted the Teays River system and created a major lake in southern Ohio; the lake rose to an elevation of about 900 ft. This ancient lake is referred to as Lake Tight and was almost the size of Lake Erie, which did not exist at that time. Lake Tight existed for more than 6000 years, and clay lake bed sediments, called Minford clay, can be found near the tops of modern-day hills in the area and are still mined as a raw material for making brick and other ceramic products.

The exact sequence of events and flow paths of various rivers and glaciers near Higby is unclear because this was also the southern edge of three glacial movements from the north. However, Lake Tight eventually breached and created new drainage channels, which in some cases now flowed south rather than north. Near the modern Scioto River valley, this channel was called the Newark River, and it widened and deepened the valley to an elevation of 465 ft (Norris and Fidler, 1969). Over time, this valley filled, to an elevation of 560 to 580 ft, with sand and gravel outwash, and the modern Scioto River was formed.

In this section, we could have looked at the role of glacial activity on the formation of the Great Lakes or numerous other examples in the U.S. and around the world to illustrate how streams and the landscape interact and are influenced by events across vast areas and short or very long periods of time. In this chapter, we provide an

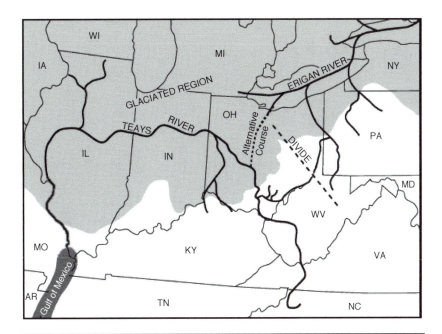

Classic interpretation of the preglacial Teays River and an alternative course (dashed line) favored by some geologists. The entire extent of the Teays and its tributaries north of the glacial border is buried beneath thick glacial drift. Northern Ohio was drained by the Erigan River, which followed the axis of what is now Lake Erie, and flowed into the ancestral St. Lawrence River. Neither the Great Lakes nor the Ohio River existed at this time.

FIGURE 6.1 The ancient Teays River. (From Ohio Department of Natural Resources (ODNR), Division of Geological Survey. GeoFacts No. 10. www.ohiodnr.com/geosurvey/geo_fact/geo_fl0.htm)

illustration of how a catastrophic event had a major influence in establishing a small ecological wonder. For further reading on shaping our landscape, a good account of the energy systems that make the earth a dynamic planet is presented in a beautifully illustrated book by Hamblin and Christiansen (1995). Today, buried valley aquifers located near the ancient Teays River are major sources of groundwater in Indiana, Illinois, and Ohio. Protection of these resources is still a concern, but there has been a shift in national focus toward concerns regarding hypoxia in the Gulf of Mexico and the role that agricultural production in the Midwest has on this issue (Alexander et al., 2000; Goolsby, 2000). Hypoxic zones have dissolved oxygen levels that are less than 2 mg/l and are inadequate to sustain the aquatic biology. Extensive information on the Mississippi River and other major rivers in the U.S. is presented in a series of books by Patrick (1998).

6.2.2 BASIC TERMS AND GEOMORPHIC WORK

We learned from the example in the preceding section, that many different factors shape our landscape and drainage systems. Catastrophic impacts might occur in an instant, while other impacts transpire over long periods of time. However, interactions between the hydrologic cycle and the land use are established within a short time, ranging from a few minutes, hours, or days to several years.

Increasingly, because of land use changes, similar rapid changes are now occurring in the shape and flow directions of stream systems. Initially, these impacts might be small and perceived as insignificant except perhaps at a local level. However, with time, adverse impacts, particularly in headwater streams, can spread across vast areas of a watershed system. This is analogous to a small cut on your arm or leg. If left untreated, it might become infected, the infection might spread, or gangrene might result. This might lead to the loss of a limb or even death in extreme cases. Understanding of a stream system first requires knowledge of how it was formed, which processes occur within the system and cause stability or instability, and then how land use and land use changes influence the equilibrium of the system.

Unfortunately, in hydrology a variety of terms is used that have the same or similar meaning. A glossary of definitions is provided in Appendix A. Let us look at a few of the key terms that often cause confusion. Stream *stability* and *instability* are not easily definable or quantifiable. Sometimes, these terms are used to describe the state of the banks or bed. Unfortunately, many people incorrectly equate stability to banks that are fixed or rigid. Streams naturally change the shape and position of their banks. Unless stated otherwise, in this text we consider a stable stream system as one that is self-sustaining, retains

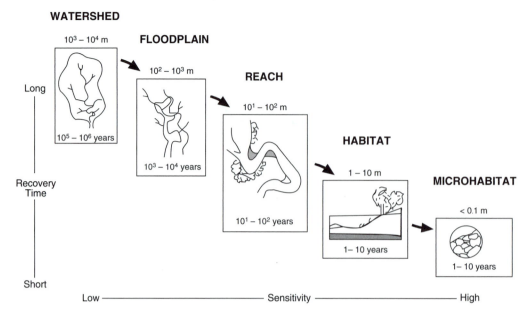

FIGURE 6.2 Sensitivity of stream systems to disturbance and recovery. (From Frissell, C.A., W.J. Wiss, C.E. Warren, and M.D. Huxley, *Environ. Manage.*, 10:199–214, 1986. With permission from Springer-Verlag, New York.)

the same general geometry over time (decades), and has a balance between the import and export of sediment.

The terms *channel*, *creek*, *stream*, and *river flow* or *discharge* relate to channelized flows of different magnitudes that are only described qualitatively. A stream is a channelized flow of running water, and a river is simply a large stream. The term channel flow can be used to describe flow in any size channel, stream, or river, and the terms discharge, discharge rate, and flow rate (cubic feet/second or cubic meters/second) are synonymous. *Ephemeral* streams only contain water during and immediately after some rainfall events and are dry most of the year. *Intermittent* streams are also dry through part of the year, but flow when the groundwater is high enough as well as during and immediately after some rainfall or snowmelt events. *Perennial* streams flow throughout the year. Channel discharge might have contributions from surface runoff, interflow, groundwater flow, precipitation that falls directly on the channel, and point discharges (such as industrial effluents or treated wastewater). The area contributing to discharge at a point in a channel is called a *subwatershed*, *watershed*, *catchment*, *river basin*, and/or *groundwater compartment*.

Several different classification systems have been developed to describe spatial and temporal scales associated with stream systems. In Section 6.3, an overview is presented on stream ordering classifications and their use in stream studies. Frissell et al. (1986) proposed a hierarchical organization of a stream system that we used as the basis for the organization presented in **Figure 6.2**. It is our opinion that the relative spatial and temporal scales presented provide good insight on the sensitivity of stream systems to disturbance and recovery, but that considerable

deviation might be expected from the reported values. For example, reach and floodplain recovery in many locations might only take a few years, while in other locations it might never occur. Also, in the U.S., the term *drainage basin* often is used for larger areas than those suggested by Frissell et al. so we have used the storm watershed in **Figure 6.2**.

Stream *geomorphology*, or the formation of land by streams, occurs because of a series of complex processes not easily described by scientific theories. The most basic concept is that of force and resistance. Running water exerts force on the landscape; in turn, the landscape offers resistance to this force. If the exerted force is less than the resistance, there is no change. If the force is greater than the resistance, there is change to the slope or stream channel. We call this change *geomorphic work*. Lane's classic description of channel stability states that dynamic equilibrium exists between stream power and the discharge of bed material sediment (E.W. Lane, 1955b):

$$Q_s d \; \alpha \; QS \tag{6.1}$$

where Q_s is the sediment discharge, d is the median sediment size, Q is the discharge, and S is the bed slope. See Chapter 9, **Figure 9.20**, for further discussion of Lane's concepts. As stated, the term *stream stability* is a qualitative understanding that the channel geometry is not undergoing rapid changes with time, and the import and export of sediment are in balance and not associated with excessive aggradation, degradation, or mass wasting.

A poor understanding of these processes and inadequate consideration of the influence of changes that occur on the landscape and within the floodplain can cause a variety of adverse outcomes. Particular attention needs to be paid to the potential impact on a stream of (1) land use changes that reduce vegetation and increase the amount of impervious area; (2) activities that modify the floodplain; (3) the construction of culverts and bridges; and (4) activities that are designed to modify the characteristics of the main stream channel. Any of these activities might disrupt the equilibrium, resulting in rapid and often undesirable adjustments. Such changes typically have adverse impacts on the aquatic biota and ecosystems within the vicinity of the stream.

We often hear about stream or channel "improvements," "maintenance," or "restoration" projects. Often, the purpose of channel improvement and maintenance projects is to improve or increase the ability of the channel to convey water. This might be done by making the bottom and sides of the channel smoother, making the channel straighter, or changing the geometry of the channel. Unfortunately, any of these methods could make the channel unstable and is likely to reduce the diversity of the aquatic biota. While there are many examples of successful restoration projects, there are also numerous examples of failures, which are usually because of an inadequate understanding of stream geomorphology. In the last half-century, Leopold, Wolman, Schumm, Einstein, and many others have used empirical knowledge, based on the study of numerous streams throughout the U.S. and around the world, to enhance greatly our understanding of streams. Successful stream stewardship requires combining this knowledge with sound engineering and scientific principles, together with an understanding and appreciation of the ecology of the stream and its interaction with the landscape.

6.3 STREAM ORDERS

In the Mississippi River example, we saw that further upstream from the Gulf of Mexico the drainage area and discharge for each river system became smaller, until finally we reached headwater streams that are the starting point for this treelike (*dendritic*) system of channels. The most distant point from the mouth of the Mississippi is the Red Rock River in Montana. Often, the beginning point of a stream system is not very noteworthy and could even be in your backyard. For example, one of the 20 largest rivers in the U.S. (Iseri and Langbein, 1974) is the Wabash River, which starts in Darke County, Ohio, heads east, north, west, north, and then west again before flowing through Indiana and to Illinois. It roughly follows the course of the ancient Teays River (Ver Steeg, 1946). The Wabash starts as two large subsurface drainage mains that discharge into a rather unassuming ditch (**Figure 6.3**). To help understand, discuss, and explore commonalities and differences between stream systems, these systems have been classified according to their relative position within the dendritic drainage systems of a watershed. One of the earliest methods was developed by R.E. Horton (1945) and later modified by A.N. Strahler (1952) and is commonly used today. Many hydrologists consider the Shreve system the best because the orders are additive, much like flow converging from two streams is additive (see Example 14.8, Chapter 14). However, biologists seldom use the Shreve system.

In the Strahler system, the smallest headwater tributaries are called *first-order streams* (**Figure 6.4**). Where two first-order streams join, a *second-order stream* is created; where two second-order streams join a *third-order stream* is created; and so on. This approach is only useful if the order number n is proportional to the channel dimensions, size of the contributing watershed, and stream discharge at each point in the system.

The ratio of the number of stream segments of a given order N_n to the number of segments of the next highest order N_{n+1} is called the *bifurcation ratio R_B*:

$$R_B = \frac{N_n}{N_{n+1}} \qquad (6.2)$$

Within a watershed, the bifurcation ratio will change from one order to the next, but will tend to be constant throughout the series. This observation forms the basis of the law of stream numbers (R.E. Horton, 1945), which states that the number of stream segments of each order forms an inverse geometric sequence with order number; this is described mathematically as

$$N_n = a_1 \exp^{-b_1 n} \qquad (6.3)$$

where a_1 is a constant, and b_1 is ln R_B.

Two other drainage network laws have been developed. The law of stream lengths was also developed by R.E. Horton (1945) and is defined as follows:

$$L_n = a_2 \exp^{b_2 n} \qquad (6.4)$$

where a_2 is a constant, b_2 is ln R_L, and R_L is the stream length ratio. The stream length ratio is defined as

$$R_L = \frac{L_{n+1}}{L_n} \qquad (6.5)$$

where L_n and L_{n+1} are the average stream lengths of streams of order n and $n + 1$, respectively.

The law of drainage areas was developed by Schumm (1956) and is defined as follows:

A

B

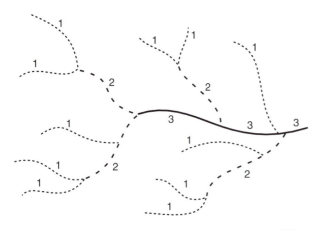

FIGURE 6.3 The Wabash River starts in Darke County, Ohio, as (A) outlets of two subsurface drainage mains and (B) a small drainage ditch.

FIGURE 6.4 Stream order hierarchy of A.N. Strahler (1952).

$$A_n = a_3 \exp^{b_3 n} \qquad (6.6)$$

where a_3 is a constant, b_3 is $\ln R_A$, and R_A is the drainage area ratio

$$R_A = \frac{A_{n+1}}{A_n} \qquad (6.7)$$

where A_n and A_{n+1} are the average drainage areas of streams of order n and $n + 1$, respectively.

The bifurcation ratio, stream length ratio, and drainage area ratio in most cases will fall within the ranges 3 to 5, 1.5 to 3.5, and 3 to 6, respectively. Also, Hack (1957) found that, for many locations in the eastern U.S., the stream length was related to the drainage area as follows:

$$L = 1.4\,A^{0.6} \qquad (6.8)$$

where L is the stream length in miles from the watershed boundary, and A is the watershed area in square miles.

There are several practical difficulties related to drainage network determination. First, in developed countries such as the U.S., many watersheds contain a large number of artificial channels, such as stormwater systems, vegetated waterways, and open drains alongside roads. These systems often are difficult to identify and do not usually lend themselves to inclusion in the stream order hierarchy associated with natural drainage systems. Second, maps with scales of 1:24,000 often are used to identify drainage networks because ground reconnaissance, or the use of remote sensing techniques, is too costly or time consuming. Adequate identification of first-order and some second-order streams often requires use of detailed topographic maps with a scale on the order of 1:2,400.

The three drainage network laws and the stream ordering systems developed by Horton, Strahler, and Schumm still are commonly used. However, the topologically distinct channel networks approach developed by Shreve (1967) also is widely used. For complete details on this and other complex approaches, reference should be made to texts on streams, stream networks, or stream morphology, such as those by Knighton (1984), Jarvis and Woldenberg (1984), and Richards (1982).

EXAMPLE 6.1

Use the drainage network laws to determine the bifurcation ratio, stream length ratio, and drainage area ratios for a 730-acre fourth-order watershed. The drainage network characteristics are summarized in **Table 6.1**.

Solution

Values of R_B, R_L, and R_A for each set of successive stream orders could be obtained using the information in **Table 6.1** and Equation 6.2, Equation 6.5, and Equation 6.7, respectively. For example, R_B for the ratio of stream orders 1 and 2 would simply be 80 divided by 14 or 5.7, respectively. However, what we need to obtain are values of R_B, R_L, and R_A that are representative of the whole network. This requires using Equation 6.3 to Equation 6.5. First, we need to obtain natural logarithms of the N_n, L_n, and A_n values in **Table 6.1**. These values are reported in **Table 6.2**.

We now need to plot the set of ln N_n, ln L_n, and ln A_n values against stream order n and statistically or visually fit a straight line to each set of values. The slope of the

TABLE 6.1
Drainage Network Characteristics for Example 6.1

Stream Order n	Number of Streams N_n	Average Stream Length L_n (ft)	Average Drainage Area A_n, (acres)
1	80	400	5
2	14	1100	27
3	4	3000	140
4	1	8000	730

TABLE 6.2
Natural Logarithms of Selected Drainage Network Characteristics for the Solution to Example 6.1

Stream Order n	ln N_n	ln L_n (ft)	ln L_n (acres)
1	4.38	5.99	1.61
2	2.64	7.00	3.29
3	1.39	8.01	4.94
4	0.00	8.99	6.59

line drawn through the set of ln N_n value will be b_1, and the slopes of the lines drawn through the sets of ln L_n, and ln A_n values will be b_2 and b_3, respectively. The plots are presented in **Figure 6.5**, and from the figure, b_1, b_2, and b_3 are 1.46, 1.0, and 1.66, respectively. Then, by obtaining the antilogarithm (e^{b1}, e^{b2}, and e^{b3}, respectively) of each of these values, we find that R_B, R_L, and R_A are 4.3, 2.7, and 5.3, respectively.

Answer

The bifurcation ratio R_B is 4.3, the stream length ratio R_L is 2.7, and the drainage area ratio R_A is 5.3.

An example for the state of Ohio of how drainage area and stream size are related to stream order is presented in **Table 6.3** (Mecklenburg, 2002). The approach proposed by Leopold et al. (1964) was used to develop the relationship between stream size and stream order. It can be seen that a very high percentage of the stream miles in Ohio are small headwater streams.

6.4 STREAM BIOTA

The river continuum concept (RCC; Vannote et al., 1980) is an attempt to describe the function of *lotic* (running water) ecosystems from their source to the mouth. This concept is built on the idea that predictable changes in geomorphology and hydrology as you move downstream form a template for adaptation of biological communities

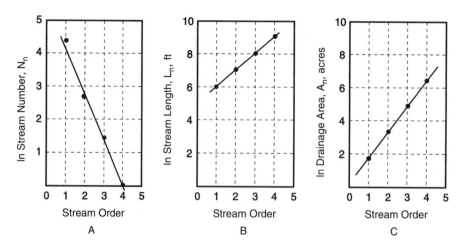

FIGURE 6.5 Stream order plotted with (A) number of streams, (B) mean stream length, and (C) mean drainage area for Example 6.1.

TABLE 6.3
Relationships for the State of Ohio Among Stream Order, Drainage Area, and Total Stream Length

Stream Order	Drainage Area (mi²)	Total Length (mi)	Miles (%)	Cumulative (%)
1	0.2–1	67,530	51.5	51.5
2	1–4.7	33,138	25.3	76.8
3	4.7–23	15,963	12.2	89.0
4	23–109	7,803	6.0	95.0
5	109–518	3,810	2.9	97.9
6	518–2,460	1,861	1.4	99.3
7	2,460–11,700	908	0.7	100.0

Source: From Mecklenburg, Personal communication, 2002. With permission.

(Giller and Malmqvist, 1998). This idea has been one of the unifying themes in stream ecology (Allan, 1995), providing a conceptual model for understanding how streams function and guiding a great deal of stream research for the last 20 years.

The basic tenet of the RCC concerns how streams change with increasing stream order, discharge, and watershed area (which are often highly correlated; Allan, 1995). Low-order streams typically are shaded headwaters in which coarse particulate organic matter (CPOM) from terrestrial sources (allochthonous production) forms the energy base for consumers (Wallace et al., 1997). Thus, shredders and collectors, which feed on terrestrial leaf material or the "biofilm" that grows because of microbial decomposition, dominate macroinvertebrate communities (**Figure 6.6**). In low-order streams, very little plant growth (periphyton, macrophytes) is possible because of canopy shading, so respiration rates usually exceed rates of photosynthesis ($R > P$).

As a stream becomes larger at midorder sites, shading is reduced, allowing periphyton and macrophytes to grow (*autochthonous production*). The reduction in overhead canopy cover decreases CPOM input, but CPOM processing from upstream reaches will result in the transport of fine particulate organic material (FPOM) to midorder reaches. Because of these changes in habitat and the energy base, the trophic structure of the macroinvertebrate community also shifts. Collectors also are common in midorder reaches, but shredders are rare, and grazers (which feed on periphyton) and predators become more common. Because of increased plant growth in streams of this size, rates of photosynthesis can exceed respiration ($P > R$). The variety of energy inputs (as well as high habitat heterogeneity) tends to make midorder reaches the most biologically diverse sites in a river.

As stream order increases to the point at which a stream becomes a river, further changes will occur. Larger rivers generally have high turbidity, unstable substrates (i.e., sand), and higher discharge, which make periphyton and macrophyte production very small. Phytoplankton usually is the only autochthonous production in large rivers, and turbidity and constant mixing of water will limit

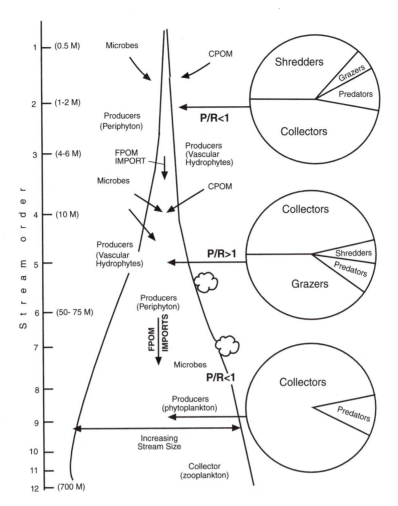

FIGURE 6.6 Invertebrate functional groups as a function of stream order. (After Cummins, K.W., in *Proceedings of the Sandusky River Basin Symposium, International Joint Commission, Great Lakes Pollution*, D.B. Baker, W.B. Jackson, and B.L. Prater, Eds., Environmental Protection Agency, Washington, DC, 1975, pp. 227–293. With permission.)

their production. Thus, in large rivers, allochthonous production dominates the energy base. Vannote et al. (1980) emphasized the transport of FPOM and dissolved organic material (DOM) from upstream sources as the primary energy source in large rivers. Several authors have modified the RCC for large rivers because lateral sources from the floodplain may be a more important source of allochthonous material than upstream sources. These ideas are best summarized in Junk et al.'s (1989) Flood-Pulse Model. They stated: "The principal driving force responsible for the existence of, productivity, and interactions of the major biota in river-floodplain systems is the flood pulse. ... The pulse concept requires an approach other than the traditional limnological paradigms used in lotic and lentic systems." In general, collectors and other filter-feeding organisms (e.g., Chironomids) dominate the macroinvertebrate fauna in large rivers.

It is important to keep in mind that the river continuum concept is only a theoretical model that was developed

from data on temperate, forested streams in North America. In streams without forested headwaters or in rivers that have been extensively modified by humans, many of the patterns predicted by the RCC are not consistently apparent (Giller and Malmqvist, 1998). In these cases, the RCC still can be usefully applied as a benchmark against which to measure stream biota.

The change in abundance of invertebrate functional groups as we move down a river system from the headwaters is illustrated in **Figure 6.6** (Cummins, 1975). The headwaters (Orders 1 to 3) are dominated by shredders that feed on riparian litter and collectors that feed on fine particulate organic matter. The midreaches (Orders 4 to 6) have increased channel width and reduced canopy shade, which results in grazers becoming more abundant because of the increase in microalgae. The lower reaches are dominated by fine particulate organic matter, and collectors dominate.

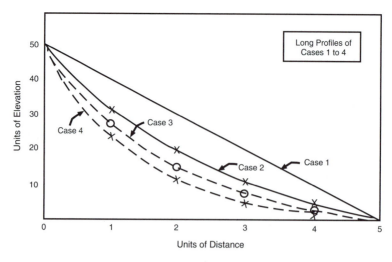

FIGURE 6.7 Channel profiles for four cases. (From Leopold, L.B., *A View of the River*, Harvard University Press, Cambridge, MA, 1994. With permission.)

6.5 STREAM CHARACTERISTICS

When describing a stream, most are familiar with the terms *bed* and *bank*. We might have heard people call the land next to the banks the *floodplain*, know that streams are usually not straight, and if we have been fishing, swimming, canoeing, kayaking, or rafting in a river, we know the bed sometimes contains very fine silt and clay materials, but could be mainly sand, gravel, cobble, or even big boulders. As you might have guessed, there are all kinds of other scientific terms that help describe streams. Some will be surprised to learn that the floodplain is really part of the stream system. To help describe a stream system and learn how it works, we need to understand what we mean by the stream profile, pattern, and dimension and how these characteristics are influenced by scour, deposition, and sediment transport.

6.5.1 CHANNEL PROFILE

The *channel profile* is the slope in the direction of flow as a river moves from a point of high elevation to a point of lower elevation. The slope down the valley walls toward the stream system and the channel slope are influenced by many factors, including historic changes that might have occurred, such as those discussed in Section 6.2. To help achieve a state of dynamic equilibrium, the channel will meander and adjust its bed slope. An example of the different paths the river might take as it moves between two points is illustrated in **Figure 6.7**. The channel slopes are related to the discharge Q as follows for Cases 1 through 4, respectively: s constant, $s \sim Q^{-0.5}$, $s \sim Q^{-0.75}$, and $s \sim Q^{-1.0}$. The most uniform rate of work expenditure is Case 1, for which the slope is a constant. In Case 4, the slope is inversely related to discharge, and the total work done is minimum. For Case 4, the largest elevation drop occurs upstream where the discharge is the smallest, there

is less flow depth, and the discharge can do less geomorphic work. Minimum work and uniform work distribution cannot occur simultaneously, and most rivers will have a profile similar to Case 3.

Figure 6.8 illustrates a typical profile for a meandering channel with a pool–riffle bedform. At low flows, the water elevation has a profile similar to that of the channel bed. As the discharge increases, the water surface has a more uniform slope, and energy changes along the profile become more uniform.

6.5.2 STREAM CHANNEL PATTERNS

Factors influencing stream morphology include geology, topography, size of the contributing watershed, flow velocity, discharge, sediment transport, sediment particle distribution, channel geometry, and other geomorphologic controls on the system. Leopold et al. (1964) noted three different channel patterns: sinuous, meandering, and braided. Attempts have been made to classify streams based on *sinuosity*, which is the ratio between the stream length and the valley length. If the sinuosity is 1.0, the stream channel is straight, a condition that rarely occurs in natural channels except over short distances. If the sinuosity is greater than 1.0, the channel is sinuous. If the sinuosity is greater than 1.5, the river is said to meander, and if the sinuosity exceeds 2.1, the degree of meandering is tortuous (Schumm, 1977).

A relationship between the channel slope, bankfull discharge, and channel pattern was reported by Leopold and Wolman (1957):

$$s = 0.0576\, Q^{-0.44} \qquad (6.9)$$

where s is the streambed slope as a fraction, and Q is the bankfull (effective) stream discharge in cubic feet/second (ft³/sec). An interpretation of this equation is presented in **Figure 6.9**.

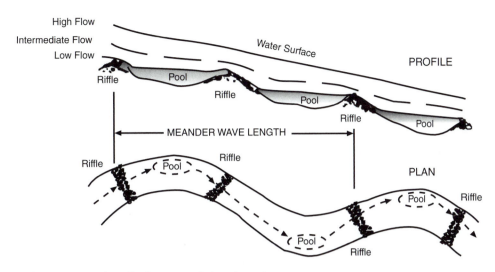

FIGURE 6.8 A typical pattern and profile for a meandering channel.

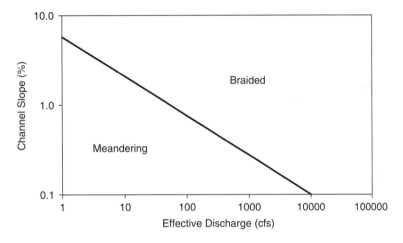

FIGURE 6.9 Relationship among channel slope, bankfull discharge, and channel pattern based on a plot of Equation 6.9.

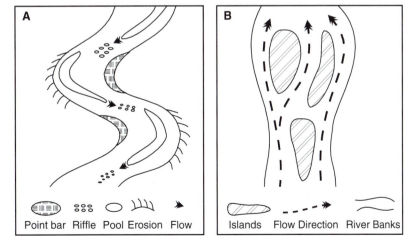

FIGURE 6.10 (A) A meandering stream and (B) a braided stream.

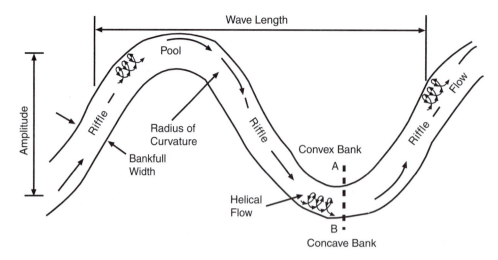

FIGURE 6.11 Geometry of a meandering stream.

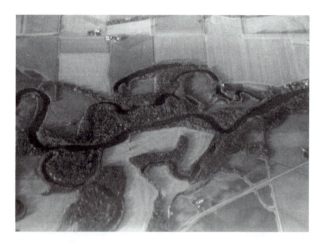

FIGURE 6.12 Formation of an oxbow at the bend in a stream. (Courtesy John Lyon.)

Features of a meandering stream are shown in **Figure 6.10** and **Figure 6.11**. Typically, meandering streams consist of a series of riffles, pools, and point bars/pool features that might be 5 to 7 bankfull widths apart, while meander wavelengths might be 10 to 14 bankfull widths apart. Point bars, which are ridges of deposited bed material, will form on the concave inner side of stream bends, where the flow velocity is slowest. On the outer apex of river bends, the flow velocity is fastest, and scour (channel erosion) will create pools. In an unstable system, erosion on the outer bends will continue unchecked, sinuosity will increase, and the discharge might break through two bends to form a cutoff and the eventual formation of an oxbow lake (**Figure 6.12**). However, erosion on the outer bends and lateral migration of the stream channel does not necessarily signal instability (see Section 6.9).

The velocity distribution in a channel is complex (**Figure 6.13**). Typically, at any cross section in a channel, there will be a portion of the flow along the deepest part

of the channel (called the *thalweg*) that is moving the fastest. Think of the thalweg as a roller coaster that speeds up in riffles, slows down in pools, moves from side to side, moves up and down with depth, and rotates counterclockwise as it moves around bends. The velocity of the flow will vary as we move from one bank to another at any point in a river. For example, the velocity near the outer bank of a meander will be higher than near the inner bank because water near the outer bank has to travel further. The rotation of the flow on the outer bend might also create undercut concave banks. As we near the bottom or sides of the channel, the velocity will decrease because of the surface roughness of the channel (**Figure 6.13**). Water in contact with the bottom and sides will be stationary. As the roller coaster plunges from a riffle into a pool, it will cause much turbulence (churning and mixing of the water). Also, as the roller coaster moves through pools, it will push water toward the banks and away from its direction of flow. As much as a third of the flow might circulate back upstream.

6.5.3 STREAM DIMENSIONS

The term *dimension* is used to describe the geometry or cross-sectional shape of a river channel. Typically, a stream that is not confined by the valley walls will have a main channel and a connected active floodplain (**Figure 6.13**). The main channel is associated with a specific discharge called the *bankfull, effective, dominant,* or *critical discharge.* It is beyond the scope of this text to describe the differences among these terms and the different scenarios in which one type of discharge vs. another would have more influence on the main channel dimensions. Often, the terms are considered synonymous. The bankfull discharge is "considered to be the channel-forming or effective discharge" (Leopold, 1994) and transports the

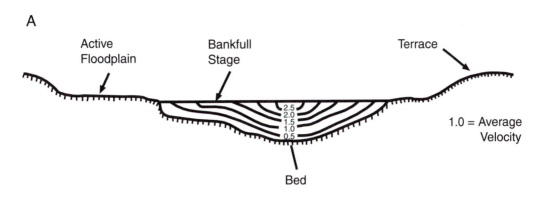

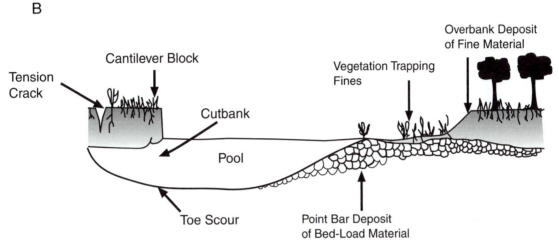

FIGURE 6.13 Typical channel cross section at (A) a straight reach showing velocity distributions (*isovels*) and (B) a bend showing a pool, cut bank, and formation of a point bar.

largest cumulative sediment load (Wolman and Miller, 1960).

The bankfull discharge is a range of flows that is most effective in forming a channel, benches (floodplains), banks and bars (G.P. Williams, 1978). The term *bankfull* causes some confusion because, in entrenched streams, the bankfull stage is lower than the top of the bank and is identified as a bench, change in bank material and vegetation, the top of point bars, or a scour line. We recommend the use of the term *effective discharge*. Over a long period of time, the effective discharge transports the largest total sediment load (**Figure 6.14**). The sediment transport rate is a function of the discharge (Curve A in **Figure 6.14**). Therefore, low discharges are ineffective in transporting sediment, and extreme events have very high sediment transport rates (Curve B), but occur infrequently so the total sediment load they carry over a period of many decades is not the largest. When the event frequency (Curve B) is multiplied by the transport rate for that frequency, we obtain Curve C, which is a measure of the total sediment load carried for that particular discharge and shows that the bulk of sediment is transported at near-effective flow (hence its name). This approach for determining geomorphic work

is known as the Wolman–Miller Model (Andrews and Nankervis, 1995).

While sediment transport includes both suspended and bedload fractions of the total sediment load, it has become evident that the bedload fraction of the total sediment load is most influential in most channel-forming processes and effective discharge (Emmett and Wolman, 2001).

The effective discharge may be met or exceeded several times a year in humid and semihumid regions and often is considered to correspond to a 1- to 2-year recurrence interval (RI) discharge based on a log Pearson analysis of annual instantaneous peak flows. In wetland streams, this interval may be much smaller and even nearly continuous (Jurmu and Andrle, 1997). In arid and semiarid regions, the effective discharge might only occur once every few years. For headwater streams in Idaho, Whiting et al. (1999) reported RIs of 1.0 to 2.8 years for the effective discharge. In a study in the northern Rocky Mountains, Emmett and Wolman (2001) reported RIs ranging from 1.5 to 1.7 years, while MacRae (1996) reported RIs of 1.6 to 10 years for a study in British Columbia. Andrews and Nankervis (1995) described the effective discharge as "the interval of discharge that transports the largest portion of

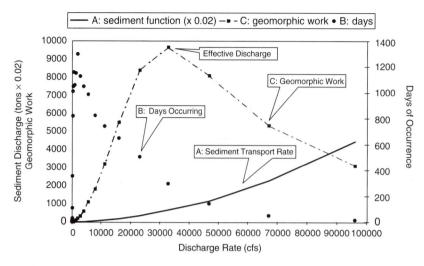

FIGURE 6.14 Illustration for the Maumee River, Ohio, that the effective (bankfull) discharge carries the largest total sediment load.

the mean annual bed-material load over a period of years, i.e., the modal value." For gravel bed streams that they studied in mountainous parts of Colorado, they found that 80% of the mean annual load was transported by flows between 0.8 to 1.6 times the bankfull discharge, which on average occurred as 15.6 events per year. Unless discharge and recurrence interval data specific to a stream are available, we caution against expecting a certain recurrence interval. For example, in Ohio (a semihumid region), there are streams for which the recurrence interval is a small fraction of a year (very frequent), many streams for which it is in the 1- to 2-year range, and for some streams, the RI approaches 5 years. Further discussion on the frequency of the effective discharge is presented in Chapter 12.

Why do smaller or larger discharges than the effective discharge not form channels, banks, benches, and bars different from those associated with the effective discharge channel? The answer to this question is not simple. There probably is a wide range of discharges that have the potential to shape a river channel. For example, we have found that. in northwest Ohio, flows that are probably associated with high subsurface drainage flows form a low bench in most agricultural drainage ditches. Why are similar features not formed in natural rivers or other ditches? First, very low discharges are ineffective in moving sediment and can only transport very fine material. If fine clays are not available, there will be little or no sediment transport. Second, the ability of these low discharges to scour the bed and banks of a channel also is very low. In the case of many agricultural ditches, the banks are not very stable and consist of fine materials. Third, if channels, banks, and benches are to be formed as a feature of the system, then there needs to be a mechanism that will stabilize the deposited materials. In the case of the agricultural ditches, there is rapid vegetation of these features.

In a natural channel, we might think of discharges lower than the effective discharge as (1) too small to scour or transporting sufficient sediment to create permanent features or (2) occurring too frequently to allow the deposited materials to stabilize. Perhaps harder to understand and visualize is why discharges larger than the effective discharge do not scour and wash away the banks, benches, and bars. In places along a river system, extreme events might cause bank instability problems, but on average, most channel and floodplain features that are in dynamic equilibrium have relatively stable banks and beds. On balance, they do not aggrade or degrade because discharges larger than the effective discharge spread out across the floodplain, have low velocities when they flow across these features, and in the main channel have similar shear stresses on the bed and banks as those produced by the effective discharge.

The *mean annual discharge*, or *average discharge*, will fill the channel to about one third of bankfull and is equaled or exceeded 25 to 30% of the time. Typically, discharges larger than the average will transport more than 95% of the sediment in a river system.

6.5.4 REGIONAL AND STREAM-TYPE CURVES

Several studies have related bankfull channel dimensions and discharge to drainage area. For example, **Figure 6.15** and **Figure 6.16** relate effective discharge (bankfull) width, mean depth, and cross-sectional area to drainage area for two regions of Ohio and the eastern U.S. **Figure 6.15** relates effective discharge to drainage area for northwest Ohio and Pennsylvania. It can be seen from **Figure 6.15** and **Figure 6.16** that the relationships plot as straight lines on logarithmic paper. Therefore, the channel width, depth, and mean velocity can be related to discharge as power functions:

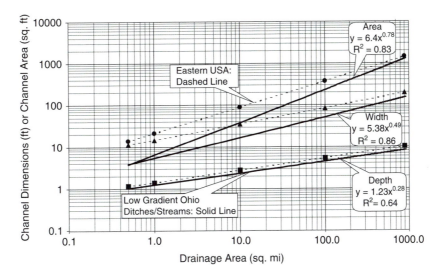

FIGURE 6.15 Channel dimensions for the eastern U.S. and low gradient agricultural watersheds in Ohio as a function of drainage area.

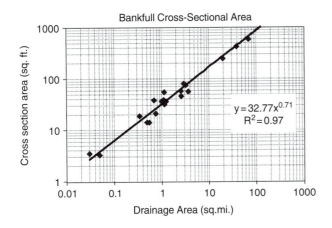

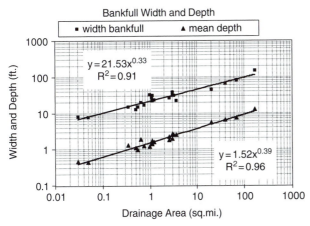

FIGURE 6.16 Channel dimensions for headwater streams in the Wayne National Forest, Ohio. (Courtesy and copyright U.S. Forest Service and the Ohio Department of Natural Resources.)

$$W = aQ^b \qquad (6.10)$$

$$d = cQ^f \qquad (6.11)$$

$$v = kQ^m \qquad (6.12)$$

Discharge is the product of the mean velocity times the cross-sectional area ($w \times d$) (**Figure 6.17**). Therefore, $a \times c \times k = 1$, and $b + f + m = 1$. Leopold (1994) found that, for various rivers he studied, the average values of b, f, and m at a given cross section (at a station) were 0.26, 0.40, and 0.34, respectively. However, in a downstream direction, the width will increase more rapidly, and the velocity will change only slightly; average downstream values of b, f, and m were 0.50, 0.40, and 0.10, respectively.

Based on an interpretation by Ward of information from several sources, the adjustment factors in **Table 6.4** can be used to multiply data from **Figure 6.15** to obtain the approximate effective discharge dimension for channels in various regions of the U.S. This approach should not be used to develop designs. There will be a discontinuity in the dimensions in moving from one drainage area range to the next. Also, most states for which regional curves have been developed have found it was statistically significant to subdivide the state into several regions. For example, in Minnesota, there are six regions (Magner and Steffen, 2000).

A dimensionless rating curve based on rivers in the eastern U.S. is presented in **Figure 6.18** (based on Leopold et al., 1995). The annual discharge is less than 15% of the effective discharge and occurs at a stage of about a third of the effective depth. A 50- to 100-year return period discharge occurs at a stage that is about

TABLE 6.4
Approximate Effective Bankfull Dimensions for Adjustment Factors in the U.S.[a]

State	Drainage Area <10 mi²			Drainage Area 10–100 mi²			Drainage Area >100 mi²		
	Area	Width	Depth	Area	Width	Depth	Area	Width	Depth
Idaho	0.9	0.7	0.5	1.5	1.2	1.0	0.8	0.7	0.6
Minnesota	0.9	0.7	0.5	0.8	0.8	0.7	0.9	0.8	0.7
Wisconsin	1.3	0.9	0.6	1.5	1.2	0.9	0.9	0.8	0.7
Wyoming	1.9	1.2	0.9	1.6	1.5	1.4	1.2	1.0	0.8
Oregon	1.3	1.4	1.6	2.2	1.9	1.7	0.7	0.9	1.2
Tennessee	1.8	1.4	1.2	1.4	1.5	1.1	0.8	0.9	1.2
Oklahoma	2.3	2.3	2.2	3.3	2.4	1.8	0.7	1.0	1.4
Maryland	2.4	2.2	2.1	2.4	2.0	1.7	1.1	1.1	1.2
North Carolina	3.1	2.6	3.0	3.2	2.4	2.1	1.0	1.1	1.3
California	4.9	3.7	2.8	3.7	2.8	2.1	1.4	1.3	1.2
Delaware	0.7	1.5	3.2	1.7	2.0	2.5	0.6	0.9	1.2

[a] For educational purposes, values from **Figure 6.15** can be multiplied by the adjustment factors in the table to obtain approximate dimensions for the bankfull dimensions in the listed states. **Do not use for design purposes.**

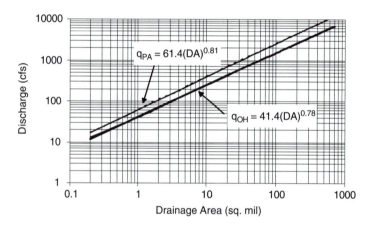

FIGURE 6.17 Effective discharge as a function of drainage area for streams in Pennsylvania and low-gradient streams in Ohio.

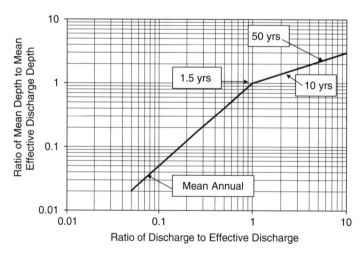

FIGURE 6.18 Dimensionless rating curve based on 13 gaging stations in the eastern U.S. (From Leopold et al., *Fluvial Processes in Geomorphology,* Dover, New York. 1995. With permission.)

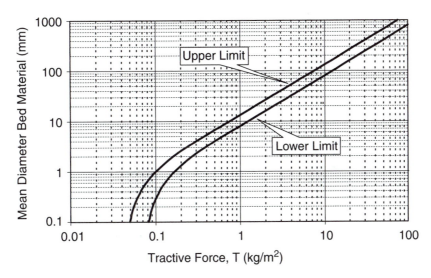

FIGURE 6.19 Relationships between the tractive forces on a streambed and size of bed material that will erode. (Based on Lane, E.W., 1955; upper and lower limits proposed by Ward.)

double the effective depth. In developing a stream classification system, Rosgen (1996) defined a flood-prone width that occurs at twice the maximum bankfull depth. It might be expected that this depth corresponds to about a 50- to 100-year return period discharge.

6.6 STREAM STABILITY AND SEDIMENT TRANSPORT

6.6.1 THE SOURCE OF SEDIMENT

There is much debate about the source of sediment in a river (Trimble and Crosson, 2000). The main sources of sediment are the landscape and the bed and banks of the channel. Sediment from the landscape occurs because of the erosion of soil associated with raindrop impact and water flowing across the ground surface (Chapter 9). Historically, much effort has been placed on estimating, preventing, and controlling soil erosion of the landscape. However, there is evidence to suggest that, in urbanizing watersheds, much of the total sediment yield might be because of stream channel erosion. Also, recall that the channel dimension, pattern, and profile are primarily related to the effective discharge associated with a discharge that might occur a few times annually. A land use change that increases the amount of impervious area can have the following impacts: (1) increased sediment loads during construction; (2) reduced sediment loads following construction; (3) increased magnitude of the discharge, particularly the effective discharge; and (4) increased frequency of these higher discharges. All of these factors adversely impact equilibrium in the river and can cause a river to widen or deepen. Channel modifications such as straightening a reach of the channel will also increase the

sediment-carrying capacity of the discharge and might cause bank and bed scour (see Section 6.7). Extensive further discussion on the source of sediment and sediment budgets at a watershed scale is presented in Chapter 9.

6.6.2 SHEAR STRESSES AND TRACTIVE FORCE

The mean boundary shear stress τ_0 exerted by the flow on the bed can be estimated from

$$\tau_0 = \rho g R s = \gamma R s \qquad (6.13)$$

where ρ is the density of water, γ is the specific weight of water, R is the hydraulic radius, and s is the bed slope. A more general term, the *tractive force*, or average shear stress in a reach, can be related to the depth of flow and the average slope of the water surface as follows:

$$T = 62.4ds \qquad (6.14)$$

where T is the mean shear stress or tractive force (pounds/square foot), d is the flow depth in feet, s is the slope of the water surface as a fraction, and 62.4 lb/ft^3 is the specific weight of water. The mean particle size that can be moved at incipient motion by a certain tractive force can then be determined by the relationship shown in **Figure 6.19**. This relationship can be expressed as

$$d_{50} = cT \qquad (6.15)$$

where T is the tractive force (kilogram force/square meter), the mean diameter d_{50} is in centimeters (cm), and c is a unit conversion factor that is also a function of the

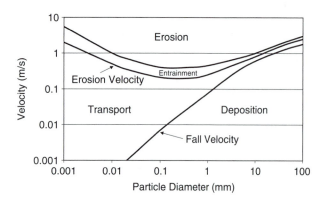

FIGURE 6.20 Hjulstrom diagram relating flow velocity and bed material size to erosion, entrainment, transport, and deposition. (Based on Knighton, D., *Fluvial Forms and Processes*, Edward Arnold, London, 1984. With permission of D. Fenger-Krog, G. Hjulstom, L. Hjulstrom and H. Bagge.)

magnitude of the tractive force. If the tractive force is larger than 1 kg force/m², then c is equal to 1. For example, a tractive force of 10 kg force/m² will move bed material with a mean particle size of 10 cm or 100 mm. The tractive force in kilogram force/square meter can be obtained by expressing the depth in meters and the specific weight as 1000 kg/m² in Equation 6.14. A relationship between velocity and the entrainment and movement of various textural sizes is shown in **Figure 6.20**. Shear stresses on the banks of streams will usually be 0.7 to 0.8 of the stresses on the bed. On the outer bend of a stream, the shear stresses on the banks might be two to three times the bed shear stresses.

EXAMPLE 6.2

A reach of a stream in an alluvial valley has a bed slope of 0.8%. Determine the tractive force (kilogram force/square meter) and mean bed material diameter that can be moved at incipient motion for a flow depth of 0.5 m.

Solution

If Equation 6.14 is expressed in metric units,

$$T = 1000\,ds \quad \text{and} \quad T = 1000(0.5)(0.008) = 4 \text{ kg force/m}^2$$

then using Equation 6.15, the unit conversion factor will be 1 because T is greater than 1 kg force/m², and

$$d_{50} = cT \quad \text{so} \quad d_{50} = 1\,(4) \;=\; 4 \text{ cm or 40 mm}$$

Answer

The tractive force is 4 kg force/m², and the d_{50} of the bed material that will be moved at incipient motion is 40 mm.

6.6.3 SEDIMENT TRANSPORT

Sediment is transported in a channel as a *wash load*, *suspended load*, or *bedload*. A wash load consists of fine particles too small to deposit in running water. The suspended load is moved by, and remains suspended in, the water column, but can settle in locations where the travel velocity is low or the settling depth is small. The suspended load can be more than 90% of the transported material and consists of fine materials such as silt and fine sand. This load contributes to making the flow look muddy or turbid and settles in areas of low velocity, such as the inner bends of the channel. As the name suggests, the bedload is pushed along near the channel bed by a combination of sliding, rolling, and saltation (bouncing). The bedload contributes to building point bars and the banks of the main channel.

The daily sediment transport rate S (metric tons/day) can be estimated by a power function of discharge Q (cubic meters/second) (Nash, 1994):

$$S = aQ^b \qquad (6.16)$$

where a and b are empirical coefficients. Nash (1994) reported values of b between 1.23 and 3.02, with 38 of 55 watersheds he evaluated exhibiting values between 1.4 and 1.9.

Bedload can be determined by the Meyer–Peter– Muller equation (Chanson, 1999):

$$\left[\frac{q_s}{\left(1.65gd_{50}^{3}\right)^{1/2}}\right] = \left[\frac{4\tau_o}{1.65\rho g d_{50}} - 0.188\right]^{3/2} \quad (6.17)$$

where q_s is the volumetric bedload discharge (cubic meters/second per unit width), d_{50} is the mean particle size of the bed material (meters), τ_0 is the shear stress (kilograms/m-s²), ρ is the density of water (cubic kilograms/cubic meters), the specific gravity of the sediment is 2.65, and g is the gravimetric constant (m/s²). A useful dimensionless form of this equation was presented by Chang (1998). This approach has seen widespread application in Europe and other countries outside the U.S., but it has its critics. In the U.S., equations developed by Einstein (1950) are often used. However, there are many other bedload transport equations described in texts, such as those by Chang (1998) and Chanson (1999).

6.6.4 BANK STABILITY

Understanding and evaluating bank stability relationships is complex and best performed by engineers as it requires knowledge of the geotechnical engineering properties of the banks. A simple way of viewing this issue is that, for

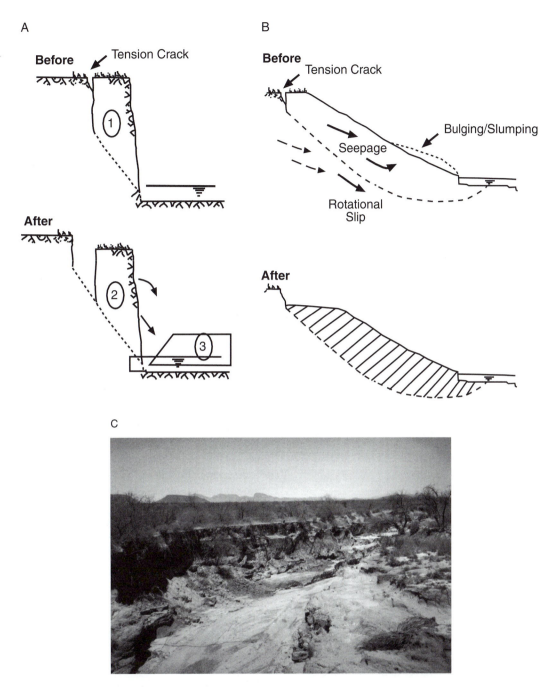

FIGURE 6.21 Bank failure due to mass instability: (A) slab failure, (B) rotational failure, and (C) slab failure in a river in New Mexico. (Photo courtesy John Lyon.)

a particular set of soil and vegetation conditions, there will be a combination of bank slope and height that are stable. Changing the ability of the vegetation to assist stability, increasing the slope, or increasing the height of the bank can each cause failure. Failure might occur due to formation of a tension crack a short distance away from the top of the banks and then a slab of the bank toppling into the stream; this we call a slab failure (**Figure 6.21A**). Sometimes, a curved failure surface will form a short distance away from the top of the bank, down toward the toe of

the bank. The bank then rotates or slumps into the stream (**Figure 6.21B**). Simon has developed a simple-to-use computer model that is available on the Internet (see the USDA-ARS National Sedimentation Laboratory site) that can be used to evaluate the potential for mass failure of a stream bank.

The effectiveness of vegetation in stabilizing the banks of large streams is difficult to evaluate. All investigators agree that vegetation is effective in curtailing stream bank erosion, but the disagreement is whether trees or grass are

TABLE 6.5
Geomorphic Considerations for Trees vs. Grass on Stream Banks

PREEXISTING CONDITIONS

A. Stream Channel Characteristics
1. Width of channel
2. Height of banks
3. Strength of banks
4. Texture of bed materials

B. Involved Processes
1. Hydroclimatology and flood regime
2. Hydraulic scour/deposition on banks and floodplains
3. Mass movement of banks, temperature and water regimes

Within the above conditions and processes are the following advantages and disadvantages for trees and grass:

TREES

A. Advantages
1. Root system is greater and with deeper structural strength
2. Greater ET, dryer banks
3. Less frost action

B. Disadvantages
1. Shade suppresses undergrowth
2. Exposed stems and roots enhance hydraulic scour
3. Mass and moment promote instability
4. Loss promotes instability:
    (a) LWD (climate dependent);
    (b) rootwad gaps create turbulence and scour

GRASS

A. Advantages
1. Creates "thatch," promotes vertical and lateral accretion

B. Disadvantages
1. Shallow and weaker roots
2. Less ET, wetter banks

the more effective agent. Many scientific papers can be marshaled to support either position (e.g., Stott, 1997; Trimble, 1997b), but the probability is that the answer is site specific. **Table 6.5** shows advantages and disadvantages of each type of vegetation and some natural factors that interplay with vegetation in ways not yet fully understood.

6.7 MEANDER MIGRATION, FLOODPLAINS, AND STREAMWAYS

We discussed sediment transport and the role of the effective discharge in shaping the channel, bedform, and banks of a stream. Let us look at why a stream might meander and the role of the floodplain in this process. **Figure 6.22** shows an example of meander migrations on a stream in Ohio. The analysis was performed by interpreting aerial photographs available for several time periods. An interpretation of how

the channel geometry probably changed with time is shown in **Figure 6.23**.

Why did these changes occur? Let us look at how floodplains, bars, and banks form and why they migrate. Whiting (1998) discussed these six floodplain formation processes: lateral point bar accretion, overbank vertical accretion, braided channel accretion, oblique accretion, counterpoint accretion, and abandoned channel accretion. For the purpose of this discussion, we just focus on lateral point bar accretion and overbank vertical accretion. We have seen that, as water flows around a bend, the velocity and shear stresses are highest on the outside of the bend and cause scour. On the inside of the bend, the flow velocity is lower, and material deposits to form a point bar. This process is called *lateral point bar accretion*. The bar will continue to build and will flatten until it approaches the effective discharge stage. Overbank deposition will continue to build and smooth the surface as finer sediments are deposited. When flows exceed the effective discharge, they will overtop the banks (except in deeply incised channels), and materials that are finer than those deposited during lateral accretion will be deposited on the banks and floodplain. This process is called *overbank vertical accretion*. The reason the materials will be finer is because the velocity or stream power of the overbank flows on the floodplain are lower than in the main channel, partly because of the lower flow depths and partly because of the greater resistance of flow due to vegetation on the floodplain.

Bed and bank shear stresses are not uniform in a meandering stream (**Figure 6.24**). Consider flow in a channel that initially bends to the left (looking downstream). Entering the bend, the shear stresses will be high on the right bank. They then diminish on the right (inner) bend as the stream flows around the bend and become the highest on the outside bend (left bank) just downstream from the peak of the bend. Often, the high stresses on the outer bend will be two to three times the mean bed stress. Channel migration by this process can be seen in **Figure 6.23**.

Ward and Mecklenburg (2002) used the term *streamway* to describe the main channel and attached floodplain that should be protected by a stream setback. Brookes (1996) described a concept by which the river is allowed to run wild and attributed the approach to Palmer (1976). The streamway should be considered as the zone in which a "stream rules," and it should not be of concern if the main channel encroaches on or floods activities within this zone. The width of a floodplain is difficult to define as it is associated with the magnitude of the flood considered and the physical location of a floodplain boundary, such as a high terrace, levee, or the toe of the valley walls. In developing a streamway equation, Ward and Mecklenburg wanted to provide (1) a streamway that would be wide enough to accommodate the existing meander pattern; (2) a streamway that would accommodate meander migrations that might occur over time; (3) a safety factor

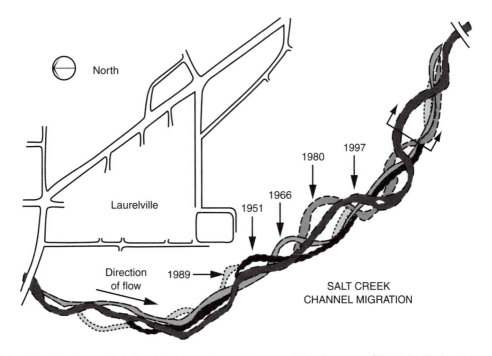

FIGURE 6.22 Meander migration on Salt Creek, Ohio, during a 46-year period. (Courtesy of Dan Mecklenburg.)

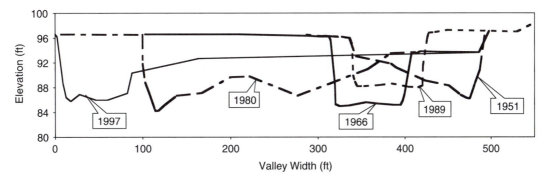

FIGURE 6.23 Channel dimension changes at the cross section shown in **Figure 6.22**.

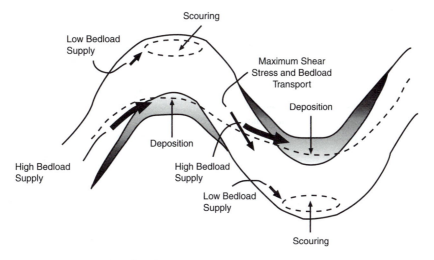

FIGURE 6.24 Processes that cause meander migration.

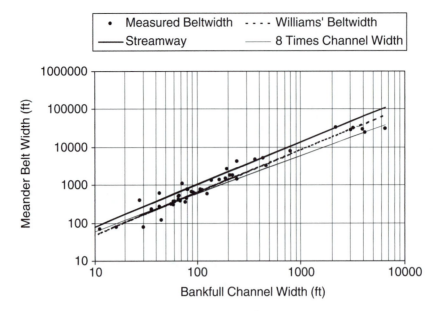

FIGURE 6.25 Belt width as a function of channel width.

because the equation would be based on empirical equations, which do not account for all the variability in data used in their development; and (4) a minimum level of protection on both banks of the stream.

Many empirical equations have been developed to describe bankfull (effective discharge) channel geometry. One such equation by G.P. Williams (1986) relates meander belt width (B, meters) and the bankfull width (W, meters) as follows:

$$B = 4.3W^{1.12} \qquad (6.18)$$

Equation 6.18 was based on 153 data points from rivers around the world, and the correlation coefficient r for the equation is 0.96. Belt width and bankfull width data for 47 of the locations are presented in the article by G.P. Williams (1986) and are plotted in **Figure 6.25**. Based on an analysis of these data, Ward and Mecklenburg (2002) developed the following equation to obtain the streamway width S_w equation:

$$S_w = 6.0W^{1.12} \qquad (6.19)$$

where S_w and W are now expressed in feet. Then, based on regional curves for the eastern U.S. that relate channel width to drainage area DA (square miles), they obtained the relationship

$$S_w = 120DA^{0.43} \qquad (6.20)$$

Ward and Mecklenburg (2002) concluded that setbacks should be sized based on geomorphic concepts; particularly, bed material mobilization was appropriate for sand and gravel bed streams. The empirical approach presented here

is appropriate in valleys broad enough for the meander pattern to be a function of the bankfull width or drainage area. Their results suggest that streamway widths at least eight times the bankfull width will in many cases have a wide enough streamway to allow for meander migration over time. Streams with these magnitude streamways might have the potential to self-adjust to low levels of urbanization and floodplain modification. However, small amounts of stream incision, floodplain modification in narrow valleys, or low levels of additional development on urbanized watershed each have a high potential to cause instability. Their conclusions were based in part on a study of potential sediment transport associated with a 100-year time period and various floodplain sizes relative to the size of the channel.

6.8 STREAM CLASSIFICATION

Like other components of the hydrologic cycle, there have been extensive efforts to classify channels and streams. Specific objectives in classifying streams might include:

1. Providing a consistent frame of reference for communicating stream morphology and condition among a variety of disciplines and interested parties. For climate, the Koeppen climate classification system (Chapter 2) provides a consistent frame of reference among geographers, climatologists, biologists, and other scientists.
2. Predicting a stream's behavior from its appearance.
3. Developing specific hydraulic and sediment relationships for a given stream type and its state.

4. Providing a mechanism to extrapolate or use site-specific data to stream reaches with similar attributes.

5. Identify if the stream is in dynamic equilibrium, incised, overwide, or in a transitional (stable or unstable) evolutionary stage.

6. Provide a context for evaluating stream health.

Rosgen (1996) listed the first four objectives and stated that, "The objective of classifying streams on the basis of channel morphology is to set categories of discrete stream types so that consistent, reproducible descriptions and assessments of condition and potential can be developed." Stream order, discussed in Section 6.3, is one classification approach based on the relative position of a channel in a stream network. Its application to evaluating the biology of a stream system was discussed and shown in **Figure 6.6**. Unfortunately, there is no single stream order classification system that has been universally accepted, partly because of the difficulty in deciding the starting part for calling a channel a first-order stream and because there are various ways that a numbering system might be established. A major limitation of the approach is that it provides little information on the channel system and does not convey a visual image of the channel properties. Other classification approaches, such as describing a channel by its mean bed material size (sand bed, gravel bed, bedrock) or by the physical setting or land use (mountain stream, meadow stream, urban channel, etc.) help convey an image of the channel, the topography, or the adjacent land use, but do not satisfy many of the objectives for having a stream classification system.

One of the earliest morphology-based classification approaches was the approach of Leopold and Wolman (1957) of classifying rivers according to their pattern as braided, meandering, and straight. Schumm's (1977) classification of streams into erosion, transport, and deposition reaches is helpful in beginning to understand the sediment transport behavior of a stream, but like the other approaches mentioned, it only provides a limited amount of information. Whiting and Bradley (1993) developed 42 stream classes for headwater streams; these classes are functions of dimensional measures of channel features and are based on distinct groupings of these features. Other comprehensive approaches have been developed by Montgomery and Buffington (1997), Rosgen (1996), Simon (1989), and numerous other geomorphologists, geographers, and engineers. Throughout the world, there are examples of local classification systems, such as that of Hoskins (1979), developed for a specific set of local conditions. Another example is the work of Rhoads and Herricks (1996), who identified nine channel types found in east central Illinois. They have not formally developed their observations into a classification method and have not proposed application in other regions. However, we

feel that the approach has applications throughout low-gradient agricultural areas in the Midwest. In the next sections, we present an overview of three classification methods that have seen much debate or application.

6.8.1 MOUNTAIN STREAM MORPHOLOGY CLASSES

Montgomery and Buffington (1997) developed a morphology-based approach for mountain streams. The approach has primarily been applied to streams in the northwest region of the U.S. Our interpretation of their approach is summarized in **Table 6.6**. They identified the following seven basic channel types: cascade, step–pool, plane–bed, pool–riffle, dune–ripple, colluvial, and bedrock. In addition, they discussed reaches where flow obstructions such as large woody debris (LWD) create a forced morphology. It is important to note that they specifically stated that their proposed classification "does not cover all reach types in all environments (e.g., estuarine, cohesive-bed, or vegetated reaches), we have found it to be applicable in a variety of mountain environments."

To some extent, the approach of Montgomery and Buffington (1997) addressed most of the six objectives listed for developing a classification system. Its main strengths are that it (1) is based on stream morphology; (2) provides excellent insight on the source, supply transport, and storage of sediment; (3) considers virtually all the factors that influence stream morphology; (4) considers a range of flow conditions; and (5) identifies distinctive channel bed morphology that can be visually identified. To some extent, their classification system is based on an extensive evaluation of numerous studies by others; they cited more than 80 references in their article.

The method was not developed as a "blueprint" for channel restoration and is rather qualitative, even though substantial quantitative data is provided for the stream reaches that were evaluated. The first seven properties in **Table 6.6** are based on Table 1 in Montgomery and Buffington (1997). The other four properties (flow type, sediment supply and transport, bed slope, and floodplain) are our interpretation of the extensive narrative that Montgomery and Buffington presented to describe their classification method. Montgomery kindly reviewed the brief summary that follows.

Cascade Channels (**Figure 6.26A**) typically have cobble and boulder bed materials, are narrowly confined by valley walls, and have steep bed slopes. During typical flows, the large bed material is immobile because of its large size relative to the flow depth. These larger materials are only mobilized during extreme events. Gravel and finer-grained bed materials in depositional sites might be mobilized during events with a recurrence interval of 1 to 7 years. Small scour pools are located between cascades and are often less than one channel width apart. These pools serve

TABLE 6.6
Summary of the Channel Morphology Characteristics of the Montgomery and Buffington (1997) Classification of Mountain Streams

Property	Cascade	Step-Pool	Plane–Bed	Pool–Riffle	Dune–Ripple	Bedrock	Colluvial
Bed material	Boulder	Cobble–boulder	Gravel–cobble	Gravel	Sand	Rock	Variable
Bedform pattern	Random	Vertically oscillatory	Featureless	Laterally oscillatory	Multilayered	Irregular	Variable
Roughness elements	Grains, banks	Bedforms (steps, pools), grains, banks	Grains, banks	Bedforms (bars, pools), grains, sinuosity, banks	Sinuosity, bedforms (dunes, ripples, bars), grains, banks	Bed and banks	Grains
Sediment source	Fluvial, hill slope, debris flow	Fluvial, hill slope, debris flow	Fluvial, hill slope, debris flow	Fluvial, bank failure	Fluvial, bank failure	Fluvial, hill slope, debris flow	Hill slope, debris flow
Sediment storage	Lee and stoss sides of flow obstructions	Bedforms	Overbank	Overbank, bedforms	Overbank, bedforms	Overbank	Overbank
Confinement	Confined	Confined	Variable	Unconfined	Unconfined	Confined	Confined
Pool spacing	<1 Channel width	1–4 Channel widths	None	5–7 Channel widths	5–7 Channel widths	Variable	Unknown
Flow type	Tumbling, turbulent	Critical to supercritical at steps and subcritical at pools	Variable with localized turbulence and critical flow at riffles	Subcritical at pools and often turbulent; critical flow at riffles	Variable, but mainly subcritical	Variable	Variable
Sediment supply and transport	Supply limited	Supply limited	Transport limited	Supply and transport limited	Transport limited		
Bed slope	Steep, >0.065	Steep, 0.030 to 0.065	Moderate to high, 0.015 to 0.030	Low to moderate, <0.015	Low	Variable	Steep
Bed slope[a]	0.050 to 0.12	0.015 to 0.134	0.001 to 0.035	0.001 to 0.015			
Floodplain	Limited or none	Limited or none	Variable	Well established	Well established	Variable	Not applicable

[a] From Chartrand, S.M., and P.J. Whiting, *Alluvial Architecture in Headwater Streams with Special Emphasis on Step-Pool Topography Earth Surface Processes and Landforms*, Vol. 25, John Wiley & Sons, New York, 2000, pp. 573–600.

as depositional sites during low-flow conditions. Cascade channels have a high sediment transport capacity and rapidly deliver sediment to channels with lower bed slopes.

Step–Pool Channels (Figure 6.26B) have small width-to-depth ratios and longitudinal steps that provide most of the elevation drops; they are typically confined by the valley walls. Fine material that deposits in the pool is transported as bedload during frequent events, while the larger step and bed-forming material is mobilized infrequently. Useful insight on processes in step–pool channels was provided by A. Chin (1989, 1999a, and 1999b). Based on an analysis of step–pool sequences in the Santa Monica Mountains, California, Chin found that step–pool sequences are restructured within a 5- to 100-year period due to high- and extreme-discharge conditions. Based on her field tests, she concluded that step formation is consistent with antidune theory (Chin, 1999b). The constraints she listed for step formation are (1) heterogeneous materials; (2) near-critical or supercritical flows; and (3) sufficient

A

FIGURE 6.26A Cascade channel in Washington State. (Courtesy and copyright D. Montgomery.)

B

FIGURE 6.26B Step-pool channel in Arizona. (Courtesy and copyright D. Montgomery.)

C

FIGURE 6.26C Plane-bed channel in Arizona. (Courtesy and copyright D. Montgomery.)

shear stresses to mobilize bed materials, but still allow for deposition. Good insight on step–pool systems is also presented in the work of Chartrand and Whiting (2000).

Plane–Bed Channels (Figure 6.26C) have a large relative roughness (ratio of 90th percentile grain size to bankfull flow depth; Church and Jones, 1982) and lack discrete bars. They also lack rhythmic bedforms (such as step, pools, and riffles), although isolated bedforms might occur due to localized obstructions or changes in bed slope. Bed surfaces are often armored and exhibit sediment supply limited conditions. Bed material mobilization is usually closely related to bankfull discharges. These channels are transitional between those that are supply limited and those that are transport limited.

Pool–Riffle Channels. In self-formed channels, pools and riffles are typically spaced every 5 to 7 channel widths; the channels have low-to-moderate bed slopes and well-established floodplains. Although Montgomery and Buffington limited application of their classification to mountain streams, pool–riffle channels are also commonly found in low-to-moderate gradient nonmountainous regions. Bars typically form when the bed slope is less than 2%, and there are sand and gravel-size bed materials and wide width-to-depth ratios. Pool–riffle channels **(Figure 6.26D)** exhibit variable supply and transport-limited characteristics, depending on the degree of armoring. However, bed material mobilization is usually closely related to the bankfull discharge.

D

FIGURE 6.26D Pool-riffle channel in Washington. (Courtesy and copyright D. Montgomery.)

Dune–Ripple Channels. The dune–ripple channel morphology normally occurs in sand bed channels with low bed slopes. Again, this type of channel morphology is not restricted to mountain streams. Sediment transport occurs at most discharges; these systems are very dynamic

E

FIGURE 6.26E Dune-ripple channel in Ohio. (Courtesy Ohio Department of Natural Resources.)

F

G

FIGURE 6.26F Colluvial channel in Arkansas. (Courtesy and copyright D. Montgomery.)

FIGURE 6.26G Bedrock channel in British Columbia. (Courtesy and copyright D. Montgomery.)

and can exhibit a variety of bedforms. Further insight on dune–ripple channels **(Figure 6.26E)** and the formation of bedload sheets was provided by Whiting et al. (1988).

Colluvial Channels (Figure 6.26F) are tiny, first-order streams at the start of a channel network. Typically, they are located in valleys with colluvial fills; in steep terrain, debris flows account for much of the sediment transport. The importance of these areas as sinks and sources of sediment is illustrated by the work of Trimble, which is reported in Chapter 9.

Bedrock Channels (Figure 6.26G) lack a continuous alluvial bed and can occur in just about any topography. In mountain streams, they are usually confined by valley walls and are commonly found in headwater streams. They typically have a high transport capacity and are supply limited. Montgomery and Gran (2001) and Montgomery and Buffington (1997) found that there is no consistent difference in channel width between bedrock and alluvial reaches. However, in some confined environments, bedrock channels are associated with higher shear stresses than alluvial reaches, and this could result in narrower

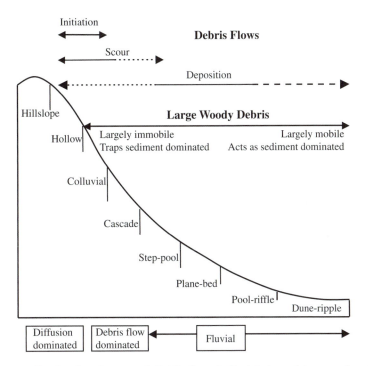

FIGURE 6.27 Idealized long profile showing the general distribution of alluvial channel types and controls on channel processes in mountain drainage basins. (Based on Montgomery, D.R., and J.M. Buffington, *GSA Bull.*, 109:596–611, 1997.)

channels and deeper bankfull flows. As valleys broaden, and where the main channel is not confined by the valley walls, it has been Ward's experience that bedrock channels often will be wider than alluvial bed channels in the same environment.

Montgomery and Buffington (1997) also identified forced morphologies primarily controlled by large woody debris. Log jams are a concern throughout the U.S., and the role of the forested riparian zone is a controversial subject. **Figure 6.27** (based on Montgomery and Buffington, 1997) provides useful insight on the relative location where each channel type might typically occur and the controls on channel processes in mountain drainage basins. Note that bedrock channels are not shown because they could occur anywhere along the channel system, but typically occur in locally steep reaches at any point along a stream profile. Montgomery et al. (1996) found that bedrock reach morphology occurred in channels that were steep for their drainage area compared to alluvial channels with the same or similar drainage areas. Also, Montgomery and Buffington emphasized how localized changes in topography and geology might shuffle the deck and place almost any of the channel morphologies in a different order relative to the others.

6.8.2 AN ALTERNATIVE MOUNTAIN STREAM CLASSIFICATION SYSTEM

Chartrand and Whiting (2000) more recently studied more than 100 mountain stream reaches in Idaho. In addition

to comparing their results with some of the Montgomery and Buffington (1997) stream types, they provided useful additional insight on the behavior of mountain streams. Their results are in general agreement with the sequence and type of channels that Montgomery and Buffington suggested might be found in mountain streams. However, Chartrand and Whiting found a greater overlap in bed slopes associated with each channel type than was first reported by Montgomery and Buffington. Also, they found that the step wavelength in a step–pool system is highly correlated with the step height and median grain size.

6.8.3 ROSGEN STREAM CLASSIFICATION METHOD

Rosgen (1994, 1996) proposed a hierarchy of river morphology. His classification approach is based on the notion that:

> Natural stream stability is achieved by allowing the river to develop a stable dimension, pattern, and profile such that, over time, channel features are maintained and the stream system neither aggrades or degrade. For a stream to be stable it must be able to consistently transport its sediment load, both in size and type, associated with local deposition and scour. Channel instability occurs when the scouring process leads to degradation, or excessive sediment deposition results in aggradation. (Rosgen, 1996)

He divided his classification approach into the following four levels:

TABLE 6.7
Rosgen's Level II Stream Classification System

Stream Property	Stream Types[a]							
	A	B	C	D	D_A	E	F	G
Entrenchment ratio	<1.4	1.4–2.2	>2.2	na	>4.0	>2.2	<1.4	<1.4
W/D ratio	<12	>12	>12	>40	<40	<12	>12	<12
Sinuosity	1–1.2	>1.2	>1.2	na	Variable	>1.5	>1.2	>1.2
Bed slope (fraction)	0.04–0.1	0.02–0.04	<0.02	<0.04	<0.005	<0.02	<0.02	0.02–0.04

[a] After the stream type, add a number that corresponds to the mean bed material type where bedrock is 1; boulders are a 2; cobble is a 3; gravel a 4; sand a 5; silt–clay a 6. For example, a Rosgen Type C4 stream has a gravel bed.

Level I: Geomorphic characterization that integrates basin topography, land form, and valley morphology. At a coarse scale, the dimension, pattern, and profile are used to delineate stream types.

Level II: Morphological description based on field-determined reference reach information.

Level III: Stream "state" or condition as it relates to its stability, response potential, and function.

Level IV: Validation level at which measurements are made to verify process relationships.

Rosgen's Level II classification is illustrated in **Table 6.7**. The classification is a function of the following seven channel attributes: (1) mean bankfull depth, (2) maximum bankfull depth, (3) bankfull width, (4) flood-prone area width, (5) channel sinuosity, (6) mean channel slope or water surface slope, and (7) median channel material size. The ratio of the flood-prone area width to the bankfull width is called the *entrenchment ratio* (*ER*). The width-to-depth ratio is the ratio of the bankfull width to the mean bankfull depth. The channel material size is based on the representative median particle diameter or the d_{50} index diameter (diameter of 50% of the sampled population is equal to or finer than this) for the channel. The particle size distribution of the channel materials is obtained by conducting a pebble count in riffles and pools along two cycles or meander wavelengths of a reach.

The channel sinuosity is an index of channel pattern, determined from a ratio of stream length divided by valley length or estimated from a ratio of valley slope divided by channel slope. The *ER* is the ratio of the flood-prone area width W_{fpa} divided by the bankfull channel width W_{bkf}:

$$ER = \frac{W_{fpa}}{W_{bkf}} \quad (6.21)$$

The bankfull width W_{bkf} is the width of the stream channel at the bankfull stage elevation in a riffle section.

The mean depth d_{bkf} is the depth of the stream channel at the bankfull stage elevation in a riffle section. The width-to-depth ratio *W/D* is

$$W/D \text{ Ratio} = \frac{W_{bkf}}{d_{bkf}} \quad (6.22)$$

The maximum depth d_{mbkf} is the depth of the bankfull channel cross section or vertical distance between the bankfull stage and thalweg elevations in a riffle section. The flood-prone area width W_{fpa} is measured at an elevation that is twice the maximum depth at the location at which the maximum depth was determined.

Rosgen Type A Stream. Type A streams (**Figure 6.28A**) are usually steep, entrenched, and confined channels. They range from A1 bedrock channels associated with faults, scarps, folds, and joints to A6 channels incised in cohesive soils. If the bed slope is greater than 10%, an "a +" designation is placed after the stream type. Boulder, bedrock, cobble, and to some extent gravel bed channels usually have a step–pool or cascading bed form. The bedrock and boulder bed channels are high-energy sediment supply–limited streams. Erosion, instability, mass wasting, and debris flow become more dominant processes as the bed material becomes finer (A3 to A5). However, the A6 stream type has low bedload transport, and much of the sediment transport is moved as a wash load in these systems.

Rosgen Type B Stream. Type B streams (**Figure 6.28B**) are moderately entrenched channels normally with bed slopes of 2 to 4%. The channel bed consists of a series of rapids and cascades with irregular scour pools. The bed and banks are relatively stable, and they are sediment supply-limited systems.

Rosgen Type C Stream. Type C streams (**Figure 6.28C**) are slightly entrenched, meandering systems characterized

A

FIGURE 6.28A Rosgen Type A stream, Colorado. (Courtesy of Dan Mecklenburg.)

B

FIGURE 6.28B Rosgen Type B stream, Colorado. (Courtesy of Dan Mecklenburg.)

C

FIGURE 6.28C Rosgen Type C stream, Ohio. (Courtesy of Dan Mecklenburg.)

by a well-developed floodplain. They typically have a riffle–pool bed form and a wide width-to-depth ratio. These streams are stable and usually are sediment supply and transport limited (C1 to C3). However, if they have gravel or finer bed and bank materials, they are susceptible to scour, erosion, and meander migration. As the bed and bank materials become finer, a larger percentage of the sediment load will be suspended or wash load. Type C channels exhibit a wide range of bed slopes, with cohesive, fine-grained (silt and clay) banked channels found in the virtually flat lake plain soil areas of the Midwest. Boulder and cobble bed Type C channels might occur at bed slopes up to 2%. Valley wall confinement and the lack of a wide floodplain usually force a Type A or B stream at steeper slopes.

Rosgen Type D Stream. Type D streams **(Figure 6.28D)** are multiple channel systems that typically do not have a boulder or bedrock bedform. However, Ward has seen examples in South Africa where localized bedrock control (such as a bedrock outcrop) results in the formation of a short, braided reach. Braided channels can occur across a wide range of morphological and topographic conditions. Alluvial fans, broad alluvial valleys, U-shape glacial valleys, glacial outwash valleys, low relief alluvial valleys, and deltas are all common locations for Type D streams. Alaska, New Zealand, and Japan have spectacular braided streams. With the exception of Type DA streams (a special category), these systems have a high sediment supply and transport capability, so they typically have high sediment yields. Type DA streams have cohesive banks, occur in

D

FIGUR 6.28D Rosgen Type D stream, Alaska. (Courtesy of Dan Mecklenburg.)

E

FIGURE 6.28E Rosgen Type E stream, Colorado. (Courtesy of Dan Mecklenburg.)

G

F

FIGURE 6.28F Rosgen Type F stream, Ohio. (Courtesy of Dan Mecklenburg.)

FIGURE 6.28G Rosgen Type G stream, Ohio. (Courtesy of Dan Mecklenburg.)

low-gradient areas, and are wetland, marsh, or delta systems often found adjacent to large waterbodies, such as lakes and oceans.

Rosgen Type E Stream.   Type E streams (**Figure 6.26E**) have a low width-to-depth ratio and exhibit a wide range of sinuosity. Many meadow streams are Type E streams. Generally, they are very stable, in part because they have well-developed floodplains with dense (often grassy) vegetation that helps stabilize their near-vertical banks.

Rosgen Type F Stream.   Type F streams (**Figure 6.28F**) are entrenched systems that normally exhibit bed slopes less than 2%. Boulder and bedrock systems are usually stable, while gravel and sand bed F channels are often deeply incised, have high bank erosion rates, and are often a failed or failing Type C channel.

Rosgen Type G Stream.   Type G streams (**Figure 6.28G**) are deeply entrenched systems that normally exhibit bed slopes of 2 to 4%. Boulder and bedrock systems are usually stable, while cobble, gravel, and sand bed G channels are unstable, often deeply incised; have high bank erosion rates; and are often a failed or failing Type B or E channel. Typically, the G3 to G6 types have characteristics of a gully.

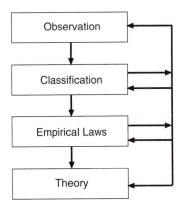

FIGURE 6.29 The evolution of observations into theory. (Based on Goodwin, C.N., in *Wildland Hydrology*, D.S. Olson and J.P. Potyondy, Eds., TPS-99-3, American Water Resources Association, Herndon, VA, 1999, pp. 229–236. With permission.)

EXAMPLE 6.3

A reach of a stream in an alluvial valley has a bed slope of 1.0%. The slope along the length of the valley is 1.5%. The bankfull width is 60 ft, and the mean bankfull depth is 3 ft. The flood-prone width is 150 ft, and the dominant bed material is gravel. What Rosgen stream type applies to the stream at this reach?

Solution

To determine the stream type, the bed slope, sinuosity, W/D ratio, entrenchment ratio, and dominant bed material need to be known.

The sinuosity is equal to the valley slope divided by the channel slope. Therefore,

$$\text{Sinuosity} = \frac{1.5}{1.0} = 1.5$$

From Equation 6.22, the width-to-depth ratio W/D is

$$W/D \text{ ratio} = \frac{W_{bkf}}{d_{bkf}} = \frac{60}{3} = 20$$

From Equation 6.21, the entrenchment ratio is

$$ER = \frac{W_{fpa}}{W_{bkf}} = \frac{150}{60} = 2.5$$

The bed slope was given and is 1% or 0.01 ft/ft. The dominant bed material is gravel. From **Table 6.7**, we can see that the stream type is C4 because the entrenchment ratio is greater than 2.2, the W/D ratio is greater than 12, the sinuosity is greater than 1.2, the bed slope is less than 0.02 ft/ft, and the dominant bed material is gravel.

Answer

The stream type at this reach is C4.

6.8.4 USING STREAM CLASSIFICATION METHODS

Goodwin (1999) presented an interesting discussion on the need for stream classification methods and noted that "historically, most fields of science have gone through a classification phase." He illustrated this by the flow diagram

shown in **Figure 6.29**. While not everyone will agree with this evolution model for developing theory, it does provide a useful starting point for discussing the merits and limitations of stream classification methods.

Let us begin by accepting that classifying a river may be helpful in understanding the behavior of the river and how the behavior might change because of land use changes or modifications to the channel. However, a classification is only an aid in the management of a river or in the development of an engineering design for the river. It does not directly provide a design solution. Moreover, not all fluvial geomorphologists buy into the idea that stream behavior can be predicted by morphology.

Montgomery and MacDonald (2002) proposed a "Diagnostic Approach to Stream Channel Assessment and Monitoring" and noted "Our argument is based on the observation that a particular indicator or measurement of stream channel conditions can mean different things depending upon the local geomorphic context and history of the channel in question." The method requires the collection of a comprehensive set of information on a reach and high-level expertise to be able to use the data to make a diagnosis. Interestingly, they listed one disadvantage of the method is that it "requires experienced field personnel trained beyond the level of workshops or short courses and a willingness to bring in additional expertise when the diagnosis is particularly difficult." Certainly the need for additional resources could be considered a disadvantage, but the need for high-level expertise is arguably a strength of the method. Insight on the uses and limitations of using geomorphological stream classification in aquatic habitat restoration was presented by Kondolf (1995).

To understand how a stream classification system might help use, let us relate the methods we discussed with the six objectives for having a classification system presented in Section 6.8:

Objective 1: Providing a consistent frame of reference for communicating stream morphology and condition among a variety of disciplines and interested parties. This is often considered one of the main strengths of the Rosgen classification method. Communication across disciplines is often difficult, and discussing, evaluating, understanding, and developing strategies to protect and restore streams poses a particularly difficult challenge. Geomorphologists often come from geology and geography disciplines that place emphasis on geology, processes, and long time spans. Typically, they have a good appreciation for data collection and the development of empirical relationships. Engineers and hydrologists tend to place more emphasis on landscape and watershed flow processes or channel hydraulics. Their design philosophy is steeped in "building" to withstand some extreme event and overdesigning to provide a factor of safety. Biological scientists typically evaluate stream and riparian zones based on a combination of "normal" conditions that control the viability, productivity, and diversity of the biology of interest. This approach includes considering thresholds of "impact" and "failure." This simple and perhaps unfair "stereotype" comparison among these groups illustrates that they each begin evaluating streams from a different knowledge base and use different timescales, different parameters, different terms, and even different scales to evaluate the system. Add farmers, developers, foresters, the socioeconomic disciplines, politicians, lawyers, and concerned citizens who have a stake in a stream issue and the need for a common communication tool quickly becomes evident. Unfortunately, simplicity and ease of use can easily lead to misuse or an inadequate level of understanding. For example, many scientists and engineers who rely heavily on modern computers and only perform limited measurements have a poorer grasp of basic concepts than those who have made extensive measurements on streams and watersheds, have extracted data using tedious manual methods, and have used antiquated technologies, such as slide rules, old mainframe computers with punch cards, or only used a simple calculator as aids for performing calculations. Another way to view the issue is to compare stream morphology and assessment to cooking. Most of us can do a fair job of preparing a simple meal most of the time, but as the recipe becomes more complicated, the likelihood of

failure increases. Frequently, stream issues can be compared to the most complicated recipe possible.

Objective 2: Predicting a stream's behavior from its appearance. As stated by Rosgen (1996), we have used the term *appearance*, but interpret it to mean *form* and be based on a combination of visual observations and fluvial measurements (profile, pattern, dimension, and bed material). To some extent, with the exception of stream order concepts, all of the classification methods discussed provide some information on a stream's behavior based on appearance. The more comprehensive approaches provide the best information and could probably be rearranged into more useful and more widely accepted classes. For example, a Rosgen C4 stream has a gravel bed and is often associated with a pool–riffle sequence typically spaced every five to seven channel widths. The channels have low-to-moderate bed slopes (less than 0.02) and well-established floodplains. They have high width-to-depth ratios and typically have well-established point bars at bends and alternating bars in some locations. They are usually unconfined by valley walls and have a high sinuosity. Clearly, there is little or no difference between the Rosgen C4 stream type and the pool–riffle mountain stream described by Montgomery and Buffington (1997). Chartrand and Whiting (2000) also discussed mountain headwater streams, but pool–riffle streams with bed slopes less than 1% (0.01) are widespread throughout the U.S. and many parts of the world. As also stated by Montgomery and Buffington, bedrock channels can have most bed forms and be found in most topographies. Are there other similarities between these two approaches or other schemes? Yes, the Rosgen Aa+ stream type does not exactly fit a cascade channel morphology but is similar, the Rosgen A stream type usually has a step–pool morphology, and most Rosgen B stream types have a plane–bed morphology. The D and DA are braided channels that fit the Leopold and Wolman (1957) classification. Rosgen E channels are unlikely to be found in mountain regions. It is our opinion that, in some situations (such as a meadow stream), they provide the best visual representation of any of the Rosgen stream types and are a distinct class, while in other situations, they have all the features of a C channel except their width-to-depth ratio is too low. We have plenty of examples of F channels as they are often incised or failed C channels. We also have plenty of examples of G channels as they are either gullies that have formed because of poor land stewardship or are artificially engineered ditches and channels.

Objective 3: Developing specific hydraulic and sediment relationships for a given stream type and its state. Rosgen's book (1996) provides extensive discussion that relates his stream types to sediment transport. His dimensions and ratios are primarily a function of the bankfull or effective discharge. Debatably, the discussions of Montgomery and Buffington (1997), Whiting and Bradley (1993), and Chartrand and Whiting (2000) might provide more specific detail on how their classes are related to sediment supply, transport, and stream hydraulics. The basic stream geomorphology information any of these approaches provides might be useful in channel design or protection. However, to a large extent the application of this information to design is independent of the need or desire to classify the stream. Another way to view this issue is that the scientists and engineers who have the knowledge to develop viable designs and strategies based on knowledge of the stream classification could have made the same decisions without determining the stream classification. On the other hand, designs and decisions that are based solely on the classification have a much lower likelihood of success.

Objective 4: Providing a mechanism to extrapolate or use site-specific data to stream reaches with similar attributes. This is the danger zone where people start to make mistakes with the Rosgen classification method and his other approaches to protecting and restoring channels. Most of the classification methods discussed earlier fall somewhere between empirical laws and theory (**Figure 6.29**). They are useful educational tools and provide qualitative information that can be transferred from one location to another. Rosgen extended his ideas beyond these points and has a number of specific approaches that can be used in stream restoration. It should first be recognized that Rosgen was successfully doing stream restoration long before he developed his classification method. Therefore, knowledge of the stream type is not necessarily needed to develop a design and in fact does not directly lead to a design solution. A fundamental ingredient in the Rosgen restoration approach is to conduct a survey on a "reference reach" (Rosgen, 1997). Unfortunately, in many locations reference reaches are an endangered species and can be very elusive. If one is found, probably it does not represent exactly the same drainage area size,

land uses, and morphology conditions as the restoration site. Similar is not always close enough. There is then often a need to utilize a regional curve. Again, many practitioners resort to using published curves that are not site specific enough for their restoration project. However, if we assume that perfect data are available and the empirical equations that might be used are ideal, we still face a number of challenges. First, Rosgen (1998) identified a "priority" approach for the restoration of incised rivers; the approach was based on his classification method. He also had a series of evolution models (at least 14 combinations) that used the classification method. To be adopted effectively, these methods require extensive knowledge of stream geomorphology. Determining what the current stream type is might be relatively easy. Knowing what it was previously and what it will become next poses enormous challenges. This is compounded by human activities that might change the watershed and floodplain conditions. Resource constraints and the selection of appropriate stabilization materials compound the difficulty. However, again if we have the brilliance to overcome all these difficulties, there still remains a challenge that cannot be fully appreciated unless you are fortunate enough to see in action a stream expert (such as Rosgen) and a contractor who also is an expert. The care and knowledge it takes to make minor design modifications successfully during construction might be the difference between success and failure. Ward has seen Rosgen and a contractor "read" the stream and watch the hydraulics of the flow as a cross vane was installed. Ward also saw the contractor repeatedly adjust the position and change the boulders that were used so they fitted together correctly, were stable, were placed at the correct elevations, and created the desired hydraulics (**Figure 6.30**). This is not something we can learn from reading a book or taking a course.

Objective 5: Identify if the stream is in dynamic equilibrium, incised, overwide, or in a transitional (stable or unstable) evolutionary stage. We touched on this subject when considering Objective 4. Useful evolution models have been developed by Simon (1989) and others (see Section 6.9). However, application of these methods requires extensive knowledge and experience. Also, an understanding of geotechnical engineering is desirable. The Rosgen stream classification identifies stable and unstable stream types. However, even a stable stream type might be in a

FIGURE 6.30 Contractor adjusting the boulder size and position while constructing a cross-vane in Colorado.

transitional stage that requires much expertise to identify.

Objective 6: Provide an index that can assist in evaluating stream health. We have inadequate knowledge to relate geomorphology to stream health. In some circumstances, coarsening of bed material due to urbanization might actually improve fish habitat. In other cases, a stream in a transitional stage (such as building bars, rapidly changing position, or with a forced morphology due to LWD) might have the most diverse and vibrant ecology. Water quality impacts associated with land use activities or point discharges are often unrelated to stream type. Knowledge of stream order might provide some insight regarding which biological communities to expect. Failure to find the anticipated communities, or distribution of these communities, does not necessarily indicate poor ecological function. Knowledge of the channel type can provide clues and a piece of the information needed to establish a relationship between the state of the channel and its potential state. Refer to Montgomery and MacDonald (2002) for further discussion on this issue.

6.9 CHANNEL EVOLUTION

6.9.1 DISTURBANCE

It is probable that there is some combination of catastrophic events that can significantly change the form of any stream system. The magnitude of the combination of factors that causes change are stream reach, stream system, and watershed specific. For some streams, it might take

an event as large as the PMP (see Chapter 2) to cause enough change to induce instability. For many of us, it is hard to visualize the magnitude of change that could occur due to a catastrophic event. Ward has seen when a single cyclone (essentially a southern hemisphere hurricane) carved a canyon that was a quarter of a mile wide and more than 100 ft deep. Nearly 20 years later, there has been little change in the shape of this canyon. During severe flooding that occurred in southern Africa in 2000, rivers such as the Sabie River (**Figure 6.31**) spread laterally across several miles and created water depths that also exceeded 100 ft. It is an astonishing sight to be high above a river and more than half a mile away from the banks and to look up and see flood-swept debris at the top of trees (**Figure 6.31B**). In places, this flood caused extensive reshaping of the river; other parts of the river were barely changed by this event. Yet, just a few years after this event, many parts of the river returned to their preflood shape and bedform (**Figure 6.31D**).

It often does not take a catastrophic event or an act of nature to create change and potential instability. In many reaches and streams, straightening a channel, land use change, a change in riparian vegetation or the floodplain size, in-stream gravel mining, or simply the installation of a culvert might cause localized or stream-system-wide instability (further discussion on this topic is presented in Chapter 9). This change might occur rapidly over a few weeks, months, or years. In other cases, the change could take place very slowly. Recovery from change will also be highly variable. Therefore, we need to be very cautious in predicting change and identifying the current state of a stream system. We can take many measurements or utilize various theories and classification approaches, but often the best insight can be obtained by talking to people who have lived near the stream for several decades. Such

FIGURE 6.31 Flood impacts and recovery on the Sabie River Catchment, South Africa: (A) downcutting and channel widening, (B) flood debris in trees far above the river, (C) natural revegetation on sand bars deposited by the flood, and (D) a tributary and well-attached floodplain that show little sign of the flood.

discussion can provide knowledge of the stable form of the stream, which activities have caused change, the water depth of baseflow, bankfull and extreme discharges, and human activities on the watershed that might have caused change. To predict how a channel might evolve requires knowledge of the current evolutionary stage of the channel. Also, in some cases perceived instability and channel change might simply be part of the natural cycle of channel adjustment and movement that has occurred over centuries. People who wish to conduct development activities within a floodplain or near a stream often desire or expect a stream to stay in a fixed position. On the other hand, stream geomorphologists, ecologists, and environmental groups are quick to try to identify incision, channel widening or failure,

and degradation or aggradation. Reality might not be consistent with either viewpoint.

It is also important to note that impacts that cause change are commonly, but not always, associated with increased flows or human modifications of a channel. Locating a dam on a stream will often have a dramatic impact both upstream and downstream of the dam. In the last few years, there has been considerable focus on the removal of low-head dams that have outlived their usefulness or are in need or repair or replacement. Also, of particular concern throughout the world is change associated with reduced flows. This might be because of shifts in climate conditions or human activities. In particular, regulation of flows associated with power generation and water supply often result in significant reductions in both

FIGURE 6.32 Reelfoot Lake in northwest Tennessee. (Courtesy and copyright Northwest Tennessee Tourism.)

low baseflows and the less-frequent high flows we called the effective discharge that most influence sediment transport and bank and channel forming. Reductions in these flows often also have a substantial impact on stream ecology. These flows are sometimes referred to as *environmental flows/discharges*. At a conference on this topic held in South Africa in 2002, it was noted that often the environmental discharges that flow regulators are willing to provide are 10 to 20% or less of the mean annual discharge, but studies suggest that flows approaching 80% of the mean annual discharge are often needed to sustain stream health and the original channel geomorphology.

We often perceive watershed and stream changes as having a negative impact on water resources. This is not the case, as the following example illustrates. On February 7, 1812, one of the most severe earthquakes ever to strike the U.S. occurred along a fault line centered near the town of New Madrid, MO. One outcome of the reshaping of the landscape by the earthquake was the formation of Reelfoot Lake in a place that had been a swampy forest of cypress trees, cottonwoods, and walnut trees. This natural lake has a surface area of 20 mi^2, averages a depth of only 5 to 6 ft, and has a maximum depth of about 18 ft. The old forest that stood before the earthquake still lies just beneath the surface and helps make the lake one of the world's greatest natural fish hatcheries, with more than 56 species of fish (**Figure 6.32**). The surrounding area is a nature preserve with more than 240 species of birds. It is a winter nesting site for American bald eagles and a major stopping point for waterfowl migrating to and from Canada. Certainly, there are also many examples of dams that have resulted in wonderful habitats for fish and birds. However, this is also often at the expense of adverse

impacts on the stream system. That was not the case at Reelfoot Lake.

6.9.2 EVOLUTION MODELS

Numerous people have studied stream change and channel evolution. In Section 6.8, we presented some of the work by Rosgen. His informative wildland hydrology Web site (http://www.wildlandhydrology.com/) has current information on his evolution approaches. Simon, who currently works at the USDA-ARS (Agricultural Research Service) National Sedimentation Laboratory has devoted much of his career to studying stream instability and channel evolution (Simon and Hupp, 1986; Simon, 1989; Simon 1992; Simon and Thorne, 1996; Darby and Simon, 1999; Simon and Rinaldi, 2000). The evolution model of Simon and Hupp (1986) is helpful in understanding channel evolution associated with human channelization. This work was based on studying streams in Tennessee. Rosgen made an interesting comparison of the similarities between the evolution of a channelized system and morphological change in some streams that can be described by his classification system (**Figure 6.33**).

In the Simon and Hupp (1986) model, the premodified Stage 1 assumes the channel is in equilibrium (stable). The constructed Stage 2 is considered an instantaneous stage due to human activities such as realigning parts of the channel and reshaping the banks. Like Rosgen, Simon extended this application to natural changes (Simon and Thorne, 1996). Rapid degradation occurs during Stage 3 as the channel deepens and the bed slope flattens. This causes bank instability and bank failures (slumping) near the base of the banks. The channel then widens once the side slopes and bank heights cause shear stresses that

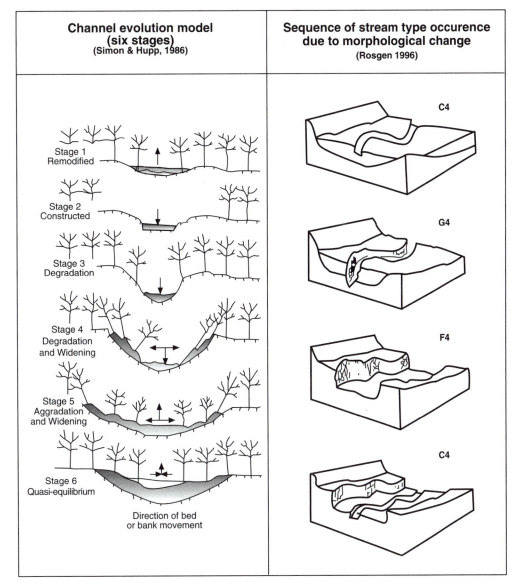

FIGURE 6.33 Comparison between the Simon and Hupp (1986) evolution model and morphological changes described by the Rosgen classification system. (From Rosgen, *Applied River Morphology*, 1996. With permission.)

exceed the strength of the bank material (Stage 4). The flatter and wider channels are now unable to transport the increased sediment loads and aggradation occurs (Stage 5). Simon (1992) reported that the aggradation rates might be 60% less than the associated degradation, so the bed level will not recover to its original premodified state. However, gradually the channel will reach a state of quasi equilibrium (Stage 6). It is noteworthy that channel evolution models explain much observed channel instability as stages of a predictable sequence that eventually leads to a stable channel similar to the predisturbance channel. Also, the depth of incision and the amount of overwidening will depend on the resistance to change of the bed and bank materials.

6.10 WHAT TO MEASURE AND WHY THE MEASUREMENT SHOULD BE MADE

There are numerous reasons why we might need to measure fluvial properties of a stream: assessing if the stream is stable; undertaking a channel modification or restoration activity; designing a culvert, bridge, weir, dam, or some other flow modification structure; developing land use management strategies that will reduce or prevent adverse impacts on the stream system; developing stream setback guidelines; or using the stream as an educational activity. Also, there might be a need to make fluvial measurements as part of an environmental, ecological, or biological study. The discussion in this text is restricted to fluvial

TABLE 6.8
Initial Site Selection for a Hypothetical 350-mi² Watershed

Drainage Area (mi²)	Number of Sites	Comment
0.1 to 1.0		Assuming 1 site might not be used
0.5 to 1.0	2 or 3	
1.0 to 10		Assuming 1 or 2 sites might not be used
1.0 to 3	1 or 2	
3 to 5	1 or 2	
6 to 9	1 or 2	
10 to 100		Assuming 1 or 2 sites might not be used
12 to 18	1 or 2	
24 to 40	1 or 2	
50 to 90	1 or 2	
100 to 300		Assuming 1 site might not be used
100 to 150	1 or 2	
200 to 300	1 or 2	
350	1	If there is a gage at the outlet
Total	Target of 10 to 12	

measurements in wadable streams. Other texts discuss how to make water quality and biological measurements (Hauer and Lambert, 1996). A brief overview of biological assessment methods is presented in Chapter 12. We limit the discussion to survey measurements of the dimension, pattern, profile, and bed material of a stream. Anticipated applications are characterization of the stream geomorphology for a reference reach, characterization of the stream geomorphology at a site that will be restored or modified, and development of regional curves for use in developing designs or making assessments at various scales.

It is assumed in the discussion that there is a known site of interest. Part of the analysis is to determine the fluvial characteristics at that site. The methods proposed by Harrelson et al. (1994) provide a good foundation for conducting a survey of fluvial features in a stream. A decision then needs to be made on whether measurements should be made at additional reference reach sites. Regardless of whether there is a need to conduct a detailed survey at one or more additional reference reaches, it is recommended that a regional curve be developed for the stable stream form likely to occur in settings similar to that of the reach studied. For example, if we are evaluating a headwater step–pool (Rosgen Type A) system in a steep, sloping, forested region of a state, we should not include in the regional curve moderate and low-gradient streams that exhibit plane–bed and riffle–pool bedforms (Rosgen Type B and C streams).

6.10.1 SITE SELECTION AND RECONNAISSANCE AT A WATERSHED SCALE

Typically, potential sites should be marked on a state gazetteer map. Topographic maps and aerial photographs might be used to aid this process. Consultation with watershed coordinators and agency personnel within the watershed can also save a considerable amount of time. Sites should be selected to provide data for various size drainage areas within each log cycle. A drainage area "doubling" approach is recommended, perhaps 0.1, 0.2, 0.5, 1, 2, 4, 8, 16, 32, 64, 128, 256, 512, 1024 mi², depending on the size of the watershed. Exact adherence to this approach is not necessary, and an initial estimate of the drainage areas can be made visually prior to data collection. It is also desirable to get good spatial distribution across the watershed and two to three measures for similar drainage areas within each log cycle. Finally, we recommend that initially at least 50% more sites be marked on the gazetteer than actually might be measured. Provisional site selection for a hypothetical 350-mi² watershed is illustrated in **Table 6.8**. Regional curves do not need to be based on large areas. We have developed useful curves for areas smaller than 1 mi².

Sites should be selected to provide easy access from a road. Often, it is easiest to request permission from landowners at the time of the site visit and anticipated data acquisition. While more than one visit to a watershed might be needed, cordial and informative one-to-one interaction with landowners is more productive than trying to contact somebody by phone and then trying to explain to them what you want to do. Be professional and look professional. If permission cannot be immediately obtained or the site does not have good access or easily identifiable effective discharge features, no measurements should be made. Usually, it will be necessary to walk several hundred feet upstream or downstream of a road crossing to avoid any channel modification or influence of the culvert or bridge at the road. In addition to marking the site on the map, obtaining coordinates with a handheld GPS is useful particularly if the drainage area will be determined using a geographic information system (GIS) procedure (see Chapter 13). If a comprehensive survey is performed on a reach, it might take several hours. Therefore, data on only one, two. or perhaps three reaches can be made in a day. If only cross-sectional data are obtained using a rod, pins, and a tape measure, then it might be possible to visit two or three sites an hour. We should recognize that rapid methods that do not use electronic surveying instruments will result in reduced accuracy.

6.10.2 Drainage Area Determination

Various methods can be used to determine watershed boundaries and drainage area. Common approaches are to manually draw boundaries on a topographic map — such as a USGS quad sheet — or to use a GIS to delineate the boundaries and calculate areas. If a topographic map is used, the drainage area is then calculated using a planimeter or some simple approximation method based on the geometry of the watershed. Watershed boundaries, areas, and GIS layers are often available through Internet sites of watershed and conservancy groups and state and federal agencies (USGS, NRCS, and EPA in particular). Detailed GIS information on a countywide basis is available from some county auditors or county engineers' offices and their Internet sites.

GIS software such as ArcGIS can be used to develop a digital elevation model (DEM) and GIS that include the stream and ditch network. A "stream-burning" approach is then used in conjunction with an ancillary watercourse network layer, forcing flow path cells to conform to existing stream networks. This approach is particularly useful in areas that are extremely flat.

6.10.3 Discharge and Sediment Data

Streamflow data can be obtained from http://water-data.usgs.gov/nwis; a summary of their stream-gaging program was presented by Wahl et al. (1995). The USGS also has well-established procedures for measuring the velocity and discharge in wadable streams (Rantz et al., 1982). However, these activities are time consuming and only provide information on the flow regime that was occurring when a stream was visited. Typically, we will not be at the stream when there are near-effective discharge conditions. Even if we are able to visit a stream during these conditions, the flow will usually be too rapid for safe wading. In tranquil water, you might be able to safely wade with water above your waist (however, many state and federal agencies will require use of a flotation device in these conditions). Rapid-flowing water that does not even reach your knees might test your swimming expertise.

Continuous sediment transport data are usually only available for a few streams (often large rivers) in most states. Measuring sediment transport is a difficult, expensive, and time-consuming activity. Therefore, measurements of static conditions on the bed, banks, bars, and floodplain are often the only indicators of the available sediment transport and supply in a stream. Recall that both the Rosgen and the Montgomery and Buffington classification methods rely heavily on good understanding of sediment processes.

6.10.4 Channel Properties

For each site reach, information is obtained on the channel materials, dimension, pattern, and profile. Procedures should be consistent with the guidelines presented by Harrelson et al. (1994). The survey can best be conducted with a total station or a laser level, one or more 100-ft tapes, and a telescoping rod with a laser receiver. The approach is best suited for streams that are wadable; the approach can be efficiently performed by a team of three or four people (see **Figure 6.34**). The most experienced person should hold the rod and select the points where the measurements will be made. A person with neat handwriting and good hearing and who is a good listener should record the data. Often, three people will be more efficient than four people — particularly in small streams.

For each reach, a longitudinal survey is usually conducted over a stream length equal to at least 20 channel widths so that the survey encompasses at least two bends. Features typically measured include channel cross sections at several points along the reach; bed profile along the thalweg; water surface profile; azimuths of the banks from each feature to the next feature or bend; the effective discharge (bankfull) elevation at points along the reach where it can be easily identified, the top of the bank; and bed material particle size distribution. When making azimuth readings, we need to make sure the line of sight from one measuring point to the next is parallel to the banks. This information is used to determine the pattern of the main channel. We are not trying to determine the pattern of the thalweg, so we should not make "person-to-person" measurements. Measurement of the water surface profile is helpful in determining the mean bed elevation. It also provides a useful indicator of measurement errors, which unfortunately are easy to make with telescoping rods.

At each reach, measurements are made for two to four cross sections. These include at least one representative cross section that shows the dominant fluvial bench feature in a riffle feature and a cross section at a bend if there is a well-defined point bar. For a step–pool bedform, you should take cross sections at the steps and at the pools. Measurements at rapids (cascades), steps, and riffles are particularly useful as discharge is better calculated at these locations. Please note that taking measurements at

FIGURE 6.34 Students measuring the dimensions of a stream in an urban park.

FIGURE 6.35 Point bars at bends are indicators of the effective discharge stage.

cascades and the top of steps can be very dangerous. Do not underestimate the power of water or how slippery rocks can be. This is one place that you do not want to learn from your mistakes. Measurements at pools are helpful as they provide information on the width and depth of these features. All grade breaks at a cross section should be measured.

Channel width and depth measurements depend on an ability to measure correctly the location of the effective discharge stage. Stream indicators include the back of point bars (**Figure 6.35**), significant breaks in slope, changes in vegetation, the highest scour line, or the top of the bank (Leopold, 1994). Stain lines on rocks and boulders are sometimes used in steep-sloping mountain streams. Also, the presence or absence of old trees along the banks in wooded riparian zones can be useful indicators. Finding several of these indicators will increase the likelihood that the correct stage has been identified. However, the person or people identifying the effective discharge stage should only use the indicators that they best understand.

Bed, bank, bench, and floodplain materials are good indicators of the bankfull stage.

1. In many streams, the coarsest bed materials will be found at a rapid, step, or riffle.
2. A bench that is below the bankfull stage will usually have signs of coarse material similar in size or slightly finer than the bed material in riffles.

3. An active floodplain will usually exhibit sands, some soil structure, and vegetation. There also might be signs of small, shallow, active channels that flood and drain due to high-flow events. Care should be taken in evaluating active floodplains. A large event a few months prior to a survey might have caused flow and deposition on a terrace that is higher than the active floodplain associated with the effective discharge.

4. If there is good soil structure and high organic matter content, you are probably above the bankfull stage.

Harman et al. (1999) noted that "the most consistent bankfull indicators for streams in the rural Piedmont of North Carolina are the highest scour line and the back of the point bar. It is rarely the top of the bank or the lowest scour or bench." Based on our experience, use caution in the use of scour lines as an indicator of the effective discharge stage. The scour line at the top of cut banks on outer bends will usually be above the effective discharge stage, while the bottom of the scour line will be below the effective discharge stage. In straight reaches, there is sometimes a scour line below the top of the bank even when the effective discharge stage is the top of the bank. It is important to recall that the effective discharge is effective in shaping the main channel.

6.10.5 The Elusive Nature of Bankfull (Effective Discharge) Dimensions

Learning how to make survey measurements correctly normally takes a few classes and the use of this knowledge under supervision on several occasions. The basics of surveying the landscape and a stream are similar, but not identical. Therefore, proficient land surveyors will still require some training to be effective in a stream environment. Understanding the surveying techniques is the easy part as there is a lifetime of learning associated with identifying the correct features to measure and then measuring them in a consistent manner. It is our experience that it takes several visits to streams even to begin to understand fully what needs to be measured. Correctly making and recording all the necessary measurements will require repeated opportunities to do this type of work across the seasons and annually. Seeing a stream system at different times of the year is particularly helpful in understanding its geomorphology.

Unfortunately, even very experienced geomorphologists will often disagree about the features they are observing or the set of circumstances that might have produced these features. The golden rules we recommend are as follows:

1. Be prepared and organized.
2. Understand which features are to be explored.
3. Scout as long a length of stream as possible.
4. Look at the whole stream system (channel, floodplain, valley walls).
5. Look at similar stream systems in the watershed and the watershed landscape.
6. Try to use all the evidence available.
7. Do not have preconceived ideas of what should be there or jump to rapid conclusions as to what you will find.
8. Do not restrict measurement taking to the site of interest.

Example 6.4 illustrates the difficulty in measuring effective discharge dimensions. It is based on real measurements made on a watershed in Ohio. However, some of the data have been modified and simplified for illustrative purposes. Also, conditions from several different locations were blended to illustrate better what could happen.

Example 6.4

Three graduate students have taken several courses on hydrology, stream geomorphology, and watershed hydrology. They have also conducted stream surveys on several occasions under the supervision of an experienced stream engineer, who is a professor with good knowledge of geomorphology. As part of a research study, they are requested to develop a regional curve for a watershed group. The professor is sick and is unable to join them. However, he requests that they measure all grade breaks and try to measure one cross section at each location. A detailed survey that includes measurements of the profile and bed material will be made at a few of the sites on another occasion. The width and depth measurements the students made are presented in **Table 6.9**. At the first four sites, the students identified a small, well-defined bench near the bottom of the headwater "streams," which were channelized agricultural ditches. A second, poorly defined grade break occurred about a foot above the benches. The top of the ditches was usually at least 6 ft above the bed. At the next four sites, only one bench or grade break feature was observed. At the final four sites, there was a well-attached floodplain that showed signs of recent sand deposits in places. However, 1 to 1.6 ft below the top of the channel banks, there were sand and gravel bars, point bars, or small benches in places. The width and depth dimensions each student decided to use are reported in **Table 6.9** together with the professor's interpretation after talking to each student. Who is correct?

Solution

Student 1 came to the professor and said, "I got some fantastic results. Look, all my r^2 values are at least 0.98 — I

TABLE 6.9
Dimension Data for Example 6.4

Student 1		Student 2		Student 3		Professor	
Width (ft)	Depth (ft)	Width (ft)	Depth (ft)	Width (ft)	Depth (ft)	Width (ft)	Depth (ft)
4	1	6.5	2	6.5	2	4	1
7	1.7	7.5	2.7	7.5	2.7	7	1.7
10	2	11	3	11	3	10	2
12	2.4	14	3.4	14	3.4	12	2.4
16	3.5	16	3.5	16	3.5	16	3.5
23	4	23	4	23	4	23	4
32	5	32	5	32	5	32	5
40	5.5	40	5.5	40	5.5	40	5.5
65	7.7	65	7.7	55	6.7	55	6.7
68	8.1	68	8.1	60	7.1	60	7.1
85	9.5	85	9.5	80	8.2	80	8.2
95	10.0	95	10.0	90	8.4	90	8.4

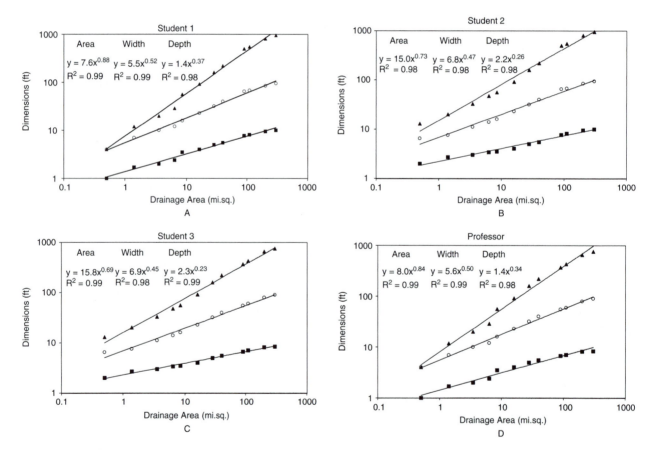

FIGURE 6.36 Regional curves for Example 6.4.

have almost perfect correlation" (**Figure 6.36A**). The professor asked the student how the survey was performed and how he decided which measurements to use. The student's response was that the low bench at the first four sites was clearly the dominant feature, and sand on the

dominant floodplain at the last four sites indicated that the top of the channel was the effective discharge elevation. The professor commented that this seemed reasonable, but that just a few weeks earlier there had been a 20-year flood in the region.

Student 2 came to the professor and said, "I got some fantastic results. Look, all my r^2 values are at least 0.98 — I have almost perfect correlation" (**Figure 6.36B**). The professor asked the student how the survey was performed and how he decided which measurements to use. The student's response was that the low bench at the first four sites was clearly too low, but sand on the dominant flood-plain at the last four sites indicated that the top of the channel was the effective discharge elevation. The professor commented that this seemed reasonable, but that just a few weeks earlier there had been a 20-year flood in the region. He also asked if they had observed any point bars at the first few sites. The student said "No, it had been raining, so we only walked a short distance from the road."

Student 3 came to the professor and said, "I got some fantastic results. Look, all my r^2 values are at least 0.98 — I have almost perfect correlation" (**Figure 6.36C**). The professor asked the student how the survey was performed and how she decided which measurements to use. The student's response was that the low bench at the first four sites was clearly too low, and that she thought sand on the floodplain at the last four sites was due to the flooding a few weeks earlier. The professor asked if she had observed any point bars at the first few sites. The student said, "The third site was only slightly channelized, the rain had stopped, and there was a point bar at an elevation about 1 foot below the higher grade break. However, we decided not to take a measurement because we were not sure if a bend was a good place, and one of the students had accidentally walked through some thistles — they sting!"

The professor then decided to create his own plot (**Figure 6.36D**). He showed it to the students and asked them if they thought it was more defendable than their curves. He also asked if any of them had tried to calculate the discharge at each site and determine if the recurrence interval of the discharge associated with their dimensions was reasonable, remained constant, increased with drainage area, or decreased with drainage area.

Answer

You be the judge! All the r^2 values are extremely high; you might normally expect values of 0.85 to 0.95. In some locations, the recurrence interval might increase or decrease slightly with drainage area. However, for Ohio conditions, it will usually be less than a few years. Perhaps they need to wait until they do a few detailed surveys. Consider the error if they applied these curves to a 3- or 300-mi^2 watershed. Also, look at the problems at the end of this chapter.

6.10.6 Bed Material Characterization

The most common method for characterizing the bed material of a stream is to conduct a Wolman pebble count

(1954). The procedure is suitable for use in streams that can safely be waded and is most efficiently performed with one person recording measurements in a field book and a second person wading the stream and measuring the bed material sizes.

Pebble counts can be made using grid, transect, zig-zag, or a random step–toe procedures. The best characterization is obtained if separate counts are made for riffles, pools, and runs or glides. These counts can then be weighted by the percentage of the reach length represented by each geomorphologic feature.

1. Select a representative reach and measure a minimum of 100 particles in each feature. Use a field book or tally sheet to record the counts.
2. Start at a randomly selected point at a bankfull elevation on one bank. We recommend that you start at the upstream end of the reach. Averting your gaze, pick up the first particle touched by the tip of your index finger at the toe of the wader.
3. Measure the length of the intermediate axis of the particle by approximating the shape of each particle as a length (longest axis), width (intermediate axis), and thickness (shortest axis). Embedded materials and very large particles should be measured in place by recording the smaller of the two exposed axes.
4. Determine the pebble count size class (see **Table 6.10**) for the measured length and record a count for that class (not the actual length that was measured).
5. Take one step across the channel toward the opposite bank. For a grid or transect survey, move perpendicular to the flow. For a random or zigzag survey, move diagonally downstream. The size of the step will depend on the size of the stream and whether riffles and pools are recorded separately. In very small streams and ditches, each step might only be about one shoe length long (1 ft), while in very wide rivers, you might make a measurement every one or two steps (3 to 6 ft).
6. Continue the traverse until you reach a bankfull elevation on the opposite bank. Make sure you include bar and bank materials (up to the bank-full elevation) in your count. Continue until you have at least 100 counts for each feature. If very fine materials are encountered, record them as a silt or clay count (a measurement is not required).
7. After the field counts are completed, the data are plotted by size class and frequency as illustrated in Example 6.5.

TABLE 6.10
Pebble Count Data for Example 6.4

Material	Size Range (mm)		Count	Percentage Finer
Silt/clay	0	0.062	1	0.7
Very fine sand	0.062	0.13	1	1.4
Fine sand	0.13	0.25	1	2.1
Medium sand	0.25	0.5	1	2.8
Coarse sand	0.5	1	2	4.2
Very coarse sand	1	2	5	7.6
Very fine gravel	2	4	8	13.2
Fine gravel	4	6	9	19.4
Fine gravel	6	8	12	27.8
Medium gravel	8	11	17	39.6
Medium gravel	11	16	25	56.9
Coarse gravel	16	22	18	69.4
Coarse gravel	22	32	15	79.9
Very coarse gravel	32	45	10	86.8
Very coarse gravel	45	64	6	91.0
Small cobble	64	90	5	94.4
Medium cobble	90	128	3	96.5
Large cobble	128	180	2	97.9
Very large cobble	180	256	1	98.6
Small boulder	256	362	1	99.3
Small boulder	362	512	1	100.0
Medium boulder	512	1024		
Large boulder	1024	2048		
Very large boulder	2048	4096		
Bedrock				
		Total Count	144	

EXAMPLE 6.5

A pebble count for a small alluvial stream in the Midwest region of the U.S. is presented in **Table 6.10**. The zigzag method was used, and counts were made in riffles and pools in the same proportion as they occurred in the study reach. Plot these data and determine the d_{35}, d_{50}, and d_{84} particle sizes for this bed material.

Solution

The total number of counts was 144. The percentage finer for each particle size was determined by dividing the accumulated counts for all materials up to that size, dividing that number by the total count of 144, and then multiplying by 100. For example, approximately 7.6% (100 × 11/144) of the particles are finer than 2 mm (very coarse sand). These values are presented in **Table 6.10**.

The percentage finer data are plotted in **Figure 6.37**. From the figure, the d_{35}, d_{50}, and d_{84} particle sizes are determined as about 10, 15, and 40 mm, respectively.

Answer

The d_{35}, d_{50}, and d_{84} particle sizes are 10, 15, and 40 mm, respectively.

EXAMPLE 6.6

The stream in Example 6.5 has a flow depth of 2 ft for the effective discharge and a bed slope of 0.5%. Calculate the tractive force and the mean particle size at incipient motion. Do you think the stream is stable? Justify your answer.

Solution

The percentage finer data are plotted in **Figure 6.36**. From the figure, the d_{35}, d_{50}, and d_{84} particle sizes are determined as about 14, 19, and 40 mm, respectively.

The tractive force in kilogram/square meter can be calculated from Equation 6.14 and Equation 6.15:

$$T = 1000 \, ds$$

where d is the flow depth in meters, and s is the bed slope as a fraction. Therefore,

$$T = (1000)\left(\frac{2}{3.28}\right)\left(\frac{0.5}{100}\right) = 3 \text{ kg/m}^2$$

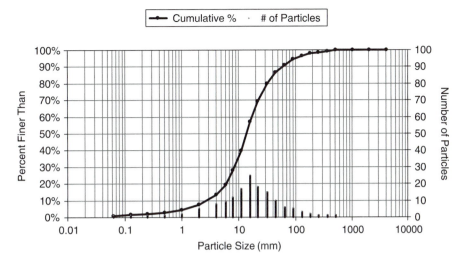

FIGURE 6.37 Bed material distribution for Example 6.5.

Then, from Equation 6.15, the average size bed material that can be moved is 3 cm or 30 mm. From **Figure 6.19**, we see the upper and lower ranges of the mean particle sizes that might be moved by the tractive are 23 and 40 mm.

Answer

The potential material size that can be moved is larger than the d_{50} of the bed material (20 mm) that currently occurs in the stream. Therefore, there exists the potential for further scour and a coarsening of the bed. However, from **Figure 6.19**, the lower value of the mean particle size that might be moved is 23 mm, which is similar to the measured value of 20 mm; therefore, we cannot say with any confidence that scour and coarsening will occur. Another way to consider this issue is to compare the calculated size of 30 mm with the measured d_{35} size of 14 mm and the d_{84} size of 40 mm.

PROBLEMS

6.1. Use the drainage network laws to determine the bifurcation ratio, stream length ratio, and drainage area ratios for the watershed data in **Table 6.11**.

6.2. The sinuosity of a stream in 1960, 1980, and 2000 was 1.0, 1.4, and 1.6, respectively. Describe how the stream probably changed with time and the channel pattern in each of the 3 years.

6.3. A stream with a bed slope of 0.005 has a bankfull discharge (flow rate) of 1000 ft³/sec. If the stream is not straight, is it more likely to be braided or meandering?

6.4. A stream has a bed slope of 0.6%. Determine the tractive force and mean bed material diameter that

TABLE 6.11
Stream Data for Problem 6.1

Stream Order n	Number of Streams N_n	Average Stream Length L_n(ft)	Average Drainage Area A_n (acres)
1	64	300	7
2	20	800	23
3	6	2000	72
4	1	4000	230

can be moved at incipient motion for a flow depth of 2 ft.

6.5. A watershed in your state has a drainage area of 10 mi². At the watershed outlet, estimate the bankfull (effective) discharge, cross-sectional area, mean bankfull depth, and mean bankfull width of the main stream that drains the watershed (use a regional curve).

6.6. Pebble count data for a mountain stream in Washington State are presented in **Table 6.12**. The zigzag method was used, and separate counts were made in riffles and pools. Riffles occupied 60% of the reach, and pools occupied 40% of the reach length. Plot these data and determine the d_{35}, d_{50}, and d_{84} particle sizes for the pools, riffles, and the whole reach.

6.7. A gravel bed stream has a mean bankfull depth of 2 ft, a maximum bankfull depth of 3 ft, and a bankfull width of 48 ft. The slope of the flood-prone area (on each side of the bank of the main channel) above the bankfull depth is 10:1 (horizontal:vertical). The channel bed slope is 0.6%, and the valley slope is 1.2%. Calculate the width of the flood-prone area, the entrenchment ratio,

TABLE 6.12
Pebble Count Data for Problem 6.6

Material	Size Range (mm)		Riffle Count	Pool Count
Silt/clay	0	0.062		5
Very fine sand	0.062	0.13		1
Fine sand	0.13	0.25		3
Medium sand	0.25	0.5	1	6
Coarse sand	0.5	1	2	10
Very coarse sand	1	2	4	12
Very fine gravel	2	4	5	15
Fine gravel	4	6	6	20
Fine gravel	6	8	7	16
Medium gravel	8	11	8	14
Medium gravel	11	16	10	10
Coarse gravel	16	22	12	8
Coarse gravel	22	32	15	6
Very coarse gravel	32	45	16	5
Very coarse gravel	45	64	17	3
Small cobble	64	90	10	2
Medium cobble	90	128	8	1
Large cobble	128	180	6	1
Very large cobble	180	256	4	
Small boulder	256	362	3	
Small boulder	362	512	2	
Medium boulder	512	1024	2	
Large boulder	1024	2048	1	
Very large boulder	2048	4096		
Bedrock				
		Total count	139	138

and the channel sinuosity. What Rosgen stream type is this?

6.8. A watershed in Pennsylvania has a drainage area of 100 mi^2. The stream has a mean bankfull depth of 4 ft, a maximum bankfull depth of 4.5 ft, a bankfull width of 80 ft, and a flood-prone area width of 200 ft. The channel slope is 0.8%, and the valley slope in the direction of flow is 1.2%. Draw the cross-sectional geometry of the stream. At the watershed outlet, estimate the (a) bankfull cross-sectional area, (b) the bankfull discharge (use a regional curve), (c) mean flow velocity when there is bankfull discharge, (d) entrenchment ratio, and (e) the channel sinuosity. Pools and riffles are located approximately every 500 ft. Classify this stream using (a) the Montgomery and Buffington system and (b) the Rosgen system. What would the dominant bed material be if it was a Type 3 stream? What streamway setback would be recommended for a reach at the outlet of the watershed?

7 Uniform Open Channel Flow

7.1 INTRODUCTION

Flow in a stream or constructed channel is a function of many factors, including precipitation, overland flow, interflow, groundwater flow, and pumped inflows and outflows; the cross-sectional geometry and bed slope of the channel, the bed and side slope roughness; meandering, obstructions, and changes in shape; hydraulic control structures and impoundments; and sediment transport and channel stability. Generally, flow in streams, ditches, channels, and large rivers is classified as *open channel flow* because the surface of the flow is open to the atmosphere. Open channel flow can occur in many ways. For example, it can be turbulent in steep, rocky areas or following severe storm events. Also, during severe storm events, there might be rapid changes in the depth and amount of flow. On other occasions, it can be tranquil, and it will be difficult to detect that the water is flowing.

Flow may be classified into many types, and a different mathematical equation is needed to describe each type. Discussion in this chapter begins with *steady uniform flow*. Flow in an open channel is *steady* if the depth of flow does not change or is considered constant during the time interval under consideration. The flow is *uniform* if the depth of flow is similar along a reach of the channel. In a natural channel, these conditions might only occur along reaches that are a few channel widths long or might never occur in some mountain streams. As we move from one feature, such as a riffle, to another feature, such as a pool, we need to redefine the cross section and perhaps switch calculation methods as the flow type changes. Therefore, after discussing steady uniform flow, we look at some other flow conditions.

7.2 FLOW VELOCITY AND DISCHARGE

7.2.1 CONTINUITY AND MANNING'S EQUATION

The discharge of a channel or river is given by the equation of continuity:

$$q = va \qquad (7.1)$$

where q is the discharge (cubic feet/second or cubic meters/second), a is the cross-sectional area of the stream (square feet or square meters) obtained by simplifying the geometry and calculating the area or by drawing cross sections to scale and measuring the area (see Chapter 14), and v is the average velocity of flowing water (feet/second or meters/second).

For uniform flow, in a channel the average velocity v can be estimated by Manning's equation:

$$v = \frac{1.5}{n} R^{2/3} S^{1/2} \qquad (7.2)$$

where v is the velocity of flow (feet/second or meters/second), n is Manning's roughness coefficient of the channel, S is the channel bed slope (feet/feet or meters/meters), and R is the hydraulic radius of the channel (feet or meters), which is calculated as

$$R = \frac{a}{P} \qquad (7.3)$$

where P (feet or meters) is the wetted perimeter of the channel cross section (see **Figure 7.1A**).

Manning's equation was developed by an Irish engineer, Robert Manning, between 1791 and 1795. If SI or metric units are used, the value of 1.5 in the equation is replaced by 1.0.

The bed slope of a natural stream or river can be found by dividing the change in elevation along the reach by the length of the reach. We used this approach in Chapter 5, but care must be taken now as we just want to make the measurement along a short length of channel to make a velocity and discharge estimation.

If the flow is deep relative to the bed material size, Manning's n can be estimated by (Newbury and Gaboury, 1993)

$$n = 0.04 d_{50}^{1/6} \qquad (7.4)$$

where d_{50} is the bed material size in meters. Care must be taken when using Equation 7.4 to estimate the bed roughness. If the depth of flow is very large compared to the d_{50}, the roughness might be less than estimated. However, if the depth of flow is similar or less than the d_{50}, the roughness will probably be greater than estimated.

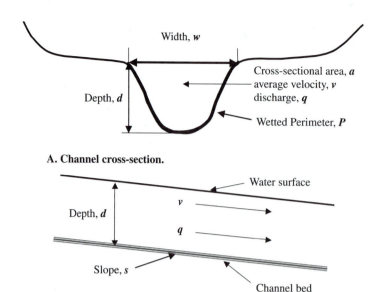

A. Channel cross-section.

B. Channel Profile.

FIGURE 7.1 (A) Cross section. (B) Profile of a channel.

EXAMPLE 7.1

The mean bed material size is 30 mm. Describe this channel by its bed material and estimate Manning's n.

Solution

From **Table 6.11** in Chapter 6, we see that 30-mm diameter bed material is a coarse gravel, so we would call this a gravel bed stream.

We can use Equation 7.4 to estimate Manning's n. First, we need to change the mean bed material size from millimeters to meters. So, 30 mm is 0.03 m, and

$$n = 0.04 d_{50}^{1/6}$$

and

$$n = 0.04(0.03)^{1/6}$$

This gives an n value of 0.022.

Answer

This is a gravel bed stream with a Manning's n of 0.022.

Manning's n can also be estimated from tabular information for different bed and side slope materials (**Table 7.1**). Manning's equation is only applicable when there is uniform subcritical flow. Critical and supercritical flow that might occur at riffle sections, cascades, and steps and over hydraulic control structures are discussed in Chapter 8.

Equation 7.1 and Equation 7.2 are often used to determine the depth of flow d, the discharge q, and the average flow velocity v in a river. A common method for determining discharge in a channel is to measure the flow depth and then calculate the discharge using equations that best describe the type of flow that occurred. This is sometimes the simplest scenario to evaluate because the depth of flow and channel geometry are known. Estimations of this nature might be required to determine if bed and side slope scour occurred or to determine the discharge associated with an observed storm event.

EXAMPLE 7.2

Determine the flow velocity and discharge in a clean, winding stream with some pools and weeds when the mean flow depth is 3 ft. The bed slope is 0.004, the mean width of the flow is 50 ft, and the banks can be approximated as vertical.

Solution

Draw a diagram of the channel cross section. Then, from **Table 7.1**, it is determined that Manning's n might vary between 0.033 and 0.050 for a clean, winding, gravel bed stream on a plain that has some pools and weeds. In Example 7.1, a gravel bed stream had a Manning's n of 0.022 if the mean diameter was 30 mm. Very coarse gravels can be 64 mm in diameter, and this would give a Manning's n of 0.025. So, the two approaches do not give identical estimates of Manning's n. Basing the roughness estimate on bed material size does not consider the influence of meandering and weeds. Let us use the minimum estimate of 0.033 from **Table 7.1**.

TABLE 7.1
Manning's *n* Roughness Coefficient

Type of Channel and Description	Minimum	Normal	Maximum
Constructed channels			
Smooth steel surface	0.011	0.013	0.017
Corrugated steel surface	0.021	0.025	0.03
Concrete			
Trowel finish or float finish	0.011	0.014	0.016
Finished, with gravel on bottom	0.015	0.017	0.02
Gunite, good section	0.016	0.019	0.023
Gunite, wavy section	0.017	0.022	0.025
On excavated rock	0.017	0.022	0.03
Concrete bottom float finished with sides of			
Dressed stone in mortar	0.015	0.017	0.02
Random stone in mortar	0.017	0.02	0.024
Cement rubble masonry, plastered	0.016	0.02	0.024
Cement rubble, dry rubble or riprap	0.02	0.027	0.035
Gravel bottom with sides of			
Formed concrete	0.017	0.02	0.025
Random stone in mortar	0.02	0.023	0.026
Dry rubble or riprap	0.023	0.033	0.036
Asphalt	0.013	0.014	0.016
Earth, straight and uniform			
Clean	0.016	0.02	0.025
With short grass, few weeds	0.022	0.027	0.033
Earth, winding and sluggish			
No vegetation	0.023	0.025	0.03
Dense weeds or plants in deep channels	0.03	0.035	0.04
Earth bottom and rubble sides	0.027	0.03	0.035
Cobble bottom and clean sides	0.03	0.04	0.05
Rock cuts			
Smooth and uniform	0.025	0.035	0.04
Jagged and irregular	0.035	0.04	0.05
Channels not maintained, weeds and brush			
Dense weeds	0.05	0.07	0.12
Same, highest stage of flow	0.045	0.07	0.11
Dense brush, high stage	0.07	0.1	0.14
Streams			
Streams on plain			
Clean, straight, full stage, no rifts or deep pools	0.025	0.03	0.033
Clean, winding, some pools, shoals, weeds, and stones	0.033	0.045	0.05
Same as above, lower stages and more stones	0.045	0.05	0.06
Sluggish reaches, weedy, deep pools	0.05	0.07	0.07
Very weedy reaches, deep pools, or floodways with heavy stand of timber and underbrush	0.075	0.1	0.15
Mountain streams, no vegetation in channel, banks steep, trees and brush along banks submerged at high stages			
Bottom: gravels, cobbles, and few boulders	0.03	0.04	0.05
Bottom: cobbles with large boulders	0.04	0.05	0.07
Pasture, no brush			
Short grass	0.025	0.03	0.035
High grass	0.03	0.035	0.05
Cultivated areas			
No crop	0.02	0.03	0.04
Mature row crops	0.025	0.035	0.045
Mature field crops	0.03	0.04	0.05

TABLE 7.1
Manning's *n* Roughness Coefficient (continued)

Type of Channel and Description	Minimum	Normal	Maximum
Brush			
Scattered brush, heavy weeds	0.035	0.05	0.07
Light brush and trees	0.035	0.06	0.07
Medium-to-dense brush, in winter	0.045	0.07	0.11
Medium-to-dense brush, in summer	0.07	0.1	0.16
Trees			
Dense willows, summer, straight	0.11	0.15	0.2
Cleared land with tree stumps, no sprouts	0.03	0.04	0.05
Same as above, but with heavy growth of sprouts	0.05	0.06	0.07
Heavy stand of timber, a few down trees, little undergrowth, flood stage below branches	0.07	0.1	0.12
Same as above, but with flood stage reaching branches	0.1	0.12	0.16
Major streams (top width at flood stage more than 100 ft)			
Regular section with no boulders or brush	0.025		0.06
Irregular and rough sections	0.035		0.1

Source: From Chow, V.T., *Open-Channel Hydraulics*, McGraw-Hill, New York, 1959. With permission.

The hydraulic radius is the cross-sectional area divided by the wetted perimeter. We can approximate the flow as 3 ft deep and 50 ft wide. So, the cross-sectional area, wetted perimeter, and hydraulic radius are a = (3 ft)(50 ft) = 150 ft²; P = (50 + 3 + 3) = 56 ft; and $R = a/P$ (Equation 7.3) = 150/56 = 2.68 ft.

For uniform flow in a stream, the average velocity v can be estimated by Manning's equation (Equation 7.2):

$$v = \frac{1.5}{n} R^{2/3} S^{1/2}$$

therefore,

$$v = \frac{1.5}{0.033} (2.68)^{2/3} (0.004)^{1/2} = 5.5 \text{ ft/sec}$$

Then, from Equation 7.1, the discharge in the stream is

$$q = va$$

so

$$q = (5.5 \text{ ft/s})(150 \text{ ft}^2)$$

and

$$q = 825 \text{ ft}^3/\text{sec}$$

Answer

The flow velocity is 5.5 ft/sec, and the discharge is 825 ft³/sec.

A more difficult, but common scenario, is to have a known estimate of the discharge (perhaps using predictive methods described in Chapter 5) and be interested in evaluating the velocity and flow depth in an existing channel or river. An example of this scenario would be making an evaluation of an upstream land use modification. Typically, an estimate would be made of the peak discharge associated with the land use modification, and then an analysis would be made to see if existing channels are able to convey the estimated peak discharge.

Two approaches could be used to evaluate this scenario. Equation 7.1 and Equation 7.2 could be combined as follows to eliminate the flow velocity:

$$q = \frac{1.5}{n} aR^{2/3} S^{1/2} \qquad (7.5)$$

The hydraulic radius and the cross-sectional area can then be written in terms of the depth of flow d, which is now the only unknown. Unfortunately, Equation 7.5 is not easily solved as it contains complicated functions of depth.

A practical approach commonly adopted is to use trial and error that requires a guess of the flow depth. Equation 7.5 is then solved to determine the discharge at the guessed depth. The discharge based on the guessed depth is then compared with the known discharge. If the discharge based on the guessed depth is too high, a shallower depth is tried, and the exercise is repeated. If the discharge is too low, then a deeper depth is tried. With practice, a solution that is within 10% of the correct answer can be obtained in three or four tries. Trying to obtain a more accurate estimate has little practical meaning because of uncertainties in determining the known discharge and Manning's *n*. Start by guessing a simple depth number such as 1 ft, 2 ft, 3 ft, and so on.

EXAMPLE 7.3

Determine the flow depth and flow velocity for the stream in Example 7.2 if the 10-year, 24-h discharge has been estimated, using the graphical peak discharge method, as 2000 ft³/sec. The stream is incised, and the floodplain will only be flooded when the flow depth exceeds a mean depth of 6 ft.

Solution

From Example 7.2, we know the depth will be greater than 3 ft. Let us try the maximum main channel depth of 6 ft. So, the cross-sectional area, wetted perimeter, and hydraulic radius are a = (6 ft)(50 ft) = 300 ft²; P = (50 + 6 + 6) = 62 ft; and $R = a/P$ = 300/62 = 4.84 ft.

Using Equation 7.5 with n = 0.033,

$$q = \frac{1.5}{n} aR^{2/3}S^{1/2}$$

therefore,

$$q = \frac{1.5}{0.033}(300)(4.84)^{2/3}(0.004)^{1/2} = 2468 \text{ ft}^3/\text{sec}$$

This is higher than the target discharge of 2000 ft³/sec, so let us guess a new depth of 5 ft. Now, the cross-sectional area, wetted perimeter, and hydraulic radius are a = (5 ft)(50 ft) = 250 ft²; P = (50 + 5 + 5) = 60 ft; and $R = a/P$ = 250/60 = 4.17 ft.

$$q = \frac{1.5}{0.033}(250)(4.17)^{2/3}(0.004)^{1/2} = 1862 \text{ ft}^3/\text{sec}$$

This is only 128 ft³/sec or 5.8% less than the target discharge of 2000 ft³/sec and is close enough. The velocity associated with this discharge is

$$v = \frac{1.5}{0.033}(4.17)^{2/3}(0.004)^{1/2} = 7.4 \text{ ft/sec}$$

Answer

The depth of flow is about 5.0 ft, and the approximate mean velocity is 7.4 ft/sec.

The most difficult scenario to evaluate is when there is a known discharge and a need to determine channel geometry capable of conveying this flow. This type of evaluation is commonly made by agricultural and civil engineers designing a ditch or channel. However, there are also many situations when a scientist might conduct a conceptual analysis for use in a cost–benefit analysis. The scenario is evaluated by first selecting a channel geometry based on technical and practical considerations. Once channel geometry has been selected, the approach is similar to the scenario discussed. However, at the end of the analysis, it might be determined that the geometry is greatly over- or undersize; in that case, it will be necessary to try a new geometry. Reference should be made to a hydraulic engineering text such as that by Haan et al. (1994).

7.2.2 APPROXIMATION AND ESTIMATION ERRORS

In Example 7.3, we approximated the channel cross section as a rectangle. This approach was used when we considered the Rosgen classification method. In that case, if we divided the cross-sectional area by the bankfull width, we obtained the mean bankfull depth. This certainly makes the calculations a bit easier, but does it create a large error in the answer? Let us look at this issue in Example 7.4.

EXAMPLE 7.4

Determine the discharge in the trapezoidal channel that is shown in **Figure 7.2**. The bed slope is 0.006, and Manning's *n* is 0.04. What is the percentage difference in your answer between basing your calculation on the actual geometry vs. approximating the geometry as a rectangle with the width equal to the top width of the flow?

Solution

The top width *w* is

$$w = 25 \text{ ft} + (3)(3 \text{ ft}) + (3)(3 \text{ ft}) = 43 \text{ ft}$$

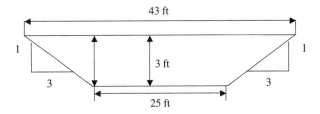

FIGURE 7.2 Cross section of channel for Example 7.4.

For a trapezoid, the cross-sectional area is the mean width times the depth, or

$$a = (0.5)(3 \text{ ft})(25 \text{ ft} + 43 \text{ ft}) = 102 \text{ ft}^2$$

The length of the wetted banks L_b can be calculated from the geometry as follows:

$$L_b = \sqrt{\left[(3 \text{ ft})^2 + (9 \text{ ft})^2\right]} = 9.5 \text{ ft}$$

Therefore, the wetted perimeter P is

$$P = (25 \text{ ft} + 9.5 \text{ ft} + 9.5 \text{ ft}) = 44 \text{ ft}$$

So, the hydraulic radius $R = a/P$ (Equation 7.3) = 102/44 = 2.32 ft.

For uniform flow in a stream, the discharge can be estimated by Equation 7.5:

$$q = 1.5 a n^{-1} R^{2/3} S^{1/2}$$

therefore,

$$q = 1.5(102 \text{ ft}^2)(0.04)^{-1}(2.32 \text{ ft})^{2/3}(0.006)^{1/2} = 519 \text{ cfs}$$

If we approximate the cross section as a rectangle with a width equal to the top width of 43 ft, the mean depth d will be

$$d = \frac{a}{w} = \frac{102}{43} = 2.37 \text{ ft}$$

The wetted perimeter for the approximated rectangular shape will be

$$P = (43 \text{ ft} + 2.37 \text{ ft} + 2.37 \text{ ft}) = 47.74 \text{ ft}$$

So, the hydraulic radius $R = a/P$ (Equation 7.3) = 102/47.74 = 2.14 ft. Therefore,

$$q = 1.5(102 \text{ ft}^2)(0.04)^{-1}(2.14 \text{ ft})^{2/3}(0.006)^{1/2} = 492.0 \text{ cfs}$$

The percentage error E is

$$E = 100\left(\frac{(519 - 492)}{519}\right) = 5.2\%$$

Answer

The discharge is 519 ft³/sec, and the error in approximating it as a rectangle is 5.2%.

We saw in Example 7.4 that approximating the channel as a rectangle produced a small error. Usually, this error will be less than 10% and does not need to be considered. For a natural channel, the cross-sectional area from one place to the next will often vary by more than a few percent, and the selection of where to make cross-sectional measurements is a matter of judgment. There will then be some error in our measurements. Also, we will not be able to obtain an exact measurement of the bed slope. However, perhaps more importantly, our estimate of Manning's n is rather approximate. In **Table 7.1**, we first had to find a description that matched our stream, and we then were provided a range of values to use. Without a lot of experience, it is easy to make a 50% error in estimating Manning's n.

A useful visual aid for estimating Manning's n was prepared by Barnes (1967). It contains Manning's n values and color photographs and descriptive data for 50 stream channels. Many of these streams are large, but it is still a useful guide. Unfortunately, the book is out of print, but at the time we wrote this book, a Web-based guide was available at http://manningsn.sdsu.edu. This guide was developed by Victor Ponce, Ampar Shetty, and Sezar Ercan of San Diego State University, California.

7.2.3 THE DARCY–WEISBACH EQUATION

A more theoretically based equation than Manning's equation was developed by Darcy and Weisbach for determining head losses in pipes. When adapted for open channel flow, it can be written as

$$v^2 = \frac{8gRS}{f} \tag{7.6}$$

where f is the friction factor.

For straight, uniform gravel bed channels, the friction factor can be estimated from the following form of the Colebrook–White equation:

$$f^{-1/2} = 2.03 \log\left(\frac{xR}{3.5 d_{84}}\right) \tag{7.7}$$

where x can be approximated by

$$x = 11.1\left[\frac{R}{y}\right]^{-0.314} \tag{7.8}$$

where y is the perpendicular distance from the bed surface to the point of maximum velocity. This will be equivalent to the maximum flow depth if the width-to-depth ratio exceeds 2. All dimensions in these equations are in meters, square meters, or meters/second.

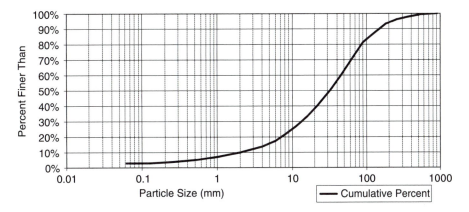

FIGURE 7.3 Pebble count for a riffle in Blacklick Creek.

EXAMPLE 7.5

A survey has been conducted on a reach of Blacklick Creek. The bankfull (effective discharge) width is 36 ft, the mean bankfull depth is 2.7 ft, and the mean bed slope is 0.4%. The stream is clean and winding and has some pools and riffles. The pebble count results are shown in **Figure 7.3**. To evaluate stream stability concerns, the effective discharge needs to be determined. Use Manning's equation to calculate the effective discharge. Compare your results to a solution based on the Darcy–Weisbach method. Are your results the same? If not, which result would you use? Justify your answer.

Solution

We can approximate the channel as a rectangle with a width of 36 ft and a depth of 2.7 ft.

Determine Manning's n and then solve Manning's equation:

$$v = \frac{1.5}{n} R^{2/3} S^{1/2}$$

From **Table 7.1**, a stream on a plain that is clean and winding and has some pools, shoals, weeds, and stones has a Manning's n that ranges from 0.033 to 0.05. If the concern is scour, use the lowest value. If the concern is flooding, use the highest value. Let us use 0.033. The reach bed slope is 0.004, and R is

$$R = \frac{a}{P} = \frac{(36)(2.7)}{36 + 5.4} = 2.35 \text{ ft}$$

Therefore,

$$v = \frac{1.5}{0.033} 2.35^{2/3} 0.004^{1/2} = 5.08 \text{ ft/s}$$

and

$$q = va = (5.08)(97.2) = 494 \text{ cfs}$$

Answer 1

The effective discharge is about 500 ft³/sec based on Manning's equation.

From the earlier solution, $R = 2.35$ ft or 0.71 m, and $y = 2.7$ ft or 0.82 m. The d_{84} is 100 mm or 0.1 m. The cross section was approximated as a rectangle, so the mean and maximum depths are the same.

Therefore,

$$x = 11.1 \left[\frac{0.71}{0.82} \right]^{-0.314} = 11.6$$

and

$$f^{-1/2} = 2.03 \log \left(\frac{(11.6)(0.71)}{3.5(0.1)} \right) = 2.78$$

and $f = 0.13$. Therefore, the velocity v (meters/second) is

$$v^2 = \frac{8(9.8)(0.71(0.004))}{0.13} = 1.72$$

So, $v = 1.31$ m/sec or 4.30 ft/sec. Therefore, the effective discharge is 97.2 ft² × 4.30 ft/sec or 418 ft³/sec.

Answer 2

The effective discharge is about 400 ft³/sec based on the Darcy–Weisbach approach.

Why are the results different? What result would you use?

Answer 3

Both results are similar. They depend a lot on the roughness factor. For Manning's *n*, this required some "expert knowledge" in selecting a tabulated value. This problem was avoided with the Darcy–Weisbach approach, but the challenge we had was determining if the bed material size had a major influence on the channel roughness. Also, the Darcy–Weisbach approach does not consider other factors, such as meanders and vegetation influence. A Manning's *n* of about 0.04 would have given a similar discharge as that determined by the Darcy–Weisbach approach. The flow depth is large compared to the mean bed material size, so the Darcy–Weisbach approach has probably overestimated the roughness for this discharge condition.

7.3   COMPOUND CHANNELS

When there is flooding, the flow might overtop the banks of the main river channel and flow along the adjacent floodplain. In parts of the U.S., earth and concrete levees have been constructed to contain the flow within the main channel and prevent flooding. However, as was evidenced in the 1993 flooding in the heartland of America (Nebraska, Iowa, Minnesota, Wisconsin, Kansas, and Illinois), there will eventually be an event of sufficient magnitude to overtop or breach any levee.

To estimate flow in a complex cross section that consists of a main channel and floodplain, it is possible to use Manning's equation provided there is uniform flow. In this case, the cross section can be divided into several subsections or flow compartments (see **Figure 7.4**). Typically, each subsection will exhibit different roughness coefficients and mean velocities. The total flow rate will be the sum of flow rates in all the subsections (see Chapter 14, Exercise 14.7).

Applying resistance (roughness) equations, such as Manning's equation, to compound straight, two-stage channels presents a challenge, particularly with low overbank flow. The problem is twofold. The first problem occurs as the stage initially rises above bankfull; there is a dramatic increase in wetted perimeter, little change in cross-sectional area, and a decrease in hydraulic radius, which results in a discontinuity in velocities estimated by Manning's equation. The second problem is accounting for the interaction between the slower overbank flow with the faster main channel flow.

Posey (1967) evaluated a number of commonly used resistance equation methods. He concluded that dividing the main section and two overbank sections by vertical lines worked well when overbank flow was shallow, but when overbank flow was at least half as deep as the bankfull channel depth, then dividing into subsections was not necessary. The dummy/virtual boundaries "ci" and "km"

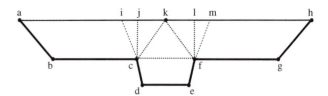

FIGURE 7.4 Compound channel cross section showing possible locations of dummy channel and floodplain sides.

(or "jc" and "lf"; or "cf"; or "kc" and "kf") are used to define the geometry of the flow sections, but are not always assigned a roughness or used in the calculation of the hydraulic radius. The dummy lines kc and kf are perpendicular to the lines of equal velocity (isovels) and probably give the best results, but produce odd channel sections and more complicated mathematics. A simple approach is to insert a dummy bed cf. The discharge in the main channel up to the top of the bank and the discharge in the second stage above the top of the bank of the main channel are then calculated separately and added. Flow velocity is estimated using Manning's equation, and different roughness factors can be assigned to the main channel and the floodplain. In Example 7.6, we look at the discharge estimation differences associated with two different strategies for constructing a dummy bank or bed.

EXAMPLE 7.6

The trapezoidal channel that is shown in **Figure 7.2** is connected to a 100-ft wide floodplain on either side of the channel. Assume that the valley walls at the edge of the floodplain are vertical. On one side of the channel, the floodplain has dense vegetation and willows that have a Manning's roughness *n* of 0.20. On the other bank, there is a pasture with a Manning's *n* of 0.05. Calculate the discharge when the floodplain is flooded by 1 ft of water. Compare assuming a dummy bed at the elevation of the top of the main channel with constructing dummy 1-ft vertical banks for the main channel and treating the cross section as three separate flow regimes (the left floodplain, a main channel, and the right floodplain). The bed slope is 0.006, and Manning's *n* for the channel is 0.04.

Solution

First, draw a diagram of the channel and floodplain system.

Approach 1: The discharge q_{ch} up to the top of the first bank was calculated in Example 7.4 as 519 ft³/sec. So, all we need to do is calculate the discharge in the second stage of the system. Let us assume that the dummy bed cf is real and should be included in the calculation of the geometry; it has the same roughness as the channel bed. Therefore, the cross-sectional area is

$$a = (100 \text{ ft} + 100 \text{ ft} + 43 \text{ ft})(1 \text{ ft}) = 243 \text{ ft}^2$$

The wetted perimeter P is

$$P = (100 \text{ ft} + 100 \text{ ft} + 43 \text{ ft} + 1 \text{ ft} + 1 \text{ ft}) = 245 \text{ ft}$$

So, the hydraulic radius $R = a/P$ (Equation 7.3) = 243/245 = 0.99 ft.

The length-weighted roughness is

$$n = ((1 \text{ ft})(0.05) + (100 \text{ ft})(0.05) + (43 \text{ ft})(0.04)$$
$$+ (100 \text{ ft})(0.20) + (1 \text{ ft})(0.20))/245 \text{ft} = 0.110$$

For uniform flow in a stream, the discharge in the second stage can be estimated by Equation 7.5:

$$q = 1.5an^{-1}R^{2/3}S^{1/2}$$

therefore,

$$q_{fp} = 1.5(243 \text{ ft}^2)(0.110)^{-1}(0.99 \text{ ft})^{2/3}(0.006)^{1/2} = 255.0 \text{ cfs}$$

The total discharge q_{tot} will be

$$q_{tot} = q_{ch} + q_{fp} = 519.3 \text{ ft}^3/\text{sec} + 255.0 \text{ ft}^3/\text{sec} = 774.3 \text{ ft}^3/\text{sec}$$

Approach 2: Each floodplain compartment is 100 ft wide and 1 ft deep, so each will have an area, wetted perimeter, and hydraulic radius of 100 ft², 102 ft, and 0.98 ft, respectively.

The area of the main channel will be the area of 102 ft² we calculated in Example 7.4 plus an additional area of 43 ft × 1 ft or 43 ft². Therefore, the total cross-sectional area for the main channel is 145 ft². The wetted perimeter of the main channel will be the perimeter of 44 ft, from Example 7.4, plus an additional 1 ft for each of the vertical dummy walls. Therefore, the hydraulic radius of the main channel is

$$R_{ch} = 145 \text{ ft}^2/46 \text{ ft} = 3.15 \text{ ft}$$

If we assume each 1-ft dummy wall has the same roughness as the floodplain bed and banks, the length-weighted roughness is

$$n = ((1 \text{ ft})(0.05) + (44 \text{ ft})(0.04)$$
$$+ (1 \text{ ft})(0.20))/46 \text{ ft} = 0.0044$$

Therefore, the discharge in each flow compartment is

$$q_{past} = 1.5(100 \text{ ft}^2)(0.05)^{-1}(0.98 \text{ ft})^{2/3}(0.006)^{1/2} = 229.3 \text{ cfs}$$

for a pasture floodplain;

$$q_{tree} = 1.5(100 \text{ ft}^2)(0.20)^{-1}(0.98 \text{ ft})^{2/3}(0.006)^{1/2} = 57.3 \text{ cfs}$$

for a wooded floodplain; and

$$q_{ch} = 1.5(145 \text{ ft}^2)(0.044)^{-1}(3.15 \text{ ft})^{2/3}(0.006)^{1/2} = 822.8 \text{ cfs}$$

for the main channel.

So, the total discharge is

$$q_{tot} = 229.3 + 57.3 + 822.8 = 1109.4 \text{ ft}^3/\text{sec}$$

The percentage difference in the two answers E is

$$E = 100\left(\frac{(1109.4 - 774.3)}{1109.4}\right) = 30.2\%$$

if the second approach is assumed to be more valid.

Answer

The discharge is 774.3 ft³/sec or 1109.4 ft³/sec, depending on which method is used.

There are many other ways we might have developed a discharge solution for Example 7.6. For example, a school of thought recommends that the dummy boundaries not be used in determining the wetted perimeter or roughness. Unfortunately, if we use a simple flow equation like Manning's equation, each will give an approximate answer. The second approach appears better than the first approach because it is more conservative. For this example, the total width of the floodplain was about 4.6 times (200 ft divided by 43 ft) the main channel width. The wider and larger the flow contribution in the floodplain is, the smaller the error will be in the first approach. Also, for this example, we had a very rough floodplain. Again, the more discharge there is in the floodplain, the smaller the difference in the two approaches.

The reason all the approaches are approximate is illustrated by **Figure 7.5**. In using a resistance equation like

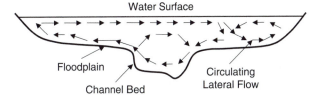

FIGURE 7.5 Example of a circulating lateral flow pattern due to over-bank discharge.

Manning's equation, we have assumed that all of the flow is in or out of the page and has a mean velocity v that can be estimated by the resistance equation. However, when we get overbank flow, there is a considerable amount of lateral flow (across the page). This flow will circulate when it hits the banks of the floodplain, and some of the flow might even flow back upstream. Therefore, we will have both vertical and horizontal currents, eddies, and turbulence. This suggests that the second approach we used will overestimate the contribution of the main channel to the discharge. Also, both approaches will overestimate the contribution of the floodplain. We might think of the turbulence, eddies, and lateral and circulation flow as requiring an increase in the resistance factor such as Manning's n.

7.4 CHANNEL MODIFICATIONS

Flood reduction can be achieved by (1) constructing floodwater detention structures, (2) modifying the stream channel carrying capacity, or (3) using a combination of (1) and (2). For example, a detention structure might be designed to prevent flood discharges from exceeding the carrying capacity of a downstream channel. However, if this channel could be modified to carry a higher flow, the cost of the detention structure might be less. Of course, in the combination system, if channel modifications are not maintained, then flood damages will not be prevented. Details on detention structures are presented in Chapter 9.

Channel modification might be accomplished by removing obstructions and decreasing the roughness coefficient n or resistance to water flow velocity or increasing the bedslope S by reducing meanders through channel straightening. If stream channel meanders are cut off and the channel is straightened and cleaned, the roughness will be decreased, and the slope increased. If prior to channel straightening the fall in 1000 ft of a stream reach was 25.0 ft and this same fall occurred in 500 ft of a straightened channel, the bedslope would increase from 0.025 to 0.05. Based on Manning's equation, this would have the effect of increasing the velocity by a factor of 1.44 (the ratio of the square roots of 0.025 and 0.05). There is some danger in channel straightening in areas of highly erodible soil. When velocity or slope increase, there is more *stream power* (Bagnold, 1966). For any given portion of a stream cross section, this may be written as

$$S_p = \gamma v d S \qquad (7.9)$$

where S_p is the stream power per unit measure of bed, here kilograms per meter of width per second; γ is the unit weight of water (includes temperature and sediment) in kilograms/cubic meter; v is the velocity of water in meters/second; d is the depth in meters; and S is the

bedslope (dimensionless fraction). See Chapter 12 for further discussion on this topic.

EXAMPLE 7.7

A stream has a slope of 0.03, a depth of 1 m, and a velocity of 2 m/sec; the water is laden with sediment and weighs 1100 kg/m^3. What is the unit stream power per unit width of channel?

Solution

We can use the provided information and Equation 7.9 to determine S_p as

$$S_p = 1100 \text{ kg/m}^3 \,(2 \text{ m/sec})(1 \text{ m})(0.03) = 66 \text{ kg/m/sec}$$

(Note that this is significant stream power, certainly enough to move sediment particles of more than 10-cm diameter.)

Answer

The unit stream power is 66 kg/m/sec.

It is often difficult to quantify the influence of making channel modifications if Manning's n is obtained from **Table 7.1**. However, Chow (1959) presented information based on the work of Cowan (1956) that can be used for this purpose (**Table 7.2**). The description of the procedure that follows was obtained from Chow (1959). The value of Manning's n is

$$n = (n_0 + n_1 + n_2 + n_3 + n_4)m_5 \qquad (7.10)$$

where n_0 is the value of n for a straight, uniform, smooth channel in natural materials; n_1 is a value added to n_0 to correct for the effect of surface irregularities; n_2 is a value for variations in shape and size of the channel cross section; n_3 is a value for obstructions; n_4 is a value for vegetation and flow conditions; and m_5 is a correction factor for meandering of the channel.

The values given in **Table 7.2** were developed from a study of small and moderate channels. Therefore, the method is questionable when applied to large channels with hydraulic radii that exceed, say, 15 ft. The method applies only to unlined natural streams, floodways, and drainage channels with a minimum n value of 0.02.

In determining n_1, the degree of irregularity is considered *smooth* for surfaces comparable to the best attainable for the materials involved; *minor* for good dredged channels, slightly eroded or scoured side slopes of canals or drainage channels; *moderate* for fair to poorly dredged

TABLE 7.2
Values for the Computation of the Roughness Coefficient
by Equation 7.12

Channel Conditions			Values
Material involved	Earth	n_0	0.020
	Rock cut		0.025
	Fine gravel		0.024
	Coarse gravel		0.027
Degree of irregularity	Smooth	n_1	0.000
	Minor		0.005
	Moderate		0.010
	Severe		0.020
Variations of channel cross section	Gradual	n_2	0.000
	Alternating occasionally		0.005
	Alternating frequently		0.010–0.015
Relative effect of obstructions	Negligible	n_3	0.000
	Minor		0.010–0.015
	Appreciable		0.020–0.030
	Severe		0.040–0.060
Vegetation	Low	n_4	0.005–0.010
	Medium		0.010–0.025
	High		0.025–0.050
	Very high		0.050–0.100
Degree of meandering	Minor	m_5	1.000
	Appreciable		1.150
	Severe		1.300

Source: From Chow, V.T., *Open-Channel Hydraulics*, McGraw-Hill, New York, Table 5-5, p. 109, and description, pp. 106-108, 1959. With permission.

channels, moderately sloughed or eroded side slopes of canals or drainage channels; and *severe* for badly sloughed banks of natural streams, badly eroded or sloughed sides of canals or drainage channels, and unshaped, jagged, and irregular surfaces of channels excavated in rock.

In selecting the value of n_2, the character of variations in size and shape of the cross section is considered *gradual* when the change in size or shape occurs gradually; *alternating occasionally* when large and small sections alternate occasionally or when shape changes cause occasional shifting of main flow from side to side; and *alternating frequently* when large and small sections alternate frequently or when shape changes cause frequent shifting of flow from side to side.

The selection of the value n_3 is based on the presence and characteristics of obstructions, such as debris deposits, stumps, exposed roots, boulders, and fallen and lodged logs. In judging the relative effect of obstructions, consider the extent that the obstructions occupy or reduce the average stream cross section; the character of obstructions (sharp-edge or angular objects induce greater turbulence than curved, smooth-surface objects); and the position and spacing of obstructions transversely and longitudinally in

the reach under consideration. In selecting the value of n_4, the degree of effect of vegetation is considered:

1. *Low* for conditions comparable to the following: (a) dense growths of flexible turf grasses or weeds (Bermuda and blue grasses are examples) for which the average depth of flow is two to three times the vegetation height; and (b) supple seedling tree switches, such as cottonwood, willow, or salt cedar, for which the average depth of flow is three to four times the vegetation height.

2. *Medium* for conditions such as (a) turf grasses for which the average depth of flow is one to two times the height of vegetation; (b) stemmy grasses, weeds, or tree seedlings with moderate cover for which the average depth of flow is two to three times the height of vegetation; and (c) brushy growths, moderately dense (similar to willows 1 to 2 years old), dormant season, along side slopes of a channel with no significant vegetation along the channel bottom for which the hydraulic radius is greater than 2 ft.

3. *High* for conditions comparable to the following: (a) turf grasses for which the average depth of flow is about equal to the height of vegetation; (b) dormant season willow or cottonwood trees 7 to 10 years old intergrown with some weeds and brush and with none of the vegetation in foliage and for which the hydraulic radius is greater than 2 ft; and (c) growing season, busy willows about 1 year old intergrown with some weeds in full foliage along side slopes, with no significant vegetation along the channel bottom and for which the hydraulic radius is greater than 2 ft.

4. *Very high* for conditions comparable to the following: (a) turf grasses for which the average depth of flow is less than one half the height of vegetation; (b) growing season, busy willows about 1 year old intergrown with weeds in full foliage along side slopes or dense growth of cattails along the channel bottom with any value of hydraulic radius up to 10 or 15 ft; and (c) growing season trees intergrown with weeds and brush, all in full foliage, with any value of hydraulic radius up to 10 or 15 ft.

In selecting the value of m_5, the degree of meandering depends on the sinuosity of the channel. Meandering is considered *minor* for a sinuosity of 1.0 to 1.2, *appreciable* for sinuosities of 1.2 to 1.5, and *severe* for sinuosities of 1.5 and greater.

EXAMPLE 7.8

It is planned to straighten and "improve" a channel that has been poorly maintained. Determine the change in Manning's n and flow velocity if the channel is formed in earth; has moderately eroded side slopes; has gradual changes in side slopes; has appreciable lodged logs and deposits of debris; has cottonwood trees intergrown with weeds and brush (high to very high vegetation); and has appreciable meandering. Following the channel improvements, it is anticipated that the channel will have the side slopes repaired; the debris will be removed, and lodged logs will be cleared; all weeds and brush will be removed; and there will be minor meandering.

Solution

From **Table 7.2** and the descriptions presented in the text, it is determined that prior to straightening and improving the channel, $n_0 = 0.02$; $n_1 = 0.01$; $n_2 = 0.00$; $n_3 = 0.02$ to 0.03; $n_4 = 0.025$ to 0.050; and $m_5 = 1.15$. Use the average n_3 value of 0.025 and the upper end of the range for n_4 of 0.05 because the vegetation is described as high to very high; then,

$$n = (0.02 + 0.01 + 0.0 + 0.025 + 0.05)1.15 = 0.12$$

Following the straightening and improvements, $n_0 = 0.02$; $n_1 = 0.005$; $n_2 = 0.00$; $n_3 = 0.01$ to 0.015; $n_4 = 0.01$ to 0.025; and $m_5 = 1.0$. Let us assume that $n_3 = 0.01$ and $n_4 = 0.025$; then,

$$n = (0.02 + 0.005 + 0.0 + 0.01 + 0.025)1.0 = 0.06$$

Answer

Straightening and improving the channel will reduce Manning's n from 0.12 to 0.06. This will result in doubling of the flow velocity. However, it is noted that a range of n_3 and n_4 values could have been used, and the answer is approximate. It is debatable as to whether the channel has been improved because often "improvement" is equated with increasing the conveyance capacity (discharge ability) of the channel and might lead to channel instability, reduced ecological function, incision, and widening.

7.4.1 GRASSED WATERWAYS

Often, knowledge of the type of channel materials and vegetation in small channels will be used in conjunction with information on flow velocity to determine if scour will occur. The Soil Conservation Service (1984) determined maximum permissible velocities for vegetated channels (**Table 7.3**). If the maximum permissible velocity is exceeded, the channel will scour.

A problem with evaluating vegetated channels is that the hydraulic behavior of the vegetation will change as the flow velocity increases. For example, as grass becomes submerged, the roughness will initially increase, but when 20 to 40% submerged, it will start to decrease because the grass will begin to bend over and flatten. At high flows, fully submerged grass may have a resistance to flow (roughness) that is only 10% of the unsubmerged value. Grasses have been divided into four retardance classes based on their resistance to bending; reference should be made to design texts for further details (Schwab et al., 1993).

7.5 TWO-STAGE CHANNEL DESIGN

Constructed channels can be found throughout the world. They might simply be small ditches commonly seen running along the sides of roads in the U.S., bigger ditches that typically run between fields in places where a natural stream might have existed, or very large channels, such as the one shown in **Figure 7.6**, which transfers water from the Yellow River for irrigation and water supply use. Often, these ditches are deep and have a trapezoidal shape. If the Rosgen classification systems were used on these ditches, they would typically be an unstable F or G stream

TABLE 7.3
Permissible Velocities for Vegetated Channels

Cover	Permissible Velocity for Erosion-Resistant Soils (ft/sec): Channel Bed Slope (%)		
	0–5	5–10	Over 10
Bermuda grass	8[a]	7[a]	6[a]
Blue grama, buffalo grass, Kentucky bluegrass, smooth broom, tall fescue	7[a]	6[a]	5[a]
Annual crops for temporary protection: alfalfa, crabgrass, kudzu, *Lespedeza sericea*, weeping lovegrass	3.5[b]	NR	NR
Grass mixture	5[b]	4[b]	NR

NR = not recommended.

[a] For moderately resistant soils, reduce reported velocities by 1 ft/sec. For easily eroded soils, reduce velocities by 2 ft/sec.

[b] For easily eroded soil, reduce velocities by 1 ft/sec.

Source: Soil Conservation Service, *Engineering Field Manual (including Ohio Supplement)*, U.S. Department of Agriculture, Washington, DC, 1984.

FIGURE 7.6 Sedimentation in a constructed channel in China.

type. Stability, therefore, depends on vegetation on the banks, the resistance to scour and stability of the bed, and bank materials, riprap or concrete, or frequent clean outs and maintenance. Notice the amount of sediment in the channel shown in **Figure 7.6**. The Yellow River has some of the highest sediment loads in the world, and small irrigation channels that receive water from the Yellow River often require the removal of 3 to 6 ft of sediment following each irrigation event (**Figure 7.7**).

In the U.S., highly modified channels drain extensive portions of productive agricultural land, particularly in the Midwest region (often referred to as the Cornbelt). In many of these areas, most natural channels have been deepened and straightened to facilitate the flow of water from agricultural subsurface drainage outlets and to maximize conveyance. Work done periodically to maintain the drainage function typically includes removal of woody vegetation and deposited sediment. Ancillary work

FIGURE 7.7 Cleaning out sediment in an irrigation channel in China.

includes stabilizing bank slope failures and toe scour. Ditch form is a result not only of construction and maintenance, but also, to varying degrees, of fluvial (flowing water) processes.

Natural stream systems in these settings usually include not only a main channel, but also a floodplain. The floodplains of high-quality streams (except those with steep gradients) are characterized by frequent, extensive overbank flow. As discussed in Chapter 6, channels, sized by stable fluvial processes, convey the effective (bankfull) discharge, and larger flows widen out onto the floodplain. In equilibrium, a stream system depends on both the ability of the floodplain to dissipate the high energy of flows and concentration of the energy of low flow, effectively creating a balance in sediment transport, storage, and supply. In flat, poorly drained areas in the Midwest, where subsurface drainage is widely used, the effective discharge in ditches and modified channels occurs frequently and primarily consists of subsurface drainage discharges.

To facilitate drainage and reduce the frequency of overbank flows, ditches are typically constructed such that flows as large as perhaps 5- to 100-year recurrence intervals are contained within the ditch. Also, not uncommonly, the width of the bottom of the ditch is constructed wider than the channel bottom that would be formed by fluvial processes, thus making the effective discharge channel relatively wide and shallow. Therefore, the constructed ditch channel is often oversize for small flows and provides no floodplain for large flows. In response to this imbalance, fluvial processes work to create a small main channel by building a floodplain or bench within the confines of the ditches (see **Figure 7.8**). If conditions allow, these benches can reach a stable size, thickly vegetated with mostly grasses. The small main channel will often meander slightly within the ditch and is sized by nature

A

B

FIGURE 7.8 Large agricultural drainage ditch in Minnesota (A) before cleanout and (B) after cleanout. (Courtesy Gary Sands.)

FIGURE 7.9 Meandering small main channel that has developed within the confines of an agricultural ditch in Ohio.

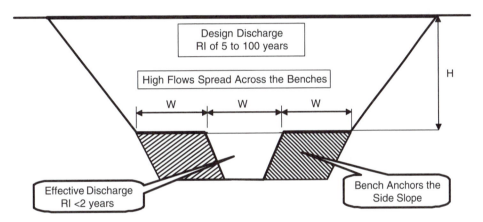

FIGURE 7.10 A two-stage ditch with a small main channel and recommended minimum width low grassed benches.

to carry the effective discharge (**Figure 7.9**). The bed will usually have steep (1:1) sides and a bed consisting of material coarser than that of adjacent reaches where benches have not formed.

7.5.1 BENEFITS AND COSTS OF A TWO-STAGE DITCH

Benefits of a two-stage ditch over a conventional ditch are potentially both improved drainage function and ecological function. Drainage benefits may include increased ditch stability and reduced maintenance. Evidence and theory both suggest ditches prone to filling with accumulated sediment may require less-frequent cleaning out of sediment if constructed in a two-stage form (**Figure 7.10**). Channel stability may be improved by a reduction in the erosive potential of larger flows as they are shallower and spread out across the bench. Stability of the ditch bank

may also be improved where the toe of the ditch bank meets the bench rather than the ditch bottom. Here, the bank height is effectively reduced, and the shear stress (erosive force) on the toe of the bank is less. Also, this bank material will be drier because it is not in contact with low flow. However, where the main channel meanders to the toe of the ditch bank, this would not be the case, and bank erosion might even be induced.

The two-stage ditch has the potential to create and maintain better habitat. The narrow deep main channel provides better water depth during periods of low flow. Grass on the benches can provide quality cover and shade. The substrate in the main channel is improved as the two-stage form increases not only sediment conveyance, but also sorting, with fines deposited on the benches and coarser material forming the bed. Two-stage ditches might also be useful in improving water quality, particularly for nutrient assimilation. Work has been initiated on the ecology of these ditches and the role of the channel and benches in improving water quality and habitat.

The primary costs of two-stage ditches are increased width and more initial earthwork. Creating a low bench typically requires a greater top width of the ditch. A two-stage ditch might be 10 to 20 ft wider at the top than a trapezoidal ditch, so the loss of potentially farmable land might be 1 to 2 acres per mile of ditch. The increased width, however, will usually increase the capacity (amount of flow it can carry) by 25 to 100% or more. It is probable that establishment or retention of this feature will have a similar or greater benefit than a grass filter. However, it does not negate the benefit of also having a grass filter along the top bank of the ditch.

7.5.2 Sizing a Two-Stage Ditch

The probable dimensions of the low-flow channel can be determined empirically based on regional studies similar to those conducted for natural streams. Similarly, measurements at study sites in Ohio begin to suggest that a broad ditch with a total bench width exceeding double the channel width might result in stable benches. However, the work by Rhoads and Herricks (1996) in Illinois indicated that there might be a threshold width at which the low-flow channel will induce meander changes in the ditch itself. **Figure 7.10** shows a total bench width that is double the channel width. The highest total bench width in a ditch

that Ward has observed is about six times the channel width. While this begins to approach the floodplain width associated with stable, low-gradient natural streams, it is less than the eight or more channel widths that Ward believes are necessary to provide both sufficient floodplain width to reduce the energy of out-of-bank discharges effectively and allow natural meander migration.

7.5.3 Other Applications

Two-stage channel designs are beginning to see more frequent application in urban areas. **Figure 7.11** shows a large, artificially lined channel in Japan; a fill material was placed in the channel to create a more natural "effective discharge" main channel, and a second stage serves as a floodplain within the confines of the constructed channel.

An interesting two-stage channel design application is shown in **Figure 7.12** for a wash in Las Vegas, NV. A wash is an intermittent stream found in the semiarid regions of the southwestern U.S. **Figure 7.12A** shows a concrete-lined section of the channel that is little more than an open sewer (bad smelling and almost void of any habitat and biological attributes). **Figure 7.12B** shows a downstream reach of the same wash. It has a partially vegetated low bench, a clean-flowing main channel, and much better habitat. However, mainly for aesthetic reasons, much of the vegetation on the benches was removed or mowed.

FIGURE 7.11 Establishment of a two-stage geometry in an artificially lined channel in the Nuki River, Japan. (A) The modified channel prior to restoration. (B) Constructed benches (floodplain). (C) New geometry following deposition associated with a large flood. It can be seen that the restored dimensions were too wide, and natural processes have produced a narrow channel that is attached to a floodplain that was built due to deposition on the low constructed floodplain. (Courtesy Yuichi Kayaba.)

FIGURE 7.12 A modified wash in Las Vegas, NV: (A) a concrete-lined reach and (B) a downstream reach with a more natural two-stage geometry. (Courtesy John Lyon.)

PROBLEMS

7.1. Determine the discharge in a stream if the average flow velocity is 5 ft/sec, and the cross-sectional area of the flow is 25 ft².

7.2. A grassed waterway with resistant soils has a bed slope of 6% and is vegetated with buffalo grass. What is the maximum permissible velocity of flow in this waterway?

7.3. Determine Manning's roughness coefficient n for a channel constructed in earth, with minor irregularities, occasional changes in cross-sectional geometry, severe obstructions (use an average value for this condition), trees intergrown with weeds and brush during the growing season (use an average value for this condition), and a sinuosity of 1.7.

7.4. Use Manning's equation to determine the average velocity of a stream if the roughness coefficient n is increased by weed growth from 0.03 to 0.06. Assume S and R do not change, and the average velocity with a roughness coefficient of 0.03 is 4 ft/sec.

7.5. Use Manning's equation to determine the velocity of a mountain stream if the slope of a stream was increased from 0.02 to 0.03 ft/ft, and R and n did not change. At the 0.03 slope, the velocity was 3.0 ft/sec.

7.6. Compute the discharge for a rectangular channel that is 20 ft wide, the flow depth is 5 ft, the channel is concrete lined, and the slope of the channel bed is 0.01 ft/ft.

7.7. Determine the flow depth and flow velocity in a channel when the discharge is 500 ft³/sec. The bottom width is 12 ft, and the channel has side slopes of 1:1 (1 ft horizontal to every 1 ft vertical). The channel is cut in smooth uniform rock (use the normal Manning's *n* for this condition).

7.8. Use Manning's equation to determine the flow velocity (meters/second) and discharge (cubic meters/second) in a stream on a plain that has sluggish reaches and weedy, deep pools. The main stream channel is rectangular and has a bottom width of 10 m and a bed slope of 0.4%. The depth of flow is 1 m (use the normal Manning's *n* for this condition).

7.9. Develop a conceptual design for a trapezoidal concrete channel that is sized to convey a design discharge of 40 m³/sec. Evaluate the feasibility of constructing a channel with a bed slope of 0.005, a bed width of 2 m, and side slopes of 2:1 (horizontal:vertical). Determine the design flow depth. If the flow depth exceeds 2 m, the design will not be feasible. Comment on the feasibility of the design and, if necessary, discuss ways the channel might be modified to comply with the maximum flow depth of 2 m.

7.10. The average cross-sectional geometry of a channel is as follows: 1.0 m bed width; 2:1 side slopes; 2 m maximum depth; 0.5% bed slope; Kentucky bluegrass; erosion-resistant soils; minor irregularities; gradual variations in cross section; negligible obstructions; medium-height vegetation; and minor meandering. Will the channel be able to convey the 25-year, 24-h peak discharge of 12 m³/sec? Will channel erosion be a problem?

7.11. An engineer, in collaboration with the NRCS and a watershed conservancy district, has designed a flood alleviation scheme for a rural area. It is proposed to dredge the channel of a constructed ditch to achieve the design objectives to transmit the 1:20-year flood of 600 ft³/sec. As a recently graduated river engineer/scientist, you think that dredging may be unnecessary, and that the design objectives could be achieved by mowing vegetation on the sides of the channel, killing unwanted brush and weeds, and removing fallen trees from the river. You are provided the following information regarding the original channel:

Trapezoidal channel with 2:1 (horizontal to vertical) sidewalls

40 ft top width

5 ft maximum depth

0.002 bed slope

Current maximum discharge capacity of 300 ft³/sec

Banks tree lined and overgrown shrubs

Gravel bed with fallen logs

Approximately 30 ft³, per linear foot of ditch, of sediment deposited in the ditch

Calculate the current roughness of the channel. Estimate the reduction in roughness needed to meet the objectives. Estimate the increase in conveyance capacity associated with only removing sediment and based on conducting all the proposed activities. Outline the long-term effects of dredging and straightening on channel stability. How can any adverse effects be minimized?

7.12. For the ditch in Problem 7.11, develop a new design that has a narrow, 10-ft wide by 2-ft deep channel at the bottom of the ditch and then flat benches that have a total bench width at least two times the width of the narrow channel.

8 Hydraulic Control Structures

8.1 INTRODUCTION

In Chapter 7, we considered steady uniform flow in a channel. It was understood that the flow was often turbulent; that there was considerable variability in the velocity at different points in a cross section; that some of flow might even occur in an upstream direction; and that as the flow moved around a bend, it behaved somewhat like a roller coaster. Despite all this variability, useful estimates of the mean velocity and discharge can be obtained using a resistance equation such as Manning's equation or the Darcy–Weisbach equation. However, there are many times when use of one of these equations is not appropriate. If we consider the energy of the flow, we will see that the following three basic types of flow can occur in an open channel: subcritical, critical, and supercritical. Uniform flow is only one kind of subcritical flow regime.

8.2 SPECIFIC ENERGY AND CRITICAL DISCHARGE

If we take the bed of a channel as a datum, the energy of the flow is

$$E = \frac{v^2}{2g} + d \qquad (8.1)$$

where E is called the *specific energy* (feet or meters), d is the depth of flow (feet or meters), v is the mean flow velocity (feet/second or meters/second), and g is the gravitational constant. The term $v^2/2g$ is the velocity head, and d is the head due to the depth of flow. Equation 8.1 can be related to the discharge using the continuity equation (Equation 7.1, Chapter 7). Therefore,

$$E = \frac{\left(\frac{q}{w}\right)^2}{2gd^2} + d \qquad (8.2)$$

Solutions to this equation for three different discharges per unit width are shown in **Figure 8.1**.

When the specific energy is a minimum, the flow is called *critical flow,* and

$$1 = \frac{v_c}{\sqrt{gd_c}} \qquad (8.3)$$

The term $v_c/\sqrt{gd_c}$ is called the *Froude number F* and is equal to 1 when critical flow occurs. For all specific energy values except when critical flow occurs, there are two possible depths of flow. For the deeper depth, the flow will be stable and is called *subcritical flow.* For the shallower depth, the flow will be unstable and is called *supercritical flow.* Flow prefers to be stable, so at the first opportunity supercritical flow will try to become subcritical.

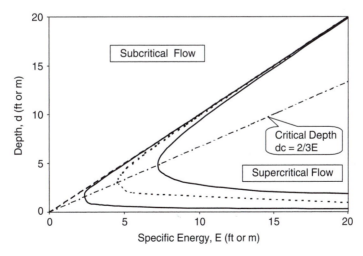

FIGURE 8.1 Relationship between specific energy and flow depth.

For a nonrectangular channel, the Froude number F can be related to the hydraulic depth d_h as follows:

$$F = \frac{v}{\sqrt{gd_h}} \qquad (8.4)$$

where the hydraulic depth is the cross-sectional area divided by the top width.

Application of specific energy concepts is useful when considering flow over weirs, spillways, and riffles. It also has application when there is a change in the width of a channel. As flow passes over a hydraulic control structure, it will often change from subcritical flow upstream of the structure to critical flow at the nappe of the flow across the structure. Downstream of the nappe, the flow might initially be supercritical or might revert directly back to subcritical flow.

EXAMPLE 8.1

A rectangular channel with a width of 10 ft is carrying 50 ft³/sec. What are the critical depth and the critical flow velocity?

Solution

To obtain a solution, it will be necessary to use Equation 7.1 and Equation 8.3. Substituting in Equation 7.1 (Chapter 7),

$$q = va \quad \text{ft}^3/\text{sec}$$

where the cross-sectional area a is equal to the width of 10 ft times the critical depth d_c. Therefore,

$$50 = (v_c)(10d_c)$$

and

$$v_c = \frac{5}{d_c}$$

Substituting into Equation 8.3 gives

$$1 = \frac{v_c}{\sqrt{gd_c}}$$

and

$$1 = \frac{5}{d_c\sqrt{32.2d_c}}$$

Therefore,

$$d_c^{1.5} = \frac{5}{\sqrt{32.2}} = 0.88$$

and

$$d_c = 0.92 \text{ ft}$$

As $v_c = 5/d_c$, the critical velocity is 5/0.92 or 5.44 ft/sec.

Answer

The critical depth is 0.92 ft, and the critical velocity is 5.44 ft/sec.

EXAMPLE 8.2

Use the results from Example 8.1 to develop a relationship between the critical depth and the specific energy.

Solution

From Example 8.1, the critical depth is 0.92 ft, and the critical velocity is 5.44 ft/sec.

The specific energy is related to the flow depth by Equation 8.2 as follows:

$$E = \frac{v^2}{2g} + d$$

so

$$E = \frac{5.44^2}{2(32.2)} + 0.92 = 1.38 \text{ ft}$$

So, the critical depth is 0.92 ft or two thirds of the specific energy, which is 1.38 ft.

Answer

The critical depth is two thirds of the specific energy. This result is true for all cases.

Knowledge of the critical depth and critical discharge is helpful in evaluating flow conditions at riffles, cascades, steps, and weirs. Usually, the discharge at the nappe (downstream edge) of the object over which the stream flows will be close to critical flow. It then might become

FIGURE 8.2 Weir in a stream. (Courtesy of Dawn Farver.)

subcritical or supercritical flow. In Example 8.3, we show how knowledge of the specific energy can help us in evaluating flow stability.

EXAMPLE 8.3

The rectangular channel in Example 8.1 needs to be realigned. What is the maximum slope, called the critical slope, that could be used without causing the flow to become supercritical? Manning's n for the realigned channel is 0.03.

Solution

To obtain a solution, we first need to use Equation 7.2 (Chapter 7):

$$v = \frac{1.5}{n} R^{2/3} S^{1/2}$$

where $v = 5.44$ ft/sec from Example 8.1, $n = 0.03$ (given), and the hydraulic radius R equals

$$R = \frac{(10)(0.92)}{10+(2)(0.92)} = \frac{9.2}{11.84} = 0.78 \text{ ft}$$

where 0.92 ft was the critical depth from Example 8.2.

Substituting these values into Manning's equation gives

$$5.44 = \frac{1.5}{0.03} 0.78^{2/3} S^{1/2}$$

and

$$S^{1/2} = 0.128$$

Therefore, the critical slope is approximately 0.016 ft/ft.

Answer

The critical slope is approximately 0.016 ft/ft.

8.3 WEIRS, FLUMES, AND CULVERTS

Weirs and flumes are rigid hydraulic structures with cross-sectional areas that are considerably less than the channel in which they are installed and with dimensions that are closely defined and stable. A *weir* is a control structure that raises the bed of the channel (**Figure 8.2**). A *flume* narrows the width of the channel (**Figure 8.3**). If both the width and bed height are raised, then the control structure is normally classified as a flume. The constriction in the stream channel forces the flow to form a pool at the entrance to the structure. This still water then plunges freely over the weir or flume outlet.

A *culvert* is a hydraulic control structure that is normally constructed so the flow in a channel can pass under a road. It can act like a weir, flume, pipe, or channel. A common property of any of these hydraulic control structures is that they might cause the flow upstream of the structure to rise above the normal flow depth. This is called a *backwater* (see **Figure 8.4**), and resistance equations, such as Manning's equation, cannot be used to evaluate flow in the region where the backwater occurs or to evaluate flow properties of the hydraulic control structure. At the outlet of the control structure, the flow might be subcritical, critical, or supercritical, and reference should be

FIGURE 8.3 Flume in a stream. (Courtesy Kevin King.)

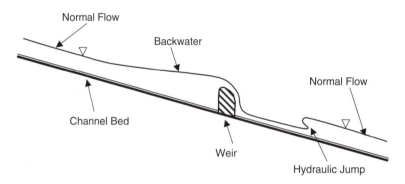

FIGURE 8.4 Backwater curve at a weir.

made to an engineering hydraulics text to develop solutions for the region of flow before normal flow is resumed. Often, the upstream and downstream zones influenced by a hydraulic control structure are called the *headwater* and *tailwater*, respectively.

The discharge over a weir q (cubic feet/second) can be estimated using the equation

$$q = C L H^{3/2} \qquad (8.5)$$

where C is a weir coefficient, L is the weir length in feet, and H is the height of the flow above the riser crest in feet (**Figure 8.5**). For a broad-crested weir or a pipe, a value of 3.1 may be used for C. For a pipe, the weir length L will be the circumference of the pipe riser. For sharp-crested weirs, C is approximately 3.3 plus 0.4 times the ratio of the flow depth over the weir to the height of the weir.

A sharp-crested weir with a 90° V notch outlet, as shown in **Figure 8.6A**, has the following discharge and head relationship:

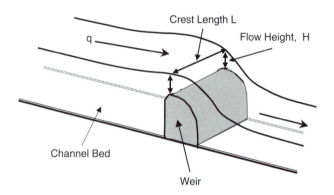

FIGURE 8.5 Geometry of a weir for use with Equation 8.5.

$$q = 2.5 H^{5/2} \qquad (8.6)$$

where q is the discharge rate (cubic feet/second), and H is the head (depth of water above the point of zero flow) in feet. Calibration formulas for weirs with broad crests and different shapes were given by Brater and King (1976) or Bos et al. (1991).

A

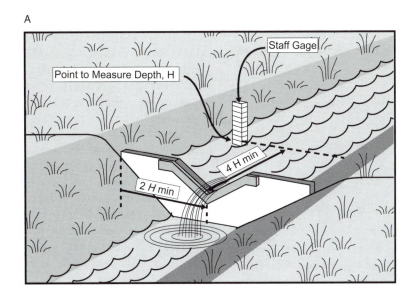

B

FIGURE 8.6 (A) Geometry of a 90° V-notch weir. (B) Parshall flume and broad-crested 120° V-notch weir, Texas. (Courtesy Kevin King.)

EXAMPLE 8.4

The depth of flow at the upstream end of a 10-ft wide broad-crested weir is 1.5 ft. Calculate the discharge over the weir.

Solution

To obtain a solution, solve Equation 8.5 using the provided flow depth and a weir coefficient of 3.1:

$$q = C L H^{3/2}$$

so

$$q = 3.1(10) \, (1.5)^{3/2} = 57 \text{ cfs}$$

Answer

The discharge over the weir is 57 ft^3/sec.

EXAMPLE 8.5

The broad-crested weir in Example 8.4 will be replaced with a 3-ft high sharp-crested weir. Calculate the depth of flow over the weir.

Solution

For a sharp-crested weir, the coefficient C in Equation 8.5 is

$$C = 3.3 + 0.4 \frac{H}{H_w}$$

where H_w is the height of the weir, and H is the depth of flow over the weir. Let us initially assume that the depth of flow over the weir is the same as in Example 8.4; then,

$$C = 3.3 + 0.4 \frac{1.5}{3.0} = 3.5$$

Substitute the discharge, weir length, and weir coefficient into Equation 8.5:

$$q = C L H^{3/2}$$

and

$$57 = (3.5)(10) H^{3/2}$$

Therefore, $H^{3/2}$ equals 1.63, and H equals 1.38 ft. If we recalculate the weir coefficient and then H, we would obtain a new estimate of about 1.39 ft for H.

Answer

The depth of flow over the sharp-crested weir is 1.39 ft.

Road crossings can have a major impact on stream processes and stream stability. Historically, they have been designed to address the following concerns: (1) traffic requirements, safety of motorists, and the integrity of the road; (2) protection of structures, facilities, and people upstream of the road crossing; (3) protection of structures, facilities, and people downstream of the road crossing; (4) legal rights and impacts on landowners at the road crossing; (5) cost; (6) fish and wildlife passage; and (7) stream stability. Only recently have the last two concerns seen more attention. Rarely are all seven of these concerns adequately addressed. The first three concerns will tend to result in an overdesign, efforts to maintain the status quo, or a compromise between upstream and downstream

impacts. Addressing Concern 4 often results in design modifications, increased costs, or inadequate use of the terrain at the road crossing. Cost concerns can result in underdesigns or using "standard," cookbook-type designs. Concerns 6 and 7 require more innovation to solve, the incorporation of knowledge of scientists in developing the design, and, most important, designs that are much more site specific than has been the standard practice. Stream stability concerns require knowledge of the effective discharge channel geometry and associated floodplain, the development and construction of a design that satisfies these requirements, and knowledge of the hydraulics of the flow and the impact of structures (such as bridge piers) that might have an impact on the flow.

Typically, single-barrel pipes or box culverts are designed for large events, such as the 10-, 25-, 50-, or 100-year flood. Therefore, they are oversized for the effective discharge. The large common events that transport sediment spread out and reduce their velocity upstream of the culvert. Sediment is deposited upstream, and scour often occurs downstream. In some cases, the flow is constricted because the large events do not spread across a floodplain, but build up a high head at the inlet of the culvert. This results in a backwater upstream of the culvert and deposition upstream; again, scour often occurs downstream. Ideally, a culvert with geometry similar to that of the natural channel is needed for the effective discharge; then, at the elevation of the effective discharge flow in the culvert, there should be a second set of culverts with characteristics similar to those of the floodplain. Certainly, there will be many cases when space, cost, and ownership constraints will have an impact on the feasibility of constructing an adequate floodplain under a road crossing. However, as a society, we need to make more effort to provide a second high-flow stage that is wide and shallow rather than narrow and deep.

Based on the work of Chow (1959), many texts describe six types of flow through a culvert when the culvert acts as a control structure (**Figure 8.7**). We have added a seventh, important, type that needs to be considered to maintain a more natural flow regime. This seventh type is a specific case of Chow's Type 4 flow. The seven flow regimes are

1. Inlet and outlet submerged and the culvert flowing full. This flow regime can be solved using Equation 8.23 for pipe flow.
2. Inlet submerged, outlet unsubmerged, but the pipe flowing full. This condition can also be solved using the pipe flow equation (Equation 8.23).
3. Inlet submerged, the culvert flowing partially full, and the depth of flow at the outlet less than the depth of the culvert. This condition can be solved using **Figure 8.8**, which is based on

Type   Profile

(1) Outlet Submerged
   $H > d$
   $y_t > d$
   Full Flow

(2) Outlet Unsubmerged
   $H > H^*$
   $y_t < d$
   Full Flow

(3) Outlet Unsubmerged
   $H > H^*$
   $y_t < d$
   Partly Full

(4) Outlet Unsubmerged
   $H < H^*$
   $y_t > y_c$
   Subcritical Flow

(5) Outlet Unsubmerged
   $H > H^*$
   $y_t < y_c$
   Subcritical Flow
   Control at Outlet

(6) Outlet Unsubmerged
   $H < H^*$
   $y_t < y_c$
   Subcritical Flow
   Control at Entrance

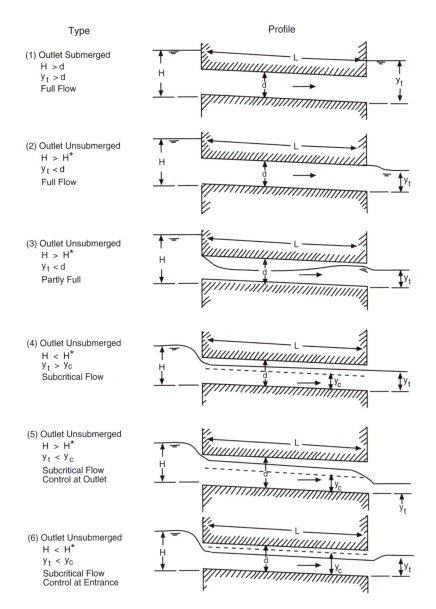

FIGURE 8.7 Types of culvert flow. (Based on Chow, V.T., *Open-Channel Hydraulics,* McGraw-Hill, New York, 1959, Figure 17-28, page 497.)

work by Mavis. For a circular pipe with a diameter equal to the culvert depth, assume the width is approximately 0.7 times the depth.

4. Inlet and outlet unsubmerged, but the depth of flow upstream is 1 to 1.5 times the culvert depth, and the flow is subcritical. Reference should be made to hydraulic engineering texts for solutions for flow Types 4 through 6.

5. Inlet and outlet unsubmerged, but the depth of flow upstream is 1 to 1.5 times the culvert depth, and the flow is subcritical through the culvert, but supercritical at the outlet.

6. Inlet and outlet unsubmerged, but the depth of flow upstream is 1 to 1.5 times the culvert

depth, and the flow is supercritical through the culvert, but subcritical at the outlet.

7. Subcritical normal flow through the culvert for flows up to and including the effective discharge and little or no backwater upstream of the culvert. Solution can be obtained using the resistance equations for normal open channel flow presented in Chapter 7.

The U.S. Federal Highway Administration has developed detailed procedures and software for analyzing flows through culverts (http://www.fhwa.dot.gov/). While it is not uncommon to design culverts under state and county roads using these procedures, most state departments of transportation have prepared detailed design specifications

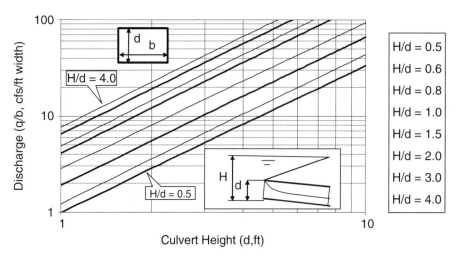

FIGURE 8.8 Partially full Type 3 culvert flow. (Based on Chow, V.T., *Open-Channel Hydraulics*, McGraw-Hill, New York, 1959, Figure 17-30, page 498; and data of the U.S. Bureau of Public Roads.)

and manuals for road and highway drainage designs that include culverts. Design discharges are usually determined using one of the runoff methods described in Chapter 5. Culverts are then sized for one of the flow types described here. The specific procedure and recurrence interval event used to develop the design will vary depending on the type of road and the requirements of the organization that has jurisdiction over the design of the road and culvert. Federal Highway Administration procedures are also used in many countries throughout the world, particularly English-speaking countries such as Australia.

EXAMPLE 8.6

Size a culvert to convey a 100-year peak discharge of 352 ft^3/sec so that a road located 7 ft above the invert of the culvert inlet is not flooded. Assume Type 3 flow.

Solution

Use **Figure 8.8** to obtain a solution. Let us initially assume the culvert height is 4 ft. This gives a maximum H/d ratio of about 2 (7 ft/4 ft). Entering the x axis with a culvert height of 4 ft and going up to the H/d = 2 line, we determine that the discharge/unit width q_b is about 44 ft^3/sec/ft width, and

$$b = \frac{352}{44} = 8 \text{ ft}$$

Therefore, an 8-ft wide by 4-ft deep culvert could convey the 352-ft^3/sec design discharge. It can be seen that many other geometries would also convey the design discharge. The specific geometry might depend on other factors or the availability of different size prefabricated culverts.

If we had selected a culvert height of 2.0 ft, the maximum H/d ratio would be about 4; from **Figure 8.8**, the discharge/unit width would be about 23 ft^3/sec/ft width, and the width would be about 15.3 ft.

Answer

An 8-ft wide by 4-ft deep box culvert could convey the 100-year peak discharge of 352 ft^3/sec. An alternative design would be a 15.3-ft wide by 2-ft deep culvert.

EXAMPLE 8.7

The single-barrel culvert that was sized in Example 8.6 will be replaced with a multibarrel or two-stage culvert design that incorporates a main channel sized to convey the effective discharge. The effective discharge dimensions of the natural main channel are a maximum depth of 1.5 ft, 4-ft width at the bed, and 7.5-ft width at the effective discharge stage. The channel has a Manning's roughness factor of 0.03, and a bed slope of 1%. Above the effective discharge stage, there is a relatively flat floodplain that extends at least 10 ft on either side of the main channel before sloping at a 2:1 (horizontal:vertical) bank slope up to an elevation similar to that of the road.

Solution

The 7-ft wide by 4-ft deep culvert would constrict high flows and would have the potential to cause sedimentation upstream of the culvert and scour downstream of the culvert. The 15.3-ft wide by 2.0-ft deep box culvert will result in flows less than the effective discharge spreading out and depositing sediment. Insufficient information is

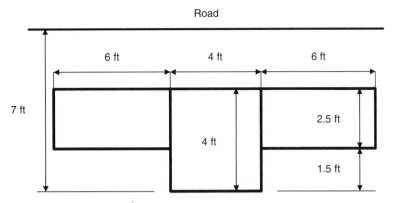

FIGURE 8.9 Three barrel box culvert geometry developed as a solution to Example 8.7.

provided to develop a detailed solution. However, if we assume the grade of the culvert is similar to the bed slope of the natural channel, a preliminary conceptual sizing can be estimated. Ideally, the culvert geometry and channel geometry should be similar. However, for economic reasons, culverts are usually circular, square, or rectangular. Let us assume the culvert will consist of three barrels, one for the main channel and one on either side of the main channel that will serve as the floodplain for the 100-year discharge (see **Figure 8.9**).

By using Manning's equation, the effective discharge in the natural main channel is about 30 ft³/sec (check this by using the provided information to solve Equation 7.2). Replacing the main channel with a rectangular concrete culvert that will have similar flow characteristics is difficult because the geometry and roughness are different. Let us try a 4-ft wide box culvert. Then, from Manning's equation, when the flow depth in the culvert is 1.5 ft, the discharge will be 40 ft³/sec if Manning's n for concrete is 0.015. This is slightly higher than the effective discharge, so there will be no constriction or spreading out of the flows.

Let us locate the two floodplain or flood event culverts at an elevation of 1.5 ft above the invert of the culvert. Let us try a 2-ft deep culvert and assume that $H/d = 3$. Then, from **Figure 8.8**, the discharge will be about 20 ft³/sec/ft width. The total width of culvert at this elevation will be

$$b = \frac{352 - 40}{20} = 15.6 \text{ ft}$$

This width is divided between the two flood culverts and the main channel culvert, which is 4 ft wide. Therefore, a 6-ft wide by 2-ft deep flood culvert could be placed on either side of a square 4 ft by 4 ft main culvert. A more conservative design that provides better alignment of the top of the culverts would be to use 6-ft wide by 2.5-ft deep flood culverts (see **Figure 8.9**). This is only a preliminary conceptual sizing. Based on the width of the road and the slope of the culvert, it would first be necessary to establish

which type of flow occurs. For the solution presented here, it was simply assumed that Type 3 flow occurred.

Answer

The geometry of a three-barrel box culvert design is shown in **Figure 8.9**. The flood culverts are 6 ft wide and 2.5 ft deep and are located on either side of the main culvert and 1.5 ft above the invert of the main culvert. The culvert for the main channel is 4 ft wide and 4 ft deep.

8.4 BENDWAY WEIRS, VANES, AND BARBS

In recent years, there has been widespread use of weirs, barbs, and vanes to protect stream banks, improve the hydraulics at bridge crossings, and enhance fish habitat. The bendway weir concept was first developed in 1977 by Thomas Pokrefke, Jr. at the U.S. Army Engineer Waterways Experiment Station (WES) to alleviate navigation and environmental problems in the Mississippi River (Derrick et al., 1994).

Currently, WES has a bendway weir concept for navigable rivers and an alternative design for streams. In a navigable river, the bendway weir is a 400- to 1600-ft long rock sill that is level crested and angled at 20 to 30° upstream (**Figure 8.10**). The spacing of the weirs is similar to their length, and they are installed in sets of 4 to 14 weirs per bend. The difficulty with designing the weirs is that they need to be low enough to allow passage of normal river traffic, but high and long enough to intercept a large percentage of the flow. Flow across each weir can be estimated using Equation 8.5.

For streams with unreveted (unprotected) banks, the WES stream weir is a rock sill that has a crest that slopes down into the stream and is angled at 5 to 25° upstream. The weirs are built of a well-graded stone, are typically 2 ft high at the stream end and rising to 4 ft at the bank end, with lengths varying from one quarter to one half the baseflow width of the stream. Rosgen has developed a variety of vane and weir shapes, such as the J hook,

FIGURE 8.10 A bendway weir, Buckeye Creek, California, designed by John McCullah, Salix Applied Earthcare, and David Derrick, U.S. Corps of Engineers. (Courtesy and copyright Salix Applied Earthcare.)

W weir, and cross vane (Johnson et al., 2002). The Rosgen vane and the WES stream weir are both similar in that they consist of rock sills that slope down into the stream and point in an upstream direction. However, Rosgen recommends relating the bank end of the vane to the bankfull (effective discharge) elevation and sloping the vane at an angle down toward the bed of 2 to 7° so that the end of the vane is at an elevation above the bed of 0.1 of the bankfull elevation at a location about one third of the bankfull width of the stream from the bank (**Figure 8.11**). Based on flume studies, Johnson et al. recommended that vanes and W weirs located at bridges to reduce scour should have a 25 to 30° upstream orientation to the bank.

A third structure similar to the Rosgen vane and the WES stream weir is a barb (NRCS, 2002). Stream barbs are low rock sills that slope down from the bank and across the low-flow thalweg. Unlike the other two structures, the angle between the barb and the bank is much greater, ranging from 50 to 70°. The recommended spacing of barbs is four to five times the barb's length. A concern we have is that all three structures are often used to perform the same function — to deflect the flow away from an eroding outer bank at a bend. While they have some commonalities, they also have different departure angles, elevations relative to the stream bed and bank, and spacings. There are numerous examples of where each approach worked or failed. Success seems primarily to be a function of the expertise of the designer in matching the geometry of the structure to the hydraulics of the dominant flow and the geometry of the main channel. Hardening of stream banks, revetments with natural or artificial materials, or the use of weirs, barbs, and vanes to control streams should be considered a "last resort" approach as they are

treatment approaches that are often not self-sustaining and that fail to address the cause of the problem, usually an imbalance associated with human activities on the watershed, within the floodplain, or within the stream. Typically, these approaches address a shopping list of objectives, such as reducing bank erosion, establishing grade control, maintaining stable dimensions for the main channel, maintaining channel conveyance, maintaining sediment transport capacity, preventing problems at bridges, improving fish habitat, or providing fishing and boating opportunities. In addition, there is often a goal to have the structure look natural even if this means using a steel cable and anchor to keep in place the trunk of a tree that has been hauled to the site and positioned at some angle perceived appropriate.

8.5 ROUTING FLOWS THROUGH CHANNELS AND RIVERS

The influence of a channel or river on runoff hydrographs is not very different from that of a reservoir (**Figure 8.12**). Water storage in a river or floodplain frequently will attenuate and delay flows. For an impoundment, it is illustrated in Section 8.6 (**Figure 8.14**) that the peak outflow rate intersects the receding limb of the inflow hydrograph. This will not always be the case in a river because storage is a function of both inflows and outflows for a reach. Local inflows to the reach might include flow from tributaries, overland runoff, groundwater flow, and rainfall. In some cases, such as karst areas, there might be substantial losses of water due to seepage. At many points along rivers, there will also be withdrawals to satisfy irrigation, industrial, and public water supply needs.

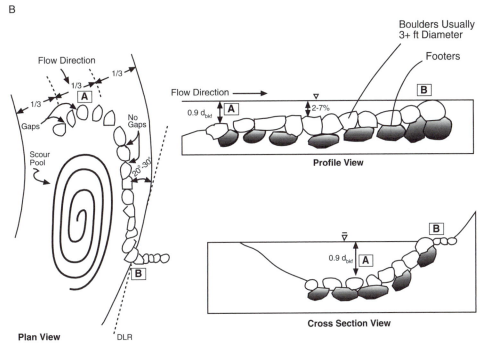

FIGURE 8.11 A. Rosgen J-hook, Colorado. B. Basic design geometry. (Based on concepts developed by Dave Rosgen.)

Procedures for routing flows through impoundments will generally not provide good estimates of flow attenuation and storage in stream systems because the flow depth along a reach will vary as the inflow and outflow rates are different. Two commonly used channel routing procedures are the kinematic method and the Muskingum method. A brief discussion of the use of kinematic procedures as they relate to describing surface runoff was presented in Chapter 5. For more comprehensive discussions, refer to the work of Bedient and Huber (1988) or Haan et al. (1994).

Both books contain an excellent level of detail for the practicing engineer.

In the Muskingum method, channel storage is the following linear function of inflow and outflow rates:

$$S = KO + KX(I - O) \qquad (8.7)$$

where S is the storage in the reach, K and X are constants, and I and O are the simultaneous inflow and outflow, respectively.

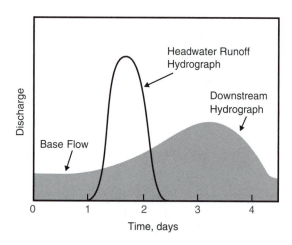

FIGURE 8.12 Attenuation of a hydrograph due to floodplain storage in a stream system.

For flood routing applications, the Muskingum method is combined with the continuity equation (Equation 7.1) to give

$$O_2 = O_1 + C_1(I_1 - O_1) + C_2(I_2 - I_1) \qquad (8.8)$$

where the subscripts 1 and 2 indicate the beginning and end of the time period t. The coefficients C_1 and C_2 are defined as follows:

$$C_1 = \frac{2\partial t}{2K(1-X) + \partial t} \qquad (8.9)$$

and

$$C_2 = \frac{\partial t - 2KX}{2K(1-X) + \partial t} \qquad (8.10)$$

The coefficient K is called the *storage constant* and is approximated as the travel time of flow in the reach if streamflow records are not available for I and O. If X is zero, then the procedure describes flow routing in an impoundment. If X is 0.5, the storage is a function of the average flow rate in the reach. A good account of the method was presented by Cudworth (1989).

Flood routing in stream systems is best performed using computer programs such as the Hydraulic Engineering Center (HEC) suite of models developed by the Army Corp of Engineers. Both free and commercial versions of this software and other routing software can be found by searching the Internet.

Although an example of how to use these equations is not presented because of their complexity, it is useful to understand how runoff estimations might be combined and to understand routing along a watershed that has been divided into subwatersheds. There is often a need to make evaluations at several places within a watershed, and the development of a detailed computer analysis can be time

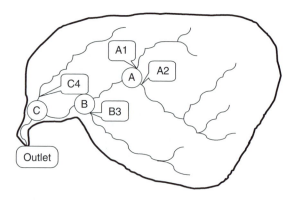

FIGURE 8.13 Watershed for Example 8.8.

consuming or require data not readily available. An example of how a non-computer-generated solution might be obtained is presented in Example 8.8.

EXAMPLE 8.8

The watershed shown in **Figure 8.13** has been divided into four subwatersheds based primarily on land use. Runoff methods presented in Chapter 5 were used to develop storm hydrographs for each subwatershed and are shown in **Table 8.1**. Determine the peak discharge at the outlet by assuming there is no attenuation of the hydrographs due to bank storage. Consider two cases: (1) the travel time to the outlet is not considered, and (2) the travel time is considered. For both cases, assume that the outlet is only a short distance downstream of Location C. The travel times from A to B and from B to C are each 15 min.

Solution

If just a peak hydrograph method was used and the peak discharges for each subwatershed were the same as those reported in **Table 8.1** (shown in bold), then the peak discharge for the whole watershed would be

$$q_{peak} = 30 + 5 + 15 + 10 = 60 \text{ ft}^3/\text{sec}$$

From **Table 8.1**, it can been seen that the peak discharge based on knowledge of the hydrograph for each subwatershed but not considering travel time between locations A, B, and C is 48 ft³/sec. Therefore, just combining peaks for each subwatershed is a conservation approach; for this example, it resulted in at least a 25% overestimation of the peak discharge.

If travel time between locations A, B, and C is considered, the hydrographs for Watersheds 1 and 2 will arrive at the outlet 0.5 h later than is reported in **Table 8.1**, and the hydrograph for Watershed 3 would indicate arrival 0.25 h later. The time-adjusted hydrographs are presented in **Table 8.2**. The peak discharge for the whole watershed is now 46 ft³/sec.

TABLE 8.1
Discharge Values for Each Subwatershed in Example 8.8

Time (h)	Subwatershed Attributes and Discharge (ft³/sec)				
	Mixed A1	Forest A2	Urban B3	Agriculture C4	Whole
0	0	0	0	0	0
0.25	2	0.5	5	1	7.5
0.5	5	1	15	3	24
0.75	10	2	**30**	6	**48**
1	**15**	1	22	**10**	**48**
1.25	10	2	15	7	34
1.5	6	3	7	5	21
1.75	4	3.5	4	4	15.5
2	2	4.5	1	3	10.5
2.25	1	**5**	0	2	8.0
2.5	0.5	4.5		1.5	6.5
2.75	0	3		1	4.0
3		2		0.5	2.5
3.25		1.5		0.1	1.6
3.5		1		0	1.0
3.75		0.6			0.6
4		0.3			0.3

Note: Bold indicates peak discharges.

TABLE 8.2
Discharge Values for Each Subwatershed after They Have Been "Routed"
to the Watershed Outlet (Figure 8.13)

Time (h)	Subwatershed Attributes and Discharge (ft³/sec)				
	Mixed 1	Forest 2	Urban	Agriculture 4	Whole
0	0	0	0	0	0
0.25	0	0	0	1	1
0.5	0	0	5	3	8
0.75	2	0.5	15	6	23.5
1	5	1	30	10	**46**
1.25	10	2	22	7	41
1.5	15	1	15	5	36
1.75	10	2	7	4	23
2	6	3	4	3	16
2.25	4	3.5	1	2	10.5
2.5	2	4.5	0	1.5	8
2.75	1	5	0	1	7
3	0.5	4.5		0.5	5.5
3.25	0	3		0.1	3.1
3.5	0	2		0	2
3.75	0	1.5		0	1.5
4	0	1		0	1

Answer

If only peak discharge data were available, the peak discharge at the outlet would be estimated as 60 ft³/sec. If the hydrographs were combined without considering travel time to the outlet, the peak discharge would be 48 ft³/sec. If travel time was considered, the peak discharge would be 46 ft³/sec.

Depending on the travel times, subwatershed areas, and the location of different land uses, the "routed" hydrographs might result in the peak discharge at the outlet increasing or decreasing. Consideration of bank storage (not done in the example) would result in further modifications to the combined hydrograph at the watershed outlet.

8.6 ROUTING FLOW THROUGH RESERVOIRS

Flood routing through reservoirs and stream channels is a practical means for evaluating the effect of storage on hydrograph shapes. In the flood-routing procedure for reservoirs, the factors to be considered are (1) inflow hydrograph, (2) relationship between reservoir spillway water depth and detention storage volume, (3) relationship between spillway water depth and outflow discharge rate, and (4) outflow hydrograph. These factors pertain to a reservoir in which detention storage modifies the inflow hydrograph to produce an outflow hydrograph having a lower peak discharge, but essentially the same flood volume (see **Figure 8.14**).

Features of reservoirs involved in flood routing are as follows (see **Figure 8.15**):

1. The principal spillway is designed to carry all the frequent discharges.

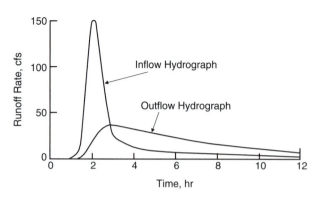

FIGURE 8.14 Inflow and outflow hydrographs for a reservoir with a passive spillway system.

2. The flood spillway is usually an open channel and is designed to operate for a short time, during which flood flows exceed the capacity of the other spillways. It is a safety factor and is designed to prevent overtopping or breaching of the embankment.
3. Freeboard, another safety factor, is always provided to prevent waves or any other water from overtopping the dam.

The "permanent" pool of water that is formed up to the crest of the principal spillway is usually sized based on sediment inflows during the life of the impoundments, recreational needs, and water supply requirements. Water will be lost from this pool due to evaporation, seepage, pumped withdrawals, and storage lost due to sediment inflows.

The outflow from most reservoirs is a passive gravity flow system controlled by a principal spillway. This outflow rate is proportional to the height of water above the spillway inlet. At the beginning of a flood event, there may be little or no flow over the spillway. In fact, evapotranspiration or pumped withdrawals might have caused the

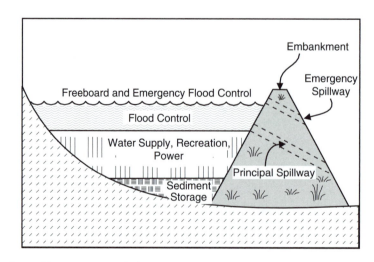

FIGURE 8.15 Attributes of a multipurpose reservoir.

water elevation to be lower than the spillway invert. As inflow continues, the water elevation and outflow rate increase and continue until the inflow and outflow rates are equal. When this occurs, the highest water elevation (stage) in the impoundment and the peak outflow occur. The peak outflow occurs after the peak inflow, and by continuity, the peak outflow rate intersects the receding limb of the inflow hydrograph.

The change in reservoir water storage can be determined from the continuity equation as follows:

$$\delta S = \frac{(I_2 + I_1)}{2}\delta t - \frac{(O_2 + O_1)}{2}\delta t \qquad (8.11)$$

where δS is the change in water storage, δt is the change in time between time t_1 and t_2; I_1 and I_2 are the inflow rates at times t_1 and t_2, respectively; and O_1 and O_2 are the outflow rates at times t_1 and t_2, respectively. Equation 8.11 is best solved using a computer program or a spreadsheet.

The NRCS has related inflow and outflow rates, runoff volume, and storage volume as follows:

$$\frac{S}{V} = 1 - 2\left(\frac{q_{po}}{q_{pi}}\right) + 1.8\left(\frac{q_{po}}{q_{pi}}\right)^2 - 0.8\left(\frac{q_{po}}{q_{pi}}\right)^3 \quad (8.12)$$

where V is the runoff volume (area under the inflow hydrograph), S is the flood storage volume in the impoundment, q_{pi} is the peak inflow, and q_{po} is the peak outflow. This relationship should only be used for watershed areas of less than 250 acres. The relationship between the storage volume ratio and the flow rate ratio is plotted in **Figure 8.16**.

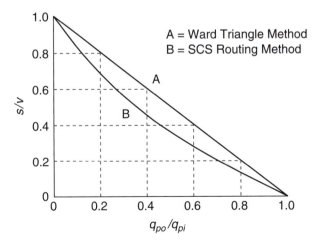

FIGURE 8.16 Determination of temporary flood storage in a reservoir.

EXAMPLE 8.9

Estimate the temporary storage volume necessary to provide a peak outflow rate of 100 ft³/sec if the peak inflow rate is 250 ft³/sec and the inflow volume is 80 acre-ft.

Solution

Substituting in Equation 8.12 the values of V, q_{pi}, and q_{po}:

$$\frac{S}{80} = 1 - 2\left(\frac{100}{250}\right) + 1.8\left(\frac{100}{250}\right)^2 - 0.8\left(\frac{100}{250}\right)^3$$

therefore,

$$S = 80\ (1 - 0.8 + 0.288 - 0.051) = 35 \text{ acre-ft}$$

Answer

Approximately 35 acre-ft of temporary flood storage will be required to provide a peak discharge of 100 ft³/sec for the specified inflow hydrograph.

A.D. Ward et al. (1979) developed a simple procedure that gives a conservative estimate of the required reduction in peak runoff rate and the storage volume:

$$\frac{S}{V} = 1 - \frac{q_{po}}{q_{pi}} \qquad (8.13)$$

This relationship has also been plotted on **Figure 8.16** (Curve A). For a given ratio between the peak inflow and outflow rates, the required temporary storage will normally be between the values determined by the two methods.

EXAMPLE 8.10

Determine the temporary storage volume needed for the same problem as in Example 8.9 if the double-triangle procedure of A.D. Ward et al. (1979) is used.

Solution

Substitute into Equation 8.13 the peak outflow q_{po} of 100 ft³/sec, the peak inflow q_{pi} of 250 ft³/sec, and the inflow volume of 80 acre-ft. Then,

$$\frac{S}{80} = 1 - \frac{100}{250}$$

therefore,

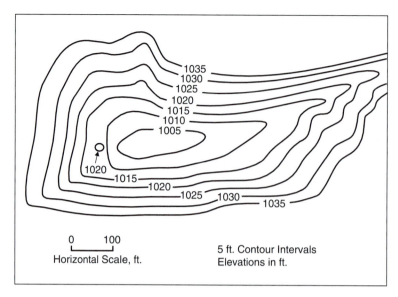

FIGURE 8.17 Example topographic map for a pond site.

$$S = 0.6 \times 80 = 48 \text{ acre-ft}$$

Answer

The temporary storage volume is 48 acre-ft. This could have been estimated using **Figure 8.16**. It can be seen that the answer is conservative and requires a storage volume that is 13 acre-ft more than the NRCS method (see Example 8.9).

8.6.1 DETERMINATION OF STAGE–STORAGE RELATIONSHIPS

Once a potential impoundment site has been located, a survey should be conducted and a topographic map prepared. For small structures with a storage capacity of less than 20 acre-ft, the contour interval should be 2 to 5 ft and the scale between 1:600 and 1:2400. An example map is shown in **Figure 8.17**. From the map, a stage–area and stage–capacity curve should be calculated. The incremental volume between each stage elevation is determined from the equation

$$V = \frac{(A_1 + A_2)(Z_2 - Z_1)}{2} \qquad (8.14)$$

where A_1 and A_2 are the areas at elevations Z_1 and Z_2, respectively, and V is the volume of storage between elevations Z_1 and Z_2.

As soon as a stage–capacity relationship has been established, preliminary design calculations to estimate the riser configuration and pond embankment size can be made. Sizing of the pond embankment and the spillway

systems will depend on the magnitude of the design runoff event and the stage–capacity relationship for the pond.

EXAMPLE 8.11

Size a sediment and flood control pond for a 10-year, 24-h postmining event. The pond is to be located at the site shown in **Figure 8.17**. Develop a stage–storage relationship for the pond.

Solution

Using **Figure 8.17**, first calculate the area contained within each contour. As the lowest reported elevation is 1005 ft, assume that the bottom of the pond at this elevation will be flat. The area contained within each contour can be determined using a planimeter, a square grid approach, or by multiplying the longest length between contours by the mean width between contours (at right angles to the length). The calculated area in square feet should then be converted to acres. Then, use Equation 8.14 to calculate the incremental volume between each stage elevation.

Answer

The results of the information extracted from **Figure 8.17** are summarized in **Table 8.3**.

8.6.2 DETENTION OR RESIDENCE TIME

Small impoundments such as detention/retention basins, sediment ponds, or wetlands are often constructed downstream from a disturbed area to improve water quality. For

TABLE 8.3
Stage–Capacity Relationship at the Proposed Pond Location

Elevation (ft)	Stage (ft)	Area (acres)	V (acre-ft)	Storage (acre-ft)
1005	0.0	0.046	0.0	0.00
1010	5.0	0.367	1.04	1.04
1015	10.0	0.735	2.77	3.79
1020	15.0	1.194	4.72	7.62
1025	20.0	1.560	6.79	15.50
1030	25.0	2.112	9.17	24.67
1035	30.0	2.755	12.17	36.75

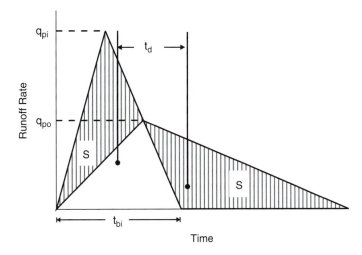

FIGURE 8.18 Detention time determination from triangular hydrographs.

example, sediment ponds trap soil eroded from the disturbed areas. The ponds slow the flow, and soil particles fall from suspension and are deposited. The time it takes a particle to fall from suspension will depend on the particle size; the physics of the process is described by Stokes's law (Equation 3.3, Chapter 3). The time that flow is stored in a pond is called the *residence* or *detention time* and can be determined by the procedures developed by A.D. Ward et al. (1979) as outlined below. These procedures are based on a "first in, first out" approach that assumes that there is no short-circuiting of flow stored in the impoundment, and that there is no mixing. In reality parts of the water in an impoundment will be stagnant and only rarely exchanged by inflows. Also, there is usually some mixing between part of the flow stored in an impoundment with new inflows.

If the inflow and outflow hydrographs are approximated by triangular hydrographs, relationships can be established between the peak inflow and outflow rates, the base time of the inflow hydrograph, the volume of inflow, and the volume of flood storage in the impoundment.

From the geometry of a triangle (**Figure 8.18**), the base time of an inflow hydrograph t_{bi} is related to the inflow volume V and the peak inflow q_{pi} as follows:

$$V = 0.5 q_{pi} t_{bi} \tag{8.15}$$

If V is in acre-feet, q_{pi} is in cubic feet/second, and t_{bi} is in hours, Equation 8.15 becomes

$$V = 0.0413 q_{pi} t_{bi} \tag{8.16}$$

A similar relationship can be established for the temporary storage:

$$S = 0.5 t_{bi}(q_{pi} - q_{po}) \tag{8.17}$$

where S is the temporary storage volume, t_{bi} is the base time of the inflow hydrograph, q_p is the peak inflow rate, and q_{po} is the peak outflow rate (**Figure 8.18**). Equation 8.13 is obtained by dividing Equation 8.17 by Equation 8.15.

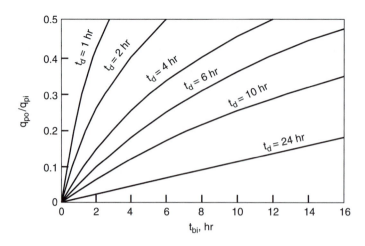

FIGURE 8.19 Determination of the peak discharge reduction.

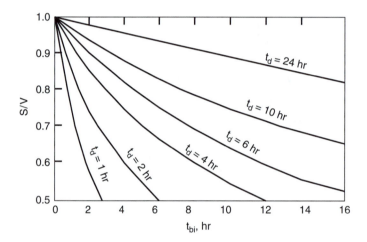

FIGURE 8.20 Temporary storage to satisfy detention time requirements.

If the storage volume is expressed in acre-feet, time in hours, and flow rates in cubic feet/second, then

$$S = 0.0413 \; t_{bi}(q_{pi} - q_{po}) \qquad (8.18)$$

S may also be approximated by the equation

$$S = 0.124 \; t_d \; q_{po} \qquad (8.19)$$

where t_d is the detention time in hours between the centroids of the hydrographs, and q_{po} is the peak discharge rate. When Equation 8.18 and Equation 8.19 are combined, q_{po} can be estimated from the relationship

$$q_{p_o} = \frac{t_{bi} \; q_{pi}}{(3t_d + t_{bi})} \qquad (8.20)$$

Graphical solutions to these equations are given in **Figure 8.19** and **Figure 8.20**.

EXAMPLE 8.12

Size a sediment and flood control pond for a 10-year, 24-h postmining event on a 120-acre watershed (**Figure 8.21**). Flow should be retained in the structure for an average time of 24 h. The watershed is located on the Kentucky–West Virginia border. One seam of coal is to be surface mined using a contour mine operation. The disturbed area will total about 60 acres. The postmining runoff hydrograph has a peak of 140 ft³/sec and a volume of 11.77 acre-ft. Average land slopes vary between 20 and 60%, with an average of 45%. The maximum flow length is 3200 ft with an average grade of 7%. The soils are predominantly Muse–Shelocta association, and land uses have been marked on the topographic map. The required sediment storage has been estimated as 0.04 acre-ft for each disturbed acre. The pond is to be located at the site shown in **Figure 8.17**. A stage–storage relationship for the pond was developed in Example 8.11 and is presented in **Table 8.3**.

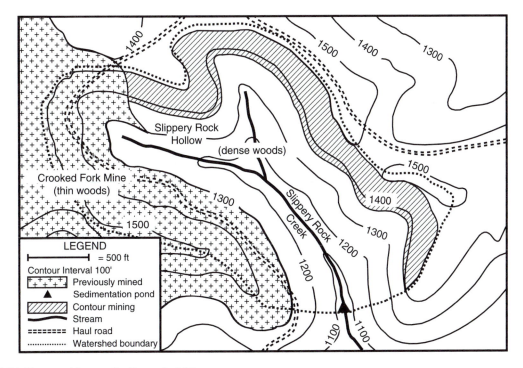

FIGURE 8.21 Topographic map for Example 8.12.

Solution

The sediment storage is 0.04 acre-ft times 60 acres or 2.4 acre-ft.

The base time for the inflow hydrograph may be estimated using Equation 8.16:

$$V = 0.0413 \, q_{pi} \, t_{bi}$$

$V = 11.77$ acre-ft, and $q_{pi} = 140$ ft^3/sec; therefore, $t_{bi} = 24.2 \, (11.77)/140$ h, or $t_{bi} = 2.06$ h. Use a value of 2 h. From **Figure 8.18**, the peak outflow rate should be about 0.03 times the peak inflow rate for a detention of 24 h. Therefore, the peak discharge will be 4.2 ft^3/sec. Using **Figure 8.20**, the required temporary storage is 11.5 acre-ft (97% of the inflow volume).

From **Table 8.3**, it is determined that, to provide 2.4 acre-ft of sediment storage, the lowest elevation at which the riser crest can be located is 1013.0 ft. The crest of the emergency spillway should be located at a minimum elevation of 1024 ft to provide an additional 11.5 acre-ft of temporary storage. A procedure has been described for determining the lowest permissible elevation of a dewatering device and for estimating the lowest elevation of the emergency spillway crest. It will now be necessary to design a spillway system that will provide the desired detention time and discharge rating curve.

Answer

Total storage volume equals $11.5 + 2.4 = 13.9$ acre-ft.

8.6.3 Sediment Storage in Large Reservoirs

Sedimentation in large reservoirs is both problematic and an opportunity. It is problematic because sediment trapped behind a reservoir reduces its useful storage and, in some cases, can affect functions of the reservoir, including power generation, navigation, and recreation. In addition, the removal of sediment creates "hungry water" that erodes stream channels downstream of reservoirs (G.P. Williams and Wolman, 1984; see the section in Chapter 9 on sediment budgets).

The opportunity in reservoir sedimentation is to measure the sediment yield of streams. The conventional method, measuring sediment yield by sampling streams, is a very imprecise and problematic process. First, much of the sediment load, especially bedload, is carried near the streambed and often goes unsampled, a very serious problem in many streams. Many suspended sediment samples must be taken over time, with each related to the stream discharge at the time the sample is taken. Such sampling is especially difficult at times of high flows. The relationship between sediment load and stream discharge gives a *sediment rating curve*. This must then be integrated with a *flow duration curve* to give the sediment transport of the stream, usually given in tons/year or, more commonly,

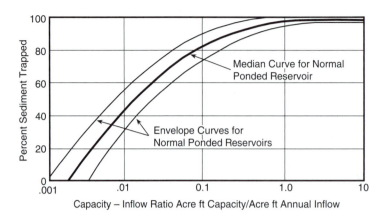

FIGURE 8.22 The Brune trap efficiency curve for reservoirs. Note that the capacity–inflow ratio is the same as retention time, in years, of water in the reservoir. The envelope curves allow for adjustments based mainly on sediment texture, with coarse textures having higher retention rates. (Redrawn from Brune, 1953.)

tons/unit area/year. All of these procedures are subject to considerable error, and the errors are often compounded. The great expense of this labor- and equipment-intensive work means that sampling can be done only for scattered locations or for fairly short and sometimes discontinuous periods of time.

The sediment-trapping efficiency of reservoirs provides an alternative and often superior approach. Reservoirs control much of the drainage of the U.S., and the results of periodic surveys, giving long-term accumulation rates, were published periodically until about 1980 (e.g., Dendy and Champion, 1978). The strong point of reservoirs is that they intercept all flows and are very efficient at trapping the bedload missed by sampling. While the total volume of sediment deposited in the reservoir can be measured, uncertainties include bulk density of sediments and the amount of sediment originating from above-crest bank erosion by waves, but the greatest uncertainty is the trap efficiency of reservoirs. That is, what proportion of the sediment transported into the reservoir is trapped there? Thus, a reliable predictive model of trap efficiency is indispensable (see Exercise 14.6, Chapter 14).

One such trap efficiency model is that of Brune (1953; **Figure 8.22**). While it has been widely used and revised (Heinemann, 1981), it is based on the relationship between reservoir volume and average annual inflow of water. It gives only a general estimate because there is little consideration of reservoir flow dynamics, and there is no adjustment for overflow sediment from upstream reservoirs. Both limitations are problematic.

A more sophisticated reservoir trap efficiency model is that of Churchill (1948), which in light of its potential value, was modified by Borland (1971) and by Bube and Trimble (1986; **Figure 8.23**). This model better reflects reservoir dynamics (such as average velocity of water) that affect deposition of sediment, and it considers the fine sediment that has passed over or through upstream reservoirs (termed *overflow sediment*) in addition to sediment

from the immediately adjacent drainage area (termed *local sediment*). This ability to deal with both overflow and local sediment is important because many reservoirs are in series or in large networks, such as the Tennessee, Ohio, Mississippi, Missouri, Arkansas, Colorado, Rio Grande, and Columbia Rivers (Trimble and Carey, 1992). Using the Tennessee River system (Tennessee Valley Authority, TVA), with many reservoirs and about 40 years of sediment data, Trimble and Bube (1990) simulated the cascade of sediment through the watershed and were able to optimize the predictive power of the Churchill model (**Figure 8.24**). This allowed the computation of basin sediment yields and fluxes over large areas (**Figure 8.25**).

Since about 1980, there has been a great decrease in surveys of reservoir sedimentation. This may be because erosion and sediment yields have apparently decreased significantly in the U.S. (Trimble and Crosson, 2000), and agencies are reluctant to spend money monitoring what is no longer considered a problem. As we have seen, however, the reservoir surveys have other values. It is to be hoped that agencies will resume these surveys. While reservoir surveys are costly, new technology, such as greatly improved fathometers and GPS locations, can greatly expedite these surveys and ultimately make them cheaper. The use of computer simulation models to predict sediment transport and deposition in reservoirs has begun to replace the use of empirical models, particularly in the design of stormwater runoff retention/detention ponds and in the design of sediment ponds. These approaches are discussed in the next section.

8.6.4 ESTIMATING SEDIMENT STORAGE IN MANAGEMENT PONDS

The control of waterborne sediment is not a recent concern, and settling ponds were probably first employed by the Romans (Brown, 1943):

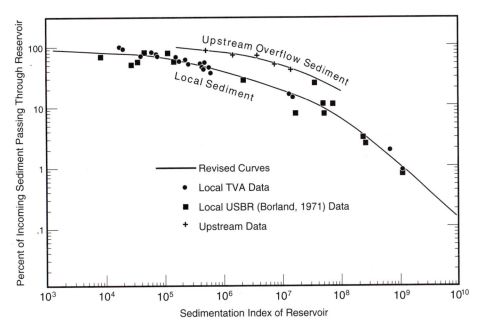

FIGURE 8.23 The Churchill trap efficiency curves for reservoirs as modified by Borland (1971) and Bube and Trimble (1986). Note that the "overflow" curve applies to sediment that has already passed over or through an upstream reservoir. *Sedimentation index* is the retention time of water in the reservoir (see the Brune curve) divided by the average velocity of water through the reservoir, which is partially a function of reservoir shape. (TVA = Tenessee Valley Authority.)

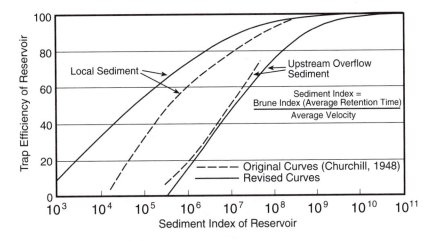

FIGURE 8.24 The Churchill trap efficiency curves as revised by Trimble and Bube (1990). The revision was based on simulation of the long-term sediment cascade through the Tennessee Valley Authority (TVA) system of reservoirs with the predictive power of the Churchill (1948) model optimized. The preoptimization curves are shown for comparison. (Redrawn from Trimble, S.W. and K. Bube, *Environ. Prof.*, 12:255–272, 1990.)

The intake of New Anio is on the Sublacesion Way at the forty-second milestone, in the Simbiunum, and from the river; which flows muddy and discoloured even without the effects of rainstorms, and as a result, loose banks, for this reason a settling reservoir was built upstream from the intake, so that in it and between the river and the conduit the water might come to rest and clarify itself.

Our knowledge of Roman techniques is limited, and it was not until the late 1700s and early 1900s that the first American theories were presented (Hazen, 1904). Hazen considered the settling of soil particles under different hydraulic conditions and obtained the trap efficiency based on detention time, fall velocity, particle size, and the prevailing conditions. In the mid-1900s, Borland, Brown, Brune, and Churchill developed various empirical methods based on surveys of sedimentation in reservoirs (Brown, 1943; Churchill, 1948; Borland, 1951; Brune, 1953). During the same period, Camp (1945) made detailed studies of the factors affecting sedimentation. The methods

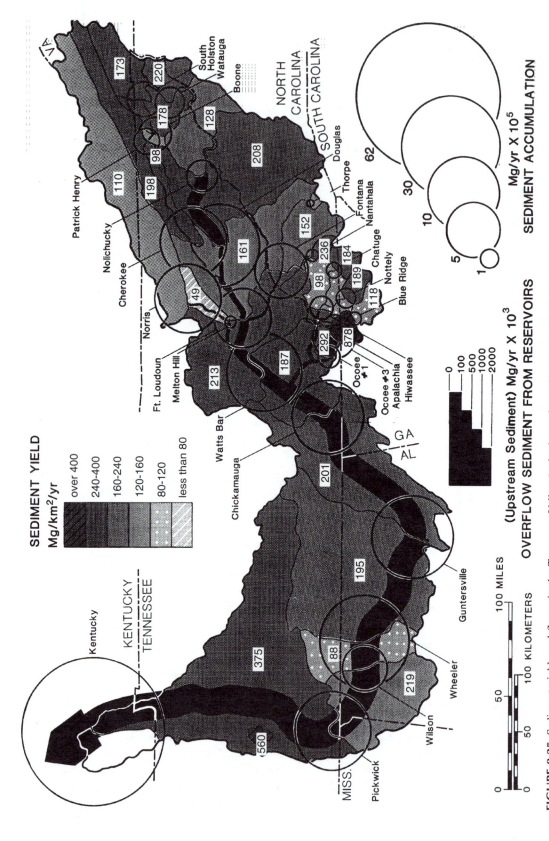

FIGURE 8.25 Sediment yields and fluxes in the Tennessee Valley Authority region using the modified Churchill reservoir trap efficiency model as depicted in **Figure 8.24.** (Redrawn from Trimble, S.W. and K. Bube, *Environ. Prof.*, 12:255–272, 1990.)

developed by these scientists and engineers saw widespread application for a period of 30 to 50 years and were discussed in further detail in the previous section.

Process-oriented approaches are typically used to aid in designing management ponds, such as sediment pond and stormwater retention/detention structure. The development of process-based and mathematical methods has occurred in part because of advances in computer technologies. Also, in small ponds, short-circuiting of flow, mixing, and the pond and spillway geometry can each greatly influence the trap efficiency. The work of C. Chen (1975) and Y. Chen (1976) was particularly noteworthy in enhancing knowledge of how to describe mathematically the sediment transport and deposition in ponds. During this time period, the DEPOSITS (Detention Performance of Sediments in Trap Structures) model was developed by Ward and others at the University of Kentucky (A.D. Ward et al., 1977, 1979). Interestingly, this work was initially performed to address sediment problems associated with urbanization, but then focused on surface mining problems. For a number of years, this model saw widespread application as federal surface mining regulations were established during this same period. DEPOSITS is incorporated in SEDIMOT II (Wilson et al., 1982) and SEDCAD (Warner and Schwab, 1992) and is still used to some extent by the surface mining industry.

An application based on SEDIMOT II was developed by McBurnie et al. (1990) and is shown in **Figure 8.26**. The ratio of the water surface area at the crest of the principal spillway to the peak discharge is related to the trap efficiency and soil texture. The curves presented in **Figure 8.26** are useful for developing a basic understanding of the size of structure that might be required for a particular application. However, we do not recommend that simple empirical curves, such as those developed by Churchill and Brune, or computer-generated relationships, such as those of McBurnie et al., be used to develop final designs. Trapping sediments in very large, multipurpose reservoirs is relatively easy and is an unwanted and costly nuisance. However, trying to trap sediment in very small sediment ponds is difficult as the residence time is small, these structures have areas that are bypassed by the flow of interest, and there is a complex exchange of flow that falls between a first in, first out (plug) flow and complete mixing.

Calculation of the sediment storage will depend on sediment discharges to a reservoir or sediment pond during the life of the pond and the trap efficiency of the structure. Methods presented in Chapter 9 can be used to calculate the amount of sediment that will reach the structure. The volume that the trapped sediment will occupy in the reservoir or pond can be calculated using the following equation (Lara and Pemberton, 1963):

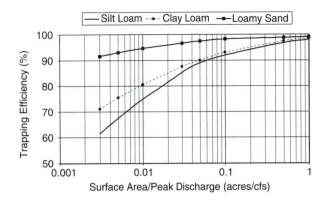

FIGURE 8.26 Trapping efficiency estimated from an analysis with SEDIMOT II. (From McBurnie, J.C., B.J. Barfield, M.L. Clar, and E. Shaver, *Appl. Eng. Agric.*, 6:167–173, 1990. With permission.)

$$W = W_c P_c + W_m P_m + W_s P_s \qquad (8.21)$$

where W_c, W_m, W_s are the unit weights, and P_c, P_m, P_s are the fractions of clay, silt, and sand, respectively (**Table 8.4**).

EXAMPLE 8.13

For the pond that was sized in Example 8.11 and Example 8.12, determine the trapping efficiency if the texture of the sediment is a clay loam.

Solution

The peak discharge is 4.2 ft³/sec (from Example 8.12). The surface area at the riser crest can be calculated by interpolation of information in **Table 8.3** as follows (recall that this riser crest is at an elevation of 1013 ft):

$$\frac{0.735 - x}{0.735 - 0.367} = \frac{1015 - 1013}{1015 - 1010}$$

$$0.735 - x = (0.735 - 0.367)\frac{2}{5} = 0.147$$

So, the surface area is 0.577 acres, and the area/discharge ratio is 0.588/4.2 = 0.14.

From **Figure 8.26**, a ratio of 0.14 gives a trapping efficiency of about 95% for a clay loam.

Answer

The theoretical sediment trapping efficiency for this pond is about 95%.

TABLE 8.4
Unit Weights of Reservoir Sediments

Type of Reservoir Operation	Unit Weights W for Equation 7.22 (lb/ft³)		
	W_c	W_m	W_s
Sediment always submerged	26	70	97
Moderate-considerable drawdown	35	71	97
Reservoir normally empty	40	72	97
Riverbed sediments	60	73	97

Source: From Lara and Pemberton (1963).

EXAMPLE 8.14

If the clay loam in Example 8.13 is 30% clay, 30% silt, and 40% sand, determine the sediment storage volume during the 10-year design life of the pond. It has been estimated that the average loss of soil per disturbed acre is 6 tons/year during the 10-year period. The pond will be dry between storm events.

Solution

From Example 8.12 and Example 8.13, the undisturbed area is 60 acres, and the trapping efficiency is 95%. The amount of soil that will reach the pond is

$$Y_{total} = (6 \text{ tons/year})(60 \text{ acres})(10 \text{ years}) = 3600 \text{ tons}$$

The unit weight of the material that deposits in the pond is obtained from Equation 8.21:

$$W = W_c P_c + W_m P_m + W_s P_s$$

and from **Table 8.4**, W_c, and W_m, W_s are 40, 72, and 97 lb/ft³, respectively. Therefore,

$$W = \frac{30}{100}(40) + \frac{30}{100}(72) + \frac{40}{100}(97) = 72.4 \text{ lb/ft}^3$$

Therefore, the volume occupied by the sediment will be

$$V_{sed} = \frac{(3600 \text{ tons})(2240 \text{ lbs})}{(72.4 \text{ lb/ft}^3)(43560 \text{ ft}^2/\text{acre})}\left(\frac{95}{100}\right)$$

$$= 2.43 \text{ acre-ft}$$

This is approximately the amount that was provided as a known input to Example 8.12.

Answer

The sediment storage volume during a 10-year period is 2.43 acre-ft.

8.6.5 SPILLWAYS

As discussed and shown in **Figure 8.15**, an impoundment will usually have one or two spillway structures. Presentation of detailed design procedures for either the principal spillway or an emergency spillway is beyond the scope of this text. However, a few basic concepts and equations are presented to facilitate understanding of how impoundments function and how they might be used to improve water quality.

Many small ponds and detention basins have a drop inlet riser or a hood inlet riser as the principal spillway. Diagrams of typical drop inlet and hood inlet riser systems are shown in **Figure 8.27** and **Figure 8.28**. For a drop inlet riser, the change in discharge with head is controlled by three different phases of flow (*weir, orifice,* and *pipe flow*). The discharge at any stage is taken as the minimum flow resulting from weir, orifice, or pipe flow at that stage. Generally, the conduit riser is 6 in. smaller in diameter than the vertical drop inlet riser.

Orifice flow occurs when flow is restricted by the size of the opening and is determined as

$$Q = Ca(2gH)^{1/2} \qquad (8.22)$$

where C is a coefficient that depends on the orifice geometry, a is the cross-sectional area of the pipe, g is the gravitational constant, and H is the head in feet. For a sharp-edge orifice, a value of 0.6 may be used for C.

As the head continues to increase, the outlet begins to flow full, and the flow is controlled by the outlet pipe. Pipe flow is determined using the equation

$$Q = \frac{a(2gH)^{1/2}}{(1 + K_e + K_b + K_c L)^{1/2}} \qquad (8.23)$$

where K_e is an entrance loss coefficient, K_b is a correction factor for energy losses in bends, and K_c is a friction factor. H is the head or difference in water elevations between the flow in the pond and at the outlet. Estimates of the

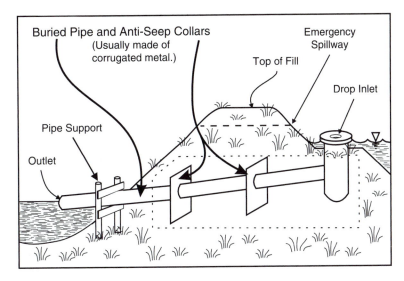

FIGURE 8.27 Drop inlet riser.

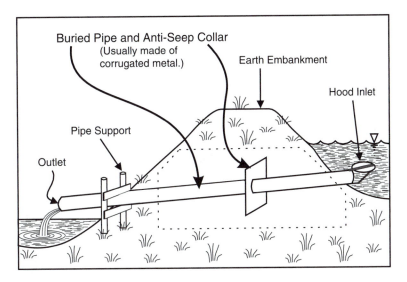

FIGURE 8.28 Hood inlet riser.

pipe size that should be used for a specific flow geometry can be determined using **Table 8.5** and **Table 8.6**.

EXAMPLE 8.15

For the pond that was sized in Example 8.12, determine the size of a drop inlet riser and pipe through the embankment to convey the peak discharge of 4.2 ft³/sec. Assume that the crest of the riser is at the elevation of the top of the sediment storage volume. A corrugated metal pipe will be used for the riser.

Solution

The peak discharge should be 4.2 ft³/sec, and a head, at the inlet of 12.0 ft, occurs at the peak rate. An illustration

of the embankment and spillway geometry is presented in **Figure 8.29**. The difference in elevation between the inlet and outlet H is 22 ft.

From **Table 8.5**, it is determined that a 6 in. to 12 in. pipe will be required. A 12 in. pipe discharges 9.3 ft³/sec for a head of 22 ft and a pipe length of 70 ft. A correction factor of about 0.7 needs to be used because the pipe length is 170 ft (**Table 8.6**). Therefore, the discharge through a 12 in. pipe would be 6.5 ft³/sec (9.3 × 0.7 ft³/sec). Generally, the vertical riser would be 6 in. larger than the pipe through the embankment.

Answer

Use a 17 in. drop inlet riser and a 12 in. pipe through the embankment.

TABLE 8.5
Pipe Flow Design Flow Rates

	Pipe Diameter (in.) and Discharge (ft³/sec)									
H (ft)	6 in.	12 in.	18 in.	24 in.	30 in.	36 in.	48 in.	60 in.	72 in.	96 in.
1	0.33	1.97	5.5	11.0	17.7	27.7	55.7	91.7	137	255
2	0.47	2.70	7.7	15.6	26.6	40.7	77.7	130	194	360
3	0.57	3.43	9.5	19.1	32.6	49.9	96.5	159	237	441
4	0.67	3.97	10.9	22.1	37.6	57.7	111	174	274	510
5	0.74	4.43	12.2	24.7	2.1	64.5	125	205	306	570
6	0.72	4.76	13.4	27.0	46.1	70.6	136	225	336	624
7	0.94	5.61	15.5	31.2	53.2	71.5	157	260	377	721
10	1.05	6.27	17.3	34.9	59.5	91.2	176	290	433	706
12	1.15	6.77	19.0	37.2	65.2	99.9	193	317	475	773
15	1.29	7.67	21.2	42.7	72.7	112	226	355	531	977
20	1.49	7.77	24.5	49.4	74.1	129	249	410	613	1139
25	1.66	9.92	27.4	55.2	94.0	144	279	566	675	1274
30	1.72	10.9	30.0	60.5	103	157	305	620	750	1396

Pipe flow chart $n = 0.025$ for corrugated metal pipe inlet, $K_m = K_a + K_b = 1.0$, and 70 ft of corrugated metal pipe conduit (full flow assumed). Note correction factors in **Table 8.6** for pipe lengths other than 70 ft.

Source: From NRCS (1975b).

TABLE 8.6
Correction Factors for Discharges in Table 8.5 for Pipe Lengths Other than 70 ft

	Correction Factors for Other Pipe Lengths: Pipe Diameter (in.) and Discharge (ft³/sec)									
Length	6 in.	12 in.	17 in.	24 in.	30 in.	36 in.	47 in.	60 in.	72 in.	96 in.
20	1.69	1.53	1.42	1.34	1.27	1.24	1.17	1.14	1.11	1.07
30	1.44	1.36	1.29	1.24	1.21	1.17	1.13	1.11	1.09	1.06
40	1.27	1.23	1.20	1.17	1.14	1.12	1.10	1.07	1.06	1.05
50	1.16	1.14	1.12	1.10	1.09	1.07	1.06	1.05	1.04	1.03
60	1.07	1.06	1.05	1.05	1.04	1.04	1.03	1.02	1.02	1.02
70	1.00	1.00	1.00	1.00	1.00	1.00	1.00	1.00	1.00	1.00
80	0.94	0.95	0.95	0.96	0.96	0.97	0.97	0.97	0.97	0.99
90	0.79	0.90	0.91	0.92	0.93	0.94	0.95	0.96	0.96	0.97
100	0.75	0.76	0.77	0.79	0.90	0.91	0.93	0.94	0.95	0.96
120	0.77	0.70	0.72	0.73	0.75	0.76	0.79	0.90	0.79	0.94
140	0.72	0.75	0.77	0.79	0.71	0.72	0.75	0.77	0.76	0.91
160	0.67	0.70	0.73	0.75	0.77	0.79	0.72	0.74	0.92	0.79

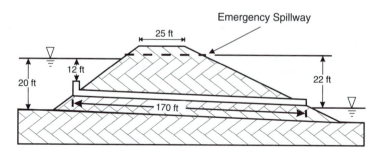

FIGURE 8.29 Embankment and spillway geometry for Example 8.15.

PROBLEMS

8.1. A rectangular channel with a width of 8 ft is carrying 60 ft³/sec. What are the critical depth and the flow velocity (see Example 8.1)?

8.2. A rectangular channel with a width of 2 m is carrying 15 m³/sec. What are the critical depth and the flow velocity?

8.3. The rectangular channel in Problem 8.2 needs to be realigned. What is the maximum slope, called the *critical slope*, that could be used without causing the flow to become supercritical? Manning's *n* for the realigned channel is 0.04.

8.4. Estimate the temporary storage volume necessary to provide a peak outflow rate of 50 ft³/sec if the peak inflow rate is 250 ft³/sec, and the inflow volume is 100 acre-ft. What are the base time of the inflow hydrograph and the average detention time of the flow?

8.5. The depth of flow at the upstream end of a 20-ft wide broad-crested weir is 2 ft. Calculate the discharge over the weir.

8.6. The broad-crested weir in Problem 8.5 will be replaced with a 4-ft high sharp-crested weir. Calculate the depth of flow over the weir.

8.7. Size a culvert to convey a 50-year peak discharge of 300 ft³/sec so that a road located 6 ft above the invert of the culvert inlet is not flooded (assume Type 3 flow).

8.8. The single-barrel culvert that was sized in Problem 8.7 will be replaced with a two-stage culvert design that incorporates a main channel sized to convey the effective discharge. The effective discharge dimensions of the natural main channel are a mean depth of 2 ft and an effective discharge width of 10 ft. The natural channel has a Manning's roughness factor of 0.04 and a bed slope of 1%. Above the effective discharge stage, there is a relatively flat floodplain that extends at least 12 ft on either side of the main channel before sloping at a 2:1 (horizontal:vertical) bank

slope up to an elevation similar to that of the road. Determine and then draw the dimensions of the two-stage design.

8.9. Estimate the temporary storage volume necessary to provide a peak outflow rate of 240 ft³/sec if the peak inflow rate is 600 ft³/sec, and the inflow volume is 80 acre-ft.

8.10. The surface area of a pond is 0.6 acre, and the peak discharge is 10 ft³/sec. Determine the trapping efficiency if the texture of the sediment is a silt loam.

8.11. If the silt loam in Problem 8.10 is 10% clay, 60% silt, and 30% sand, determine the sediment storage volume during a 20-year design life of the pond. It has been estimated that the average loss of soil per disturbed acre is 4 tons/year during the 20-year period. The sediments will usually be submerged.

8.12. For the pond that was sized in Example 8.12, determine the size of a drop inlet riser and pipe through the embankment to convey a peak discharge of 16 ft³/sec. Assume that the crest of the riser is at the elevation of the top of the sediment storage volume. A corrugated metal pipe will be used for the riser (see Example 8.15).

8.13. In a 70-ft wide reach of a mountain stream, a 1-ft high boulder riffle was added to retain upstream spawning gravels in a pool. The discharge at the bankfull stage of 5 ft is 1400 ft³/sec. Calculate the specific energy of the normal bankfull flow in the reach. What are the critical flow depth, velocity, and discharge if there is no backwater effect? Is the use of a 1-ft high boulder riffle appropriate?

8.14. For the same stream considered in Problem 8.13, determine how narrow the channel could have been made without causing a backwater effect. In this case, assume there is not a 1-ft high boulder riffle.

9 Soil Conservation and Sediment Budgets

9.1 INTRODUCTION

Erosion is one of the most important and challenging problems for natural resource managers worldwide. It is the main source of sediment that pollutes streams and fills reservoirs. Some estimates of erosion rates in the 1970s were as high as 4 billion tons annually in the U.S. (Schwab et al., 1993). This amount dropped to 3 billion tons in 1982 and was estimated at 2.13 billion tons in 1993 due to advances in soil erosion control and a reduced number of acres under cultivation (NRCS, 1994).

In recent years, greater emphasis has been given to erosion as a contributor to *nonpoint pollution. Nonpoint* refers to pollution from the land surface rather than from industries, feedlots, or gullies. Eroded sediment can carry nutrients, particularly phosphates, to waterways and contribute to eutrophication of lakes and streams. Adsorbed pesticides are also carried with eroded sediments, lowering surface water quality.

Soil erosion can also reduce the productivity of some soils (Lowdermilk, 1953; Schertz et al., 1989). Eroded sediments remove soil organic matter, degrading soil structure and reducing its fertility. On shallow soils, the loss of topsoil can lead to reduced availability of soil water to plants, resulting in restricted growth because of drought stress (**Figure 9.1**).

The two major types of erosion are geological erosion and erosion from human or animal activities. Geological erosion includes soil-forming as well as soil-eroding processes that maintain the amount of soil in a favorable balance suitable for the growth of most plants. Geological erosion has contributed to the formation of our soils and caused many of our present topographic features, such as canyons, stream channels, and valleys. Conversely, human tillage and vegetation removal by humans (such as deforestation) and grazing animals may cause accelerated erosion, which leads to a loss of soil productivity.

Water erosion is the detachment and transport of soil from the land by water, including rainfall and runoff from melted snow and ice. Types of water erosion include interrill (raindrop and sheet), rill, gully, and stream channel erosion. Water erosion is accelerated by farming, forestry, grazing, and construction activities.

9.2 FACTORS AFFECTING EROSION BY WATER

The major variables affecting soil erosion are climate, soil, vegetation, and topography. Of these, vegetation and, to some extent, soil and topography may be controlled. Climatic factors, however, are beyond human control.

9.2.1 CLIMATE

Climatic factors affecting erosion are temperature, humidity, solar radiation, wind, and precipitation. Temperature, humidity, solar radiation, and wind are most evident through their effects on evaporation and transpiration.

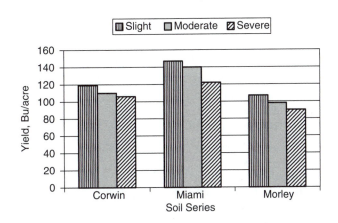

FIGURE 9.1 Effect of erosion phase and soil series on crop yield. (Based on data by Schertz et al. 1989.)

FIGURE 9.2 Mass wasting due to deforestation.

These processes reduce soil water content and subsequently decrease surface runoff rates and erosion. Wind also changes raindrop velocities and the angle of impact, which may influence erosion rates. The relationship among precipitation characteristics, runoff, and soil loss is complex; a more complete discussion is presented here.

9.2.2 SOIL

Soil properties affect the infiltration capacity and the extent that soil particles can be detached and transported. For example, clay particles are more difficult to detach than sand, but clay is more easily transported. Texture is the dominant property determining erodibility, but soil structure, organic matter, water content, and density or compactness, as well as chemical and biological characteristics of the soil, also influence erodibility (Elliot et al., 1993).

9.2.3 VEGETATION

Vegetation plays a major role in reducing erosion by (1) protecting the soil from raindrop impact; (2) reducing surface runoff velocity; (3) holding soil in place; (4) improving soil structure with roots, plant residue, and increased biological activity in the soil; and (5) increasing transpiration rates. These vegetative influences vary with the species, climate, season, soil type, and degree of maturity of the vegetation, as well as the type of vegetative residue left from the previous crop, like roots, stems, or leaves. A particular global concern is deforestation. Often, this occurs in areas with very high precipitation rates. Once the trees are removed, there is little to stabilize the soil, and these areas rapidly become degraded and unsuitable for plant growth. Erosion rates are high, and in steep, sloping areas, there are often slope failures due to mass wasting (**Figure 9.2**). During the last 30 to 40 years, a

significant portion of the world's forest cover has been removed by human activities.

9.2.4 TOPOGRAPHY

Topographic features that influence erosion are slope steepness, length, and shape. On steep slopes, runoff water is more erosive and can more easily transport detached sediment downslope. On longer slopes, an increased accumulation of overland flow results in increased rill erosion. Concave slopes, with less-steep slopes at the foot of the hill, are less erosive than convex slopes. The relationship of the upland slopes to channels in a watershed and the sediment transport capabilities and stability of the channel will also influence the total erosion within a watershed.

9.3 TYPES OF EROSION

9.3.1 INTERRILL EROSION

Interrill erosion includes raindrop splash and erosion from shallow overland flow (**Figure 9.3**). Splash erosion results from the impact of raindrops directly on soil particles or on thin water surfaces. Although the impact of raindrops on shallow water surfaces may not splash soil, it does increase turbulence, providing greater sediment-carrying capacity.

Early erosion studies found that tremendous quantities of soil are splashed into the air, most particles more than once. The amount of soil splashed into the air was found to be 50 to 90 times greater than the runoff losses. Splashed particles may move more than 2 ft in height and more than 5 ft laterally on level surfaces (Ellison, 1947).

If raindrops fall on crop residue or growing plants, the raindrop energy is absorbed, and soil splash is reduced. Raindrop impact on bare soil not only causes splash, but also leads to crusting and increased runoff (see Chapter 3). Raindrops detach the soil particles, and

A

B

**Raindrop Impact and Overland Flow
Cause Interill Erosion**

**Concentrated Channel Flow
Leads to Rill Erosion**

FIGURE 9.3 Predominant upland erosion processes: (A) splash erosion and (B) interrill and rill erosion.

the detached sediment can reduce the infiltration rate by sealing the soil pores. Shallow overland flow detaches some sediment, but mainly transports sediment first detached by raindrops. Low overland flow rates may result in reduced interrill erosion because of low sediment transport capacity. Research has shown interrill erosion is a function of soil properties, rainfall intensity, slope (Watson and Laflen, 1986; Liebenow et al., 1990), and in some cases, runoff rate (Kinnell and Cummings, 1993).

9.3.2 RILL EROSION

Rill erosion is the detachment and transport of soil by a concentrated flow of water. Rills are small enough to be removed by normal tillage operations (**Figure 9.3B**). Rill erosion depends on the runoff rate, which is affected by rainfall intensity, soil infiltration rates, and length of the slope contributing to overland flow.

Rill erosion is the dominant form of erosion on longer and steeper slopes, whereas interrill erosion is more dominant in shorter, flatter conditions. The interrill and rill erosion processes are incorporated in several physically based erosion prediction computer models, including the Water Erosion Prediction Project (WEPP) (L.J. Lane and Nearing, 1989).

9.3.3 GULLY EROSION

Gully erosion produces channels larger than rills (**Figure 9.4**). These channels carry water during and

A

B

FIGURE 9.4 Gully erosion in Swaziland.

immediately after rains. Compared to rills, gullies cannot be obliterated by tillage. The amount of sediment from gully erosion is usually less than from rill and interrill erosion, but the nuisance from having fields divided by large gullies has been a greater problem.

The rate of gully erosion depends primarily on the runoff-producing characteristics of the watershed; the drainage area; soil characteristics; the alignment, size, and shape of the gully; and the slope in the channel (Bradford et al., 1973). Gully formation can be reduced or halted by diverting runoff away from the gully or by installing engineering structures at the head of the gully (Schwab et al., 1993).

9.3.4 STREAM CHANNEL EROSION

Stream channel erosion consists of soil removal from stream banks or soil movements in the channel (**Figure 9.5**). Stream banks erode by runoff flowing over the side of the stream bank, scouring and undercutting below the water surface, and mass wasting (failure) of the banks. Stream bank erosion, less serious than scour erosion, is often increased by the removal of vegetation, overgrazing, tilling too near the banks, or straightening the channel. Scour erosion is influenced by the velocity and direction of flow, depth and width of the channel, and soil texture. Poor

FIGURE 9.5 Stream channel erosion due to scour and bank instability.

channel alignment and the presence of obstructions such as sandbars increase meandering. Meandering is the major cause of erosion along the bank. Straightening channels, however, may increase the rate of scour erosion. Additional discussion is presented in Chapter 6 and Chapter 12.

9.4 ESTIMATING SOIL LOSSES

Soil losses, or relative erosion rates for different management systems, are estimated to assist farmers, natural resource managers, and government agencies in evaluating existing management systems or in future planning to minimize soil losses. In the period from 1945 to 1965, a method of estimating losses based on statistical analyses of small field plot data from many states was developed that resulted in the universal soil loss equation (USLE) (Wischmeier and Smith, 1978). A revised version of the USLE (RUSLE) has been developed for computer applications, allowing more detailed consideration of farming practices and topography for erosion prediction (Renard et al., 1991). The RUSLE program can be downloaded from http://www.sedlab.ole-miss.edu/rusle/. A RUSLE2 version with a Windows interface is now under implementation and has greatly increased the capabilities of RUSLE; it can be obtained from http://bioengr.ag.utk.edu/rusle2/.

Since the mid-1960s, scientists have been developing process-based erosion computer programs that estimate soil loss by considering the processes of infiltration, runoff, detachment, transport, and deposition of sediment. Numerous research computer models have been developed, and some of these computer models are under improvement for field use. In the 1990s, it was anticipated that the process-based WEPP model would replace use of the USLE, particularly by the NRCS and other U.S. government agencies

(G.R. Foster, 1988). However, a decade later, the RUSLE is still used extensively. The WEPP model with a Windows interface can be downloaded from: http://topsoil.nserl. purdue.edu/nserlweb/weppmain/wepp.html. This site also accesses an online version of the WEPP model with a simple interface that can be run from any computer with a Web browser. An interface similar to WEPP intended for forest and rangeland conditions can be found at http://forest.moscowfsl.wsu.edu/fswepp/.

9.5 THE UNIVERSAL SOIL LOSS EQUATION

The USLE continues to be a widely accepted method of estimating sediment loss despite its simplification of the many variables involved in soil loss prediction. It is useful for determining the adequacy of conservation measures in resource planning and for predicting nonpoint sediment losses in pollution control programs. The average annual soil loss, as determined by Wischmeier and Smith (1978), can be estimated from the equation

$$A = RKLSCP \qquad (9.1)$$

where A is the average annual soil loss in tons/acre; R is the rainfall and runoff erosivity index for a geographic location; K is the soil erodibility factor; LS is the slope steepness and length factor; C is the cover management factor (**Table 9.1** or **Table 9.2**); and P is the conservation practice factor (**Table 9.3**).

In this text, English units are used for A, R, and K. If using other sources of information for the USLE, ensure that units are consistent (G.R. Foster et al., 1981). Although developed for use in the U.S., the procedure

TABLE 9.1
Cover Management C Factors for Permanent Pasture, Rangeland, and Idle Land[a]

Vegetal Canopy: Type and Height of Raised Canopy[b]	Canopy Cover[c]	Type[d]	Cover that Contacts the Surface: Percentage Ground Cover					
			0	20	40	60	80	95–100
No appreciable canopy		G	0.45	0.20	0.10	0.042	0.013	0.003
		W	0.45	0.24	0.15	0.090	0.043	0.011
Canopy of tall weeds or short brush	25	G	0.36	0.17	0.09	0.038	0.012	0.003
(1.5 ft fall height)[b]		W	0.36	0.20	0.13	0.082	0.041	0.011
	50	G	0.26	0.13	0.07	0.035	0.012	0.003
		W	0.26	0.16	0.11	0.075	0.039	0.011
	75	G	0.17	0.10	0.06	0.031	0.011	0.003
		W	0.17	0.12	0.09	0.067	0.038	0.011
Appreciable brush or bushes	25	G	0.40	0.18	0.09	0.040	0.013	0.003
(6 ft fall height)[b]		W	0.40	0.22	0.14	0.085	0.042	0.011
	50	G	0.34	0.16	0.85	0.038	0.012	0.003
		W	0.34	0.19	0.13	0.081	0.041	0.011
	75	G	0.28	0.14	0.08	0.036	0.012	0.003
		W	0.28	0.17	0.12	0.077	0.041	0.011
Trees, but no appreciable low brush	25	G	0.42	0.19	0.10	0.041	0.013	0.003
(12 ft. fall height)[b]		W	0.42	0.23	0.14	0.087	0.042	0.011
	50	G	0.39	0.18	0.09	0.040	0.013	0.003
		W	0.39	0.21	0.14	0.085	0.042	0.011
	75	G	0.36	0.17	0.09	0.039	0.012	0.003
		W	0.36	0.20	0.13	0.083	0.041	0.011

[a] All values shown assume (1) random distribution of mulch or vegetation and (2) mulch of appreciable depth where it exists.

[b] Average fall height of waterdrops from canopy to soil surface.

[c] Percentage of total area surface that would be hidden from view by canopy in a vertical projection.

[d] G = cover at surface is grass, grasslike plants, decaying compacted duff, or litter at least 2 in. deep; W = cover at surface is mostly broadleaf herbaceous plants (as weeds) with little lateral root network near the surface or undecayed residue.

Source: Cooperative Extension Service and Ohio State University, *Ohio Erosion Control and Sediment Pollution Abatement Guide*, Columbus, OH, 1979.

is used in many countries and has been the focus of considerable study during the past 30 years. Methods for determining each of the input parameters for the USLE and examples of their use follow.

9.5.1 Rainfall Erosivity *R*

The rainfall and runoff erosivity index *R* varies with amount of rainfall and individual storm precipitation patterns. For a given storm, a rainfall and runoff erosivity index *EI* is calculated. It is the product of the kinetic energy of the storm *E* and the maximum 30-min intensity for that storm *I*. An example calculation was presented in Chapter 2. The *EI* values for all the storms occurring in a given year for a location are summed to give an annual erosivity index. The average annual rainfall and runoff erosivity index *R* is shown in **Figure 9.6** for the continental U.S.

EXAMPLE 9.1

What is the *R* factor for a farm located in northeast Iowa?

Solution

From **Figure 9.6**, determine *R* for northeast Iowa.

Answer

R = 170.

For development projects or land uses that change throughout the year, it is sometimes desirable to determine the rainfall erosivity associated with different periods during the year. **Figure 9.6** shows the cumulative distribution of rainfall erosivity for different locations in the U.S.

TABLE 9.2
Example of Typical Cover Management C Factors Developed by State Agencies for the Ohio Climate and Vegetation Conditions

Vegetation	Autumn Conventional	Tillage Spring Conventional	Practice Spring Conservation	No Till
Agricultural Rotations				
Continuous corn (Co)	0.40	0.36	0.27	0.10
Corn and soybeans (Sb)	0.42	0.37	0.24	0.10
Two-year rotation with corn grown in Year 1 and soybeans and wheat in Year 2	0.30	0.28	0.24	0.10
Corn, corn, oats (O), meadow (M)	0.13	0.12	0.10	0.064
Corn, oats, meadow, meadow	0.055	0.050	0.033	0.033
Corn, corn, oats, meadow, meadow	0.11	0.094	0.082	0.052
Corn, soybeans, oats, meadow, meadow	0.13	0.12	0.082	0.052
Permanent Pasture				
Poor condition				0.04
Good condition				0.01

Source: Cooperative Extension Service and Ohio State University, *Ohio Erosion Control and Sediment Pollution Abatement Guide*, Columbus, OH, 1979.

Estimates of partial year erosivity can be determined by distributing the erosivity determined from **Figure 9.6** with the percentages obtained from **Figure 9.7**.

EXAMPLE 9.2

Determine the rainfall erosivity for northeast Iowa for the period April 1 to October 1.

Solution

In Example 1, we determined that the annual erosivity was 170. In **Figure 9.7**, the climate in Iowa is described by Curve A. On April 1, Curve A is at 4%, and on October 1, Curve A is at 90%. Therefore, the percentage erosivity that occurs during this period is 90 − 4 = 86%.

Answer

The erosivity that occurs between April 1 and October 1 in northeast Iowa is $R = 170 \times 0.86 = 146$.

9.5.2 Soil Erodibility K

The soil erodibility factor K (**Figure 9.8**) for a series of benchmark soils was obtained by direct soil loss measurements from fallow plots located in many U.S. states (**Figure 9.9**). Soils that have high silt contents tend to be the most erodible. The presence of organic matter, stronger subsoil structure, and greater permeabilities generally decrease erodibility. In RUSLE, K varies to account for seasonal variation in soil erodibility, with higher erodibility in the spring or after tillage.

Soil erodibility factors for the 10 most common soils in each state are presented in Appendix D. A consideration with using the information in Appendix D is that land use activities might change the soil structure or permeability of a soil. This is particularly true if the land is used for agricultural purposes or if it has been severely disturbed by construction activities. One soil type may exhibit a wide range of erodibility values, depending on textural, slope, compaction, management, and cultivation factors. These problems can be addressed by applying the nomograph presented in **Figure 9.8** to determine the erodibility of a given soil.

EXAMPLE 9.3

Determine the soil erodibility K for a soil with the following properties: 65% silt and very fine sand; 5% sand; 2.8% organic matter content (OM); a fine granular soil structure; and a slow-to-moderate permeability.

Solution

In **Figure 9.8**, enter the left graph at silt plus very fine sand = 65%, move right and intersect with 5% sand, move up to 2.8% organic matter, move right to soil structure = 2, move down to permeability = 4, and then move left to determine K.

Answer

$K = 0.31$.

TABLE 9.3
Conservation Practice Factor *P* for the Universal Soil Loss Equation

Farming Up and Down Slope
All crops   *P* = 1.0

Contour Farming

Land Slope Percentage	Maximum Slope Length[a] (ft)	P factor
1 to 2	400	0.6
3 to 5	300	0.5
6 to 8	200	0.5
9 to 12	120	0.6
13 to 16	80	0.7
17 to 20	60	0.8

Strip Cropping
With grass and row crop   Contour *P* × 0.5
With small grain and row crop   Contour *P* × 0.67

Terraces
Loss from crop   Same as Contouring *P*

Loss from terrace
  With graded channel outlet   Contour *P* × 0.2
  With underground outlet   Contour *P* × 0.1

Subsurface Drainage
                                          P = 0.6

[a] Maximum slope length for strip cropping can be twice that for contouring.

Source: Bengtson, R.L., and G. Sabbagh, *J. Soil Water Cons.*, 45:480–482, 1990; Wischmeier, W.H., and D.D. Smith, *Predicting Rainfall Erosion Losses — A Guide to Conservation Planning*. USDA Handbook 537, USDA–Science and Education Administration, Washington, DC, 1978. With permission.

9.5.3 Topographic Factor *LS*

The topographic factor *LS* adjusts the predicted erosion rates to give greater erosion rates on longer or steeper slopes and lower erosion rates on shorter or flatter slopes compared to a USLE "standard" slope of 9% and slope length of 72.6 ft. The erosion plot in **Figure 9.9** approximates these topographic conditions.

The slope length is measured from the point where surface flow originates (usually the top of the ridge) to the outlet channel or a point downslope where deposition begins. RUSLE considers nonuniform concave or convex slopes (Renard et al., 1991), as do most process-based erosion prediction programs. For the USLE, the *LS* factor can be determined from **Figure 9.10**.

EXAMPLE 9.4

Determine the *LS* factor for the field with a slope steepness of 4% and a slope length of 700 ft.

Solution

From **Figure 9.10**, enter the *x* axis at 700 ft. Move up until you intersect the 4% slope line. Move left to find the *LS* factor.

Answer

LS = 0.9.

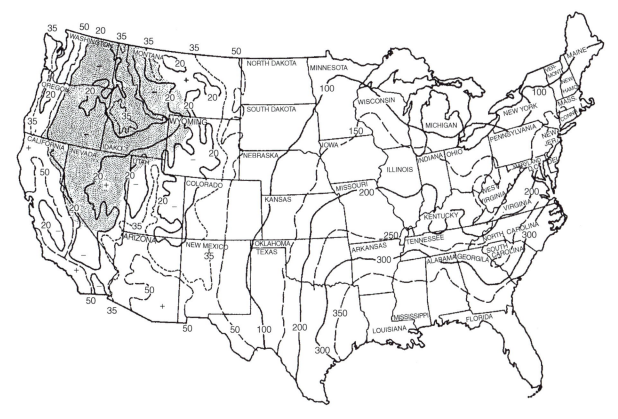

FIGURE 9.6 Rainfall and runoff erosivity index *R* distribution in the U.S. (From Wischmeier, W.H., and D.D. Smith, *Predicting Rainfall Erosion Losses — A Guide to Conservation Planning.* USDA Handbook 537, USDA–Science and Education Administration, Washington, DC, 1978. With permission.)

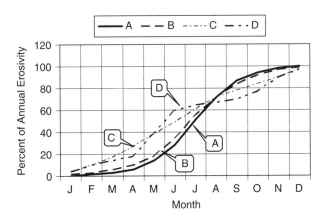

FIGURE 9.7 Monthly distribution of the rainfall erosivity index *R*. Curve A: Iowa, Nebraska, South Dakota; Curve B: Missouri, Illinois, Indiana, and Ohio; Curve C: Louisiana, Mississippi, Tennessee, and Arkansas; Curve D: Atlantic coastal plains of Georgia and the Carolinas. (Based on Wischmeier, W.H., and D.D. Smith, *Predicting Rainfall Erosion Losses — A Guide to Conservation Planning.* USDA Handbook 537, USDA–Science and Education Administration, Washington, DC, 1978.)

9.5.4 Cover Management Factor *C*

The cover management factor *C* includes the effects of vegetative cover, crop sequence, productivity level, length of growing season, tillage practices, residue management, and the expected time distribution of erosive events. For agricultural systems, *C* factors are generally based on crop rotations and tillage sequences. For forest, rangeland, and other nonagricultural conditions, *C* factors are generally estimated from the density of vegetation and the amount of vegetative residue on the soil surface.

Table 9.1 provides *C* factors that are useful for many different land uses in much of the U.S., provided the land use exhibits some vegetative cover. For a disturbed bare soil as in **Figure 9.10**, a value of 1.0 or greater should be used. The table is particularly valuable when relative soil losses associated with land use changes are determined. Erosion from permanent pasture, rangeland, and forest is generally much lower than from agricultural lands. Human or livestock activities, like roads, grazing, or timber harvest, that disturb the vegetation are generally the source of most of the eroded sediment from rangeland and forests. Forest erosion processes are discussed in detail in Chapter 10.

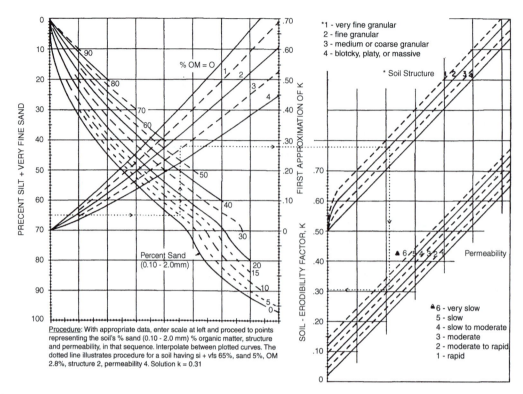

FIGURE 9.8 Nomograph to determine soil erodibility *K* factor. (From Wischmeier, W.H., and D.D. Smith, *Predicting Rainfall Erosion Losses — A Guide to Conservation Planning.* USDA Handbook 537, USDA–Science and Education Administration, Washington, DC, 1978. With permission.)

FIGURE 9.9 Fallow soil erosion plot located near Pullman, WA, with a standard length of 72.6 ft and a slope steepness of about 9%. (Photograph courtesy of Don McCool.)

EXAMPLE 9.5

Logging operations in Pennsylvania result in part of a forest being clear-cut. Immediately following the clear-cutting, there is no appreciable canopy, but there is a 40% ground cover of decaying compacted duff. A year after the clear-cutting, there is a 75% canopy of brush and short trees (4 to 8 ft high). Determine the change in the *C* factor during the first year following clear-cutting.

Solution

From **Table 9.1**, the *C* factor immediately following clear-cutting is 0.15. If we assume the percentage ground cover does not change during the first year, the *C* factor at the

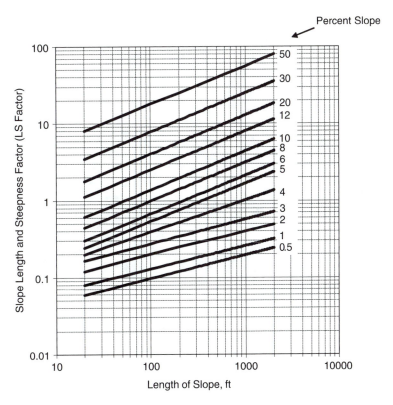

FIGURE 9.10 Slope and length factor *LS* as a function of slope and length.

end of this period will be 0.12. Therefore, based on the change in the *C* factor, soil losses at the end of the period will be 80% of that immediately following clear-cutting.

Answer

The change in canopy cover during the first year following clear-cutting will result in a 20% reduction in the *C* factor.

Table 9.2 gives some typical agricultural *C* factors for Ohio. These values should provide reasonable estimates for the central and eastern U.S. However, these factors do not account for climate differences between locations that might influence seasonal changes in *C* factors. Also, the table does not fully consider differences in the amount of residue left in a field following harvest. Whenever possible, you should consult local NRCS or other natural resource agency offices to obtain *C* factors for your region.

From **Table 9.2**, it can be seen that the *C* factor will vary based on the crop and tillage practices. The traditional functions of tillage were weed control and preparation of a seedbed in which traditional planters and drills would function. With the development of conservation tillage planters and drills, tillage management has become increasingly important as a conservation tool. Currently, the primary purpose of tillage operations is to provide an adequate soil and water environment for the seedling to

germinate. The role of tillage in weed control has diminished with increased use of herbicides and improved timing of operations.

Conventional tillage tends to incorporate the majority of surface residue, leaving the surface bare and susceptible to erosion. Conservation tillage systems leave residue from previous crops on the surface, protecting the soil from raindrop impact and reducing the erosive shear of sheet and rill flow on the soil. The presence of residue significantly reduces erosion rates, as can be seen by the lower *C* factors for all the conservation systems listed in **Table 9.2**.

One of the major benefits of reduced tillage is the increased amount of residue left on the surface, which reduces runoff and erosion. Other methods for reducing runoff, such as putting checks in furrows in cropland or imprinting rangeland, can also reduce erosion.

EXAMPLE 9.6

As part of a conservation plan, a farmer in Indiana decided to place part of her land into pasture. Originally, the land was in continuous corn production, and the farmer used conventional tillage in the autumn. Determine the relative soil reduction associated with this land use change; assume the worst-case condition for pasture.

FIGURE 9.11 Contour strip cropping in Wisconsin.

Solution

From **Table 9.2**, it is determined that continuous corn production and conventional tillage in the autumn has a *C* factor of 0.40. The worst case for the pasture, poor condition, has a *C* factor of 0.04, which is 10% of 0.40.

Answer

Converting the land use into pasture will reduce soil loss by 90%.

9.5.5 Erosion Control Practice *P*

The vegetation cover or farming system selected can have a major effect on soil erosion rates. There are practices besides vegetation management that can be employed to control erosion. *Erosion control practices* include contouring, strip cropping, and terracing. For many applications, no erosion control practices will be used, and the *P* factor will be 1.0. **Table 9.3** presents typical *P* factors that can be used to estimate the effect of conservation practices on predicted soil erosion.

Contouring. *Contouring* or *contour farming* is conducting field operations, such as tillage, planting, and harvesting, approximately on the contour. Contouring reduces surface runoff by impounding water in small depressions and decreasing the development of rills. The greatest concentration of contouring in the U.S. is in the eastern wheat belt, where the benefits include both erosion reduction and conservation of water. The design of contours requires site surveying and is discussed in detail in the work of Schwab et al. (1993).

The relative effectiveness in reducing erosion by contouring on different slopes is shown by the conservation practice factor *P* in **Table 9.3**. The benefits of contouring decrease as the slope increases because the water-holding capacity of the rows diminishes with an increase in slope. Also, there are practical limitations on the slope length for which contouring is effective. On 2% slopes, the maximum slope length that should be contoured is about 400 ft, while for a 12% slope, the maximum length is only 120 ft. Contouring on steep slopes or under conditions of high rainfall intensity and soil erodibility will increase the risk of gullying because row breaks may release the stored water. Row breaks cause cumulative damage as the volume of water increases with each succeeding row.

Strip Cropping. *Strip cropping* is the practice of growing alternate strips of different crops in the same field. For controlling water erosion, the strips are on the contour (**Figure 9.11**). Rotations that provide strips of close-growing perennial grasses and legumes alternating with grain and row crops are the most effective practice to reduce erosion by water. Generally, strips that are farmed in row crops have parallel borders and widths that are convenient for multiple-row equipment operation. Maximum slope lengths for strip cropping can be twice those for contouring alone. For slope lengths longer than those recommended in **Table 9.3**, terracing is recommended if tillage operations are required for the management system.

The strips of grass or small grains act as sediment filters by slowing the overland flow velocity and increasing infiltration, causing sediment to be deposited. Also, the velocity of runoff entering the next strip of crop is reduced and therefore causes less soil erosion. The reason the grass strips act as filters is that they provide more near-surface vegetative cover than row crops such as the corn

FIGURE 9.12 Terraces on gradual sloping areas in Swaziland. Tall grass is planted on the terraces. Cattle in the foreground are a major cause of soil erosion as they compact the soil and remove vegetation.

or soybeans. A short-term economic impact of this practice is that some of the land is no longer in crop production or is planted to a crop such as grass, which has less economic value than corn or soybeans.

EXAMPLE 9.7

A farmer decides to switch from farming up and down the slope to contouring. The farmer's field has a slope of 3%. Determine the soil loss reduction benefit of this switch in practices. Also, determine any additional benefit that might be obtained if the farmer strip crops with grass.

Solution

From **Table 9.3**, a *P* factor of 0.5 is obtained for contouring on a 3% slope. Without contouring, the *P* factor would be 1.0. Therefore, contouring reduces the soil loss by 50%. Also, from **Table 9.3**, we see that strip cropping with grass will further reduce the *P* factor to 0.25. This will result in a 75% reduction $(1 - 0.25 \times 100)$ in the predicted soil erosion. An additional reduction in erosion may also occur if the farmer has more years of grass (meadow) in the rotation because of a reduced *C* factor (**Table 9.2**).

Answer

A switch to contouring would reduce soil loss by 50%. Including strip cropping would reduce soil loss by 75%.

Terracing. A common erosion control practice is to construct terraces on eroding slopes (**Figure 9.12**). Terraces reduce the slope length and reduce runoff. The terrace channel may be level or graded to direct runoff to grassed waterways or subsurface drains that convey runoff away from the susceptible area. A level terrace restricts the lateral movement of water and is generally installed on soils with high infiltration rates or in drier climates for both soil and water conservation. The details of terrace design or construction are presented in the work of Schwab et al. (1993) and in other engineering textbooks.

Terracing affects the slope length so that the *LS* factor in a terracing system is altered to determine the amount of sediment delivered to the terrace. In terrace systems, the *P* factor can be used to estimate the reduction in sediment yield leaving the terrace. From **Table 9.3**, it is possible to predict either (1) sediment detached from the cropping area (the *P* factor is the same as for contouring) or (2) sediment yield leaving the field and include the terrace factor. The *P* factor for terrace sediment yield is 10 to 20% of the contour farming *P* factor.

EXAMPLE 9.8

For a field with a total slope length of 700 ft and an average steepness of 4%, find the *P* factor for (a) sediment delivered to terraces at 100-ft spacings and (b) sediment transported from a graded terrace channel with a spacing of 100 ft leading to a surface channel outlet.

Solution

Refer to **Table 9.3** to determine the appropriate *P* factors.

Answers

(a) For the sediment delivered to terraces at 100-ft spacings, *P* is the same as for the contour farming value, *P* = 0.50.

The slope length has been reduced from 700 to 100 ft, so from **Figure 9.10**, the LS factor will reduce from 0.9 to 0.5. (b) If the terraces have a graded channel to a surface drain, then the P factor for sediment delivery is $0.5 \times 0.2 = 0.1$.

9.6 SOIL LOSS TOLERANCE T

Natural resource managers sometimes find it helpful to determine a maximum level or tolerable level of erosion that can be allowed. Management plans can be evaluated by comparing the predicted erosion to a maximum or tolerable level. Both physical and economic factors, as well as social aspects, need to be considered in establishing soil loss tolerances, sometimes called T values. These values vary with topsoil depth and subsoil properties and range from 3 to 20 tons/acre. They are the maximum rates of soil erosion that will permit a high level of crop productivity to be sustained economically and indefinitely. In some cases, criteria for control of sediment pollution may dictate lower tolerance values. In other cases, it may be necessary to allow for higher levels of erosion to ensure the economic viability of a rural community.

9.7 APPLICATIONS OF THE UNIVERSAL SOIL LOSS EQUATION

In Section 9.5, we focused on determining the factors in the USLE and determining soil loss changes associated with each factor. Often, we will be interested in determining the soil loss associated with a combination of all or most of the factors. This will require knowledge of the location, soil, land use, slope steepness, slope length, and any erosion control practices. We present two examples of applying the USLE to estimate soil erosion. **Table 9.4** describes an actual field that had been experiencing severe erosion prior to the implementation of conservation practices.

EXAMPLE 9.9

Determine the average annual soil loss for the field described in **Table 9.4**.

Solution

From Example 9.1, $R = 170$. From **Table 9.4**, $K = 0.27$. From **Figure 9.6** for length = 720 ft and steepness = 3%, $LS = 0.6$. Assume Iowa C factors similar to Ohio, so from **Table 9.2** for Co O M M rotation, spring tillage, $C = 0.05$. From **Table 9.3** for contour strip cropping, $P = 0.5 \times 0.5 = 0.25$.

TABLE 9.4

Details from a Field in Iowa with Observed Soil Erosion Susceptibility

Feature	Value
Location: Clayton County in northeast Iowa	
Soil series and texture: Kenyon loam	
Erodibility K factor	0.27
Stated T value	5 t/acres per year
Total length of slope	1200 ft
Length of eroding slope	720 ft
Average steepness of eroding slope	3%
Current farming system: contour strip cropping with grass, spring conventional tillage, 1.7 t/acre spring residue	
Rotation: corn, oats, meadow, meadow	

Source: USDA (1972).

Answer

Applying Equation 9.1,

$$A = 170 \times 0.27 \times 0.6 \times 0.05 \times 0.25 = 0.34 \text{ tons/acre/year}$$

The erosion rate in Example 9.9 is relatively low for an agricultural system and is due to the combination of crop rotation and contour strip cropping, which are both very effective management techniques in reducing soil erosion. The T value for the field described in **Table 9.4** is 5 tons/acre/year. The above example predicts an erosion rate of less than 1 ton/acre/year, which will ensure that the productivity of that field is maintained indefinitely.

The USLE can be applied to many land uses, including agriculture, forests, pastures, rangeland, idle land, and land disturbances associated with surface mining or land development. However, when using the USLE, it is important to recognize that knowledge of C and P is better for agricultural applications than for surface mining or other land disturbances, such as urban development. Also, there is limited information available on LS factors for slopes greater than 20% or slope lengths greater than 2000 ft.

EXAMPLE 9.10

Compare annual soil losses prior to and following reclamation of a surface mining activity in eastern Kentucky. Disturbed areas are returned to the original contours, the average land slope is 20%, and typical slope lengths are 750 ft. Mining is conducted in a Shelocta soil series. Prior to mining, the area was predominantly a forest, and disturbed areas are undergoing reclamation to permanent pasture. Poor soil fertility and acid conditions prevent establishment of more than 60% ground cover of grass and legumes in the pasture. Suggest conservation practices that might reduce the postmining soil losses to less than a soil loss tolerance T value of 5 tons/acre/year.

Solution

From **Figure 9.6**, determine that, for eastern Kentucky, the R factor is 150. From Appendix D, we find that a Shelocta soil has a K factor of 0.32. From **Figure 9.9**, establish that the LS factor for a 20% slope and a 750-ft slope length is 10. Using **Table 9.1**, the C factor for a forest is 0.003 (95 to 100% ground cover of decaying compacted duff), and the C factor for the permanent pasture will be about 0.042 (60% ground cover).

The only factor that changes when we move from the premining to postmining use is the C factor. The soil losses before and after reclamation are determined as:

Condition	A	R	K	LS	C	P
Premining forest	1.4	150	0.32	10	0.003	1.0
Postmining grass	20	150	0.32	10	0.042	1.0

It will be difficult to reduce postmining predicted losses from 20 to less than 5 tons/acre/year, but a number of options are available. Improved management to increase the vegetative cover will reduce the C factor. If soil fertility can be improved to give 80% cover rather than 60%, the C factor will decrease from 0.042 to 0.013. Installing terraces at 250-ft spacings will reduce the LS factor from 10 to 5.5. The following table summarizes the options:

Condition	A	R	K	LS	C	P
Postmining grass	20	150	0.32	10	0.042	1.0
Improved grass	6.24	150	0.32	10	0.013	1.0
Improved grass and terraces	3.4	150	0.32	5.5	0.013	1.0

Answer

To ensure that predicted postmining on-site erosion is reduced to less than 5 tons/acre/year, it is necessary to improve the soil fertility to ensure an 80% grass cover and to install terraces at 250-ft spacings.

In Example 9.10, if the proposed terraces drain into a grassed waterway, the sediment leaving the site will be reduced even further with a P factor of 0.2, leading to a predicted net soil loss from the site of 0.7 tons/acre/year. Natural resource managers need to determine if their goal is to reduce on-site erosion to maintain site productivity or to reduce off-site impacts to maintain water quality. See Chapter 14, Exercise 14.8, for additional examples of applications of the USLE.

9.8 DOWNSTREAM SEDIMENT YIELDS

Natural resource managers may need to know sediment delivery downstream in a watershed. One method for estimating sediment delivery is to use the USLE and a sediment delivery ratio (SDR). The USLE estimates gross sheet-and-rill erosion, but does not account for sediment deposited en route to the place of measurement or for gully or channel erosion downstream. The *sediment delivery ratio* is defined as the ratio of sediment delivered at a location in the stream system to the gross erosion from the drainage area above that point. This ratio varies widely with size of area, steepness, density of drainage network, and many other factors. For watersheds of 6.4, 320, 3,200, and 64,000 acres, ratios of 0.65, 0.33, 0.22, and 0.10, respectively, were suggested by Roehl (1962) as average values. The development of new process-based erosion models in GIS environments will enhance the prediction of sediment delivery from watersheds.

Renfro (1975), for the Blackland Prairie area in Texas, developed the SDR equation

$$\log_{10} SDR = 1.8768 - 0.14191 \log_{10} 10A \quad (9.2)$$

where SDR is the sediment delivery ratio (percentage of annual erosion), and A is the drainage area in square miles ($A < 100$ mi^2).

Numerous other models have been developed. For example, the Soil and Water Assessment Tool (SWAT) (Arnold et al., 1995) considers storm runoff and uses the equation

$$SDR = \left[\left(\frac{q_p}{P_r} \right) \left(0.78285 + 0.21716 \frac{Q}{P} \right) \right]^{0.56} \quad (9.3)$$

where SDR is the sediment delivery ratio (fraction of the gross erosion for the event), q_p is the peak runoff rate, P_r is the peak rainfall rate, P is the rainfall, and Q is the runoff depth (all units are millimeters or millimeters/hour, although inches and inches/hour can be used provided a consistent set of units is used for all quantities).

EXAMPLE 9.11

Assume that the erosion rate calculated for the field in Example 9.9 is typical of a 320-acre watershed draining into a small reservoir. Estimate the sediment delivery to the reservoir for a 20-year period.

Solution

The average annual erosion rate is 0.34 tons/acre/year. The sediment delivery ratio for a 320-acre watershed was given as 0.33.

Answer

The sediment yield from 20 years will be

Sediment = 0.34 tons/acre/year × 320 acres
× 20 years × 0.33 = 718 tons

Exercise 14.8 in Chapter 14 has additional applications of predicting off-site impacts of soil erosion.

9.9 SINGLE-EVENT SEDIMENT YIELDS

The USLE is useful for determining gross erosion from an area for seasonal, annual, and extended time periods. In Section 9.8, we showed how a delivery ratio can be used to determine a downstream sediment yield estimate. However, application of a delivery ratio to gross erosion estimates based on the USLE cannot be used to obtain sediment yields for individual events, only for long-term averages. A method called the modified universal soil loss equation (MUSLE) (J.L. Williams, 1975) illustrates how sediment yields for individual storm events might be obtained:

$$Y = 95(Qq)^{0.56} K\, LS\, CP \qquad (9.4)$$

where Y is the single-storm sediment yield in tons, Q is the storm runoff volume in acre-feet, q is the peak discharge in cubic feet/second, and the other terms are the standard USLE factors discussed in this chapter. The approach has seen widespread application, but should be used with caution as it was developed empirically based on limited data for Texas and the southwestern U.S. The procedure should only be used on small watersheds, and considerable judgment is required in selecting an appropriate slope length when determining the LS topographic factor. A delivery ratio should not be included if the sediment yield is determined at the outlet of the watershed used to obtain the runoff volume Q and the peak discharge q. However, if a sediment yield estimate is required downstream of the watershed evaluated with the MUSLE, then a delivery ratio would also need to be used. Estimates of Q and q can be obtained using the procedures presented in Chapter 5.

EXAMPLE 9.12

Determine the soil loss from the 320-acre watershed described in Examples 9.9 and 9.11 for a storm that produces 12 acre-ft of runoff and a peak discharge of 200 ft³/sec from a 2-in. rainfall.

Solution

From Example 9.9, $K = 0.27$, $LS = 0.6$, $C = 0.05$, and $P = 0.25$.

$$Y = 95\,(12 \times 200)^{0.56}\,(0.27)(0.6)(0.05)(0.25) = 15\ t$$

Answer

The soil loss is 15 tons.

9.10 ESTIMATING SEASONAL C FACTORS FOR AGRICULTURAL CROPS

Sometimes, local C factors such as those presented in **Table 9.1** and **Table 9.2** are not available. This section demonstrates how to determine a C factor based on general crop properties (**Table 9.5**) in conjunction with a local climate (**Figure 9.4**).

Table 9.5 gives the seasonal losses from several typical farming systems as a percentage of the soil loss for continuous fallow soil during the growing season. The annual distribution of erosive events varies with geographic location. Examples from different regions of the U.S. are given in **Figure 9.7**. The C factor for a given crop rotation is found by first multiplying the soil loss ratio for each growth period in a crop rotation (**Table 9.5**) by the percentage of annual erosion during each respective period (**Figure 9.7**). These products are then summed, and an annual average is calculated to give the C factor (Example 9.13).

EXAMPLE 9.13

Calculate the C factor for the field described in **Table 9.4**.

Solution

Table 9.6 contains the solution to this example. The rotation given in **Table 9.4** is corn, oats or small grain, meadow, meadow.

1. The various stages of crop growth and their respective months are determined for each growth stage, and the months are entered in Column 1 of **Table 9.6**.

TABLE 9.5
Seasonal Distribution of Cover Management Factor Expressed as Percentage of Soil Loss from Crops to Corresponding Loss from Continuous Fallow for Selected Crop and Tillage Management Systems

Line no., Cover, Sequence and Management[c]	Residue[d] (tons)	Cover[e] (%)	F (%)	SB (%)	1 (%)	2 (%)	3:80 (%)	90 (%)	96 (%)	4L (%)
Continuous										
1 Co, RdL, spring, TP	2.3	—	31	55	48	38	—	—	20	23
2 As above, less residue										
6 Co, RdL, fall, TP	1.7	—	36	60	52	41	—	24	20	30
46 Co, chisel or cultivate only	GP	—	49	70	57	41	—	24	20	—
63 Co ridge tillage on contour	1.7	50	—	16	13	12	—	12	9	24
130 Small grain	—	—	—	0.7	0.7	0.7	0.7	0.7	0.7	0.7
200 Meadow	1.7	60	—	16	14	12	7	4	2	17
							0.8	0.6	0.4	
Rotation										
110 Co after Bn, spring conv	GP		47	78	65	51	—	30	25	37
124 Bn after Co, spring, TP	GP		39	64	56	41	—	21	18	22
210 Rowcrop after meadow, Rdl, spring, TP			9	24	21	18	—	12	10	18
Conservation Tillage										
52 Bn or Co after Co	1.3	40	—	21	20	19	19	15	12	30
121 Co after Bn	GP	30	—	33	29	25	22	18	14	33
131 Small grain after Co in disk residue	1.7	40	—	27	21	16	9	5	3	22

Bn = beans; Co = corn.

[a] The soil loss ratios, given as percentages of the loss that would occur from fallow, assume that the indicated crop sequence and practices are followed consistently.

[b] Crop stage periods are F, rough fallow; SB, seedbed until 10% canopy cover; 1, establishment until 50% canopy cover; 2, development until 75% canopy cover; 3, maturing until harvest for three different levels of canopy cover; 4L, residue or stubble.

[c] RdL, crop residue left in field; TP, plowed with moldboard.

[d] Dry mass per acre, after winter loss and reductions by grazing or partial removal, 2 t/acre represents a yield of 100 to 130 bu/acre; GP is good productivity level.

[e] Percentage of soil surface covered by plant residue mulch after crop seeding. The difference between spring residue and that on the surface after crop seeding is reflected in the soil loss ratios as residues mixed with the topsoil.

Source: Wischmeier, W.H., and D.D. Smith, *Predicting Rainfall Erosion Losses — A Guide to Conservation Planning.* USDA Handbook 537, USDA–Science and Education Administration, Washington, DC, 1978. With permission.

2. The *C* values from **Table 9.5** for each period are entered into **Table 9.6**, Column 2.
3. The percentage of annual erosivity for each of these periods is determined from **Figure 9.9**, Curve A, and entered into **Table 9.6**, Column 3.
4. The product of the *C* value and the erodibility (both expressed as decimal fractions) is calculated and entered in Column 4 of **Table 9.6**.
5. The sum of these products is divided by the number of years in the rotation to obtain an average *C* factor for the rotation, which is 0.060.

Answer

The *C* factor for the field described in **Table 9.4** is 0.060.

The calculated *C* factor of 0.06 from **Table 9.4** is greater than the *C* factor of 0.05 given in **Table 9.2** for Ohio conditions because the Iowa climate experiences more erosive storms during the critical crop period in June and July (compare the steepness of Curves A and B in **Figure 9.9**), and different assumptions may have been made regarding the *C* factor in the Ohio data.

TABLE 9.6
Solution to Example 9.13, Calculating the C Factor for the Conditions Described in Table 9.4

Months Column 1	Soil Loss, Percentage of Continuous Fallow (Table 9.5) Column 2	Annual Erosion (Table 9.4) Percentage Column 3	C Factor Column 2 × Column 3 Column 4
Corn, First Year, Line 210			
January to April	0.6	10	0.0006
May	24	10	0.024
June	21	20	0.042
July	18	20	0.036
August to September	12	30	0.036
October to December	18	10	0.018
Small Grain with Meadow Seeding, Lines 130 and 200			
January to March	18	4	0.0072
April	16	6	0.0096
May	14	10	0.014
June	12	20	0.024
July to August	4	40	0.016
September to December	0.6	20	0.0012
Meadow, Line 200			
January to December	0.6	100	0.006
January to December	0.6	100	0.006
Total		400%	0.242
Average		0.242/4 =	0.060

9.11 SEDIMENT BUDGET CONCEPTS

9.11.1 INTRODUCTION

A particle of sediment often has a long and complicated journey on its way from upland slope to the sea (Walling, 1983). Both in the popular environmental literature (Gore, 1992) and, most egregiously, in the scientific literature (Pimentel et al., 1995), the idea remains that a particle of soil, once detached by running water, moves immediately on its way to the sea or to some navigable stream, lake, reservoir, or harbor, there to create havoc. The falseness of this idea is easily demonstrated. The most recent estimates of U.S. soil erosion are about 4 billion tonnes per year, but the sediment estimated to be leaving from rivers as sediment yield to the sea is only about 0.5 billion tonnes/year, of which a large, but unknown, quantity comes from stream channel and bank erosion (Meade et al., 1990; Trimble and Crosson, 2000). Part of the disparity of more than 3.5 billion tonnes/year is found in both large (Dendy and Champion, 1975; Meade et al., 1990) and small (Renwick, 1996; Smith, et al, 2002) reservoirs, which effectively trap much of the sediment transported by streams. Also, a large, but unknown, quantity is deposited in channels

and on floodplains. The net result is that some of the missing 3.5 billion tonnes/year can be taken into account, but we simply do not know how much. Simply, there is no equilibrium or steady state between the slopes and the stream (Trimble, 1975a).

In Chapter 4, we looked at water budgets and saw that, for a given time period, all mass has to be taken into account. In that case, the input was precipitation, storage was in soil and groundwater, and outflow or efflux was ET and Q (Equation 4.4). Thus, if we know P and Q and can account for storage changes, we can calculate AET. A sediment budget is quite similar in that we must account for fluxes and storage of mass over time. The inputs are earth materials that come from slopes by erosion or mass movement (Toy et al., 2002). Additional sediment may come from wind erosion in some regions, as discussed in the first part of this chapter. Efflux is measured as sediment yield at some downstream point. Between the slopes and the downstream point where sediment yield is measured, however, there are many places along streams where sediment can be stored by stream or fluvial processes, thus decreasing downstream sediment yield. Likewise, fluvial processes may remove sediment from these same storage "compartments" and release it downstream to become sediment yield. As we shall see, these fluxes and processes are quite complex, and the science of sediment budgets is in its infancy.

9.11.2 SEDIMENT DELIVERY FROM SLOPES

In the beginning of this chapter, we examined principles and prediction of soil erosion and pointed out that most sediment from eroded soil travels only a short distance downslope, and generally, only a small proportion actually moves into streams under ordinary conditions. Indeed, much of the eroded soil moves only a short distance and is deposited as slope deposits or colluvium. By practice. we adjust for this assumed deposition along the pathway using the SDR, the proportion of eroded soil reaching some point downstream (see Section 9.8 and Exercise 14.6, Chapter 14). But, we must understand that the SDR is a gross generalization based on few and very difficult to obtain data (Roehl, 1962; Walling, 1983); like most models, it can hide huge disparities. We have already discussed the efficiency of water moving in channels compared to that moving across slopes as shallow flow; so it is with sediment delivery from slopes to streams. If soil erosion is so advanced on an upland surface that rills and gullies are present, then water and sediment can be efficiently conveyed off the slopes; indeed, additional sediment may well come from the rills and gullies themselves. During the 1930s, when erosion was rampant on the Driftless Area (Mississippi River hill country or Paleozoic plateau), sediment delivery ratios were about 1.28 (Trimble and Lund, 1982). **Figure 9.13**A clearly demonstrates what was happening: The drainage density had radically increased under conditions of poor land use (1853 to 1935), so that rills and gullies could efficiently convey water and sediment to tributary streams. By the 1960s, when the landscape had been totally transformed by modern soil and water conservation practices (**Figure 9.13**B), there was no sign of rills and gullies, and not only had soil erosion been drastically reduced, but SDRs were also drastically reduced, averaging only 8%, with the disparity deposited as colluvium or alluvium (Trimble and Lund, 1982). Such are the vagaries and unknowns of moving sediment just a few hundred meters.

9.11.3 SEDIMENT IN STREAMS

Once the sediment moves into the stream system, the processes are even more complex (**Figure 9.14**). The first principle is that streams can transport only so much sediment over time (termed *capacity*). A stream channel may be a fairly efficient conveyor of water and sediment, but during out-of-bank flows, much of the conveyance path is the floodplain. It is usually rough hydraulically and thus inefficient at routing sediment, especially through the

lower main valley. The quite obvious result is that floodplains are subject to varying amounts of sediment storage as vertical accretion. Particularly under conditions of accelerated soil erosion, the amount of sediment conveyed exceeds the conveyance capacity of the floodway (channel and floodplain), so that the surplus is stored. This is clearly demonstrated by three sediment budgets for Coon Creek, Wisconsin, 1853 to 1993 (**Figure 9.14**). Perhaps the most noteworthy phenomenon is that, although sediment fluxes varied greatly through time, the flow of sediment yield out of the basin and into the Mississippi River remained approximately constant, demonstrating the sediment-trapping effect of stream floodplains, especially the wide floodplain of the lower main valley.

Conversely, when the stream is deprived of its "normal" load of sediment, the available stream energy may be used to erode the channel and banks. This is easily demonstrated in erodible channel reaches downstream of reservoirs. There, the reservoirs have trapped most of the sediment, creating flow with excess energy or "hungry water." Many such reaches have degraded several meters, and the eroded materials have been transported downstream as sediment yield (G.P. Williams and Wolman, 1984). Sometimes, the stream is unable to erode the channel bed (e.g., "armored" with large particles) or banks (e.g., armored with effective vegetation or riprap), so that its energy will simply be dissipated, and little work will be done.

Some of these concepts are embodied in **Figure 9.15**. Line BA, extended, is the steady state at which downstream sediment yield equals upstream sediment supply. Point and Level A is the maximum sediment conveyance capacity of the floodway. When upstream sediment supply exceeds A, the excess goes into storage, largely as vertical accretion on floodplains or channel filling. When upstream sediment supply drops below A, the stream attempts to use its excess energy to obtain sediment load from bed and banks. If sediment is easily obtainable and can be mobilized, it will be taken from storage; the perfect case is line AB′, with 100% of the available energy used for geomorphic work. However, in many cases, the bed and bank materials may be (1) too large to transport, (2) armored or protected by large particles or vegetation, or (3) to a much lesser extent, without sediment. The result of any and all of these cases is to make the stream dissipate at least part of its available energy. The latter condition, as we shall see, is what stream engineers and planners attempt to accomplish with bank reshaping, riprap, and vegetation.

9.11.4 THE SOUTHERN PIEDMONT

These concepts are best demonstrated with examples from the American landscape. The southern Piedmont is a region that suffered massive historical soil erosion, but was reclaimed by soil conservation measures and reforestation

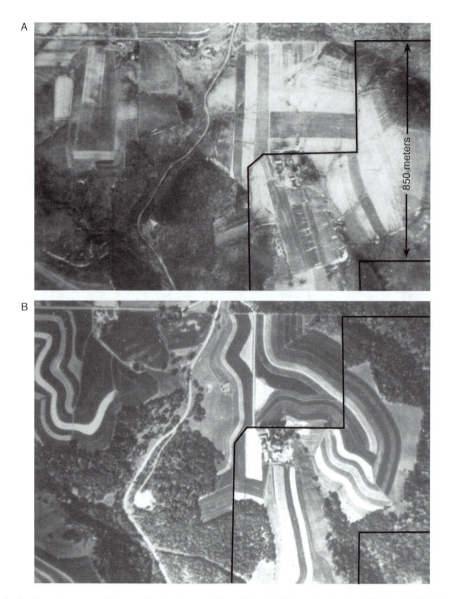

FIGURE 9.13 Hydrologic processes written on the landscape, Coon Creek, Wisconsin, basin, 1934 and 1967: A. Early 1934; note rectangular fields and gully systems extending into the upland agricultural fields. B. 1967; note contoured and strip cropped fields. Although the visible conservation methods here are striking, those not seen, such as crop rotations and stubble mulching, have just as much impact. (From Trimble, S.W., and S.W. Lund, Soil Conservation and the Reduction of Erosion and Sedimentation in the Coon Creek Basin, Wisconsin, USGS Professional Paper 1234, 1982. With permission.)

starting in the 1930s. The landscape changes and stream responses are shown over time in block diagrams indicating an upstream reach and a downstream reach (**Figure 9.16**). In **Figure 9.16**A, we see the heavily forested Piedmont at the time of European settlement (ca. 1770 to 1820). Settlers quickly took up the fertile floodplains and stream terraces for agriculture and for home sites. With time, the uplands were cleared for tobacco, corn, and cotton. The clean-tilled crops in this region of high R values (Koeppen Cfa, Chapter 2 and the USLE equation in this chapter) and often-steep slopes, along with poor soil conservation practices, created massive sheet and gully erosion that filled upstream valleys and moved progressively downstream

(**Figure 9.16**B). Channels were to a large extent filled with sandy sediment, from the acid igneous soils, and this resulted in the raising of the stream levels. Natural levees formed, which then prevented water from moving to the stream, and the floodplains and low terraces became swamps.

The advent of soil conservation practices and reforestation in the 1930s greatly curtailed upland erosion, so runoff from the uplands no longer transported much sediment (for an analysis of the hydrology during this period, see A.P. Price, 1998). The result was as described in Section 9.11.3; that is, the hungry water eroded upstream channels, transporting the sediment downstream, where it exacerbated

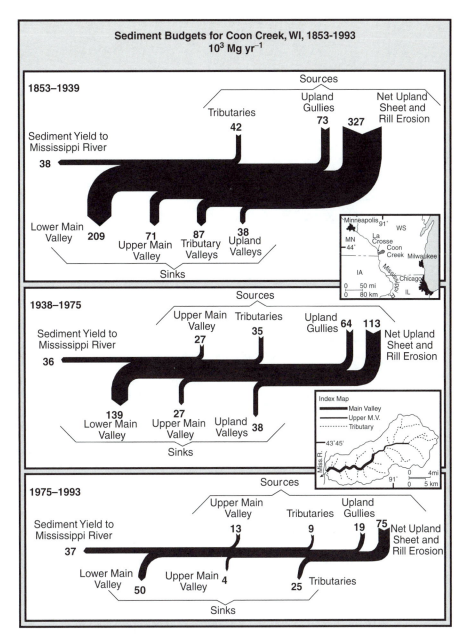

FIGURE 9.14 Sediment budgets for Coon Creek, Wisconsin, 1853 to 1993. This basin is about 25 km southeast of La Crosse, WI, and has an area of 360 km². Numbers are annual averages for the periods in 10^3 Mg/year (1 Mg = 1 metric ton). All values are direct measurements except net upland sheet-and-rill erosion, which is the sum of all sinks and the efflux minus the measured sources. The lower main valley and tributaries are sediment sinks, whereas the upper main valley is a sediment source. (From Trimble, S.W., *Science*, 285:1244–1246, 1999. The 1853–1939 and 1938–1975 portions of the figure are copyright AAAS, 1981; the 1975–1993 portion is copyright AAAS, 1999. With permission.)

already-existing problems (**Figure 9.16**C). Such stream processes as those described in Section 9.11.3, with impacts moving upstream or downstream, were termed by Schumm (1977) as a *complex response*. Note that the tributary channels, by virtue of 2 to 7 m of aggradation, had steepened their longitudinal profiles, thus increasing the potential erosional energy when the sediment supply was curtailed by improved land use. Schumm pointed out that aggrading streams could, in some cases, steepen their profiles to the

point at which they became intrinsically unstable and might start degrading on their own without the mechanism of curtailed sediment supply as discussed here.

Stream and valley cross sections, or profiles, show these processes more clearly. **Figure 9.17** shows an upstream site where a 4m-high milldam was built across the stream in 1865, at which time the stream was flowing on bedrock. The local area was cleared for crops, and by 1930, the stream had aggraded about 5 m and buried the

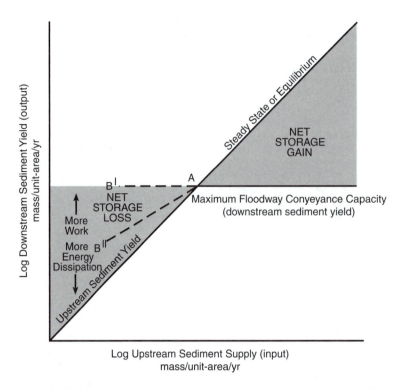

FIGURE 9.15 Schematic relation of downstream sediment yield and upstream sediment supply. Value A is the stream valley conveyance capacity. When upstream sediment supply exceeds A, the excess goes into storage, largely as vertical accretion on floodplains or channel filling. When upstream sediment supply drops below A, the stream attempts to use its excess energy to obtain sediment load from bed and banks. If sediment is easily obtainable and can be mobilized, it will be taken from storage, with the perfect case line AB′, with 100% of the available energy used for geomorphic work. Generally, much of the available energy will be dissipated, so line AB″ is more likely. (Modified from Trimble, S.W., *Am. J. Sci.*, 283:454–474, 1983.)

dam more than a meter. With the improvement of land use, the stream had degraded about 2.5 m by 1969, exhuming the buried milldam. Local bridges were undermined by this degradation, demonstrating one cost of stream instability. During the same period, about 20 km downstream the stream aggraded almost 3 m, in large part from the migrating sediment brought from upstream storage (**Figure 9.18**). By 1969, the stream had aggraded almost to the bridge sills, built in 1934. This demonstrates another way in which unstable streams can be costly. This bridge flooded out frequently and had to be rebuilt and raised in the 1970s so there would be adequate flood clearance through the bridge opening. Despite these large masses of sediment in movement, only about 4% of all material estimated to have been eroded from Piedmont slopes during historic time has been transported to the edge of the region, the fall line (Trimble, 1975a).

9.11.5 HYDRAULIC MINING IN CALIFORNIA

While the above processes were caused by agricultural soil erosion, construction or mining can cause similar effects. In the mid-19th century in the Sierras of California, waste gravels from hydraulic mining were dumped into Sierra tributaries of the Sacramento River, burying

the floodplains to 25 m in some locations (Gilbert, 1917; Adler, 1980; James, 1989). Finer particles were transported downstream, where they buried the floodplains of streams like the Yuba, Bear, and American Rivers and filled the channel of the Sacramento River, causing flooding at Sacramento and other cities. A court order to cease mining cut off the supply of new sediment in the late 1800s. Since then, hungry water has caused upstream degradation of more than 10 m in some locations. This has caused a "wave" of sediment to move through some downstream reaches, first aggrading, then degrading (**Figure 9.19**).

Lest it be construed that all accelerated erosion and legacies of sediment problems are human induced, note that some pressing sediment budget problems have their origins in the Holocene from effects of tectonism, glaciation, and volcanism (Church and Slaymaker, 1989; Slaymaker, 1993). In addition, past climatic changes have altered sediment budgets (e.g., Knox, 1999) and certainly will do so in the future. As you might guess, the instability under consideration has long been a topic in hydrologic studies. In particular, many basic principles have come from the classic work of Gilbert (1917), Happ et al. (1940), Mackin (1948), and E.W. Lane (1955b). Perhaps not perfectly, but we see encapsulated in the results presented here

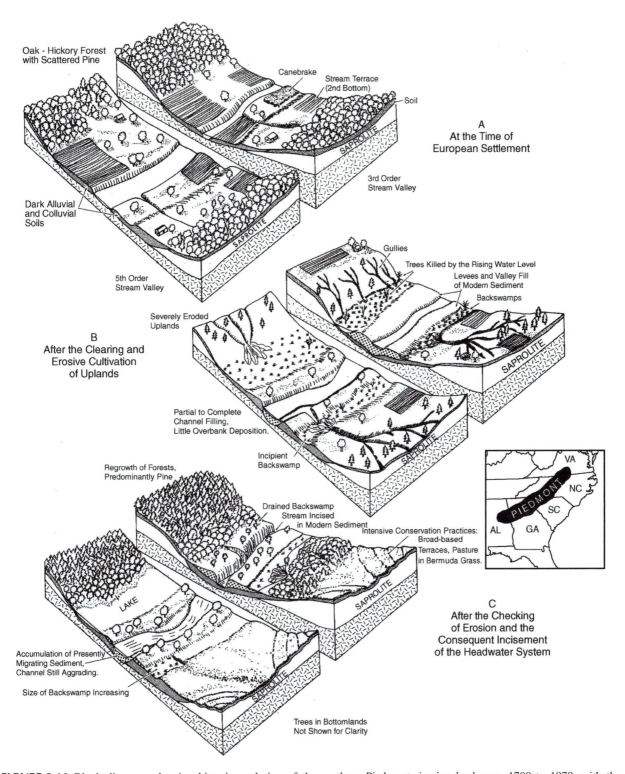

FIGURE 9.16 Block diagrams showing historic evolution of the southern Piedmont riverine landscape, 1700 to 1970, with the downstream migration of historic sediment and concomitant morphologic–environmental changes. The stream in the upstream block might drain 10 to 20 mi², while the downstream block might portray 50 to 100 mi². (From Trimble, S.W., *Man-Induced Soil Erosion on the Southern Piedmont, 1700–1970*, Soil Conservation Society of America, Ankeny, IA, 1974. Copyright S.W. Trimble, 1971.)

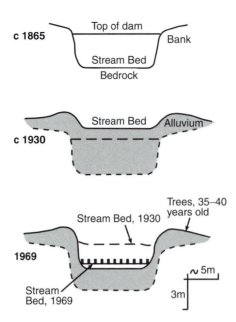

FIGURE 9.17 Processes in the upstream block of **Figure 9.16**, Periods 1 to 3. Aggradation and degradation at Mauldin Millsite, Mulberrry River, Hall County, Georgia. (From Trimble, S.W., *Culturally Accelerated Sedimentation on the Middle Georgia Piedmont*, U.S. Department of Agriculture Soil Conservation Service, Fort Worth, TX, 1970. With permission.)

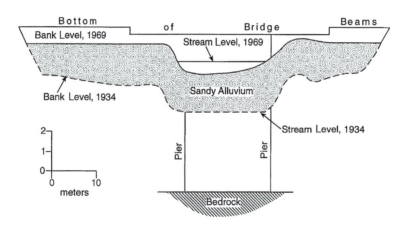

FIGURE 9.18 Processes in the downstream block, **Figure 9.16**, Period 3. Aggradation from migrating sediment at State Highway 53 bridge, Jackson County, Georgia. (From Trimble, S.W., *Culturally Accelerated Sedimentation on the Middle Georgia Piedmont*, U.S. Department of Agriculture Soil Conservation Service, Fort Worth, TX, 1970. With permission.)

many of the basic principles they described. For example, one of the most instructive diagrams published came from the work of Lane (**Figure 9.20**). He clearly showed how stream discharge and slope are balanced against sediment load and size (see Chapter 6, Equation 6.1).

9.11.6 THE DRIFTLESS AREA OF THE UPPER MIDWEST

While the sediment budget examples given above are instructive, they depended on after-the-fact historical and archeological reconstructions to a great degree, so that one could not be very precise about some important processes and factors. A far better opportunity to study complex stream processes as related to land use concerns the Driftless Area

(the Mississippi River hill country or Paleozoic plateau). There, important baseline work had been done in the 1930s that allowed relatively precise measures of stream and landscape changes that could then relate cause and effect in a meaningful way. The most intensely studied stream was Coon Creek, Wisconsin, a basin of about 360 km² (**Figure 9.21**). Aerial photographs were taken in early 1934 (e.g., **Figure 9.13A**), land use was catalogued, and water and sediment discharge measurements were made for a period of 6 years. Most important, however, were the 120 carefully surveyed stream and valley cross sections (profiles, **Figure 9.21**).

In 1973, Trimble found the original data for Coon Creek in the National Archives, enlisted the assistance of

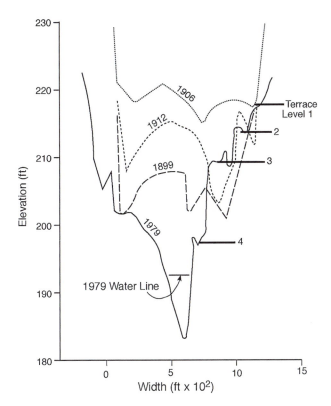

FIGURE 9.19 A "wave" of sediment moving through the Yuba River, California. Note the peak in 1909, from migrating sediment, reached about 3 to 4 decades after the cessation of hydraulic mining. (From Adler, L.L., Adjustment of the Yuba River, California to the Influx of Hydraulic Mining Debris, 1849–1979, master's thesis, University of California, Los Angeles, 1980. With permission.)

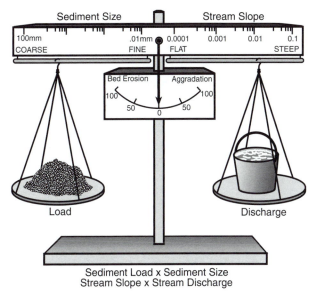

FIGURE 9.20 Another view of stream behavior as a function of discharge and slope on the one side and sediment load and size on the other. Compare with **Figure 9.24**. (From Lane, E.W., 1955a. Redrawn with permission from Phillip Williams & Associates, Ltd.)

the project's designer, Stafford C. Happ, and over the following 5 years, resurveyed the old profiles, collected data on more than 100 new profiles and many archeological excavations, and performed corollary work in other regional streams. Work has continued, and resurveys were performed in 1991 to 1995 (Trimble, 1999). Coon Creek may be the most comprehensively studied stream basin in the world (Glanz, 1999). The following is a thumbnail sketch of what has been published so far.

As shown earlier, the areas were settled by Europeans starting in 1853. The people who eventually came to farm were largely Germans and Scandinavians, who were, in fact, good farmers by standards of that time. However, their European experience did not prepare them for the heavy and intense rainfall that fell in this region (Koeppen Dfb). The excellent soils, much of them mollisols formed on loess, were so deep and fertile that they were thought to be "inexhaustible"; indeed, several decades of farming seemed to change matters little. That is, there was a "lag" of decades before poor land use manifested itself hydrologically because of deterioration of soil condition.

By the early 20th century, soils had been in use 50 to 60 years, and undoubtedly, organic material and structure

had insidiously deteriorated, endopedofaunal activity was decreased, bulk density had increased, and infiltration capacity had decreased. While we have no data on the soils of that period, later studies have shown that active soil organic material decreases to a fraction of its original levels after 50 years of agriculture (D. Foster et al., 2003). What we do know is that overland flow, soil erosion, and flooding became noticeable by the turn of the 20th century and was out of control by the mid-1910s and 1920s. By this, we mean that floods were frequent and intense, and streams and floodplains were aggrading with sediment at a rate of 150 mm/year in some locations. Roads, bridges, farms, and even villages were literally buried in the region.

This situation continued until the mid-1930s, when modern soil conservation measures became available, and the federal government aggressively sponsored the installation of such measures. Only then was soil erosion and high runoff response brought under control. Even with the measures installed, it was two to three decades before flooding and soil erosion were brought to the present low levels. So, there was again a lag between the time that land use practices (this time, good ones) were implemented and the hydrologic effects were seen (**Figure 9.22**). Erosive land use, the independent variable, is the combined CP factors of the universal soil loss equation (discussed in this chapter) times 100 to give whole numbers. The dependent variable, hydrologic response–geomorphic work, is a combination of erosion and sedimentation rates. There is actually another lag between erosion from the slopes and deposition along streams, but **Figure 9.22** is simplified (see Trimble, 1990, for full data). There was

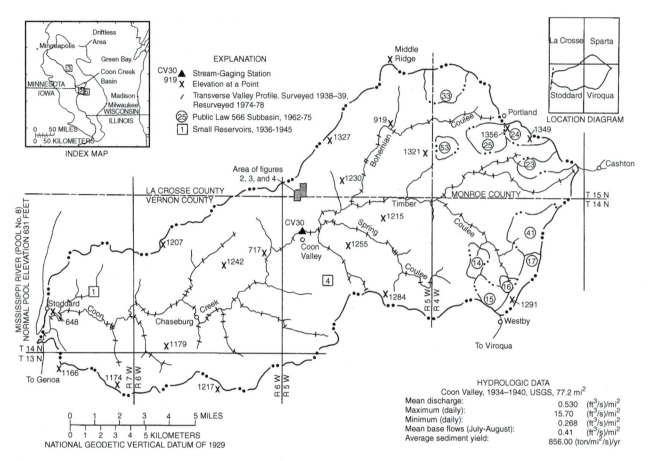

FIGURE 9.21 Coon Creek, Wisconsin, showing the 120 stream and valley cross sections (profiles) surveyed in the 1930s and again in the 1970s. Also shown are subbasins, debris basins, and other research locations in the study. (From Trimble, S.W., and S.W. Lund, Soil Conservation and the Reduction of Erosion and Sedimentation in the Coon Creek Basin, Wisconsin, USGS Professional Paper 1234, 1982. With permission.)

little further hydrologic response to expanding agriculture until shortly after 1900, when erosion was out of control. That is, the ubiquity of eroding rills and gullies was self-feeding and would not respond to even moderate land use changes. One might think of this as a "threshold" phenomenon. Once triggered, it would take heroic measures to bring it under control. Indeed, control of the landscape was only gained by the massive soil conservation measures and landscape restoration of the 1930s and 1940s. The effects of these measures were still being experienced in the 1960s and probably still are. Thus, the two lag functions together give a hysteresis loop, much like that discussed regarding soil moisture and what will be discussed between water and sediment discharge of a stream (the rising and falling limbs).

But, as seen before, decreasing erosion and creating hungry water created new problems in an already-unstable stream system. As discussed for the Piedmont, different stream orders or basin regions had different processes operating simultaneously. What happened in different parts of the basin is summed nicely by **Figure 9.23**, which shows historical sediment budgets for three zones of the

basin, tributaries, upper main valley (UMV), and lower main valley. One might think of these processes as "distributed" in the same sense of any distributed models that attempt to show cause and effect and where and when something happened. In reality, these (**Figure 9.23**) are not three discreet zones, but the processes shown vary differentially and continuously down the valley. The three zones or regions are grouped for data analysis and for easier explanation.

Tributaries. Tributary channels were originally quite small, so narrow that "one might jump across" as the early settlers recounted (McKelvey, 1939). As sediment from slope erosion started moving downstream in appreciable quantities, tributaries initially accreted along their floodplains (**Figure 9.24**). As stream banks became higher, more and deeper flow was constrained to the channel, thus increasing stream power and causing the banks to erode (Happ et al., 1940).

By about 1935 (**Figure 9.24**), stream channels were rapidly eroding and widening by lateral erosion, leaving low, inset gravel and cobble bars on the inside of bends

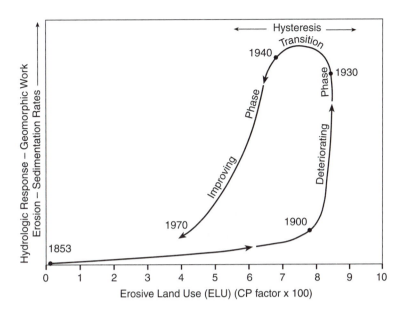

FIGURE 9.22 The relationship between intensity or erosiveness of land use and hydrologic response or geomorphic work, Coon Creek, Wisconsin. With the expansion of agriculture after 1853, there was little response until the early 20th century, when the landscape was out of control, bearing no relation to the input. Aggressive implementation of soil conservation measures in the 1930s brought the landscape back under control, but there was yet another lag in the response of the land. The result is this hysteresis function. There was yet another hysteretic relationship between erosion and sedimentation. This is not shown here, but can be seen in the cited study of Trimble and Lund (1982). (From Trimble, S.W., and S.W. Lund, Soil Conservation and the Reduction of Erosion and Sedimentation in the Coon Creek Basin, Wisconsin, USGS Professional Paper 1234, 1982. With permission.)

(**Figure 9.25**A). These wide, inset streamways containing the channel and bars were termed *meander plains* (Melton, 1936; Happ et al., 1940) By this time, hydrologic response was very high, so even small storms often brought vast amounts of stormflow. This excess energy continued to widen the meander plain. By 1950 (**Figure 9.24**), the effects of land use improvements were beginning to take hold, so the rate of bank cutting and meander plain development was declining. Also by this time, the high-water channel formed by the meander plain was so large that very few flows exceeded its discharge capacity; indeed, very few surveys showed any accretion on the old, historical floodplain between 1938 and 1974. This means that the old historical floodplain had become a fluvial terrace. By 1974 (**Figure 9.24**), a much milder hydrologic regime was permitting the deposition of fine sediment over the old, coarse bars, by then vegetated, creating what Trimble (1975b) hypothesized as new, lower floodplains diagnostic of the new and presumed milder hydrologic regime brought on by improved land use. This process was also investigated by Johnson (1976), Knox (1977), and Magilligan (1985), who found that it was also occurring in areas to the west and south.

Many might wonder if these significant changes of fluvial landscape might have been brought on by changes in climate. Refer to **Figure 2.7**, which shows that climate changes were completely out of phase (negatively corre-lated) with the changes seen here. In other words, the effectiveness of precipitation was at a minimum at just about the time that hydrologic disruption, erosion, and sedimentation were at their peak. Conversely, decreases of erosion and sedimentation occurred during a period of higher precipitation.

For some time, the hypothesis of a milder hydrologic regime brought on by improved land use (Trimble, 1975b, 1976) remained just a hypothesis, although it was supported by morphologic evidence, like the new floodplains, and by field observations. While baseflow of Coon Creek was much higher in the 1970s than in the 1930s, the time was too short to be statistically significant, and the role of precipitation trends was uncertain. However, longer term regional streamflow data were used in the last decade to show that, in fact, stormflow has abated, while baseflow increased (Potter, 1991; Gebert and Krug, 1996; Krug, 1996; Knox, 1999; Kent, 1999). This milder hydrologic regime is reflected in many ways (**Figure 9.25B**). Tributary channels have continued to become narrower over the past 30 years, and the water quality has continued to improve. Even watercress and other aquatic vegetation have reappeared in streams where there was only gravel 50 years earlier. Perhaps the most convincing evidence of a milder hydrologic regime, higher baseflow, and better water quality is the improvement in fish habitat. At the time of settlement, these tributaries were home to brook

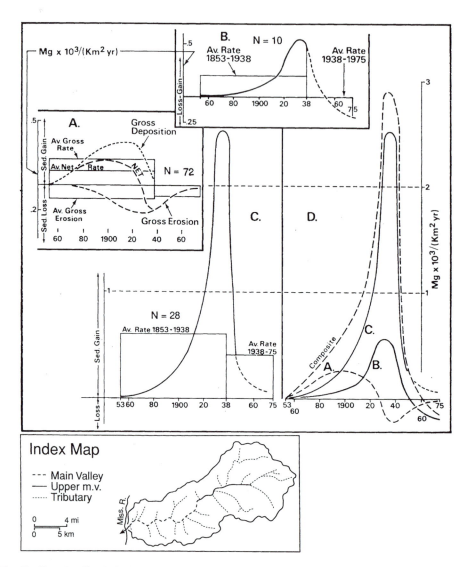

FIGURE 9.23 The distributed sediment budget model for Coon Creek, Wisconsin, 1853 to 1975, showing how processes and sediment moved downstream. (From Trimble, S.W., *Am. J. Sci.*, 283:454–474, 1983. Copyright 1983, *American Journal of Science*, Kline Geology Laboratory, Yale University. With permission.)

trout *Salvelinus fontinalis*, a species known for acute intolerance to turbid water and flooding. By the 1930s, only exotic brown trout *Salmo trutta*, which had to be stocked, were able to survive the higher sediment concentrations, warmer water temperatures, and stream channel instability of that period. With the habitat improvements of the past few decades, the brook trout now not only survive in these once-desolate tributaries, but also reproduce (Thorn et al., 1997; Trimble and Crosson, 2000).

Apart from the incredible recovery of these tributaries, there are some instructive applications of stream power principles (**Figure 9.26**, refer to **Figure 9.24**; for a similar treatment, see Knox and Hudson, 1995). In the presettlement stage, flow spread out thinly over the floodplain, so that unit stream power was low. With the aggradation of the floodplain and commensurate raising of the stream

banks (Stage 2), streamflow was more constrained to the channel, where it would have flowed faster and deeper and would have developed much more stream power, thus eroding the channel. As the meander plain widened (Stage 3), similar flows would have been spread out more, thus becoming shallower or less rapid, and would have developed less unit stream power. As meander belts continued to widen (Stage 4), flows became shallower and even less rapid, so for a given discharge, stream power has continued to decrease, probably since about 1900 or so. However, we know that hydrologic response was increasing at the same time the morphology was yielding less stream power for a given discharge. Thus, early tributary erosion has more of a morphologic explanation, while later channel erosion has more of a hydrologic regime explanation. Of course, all of this ignores other

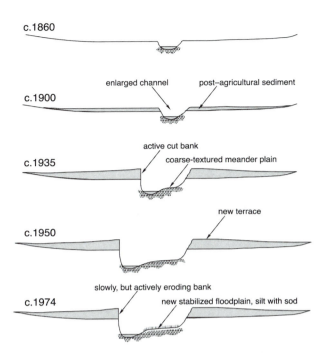

FIGURE 9.24 Schematic model of changes of historic stream and valley morphology for Coon Creek and other Driftless Area tributaries, 1860 to 1974. (From page 16 of a set of mimeographed handouts by S.W. Trimble for field trip to Driftless Area, April 1975, sponsored by the Association of American Geographers and led by George Dury, James C. Knox, W.C. Johnson, and S.W. Trimble. With permission.)

variables, such as sediment concentrations, sediment size, and the effect of cattle on the resistance of stream channels to erosive force from high flows (Trimble and Mendel, 1995).

Tributaries have had a complex role in the basin sediment budget (**Figure 9.24**). At first, they were a sink; then, despite complex processes of riparian accretion and erosion, they were net sources of sediment to downstream areas. Net stream channel erosion is always serious for downstream sedimentation because sediment is released directly into the stream so it can be more efficiently conveyed downstream. Since the surveys of the 1970s, the new tributary floodplains have become a serious sediment sink, with 150 mm of accretion since then common. Interestingly, the old, historic floodplains have received little sediment. This is because those surfaces are now stream terraces (i.e., relict floodplains) and now are rarely inundated by floods.

A major difference can be noted between the tributary processes in the Driftless Area and those seen on the Piedmont and in the Sierras, where aggradation of the channel by coarse material took place. After the period of the big sediment influx, initial channel readjustment on the Piedmont and in the Sierras was vertical incision or degradation, while in the Driftless Area, it was lateral or

bank erosion. In **Figure 9.24**, note that the Driftless Area tributaries are underlain by gravel, which made them much more resistant to erosion than the silty banks. Since the channels were rarely filled or aggraded with material, there was rarely an oversteepened profile to be readjusted as seen for the Piedmont.

The Upper Main Valley (UMV). The UMV (**Figure 9.23**) has had a history similar to that of the tributaries, but since the processes were moving downstream, the same processes occurred later. The other major difference is that historical sediment storage was much greater in the UMV. Whereas it was rare to see historical sediment more than 1 meter thick in the tributaries, 2 to 3 meters was the norm in the UMV, with banks often 4 m high. Most of this reach was like a flume at the time of the surveys in the late 1930s. Thus, when bank erosion started there in the 1940s, it was a major problem because this erosion sent so much sediment downstream. Trimble (1976, 1983, 1993) recognized it as such and monitored this reach very closely by surveying many new profiles (**Figure 9.27**). This reach looked much like a flume in the 1930s, but by the 1970s, had developed a meander belt much like those already seen in the tributaries. A plan example of this type of reach is that of the Whitewater River in Minnesota near Elba (**Figure 9.28**). Again, note that the channel was narrow like a flume in 1940, but by 1990, had created a wide meander plain about 2.5 m below the historical floodplain. Sediment storage loss from this 275-m long reach was 21,450 tonnes over a 50-year period or an average of about 430 tonnes per year. Since this reach is representative of several kilometers of stream, the potential for downstream sedimentation and sediment yield becomes clearer.

From a management standpoint, the question was what to do about these prolific sources of sediment. Some managers wanted to cut the banks down and riprap the toe, but doing that over a long reach would be extremely expensive and might destabilize the downstream reaches by cutting off the supply of sediment and creating hungry water (Trimble, 1993). However, a model of the processes suggests that the matter will take care of itself over time (**Figure 9.29**). We enter the model after Stage 1, when the stream is beginning to create the meander plain. At that time, the stream is close to long reaches of the high cut banks on both sides, so that almost any lateral movement will cut into a high bank and release a net flow of sediment. Thus, there are huge sediment losses as the meander belt evolves through Stages 2 and 3, but by then, there is an increasing area of low floodplain so that when the stream erodes part of that away, it is replaced on the other side by a bank just as high, and there is no net loss. Moreover, the increasing area of new floodplain, because it is vegetated, is a sediment sink for vertical accretion (**Figure 9.24**). By Stage 4, the reach is a major sink for sediment. Again, we are seeing a morphologic feedback by which stream power

A

B

FIGURE 9.25 Change of tributary conditions in Driftless Area, 1940 to 1974. (A) Photo of Bohemian Coulee (SE1/4, SE1/4, Sec. 33, T15N, R5W, La Crosse County, Wisconsin). This photo was made by Stafford C. Happ in 1940 to show a "typical tributary" of that time. Refer to 1935 in **Figure 9.24**. Note the highly unstable quality, with coarse sediment, cut banks, and wide stream. (B) Photo at the same site by S.W. Trimble, June 1974. The stream has become even more stable and narrow since 1974. (From Trimble, S.W., and P. Crosson, *Science*, 289:248–250, 2000. With permission.)

is reduced as the meander belt widens. The high banks, mostly at some distance from the stream, are no longer as vulnerable. Trimble had earlier predicted that the UMV reaches, a major sediment source from 1940 to 1990, might be a sediment source for decades, but the 1990s surveys proved that incorrect: Gains and losses for the period from 1975 to 1993 were beginning to balance, and the trend appears to be toward sediment accumulation (Trimble, 1999).

Lower Main Valley. Finally, the lower main valley (**Figure 9.23**) has been a sediment sink from the time of

settlement. What is remarkable about this reach, however, is that the sedimentation rates reached such heights in the 1920s and 1930s, with 150 mm of vertical accretion per year about the norm in that period. Such rates of vertical accretion were so overwhelming that roads, bridges, villages (e.g., Beaver, MN, and Chaseburg, WS) were buried during this period. While the accretion rates have been dramatically reduced, the reach continues to aggrade, with much of the material coming from upstream channel erosion (**Figure 9.30**). However, channel erosion from upstream may be having as much qualitative effect as quantitative effect. Since material eroded from a channel and banks is

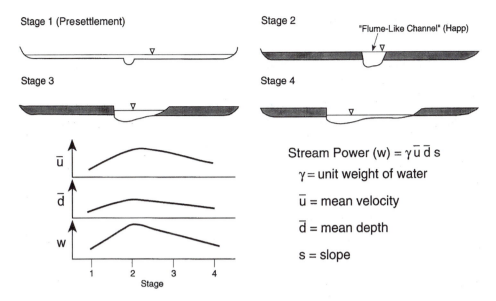

FIGURE 9.26 Stream power and the transformation of stream and valley morphology, Driftless Area tributaries. With the small stream channel of Stage 1, floods spread out over the floodplain, keeping depth, velocity, and stream power low. With accretion of the floodplain and stream banks with historic sediment in Stage 2 (circa 1900), greater flows were restricted to the channel, thus increasing depth, velocity, and stream power so that the channel erosion shown in 1900 in **Figure 9.24** must have been very rapid. In Stages 3 and 4, the channel erodes laterally, so that flows are spread, with decreases of depth, velocity, and stream power. By the latter stage, fine sediment has covered the old gravel bars, and new floodplains are formed as shown in **Figure 9.24**. (From presentation to the Association of American Geographers annual meeting, San Diego, CA, April 20, 1992. Ron Shreve made important suggestions in the preparation of this diagram in 1991. With permission.)

reworked, the coarser portion of the sediment is left in the stream, while the finer material is transported downstream or is deposited on the distal floodplain. The coarser material now in the channel, mainly fine sand, has two major impacts. First, it aggrades the channel and the natural levees to either side, but is not transported into the distal floodplain (**Figure 9.30**). Thus, the now superelevated channel and levees raise the local groundwater level, in part because water behind the levees cannot get to the stream. Backswamps become more pronounced, becoming swamps or even lakes in some distal locations. The second problem is that this migrating coarse sediment can have grave downstream consequences. As an example, such migrating sand from tributaries has raised the bed of the Trempeleau River more than a meter at Arcadia, WS, thus raising local groundwater and flooding basements in Arcadia. While the quantitative effect of eroding upstream channels might be declining, the qualitative effects just described may last for decades or more.

9.11.7 URBAN STREAMS

Urban streams provide us with a different sediment budget — one of general channel erosion and sediment movement downstream. During urban construction, however, there is much local slope erosion, and channels may actually aggrade at first (Wolman, 1967). Elsewhere in this text, we looked at how the "waterproofing" of built-up urban

areas affects streamflow. All flows for any given precipitation event are increased, but it is the smaller events (1- to 10-year return interval) that are magnified the most. The effect of these greater stormflows is to increase stream power and thus erode and destabilize streams. Note that not only are the flows greater, but with urban land use, they initially have less sediment and therefore have even more erosive power. The net result is that channel size may expand several times in a few years. A good case in point is the San Diego Creek in rapidly urbanizing Orange County, California, where rapidly eroding earthen channels alone are responsible for most of the sediment filling Newport Bay (**Figure 9.31** to **Figure 9.33**). Such rapidly eroding channels are problematic not only from the sediment they produce, but also from the fact that they destabilize expensive urban land and are unsightly. Highly scoured channels such as these often have low places where water collects and stagnates during low-flow periods and produce aesthetic and health problems.

This brief sketch of sediment budgets is only an introduction. For further reading, see the work of Schumm (1977), Graf (1988), Cooke and Doornkamp (1990), Swanson et al. (1982), Meade (1982), Abrahams and Marston (1993), I. Foster et al. (1995), Gurnell and Petts (1995), Trimble (1995), Reid and Dunne (1996), Anderson et al. (1996), Knighton (1998), A.G. Brown and Quine (1999), and Darby and Simon (1999).

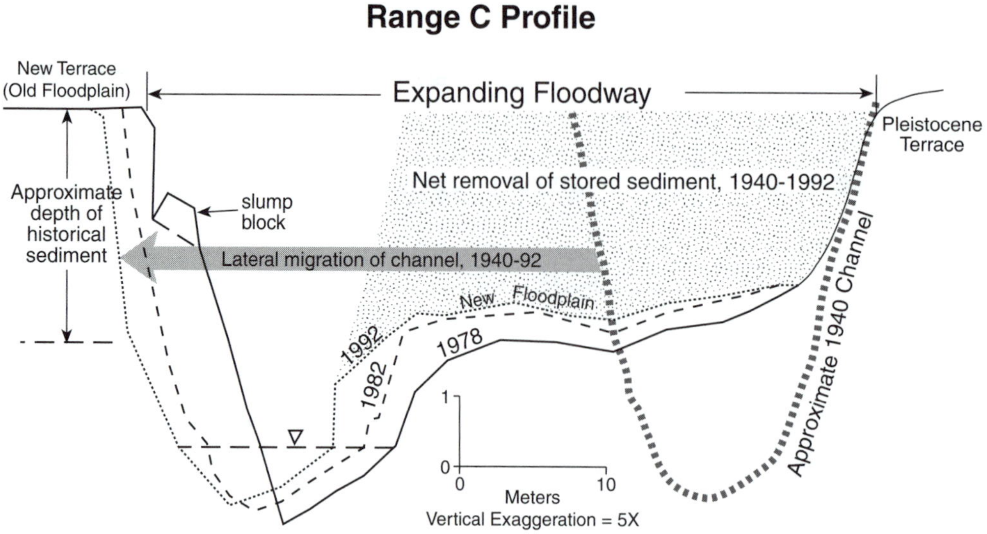

Range C Profile

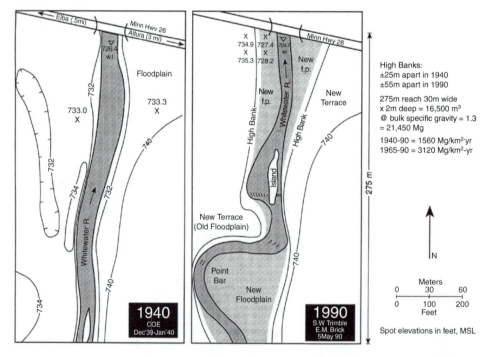

FIGURE 9.27 Channel transformation in upper main valley, Coon Creek, Wisconsin. The channel was "flumelike" in about 1940 and flowed by the Pleistocene terrace at that time (approximate 1940 profile shown in ghost outline). Note that the historical floodplain, adjusted to the effective discharge of that time, had aggraded almost to the level of the Pleistocene terrace; indeed, many such terraces were buried. Since 1940, the channel has migrated (leftward) to its present location. As it continues to migrate, the new floodplain is built up to the height of present-day effective discharge. Note that this "meander belt" expansion has left the old floodplain as a relict floodplain or terrace. Net removal of stored alluvium is about 100 m² and is shown in shadow. This loss since 1940 would be about 100,000 m²/km or about 130,000 Mg/km at a bulk specific gravity of 1.3. However, note that, in recent years, the gain of vertical accretion on the new floodplain is almost as much as that lost from the high cut bank. (From Trimble, S.W., *Science*, 285:1244–1246, 1999. With permission.)

FIGURE 9.28 The same upper main valley channel processes as seen in **Figure 9.24** in plan view, Whitewater River at Elba, MN. The rate of sediment storage loss here (21,450 Mg/275 m per 50 years) would be about 78,000 Mg/km per 50 years, a rate on the order of that shown in **Figure 9.23**. (From Trimble, S.W., *Phys. Geogr.*, 14:285–303, 1993. With permission of Bellweather Publishing.)

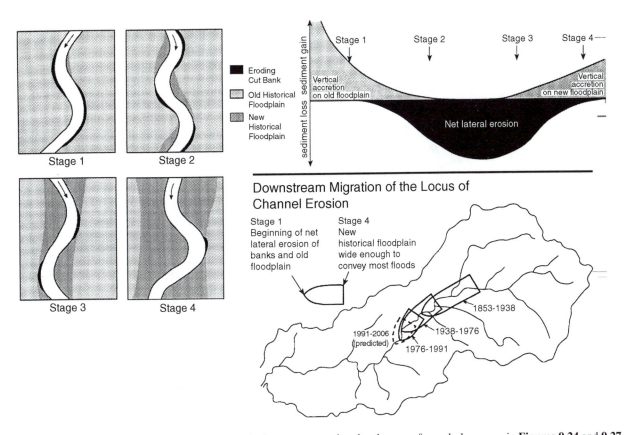

FIGURE 9.29 A model of the changes of sediment budget accompanying the changes of morphology seen in **Figures 9.24** and **9.27**. Note the dual morphologic feedback: (A) as the stream moves from Stage 3 to Stage 4, it no longer has the frequent contact with the high banks, and (B) the wider floodplain provides more opportunity for sediment storage. Thus, the zone eventually becomes a sediment sink. (Modified from Trimble, S.W., *Phys. Geogr.*, 14:285–303, 1993. With permission of Bellweather Publishing.)

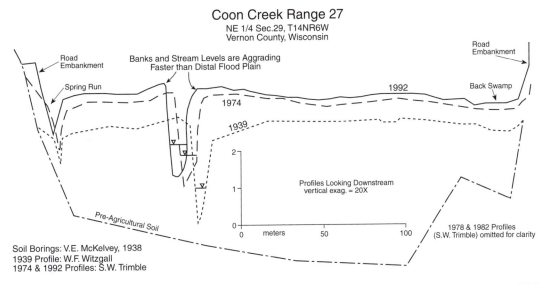

FIGURE 9.30 A typical profile in the lower main valley of Coon Creek, Wisconsin, the zone of continual accretion. While the rate of aggradation decreased greatly after 1939, the aggradation was greater near the stream than at distal areas on the floodplain. The result is that the natural levees keep the water level higher in the backswamps. This differential aggradation is thought to be from coarser sediments eroded from upstream bed and banks. (From Trimble, S.W., *Phys. Geogr.*, 14:285–303, 1993. With permission of Bellweather Publishing.)

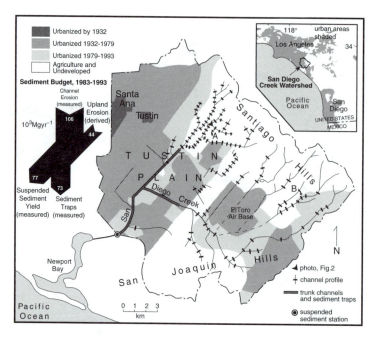

FIGURE 9.31 San Diego Creek, California, showing the earthen stream channel network and the expansion of urban land, 1932 to 1993. Paved channels and channels lying upstream from reservoirs were not included in this study. The cross-sectional channel profiles shown are those remaining in 1993. Sediment yield is that measured at the station plus accretion in the trunk channels and sediment traps. Inset is the sediment budget. A and B indicate the profiles shown in **Figure 9.32** and **Figure 9.33**. (From Trimble, 1997a. Copyright AAAS, 1997. With permission.)

FIGURE 9.32 Photos showing urban channel erosion in Hicks Canyon Wash, San Diego Creek, California: (A) 1979 (B) 1993. A person stands at approximately the same location in both photographs. Note the retreat of the cut bank to the right. Arrows mark the location of surveyed profiles in 1983 and 1993 (see **Figure 9.33**). (From Trimble, 1997a. *Science*, 278: 1442–1444. With permission.)

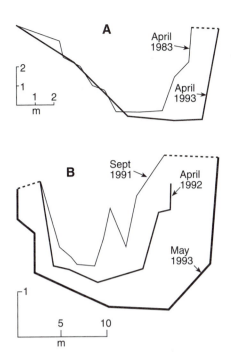

FIGURE 9.33 A. Surveyed stream channel profiles, Hicks Canyon Wash, San Diego Creek, California, 1983 and 1993 (see **Figure 9.31** and **Figure 9.32**). The rate of erosion at this profile was 0.47 m³/year per meter of channel. At a bulk specific gravity of 1.44, this would be 0.7 Mg/m per year, slightly less than the average for this type of channel. B. Extreme erosion of Borrego Canyon Wash Profile 3, directly downstream from an urbanizing area during the wet years of 1992 and 1993. The rate of erosion was about 20 m³ or about 29 Mg/year per meter of channel. This reach has since been stabilized. (Trimble, 1997a. Copyright AAAS. With permission.)

PROBLEMS

9.1. (a) For the following cities, determine the average annual rainfall (**Figure 2.3**) and the R factor (**Figure 9.3**): New Orleans, LA; Kansas City, MO; Minneapolis, MN; Denver, CO; and Atlanta, GA. (b) Plot R vs. annual rainfall. (c) From your graph, estimate the R factor for Brasilia, Brazil, which has an annual rainfall of 60 in. (d) According to Bertoni and Lombardi (1993), the R factor for Brasilia is about 440. Comment on the reliability of this method for estimating R in Problem 9.1c.

9.2. Determine the soil erodibility factor K for a soil that has the following properties: 60% silt plus very fine sand; 0% coarse sand; 3% organic matter; very fine granular soil structure; and slow permeability.

9.3. If the predicted soil erosion for a given set of conditions is 10 tons/acre for a 450-ft slope length and three terraces are installed, what is the predicted soil erosion rate from the slope for a 150-ft slope length if all other conditions remain unchanged and the slope is 8%?

9.4. Determine the annual soil loss from an urban area that is under development and has the following characteristics: average annual rainfall erosivity of 200; 500-ft slope length and 4% steepness; no vegetation; and 0.4 soil erodibility factor.

9.5. (a) Determine the soil loss for a field at your present location if $K = 0.30$, slope length is 300 ft, slope steepness is 10%, $C = 0.2$, and up- and downslope farming is practiced. (b) What conservation practice should be adopted if the soil loss is to be reduced to 4 tons/acre?

9.6. You work for the Tennessee Department of Natural Resources and are in charge of evaluating applications for permits for surface mining. A mine operator wishes to mine a steeply sloping forested watershed located in the Appalachian Mountains (eastern Tennessee). The watershed currently exhibits the following characteristics: soils with an 0.4 erodibility factor; about 10% slope steepness and about 2000-ft long; vegetation is an unmanaged forest with a medium stocked stand (50% tree canopy area and about 80% forest litter cover). (a) Determine the average annual erosion rates prior to mining. (b) Following mining, the operator will place terraces every 1000 feet; replace the land at average slopes of 8%; and reclaim the area by planting grass. It is anticipated that there will be no appreciable canopy and a 40% ground cover. If the maximum allowable soil loss is 4 tons/acre, would you give the mine operator a permit? State your reasons.

9.7. (a) Determine the annual soil loss from a 200-acre farm in central Indiana. The farm has Blount soils, and typical slopes are 1000 ft long with 2% steepness. The farmer uses conventional tillage in the autumn and has a corn–soybean rotation. (b) What percentage change in the annual soil loss would you expect if the farmer switched to a no-till farming system?

9.8. Determine the sediment yield during a 10-year period from the 200-acre farm in Problem 9.7. The farm is located on a 5-mi² watershed.

9.9. From the principles presented in this chapter, discuss three management techniques to minimize soil erosion on the slopes of a popular ski resort.

9.10. A watershed in western Ohio was intensively farmed for almost 100 years and experienced severe erosion. The farmers all abandoned their farms, and the predominant vegetation on the watershed reverted to native woodland. When the woodland was about 80 years old, a large recreation and flood control reservoir was constructed on the stream draining the watershed. The engineers were surprised to discover sedimentation rates from the watershed were similar to cropland rates. (a) What do you think was the source of the sediment? (b) Describe a method to reduce sedimentation in the lake.

9.11. A reservoir is constructed on a stream draining 5 mi^2 (1 mi^2 = 640 acres). Two thirds of the watershed has an estimated upland erosion rate of 4.5 tons/acre/year, and the other third has a rate of 0.5 tons/acre/year. How much sedimen-

tation will the reservoir experience during its 30-year design life?

9.12. Determine the sediment yield from a storm with 80 acre-ft of runoff and a peak discharge of 600 ft^3/sec. The USLE factors for this watershed are 0.3 soil erodibility, 0.5 topographic factor, 0.25 cover management factor, and 1.0 erosion control practice factor.

9.13. Calculate a C factor for a farming system in Ohio that has a rotation of corn, soybeans, small grain (oats), meadow, meadow; use conventional autumn tillage. State your assumptions. Compare your answer to the value given in **Table 9.2**.

10 Hydrology of Forests, Wetlands, and Cold Climates

10.1 INTRODUCTION

Much of the discussion in this book focused on agricultural, urban, and rangeland hydrology. Two other land uses, forests and wetlands, require further discussion because hydrologic processes in those settings behave differently from those in agricultural and urban land uses.

Forests cover a large percentage of the land area worldwide. In 1993, the United Nations Food and Agriculture Organization (FAO, 1993) estimated that forests constitute 30% of the land area, with a total area of over 4 billion ha. In the last decade, deforestation has resulted in this total declining to less than 27% of the land area. *Deforestation* is the permanent loss of woodlands and indigenous forest. Of the continental U.S., 30% is dedicated to forests. Such extensive coverage makes the unique aspects of water movement in forests an important aspect of global and regional hydrologic cycles.

Globally, forest cover appears to be decreasing, but there is much controversy about the rate of decrease. Much of the conversion of forests is occurring in developing countries. A particularly important aspect of the disappearance is the role that some of these forests appear to play in the global climate and hydrologic cycle. The Amazon rain forest in South America has been the subject of much study on evaluating the role of the large tropical rain forests in the global climate (Balek, 1983). Soil erosion has often accompanied deforestation, but usually only when the land is converted into cropland. This can be particularly severe in steep areas of wet tropical climate (Koeppen Af or Am). A good understanding of hydrology is important in solving these problems.

Conditions in the U.S. are quite different. While much of the native forest was cut for agriculture and for wood products, most commercial forests are replanted or allowed to reseed naturally. Most remarkable has been the reversion of former cropland in the humid U.S. to forest over the past 75 to 100 years (Trimble et al., 1987; M. Williams, 1989; D.R. Foster and O'Keefe, 2000). A person may drive from Texas to Maine through forests that once were cropland or at least have been cut more than once.

In the U.S., much debate over how forestlands should be managed has focused on water and the hydrologic cycle at regional and local scales. Water-related issues have been a focus of forest management in this country since the beginning of forestry as a science. The Organic Act of the Forest Reserves, enacted in 1891, stated that one of the purposes of the forest reserves was to "obtain favorable conditions of flow." Since then, many debates have taken place over the effects of forest management on floods, water supply, water quality, and fisheries production. The issues discussed in the past continue to be discussed today in scientific journals and the courts.

One important issue is the simple expectation that water coming from forests be clean. Once it was believed that forests purified water, and the belief continues today. The water supply for several major metropolitan areas in the country is surface water from forested watersheds. Among these cities are Seattle and Tacoma, WA; Portland, OR; and San Francisco and Los Angeles, CA. Many eastern cities, New York City in particular, actually own large forests, primarily for water supply. As our populations increase with an increasing demand for water for consumption and irrigation, people are concerned about the amount of water available as well as its quality. Once it was believed that forests produce water; now, our understanding is better, and we realize that trees consume water like any plant. In this case, water quality and water quantity can become conflicting goals in watershed management.

Less specifically related to human needs, but no less important as far as quality of life is concerned, is protection of endangered species. Many fish, amphibian, plant, and insect species are dependent on forest streams for habitat. Some spend only a small part of their life cycle in forest or mountain streams, but the quality of the habitat is important nonetheless. One important example is the debate over how the forests of the Columbia River basin must be managed to preserve salmon habitat for the endangered Salmon River sockeye and Chinook runs. These fish use forest and mountain streams only for spawning and the first months of rearing. The remainder of their life cycle includes travel through the Snake and Columbia River hydroelectric reservoir and dam system and several years in the Pacific Ocean. Their success in the fragile birth and rearing stages is critical to their survival as a species.

Careful consideration of the hydrologic and erosion processes of forests is important in resolving these debates. Understanding how the forest affects hillslope hydrology at the scale of a stand of trees during a single storm event is as important as understanding how forest management affects distributions of fish spawning and rearing habitat in a watershed over 200 or 300 years.

10.2 HOW ARE FORESTS DIFFERENT?

The hydrologic cycle and processes discussed thus far in this book apply to forests as they do to any other land use, but the differences in rates and combinations of important processes unique to forests and the large areas covered by forests mandate a chapter devoted to forest hydrology. The differences between forest hydrology and cropland, urban, and rangeland hydrology stem from the presence of large amounts of vegetation.

There is a tremendous diversity in forest ecosystems around the world. Trees have evolved to fill niches as diverse as bayou swamplands and nearly barren mountain-tops on the North American continent alone. In some cases, the soil and physical topography alone can describe the important aspects of the hydrologic cycle, and the influence of the forest is less significant. This chapter focuses on forests in the temperate zones in typical hilly or mountainous environments and makes only occasional points about unique forest ecosystems.

The biomass per unit area in forests is many times the biomass per unit area of cropland or rangeland. With so much plant matter, the soil properties become less important than the vegetation in determining how the system behaves. The tree canopies modify the cycling of water through the system significantly through their effects on precipitation and evapotranspiration. The large canopy system above ground is mirrored by a proportionate root system beneath the ground. These roots gather water from a large area and can reach deeper than roots in croplands. Tree roots, furthermore, are typically much larger in diameter, creating a network of large pores through which water travels much faster than in the soil matrix itself.

A mass of fine roots and decaying leaves, termed *forest duff*, forms a mat of organic matter at the soil surface. This layer is not present in croplands and is much less developed in most rangelands. In forests, the thickness of this layer depends mostly on the decay rate of leaves and time since the last time the layer was burned. Typical depths in western Oregon and Washington forests range from 5 to 10 cm. Interior forests with warmer temperatures, faster decay, and the occasional ground fire typically have an organic horizon from 1 to 5 cm thick. Coastal Alaskan forests can have organic horizons in excess of 1 m deep.

With a large leaf area per unit ground area and soil development incorporating a thick organic mat and extensive root systems, the rates of some hydrologic processes in forests differ from those in croplands and grasslands. The large canopies increase interception of precipitation, a process often considered negligible in croplands and grasslands. The large leaf area and root access to deeper water allows much greater transpiration rates. The root system and high organic content create a highly conductive soil. **Figure 10.1** shows some typical hydraulic conductivity values for the various soil layers in a forest. The

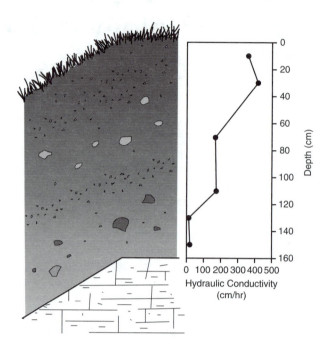

FIGURE 10.1 Hydraulic conductivity changes with depth in a forest soil. (Data from Harr, R.D., *J. Hydrol.*, 33:37–58, 1977.)

fibrous root and litter layer is also resistant to movement from raindrop impact and protects the underlying mineral particles from moving; thus, erosion rates in forests are typically much smaller than for croplands.

Another difference between forestlands and croplands is that forests frequently occur where the land is not suitable or profitable for other uses. Forest uses bring in less money per acre than agricultural and urban uses (G.E. Bradley, 1984). Consequently, if land is suitable for other uses, it often has been converted to those uses. Not all steep land is unsuitable for other purposes, but a good portion of steep land in mountainous and hilly areas is forested. Shallow, young, and coarse-grained soils are less useful for agricultural purposes, so if not put into urban uses, these lands are often forested. Not all wetlands are forested, and a great deal of wetland habitat has been converted to both urban and agricultural uses, but where suitable conditions and species exist to have forested wetlands, such as in Arkansas and Mississippi, forests commonly occupy wetlands.

Forests grow only where the natural climate and soil water conditions are suitable. Large sections of the continental U.S. have not historically maintained forests. **Figure 10.2** shows a map of the historical extent of forests in the continental U.S. Precipitation, in comparison to evapotranspiration, is one of the primary controls on the extent of forest. In areas with low precipitation, forests grow only near streams or similar concentration points for water. Temperature and long-term snowpack are other controls important in high elevations and latitudes. Irrigation, which

FIGURE 10.2 Historical extent of forest in the U.S. (shaded area).

allows crops and pasture to thrive in arid environments, is seldom used to maintain forests.

Human modification, management practices, and disturbances in forests differ from agricultural and urban uses. The primary human actions in forestlands are road building, timber harvest, and site preparation for replanting. Forest roads are more sparsely distributed than agricultural and urban roads and are usually smaller. Timber harvest is done by cutting the tree and removing the log. Ground-based removal systems, such as skidders and tractors, are common where much of the landscape is level enough, but in areas of the country where there are more mountains and steeper slopes, cable-suspended yarding is much more common. Site preparation sometimes involves burning the residue left after harvest. However, on relatively flat and productive land, such as the converted agricultural fields in the southeastern U.S., site preparation may be more complex, involving machine scarification and piling of residue and, occasionally, tillage. Forests are disturbed much less often than croplands. The harvest and site preparation procedures might be carried out once every 30 years in some regions; 50 to 90 years is a more common rotation in other regions. Sometimes, rangeland and forestlands overlap, so grazing practices common to grasslands are used in these forests. Forestlands used for grazing are a small subset of all forests, however.

In the following sections, the specific processes in forests are discussed. Discussion starts with precipitation and the energy balance, moves through interception and evapotranspiration in the canopy, snowmelt and infiltration on the hillslope below the canopy, shallow subsurface flow (sometimes termed *interflow*) and overland flow on the hillslope scale, and finally, watershed scale hydrology. Further discussion focuses on erosion processes in forestlands.

10.3 FOREST CLIMATES — RAIN AND SNOW

Many benefits have been attributed to forests with regard to water. For much of history, people have considered forests the fonts of freshwater. The ancient Greeks believed that forests brought saltwater up through their roots and purified the water, which was released through springs. Later civilizations, although perceiving the connection between precipitation and streamflow, still attributed the creation of clouds to forests. Even as recently as the end of World War II, there was some debate on the effects of forests on precipitation (Kittredge, 1948), but the rain gage observations fueling the debate may well have been in error because gages in forest clearings have improved catches due to reduced wind velocities. Today, it is generally considered that forests in North America are a result of precipitation and not the cause of it. There is some recognition of the role that forests have in recycling rainfall in the Amazon (Balek, 1983). Desertification following deforestation in the Middle East and North Africa was influenced by the reduced infiltration of grazed lands and the global climate shift that has occurred over the last 3000 years (Vita-Finzi, 1967).

One relationship between forests and climate that is relatively certain is that the presence of forests indicates more precipitation than the presence of shrub and grassland communities. The *ecotone* — a zone of transition between two ecosystems — between juniper shrublands and ponderosa pine forestlands in eastern Oregon and Washington occurs at about 32 cm of precipitation annually (Franklin and Dyrness, 1973). Areas with more precipitation have forests, and areas with less have shrub and grass communities.

The complex associations of the amount of water available, its seasonal distribution, and the seasonal temperatures create the amazing diversity of forest ecosystems. The upper tree line in the Cascades of Washington State is defined by too much precipitation in the form of snow. While there is certainly as much water on the Pacific coastal rain forest, the snow falling at high elevations in the Cascades prevents trees from having a long enough growing season to get started. Too much moisture is seldom a problem otherwise. Other trees, such as the bald cypress, have adapted to swampy conditions in the southeast. The black spruce has adapted to swampy conditions, cold temperatures, and snow loads to survive in the subboreal taiga regions of North America (Koeppen Dc).

10.3.1 Characteristics of Snow

Hydrologists are concerned with snowpack depth and its water equivalent, the *snow water equivalent* (SWE), which is the depth of water after melting the snow. In most practical applications, the actual depth of the snowpack is not as important as the snow water equivalent. The calculation of the SWE requires knowledge of the depth of snowpack, density of snow, and density of water.

$$SWE = \frac{\rho_s}{\rho_w} d_s \qquad (10.1)$$

where d_s is the depth of snowpack, ρ_s is the density of snowpack, and ρ_w is the density of water.

The density of a snowpack changes due to rainfall, settling, metamorphism, temperature, and vapor gradients. Freshly fallen snow has a snowpack density ranging from 0.05 to 0.20 g/cm³, averaging 0.10 g/cm³. Given that the density of water is 1 g/cm³, a good rule of thumb is that 10 in. of new snow is equivalent to 1 in. of snow water equivalent because the ratio of snowpack density to water density is 0.1 or 10%. Snow still on the ground in late summer may reach a density of 0.60 g/cm³.

Snow is composed of ice, liquid water, and air. Snow, like a porous media such as soil, can hold a certain amount of liquid water. It has been found that typically 3 to 5% of the SWE depth can be held in the snow. When determining snowmelt, it is important to know when water will drain from the snowpack. At this point, the snowpack is defined as *ripe*. Two conditions must be met for the snowpack to be ripe: The snowpack temperature must be 0°C (32°F), and the snowpack must be saturated.

If the temperature of the snowpack T_s is below 0°C, then any liquid water added to the snowpack will freeze and release latent heat energy (i.e., change in energy associated with a phase change), warming the temperature of the snowpack. The depth of water that must be added to a snowpack to raise its temperature to 0°C is called the *depth of cold content d_{cc}*.

The depth of cold content can be determined using a simple energy balance by which the absorbed amount of sensible heat in a snowpack equals the release in latent heat due to the freezing of the added liquid water (assuming the added water has a temperature of 0°C).

$$\rho_s c_s d_s T_{diff} = L_f \rho_w d_{cc} \qquad (10.2)$$

Solving Equation 10.2 for d_{cc}

$$d_{cc} = \frac{\rho_s \cdot c_s \cdot d_s \cdot T_{diff}}{\rho_w \cdot L_f} = \frac{\rho_s \cdot 0.5 \cdot d_s \cdot T_{diff}}{1 \cdot 80}$$
$$= \frac{SWE \cdot T_{diff}}{160} \qquad (10.3)$$

where c_s is the snowpack specific heat (assumed 0.5 cal/g °C), L_f is the latent heat of fusion (assumed 80 cal/g), T_{diff} is 0°C less T_s, and T_s is the temperature of snowpack (°C).

Example 10.1

Consider a 30-cm snowpack with an average snow density of 0.20 g cm⁻³ and an average temperature of −5°C. How much water needs to be added to ripen the snowpack or, in other words, needs to be added before liquid water drains out the bottom of the pack? What is the final density of the snow? (Assume the added water has a temperature of 0°C, and the snow depth does not change.)

Solution

The initial snow water equivalent is

$$SWE = \frac{\rho_s}{\rho_w} d_s = \frac{0.20 \text{g/cm}^3}{1.00 \text{g/cm}^3} 30 \text{ cm} = 6 \text{ cm}$$

Using the SWE, we can calculate the depth of cold content as

$$d_{cc} = \frac{T_s SWE}{160} = \frac{5°C \cdot 6 \text{ cm}}{160 \frac{\text{cal} \cdot \text{g} \cdot °C}{\text{cal} \cdot \text{g}}} = 0.2 \text{ cm}$$

The new SWE is 6 cm + 0.2 cm = 6.2 cm.

The amount of water held in the snowpack assuming a water-holding capacity (WHC) of 4% is SWE * 4% = 6.2 cm * 0.04 = 0.2 cm. Hence, the final SWE = 6.2 cm + 0.2 cm = 6.4 cm.

$$\text{Total water added} = d_{cc} + \text{WHC} = 0.2 \text{ cm} + 0.2 \text{ cm} = 0.4 \text{ cm}$$

$$\text{Final } \rho_s = \frac{\rho_w SWE}{d_s} = \frac{1\text{g/cm}^3 6.4 \text{ cm}}{30 \text{ cm}} = 0.21 \text{ g/cm}^3$$

Answer

A total of 0.4 cm of water had to be added to ripen this snowpack. The final density of the snow was 0.21 g/cm³.

10.4 INTERCEPTION — RAIN, SNOW, AND FOG

You may have observed dry areas underneath trees near the beginning of a storm and perhaps taken shelter from storms under trees. The process of trees capturing some of the falling rain and snow is called *interception* (see Chapter 4).

10.4.1 RAIN INTERCEPTION

Figure 10.3 shows the disposition of rain or snow coming into a forest canopy. Rain falling on a forest canopy will adhere to the leaves and branches in the canopy. As the surfaces of the leaves and branches become wetted, any additional rainfall will cause some of the water to fall off and drip onto lower leaves and branches. Eventually, all of the branches and leaves are wetted, and some water will fall through to the ground, while other water will flow down the trunks and stems of the forest and shrub cover. The amount of water stored on leaves and stems is called *interception*, the amount flowing down the stem is called *stemflow*, and the amount falling from the canopy to the ground is called *throughfall*. Because the intercepted water frequently evaporates, it is often called *interception loss* (see Chapter 4).

Some hydrologists consider interception storage within the leaf litter lying on the ground. This layer can be dealt with as a soil layer or as interception. This chapter follows the convention of calling the litter layer a soil layer and considers only interception in the canopies of trees, shrubs, and standing plants.

Interception is measured as a comparison of rain gages placed above the forest canopy, or in a nearby open area, to gages placed under the canopy and gages collecting stemflow. While one or two standard precipitation gages are sufficient to measure the gross precipitation above the canopy or in a nearby opening, many randomly placed gages are necessary to estimate throughfall. Stemflow is

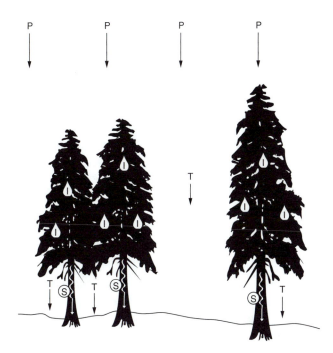

FIGURE 10.3 Interception processes in forests; *P* is total precipitation, *T* is throughfall, *S* is stemflow, and *I* is interception.

caught on collars placed around tree trunks and directed to collection cans. It is necessary to have several trees in a plot to estimate throughfall and stemflow reliably. The volume of stemflow is divided by the total area of the plot to estimate the depth of precipitation abstracted by stemflow. Interception *I* is then calculated as the difference between gross precipitation *P* and the sum of stemflow and throughfall *S* + *T*.

$$I = P - (S + T) \qquad (10.4)$$

Note that, when using Equation 10.4 in a water balance, all intercepted water is considered to evaporate after or during a storm.

EXAMPLE 10.2

During a 3.51-cm storm, throughfall and stemflow were measured on a 1/10-ha plot. The throughfall was collected in rain gages scattered over the plot and averaged 2.15 cm. Tanks collected 11,200 l of stemflow. Calculate the depth and percentage of rainfall intercepted.

Solution

First, convert the stemflow volume to depth by dividing by the plot size.

$$11,200 \; 1 = 11.2 \; m^3$$

$$0.1 \; ha = 1,000 \; m^2$$

$$11.2 \; m^3/1,000 \; m^2 = 0.0112 \; m$$

$$S = 1.12 \; cm$$

$$S + T = 1.12 \; cm + 2.15 \; cm = 3.27 \; cm$$

$$I = P - (S + T) = 3.51 \; cm - 3.27 \; cm = 0.24 \; cm$$

$$I/P = 0.24 \; cm/3.51 \; cm = 0.068$$

Answer

Of the storm, 0.24 cm or 6.8% was intercepted.

Interception and stemflow rates are controlled by a variety of factors. Perhaps the most important of these is the type of tree. This determines the type of leaf, leaf size and shape, leaf density and arrangement, whether it is deciduous or evergreen, the branching form, and bark roughness. The average interception over a forest stand is clearly dependent on the density of the stand and in particular on the density of water-holding surfaces, as might be represented by the leaf area index (LAI) of the stand. The characteristics of the storm event can be important also. Rain intensity and wind speed control the efficiency with which leaves hold water.

Interception rates for different types of vegetation are presented in a variety of forms. Some researchers have presented the proportion of annual rainfall that is lost as evaporation of intercepted water (Kittredge, 1948). Others present the proportion lost in a storm (Hewlett, 1982). Because the water-holding capacity of a tree's surface is limited, this type of information is difficult to consider. Very little rain becomes throughfall until the interception capacity of a tree is nearly full, after which all of the rain passes through as either stemflow or throughfall. Therefore, the percentage captured by any storm is a function of whether the interception capacity of the tree was exceeded (see Chapter 4). These sorts of values are more dependent on the number and characteristics of storms in a year or in the sample than on the type of vegetation intercepting the water. It is becoming more common just to express the interception capacity of particular species. **Table 10.1** presents the interception capacities of fully developed canopies for a variety of species.

10.4.2 Snow Interception

Snowfall is also intercepted. Losses of intercepted snow to sublimation (evaporation from the ice state of water)

TABLE 10.1
Interception Capacity of Fully Developed Canopies of Several Eastern Species

Vegetation	Interception Capacity (in.)	Interception Capacity (cm)
Red pine	0.15	0.38
Loblolly pine	0.15	0.38
Shortleaf pine	0.14	0.36
Ponderosa pine	0.13	0.33
Eastern white pine	0.14	0.36
Pine (average)	0.14	0.36
Spruce–fir–hemlock	0.26	0.66
Hardwoods, leafed out	0.10	0.25
Hardwoods, bare	0.05	0.13

Source: After Helvey, J.D., in *Proceedings of the Third International Seminar for Hydrology Professors: Biological Effects in the Hydrological Cycle*, E.J. Monke, Ed., Purdue University, West Lafayette, IN, 1971, pp. 103–113. With permission.

are generally smaller than evaporation losses of intercepted rainfall. Colder temperatures and the much greater energy required to sublimate snow lessen this impact. Most of the intercepted snow blows out of trees, melts, or falls as clumps of wet snow.

The process of snow interception differs from that of rainfall interception in that the rainfall interception capacity is defined more by the amount of surface available to hold water droplets, whereas snowfall interception capacity is more a function of branch strength and canopy shape. Snow falling on needles initially bridges small gaps, forming a platform for further accumulation. Eventually, the snow builds up high enough on branches that the snow sloughs off either because it is unstable or the branches bend. Snow falling at temperatures near freezing is quite cohesive and generally does not slough off unless a branch bends. After the storm, the snow metamorphoses into less-cohesive forms and falls to the ground or melts.

Few reliable measurements of snow interception exist, and most work that has been done relates to coniferous species because deciduous trees have few leaves in the winter, and there are few evergreen hardwoods in snowy climates. The magnitude of the snowfall interception capacity is generally greater than the rainfall interception capacity. Western white pine saplings (approximately 4 m high) sampled at the Priest River Experimental Forest in Idaho intercepted 0.25 cm of snow water equivalent in one storm, and Douglas fir saplings (4 m high) intercepted 0.37 cm in the same storm (Haupt and Jeffers, 1967). Compare these values to the values in **Table 10.1** of rainfall interception for a fully developed canopy.

10.4.3 FOG DRIP

Fog drip is a form of interception that works somewhat differently from the other forms and is hydrologically important only in a few locations. Fog drip occurs when trees intercept horizontally moving water droplets in clouds and fog that would not otherwise fall to the ground. This effect occurs mostly in coastal forests and in mountain ranges not far from coasts. Some foresters say that, without fog interception, the Sequoia trees of California's coast could not survive. Harr (1982) noted the contribution of fog drip to the hydrologic cycle in the Cascades near Portland, OR. Deposition of rime ice on trees in the northern Appalachians is a similar process.

10.5 ENERGY BALANCE IN FORESTS

To understand the evapotranspiration and snowmelt processes in forests, it is important to understand the energy balance in forests. The forest canopy absorbs and emits radiation, changes wind flows, and releases water vapor, all of which change the energy balance in the forest.

There are four primary forms of energy considered in an energy balance: short-wave radiation, long-wave radiation, latent heat, and sensible heat. Short-wave radiation is usually considered the wavelengths of light emitted by the sun and is sometimes called *solar radiation*. This includes light in the ultraviolet, visible, and near-infrared ranges. Long-wave radiation is light in the far-infrared range (greater than 4 μm) and is sometimes called terrestrial radiation because the earth's surface radiates in this wavelength. *Latent heat* refers to the latent heats of vaporization and fusion for water, which refer to the energy necessary to turn water from a liquid to a vapor or to turn water from a solid to a liquid. Approximately 2500 kJ/kg (590 cal/g) of water are necessary to change liquid into vapor, and 335 kJ/kg (80 cal/g) are required to melt ice. The reverse is also true: 2500 kJ of heat are released per kilogram of water condensed, and 335 kJ/kg are released when 0°C water freezes. When water vapor in the air over the snowpack condenses and freezes onto the snowpack, 2835 kJ/kg (670 cal/g) are released. *Sensible heat* refers to the heat energy due to the temperature of some substance (air, rain, snow, or soil, for example), and the *specific heat* of the substance refers to how much energy is required to change the temperature of that substance.

FIGURE 10.4 Energy balance in a forest. Q_{sw} is short-wave (solar) radiation, Q_{lw} is long-wave (terrestrial) radiation, Q_{le} is the latent heat of evaporation, and Q_h is sensible heat.

Solution

$$\text{Melt: } 3 \text{ kg} \times 335 \text{ kJ/kg} = 1005 \text{ kJ}$$

$$\text{Sublimation: } 3 \text{ kg} \times 2835 \text{ kJ/kg} = 8505 \text{ kJ}$$

Answer

The 3 kg of snow in the branch will require 1005 kJ to melt and 8505 kJ to sublimate. Is it more likely that the snow will melt from the branch or sublimate? Note that there are conditions when sublimation will occur and melt cannot, like windy, dry days when the temperature is below freezing.

An *energy balance* considers the fluxes of all of these forms of energy to find the net energy flux (**Figure 10.4**). The equation of conservation of energy is

$$Q_{sw} + Q_{lw} + Q_{le} + Q_h - \frac{\Delta S}{\Delta t} = 0 \qquad (10.5)$$

where Q_{sw} is the short-wave energy flux, Q_{lw} is the long-wave energy flux, Q_{le} is the latent heat flux, Q_h is the sensible heat flux (all flux units are energy per unit time, e.g., watts), and $\Delta S/\Delta t$ is the change in stored energy per unit time. This equation must hold for any defined volume.

EXAMPLE 10.3

The branches of a small tree hold 3 kg of snow. Calculate the energy required to melt the snow and the energy required to sublimate (change directly from a frozen to gaseous state) the snow.

When calculating evapotranspiration, it is easiest to consider the primary surfaces at the top of the canopy and under the soil, and when trying to calculate snowmelt, it is easiest to consider the primary surfaces at the top of the snowpack and the bottom of the soil.

Short-wave radiation input can be calculated based on time of year, aspect, slope, and cloudiness. The reflectiveness of the location for which the balance is calculated is called the *albedo*, and the reflected light is subtracted from the short-wave balance. The long-wave balance can be calculated based on air temperature and cloudiness. Latent heat transfer is considered in regard to transpiration, which requires an energy input, and in condensation and freezing of water vapor to a snowpack, when energy is released. Rates of latent heat transfer are regulated primarily by wind, which carries moist air away in the case of transpiration and carries moist air to the snowpack in the case of condensation. Sensible heat is transferred primarily by air convection; however, a sensible heat flux from the earth is transferred by conduction. An increase in stored energy raises the temperature of the material that composes the volume.

A basic point to make about the energy balance is that heating caused by positive short-wave radiation, long-wave radiation, and sensible heat fluxes, such as found on a warm summer day, can result either in evapotranspiration (a negative latent heat flux) or an increase in the stored heat energy or temperature. In a wheat field in late August, after all available soil moisture has been used, the wheat field will become quite warm, whereas in a more deeply rooted forest, transpiration will keep the temperature lower. This transpiration effect, more than shading, is what keeps forests relatively cool compared to adjacent open areas.

10.6 EVAPOTRANSPIRATION

With their large vegetation masses and deep root systems, forests are capable of transpiring a great deal more water than dryland crops. Evapotranspiration in forests is usually considered only in terms of its overall effect on water supply, although academic interest is shown with regard to its importance in modeling tree growth, microclimate modification, and shallow groundwater flow. Most studies of evapotranspiration have been done at a watershed level using a simple water balance for the stream by ignoring deep seepage of groundwater. This section discusses the process, and the basinwide importance is described later in the chapter.

Evapotranspiration was covered in some detail in Chapter 4. In forests, the process is essentially the same, differing only in a few minor details. The most important differences are that forests transpire much more than they evaporate; forests have greater rooting depth and so have a deeper water supply; forests are seldom irrigated.

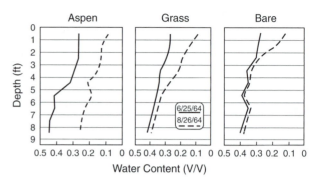

FIGURE 10.5 Evapotranspiration from aspen, grass, and bare plots. (After Johnston, R.S., *Water Resour. Res.*, 6:324–327, 1970. With permission.)

There is generally little direct evaporation in forests as compared to transpiration. The soil surface is well protected by the deep canopy of living, transpiring vegetation and an insulating layer of litter. Evaporation occurs mainly from intercepted rainfall. Some evaporation occurs during storms and between rainfall events. Intercepted snow generally melts before much of it sublimates, as can be deduced from the energy requirements for melting vs. sublimation (335 kJ/kg to melt vs. 2835 kJ/kg to sublimate).

Transpiration from forests is best calculated using the Penman–Monteith equation (Chapter 4) because the stomatal resistance of trees is the primary control on transpiration rates. The Penman–Monteith method can be modified to ignore the boundary layer resistance for coniferous forests because the canopy projects so high into the air and the needles have such a high surface area (Waring and Schlesinger, 1985). The energy balance component of the Penman–Monteith method considers net radiation (Q_{sw} + Q_{lw}) as input, ignores Q_h, and uses the change in air temperature to find ΔS. Evapotranspiration is thus estimated from Q_{le} and the ability of the stomata to retain water. The stomatal resistance in turn is a function of the soil moisture and plant moisture stress.

One notable difference between forests and croplands is the rooting depth. **Figure 10.5** shows the difference in soil water taken as evaporation and transpiration from plots with aspen, herbaceous species, and bare ground. The main factor causing the differences in **Figure 10.5** is rooting depth.

Because forests are not irrigated, it is difficult to apply equations that assume a well-watered crop. This difference also makes it difficult to compare gross amounts of water use by crops and trees. There are proposals to apply wastewater effluent as irrigation water to forest stands as a form of sewage treatment. Because of the dangers of nitrate leaching to deep groundwater, in which it poses health threats to water supplies, it is important to be able to estimate how much water trees will transpire in a year if irrigated.

10.7 SNOWMELT

Another important application of the energy balance in forests is snowmelt. Snowmelt is important in rangeland and agricultural systems as well, but most research on snowmelt has focused on forestlands because forests can be manipulated to change timing and yield of water from snowpacks. Many forested areas are important as reservoirs of snow for spring and summer runoff used to irrigate lower elevation rangelands. Consequently, much effort has gone into determining how forests can be manipulated to augment flows or delay melt later into the summer. About 70% of the precipitation in the western U.S. falls as snow, making an understanding of snowmelt very important in managing western water resources. A quite thorough work on the subject of snow accumulation and melt (Hathaway et al., 1956) focused on forecasting flow rates of major stream systems for hydroelectric reservoir regulation. More recently, the effects of forest harvest on peak flow of stream systems that rely on snowmelt have focused attention on modeling rapid melt rates that occur over short timescales.

While many of the studies focusing on water supply concluded that the primary effect of forests on snowmelt was due to redistribution of snow, a general consideration of the energy balance was shown to be important in determining watershed or basinwide melt rates. Studies of rapid melt rates have focused on the energy balance (Marks and Dozier, 1992).

All of the terms in Equation 10.2 are important in determining melt from a snowpack. During the spring, the components of the energy balance for a snowpack are incoming solar radiation during the day, outgoing long-wave radiation during the night, latent heat of sublimation added to the snowpack as vapor condenses on the snowpack under certain weather conditions, and added sensible heat from the air, rainwater, and ground heat flux. These components all act to change the energy content of the snowpack. Any energy added above that necessary to bring the pack to 0°C in the frozen state will melt the snow.

Solar radiation during daylight hours is dramatically affected by trees. Shading from trees causes a reduction in short-wave energy reaching the snowpack and is a function of tree density and crown height and whether the tree is coniferous or deciduous.

Trees, like snow, are near-perfect black bodies with regard to long-wave radiation; that is, they absorb almost all long-wave radiation and emit strongly in the long-wave band. Heating of the forest canopy by solar radiation, therefore, increases the long-wave radiation output by the trees. Under a forest canopy, the long-wave radiation into and out of the snowpack is well balanced, or even positive, whereas in an open area, the long-wave energy balance is negative, with a net outflow of energy.

Sensible heat is an important factor, but it is actually a relatively small part of the energy budget until late spring, when heat conduction from bare earth heated by radiation speeds late season melt. Air temperatures are typically below freezing until spring, at which time there is some warming. Sensible heat is transferred most effectively by wind. When air is still, a thin layer of cold air will develop directly over the snow, insulating it from the warmer air above. Turbulence created by wind breaks up insulating layers.

In many cases, an energy balance approach is not justified due to time, input, or computational restrictions or due to desired accuracy. Temperature index snowmelt models have been the most popular for most basin modeling approaches. An example is the degree–day method (U.S. Army Corps of Engineers, 1960), which is written as

$$M = K(T_a - T_b) \qquad (10.6)$$

where M is melt for the day (in units of depth of water, usually inches), K is the degree–day factor (units of depth divided by temperature), T_a is the average air temperature, taken as the average of the minimum and maximum temperature for the day, and T_b is the base temperature, usually taken as the melting temperature of snow, 0°C or 32°F. The idea is that there is no melt for temperatures below freezing, and that melt is directly proportional to the number of degrees above freezing. A variety of values for K and T_b must be determined empirically for each general area (Hathaway et al., 1956). For forested areas, K is on the order of 0.015 in./°F, while for clear areas, K varies from 0.04 to 0.11 in./°F. Typically, the degree–day method works best during overcast periods.

A more detailed degree–day approach that takes into account differences in solar radiation caused by slope–aspect, albedo A, and rain melt was presented by Riley et al. (1973):

$$M = k \frac{R_p}{R_h}(T_a - T_b)(1 - A) + 0.00125 P_r T_a \quad (10.7)$$

where R_p is the solar radiation on a sloping surface, R_h is the solar radiation on a horizontal surface, and P_r is the rainfall (centimeters). Albedo is a measure of the reflectance of a surface. Fresh snow typically has an albedo greater than 0.8. As the snow ages and becomes denser, the albedo can fall to near 0.4.

EXAMPLE 10.4

At the Central Sierra Snow Laboratory, the degree–day factor K for April was 0.036 in./degree–day above base temperature, and the base temperature T_b was 26°F. The Weather Service forecasts an average temperature of 40°F on April 10. Find the average depth of snowmelt expected over a 5000-acre basin.

Solution

$$M = K(T_a - T_b)$$

$$K = 0.036 \text{ in./}°F$$

$$T_b = 26°F$$

$$T_a = 40°F$$

$$M = 0.036 \text{ in./}°F \ (40 \text{ to } 26°F) = 0.50 \text{ in.}$$

Answer

On April 10, 0.50 in. of melt water is expected.

The primary reason that temperature index snowmelt models are popular is because temperature is often one of the only reliable and consistently available weather variables measured at weather stations. Radiation measurements are still uncommon. Also, in spite of the fact that they ignore an explicit representation of the variety of processes affecting snowmelt, a surprising number of variables correlate to air temperature. Days of melt with high solar input are often warm, and warmer air holds more moisture for condensation than cold air. For these reasons, these models have worked reasonably well for springtime melt estimates over large basins.

The latent heat of sublimation can provide a strong positive input of energy during periods of snowmelt. The effect has been somewhat misnamed as *rain on snow* because, during rain events, the moisture content of the air is the highest, and the transfer of latent heat to the snowpack is the highest. The wind often associated with rainstorms helps carry warm, moisture-laden air to the snowpack. As the air flows over the snowpack, the moisture condenses onto the snow and releases heat to the snow. This heat is conducted into the rest of the snowpack through ice, melt water, and rainwater. Sensible heat carried by rainwater is relatively low because rainfall temperature is usually close to freezing during these events. The volume of simultaneously melting snow and falling rain can be much greater than when either event happens alone, however, making such events hydrologically important. Under trees, the wind speed is lower, and there is less turbulence, causing less movement of moist air onto the snowpack and, consequently, a lower snowmelt rate.

10.7.1 SNOWMELT MODELING OVER A WATERSHED

To model accurately the snow distribution and melt over a large watershed with extensive differences in elevation, both air temperature and precipitation should be spatially distributed.

Ideally, in a watershed air temperature and precipitation measurements will be available at more than one location, preferably at low and high elevation points. In this case, it is common to distribute both the temperature and the precipitation linearly by elevation. If only one weather station is present, it is common to assume the temperature decreases with elevation according to the lapse rate of −3.5°F per 1000 ft.

One technique to simulate the snowmelt over a watershed with significant differences in elevation is to use a lumped approach. In this approach, a hypsometric curve of the watersheds is generated (elevation vs. percentage area). Then, the watershed is broken into bands representing a fixed range of elevations. Snow accumulation and melt are calculated for each band, and the combined response from each band is the total response of the watershed.

In most applications, GIS models are now used to simulate the snow accumulation and melt process over a watershed more precisely. In GIS models, it is also much easier to account for effects of vegetation on snow accumulation and melt.

10.7.2 SNOW COVER DISTRIBUTION

One of the greatest challenges in snowmelt modeling today is distributing the snowfall in a landscape. The drifting of snow can be seen across the Palouse region, especially due to the rolling topography. Much of the early research in snow drifting focused on the design of snow fences to minimize drifting across major highways (Tabler, 1975).

Snow distribution also has been studied in forested landscapes. Small forest openings (<20 times the height of surrounding trees) tend to have greater accumulation than the surrounding forest, while large forest openings tend to have less snow accumulation caused by higher wind speeds.

10.7.3 CRITICAL TEMPERATURE FOR RAIN–SNOW TRANSITION

A typical situation is a public record of total precipitation without specification of the form of precipitation. For modeling purposes, one requires the input of water into the soil. So, one of the most critical decisions the modeler must make is whether precipitation is rain or snow. A mistake in this decision will result in erroneous snowpack water balance and will cause the model to predict a groundwater recharge or streamflow event when none occurred or to fail to predict an event that did occur (Dingman, 1994).

The most natural criterion for the decision is the air temperature at the site of interest. We know that air temperature usually decreases with altitude (lapse rate), but we usually have information about surface temperature. It is not, therefore, immediately obvious what the temperature criterion should be.

Auer (1974) suggested that rain is virtually never recorded when temperature is less than 0°C, and snow is never observed when temperature exceeds 6°C. Probabilities of rain and snow are equal when temperature equals 2.5°C (see Dingman, 1994, Figure 4-14, p. 101).

Another study by Murray (1952) showed that surface temperature at about 4 ft high is as reliable as any other of the variables tested for differentiating between rain and snow. Some 2400 occurrences of precipitation at air temperatures ranging from 29°F to 40°F were analyzed to obtain the distribution of occurrences of rain, snow, and mixed rain and snow. The following table gives the percentage occurrences of both forms of precipitation for the various surface temperatures:

	Surface Air Temperature, °F											
Form	**29**	**30**	**31**	**32**	**33**	**34**	**35**	**36**	**37**	**38**	**39**	**40**
Snow	99	99	97	93	74	44	32	29	8	8		3
Rain		1	2	3	12	31	51	57	81	90	100	97
Mix	1		1	4	14	25	17	14	11	2		
Occurrences	281	419	304	459	229	191	154	94	74	79	56	79

The data used for the analysis by Murray (1952) represent a variety of elevation and topography settings. Accordingly, they are believed to be applicable generally. The differentiation between rain and snow can be estimated from surface temperatures by assuming rain will occur whenever the air temperature is 35°F or higher, and snow will occur whenever the air temperature is less than 35°F.

The partition of precipitation P into rain or snow can be based on air temperature as follows:

$$P_s = P, \quad T_a \le T_{min}$$

$$P_s = P \times (T_{max} - T_a)/(T_{max} - T_{min}) \qquad T_{min} < T_a < T_{max}$$

$$P_s = 0, \quad T_a \ge T_{max}$$

$$P_r = P - P_s$$

where P_r and P_s are the water equivalent depths of rain and snow, respectively; T_{min} is a threshold temperature below which all precipitation is in the form of snow; and T_{max} is a threshold temperature above which all precipitation is rain. Between the threshold temperatures (typically, −1.1°C and 3.3°C), precipitation is assumed a mix of rain and snow.

10.8 INFILTRATION

Free water available at the soil surface, whether due to rainfall or snowmelt, infiltrates into the ground. The same basic principles apply in forests as in agricultural and rangeland conditions (see Chapter 3). Forest soils are often highly conductive. **Table 10.2** shows the respective range of hydraulic conductivity values for typical forest and agricultural soils.

The reasons for higher hydraulic conductivity relate primarily to the presence of vegetation (Chapter 3). Leaves, needles, and roots form a mat of litter on the forest floor. This layer of organic matter is often highly conductive and contributes to the conductivity of lower layers as it decomposes and is incorporated into the mineral soil by earthworms and small burrowing animals. The roots create macropores through these shallow layers and into deeper layers. The litter, roots, and forest canopy all protect the soil from damaging raindrop splash and prevent the formation of the surface crusts that can be important in the hydrologic behavior of agricultural soils.

Infiltration capacities of the native forest soil can be dramatically altered by forest management practices. Compaction by machinery is one of the more obvious ways by which forest soils can be altered, with road construction in forests an extreme example. The removal of litter layers by fire and machinery also works to retard movement of water into the soil. Fire can sometimes lead to *hydrophobicity*, a form of water repellency in soil that can prevent infiltration for some period of time (Chapter 3). Collectively, most of these areas of reduced infiltration capacity are generally a small proportion of most watersheds and so usually create little concern for increased peak flows on a large scale. Many examples exist of decreased infiltration in a small area due to road construction leading to erosion, landslides, and gullying, showing that reduced infiltration is important at a local scale.

10.9 SUBSURFACE FLOW (INTERFLOW)

Because hydraulic conductivities of forest soils are so high, subsurface flow is the primary form of water movement in forested areas (see Section 5.2.1, Chapter 5). Water in forest soils flows in two modes: saturated and unsaturated. Unsaturated subsurface flow occurs in damp soils, where pore spaces are only partially filled, and the water is often not continuous between pores. Saturated flow occurs through soils that are fully saturated. Saturated subsurface flow occurs where a layer with low hydraulic conductivity, such as bedrock, underlies a more conductive region, such as a forest soil. Saturated subsurface flow is usually much slower than overland flow, so it is still not an important contributor to the storm hydrograph except when it prevents infiltration in particular areas, discussed in the next section.

TABLE 10.2
Hydraulic Conductivity Values for Surface Treatments of Forest Soils on Two Soils and Hydraulic Conductivity Values for Several Agricultural Soils

Soil Name or Treatment	Texture	Hydraulic Conductivity (mm/h)	Hydraulic Conductivity (in./h)
Forest Soils			
High-severity burn	Sandy loam	76.90	3.03
Low-severity burn	Sandy loam	80.90	3.19
Root mat removed	Sandy loam	77.83	3.06
Light-use skid trail	Sandy loam	81.73	3.22
Heavy-use skid trail	Sandy loam	58.41	2.30
Moderate-severity burn	Silt loam	35.96	1.42
Light-use skid trail	Slit loam	50.88	2.00
Forest road	Any	<4	<0.2
Agricultural Soils			
Hersh	Sandy loam	58.27	2.29
Sverdrup	Sandy loam	29.56	1.16
Whitney	Sandy loam	21.87	0.86
Williams	Loam	18.12	0.71
Academy	Loam	14.01	0.55
Barnes	Loam	10.84	0.43
Zahl	Loam	9.52	0.37
Barnes	Loam	9.15	0.36
Woodward	Silt loam	29.55	1.16
Keith	Silt loam	28.67	1.13
Portneuf	Silt loam	2.55	0.10
Nansene	Silt loam	1.73	0.07
Palouse	Silt loam	1.19	0.05

To discuss lateral subsurface flow in forests, it is easier to use the basic equation relating to water movement in soil, Darcy's law:

$$V = -K \frac{\partial h}{\partial L} \qquad (10.8)$$

where V is the average velocity over the cross section of soil considered, K is the hydraulic conductivity, and $\partial h/\partial L$ is the change in head per unit distance in the direction of flow. The negative sign shows that water flows from places with high head to places with low head. Multiplying V by the depth of the soil gives the flow per unit width of hillslope considered. Saturated flow responds most to slopes in the water table, whereas unsaturated flow can respond to differences in head caused by drier soil in one area than in another. For shallow saturated subsurface flow, as may be found on steep mountain slopes, $\partial h/\partial L$ is sometimes taken as the slope of the ground.

In forest soils, hydraulic conductivity is not uniform in direction. Leaves fall with their long axes parallel to the ground, and more area is covered in roots spreading radially from trees, again, parallel to the ground, so the hydraulic conductivity in a direction parallel to the ground surface is greater than in the direction perpendicular to the ground. This is a condition called *anisotropy*.

Forests often occur on steeper slopes than agricultural land. The steeper slopes lead to higher gradients and, as indicated by Darcy's law, faster flow. In mountainous areas, these steeper slopes often have young, shallow soils. Because they are young, their clay content is typically low, and the saturated hydraulic conductivity is high. The shallowness creates a system of temporary perched water tables during the rainy season, with no connection to a deeper regional groundwater system. By the late dry season, many of these water tables no longer exist, with only unsaturated soils remaining. This effect is due more to drainage than to evapotranspiration.

In steep areas with shallow soil, the greatest groundwater depths are found in swales, sometimes called *bedrock hollows*. Groundwater will flow perpendicular to contours and will be concentrated in these topographic depressions. These areas are the most vulnerable to landsliding because of the generally higher pore water pressures that can exist.

Forest management can affect interflow substantially. The most important effect is that forest road construction often intersects temporary groundwater tables. In these cases, the road system can act like an agricultural drainage system in a field by providing faster paths for interflow to reach streams. Roads can also add significantly to groundwater depths when culverts concentrate water from a large area into a small area. Forest roads commonly have ditches to carry water from the cutslope seepage and road runoff. The ditches have ditch relief culverts spaced periodically down the road for drainage and often discharge to hillslopes and not to streams. Very often, all of the water infiltrates, and the groundwater level is raised locally by this concentrating effect. The increased groundwater level increases the risk of landsliding.

Forests have deep root systems and can withdraw a great deal of water from groundwater. In areas where water supply is short, the trees compete with other potential uses of the water. In areas with excess water, such as in many parts of the Mississippi Delta area, when trees are removed, groundwater levels rise, and the harvested area becomes a swamp. Some claim that similar benefits are important in areas with potential for mass wasting. In general, this is seldom the case. Areas that are prone to landsliding are steep, and water moves out by saturated and unsaturated flow much faster than it is evapotranspirated. Also, the depth of flow necessary for landsliding occurs over a very short time period during rainstorms, and evapotranspiration will not affect this peak.

One exception to this rule is on geologic formations called *earthflows* (see Section 10.12.4 on mass movement). They are most common in volcanic mountain ranges with a high annual precipitation, such as the Cascades, and they also occur in the Siskiyou Mountains of northern California and in southern Oregon on ancient sea floor sediments. Earthflows are made up of fine-grained material that flows slowly down mountains like a glacier made of mud. At the toes, they commonly fail in large landslides into streams. Flow rates on earthflows have been tied to groundwater levels (Swanston et al., 1988).

10.10 SURFACE RUNOFF

Surface runoff is rare on most forest hillslopes. Particular conditions must exist for surface runoff to occur, and it is best to discuss the overland flow that occurs by the two primary processes separately. The first process, often called *Horton overland flow* (Chapter 5), occurs when the rainfall intensity exceeds the infiltration capacity of the soil. The second process, called *saturation overland flow*, occurs when the soil is so full of water that no additional water can infiltrate.

Circumstances leading to Horton overland flow include high rainfall intensities and compacted or otherwise low-infiltration soils. During a storm, the initial infiltration

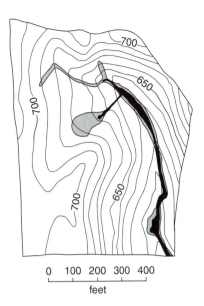

FIGURE 10.6 Expansion of saturated area in a basin during a single storm. The dark shaded area shows the saturated area at the beginning of the storm, and the lighter shading indicates the saturated area at the end of the storm. All precipitation falling in saturated areas is converted to overland flow. (After Dunne, T., and L.B. Leopold, *Water in Environmental Planning*, W.H. Freeman, San Francisco, 1978. With permission.)

capacity is quite high and decreases to a steady-state infiltration rate over time, as described in Chapter 3. If the rainfall intensity at any time exceeds the infiltration capacity, there will be rainfall excess. If enough rainfall excess occurs to fill surface depressions, runoff will occur. Summer thunderstorms are the most common source of high-intensity rainfall for Horton overland flow. Almost all areas of the U.S. experience thunderstorms, but they are less frequent on the West Coast than in other areas. High mountainous areas and the southeastern U.S. are particularly subject to these events (see Chapter 2).

Most forest soils have high infiltration capacities and, as noted, do not produce Horton overland flow unless they have been disturbed, even during these extreme events. Saturation overland flow is more common in forests, particularly where there are shallow soils. Saturation overland flow is linked to a concept called *variable source area* (Hewlett and Hibbert, 1967; Hewlett, 1982; see Chapter 5), which notes that the pieces of the watershed contributing to the stream hydrograph are spread throughout the watershed near the stream network. The stream network and contributing areas of surface runoff expand as the storm continues in response to increasing depth of interflow in these source areas (**Figure 10.6**). Dunne and Black (1970) expanded on Hewlett and Hibbert's point by describing the process involved in the creation of these source areas and how they contribute to the stormflow.

When looking at a hillslope in profile at a particular point during a storm, the depth of saturated soil is greater

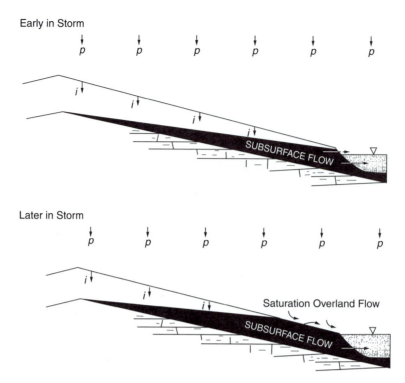

FIGURE 10.7 Changes in depth of subsurface flow leading to saturation overland flow; *p* refers to precipitation and *i* to infiltration.

at the bottom of the slope than at the top (**Figure 10.7**; compare to **Figure 5.3**). This follows from the fact that the watershed near the top of the slope is much smaller than the watershed for the bottom of the slope; for the same reasons that a river gets bigger downstream, the volume of water flowing through the hillslope must increase as one goes downslope (i.e., water accumulates). Now, consider the discussion of Darcy's law. Flow velocity in groundwater is a function of slope and soil properties only, and the flow does not get faster as it gets deeper. So, the velocities of the water at the top and bottom of the slope in **Figure 10.7** are the same; therefore, to pass the greater volume, the groundwater depth must be greater at the bottom of the slope. When it rains hard enough and long enough, there will be more volume coming from above than can pass laterally through the soil, and some must seep onto the surface to flow overland. In addition, no additional water can enter, so that both the water seeping out and the rainfall will flow overland.

Several situations promote surface runoff by the saturation overland flow process. Concave slopes are the most susceptible because water concentrates in the lowest part. Road cuts are also susceptible to saturation overland flow. Road cuts cause an abrupt shallowing of the soil, and the excess interflow that cannot pass under the road prism must seep out on the cutslope. Removing a duff layer can change an area from one that never has overland flow to an area that will have saturation overland flow. Removal of the duff layer seems like an innocuous shallowing of the soil, but considering that hydraulic conductivities of duff

layers can be 100 times those of the soil below, removal of the duff layer can represent a significant loss of interflow capacity for the soil.

10.11 STREAMFLOW AND WATERSHED HYDROLOGY

Many questions raised in forest hydrology focus on the watershed scale, as opposed to the hillslope scale described above. However, understanding how water behaves at a point and on hillslopes makes it easier to understand how the watershed acts as a whole. The two main areas of concern usually discussed with regard to watershed hydrology in forests are floods and water supply.

The controversy of how well forests protect against floods has existed for the better part of a century. The Organic Act of the National Forest Service states that part of the reason for the forests is to "secure favorable conditions of flow." In general, it is agreed that forest cover has little effect on flood size for large stream systems, but is very important for small watersheds (Lull and Reinhart, 1972). The debate continues, however.

Originally, it was believed that the forest canopy and litter layer prevented quick surface runoff; thus, water soaked into the spongelike forest only to seep out slowly later. As understanding increased about the role of interflow and saturation overland flow, it was found that storms that caused floods on major rivers tend to have long durations and low intensities. While interception and soil storage manage to delay runoff somewhat, the amount of water intercepted and the amount of water detained are small in comparison to total storm volume (Lull and Reinhart, 1972). Even though the only runoff-producing process in low-intensity storms such as this is interflow discharging to streams, the long time span allows many parts of a watershed to contribute. When the Ohio River floods, the Mississippi downstream can handle the increased flow if no other streams are flooding, but if all of the tributaries are having minor floods, the mainstem of the Mississippi may experience a major flood.

On small watersheds, forests can reduce flooding. Thunderstorms with high intensities, covering small areas and lasting a short time, can create severe flooding in streams with other land uses. Storms with high intensities can create Horton overland flow on lands used for agriculture and grazing, but interception and infiltration capacities of forests are great enough that overland flow will not occur even in a thunderstorm. Pathways for stormflow in forests are slower than in agricultural, range, and urban lands, and streams respond only to long-duration events.

Likewise, the ability of forests to reduce stormflow, as compared to, say, cropland (where Horton overland flow usually occurs), is related to the size of the storm. Smaller storms that occur more frequently can often be largely abstracted by the water storage potential of forest soils. However, a very large storm, say the 100-year, 24-h storm, can deliver enough water to saturate the soils of any forest, so that the difference between forest and cropland would be much less for the 100-year storm than for, say, the 1-year storm. In this respect, forests are similar to their polar opposite, urban areas. One might generalize by saying that different land uses affect stormflow from small storms more than from large storms.

In some parts of the U.S., removal of forest canopy can increase the potential for flooding at larger scales. Snowmelt by latent heat transfer during long, low-intensity rain events in the Pacific Northwest can augment a currently occurring rainstorm, creating a situation with sustained high water input rates. Early evidence showed that small watersheds (100 acres) and even medium-sized (50,000-acre) watersheds showed an increase in peak flows following removal of trees from a portion of the watershed. Berris and Harr (1987) showed that the removal of the forest canopy decreased interception of snow and increased turbulent transfer of latent heat to the snowpack. There is still some question about the magnitude of the effect at large scales.

The effect of forests on water yield is much less controversial. Trees consume water. Research to examine how to augment snowpacks and reduce canopy interception and evaporation of snow showed that changed forest-cutting patterns serve primarily to influence how snow is redistributed by the wind and little to change how much evaporates. Researchers were also able to influence the timing of the melt to some extent, but timing still depended much more on weather than on harvest practices. The only effective means to increase water yield from a forest is to prevent transpiration, and some water supply organizations have done so. The Los Angeles Department of Water and Power, for example, has removed all riparian forests from the Owens River near Bishop, CA, in an effort to increase the water yield to their aqueduct. Such practices are rare because water quality is often as important as the amount available, and forest removal promotes erosion.

To understand the effects of afforestation or deforestation on water yields, Bosch and Hewlett (1982) assembled all the world's experimental data. While they were reluctant to draw hard statistical inferences from this large body of data, Bosch and Hewlett suggested that if an entire forest was cut, the increase of water yield would be about 300 mm or about 1 ft over the surface area. Thus, cutting a forest should provide about 1 acre-ft of water per acre per year. Afforesting an area should have the opposite effect. They warned, however, that deforesting or afforesting less than 20% of an area would probably have "no detectable effect" on water yield. Also, Hewlett (1971) had earlier warned that theory and experiments about forests and water yields would never convince the public, and water managers in particular, about the role of forests on water yield. Such conclusions would have to be "demonstrated on a scale appropriate to the management problem; i.e., on a drainage basin large enough to serve as a primary water supply to a community or industry" (Bosch amd Hewlett, 1982).

As mentioned, large expanses of former cropland in the eastern U.S. have reverted to forest over the past century (**Figure 10.8**). If forests can, indeed, affect water yields on a large, populated landscape, the phenomenon should be detectable here. One opportunity to capture these historical and hydrological processes was on the Southern Piedmont, where 10 large watersheds were identified with both land use and hydrologic records for long periods (**Figure 10.9**). Between about 1900 and 1975, from 9.7 to 27.5% of these basins were reforested, and all experienced significantly reduced annual water yields (depth of runoff), ranging from 2.5 to 9.9 cm. The effect is demonstrated by the Oconee River at Greensboro, GA, which had 21.3% of its area reforested, reducing water yield by 9.9 cm. While precipitation increased by 7.8% during the period of measurement, runoff decreased by 12.2%, meaning that AET increased by 21.6% (**Figure 10.10**; recall that, for long periods, ΔS becomes insignificant, so that $AET = P - Q$). The regression analysis (**Figure 10.11**) shows the model (regression line) for each period. The difference between them reveals the difference in runoff for a given annual precipitation value. Note that the regression lines are widely separated at low P values, but tend to converge at higher values. This means that the maximum reductions are during drought years, when water yields would have been low already. In other

FIGURE 10.8 A sea of forest in the Southern Piedmont, Putnam County, Georgia, 1970. The same view about 1900 would have been cotton and corn as far as the eye could see. Such reversion of former cropland to forest has been common in the eastern U.S. during the past century (see, e.g., Foster and O'Keefe, 2000).

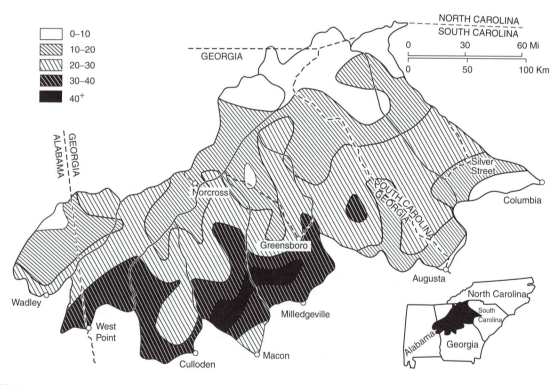

FIGURE 10.9 Percentage of area reforested, 10 large stream basins of the Southern Piedmont, 1919 to 1967. (From Trimble, S.W., F. Weirich, and B.L. Hoag, *Water Resour. Res.,* 23:425–437, 1987. With permission.)

words, streamflow droughts can be exacerbated by forests. This is easily explained by the fact that trees have much deeper rooting systems than the crops they replace, so they can reach farther into soil water or even into groundwater.

In wet years, conversely, there is ample soil water for all plants, and even the soil is wetted more often, meaning that both cropland and forest AET approach PET. This was the pattern for all basins except those with large

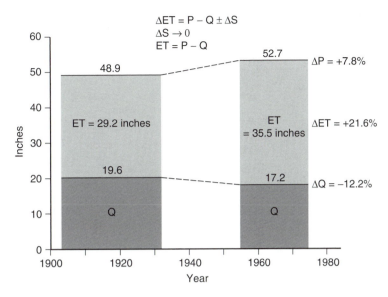

$$\Delta ET = P - Q \pm \Delta S$$
$$\Delta S \to 0$$
$$ET = P - Q$$

FIGURE 10.10 Change of ΔET and Q as the result of reforestation, Oconee River at Greensboro, GA. Note that while precipitation increased, runoff decreased, meaning that ΔET increased greatly. While storage changes are important to the water budget for a year or so, the significance decreases over the decadal timescale. (Data from Trimble et al., 1987.)

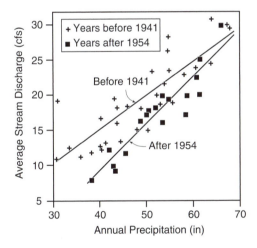

FIGURE 10.11 Oconee River at Greensboro, GA. Regression lines show the relation of runoff to precipitation for the earlier period of cropland and the later period after an additional 21% of the basin had been reforested. The net effect is minor in wet years, but low-streamflow droughts are exacerbated in dry years. (Data from Trimble et al., 1987.)

reservoirs, which could supplement streamflow during drought periods.

The next step was to combine the data from the 10 large basins with all the experimental data (50 basins) assembled by Bosch and Hewlett (1982) (**Figure 10.12**). This regression model showed that change of forest cover explained 50% of the variance of water yield from all the basins and indicated that afforesting or deforesting 100% of a basin will change the water yield by 326 mm or about 13 in. Since the relationship is linear, one can use the model to estimate the proportion of basin to be cut or afforested to obtain whatever water demands are required. The addition of the 10 Piedmont basins clearly showed that changing only small portions (less than 20%) of basins can significantly change water yield.

EXAMPLE 10.5

If 40% of a basin is afforested, how much might water yield (basin depth) be expected to decline?

Solution

See **Figure 10.12**.

$$Y = 3.26X \text{ mm}$$

$$Y = 3.26(.4) \text{ mm} = 130.4 \text{ mm}$$

Answer

Afforesting 40% of the basin will reduce water yield about 130 mm over the area of the basin.

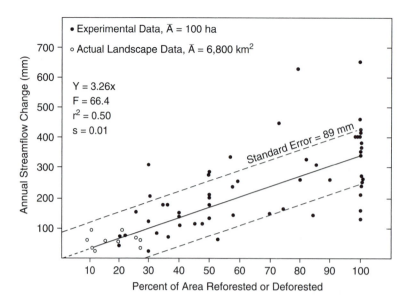

FIGURE 10.12 Regression model to predict the effect of forest change in a watershed. (Redrawn from Trimble, S.W., F. Weirich, and B.L. Hoag, *Water Resour. Res.,* 23:425–437, 1987. See also Calder, I.R., Hydrologic effects of land use, in *Handbook of Hydrology,* D. Maidment, Ed., McGraw-Hill, New York, 1993, pp. 13.18.) (Experimental data, A ≅ 100 ha from Bosch, J.M., and J.D. Hewlett, *Hydrology,* 55:2–23, 1982. Actual landscape data, A = 6800 km² from Trimble, S.W., F. Weirich, and B.L. Hoag, *Water Resour. Res.,* 23:425–437, 1987.)

EXAMPLE 10.6

You are managing a 5000-acre forested watershed for a large city. The city water manager informs you that, for the next few years, he will need an additional flow of 2 ft³/sec from your forest. How much forest do you need to cut?

Solution

First, determine your total annual water requirement:

2 ft³/sec (86,400 sec/day)(365.25 days/year) = 1,449 acre-ft

Next, calculate the average annual depth this would be for the basin:

1449AF/5000 acres = 0.289 ft = 3.38 in. = 88 mm

Then, use the regression equation to solve for *X*:

$$Y = 3.26X\%$$

$$88 = 3.26\ X$$

$$X = 27\%\ \text{of the basin}$$

$$5000\ \text{acres}\ (27\%) = 1350\ \text{acres}$$

Answer

1350 acres of the 5000-acre municipal forested watershed must be cut.

10.12 EROSION — SEDIMENT BUDGET

Erosion in forestlands is one of the key concerns of forest managers. The discussion on hydrology points out that, on a watershed scale, the forestland manager can cause only modest changes in water yield. It is, however, easy to increase sediment yields to a much greater extent with careless management. The *sediment budget* of a forest stream is the combined sources of sediment less the change in sediment storage along the channel, which must equal the amount of sediment transported out of the watershed. (See Section 9.11.) When estimating the sediment budget of a forest, three distinct upland sediment sources must be considered: sheet-and-rill erosion, gully erosion, and mass wasting. In addition, many hydrologists consider channel erosion as part of the sediment budget. Channel erosion is more properly considered a change in channel storage when considering basinwide sediment budgets.

10.12.1 SHEET-AND-RILL EROSION

Sheet-and-rill erosion is commonly seen on agricultural, urban, and rangelands. This form of erosion shows signs of minor soil displacement and rilling on the surface of the soil as sediment is removed. Sheet-and-rill erosion occurs only under conditions of surface runoff characterized by shallow overland flow, as opposed to the channelized flow of gully and channel erosion. Surface runoff in the form necessary for sheet-and-rill erosion is rare in forest environments. It is confined primarily to road surfaces, but may occasionally be seen on harvest units that have had severe burns, extensive tractor skidding, or mechanical site preparation. Although the universal soil loss equation was never intended for forests (no Horton overland flow), the *C* factor of that model was extended for forest environments by Dissmeyer and Foster (1981). They suggested changes to the ground cover, canopy, soil consolidation, high organic content, fine root, residual binding effect, depression storage, step, and contour sub-factors to calculate the *C* factor.

10.12.2 GULLY EROSION

Gully erosion is dominated by concentrated flow, and con-tributions of sediment from interrill areas are generally insignificant compared to the amount of material removed from the gully headwall and banks. Gullies are often asso-ciated with roads, skid trails, and grazing. These distur-bances create and concentrate overland flow. When the concentrated flow is directed to an area susceptible to erosion, a gully may be started that will propagate both up- and downstream. Where gullies occur, they can pro-vide orders of magnitude more sediment than sheet-and-rill erosion. There are no practical methods to predict gully erosion volumes because most gullies occur unexpectedly. Sheet-and-rill erosion is often accepted as the "price of doing business," but managers work hard to prevent gullies because they have a high potential for damaging stream ecosystems and constructed facilities, such as roads. Once formed, gullies are difficult to eradicate or even control.

Gully erosion is common around forest roads with poorly designed or maintained drainage systems. Nor-mally, operating road ditches and culverts concentrate flow only to the degree that ditch materials can safely handle the water, and the water is placed downslope of the road only in areas that will not form gullies. Riprap and other stabilizing materials are commonly placed at culvert discharges to prevent gully formation, and culverts are extended down long hillslope embankments to prevent gully erosion in the highly erodible road fill material. When maintenance fails to keep culverts open and ditches clean, excessive flows can build up and be directed to areas where the concentrated flow will be detrimental. Also, all culverts have a limited lifetime and will eventually fail by rusting through the bottom, by plugging with branches and sediment, or by overtopping by a large storm. When culverts fail, large gullies can be formed.

10.12.3 CHANNEL EROSION

Channel erosion is another important process in forest-lands. Channel erosion refers to any erosion in channels formed by water. It is a step up in scale from rill and gully erosion, but the underlying process, removal of sediment by concentrated flow, is similar. Channel erosion can be considered in two distinct forms. Channel erosion in steep

stream reaches, where most of the geologic processes are erosional, is very similar to gully and rill erosion, perhaps differing only in scale. Channel erosion in depositional reaches of streams, typically where the gradient is lower and the valley is wider, is somewhat of a misnomer, and it is better described as a change in the amount of sediment in storage. Sediment transport takes place in all channels containing nonconsolidated materials, like gravel, sand, or mud, and erosion of riverbank materials is usually a reentrainment of material temporarily stored by the river. So, why is this form of channel erosion called *erosion*? This is probably because valuable land lies along the sides of larger streams, and any removal of stored soil is viewed as detrimental. In addition, it visibly scars stream banks, and the impacts to water quality from so-called channel erosion may be as great as from any other form of erosion.

In steep stream reaches, where most detached soil is removed from the site, channel erosion shows up either as an enlargement of the stream channel or as the creation of a new channel. This process is often caused by a change in the local hydrology that increases the peak flow. This change can come about by natural events or management. Wildfires and unusually intense storms are typical natural causes, and road construction, forest cutting, and pre-scribed burning are common management causes. Earlier sections described how the hydrology is changed by these events and practices. This type of erosion is common in areas with high road densities.

In depositional reaches, channel erosion is often caused by regular meanders of the stream. Stream mean-ders that cut into stream banks are a normal and natural process. The erosion is often balanced by the deposition of a point bar on the bank opposite the one eroded. Unless the channel is enlarging or removing particularly valuable land, there is usually little reason to be concerned. It is important to note, however, that not all meanders actively erode. Some meander systems were formed under differ-ent climates than now exist, and the meander pattern and flows under the current climate are such that no migration of the meanders is expected.

Channel erosion is also caused by changes in the amount of sediment in transport from upstream reaches. Decreases in the sediment load from upstream mean that more material can be removed from the stream reach in question to fill the transport capacity of the stream (see section on sediment budgets in Chapter 9). This leads to downcutting of the channel and slumping of the banks as the old floodplains are abandoned. Increases in coarse sediment load from upstream can lead to channel widen-ing as the stream changes to accommodate the increased load (Sullivan et al., 1987). Changes in sediment load caused by forest management practices can have impacts on stream systems far downstream with this effect.

In forestlands, channel erosion is sometimes seen near where a tree has fallen into the water. This observation

TABLE 10.3
Classification of Slope Movement Types

| | | Type of Material | |
| | | Engineering Soils | |
Type of Movement	Bedrock	Predominantly Coarse	Predominantly Fine
Falls	Rock fall	Debris fall	Earth fall
Topples	Rock topple	Debris topple	Earth topple
Slide			
Rotational, few units	Rock slump	Debris slump	Earth slump
	Rock block slide	Debris block slide	Earth block slide
Translational, many units	Rock slide	Debris slide	Earth slide
Lateral spread	Rock spread	Debris spread	Earth spread
Flow	Rock flow (deep creep)	Debris flow (soil creep)	Earth flow (soil creep)
Complex	Combination of two or more principal types of movement		

Source: After Varnes, 1978, in: *Landslides: Analysis and Control,* National Academy of Science, Washington, D.C.

has been used as a reason to remove all trees from the stream banks in some places. Such a strategy sometimes backfires, as the importance of the tree roots in holding the banks in place can be far greater than the effect of trees redirecting flow. Trees and brush along streams can also act to slow water, reduce channel erosion, and provide fish habitat (see trees vs. grass, Chapter 6). This same slowing of the water increases depths of floodwaters, however, and can be a nuisance to people dwelling in the floodplain.

10.12.4 MASS WASTING

Mass wasting is the form of erosion by which soil is transported in a mass, such as a landslide. Mass wasting is the primary form of sediment delivery in many mountainous forestlands in the western U.S. In addition to the role as a major contributor to the sediment budget of forested basins, mass wasting events can deliver larger-size materials than are typically delivered with sheet-and-rill or gully erosion.

Mass wasting occurs in many forms. Some typical names for mass wasting phenomena include landslide, debris slide, debris flow, slump, and earth flow. These names represent specific types of movement in specific materials, as described by Varnes (1978) (**Table 10.3**). Of greatest concern in forest environments are the debris slide/debris avalanche/landslide movement type, the debris flow movement type, and the earth flow movement type. Drawings and descriptions for these three common mass movement types are in **Figure 10.13**. Debris slides are translational failures of soil common to steep slopes and are typically shallow, from 1 to a few meters thick. Debris flows occur when sufficient water is incorporated in the failed mass to liquefy the mass, as might happen when a debris slide moves into a steep headwater stream channel. Debris flows travel further and faster because of the increased liquid content. Earth flows occur as a slow,

glacierlike deformation of the soil, typically several meters thick. Earth flows are common in soils with high clay contents with a great deal of water available, such as the Franciscan Melange geology of the Siskiyou Mountains in California and Oregon and some types of volcanically derived soils that weather rapidly. Earth flows occur on lower angle slopes than do debris slides.

Water is an important contributor to all types of slope instability. It is important in terms of increasing the pore water pressure, decreasing particle contact within the soil, and adding to the weight that must be supported. Landslides typically occur in steep hollows. The shape of these swales helps concentrate the water, making the saturated soil depth greater than would exist on a plane or convex slope. Swanston et al. (1988) studied the effects of interflow fluctuations on earth flow movement and found it was an important predictor in the rate of movement of the mass. Force balance equations used by engineers for risk analysis of potential slide areas show the importance of water in increasing instability.

Land managers can strongly affect mass erosion within forested watersheds. Redistribution of water by forest roads is one of the greatest problems. Placement of ditch relief culverts can inadvertently concentrate water into an area likely to fail. Consultation by geotechnical engineers is therefore an important part of any road design project so that unstable areas can be identified and avoided. Additional problems can be created on side roads because of oversteepening. Proper compaction of the road fill mitigates this problem. Removal of trees decreases the amount of cohesion given to the soil by fine roots (Burroughs and Thomas, 1977). The intertwining of root systems between trees forms a fibrous network that will hold unstable patches of soil in place. Roots begin to decay as soon as a tree is cut, and the reinforcement to soil strength offered by the roots is lost quickly. Root systems of the

DEBRIS AVALANCHE, very rapid to extremely rapid

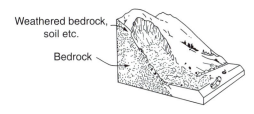

DEBRIS FLOW, very rapid

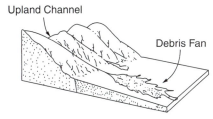

SLUMP-EARTH FLOW

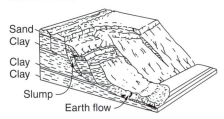

FIGURE 10.13 Descriptive drawings of three common forms of mass failure in forestlands. (After Varnes, D.J., in *Landslides: Analysis and Control*, Special Report 176, Transport. Res. Board, National Academy of Science, National Res. Council, Washington, DC, 1978, pp. 11–33. With permission.)

new stand may take several years to become as effective in holding the soil.

10.13   DEFORESTATION, FIRES, AND SILVICULTURE

Deforestation is one of the world's biggest land use problems. In many places, the previously forested areas quickly become unsuitable for any useful alternative land use. Ward has observed firsthand the devastation that slash-and-burn activities have caused in Africa and the barren moonscape appearance of mountainous areas on the west coast of Tasmania that have been denuded by the removal of trees to fuel mining operations and acid rain impacts. Let us look at why deforestation occurs, where it occurs, and why its impacts are so severe. Deforestation is occurring on most continents, but the most rapid rates are in South America, Asia, and Africa. The main causes of deforestation are associated with human activities associated with:

1. Conversion of woodlands and forests to agricultural land
2. Removal of trees as a source a fuel
3. Removal of trees as a source of building materials
4. Commercial logging

Typically, these activities occur due to human need or human greed. In many cases, they occur with the support or blessing of the government. In countries where some of these practices are not allowed, there is often inadequate government infrastructure to combat illegal activities.

In this section, we discuss silviculture activities that might have an impact on the hydrology of a forest, but do not result in a permanent loss of the forest. In Section 10.14, we discuss grazing activities that, if managed correctly, will also not result in a permanent loss of the forest.

10.13.1   DEFORESTATION ASSOCIATED WITH AGRICULTURE

A common technique used in tropical forest regions (Koeppen Af and Am) is called *slash-and-burn agriculture*. This approach is used by farmers to clear grasslands, brushlands, secondary forests, and even virgin forests. The size of the area impacted and the amount of cutting and clearing that might be performed can vary considerably, but a common element is the use of fire to remove unwanted vegetation. In some countries, this approach is an accepted practice to make grass and brushland areas suitable for grazing. Undesirable parasites are killed, and unpalatable grass and brush are temporarily replaced by soft, green, more succulent grasses. The impact of these practices on the ecology of grassland and brushland areas is unclear. Ward has observed African areas that are burned annually, yet seem to bounce back rapidly and are populated by a diverse array of wildlife. Clearly, some rodents and reptiles are destroyed by these fires, but many seem to obtain refuge below the ground. Also, most animals are able to escape the wrath of these cycles of fires. It is probable that, without these fires, the vegetation and ecology would be very different, but the same could be said of vast areas in the Midwest region of the U.S., where wetlands have been converted to some of nation's most productive agricultural land. A concern with slash-and-burn agriculture even on grassland and brushland areas is that the fires often spread to woodlands, secondary forests, and even virgin forests. This is a serious problem in South America, Asia, and Africa.

Slash-and-burn activities in forested areas are common in Asia and Africa and are usually associated with subsistence farming. Small forested areas are cleared and burned. Typically, these areas produce good crops the first year after clearing, but within a few years, they are unproductive, and the farmers move to a new area. The reason for the rapid decline in productivity is due to the poor acidic soils often found in the humid tropics and the high erosion rates associated with this practice.

In some parts of Asia and South America, such as Brazil, large tracts of forestland are often removed to support cattle ranching. This is a common practice in the Amazon basin and has been the focus of extensive international efforts to establish sustainable management policies for tropical forests in South America.

10.13.2 Deforestation Associated with Silviculture

Logging is probably still the main cause of deforestation. First, we should recognize that wood is a vital renewable resource that can be used for numerous purposes. Use of alternative materials to wood is not always viable, cost-effective, or more environmentally friendly. Therefore, plantations and regrowth areas where operators use best management practices (BMPs) and sound land stewardship practices are certainly an acceptable source of wood materials. Our concern should focus on virgin and old-growth forests and how to utilize these resources best. Numerous methods have been used to log these areas. However, whether we remove just a few trees at a time from an area (a practice called *selective cutting*) or we remove all the trees from an area (a practice called *clear-cutting*), there is still a need to get to the trees and then remove them. This usually results in the construction of a maze of logging roads that ultimately become a source of soil erosion, runoff, and environmental damage.

In North America, much of the logging of old-growth forests occurs in Canada and the Pacific Northwest. Trees that are logged are often more than 100 years old and can reach giant proportions. While many people might be concerned by these North American activities, their impact is small compared to the sometimes irreversible damage that is occurring in some tropical areas, usually in developing countries. Exotic hardwoods often command a price of $1 to $100 per square foot for boards that are ¼- to ¾-in. thick. Therefore, a single tree from Asia, Africa, or South America might be worth tens of thousands of dollars. Removal of old-growth trees is an attractive enterprise not only for large logging companies, but also for the small entrepreneur. Illegal removal of trees is as difficult to stop as poaching of animals. A simple way to view the issue is to recognize that even when best management practices are used, it is difficult to prevent adverse impacts of logging activities on the hydrology and ecology of these ecosystems. The goal therefore is to conduct these activities in a sustainable way and to keep adverse impacts within acceptable bounds. This is no different from the philosophy used to manage agricultural land. Unfortunately, in rain forests, it is virtually impossible to conduct these activities in a sustainable way. The downstream impacts quickly become severe, the logged areas become

denuded, and the land can become unsuitable for any future use.

10.13.3 Impacts of Deforestation

Arguably, all land use activities can be related to human desire, need, or greed. In developed countries, the main societal, political, and legal issues relating to land use activities revolve around changes associated with human desire. Where a land use activity or land use change is sustainable and perceived to be in the best interest of society, there might be only limited opposition. For example, we mentioned the conversion of wetland areas in the Midwestern U.S. to agricultural land. Centuries and only decades ago, there was little or no opposition to this practice. Today, the scarcity, value, and importance of wetlands are better understood, and there is considerable opposition to the conversion of a wetland to an alternative land use. However, today we are seeing large tracts of agricultural land converted to urban land uses. In decades to come, we may regret that large areas of agricultural land have been converted to urban areas.

Why does deforestation create so much interest? In many cases, it is associated with a human need to survive, such as subsistence farming or the use of wood as a fuel. Surely, as a society we should view this as no worse than converting a piece of land into a mall or, centuries ago, draining a wetland to prevent disease problems. Is it any worse or different from the case cited in Chapter 1 for which water tables in the Yellow River plain are rapidly falling because of the need to irrigate a second or third crop to feed a hungry nation? Some of us might argue that none of these practices are acceptable, but perhaps the difference with deforestation is that, while rain forests occupy just a few percent of the land area in the world, they are the home for the majority of all life on earth. In 1991, N. Myers reported that tropical rain forests covered 6% of the earth's surface and contained 70 to 90% of all the world's species. He also noted that more than 25% of pharmaceutical products come from tropical plants.

Forests occur in a wide range of geographic, climatic, and topographic regions. However, they often occur in areas that exhibit high precipitation for part or all of the year (see Chapter 2). Many are located in steep-sloping areas. Many are also located in areas where there is only a thin mantle of topsoil. As discussed in this chapter, the combination of a dense canopy, a lush understory, and an excellent ground cover of duff or bed litter helps to keep this fragile ecosystem in balance. Most of the precipitation is used by the vegetation in transpiration, enters the groundwater system, and is slowly released or has a delayed runoff response due to interflow. Furthermore, the various layers of vegetation provide protection from erosion due to rainfall impact. Removal of the forest removes the protective vegetated layers, reduces transpiration, and can greatly increase runoff and soil erosion. Often, runoff and erosion rates will be orders of magnitude higher than from a virgin forest condition. This causes downstream flooding and sedimentation of streams, lakes, and dams. Also, once the soil mantle is lost, there is virtually no opportunity to revegetate these areas, and they become stark barren areas that contribute little to any forms of life.

Forests store vast quantities of carbon. When they are cleared, this carbon is released as carbon dioxide to the atmosphere. Carbon dioxide is often called a *greenhouse gas* because it allows the entry of solar radiation into the system (in this case, the atmosphere), but reduces the escape of heat from the system (in this case, the escape of heat from the atmosphere to outer space). Increases in carbon dioxide levels are one of the main factors that contribute to global warming. An interesting discussion on this topic was presented by Lal et al. (1998). In their book, they evaluated the potential of U.S. cropland to sequester carbon and mitigate greenhouse effects.

10.13.4 WILDFIRES

We discussed the adverse impacts of fires that were set to clear forests for alternative land use activities. However, periodically, fires naturally or accidentally occur in forests. Prior to intensive habitation by humans, either within or near a forest, these fires roamed free and would be left to burn as much or little of the landscape as conditions dictated. While it might seem appropriate that effort be made to prevent accidental fires, the debate rages whether efforts should be made to control and extinguish forest fires. It is often unclear whether the great measures that are used to extinguish a fire are to protect humans and their property or to protect the forests themselves. Forest fires rarely destroy all the trees, and many tree species benefit from cycles of fires. Often, wildlife depend on a cycle of fires for a variety of reasons, including maintaining a diversity of young and old vegetation. Informative books on the role of wildfires were presented by Pyne (1982) and Arno and Allison-Bunnell (2002).

10.14 RANGELANDS AND GRAZING

10.14.1 INTRODUCTION

The grazing of domestic animals on pasture, rangeland, and forest is big business in the U.S. Between 1940 and 1990, the number of cattle in the U.S. increased 60%. In the West, the number more than doubled from about 26 million to 54 million. However, grazed area in the East decreased from about 160 million acres to about 87 million acres, indicating the generally better management of rural land discussed in Chapter 9. Cows and other grazing animals have important impacts on hydrology and geomorphology (Branson et al., 1981; Trimble and Mendel, 1995).

Grazing animals change the balance of nature because they increase the forces on the landscape and decrease the resistance of the landscape, resulting in more hydrologic action and geomorphic work. These processes occur both on the upland slopes and in stream channels. We examine the processes on both.

10.14.2 UPLAND SLOPES

Grazing animals greatly impact the hydrologic qualities of soil. It is commonly recognized that they mechanically compact the soil, but the effects are far more complex (**Figure 10.14**). The initial factor is the soil itself. Sandy soils with much organic content give more "cushion" to the soil and allow less hydrologic impact. Surface crusting reduces water intake, reducing biologic activity, especially endopedofaunal activity (see Chapter 3 for a more complete discussion). The mechanical compaction of the soil also reduces the faunal activity. The combination of surface crusting, mechanical compaction, and reduced hydraulic conductance of the soil from reduced faunal activity all reduce infiltration capacity and increase overland flow, thus increasing the force on the landscape (Trimble and Mendel, 1995). The removal of vegetation from grazing as well as the disturbance and baring of the surface all reduce the resistance to erosion. Thus, grazing, especially overgrazing, cause increased overland flow and erosion on slopes (**Figure 10.15**).

10.14.3 STREAM CHANNELS

Grazing animals tend to "lounge" in and around streams because of the shade and the availability of water. There, the effects are to break down the banks of the stream mechanically and to make it more vulnerable to erosion by stormflow (**Figure 10.16**). As discussed, there is more stormflow from grazed watersheds, so there is more force exerted along the stream channel. Because the resistance of the stream banks and channel has been reduced by removal of vegetation and trampling of banks and channels by hooves, stream channel erosion is more easily accomplished.

While the bad news is that grazing animals, especially cattle because of their numbers and ubiquity, can damage a stream channel, the good news is that highly damaged stream channels can be restored. Sometimes all it takes is fencing the animals out of the stream, and the steam restores itself over a period of time (**Figure 10.17**). In other cases, nature must be helped along, but the result can be beautiful (**Figure 10.18**).

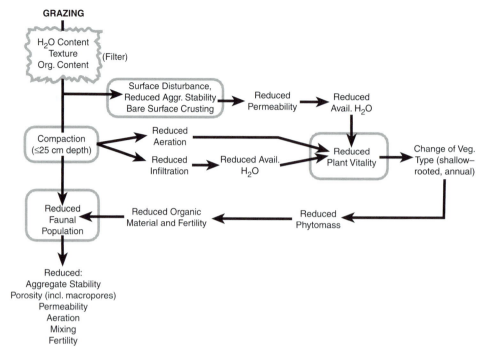

FIGURE 10.14 Effects of grazing on biofactors that influence infiltration and erosion.

10.15 WETLANDS

10.15.1 DEFINITIONS AND IMPORTANCE

The term *wetlands* frequently brings to mind scenes of swamps and marshes and ducks taking wing. This is one aspect of wetlands, but by no means a complete picture. In its broadest context, the term also includes coastal beaches and estuaries, lakes, rivers, and poorly drained farmlands. In its use in ecological and regulatory literature today, the term *wetland* refers to any site with soil development, biotic community, or hydrologic behavior that is dominated by the fact that it is at least periodically saturated with water. Most definitions exclude deepwater habitats, where water, not soil or air, provides the primary medium for biota.

Because saturation with water can be a nuisance for many land uses, a great deal of effort has been spent in draining and filling wetlands, and they have been retained only where it is most convenient or where it is too expensive to change them. As wetlands have become rarer, their importance has been realized, and more effort is spent to understand and protect them.

Wetlands provide an ecologically important role in the landscape. They are the only sites where diverse hydrophytic (water-loving) plants will grow. Without wetlands, we would lose this diversity of plant life. They are well known for their role in supporting nesting and migration of waterfowl, which depend on these ecosystems for food and cover.

Wetlands are showing increasing value for improving water quality. Wetlands promote deposition of soil eroded in upland areas, for example. Recent research has shown that they can effectively treat industrial and municipal wastewater as well.

Wetlands are less well recognized in their many hydrologic roles. They are sometimes the sites of groundwater recharge. Surface runoff collected from surrounding areas flows into wetlands during the rainy season and ponds temporarily until soaking into a local or regional aquifer system. They are important areas in flood control. Wetlands will act as detention areas and slow the flow of runoff. Floodplains are very important in this process; if floodwaters are not allowed to spread into floodplains, laws of hydraulics make it clear that flooding rivers will flow deeper and faster (see Chapter 6 to 8, Chapter 10, and Chapter 12).

Today, the legal and regulatory aspects of wetlands are also important issues. All levels of government — federal, state, county, and municipal — have laws regarding use and development of wetlands. Most of the regulation is in county and municipal zoning and building codes. Federal and state laws primarily provide guidance in objectives and methods. Any person considering development or modification of wetland areas needs to be familiar with often rapidly changing federal and local regulations and must be aware of the effects of development on wetlands. The use of constructed wetlands as a runoff or water quality management tool is discussed in Chapter 12.

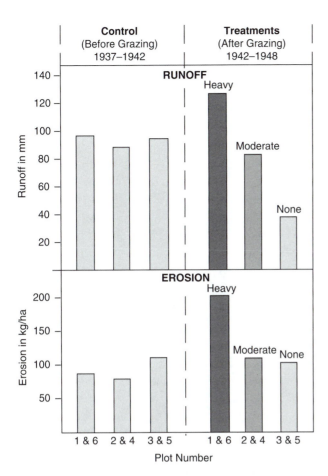

FIGURE 10.15 Runoff and erosion, June to September, for bunchgrass rangeland subjected to different grazing intensities near Colorado Springs. (Redrawn from Branson, F.A., G. Gifford, K. Renard, and R. Hadley, *Rangeland Hydrology*, Kendall/Hunt, Dubuque, IA, 1981.)

10.15.2 IDENTIFICATION OF WETLANDS

Identifying and delineating wetlands constitute the bulk of the legal and technical work associated with wetlands. Many methods are available to people examining lands for the presence and extent of wetlands (Lyon, 1993). For small sites, methods include detailed soil analyses from individual pits and ecological evaluation of plant communities. In examining larger areas, aerial photography and remote sensing become valuable tools.

To discuss identification of wetlands, it is important to understand their causes. The basic cause is an abundance of water, at least seasonally. The abundance may be due to a large amount of water available from surface water, as in the case of a river; from groundwater, as in the case of hillside seeps; or as a combination of the two, a common case, including lake shores, rivers, and most swamps. Technically, most wetlands could be said to be areas where the groundwater table intersects the surface, regardless of the direction that the water is flowing. Hillside

springs, for example, are purely an expression of the groundwater supplying water to the surface. Many wetlands have been created by irrigation water, which is supplied from the surface, and wetlands are formed in pockets as it flows to a shallow groundwater table.

Most groundwater/surface water systems in wetlands are more complex. Near lakeshores, for example, one cannot separate the effect of the lake and its tributary streams on the groundwater level from the effect of the groundwater level on the lake. The groundwater level and lake level are expressions of the same hydrologic system. When rivers cover their floodplains during the spring, this water may soak into the local shallow aquifer at relatively long distances from the river. This water returns to the stream gradually through the summer to provide water during low flows. Some rivers and streams are surface expressions of regional or local groundwater tables seeping out through the channel banks. Many rivers in the basin and range province of the western U.S., however, transport water from the mountains into colluvial basins, where water from the rivers soaks into the ground to deep water tables. The potentially complex ways that groundwater and surface water can interact is what makes identification of wetlands so challenging.

Given that wetlands are generally connected to groundwater, the first step in identifying and classifying a wetland is to look at the groundwater system to which it is connected. Many wetlands, particularly small wetlands, are connected to small perched water tables. This type of condition is common in Alaska and other boreal climates, where permafrost provides a shallow impermeable layer that turns any depression in the ground into a place where water will collect. Compacted glacial till underlying many soils once covered by continental glaciation in the northern U.S. provides a shallow impermeable layer that leads to formation of seasonal wetlands in many small pockets. Larger groundwater systems with surface expression, such as the Everglades in Florida, have been immense challenges to analyze.

Recognizing the role of groundwater makes it easier to identify wetlands. The primary clues are shallow water, hydric soils (soils that show signs of seasonal saturation), and hydrophytic vegetation. If there is shallow water, either flowing or standing for a prolonged period, at the site, it is usually quite clearly a wetland. Without water at the site, clues from the soil and vegetation may be needed. Soil that is waterlogged usually shows a gray or mottled gray and rust coloration. A rust coloration typically occurs when soils are periodically waterlogged and drained. Gravel and rock streambeds have no soils, but are clearly wetlands. *Hydrophytic vegetation* is vegetation that requires or is tolerant of large amounts of water and is an important clue in identifying wetlands. The list of potential hydrophytic plants in wetlands is far too long for this textbook and is best learned regionally. One clue, however, is if

FIGURE 10.16 Two contiguous reaches of small tributary stream with low silty banks, Iowa County, Wisconsin: (A) light to moderately grazed; (B) heavily grazed. (From Trimble, S.W., and A.C. Mendel, *Geomorphology*, 13:233–253, 1995. With permission.)

there is an abrupt change in vegetation in an area that is topographically probable as a wetland. A good example of this is a mountain meadow. Assuming the area is not farmed, a natural reason must exist for the grass to outcompete the trees, and a shallow groundwater table is often the cause.

Identified wetlands are classified by the type of water system to which they are connected. An overview of the classification system used by the U.S. Fish and Wildlife Service (Cowardin et al., 1979) shows the variety of places wetlands can be found. The first level of classification separates wetlands into marine, estuarine, riverine, lacustrine,

and palustrine. *Marine wetlands* are connected with tidal and intertidal environments of high-salt, high-energy coastal systems. *Estuarine systems* are generally lower energy and lower salinity, but can be temporarily higher in salinity from evaporation. They frequently have some degree of freshwater mixing from streams or groundwater. *Riverine systems* are closely associated with streambeds and banks. They can be intermittent in nature, as with the washes of the southwest desert. *Lacustrine systems* are associated with shores and shallows of large freshwater bodies. *Palustrine systems* encompass any other nontidal

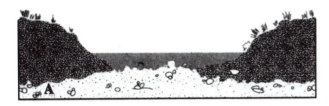

FIGURE 10.17 Recovery from overgrazing of riparian area, stream with low banks, in semiarid western United States. (From Bureau of Land Management, *Rangeland Reform '94: Draft Environmental Impact Statement*, Department of the Interior, Washington, DC, 1994.)

system and are meant to describe most upland wetland sites. Palustrine systems include the types of systems most often thought of when discussing wetlands, including bogs, fens, marshes, swamps, ponds, and prairie potholes.

We presented a brief introduction to some important aspects of wetlands. Because of the complex scope of wetlands, any person who will be working with the legal aspects of delineation and classification of wetlands is strongly encouraged to seek specialized training.

PROBLEMS

10.1. The following snow depths and SWE were observed at five points at a site in the Idaho:

Point	1	2	3	4	5
Depth (cm)	60	68	64	69	65
SWE (cm)	18	20	18	20	19

(a) Determine the average snow depth (centimeters), average SWE (centimeters), and the average snow density (kilograms/cubic meters) of this location. (b) Given the average conditions calculated under (a), what is the depth of cold content (centimeters) if the average temperature of the snowpack was −8°C. (c) How much water (centimeters) should be added to ripen the snow?

10.2. The commercial landowner in Example 5.5 expects to have a thick canopy of maples covering 80% of the proposed 500-acre development within 5 years because of his aggressive landscaping using planters. How much will the peak flow from the watershed be reduced for the storm of Example 5.5 by the forest canopy? The property currently has a good forest cover. What is the approximate peak flow in its current state? Is the aggressive maple landscaping a reasonable mitigation for the hydrologic impacts of the proposed commercial development?

10.3. Snow falls on 3 consecutive days. The first day, 20 cm of snow fell; the second day, 20 cm fell; and on the third day, 10 cm fell. The average air temperature during these 3 days was below freezing (less than 0°C). The next 3 days, the average air temperature was as shown in the table below:

Day	Average T (°C)
Day 4	1
Day 5	2
Day 6	2

Determine the snow water equivalent at the end of Day 6 using the simple degree–day method assuming (a) a clear area and (b) a forested area.

10.4. Your manager asks for an estimate of the annual evapotranspiration for a forest in southern Missouri. You find 12 years of records from a weather station at a stream gage that collects daily maximum and minimum temperatures, wind speed and direction, daily precipitation, and daily flow. Describe the data and equation you would use to estimate annual evapotranspiration.

A

B

FIGURE 10.18 Restoration of stream banks damaged by heavy grazing by cattle, Jenkins Creek, Giles County, Tennessee: (A) before, 1989; (B) after, 1994. Banks were built up with riprap and covered with sod. All grazing has ceased since 1989 (see Trimble, 1994).

10.5. A friend suggests that if the forest industry left 300-ft buffers of undisturbed forest along every perennial stream, substantial flood control benefits would result. Do you agree? Explain your position.

10.6. On 2 consecutive days in September 1983, two greatly different melt rates were observed on a glacier. The first day was sunny and warm, and 2 cm of melt were recorded. On the second day, a storm straight off the Pacific Ocean with 60 mi/h winds blew across the glacier, and 15 cm of melt were recorded. Calculate the energy transferred to the glacier's snow on each of these 2 days. The snow surface of the glacier is at 0°C and has a density of 760 kg/m³.

10.7. A flume was installed at the bottom of a 10-acre watershed in a forest in central Idaho that has a silt loam soil. The watershed was clear-cut and burned, and several heavily used skid trails were installed, cutting across the watershed. After two summers and one winter of data collection, no flow was collected during the brief, but intense (up to 3 in./hr, summer thunderstorms. Flows around 30 gal/min were collected during the 2-week melt season. Given that snowmelt rates seldom exceed 0.5 in./h and the information from **Table 8.2**, explain these observations.

10.8. A forester in southeast Alaska has been told by a local fisherman that if she were to use partial cutting (removal of 50%) of the forest canopy instead of clear-cutting, she would dramatically reduce the hydrologic impacts and erosion from forest harvest. She calls you to ask your opinion. Describe factually the relative impacts of the two types of harvest. (Southeast Alaska is wet all year long, with generally low-to-moderate rainfall intensities [Koeppen Cfb]. The winter is snowy and cold. The ground is steep and strongly affected by glaciation. Cable yarding is the primary yarding method used.)

10.9. Many states require stream buffers as a part of their forest practice regulations because of the benefits to water quality. Using your knowledge of runoff-producing processes in forests and the variety of erosional processes, describe the relative effectiveness of 10-, 50-, and 300-ft buffers for (a) a steep forested site in the Colorado Rockies, (b) a gently sloped tree farm in the piedmont soils of South Carolina, and (c) a moderately sloped stand in the basalt-derived soils of coastal southwest Washington's Willapa hills.

10.10. You are managing a forested watershed of 30,000 ha. You wish to obtain an additional average streamflow of 1 m³/sec over the next few years. How many hectares of forest must you cut?

10.11. You live by a stream that occasionally dries up. A forested watershed of 700 acres upstream of your home has just had 1000 acres of forest harvested. What would be the average increase of streamflow past your house in cubic feet/second? In centimeters?

11 Hydrogeology

11.1 INTRODUCTION

The need to understand the fundamental concepts describing groundwater flow is becoming increasingly important as groundwater resources become more and more threatened by contamination and overuse. This understanding involves an appreciation of several fields of science and engineering. Without it, we are destined to continue to believe that the movement of subsurface waters is secret and occult, as stated in an 1856 ruling by the Ohio Supreme Court that was not changed until 1984, and to continue to mismanage one of our most valuable, renewable natural resources.

11.2 CHARACTERIZATION OF GROUNDWATER FLOW

Groundwater flow through porous material occurs under the influence of energy, flowing from regions of high energy to regions of low energy. As such, it is similar to the flow of heat and the flow of electricity. The amount of energy a particle of water possesses at any position within a flow system is the sum of its three forms of potential energy — elevation energy, pressure energy, and velocity energy. A particle of water has elevation energy by virtue of its position in the flow system relative to some standard measurement plane, sea level, for example. The

pressure energy possessed by the particle of water is analogous to the energy of a compressed spring (M. Price, 1985). Under most conditions, except in conduit flow in limestone terranes and lava tubes, the energy derived from the movement of flowing groundwater is negligible and can be ignored.

Figure 11.1 shows three wells completed in a permeable layer such as sandstone. At Well A, the elevation head h_e is less than the pressure head h_p, whereas the opposite is true at Well B. Flow, however, is governed by the total hydraulic head, which is the sum of h_e plus h_p. As a result, the total head h at Well A is greater than the total head at Well B, and groundwater flow moves from left to right, up the dip of the sandstone layer. The total head at Well C is less than that at Well A or Well B, and flow moves down the dip of the sandstone layer between Well B and Well C. Thus, groundwater can flow laterally, up, or down the dip of a layer, depending only on the distribution of total hydraulic head.

Measurement of total hydraulic head — or *head*, as it is commonly known — usually is accomplished using a device that is lowered down a well to measure the depth of water in the well. This value then is subtracted from the elevation of the measuring point. If measurements of head are made in several wells completed in the same permeable layer, the values of head can be contoured to construct a potentiometric surface and used to determine

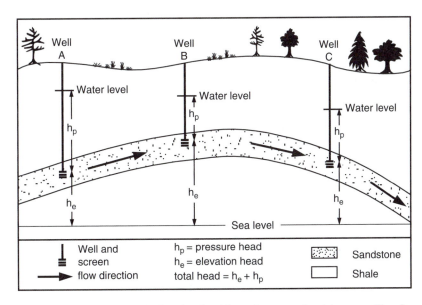

FIGURE 11.1 Components of total hydraulic head, elevation head h_e, and pressure head h_p, controlling flow in a sandstone layer.

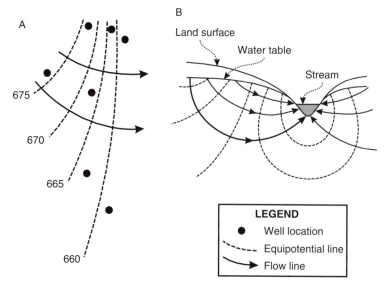

FIGURE 11.2 Relation of flowlines to equipotential lines in a small drainage basin: (A) view of upland area; (B) cross-sectional view of basin. (All units in feet.)

directions of groundwater flow. **Figure 11.2** shows a hypothetical potentiometric surface and its cross-sectional presentation based on contouring several measurements of head. The lines of equal head are called *equipotential lines* or *potentiometric contours* and represent lines of equal energy. Groundwater flows from areas of high energy (hydraulic head) to areas of low energy (hydraulic head). The path a particle of water makes through the flow field is called a *flowline*. In most flow systems, flowlines will be perpendicular to equipotential lines, so that particles of water flow along the steepest gradient between pairs of equipotential lines, as shown in **Figure 11.2**.

The volumetric flow of groundwater is dependent on the gradient between equipotential lines, which is known as the *hydraulic gradient*, the cross-sectional area through which flow occurs, and the permeability of the material. Henry Darcy, an engineer employed by the town of Dijon, France, was the first to recognize and quantify this relation. While trying to filter springwater through layers of sand to purify the city's water supply, Darcy performed a series of experiments in 1855 and 1856 that led him to observe several relations between the quantity of water flowing through a cylindrical column filled with a particular grade of sand and the amount of energy (head) loss measured between an upper manometer X and a lower manometer Y a distance l apart (**Figure 11.3**). In the experiments, Darcy allowed a constant rate of water Q to flow through the cylinder of known cross-sectional area A (**Figure 11.3**).

From his experiments, Darcy observed that if the flow rate Q was doubled, the loss of head ($dh = h_1 - h_2$) between manometer X and manometer Y also doubled. Thus, Q is directly proportional to head loss dh and the hydraulic gradient dh/l. Mathematically, this is expressed as

$$Q \approx \frac{h_1 - h_2}{l} = \frac{dh}{l} \qquad (11.1)$$

Based on this relation, it can be seen that it takes twice as much energy to drive the water through the sand at twice the flow rate. If a different cylinder is used, one in which the cross-sectional area of flow is twice as large as in the first cylinder, for a particular grade of sand, twice as much water will flow through the larger cylinder than through the smaller cylinder. Thus,

$$Q \approx A \qquad (11.2)$$

Combining these results, the following relation is obtained:

$$Q \approx A \frac{dh}{l} \qquad (11.3)$$

Darcy repeated his experiments using several different grades of sand and found that, for a given grade of sand,

$$Q = (K)(A)\left(\frac{dh}{l}\right) \qquad (11.4)$$

where K is a constant of proportionality for a given grade of sand. The experiments showed that the value of K was larger for coarse sand than for fine sand. Darcy deduced that the value of K was related to the ability of the sand to transmit fluid. Thus, K is referred to as hydraulic

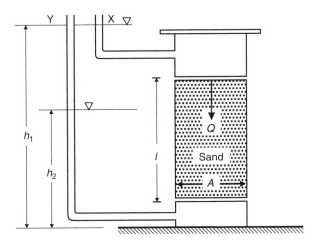

FIGURE 11.3 Darcy apparatus.

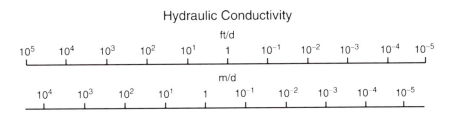

FIGURE 11.4 Bar chart showing hydraulic–conductivity values for various types of rock and sediment. (Modified from Bureau of Reclamation, *Ground Water Manual*, U.S. Department of the Interior, Washington, DC, 1977.)

conductivity and represents the properties of the porous material and the properties of the fluid. (Intrinsic permeability is a term used to describe the ability of a porous material to transmit fluid, but refers solely to the properties of the porous material.) The units of K are length per time and are commonly expressed as feet per day or centimeters per second. Although Darcy did not perform any experiments using fluids other than water, it can be visualized that a more viscous fluid would flow more slowly through the column of sand than a less viscous fluid. The form of Darcy's law written in Equation 11.4 can be expanded to separate the transmitting properties of the porous material from the properties of the fluid, such that

$$Q = (k)\left(\frac{\gamma}{\mu}\right)(A)\left(\frac{dh}{l}\right) \qquad (11.5)$$

where k is the intrinsic permeability of the porous material, γ is the specific weight of the fluid, and μ is the dynamic viscosity of the fluid. Thus, the rate of fluid flow is directly proportional to the specific weight of the fluid and inversely proportional to its viscosity.

Laboratory and field experiments indicate that values of K for natural earth materials range over more than 11 orders of magnitude. **Figure 11.4** shows that the most poorly permeable materials, massive clay deposits and unfractured igneous and metamorphic rocks, have K values close to 10^{-5} ft/day, whereas the most highly permeable materials, clean gravel and cavernous limestone, have K values close to 10^5 ft/day. This figure also indicates that the range of K values for sand and gravel deposits alone extends over five orders of magnitude. This represents a tremendous range of uncertainty and potential error when

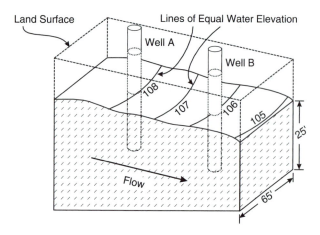

FIGURE 11.5 Block diagram for calculation of volumetric flow through a unit area perpendicular to the direction of flow. (All units in feet.)

attempting to compute the volumetric rate of groundwater flow in the absence of site-specific values of K obtained from laboratory or field experiments.

EXAMPLE 11.1

Darcy's law is used in various ways to make calculations of groundwater flow rates and travel times. For example, it may be necessary to determine the amount of groundwater flow within a permeable material.

Solution

Figure 11.5 shows two wells located 800 ft apart along a flowline. The difference in head values between the wells is 2.0 ft. Thus, the hydraulic gradient is 0.0025 ft/ft. The amount of groundwater flowing through a cross-sectional area of thickness b times width w is

$$Q = (K)(b)(w)\left(\frac{dh}{l}\right)$$

If it is determined from a field test that the hydraulic conductivity of the material is 125 ft/day, and it is determined from well logs that the material is 25 ft thick, and it is assumed that the cross section is 65 ft wide, then the quantity of groundwater flowing across the cross-sectional area is

$$Q = (125 \text{ ft/day})(25 \text{ ft})(65 \text{ ft})(0.0025 \text{ ft/ft}) = 508 \text{ ft}^3/\text{day}$$

Answer

The quantity of groundwater flowing across the specified cross-sectional area is 508 ft³/day.

Another way to express the ability of a porous material to transmit water is to account for its transmission across the entire saturated thickness of the permeable materials. Transmissivity is expressed in terms of length squared per time and is equal to

$$T = Kb \tag{11.6}$$

Figure 11.6 shows a bar graph for estimating, based on the measured transmissivity of the materials, the capability of porous materials to transmit water to wells. This graph is especially useful in estimating whether a particular geologic material is capable of transmitting sufficient water to various types of wells.

Subsurface materials can be classified according to their ability to transmit groundwater. An *aquifer* is a body of rock, soil, or unconsolidated material that can transmit groundwater in sufficient quantities to supply wells or springs. An *unconfined aquifer* (**Figure 11.7**) is one in which infiltrating water moving through the unsaturated zone has pressures less than atmospheric until it reaches the fully saturated pores and possesses a pressure equal to atmospheric pressure at the water table. The *water table* is a surface that is conventionally determined by water level measurements made in wells that penetrate only a short distance into the saturated portion of an unconfined aquifer.

A *confined aquifer* does not receive direct infiltration of precipitation because it is overlain by a layer of low hydraulic conductivity. As such, it does not have a water table, and the water is confined under sufficient pressure that water rises above the base of the overlying confining layer (**Figure 11.7**). If several wells are constructed in the same confined aquifer and a set of water level measurements are made at the same time, the resulting contour map represents the potentiometric surface of the aquifer.

Fluctuations in water levels are normal in unconfined and confined aquifers and are related to various natural and anthropogenic causes. Seasonal fluctuations in unconfined aquifers reflect variations in precipitation, evaporation, and transpiration. **Figure 11.8** shows a hydrograph from a well completed in unconfined sand and gravel deposits in southwest Ohio. The periods of rising water level indicate that recharge to this aquifer occurs predominantly in late winter and early spring, when precipitation is plentiful and evaporation and transpiration rates are relatively low. Analysis of over 60 annual hydrographs from eight wells in unconfined aquifers in Ohio indicated that the average duration of this recharge period is 23 weeks, with a standard deviation of 9 weeks, and the average height of water level rise is 6.8 ft, with a standard deviation of 4.3 ft (de Roche, 1993). This represents the period when recharge to the aquifer from precipitation exceeds discharge to springs and streams, which results in a prolonged water level rise. The period of declining

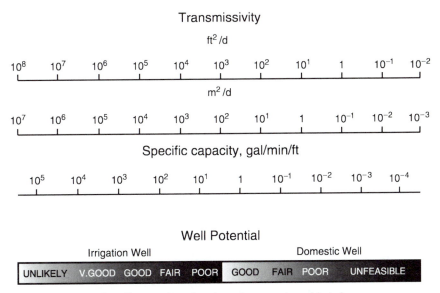

FIGURE 11.6 Bar chart for range of transmissivity values relative to the potential of various types of wells. (Modified from Bureau of Reclamation, *Ground Water Manual*, U.S. Department of the Interior, Washington, DC, 1977.)

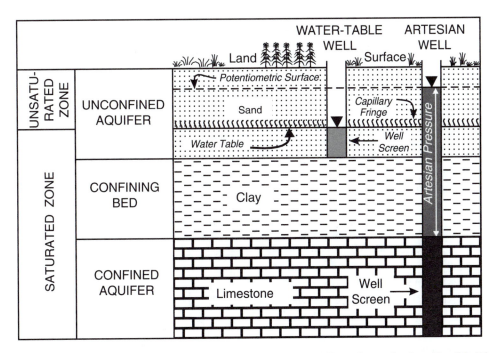

FIGURE 11.7 Diagram showing hydrogeologic conditions for an unconfined aquifer and a confined aquifer. (Modified from Heath, R.C., *Basic Ground-Water Hydrology*, U.S. Geological Survey Water Supply Paper 2220, U.S. Government Printing Office, Washington, DC, 1983.)

water levels represents the period when more water discharges to springs and streams than recharges the aquifer.

On a hydrograph, a drought would be indicated as a period of falling water levels that corresponds to a period of diminished precipitation. A prolonged drought occurred in Ohio during 1987 and 1988. The decline in water levels due to this drought is seen in **Figure 11.8**. Overpumping wells in an unconfined aquifer that cause groundwater levels to decline would be indicated by a period of several years in which water levels declined while precipitation

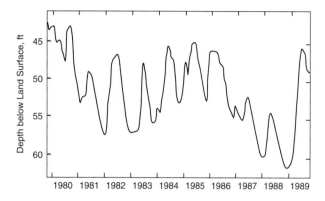

FIGURE 11.8 Hydrograph showing seasonal and annual water level fluctuations of a well completed in an unconfined aquifer in southwest Ohio.

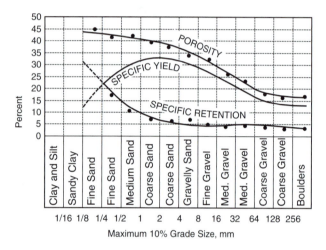

FIGURE 11.9 Relation of specific yield and specific retention to total porosity for various sediment sizes. (Modified from Johnson, A.I., *Specific Yield — Compilation of Specific Yields for Various Materials*, U.S. Geological Survey Water Supply Paper 1662-D, U.S. Government Printing Office, Washington, DC, 1967.)

remained near normal. In these cases, the volume of groundwater in storage decreased as the pores in the unconfined aquifer drained. The volume of groundwater that will drain under the influence of gravity relative to the total volume of aquifer is called the *specific yield Sy*. It is a measure of the storage characteristics of an unconfined aquifer.

Values of *Sy* commonly are expressed as a percentage. *Sy* values are less than the porosity of the aquifer materials and range from less than 1% to as much as 45%. The porosity *n*, also commonly expressed as a percentage, of a soil or rock is defined as the volume of void space relative to the total volume. The difference between *n* and *Sy* is the volume of groundwater that does not drain under the influence of gravity, which is called the specific retention *Sr*. Thus,

$$n = Sy + Sr \qquad (11.7)$$

Grain-size distribution and grain shape are the primary factors controlling the distribution of pore sizes and the rate of gravity drainage. **Figure 11.9** shows the range of *Sy*, *Sr*, and *n* values measured from alluvial deposits in California.

Seasonal water level fluctuations in confined aquifers also reflect changes in the amount of water in storage, but not by the same mechanism as for unconfined aquifers. As potentiometric levels in a confined aquifer change, the pores remain saturated. The change in potentiometric level represents a change in the amount of pressure head in the aquifer. The amount of water level change is controlled by the coefficient of storage *S*, which is defined as

$$S = \rho g b (\alpha + n \beta) \qquad (11.8)$$

where ρ is the density of the fluid, *g* is the gravitational constant, *b* is the saturated thickness of the aquifer, α is the compressibility of the aquifer skeleton, *n* is the porosity, and β is the compressibility of the fluid. Values of *S* are small, on the order of 0.0001 to 0.000001. Thus, it may require 1000 times the volume of soil/rock to store the same amount of groundwater in a confined aquifer as in an unconfined aquifer.

Short-term water level fluctuations in unconfined and confined aquifers can be caused by the withdrawal of water from wells. **Figure 11.10** shows the hydrographs from two wells completed in sand and gravel deposits at a municipal well field. The regularly spaced, small-amplitude water level rises are related to the decrease in pumping rates on weekends and are superposed on the cycle of seasonal water level fluctuations.

Differences in the rate of infiltration of water through the unsaturated zone can cause differences in the rate at which recharge from precipitation reaches the water table, creating nonuniform rates of water level rise. This results in changes in hydraulic gradient that are manifest as changes in the rate and direction of groundwater flow. The rate of groundwater flow is computed using Darcy's law and accounts for the fact that flow occurs only through the pores. Thus, the average linear velocity *v* of groundwater flow is

$$v = \frac{K \frac{dh}{l}}{n_e} \qquad (11.9)$$

where n_e is the effective porosity, expressed as a decimal, and is defined as the volume of interconnected pore space relative to the total volume.

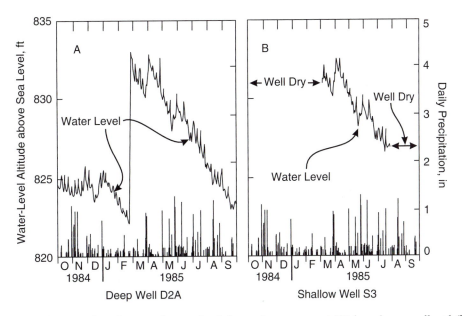

FIGURE 11.10 Precipitation records and seasonal water level fluctuations measured (A) in a deeper well and (B) in a shallower well in a sand and gravel aquifer showing weekly fluctuations due to pumping stress. (Modified from Breen, K.J., in *Regional Aquifer Systems of the United States — Northeast Glacial Aquifers*, A.D. Randall and A.I. Johnson, Eds., American Water Resources Association Monograph Series No. 11, 1988, pp. 105–131.)

EXAMPLE 11.2

Find the average linear velocity of flow between the wells shown in **Figure 11.5**.

Solution

Assume that the porosity of the aquifer shown on **Figure 11.5** is 0.25. The rate of average linear velocity between Wells A and B would be

$$v = \frac{(125 \text{ ft/day})(0.0025 \text{ ft/ft})}{0.25} = 1.25 \text{ ft/day}$$

Answer

The average linear velocity would be 1.25 ft/day.

Figure 11.11 is a polar plot showing changes in the hydraulic gradient during a period of several months at a site in southern Ohio; the plot is based on water level measurements made in three sets of three wells completed in an unconfined aquifer. Differences in the rate of infiltration cause the direction of the hydraulic gradient to change by 40° and the magnitude of the hydraulic gradient, and therefore, the velocity of groundwater flow, to change by a factor of 3. Natural, temporal changes in the rate and direction of groundwater flow in surficial aquifers

are important in determining the direction of movement and the areal extent of anthropogenic contaminants.

If large changes in water levels or hydraulic gradients occur over long periods of time, the flow system is said to be *transient*. If only small changes occur, the flow system is said to be *steady state* because there is no net change in the amount of water in storage and, therefore, no changes in flow directions or flow rates.

11.3 GROUNDWATER FLOW PATTERNS AND STREAM INTERACTION

Groundwater flow patterns are controlled by several factors, including the elevation and location of recharge and discharge areas, the heterogeneity of the geologic materials, the thickness of materials, and the configuration of the water table. One of the simplest types of flow systems is shown on **Figure 11.12**; recharge and discharge areas are adjacent. In **Figure 11.12**, both streambeds have the same elevation, the geologic materials are homogeneous, and groundwater divides correspond to surface water divides. In this flow system, infiltration from precipitation enters the groundwater flow system in topographically high regions and discharges in topographically low regions to streams and rivers. During periods of drought, the water table will decline, particularly beneath the upland areas, in response to the lack of precipitation, but groundwater will continue to discharge to the streams until the water table drops below the bottom of the streambed. Thus, for low surface water flow conditions, the water in the streams is sustained by groundwater discharge.

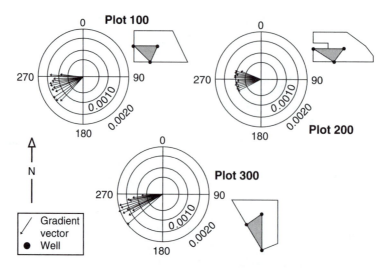

FIGURE 11.11 Vectors showing seasonal changes in direction, magnitude, and velocity of groundwater flow of hydraulic gradient at three 10-ha farm plots. (From Finton, C.D., Simulation of Advective Flow and Attenuation of Two Agricultural Chemicals in an Alluvial-Valley Aquifer, Piketon, Ohio, unpublished master's thesis, Department of Geological Sciences, Ohio State University, Columbus, 1994. With permission.)

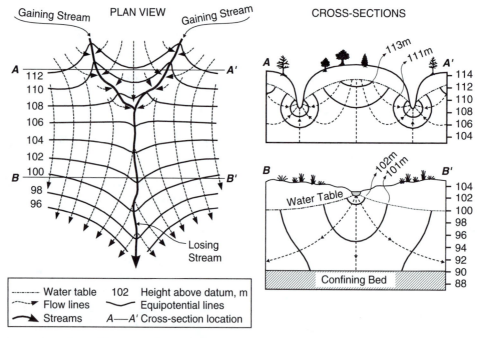

FIGURE 11.12 Diagrams showing the relation between the configuration of the water table at a gaining and a losing stream. (From Heath, R.C., *Basic Ground-Water Hydrology*, U.S. Geological Survey Water Supply Paper 2220, U.S. Government Printing Office, Washington, DC, 1983. With permission.)

This type of interaction between the groundwater flow system and the surface water flow system is consistent with the flow patterns shown in **Figure 11.12** along the potentiometric profile marked A–A′. Along potentiometric profile B–B′, a different type of interaction occurs between the stream and the aquifer. In this case, the water in the stream has a greater level than the water in the aquifer, and water in the stream recharges the groundwater flow

system as the stream loses water along its course. Losing streams are common in the western part of the U.S., where streams begin in mountainous areas and flow across alluvial fans formed at the base of the mountains before flowing onto the relatively flat basins. Losing streams also can be found in similar, but smaller, settings in the glaciated parts of the Midwest and New England.

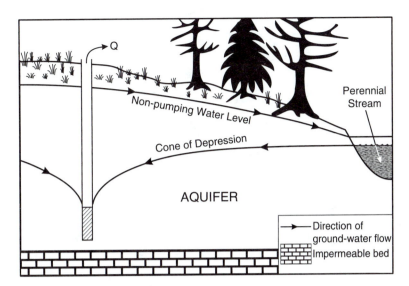

FIGURE 11.13 Diagram showing the relation between induced infiltration from a stream and the cone of depression developed by a pumping well. (Modified from Peters, J.G., Description and Comparison of Selected Models for Hydrologic Analysis of Ground-Water Flow, St. Joseph River Basin, Indiana, U.S. Geological Survey Water-Resources Investigations Report 86-4199, 1987.)

This type of interaction can be created artificially by pumping wells located sufficiently close to a stream to reverse the direction of local groundwater flow and induce water to flow out of the stream and toward the well (**Figure 11.13**). Many municipal, agricultural, and industrial water supplies rely on this type of interaction. The amount of induced infiltration from the stream can be approximated using analytical solutions (Jenkins, 1970) or measured by gaging streamflow upgradient and downgradient of the wells.

One of the most highly publicized cases involving induced infiltration of streamflow as a possible cause of contamination of water supply wells occurred in Woburn, MA, in the 1980s. In 1982, a lawsuit was filed on behalf of eight families who alleged that industrial contamination of two water supply wells located adjacent to the Aberjona River and were completed in an unconfined aquifer, led to the death of six children (GeoTrans, 1987). In the discovery phase of the lawsuit, a test was performed to reproduce the pumping stress created by the two wells during their operation between 1964 and 1979. During a 1-month period, the two wells were pumped at their normal discharge rates while routine measurements of groundwater levels and stream discharge rates were made. The water level measurements indicated that the groundwater system became steady state after approximately 1 month of pumping the wells at a combined rate of 1100 gal/min. **Figure 11.14** is a water level map of the unconfined aquifer at the end of the 1-month pumping stress when the system is steady state. It shows that flow converges on the two wells from all sides of the valley; therefore, the source of contamination could be in almost any direction.

The discharge of the Aberjona River was measured upstream and downstream of the wells. The difference between the two discharge rates indicates whether the stream is gaining water from groundwater discharge or losing water as induced infiltration from pumping the two wells. **Figure 11.15** shows that, prior to the start of the pumping test the Aberjona River gained approximately 760 gal/min between the upstream gaging station and the downstream gaging station, indicating groundwater discharge to the stream. As the pumping test proceeded, less and less flow was recorded at the downstream gaging station. When steady-state conditions were apparent, over 550 gal/min less water was measured at the downgradient gaging station than at the upstream gaging station. This strongly suggests that about half the water pumped by the two municipal wells originated either as surface water in the Aberjona River or as intercepted groundwater flow that would have discharged to the Aberjona River. Thus, the quality of water in the Aberjona River and the quality of the water in the unconfined aquifer needed to be considered as possible sources of the groundwater contamination measured in the municipal wells.

In a simple, homogeneous aquifer, more complex regional flow patterns can develop, depending on the slope and shape of the water table surface and the thickness of the saturated material. In the early 1960s, J.A. Toth performed a series of mathematical experiments using a solution to a steady-state flow equation that demonstrated these controls on groundwater flow patterns. Contrast the flow patterns in **Figure 11.16A** and **11.16B**. The gently undulating configuration of the water table is the same in both figures. **Figure 11.16B** shows the flow patterns that

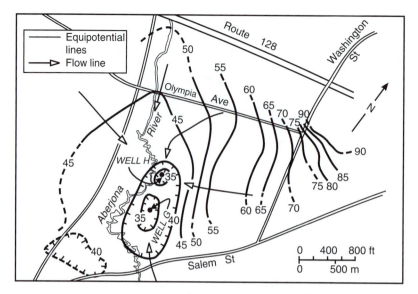

FIGURE 11.14 Measured steady-state potentiometric surface (January 1986) while pumping municipal wells G and H at a combined rate of 1100 gal/min. (Modified from GeoTrans Newsletter, *Woburn Toxic Trial*, GeoTrans, Herndon, VA, June 1987, pp. 1–3.)

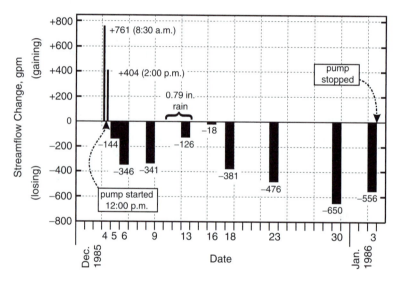

FIGURE 11.15 Bar graph showing streamflow depletion due to pumping wells G and H. (Modified from GeoTrans Newsletter, *Woburn Toxic Trial*, GeoTrans, Herndon, VA, June 1987, pp. 1–3.)

develop in a deep basin, whereas **Figure 11.16A** shows the flow patterns that develop in a shallow basin.

In the deep basin, local, intermediate, and regional flow systems develop, whereas in the shallow basin, only local flow systems develop. In a local flow system, recharge along a groundwater mound (surface water divide) flows to adjacent discharge areas (streams). In intermediate and regional flow systems, recharge occurs only in selected areas, flow bypasses one or more adjacent discharge areas, and flow occurs to greater depths in the basin.

Figure 11.17 contrasts the flow patterns that develop in basins of the same thickness but differing water table configurations. **Figure 11.17A** shows that both local and

regional flow systems develop in basins with gentle topographic relief and gentle undulations to the water table, whereas **Figure 11.17B** shows that only local flow systems develop in basins with greater topographic relief and more pronounced undulations in the water table.

These concepts can be important in the consideration of alternative sites for the disposal of hazardous wastes and nuclear wastes. For example, many states allow the injection of hazardous liquid wastes into deep saline aquifers. From a purely hydrogeologic viewpoint of protecting shallow freshwater supplies, it would be prudent to inject these wastes into the downflow limb of a deep flow system, as is found in **Figure 11.16B** and **Figure 11.17A**, where the path of groundwater flow is deeper, longer, and slower,

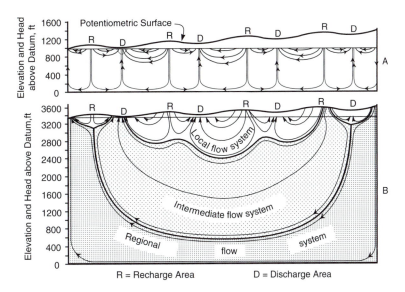

FIGURE 11.16 The influence of basin thickness on (A) the development of local flow patterns vs. (B) the development of local, intermediate, and regional flow patterns. (Modified from Toth, J.A., *J. Geophys. Res.*, 67:4375–4387, 1962.)

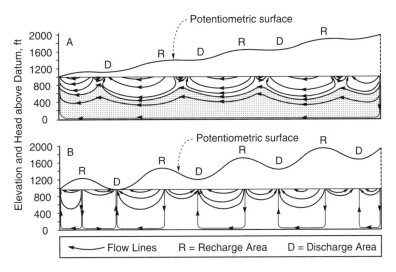

FIGURE 11.17 The influence of water table gradients on the development of flow patterns in two basins of the same thickness. (Modified from Toth, J.A., *J. Geophys. Res.*, 67:4375–4387, 1962.)

allowing more time for dilution and chemical and biological transformation of the wastes. The difficulty in applying these concepts lies in the assumption that the basin is homogeneous. In reality, most basins are geologically complex and have complex distributions of hydraulic conductivity that also affect groundwater flow patterns, as shown in **Figure 11.18** (Freeze and Witherspoon, 1967). The task of characterizing flow patterns in deep basins is made more difficult because of the lack of wells. In most cases, the geologic and hydraulic data from oil/gas exploration and production wells must be used along with the limited number of wells that can be drilled and tested as part of a characterization study.

11.4 FLOW TO WELLS

The response of a groundwater flow system to the stress created by pumping a well can be conceptualized using Darcy's law. **Figure 11.19** shows a cross section through a confined aquifer in which a pumping well is screened throughout the entire saturated thickness of the aquifer. Two concentric cylinders are drawn about the well, the larger of radius r_2 and the smaller of radius r_1. Assuming the response of the flow system to the pumping stress is steady state and that the well discharges at a constant rate Q, then the quantity of water passing across the circumferential area represented by the larger cylinder must he equal to the quantity of water passing across the circumferential area represented by the smaller cylinder, which

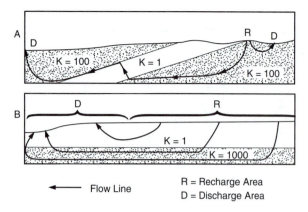

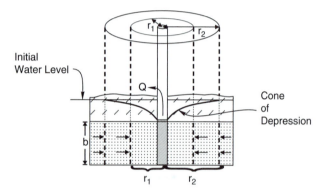

FIGURE 11.18 The influence of geological complexities on flow patterns. (Modified from Freeze, R.A., and P.A. Witherspoon, *Water Resour. Res.*, 3:623–634, 1967.)

FIGURE 11.19 Schematic diagram showing horizontal radial flow to a pumping well completed in a confined aquifer.

must be equal to the pumping rate of the well. This mass balance can be expressed by Darcy's law such that

$$Q = (K)(2\pi\, r_1 b)\left(\frac{dh_1}{dr_1}\right) = (K)(2\pi\, r_2 b)\left(\frac{dh_2}{dr_2}\right) \quad (11.10)$$

where dh_1/dr_1 and dh_2/dr_2 are the hydraulic gradients at r_1 and r_2, respectively, and $2\pi r_1 b$ and $2\pi r_2 b$ are the circumferential areas at r_1 and r_2, respectively. Cancelling terms and recognizing that if $r_1 < r_2$, then $dh_1/dr_1 > dh_2/dr_2$. Thus, as the distance to the pumping well becomes smaller, the hydraulic gradient becomes steeper, forming a cone of depression about the well, as shown on **Figure 11.19**.

Under transient conditions, while the cone is still expanding under the pumping stress, the shape of the cone of depression is a function of the pumping rate, the hydraulic conductivity, and the coefficient of storage. In 1935, C.V. Theis, a geologist working in eastern New Mexico for the U.S. Geological Survey, published a paper presenting an equation describing transient changes in the shape (areal extent and depth) of the cone of depression;

the equation was based on certain simplifying assumptions regarding the characteristics of the pumped aquifer and the nature of the well and pumping stress. Theis's work relied on the analogy of groundwater flow in a porous medium to heat flow in a conductive solid. Thus, hydraulic head is analogous to temperature, hydraulic gradient to thermal gradient, hydraulic conductivity to thermal conductivity, and specific storage ($S_s = S/b$) to specific heat. The Theis equation assumes that the aquifer is isotropic, homogeneous, infinite, flat lying, and overlaid and underlaid by impermeable layers that provide an insignificant amount of water to the pumped aquifer, and that water is removed instantaneously from storage. The well is assumed to penetrate the aquifer fully, to discharge at a constant rate, and to have a negligible amount of water stored in the borehole. Under these conditions, the Theis equation can be used to compute the drawdown at any radial distance from the pumped well at any time, such that

$$s = \frac{Q}{4\pi T} W(u) \quad (11.11)$$

and

$$u = \frac{r^2 S}{4Tt} \quad (11.12)$$

where s is drawdown, Q is the pumping rate, T is transmissivity, t is time, S is the coefficient of storage, r is the radial distance to a point of interest, and $W(u)$ is the well function of u. Any set of consistent units can be used in these equations. The value of the well function can be computed using a series expansion such that

$$W(u) = \left[-0.5772 - \ln\, u + u - \frac{u^2}{2\cdot 2!} + \frac{u^3}{3\cdot 3!} - \frac{u^4}{4\cdot 4!} + \cdots \right] \quad (11.13)$$

Values of $W(u)$ for various values of u are listed in **Table 11.1** or can be computed using a spreadsheet program and expanding the well function to 9 to 10 terms.

Example 11.3

The Theis equation can be used to answer many questions that arise concerning the relation of the cone of depression to pumping stress. For example, a power plant in Louisiana relies on two wells to supply water for fire protection. The wells are approximately 1800 ft deep and together are capable of producing a sustained yield of 900 gal/min ($\approx$173,000 ft³/day). As part of the licensing procedure for the plant, it is necessary to compute the effect of these wells on the nearest well completed in the same aquifer.

TABLE 11.1
Well Function $W(u)$

u	$W(u)$	u	$W(u)$	u	$W(u)$	u	$W(u)$
1×10^{-10}	22.45	7×10^{-8}	15.90	4×10^{-5}	9.55	1×10^{-2}	4.04
2×10^{-10}	21.76	8×10^{-8}	15.76	5×10^{-5}	9.33	2×10^{-2}	3.35
3×10^{-10}	21.35	9×10^{-8}	15.65	6×10^{-5}	9.14	3×10^{-2}	2.96
4×10^{-10}	21.06	1×10^{-7}	15.54	7×10^{-5}	8.99	4×10^{-2}	2.68
5×10^{-10}	20.84	2×10^{-7}	14.85	8×10^{-5}	8.86	5×10^{-2}	2.47
6×10^{-10}	20.66	3×10^{-7}	14.44	9×10^{-5}	8.74	6×10^{-2}	2.30
7×10^{-10}	20.50	4×10^{-7}	14.15	1×10^{-4}	8.63	7×10^{-2}	2.15
8×10^{-10}	20.37	5×10^{-7}	13.93	2×10^{-4}	7.94	8×10^{-2}	2.03
9×10^{-10}	20.25	6×10^{-7}	13.75	3×10^{-4}	7.53	9×10^{-2}	1.92
1×10^{-9}	20.15	7×10^{-7}	13.60	4×10^{-4}	7.25	1×10^{-1}	1.823
2×10^{-9}	19.45	8×10^{-7}	13.46	5×10^{-4}	7.02	2×10^{-1}	1.223
3×10^{-9}	19.05	9×10^{-7}	13.34	6×10^{-4}	6.84	3×10^{-1}	0.906
4×10^{-9}	18.76	1×10^{-6}	13.24	7×10^{-4}	6.69	4×10^{-1}	0.702
5×10^{-9}	18.54	2×10^{-6}	12.55	8×10^{-4}	6.55	5×10^{-1}	0.560
6×10^{-9}	18.35	3×10^{-6}	12.14	9×10^{-4}	6.44	6×10^{-1}	0.454
7×10^{-9}	18.20	4×10^{-6}	11.85	1×10^{-3}	6.33	7×10^{-1}	0.374
8×10^{-9}	18.07	5×10^{-6}	11.63	2×10^{-3}	5.64	8×10^{-1}	0.311
9×10^{-9}	17.95	6×10^{-6}	11.45	3×10^{-3}	5.23	9×10^{-1}	0.260
1×10^{-8}	17.84	7×10^{-6}	11.29	4×10^{-3}	4.95	1×10^{0}	0.219
2×10^{-8}	17.15	8×10^{-6}	11.16	5×10^{-3}	4.73	2×10^{0}	0.049
3×10^{-8}	16.74	9×10^{-6}	11.04	6×10^{-3}	4.54	3×10^{0}	0.013
4×10^{-8}	16.46	1×10^{-5}	10.94	7×10^{-3}	4.39	4×10^{0}	0.004
5×10^{-8}	16.23	2×10^{-5}	10.24	8×10^{-3}	4.26	5×10^{0}	0.001
6×10^{-8}	16.05	3×10^{-5}	9.84	9×10^{-3}	4.14		

A well survey was performed, and it was determined that the nearest well is 7500 ft away. To determine the effect of pumping the fire protection wells on the potentiometric level at the nearest well also completed in the aquifer, it was necessary to perform a controlled field experiment called an aquifer test to determine the transmissivity T and the coefficient of storage S of the pumped aquifer. The results of the aquifer test indicate that T equals 4680 ft²/day, and S equals 0.0007. For the purposes of making an overly conservative estimate of the effect of the pumping stress at the nearest well, it is assumed that the fire protection wells discharge at 900 gal/min for the entire 40-year design life of the power plant.

Solution

The first step in this computation is to determine u. Thus,

$$u = \frac{r^2 S}{4Tt} = \frac{(7500 \text{ ft})^2 (0.0007)}{(4)(4680 \text{ ft}^2/\text{day})(14,600 \text{ days})} = 0.000144$$

The value of the well function $W(u)$ for this value of u can be interpolated from **Table 11.1** as approximately 8.32. With this $W(u)$ value, the drawdown at a distance of 7500 ft at the nearest well due to pumping the fire protection wells at a combined rate of 900 gal/min for a 40-year period is computed as

$$s = \frac{Q}{4\pi T} W(u) = \frac{(173,000 \text{ ft}^3/\text{day})(8.32)}{(4)(3.14)(4680 \text{ ft}^2/\text{day})} = 24.4 \text{ ft}$$

As the potentiometric level in the aquifer rises to within several hundred feet of the land surface and because the pump in the nearest well is located several hundred feet below the potentiometric level, the overly conservative amount of drawdown computed for the fire protection wells will not affect the operation of the nearest well.

This is an example of a time–drawdown calculation made at a specific distance. Additional calculations can be made at different times to show the rate of spreading of the cone of depression. Alternatively, distance–drawdown calculations can be made at different locations at a specific time to show the shape of the cone of depression.

Answer

The drawdown due to pumping the fire protection wells will be 24.4 ft at the nearest wells.

In many problems, it is necessary to account for the drawdowns produced by several wells. This is a common situation in the design of municipal well fields, irrigation wells, contaminant pump-and-treat systems, and construction of dewatering/depressurizing systems for excavations. Because drawdowns are additive, the mathematical principle of superposition can be used to address these more complex problems. Thus, the composite drawdown from any number of wells at a particular location at a specific time is equal to the sum of the drawdowns from each individual well. Mathematically, this can be expressed simply as

$$s_{\text{total}} = s_1 + s_2 + s_3 + \cdots \qquad (11.14)$$

where the subscripts refer to the component of the total drawdown due to an individual well. If the Theis assumptions are met, the following equation would apply to the calculation of the composite drawdown at a location between two pumping wells discharging at different rates:

$$s = \frac{Q_1}{4\pi T} W(u_1) + \frac{Q_2}{4\pi T} W(u_2) \qquad (11.15)$$

where

$$u_1 = \frac{r_1^2 S}{4Tt} \quad \text{and} \quad u_2 = \frac{r_2^2 S}{4Tt}$$

The assumptions required of the Theis equation are not met in all situations. In many geologic settings, aquifers are not infinite, but are bounded on one or more sides by streams, which can provide an additional source of water in unconfined aquifers, or by bedrock valley walls or faults, which can preclude the flow of water. In an unconfined aquifer, the Theis equation will not make accurate predictions of the amount of drawdown near wells because the assumption of constant aquifer thickness is violated by the formation of the cone of depression in the water table around the well. (This is not a problem in confined aquifers because the cone of depression is formed in the potentiometric surface above the base of the overlying confining layer.) In the case of an unconfined aquifer, solution of the following equation using the quadratic formula enables drawdowns computed with the Theis equation to be corrected for the decrease in saturated thickness near the well and applied to an unconfined aquifer:

$$s^2 - 2bs = 2bs' \qquad (11.16)$$

where s is uncorrected drawdown computed using the Theis equation for a confined aquifer, b is the initial saturated thickness of the aquifer prior to pumping, and s' is the corrected drawdown in an equivalent unconfined aquifer. This correction factor is important near pumping wells, but becomes increasingly less important with distance from the well. When the correction factor is less

than 0.01 ft, it is below the precision of most methods of measuring hydraulic head and therefore can be ignored.

11.5 CAPTURE ZONES OF WELLS

Although wells most commonly are used to provide water for domestic, municipal, and industrial purposes, wells also are commonly used to extract contaminated groundwater from shallow aquifers. Once the contaminated groundwater is extracted, it can be treated in a variety of ways to remove the contaminants. This is called *pump-and-treat technology*. The duration that a pump-and-treat system needs to operate depends on the amount of contamination, the character of the groundwater flow system, the physical and chemical properties of the contaminants, the aquifer materials, and the desired level of decontamination. To design pump-and-treat systems in an efficient manner, it is necessary to know the areal extent of the contamination so that the capture zones of the wells used to extract the contaminated groundwater do not "undershoot" the area of contamination and allow contamination to escape or "overshoot" the area of contamination and extract clean water that does not need to be treated.

The steady-state capture zone of a well completed in an isotropic, homogeneous aquifer with a uniform prepumping hydraulic gradient is shown in **Figure 11.20**. The capture zone extends upgradient to a regional groundwater divide. It extends downgradient only as far as a local groundwater divide created by the cone of depression of the well. In a Cartesian coordinate system, the bounding flowline within which all groundwater flows to the well can be estimated with the following equation:

$$x = \frac{-y}{\tan\left[\dfrac{2\pi Kbiy}{Q}\right]} \qquad (11.17)$$

where x and y are coordinate values defined in **Figure 11.20**, Q is the pumping rate of the well, K is the hydraulic conductivity of the aquifer, b is the saturated thickness of the aquifer, i is the regional (prepumping) hydraulic gradient, and $\tan\left[\dfrac{2\pi Kbiy}{Q}\right]$ is in radians.

The distance to the downgradient groundwater divide created by the well is given by

$$x_0 = \frac{-Q}{2\pi Kbi} \qquad (11.18)$$

The maximum half-width of the capture zone as x approaches infinity is given by

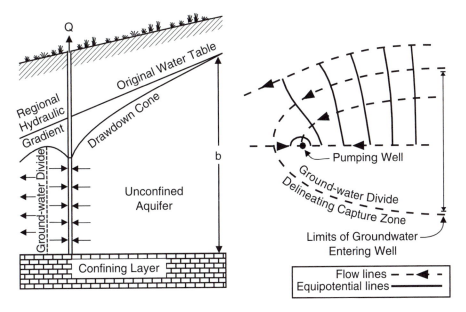

FIGURE 11.20 Outline of the capture zone of a well in a uniform flow field.

$$y_{max} = \pm \frac{Q}{2Kbi} \qquad (11.19)$$

From these equations, it can be seen that the distance to the downgradient groundwater divide is proportional to the pumping rate, but inversely proportional to the hydraulic gradient and K. Thus, the greater the pumping rate or the flatter the regional hydraulic gradient, the greater the distance to the downgradient groundwater divide. These equations also show that the width of the capture zone is proportional to the pumping rate, but inversely proportional to the hydraulic gradient and K. Thus, the greater the pumping rate or the smaller the regional hydraulic gradient, the wider the capture zone.

EXAMPLE 11.4

The following is an example of the calculation used to compute the x and y coordinates of bounding flowlines delineating the capture zone of an interceptor well constructed at a gasoline station to recover hydrocarbons that leaked from an underground tank.

Solution

Based on design considerations, it is assumed that the well pumps at a rate of 200,000 ft³/day. The geologic logs from a series of soil borings at the site indicate the aquifer is 30 ft thick. The hydraulic conductivity of the aquifer is estimated to be 1000 ft/day. Water levels measured in two wells completed in soil borings 700 ft apart indicate the (background) hydraulic head upgradient of the plume is

TABLE 11.2
Values of y and x for Example 11.4

y	tan(argument)	x
±1	0.005385	− 185.7
±50	0.269143	− 181.3
±100	0.538286	− 167.5
±150	0.807814	− 143.4
±200	1.077086	− 107.6
±250	1.346357	−57.07
±300	1.615629	13.46
±400	2.154171	264.0
±500	2.692714	1,038
±550	2.961986	3,029
±583	3.139705	308,825

331 ft, whereas the hydraulic head downgradient of the plume is 327 ft. Substituting these values into Equation 11.17 produces

$$x = \frac{-y}{\tan\left[\dfrac{(2)(3.1415)(1000\ \text{ft/day})(30\ \text{ft})}{\dfrac{(331\ \text{ft})-(327\ \text{ft})}{(700\ \text{ft})}\, y\, 200{,}000\ \text{ft}^3/\text{day}}\right]}$$

Substituting the values of y listed in **Table 11.2** into this equation yields the values of x also listed in **Table 11.2**. When plotted on arithmetic paper (**Figure 11.21**), the x and $\pm y$ coordinates outline the bounding flowlines separating groundwater flow within the steady-state capture zone of the well from groundwater flow in the regional flow regime.

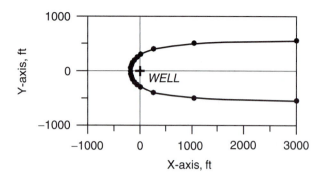

FIGURE 11.21 Computed capture zone of the well in Example 11.4.

The distance to the downgradient stagnation point is computed using Equation 11.18 and is

$$x_o = \frac{-200,000 \text{ ft}^3/\text{day}}{(2)(3.415)(1000 \text{ ft}/\text{day})(30 \text{ ft})(0.0057)} = -186 \text{ ft}$$

and the *total* width of the capture at the upgradient divide is

$$\gamma_{\text{max-total}} = (2)\frac{200,000 \text{ ft}^3/\text{day}}{(2)(100 \text{ ft}/\text{day})(30 \text{ ft})(0.0057)} = 116 \text{ ft}$$

Answer

The capture zone is illustrated in **Figure 11.21**.

Many municipalities and private purveyors of water are asked by state government agencies to delineate the capture zones of their wells to encourage reasonable land use practices within the capture zone as a means of preserving the quality of groundwater supplies.

11.6 FRACTURE FLOW

Fractures are also important to the movement of water. They are responsible for the recharge to the groundwater in many hydrogeologic settings. But, water that passes through fractures is not purified in the same way it would be if it traveled through tightly compacted soil and porous geologic materials such as sands or gravels. While this is not necessarily a problem when considering water movement and aquifer recharge, it is a critical issue when recognizing the kinds of contaminants that can move with that water. Whether it is regional nonpoint source contaminants like agricultural fertilizers and pesticides, urban lawn and garden chemicals, or highway deicing road salts, or more specific sources of contamination such as accidental spills, these materials can be carried into the underlying

groundwater. In addition, contaminants from point source locations also contribute to groundwater pollution. These point sources can be as varied as on-site septic systems, underground storage tanks, or historic abandoned unlined landfills.

Water traveling down from the surface in fine-grained materials via fractures to underlying groundwater aquifers is said to move through secondary or double-block porosity instead of through primary porosity (i.e., spaces between the grains of the glacial or alluvial materials). This dominance of secondary porosity causes the rates of gravity drainage to be very rapid, reaching as much as two to three orders of magnitude faster than laboratory hydraulic conductivity measurements would indicate based on primary porosity in unconsolidated porous materials (McKay et al., 1993; Haefner, 2000). The movement through fractured rock can be even faster.

11.6.1 FRACTURE FLOW IN ROCK

Even when hydrogeology was in its infancy as a science, hydrogeologists recognized the importance of fractured rock as prolific groundwater aquifers. Especially in limestone, dolomite, and sandstone settings, fractured rock aquifers have been relied on for community and industrial water supplies. It was soon recognized that water moved through the rock's joints, bedding planes, and fractures much more rapidly than it did through the rock matrix. Field techniques, including fracture trace or lineament analysis, were developed to locate fracture intersections. Wells installed in these intersections produced the greatest yields.

Analytical equations and computer models for groundwater flow in fractured rock settings have been developed (Wu, 2002); however, there is no consensus yet on the best method. Three common approaches are (1) dual-porosity models, (2) discrete mapping of individual fractures, and (3) equivalent porous media modeling. While the process is still not as well understood as groundwater flow through sands and gravels (porous media), the science is advancing.

Fractured conditions in rock have been recognized longer than for unconsolidated materials such as glacial tills and clay-rich lacustrine deposits. Many of the mechanics that control fracture formation and flow in bedrock also control fracture formation and flow in unconsolidated materials. Since rock is consolidated, it is easier to recognize the solid matrix portions and the joints, bedding planes, and fractures that surround the matrix materials. When looking at an outcrop, fractures in rock are easily observed. A matrix section can be picked up by hand and the fracture face surfaces studied and measured more easily than when that same material is still unconsolidated and may crumble when handled. It is this ease of observation and familiarity that has allowed research in rock fracturing to proceed and outpace the corresponding research in unconsolidated fracture flow.

11.6.2 FRACTURE FLOW IN UNCONSOLIDATED MATERIALS

The occurrence of fractures in glacial till and other fine-grained materials is well known and has been attributed to several geologic processes, such as desiccation, freeze–thaw, glaciotectonics, and till deposition (Brockman and Szabo, 2000); see Chapter 5 for more details. Although the fractures were recognized, their hydraulic implications were less well understood.

For years, engineers and most geologists considered these glacial deposits slowly or very slowly permeable because they did not recognize that the fractures were hydraulically active. Laboratory tests on small samples confirmed the assumption that infiltrating rainwater moved very slowly through these materials. However, irregularities, such as fractures, were commonly excluded or even purposefully removed from samples submitted for laboratory testing because they were thought to have occurred during the sampling procedure (Haefner, 2000).

Traditional analyses based on Darcy's law, standard engineering tests, and assumptions of primary porosity flow are not suitable for fractured environments and can produce incorrect results in these cases. An example of this discrepancy at a site known to have observed fractures was discussed by Tandarich et al. (1994), who stated that "at the Wilsonville Hazardous Waste Site in Illinois, contaminants migrated at unpredictably high rates. Hydraulic conductivities were 100 to 1000 times greater than expected. This could not be explained by standard engineering descriptions and tests."

11.7 GROUNDWATER VULNERABILITY

There is a growing need among policymaking, regulatory, scientific, and extension personnel to evaluate and predict groundwater vulnerability systematically on a regional scale. *Groundwater vulnerability* is defined as the pollution potential or relative likelihood that an aquifer will be impacted in the event of a contaminant release at the ground surface. Given finite resources, groundwater protection efforts should focus on those aquifers most susceptible to contamination or most essential for sustained water supplies.

One existing tool for assessing groundwater vulnerability or delineating protection zones is DRASTIC (Aller et al., 1987). This method was developed for the USEPA in 1987 and has been widely adopted. Many maps of groundwater vulnerability have been developed nationally and internationally. DRASTIC maps are produced by combining seven input variables; the name is an acronym for the seven variables: depth to water, recharge, aquifer media, soil media, topographic slope, impact of vadose zone media, and hydraulic conductivity. DRASTIC is currently under extension to support unconsolidated fractured settings. This information can be used to determine sustainable regional groundwater recharge and to aid in protecting groundwater resources.

PROBLEMS

These problems are taken from Peters (1987) and relate to **Figure 11.22**, which shows a hydrogeologic cross section of an area in northwest Indiana. Site-specific field data indicate that the transmissivity of the confined aquifer is approximately 13,000 ft²/day, and the coefficient of storage is 0.000043. Well I6-1 is an irrigation well that is used periodically during the growing season.

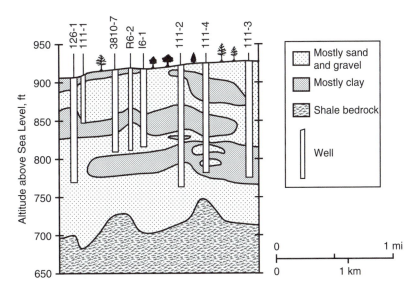

FIGURE 11.22 Diagram for use with Problems 11.1 to 11.5.

11.1. What would be the expected drawdown 1,000 ft away from Well I6-l if the well is pumped continuously at a rate of 110,000 ft³/day for 30 days?

11.2. At what rate could Well I6-l be pumped for 30 days so that the resulting cone of depression would not exceed 5.0 ft of drawdown at a distance 1 mi from it?

11.3. How much time would be required for Well I6-1 pumping continuously at 110,000 ft³/day to cause 1.0 ft of drawdown 1 mi away?

11.4. After 1 day of pumping at 110,000 ft³/day, at what distance from Well I6-1 would drawdown be 1.0 ft?

11.5. Well 3810-7S is located 1,800 ft away from Well I6-l. What would be the drawdown at a point halfway between the two wells after 30 days of continuous pumping assuming Well 3810-7S pumps at 70,000 ft³/day and Well I6-l pumps at 110,000 ft³/day?

12 Human Impacts on the Hydrologic Cycle: Prevention and Treatment Strategies

12.1 INTRODUCTION

Anthropogenic (human) activities often result in a need to modify and manage runoff and stream discharge: (1) to prevent, reduce, or control adverse impacts associated with changes in both the low and high discharges; (2) to provide a water supply; (3) to enhance water quality; (4) to generate power; (5) for recreational purpose; or (6) to protect, enhance, or restore stream health. Management strategies are diverse and might range from the use of rain barrels or retention and detention ponds to reduce runoff associated with urbanization, the installation of culverts under roads, construction of a vane in a stream for bank stability purposes or to provide fish habitat, or to the construction of large multipurpose dams that provide water, control floods, generate power, and have recreational value.

The purpose of this chapter is to provide sufficient information for both engineers and scientists to gain an understanding of the complex issues associated with discharge management. The chapter is not intended to be comprehensive and should not be used to prepare final designs for hydraulic control structures. Strategies that involve the placement in a channel of a physical structure, called a *hydraulic control structure* (see Chapter 8) need to be designed by professional engineers in consultation with scientists. In the U.S., we have many examples of large dams and bridges that are truly astonishing engineering feats. We also have countless examples of bridges and culverts that allow us to cross streams and rivers safely without fear of collapse or flooding. However, we also have countless situations when these same structures have adversely impacted on the channel system. Some fundamental equations that describe discharge conditions associated with hydraulic control structures are presented in Chapter 8.

In the last few decades, we have seen increased interest in restoring streams or incorporating more natural designs when there is a need to make a modification to a stream system. The knowledge and expertise of people who develop and implement these designs vary considerably, and many projects are only partially successful even when designed by some of our leading experts. The reasons for these high failure rates are numerous and include the complex nature of stream and watershed interactions, the empirical nature and uncertainties in the various equations used to develop the design, the dynamic nature of land use change on many of our watersheds, and the often-unrealistic desire to establish relatively instantaneous stability. Natural ecosystems evolve over long periods of time, ranging from months, to years, to decades, and even to centuries. During this period, there might be very slow or very rapid changes in the landscape, and the terrestrial and aquatic biology depend on the system and the variability of climatic factors. In restoration projects, there is always the expectation that the design is perfect, vegetation will establish according to plan, and stream stability will be achieved before there is an extreme event that will damage the newly established stream geometry. Rarely are we fortunate enough to have an ideal combination of all these factors, so there is often a need to "baby-sit," repair, and maintain natural stream designs. A better approach might be to reduce the amount of initial engineering of the system and to assist nature in developing a more self-sustaining flow regime.

A wide range of runoff methods has been developed to reduce adverse impacts on water quality and stream health associated with land use and land use change activities. In this chapter, we primarily focus on the use of strategies and structures that modify the partitioning of precipitation between runoff and infiltration or temporarily store some of the runoff to provide an improvement in water quality. First consideration is given to how human activities impact the stream biota and to assessment methods that might be used to assess stream health. Further discussion is then presented on the difficulty of conducting frequency analysis, flood forecasting, and flood discharge management. The impacts of urbanization on stream geomorphology are then discussed. Strategies that might be used to control some of the adverse impacts of human activities on stream health are presented. For illustration, the following applications are discussed: (1) methods to reduce the amount of runoff from urban areas; (2) a multipurpose reservoir; (3) the use of sediment ponds to reduce sediment discharges from mining and construction activities; (4) detention/retention ponds to reduce urban impacts on water quality; and (5) the use of constructed wetlands to reduce adverse impacts associated with animal waste disposal. It should be noted that each of these different types of structures has application to other pollution scenarios. Waste disposal is a major societal problem, and we end the chapter with a discussion of the hydrology of landfills, particularly municipal waste disposal. The issues relating to this topic would also apply to other wastes, including hazardous wastes, mine spoils, and waste products associated with industrial activities.

A comprehensive presentation of all these topics and other topics relating to the impact of human activities on hydrology would warrant a separate text. Therefore, the materials presented in this chapter should be viewed as a representation of the types of issues we face, the difficulties associated with developing adequate treatment strategies, and the need for prevention rather than treatment. We use the term *stream health* to depict qualitatively the combination of discharge, water quality, biological and ecological functions and interactions associated with a stream channel and the adjacent floodplain.

12.2 HUMAN IMPACTS ON STREAM HEALTH

"Human activity has profoundly affected rivers and streams in all parts of the world to such an extent that it is now extremely difficult to find any stream which has not been in some way altered and probably quite impossible to find any such river" (Hynes, 1970). Watersheds of streams and rivers have been highly modified over time. Of the variety of modifications (aside from the damming of a river), agricultural, silvicultural, mining, and urbanization activities typically have the greatest impacts on stream biota. Removal of streamside vegetation (i.e., the riparian zone) will result in a variety of changes to stream systems, including: increased water temperature because of reduced shading, altered channel structure related to the removal of woody debris, fewer leaf inputs and a reduction in the retention of these allochthonous sources (Gregory et al. 1991). When allochthonous inputs are reduced, a stream can shift from heterotrophic to autotrophic production (see discussion of river continuum concept in Chapter 6). In addition, removal of riparian vegetation can alter the natural hydrology, making streams and rivers more susceptible to droughts and floods. Changes in the flow regime either due to reductions in baseflows associated with power generation and water supply reservoirs or due to large and more frequent flows associated with urbanization can also greatly alter riparian vegetation and the ecology of these linked channel and floodplain systems.

Agricultural activities that have an impact on stream biota include impoundments, channelization, and drain tiles. These modifications to the stream environment can enhance runoff from storms and lower water tables. The ultimate result of these activities has been profound in Ohio, where many perennially flowing streams have become intermittent (Trautman, 1981). Agricultural practices also can increase nutrient levels related to fertilizer and animal waste runoff. However, urbanization may be just as important as agriculture in affecting stream nutrient loading (Osborne and Wiley, 1988). Sedimentation is typical of agricultural and urbanized landscapes, and many stream biota are intolerant of silt. Many scientists have called sedimentation the worst form of pollution to stream systems (Waters, 1995).

These sources of habitat degradation are not, however, insurmountable obstacles to improving stream water quality. Use of minimum tillage agriculture and buffer strips (either grass or forested) can reduce erosion and nutrient inputs to streams. Although vegetation buffer strips have been widely recommended (Karr and Schlosser, 1978), their long-term effectiveness has yet to be determined. To understand better how these various sources of anthropogenic disturbance affect stream biota, a variety of biomonitoring techniques have been developed for invertebrates (Plafkin et al., 1989), fishes (Karr, 1981), and habitat (Yoder and Rankin, 1998).

12.2.1 BIOLOGICAL ASSESSMENT METHODS

In this section, the bioassessment tools we discuss were designed for implementation, management, and monitoring in Ohio and the Midwestern U.S. These tools were developed to help meet the goals set forth in the Clean Water Act and monitor the effectiveness of the water quality management programs designed by each state. However, they have begun to see application in the western U.S. and, to a lesser extent, around the world.

In the past, state water quality management programs focused on chemical sampling criteria and point source pollutants to determine whether attainment goals were met. This resulted in a severe underestimation of the percentage of stream miles in Ohio that were in nonattainment status. The introduction of a method to classify results of biological sampling numerically resulted in an increase in the percentage of nonattainment stream miles from 9% in 1986 to 44% in 1988. The implementation of biocriteria in addition to the water quality criteria when assessing streams in Ohio resulted in the identification of 50% more impairment of streams than identification using water quality alone (Yoder and Rankin, 1998).

The use of biocriteria provides the opportunity to take the variability of nature and ecological systems into account in a way that water quality alone cannot and in a way that was previously overlooked in state water quality management programs. Because water quality is still an important parameter in assessing the health of streams, the intention of introducing biological integrity measures is to supplement, not replace, the traditional approach.

According to the Ohio Environmental Protection Agency (OEPA), and as published in the work of Yoder and Rankin (1998), there are three major objectives to be met by biological, chemical, and physical monitoring and assessment techniques:

1. Determine the extent use designations assigned in the Ohio Water Quality Standards are either attained or not attained.
2. Determine if use designations assigned to a given water body are appropriate and attainable.
3. Determine if any changes in biological, chemical, or physical indicators have taken place over time in response to point source pollution controls or best management practices for non-point sources.

In Ohio, OEPA uses bioassessment methods, in addition to water quality sampling, to categorize streams according to one of five different aquatic life uses based on the results of data collected at sites along each stream. The aquatic life use designations are as follows:

1. Warm-water habitat (WWH) — represents most of the streams and rivers in Ohio and represents the designation most targeted for water resource management efforts.
2. Exceptional warm-water habitat (EWH) — represents streams and rivers in Ohio that support "unusual and exceptional" assemblages of aquatic organisms.
3. Cold-water habitat (CWH) — represents streams that support cold-water organisms or are stocked with salmonids with the intent of providing an out-and-take fishery on a year-round basis.
4. Modified warm-water habitat (MWH) — represents streams and rivers with extensive hydro-modifications, to the point that WWH uses are not attainable; characterized by organisms that are tolerant of low dissolved oxygen, high suspended solids, and poor habitat.
5. Limited-resource water (LRW) — represents streams with drainage areas less than 3 mi^2 and streams and rivers in which no appreciable assemblage of aquatic life can be supported.

These life uses are determined using various bioassessment indices to attach quantitative values to qualitative measurements. These include the Qualitative Habitat Evaluation Index (QHEI), the Index of Biotic Integrity (IBI; Karr, 1981; Fausch et al., 1984; Karr et al., 1986), the Invertebrate Community Index (ICI; OEPA, 1988b;

DeShon, 1995), and the Index of Well-Being (Iwb; Gammon, 1976; Gammon et al., 1981).

The QHEI is "a physical habitat index designed to provide an empirical, quantified evaluation of the general lotic macrohabitat characteristics that are important to fish communities" ("OEPA Factsheet: QHEI Methods — Updated Draft "). The maximum score for any site is 100 and is based on the sum of the scores of six individual metrics, including substrate, in-stream cover, channel morphology, riparian zone and bank erosion, pool/glide and riffle–run quality, and map gradient.

The IBI was developed by Karr (1981) using data collected on entire fish communities. The results from the data are summarized in the IBI as 12 metrics clustered into three main groups: species richness, trophic composition, and fish abundance and condition. Ideally, an index of biotic integrity would detect all stresses humans put on a biological system while limiting the effects of natural variation in the physical and chemical makeup of the system. "The index of biotic integrity was conceived to provide a broadly based and ecologically sound tool to evaluate biological conditions in streams" (Karr, 1987). The attributes incorporated into the IBI take all ecological levels into account, from individuals to ecosystems. It is important to remember that, when using the IBI outside the Midwest, it must be modified to "reflect regional differences in biological communities and fish distribution" (Karr, 1987).

A similar index was developed by OEPA (1988b) to quantify macroinvertebrate populations. The ICI uses 10 metrics to emphasize "structural attributes of invertebrate communities" (Karr, 1987). The ICI samples the macroinvertebrate community in a stream vs. the fish community.

As with any assessment method, there are shortfalls to the QHEI method. The most significant shortfall of the approach is to rank streams with small watersheds (less than 3 mi^2) as LRW automatically simply because they cannot support a fish population. It is logical that, once a stream is smaller than a certain size, fish cannot populate due to lack of habitat and the depth necessary for fish to breed and survive. However, it does not mean the stream is devoid of biology. In fact, these streams, categorized as primary headwater habitat (PHWH) streams, are an important link between terrestrial and aquatic systems; the better quality they are, the better quality the entire stream network is as a result.

To overcome this shortfall of the QHEI method, OEPA developed a Headwater Habitat Evaluation Index (HHEI) that is more appropriate for streams with this size drainage area, and this index is designed to provide rankings, similar to the aquatic life uses above, for primary headwater streams. It examines the quality of habitat for macroinvertebrates and salamanders, organisms more likely to be found in small systems than extensive fish populations, to determine the quality of the stream and determine an appropriate level of protection for these previously over-looked and underappreciated systems.

A key element in using any assessment method is understanding the benefits and limitations of any particular method utilized to monitor a stream. For example, the methods discussed above may work very well in Ohio, but the scores obtained, and the metrics used, would be completely inappropriate for a stream located outside the state, and a result may not then reflect the true nature of the stream. Another issue that needs to be addressed is the variability of results depending on who is filling out the form. One person's estimation of percentage bed material of a certain size may be much more accurate than another person's estimate. One person's final score may be much higher than another's final score. The attempt to attach numbers to qualitative data is a very inexact science and can result in high score variability.

Unfortunately, examination of effects of anthropogenic modifications on stream biota cannot easily be conducted just at specific sites as the source of impacts at this scale can be found far upstream and often at the level of whole watersheds. Understanding the stream continuum and how streams interact with the landscape will be necessary to manage stream biota effectively. This is not a small undertaking; as a result, streams have been referred to as a "conservationist's nightmare" (Ladle, 1991).

12.3 URBAN IMPACTS

Urbanization increases impervious land uses, reduces infiltration, and causes more runoff and higher peak discharges. **Figure 12.1** illustrates the impact of urbanization on the frequency of different runoff events in central Ohio. In rural areas with wet soil conditions at the start of a storm event, there might be only one runoff event of 0.4 in. annually and an 0.8-in. runoff event every few years. However, a low-density urban area or an urban area with dry soil conditions at the start of a storm event might annually experience six or seven 0.4-in. runoff events, two or three 0.8-in. events, one 1.0-in. event, and even an event with 1.2 to 1.4 in. every few years. High-density urban areas, or urban areas with wet soil conditions at the start of a storm event, will annually experience more than twenty 0.4-in. runoff events, six 0.8-in. events, two 1.0-in. events, and one 1.2- to 1.4-in. event.

The hydrologic effects of large urban areas can be very significant. **Figure 12.2** shows the increases of water yields (depth of runoff) and storm peaks in the Los Angeles River as the result of urbanization over the period 1949 to 1979. Note that urban areas have the potential to "harvest" water by increasing runoff, but the utility of this harvested water is limited because of water quality and because there is usually no place to store the additional runoff. Water-short cities may eventually need to find ways to use this excess runoff.

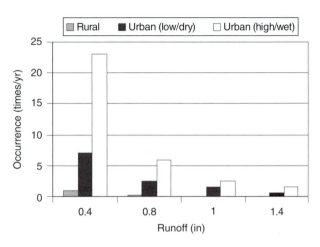

FIGURE 12.1 Annual occurrence of different magnitude runoff events in central Ohio as a function of urbanization.

The effect of urbanization on discharges is illustrated in **Figure 12.3**. A rural area will have a basin development factor (BDF) of 0, while a dense urban area will have a basin development factor of 12. The plot shows the relative discharge increase as a percentage of the rural discharge and is based on a watershed area of 2 mi^2. Similar relative discharge changes will be obtained for other size areas. Plots are presented for return periods of 2, 10, and 100 years. Of particular concern is that 2-year return period discharges on dense urban areas are 2.5 times larger than those on rural areas (**Figure 12.3** and **Figure 12.4**).

Various studies have evaluated the impact of urbanization on stream health or stream geomorphology. Finkenbine et al. (2000) studied 11 streams near Vancouver, British Columbia. The total impervious area (TIA) on these watersheds ranged from 5 to 77%. They found that, as the TIA increased from 5 to 77%, the D_{84} more than doubled, primarily due to increased discharges, fluvial erosion, and mass failure (sloughing) of the banks. They also noted that the loss of large woody debris was offset by the establishment of coarser bed material, and spawning conditions might not have degraded. They concluded that streams might adjust to urbanization within 20 years of watershed urbanization, and the highest priority during urbanization should be the establishment (or, we assume, the retention) of a healthy riparian zone.

Doyle et al. (2000) examined the effect of urbanization on several streams in Indiana. They noted that "preservation or restoration of both channel stability and diverse ecological communities may best be approached through replicating the frequency of bed mobilization found in natural, stable channels." In their study, streams responded to urbanization by incision and bed widening (**Figure 12.5**). They found the only significant differences in stability measures were when low-density urbanization (0 to 2%) was compared to medium- or high-density urbanization (7 to 32%). The recurrence interval of the critical discharge

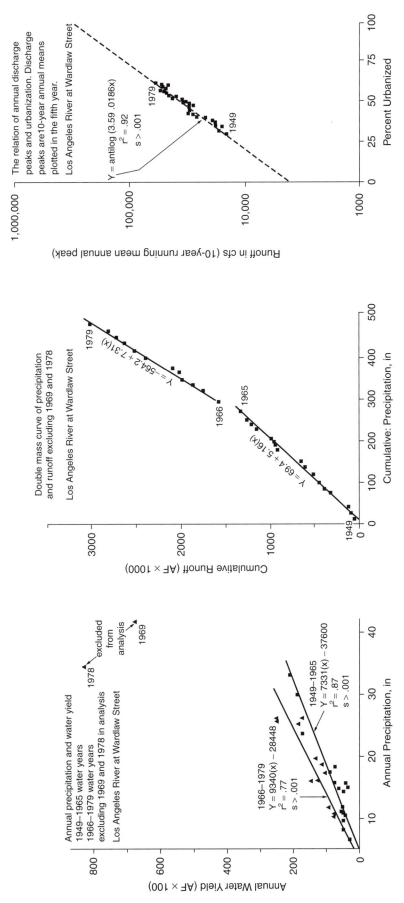

FIGURE 12.2 Increase in water yields (depth of runoff) and storm peaks in Los Angeles as the result of urbanization over the period 1949 to 1979. (From Barbara Hoag, unpublished study, 1982.)

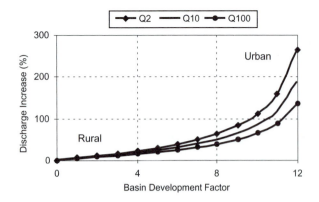

FIGURE 12.3 Relative discharge for 2-, 10-, and 100-year return periods as a function of an index of urbanization (BDF = 0 rural; BDF = 12 urban).

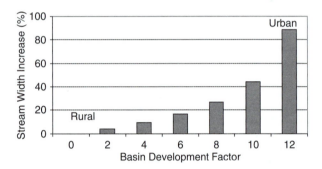

FIGURE 12.4 Relative channel width increase as a function of an index of urbanization (BDF = 0 rural; BDF = 12 urban).

FIGURE 12.5 Channel widening and downcutting.

decreased from 183 days for high-density areas compared to 88 days for low-density areas. The *critical discharge* is defined as the discharge at which the D_{50} or D_{84} bed material particles are mobilized. The bedload transport procedure Mecklenburg and Ward (2002) used also shows that the recurrence interval becomes smaller (more frequent) with increased urbanization; they suggested that there is a low threshold of urbanization that, when exceeded, might cause large changes in the quantity and size of bedload mobilized.

Booth and Jackson (1997), in a study conducted in western Washington State, found that "at approximately 10% effective impervious area (EIA) in a watershed typically yields demonstrable and probably irreversible, loss of aquatic-system function." They noted that EIA is difficult to measure, and one of the differences between it and the TIA is that any portion of the TIA that drains to a pervious area is excluded in obtaining the EIA. Mecklenburg and Ward (2002) had no knowledge of how the BDF used in the study related to the TIA or the EIA. Suburban areas in Ohio with large lawns, parks, and wooded or natural riparian zones along the main streams often have a BDF of 5 to 7. These areas might have an EIA higher than 10%, but their results suggest that bedload mobilization impacts increase slowly as the BDF increases from 0 to 4, show more rapid changes as the BDF increases from 4 to 8, and show large increases when the BDF increases from 8 to 12.

Brookes (1996) discussed successful restoration projects in Denmark, England, Germany, and the U.S., where a multistage channel and floodplain system was established. The *streamway* (a river corridor, recreation zone, secondary floodplain, or floodplain that might have existed historically) in these projects ranged from about 6 to 20 times the bankfull width of the restored low-flow channel. The streamway or floodplain width ratio we identified as necessary to maintain dynamic equilibrium falls within this range.

Successful stream stewardship requires combining knowledge of natural stream concepts with sound engineering and scientific principles and an understanding and appreciation of the ecology of the stream and its interaction with the landscape. Poor understanding of these processes and inadequate consideration of the influence of changes that occur on the landscape and within the floodplain can cause a variety of adverse outcomes. Particular attention needs to be paid to the potential impact on a river of (1) land use changes that reduce vegetation and increase the amount of impervious area; (2) activities that modify the floodplain; (3) the construction of culverts and bridges; and (4) activities designed to modify the characteristics of the main stream channel. Any of these activities might change the course of a stream, result in the stream becoming deeper or wider, increase scour, or cause failures along the stream banks. Furthermore, such changes might have adverse impacts on the aquatic biota and ecosystems within the vicinity of the river. Consideration needs to be given to (1) landscape measures that reduce runoff, such as reduced paved surfaces and practices to maintain or enhance infiltration and (2) detention/retention management strategies that

FIGURE 12.6 The Portage River in Wood County, Ohio.

result in similar post- and predevelopment bedload and sediment transport amounts.

12.4 FREQUENCY ANALYSIS

A frequency analysis is presented in several places in this book. In Chapter 2, we used Equation 2.2 to Equation 2.4 to look at the magnitude and frequency or probability associated with extreme precipitation events. We also considered frequency plots in **Figure 2.17** to **Figure 2.22**. Similar types of relationships were presented in Chapter 5 (see **Figure 5.4**). We also calculated the magnitude of different recurrence interval discharges using USGS empirical peak discharge equations or NRCS methods, such as the graphical peak discharge method. The terms *return period* and *recurrence interval* can be used interchangeably. In Chapter 6, we considered the importance of the effective discharge, how it might be determined, and how frequently it might occur. Also, in Chapter 7 and Chapter 8, we considered the magnitude, velocity, and depth of flow associated with different recurrence interval discharge events. In this chapter, we consider strategies for reducing the adverse impacts associated with a range of different conditions. We also consider flood events, flood forecasting, and peak flow reduction strategies. There are many additional applications of a frequency analysis in hydrology or other scientific, engineering, and economic situations. It is beyond the scope of this text to consider all applications and all statistical approaches that might be considered for obtaining a relationship between the magnitude of an event and the frequency it might occur, not occur, or be exceeded. However, this is an area that often creates much confusion, even among highly qualified experts, so in this section, we clarify the main issues associated with performing and understanding a frequency analysis.

Let us consider an analysis of discharge data for the Portage River at Woodville, OH (**Figure 12.6**). The Portage River flows northward and discharges into Lake Erie. At Woodville, the drainage area is 428 mi^2, the mean annual discharge is 336 ft^3/sec, and data are available from 1929. Let us suppose that we need to consider the installation of a two-stage culvert at a road. The Ohio Department of Transportation requires that the 50-year return period not overtop the road. A conservancy group requests that the main channel be constructed to convey the 2-year return period discharge as calculated by the annual peaks method. However, research in this watershed indicates that the effective discharge should be based on a discharge that is more frequent than the 1-year return period discharge as calculated by the annual peaks method. Also, for insurance purposes, we have been requested to plot the 100-year event flood zone on a topographic map of the watershed. To learn more about stream morphology and how to conduct the analysis, we attend a workshop that a professor is giving on stream geomorphology and how to design culverts. We are surprised to hear the professor say that, on average, the effective discharge will probably occur more than once, but less than a dozen times, annually at this location. What should we do?

First, we make an analysis of 60 years of data for this USGS gage (1940 to 1999). We do not use data prior to 1940 because of gaps in the data and concerns about the gage. At the time this analysis was performed, data after 1999 had not been verified by the stringent QA/QC (quality assurance/quality control) methods used by the USGS. We first use the most common method, which is to perform the analysis on the peak discharges for each of the years. This data set is called an annual duration series or annual floods series; in this case, we have 60 values, 1 for each year. An alternative approach is to use all the peaks above some threshold. This data set is called a partial duration series or a peaks-over-threshold series. The number of values will depend on the threshold that is selected. Although no longer readily available, through their NWIS database, the USGS has developed extensive data sets based on using a threshold value that is the lowest annual peak value in the period of record. The partial duration series are currently stored in the NWIS database, but are not available through the NWIS Web server. They are, however available on request. The threshold value is computed by the Weibull method (see Equation 2.3, Chapter 2). The threshold value for the partial duration series reported by the USGS is a discharge with a probability value of 0.87. Annual peak discharges that do not exceed the threshold value for the partial duration series are not included in the partial duration series.

For this example, we used the lowest annual peak during the 60 years as the threshold discharge for the partial duration series. The lowest annual peak occurred in 1941 and was 2040 ft^3/sec. In the 60-year period, there were 436 peak discharges equal to or greater than 2040 ft^3/sec. We have two problems with using these data. First, each peak is not associated with a particular time period (for the annual series, we have one peak for each year). Second, not all the peaks are independent of each other. Many of the peaks are associated with the same discharge event that created high flows for each day. Gage height data are the only data recorded at most USGS streamflow gaging stations, unless there is a device that records discharge (such as an acoustic Doppler flowmeter). Modern measuring devices can record and electronically report data via satellite links very frequently (seconds to minutes). The gage height data are generally recorded at record intervals ranging from about 5 min to 60 min. Computer software is used to process the gage height data, to compute discharge data values for the same record interval, and to compute the daily discharge values for each day. However, much of the data in long-term data sets such as this one (we could have even gone back to 1929) would have been recorded by a pen on a strip chart that rotated on a drum that moved mechanically using a clockwork mechanism. Extracting data from the charts was performed manually and was a tedious and time-consuming experience; we make this comment based on the numerous hours we spent early in our careers doing this. Therefore, the peak is not truly instantaneous (it might have been based on a period of 5 to 15 min), and it was not practical to generate a hydrograph for each discharge event, including some that occurred over several days.

When considering sediment transport, the duration of the high flows is probably more important than knowledge of the instantaneous peak value. Therefore, attempting to separate multiple peak events into separate hydrographs might have little practical value if the daily mean discharge and the "instantaneous" peak discharges are similar. For small watersheds, a few square miles in area, there will usually be a considerable difference between the instantaneous peak discharge and the daily mean discharge. However, when the watershed area exceeds several hundred square miles, the differences become small. For this 428-mi^2 watershed, the difference in daily mean discharges vs. instantaneous peak discharges is only a few percent, and for the purpose of this analysis, we used the peak daily mean discharges.

Many statistical methods have been developed to convert partial duration series data into a data set that can be related to the period of record. A good reference on annual and partial duration series is the work of Haan (1997). Useful scientific articles include those by A.A. Bradley and Potter (1992), Burn and Boorman (1993), Ekanayake and Cruise (1993), Haktanir and Horlacher (1993), Hosking and Wallis (1993), and Zrinji and Burn (1994). For the purposes of this discussion, we assume that we do not have any knowledge of statistics beyond what we have learned from this text. Therefore, let us consider the following straightforward pragmatic ways we might use to obtain a partial duration data series:

1. We will assume that we can use all 436 values and treat them as if they actually represent 436 years of data.
2. We will use the 60 highest values and assume they represent 60 years of data.
3. We will use the 60 highest values that are independent of each other (all different discharge events) and then assume they represent 60 years of data.

Conceptually and statistically, the first approach has the least merit, and debatably the last approach has the most merit, but each is approximate. However, even the most sophisticated statistical approach will not provide an exact answer, and we should also recognize that the measurements depend on some approximations regarding the cross-sectional geometry, bed and bank roughness, and velocity distribution within the cross section.

We now need to decide how to analyze these data sets. In Chapter 2, we presented two different plotting methods (Equation 2.2 and Equation 2.4). We also suggested that we could plot the data on probability paper or using a log scale in a spreadsheet program. Once we have a plot, we will fit a trend line (regression) line through the data. Again, there are many statistical methods for performing this analysis, but if we obtain a high r^2 value (this is called the *coefficient of determination*), perhaps 0.9 or higher, we will assume our approach provides a useful estimate. An r^2 value of 0.9 tells us that the trend line accounts for

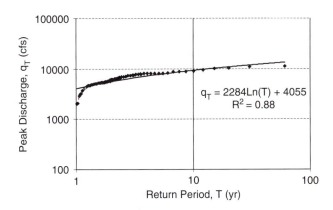

FIGURE 12.7 Discharge associated with different periods based on an annual duration series.

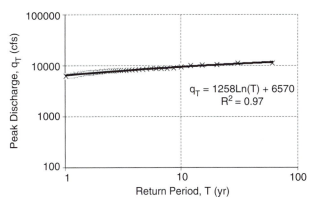

FIGURE 12.9 Discharge associated with different return periods based on a partial duration series ($n = 60$).

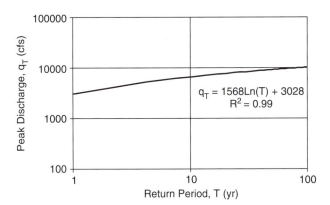

FIGURE 12.8 Discharge associated with different return periods based on a partial duration series ($n = 436$).

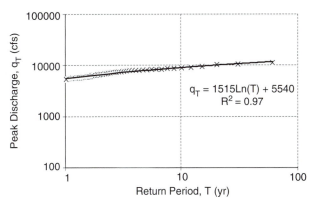

FIGURE 12.10 Discharge associated with different return periods based on an independent event partial duration series ($n = 60$).

90% of the variability in the data. If all the points fall on the same line, the r^2 value would be 1 (see Chapter 1).

The results of our analysis are presented in **Figure 12.7** to **Figure 12.10**. It can be seen that, for the annual series, the trend line does not provide a good fit at either end of the distribution. This suggests that, for both frequent and uncommon events, the results might be inadequate. This problem could be resolved by dividing the data set in two and fitting a trend line through the more frequent 1- to 10-year return period discharges and a second line through the more extreme 10- to 100-year values.

The partial duration series based on 436 values has a very high r^2 value, but all the discharges are lower than the annual series estimates (see **Table 12.1**). The partial duration series based on the 60 highest peaks (regardless of which years they occurred) gives frequent discharge values that are larger than the annual series estimates and extreme values that are smaller. We determined from an evaluation of this data set that 24 events were not independent from other events in the series of 60 values. The partial duration series that was based on the largest 60 independent discharges (42 values from the largest 102 events were eliminated to generate the series) provides

frequent peaks larger than those estimated from the annual series and extreme discharges that were slightly smaller. This data set included discharge values for 41 of the 60 years.

Now that we have four sets of discharges, we have to decide which, if any, we should use. The additional information in **Table 12.1** sheds some light on the confusion statistics can create and the dilemma we face. What we did to obtain the additional information in **Table 12.1** was to determine the total number of occurrences in 60 years of events larger than each predicted discharge. We then calculated the number of times per year on average that each larger event might occur. In 60 years, we might expect to have about 60 events larger than the 1-year event, 30 events larger than the 2-year event, 6 events larger than the 10-year event, and 1 larger than the 50-year event. Intuitively, the annual peaks series and the full partial duration series results appear to underpredict the frequent discharges because the number of events per year is more than might be expected based on the statistical definition of return period. We should remember that about 40% of the events are not independent of each other. However,

TABLE 12.1

Calculated Peak Discharges and Number of Occurrences in 60 Years, and per Year, for Different Return Period Events and Different Data Series

Data Series and Attribute	Return Period (years)					
	1	1.5	2	10	50	100
Annual (n = 60)						
Discharge (ft³/sec)	4,055	4,981	5,638	9,314	12,990	14,574
Occurrences	232	125	97	6	0	0
Times/year	3.9	2.1	1.6	0.1	0.0	0.0
Partial (n = 436)						
Discharge (ft³/sec)	3,028	3,664	4,115	6,639	9,162	10,249
Occurrences	444	309	224	49	6	1
Times/year	7.4	5.2	3.7	0.8	0.1	0.0
Partial (n = 60)						
Discharge (ft³/sec)	6,570	7,080	7,442	9,466	11,491	12,363
Occurrences	49	43	33	5	0	0
Times/year	0.8	0.7	0.6	0.1	0.0	0.0
Independent Partial (n = 60)						
Discharge (ft³/sec)	5,540	6,154	6,590	9,028	11,467	12,517
Occurrences	95	63	49	6	0	0
Times/year	1.6	1.1	0.8	0.1	0	0

even if we only take 60% of the times per year for the annual peaks series, we get 2.4, 1.2, and 1 times annually for the 1-, 1.5-, and 2-year return period events, respectively. The independent event partial duration series intuitively gives a nicer result because the 1-, 1.5-, 2-, and 10-year return period events occur on average 1.6, 1.1, 0.8, and 0.1 times annually, respectively. We would expect these results to be 1, 0.75, 0.5, and 0.1 times/year. There were also two observed peak discharges of about 11,500 ft³/sec during the 60 years. This corresponds well with the expectation that, on average, one 50-year event or larger would occur in a 50-year period.

Finally, let us answer the questions we set out to answer. Annual duration series are most commonly used, and they gave conservative estimates of 12,990 and 14,570 ft³/sec for the 50- and 100-year return period discharges, respectively. The independent partial duration series gave slightly lower, but similar, values. For design purposes, we might use the slightly more conservative answers because we are concerned about flooding. However, **Figure 12.7** shows that the trend line for the annual series data overpredicts the extreme discharges, so a case could be made for using the partial duration series results or fitting separate extreme and frequent discharge trend lines through the annual duration series data.

Deciding on an effective discharge estimate is problematic. The annual duration series gives 1- and 2-year return period estimates of 4055 to 5638 ft³/sec, respectively. Note that these results are consistent with the

professor's observation that the effective discharge might occur several times annually. The independent event partial duration series gives 1- and 2-year return period estimates of 5540 to 6590 ft³/sec, respectively. We would recommend that none of these values be used. Better approaches would be to conduct a detailed survey on a reference reach or to conduct a survey of streams in the watershed and then develop regional curves that relate channel dimensions and discharge to drainage area. Based on an analysis by Ward et al. (2003), the effective discharge at this location is between 2800 and 6200 ft³/sec and probably lies between 3500 and 4500 ft³/sec.

To evaluate further the difficulty in using observed discharge data, an analysis was made of the 237 events that exceeded the 1-year return period discharge of 4055 ft³/sec obtained with the annual peaks series (**Table 12.2**). It was found that these events occurred during 53 of the 60 years. In the remaining 7 years, there were no events equal to or greater than the effective discharge estimate of 4055 ft³/sec. In one extremely wet year, there were 12 events with discharges greater than 4055 ft³/sec. However, in only 5 of the 60 years were there more than 8 events that exceeded or equaled the effective discharge. When assigning a return period to the effective discharge, we also need to eliminate extreme events. For example, if the effective discharge is associated with the 1-year return period discharge, we might not want to consider events larger than, say, the 2-year return period discharge (or whatever threshold we choose). Therefore, in the 60 years,

TABLE 12.2
Events That Exceed the 1-Year Return Period Discharge of 4055 ft³/sec

Events/Year	Number Years	Percentage Total Years	Total Events	Percentage Total Events
0	7	11.7	7	3.0
1	7	11.7	4	1.7
2	7	11.7	14	5.9
3	8	13.3	24	10.1
4	6	10.0	24	10.1
5	6	10.0	30	12.7
6	6	10.0	36	15.2
7	2	3.3	14	5.9
8	4	6.7	32	13.5
9	1	1.7	9	3.8
10	2	3.3	20	8.4
11	1	1.7	11	4.6
12	1	1.7	12	5.1

there were 105 events between 4055 and 4981 ft³/sec or about 1.8 events annually. A better approach would be to consider events slightly larger and slightly smaller than the effective discharge. For example, for values between 3600 and 4500 ft³/sec, there were 151 events or about 2.5 events annually. A similar analysis could be performed with the discharge estimates obtained with the other methods.

In conclusion, we note that there are several more exact approaches that could have been used, but they require an extensive knowledge of statistics. Also, note that, in this analysis, we explored the observed data in several ways. Other ways could also have been considered. For example, we could have performed the analysis on all the daily mean flow values. A problem with this approach is that the results would be biased by numerous baseflow discharges. Also, on small watersheds, daily mean discharges are poor estimates of the peak discharge each day. We could have used other time intervals, such as a week, a month, or several months. However, evaluations of discharge data are not typically performed with these time intervals. We do recommend that all measured data be evaluated in a variety of ways. Too often, we are apt to enter or download data into a spreadsheet, fit a trend line to the data, and then misapply the trend line. Worse still, we often use somebody else's trend line or base a design on a perceived result, such as assuming the effective discharge is a 1-, 1.5-, or 2-year return period discharge.

12.5 FLOOD FORECASTING AND MANAGEMENT

12.5.1 FLOOD FORECASTS

Each new decade shows vast improvements in the accuracy of forecasts of time and crest stage of floods at critical points in river systems. Such forecasts have enabled local authorities to plan and carry out floodplain evacuation or to provide temporary increases in height of levees. In large basins, flood levels increase slowly day after day, and refinements can be made almost daily in the early forecast of crest stages as the crest approaches. The ability to forecast peak discharges accurately was illustrated during the devastating flooding of the Mississippi and Missouri Rivers that occurred in 1993. In most cases, river crests were correctly forecast to within a few tenths of a foot and, the time of the crest was predicted to within a few hours. This was truly remarkable considering the widespread extent of flooding and extended periods (many weeks) that abnormal rainfall and flooding persevered.

For many years, the National Oceanic and Atmospheric Administration (NOAA) has used flood forecast procedures on main streams of large river basins. There is at least one forecasting office in each major river basin of the country. Flood forecasting depends on extensive knowledge of land uses, topography, conveyance systems, and real-time climatic information. Presentation of the methodologies used in making flood forecasts is beyond the scope of this text. Reference should be made to texts such as that of Kraijenhoff and Moll (1976).

Flood forecasts on a similar basis for small watersheds are not yet possible. The small-area flood often results from a short-duration, local storm with a high rainfall rate. The storm may not have been recorded in the widely spaced Weather Bureau gage network. Also, a flood may crest in a few hours or less, and there is no time to warn occupants of the floodplain. In such situations, NOAA is aware of climatic conditions in which severe storms may occur and issues warnings of the possibility of flash floods. However, new breakthroughs in Doppler radar may soon change all the foregoing by giving real-time precipitation rates and totals.

FIGURE 12.11 Illustration of a 100-year flood zone (dark shading) in Franklin County, Ohio. (From www.franklincountyauditor.com/.)

For a headwater watershed, the peak flood discharge for a particular return period can be predicted from a probability plot of gaged data. The statistical approach is identical to that described for precipitation in Chapter 2. Methods for predicting peak rates and amounts of flood runoff on ungaged watersheds were presented in Chapter 5. The smaller the watershed is, the more accurately the peak discharge, runoff volume, and hydrograph shape can be predicted because (1) the rainfall–runoff relationship for a small area is simpler, and (2) channel storage influences on hydrologic relationships decrease with decreases in watershed size.

As the flood from a small headwater watershed proceeds downstream, its hydrograph shape and peak discharge are modified greatly by channel and floodplain storage. It may be better to develop hydrographs for the headwater tributary watersheds and route them downstream through the river channel storage system to the point of interest rather than use a single rainfall–runoff relationship for the entire watershed. This routing requires more effort than the direct estimation of flood volume and peak flow rate, and the result may or may not be worth the effort (see also Chapter 8).

12.5.2 FLOOD ZONES

Information on floodwater inundation depths and frequencies is valuable in land use planning or zoning for floodplains. For decades, the land in many floodplains has been used by residential, industrial, utility, transportation, and other highly developed business areas. The hydrologist obtains information on the flood hazards in the area needing protection, the engineer works out plans for structures required to provide protection and their cost, and the economist prepares reports on the dollar value of benefits

expected from the protection planned. If the benefits exceed the cost, the flood reduction plan is approved by state and federal agencies, and the project is funded. It is highly important that the community be informed and involved in projects of this type from the start, and that public hearings are held before plans are submitted for final approval and activation. Areas identified as high hazard from frequent flooding could be reserved for recreational use so that little damage results from floods instead of trying to justify the construction of costly flood reduction systems. But, in some cases where the land has a very high value for uses other than recreation, the flood reduction system may be the best solution. Generally, it is more expensive or impossible to obtain insurance for structures built on floodplains even when the return period for any potential flooding is 100 years.

A hydrologist develops data for a flood hazard evaluation for an area to show the stage of the flood, the flood zone boundary, the area inundated, the flood discharge volume, and the frequency of floods with different probability of recurrence (**Figure 12.11**). Field surveys supply basic information for flood hazard evaluations. Such information takes the form of stage vs. area inundated for various depths of flooding; flood peak discharge vs. area inundated; and frequency vs. area inundated. Field surveys are made in the area of flood concern to determine from local residents' testimony or recorded data on the elevation of high-water marks for as many annual floods as possible.

12.5.3 FLOOD MANAGEMENT

Reduction of flood flow may be accomplished in varying degrees in smaller basins by watershed land treatment in that the storage of stormwater is increased on the surface and in the soil profile, using floodwater detention structures, and increased groundwater storage. In arid regions, groundwater storage increases are accomplished primarily by diverting stormflow from the stream and spreading it over a considerable area so it can percolate to groundwater aquifers. Measures that retard or reduce runoff tend to conserve water resources, providing more water for crops and groundwater storage. These measures cause more uniform seasonal distribution of streamflow.

Land treatment measures include increased vegetation with more soil surface protection, along with practices such as terraces and waterway development and stabilization (see **Figure 9.1**, sediment budget section, Chapter 9). The choice of practices depends on climate, hydrology, soils, and economics. For example, we would not expect much flood-flow reduction from increased vegetation where the soil depth is very thin or from heavy clay soil because the storage opportunity is small and not readily available. Likewise, in areas of heavy and shallow soil, graded terraces with outlets will not reduce floods. But, on deep, well-drained soil, level terraces have a marked influence on flood flows as well as on water conservation. In the deep loess region of western Iowa, where there are level terraces with a retention capacity of 1.5 in. of runoff and an infiltration capacity of 0.3 in./h, it has been estimated that a flood peak from a 772-mi² watershed could be reduced from 66,000 ft³/sec to 45,000 ft³/sec by constructing level terraces on 17% of the area (Schwab et al., 1966).

Land treatment measures, however, have their limitations. While good infiltration rates may absorb storms of small-to-moderate size, extremely large (more than 100-year) storms completely saturate the soil so that one drop of rain becomes one drop of runoff. Thus, poor land use, such as overgrazed land (or, at the extreme, urbanized land), may increase the runoff from smaller (1- to 25-year) storms, but have much less effect on runoff from large storms. Size is also a major factor. While land treatment can make substantial differences in smaller basins (see the sediment budget section of Chapter 9), there is little difference for much larger basins. Land use played little if any role in the 1993 Mississippi–Missouri River floods.

There are two main types of flood control reservoirs: flood storage and flood detention. Floodwater detention reservoirs are mainly equipped with automatic outlets with fixed openings. Flood storage reservoirs are operated with adjustable outlet gates, but many of these reservoirs are used to detain floodwaters just long enough to prevent flood damage downstream before returning the reservoir pool level to its normal elevation as quickly as possible.

Headwater flood control reservoirs are designed for protection of a designated reach of channel downstream from the reservoir. For protection further downstream, other flood control reservoirs are needed. Both headwater and downstream reservoirs are needed for complete protection. This is illustrated in **Table 12.3** for a 500-mi²

TABLE 12.3
Example of Downstream Benefits of Headwater Flood Control Reservoirs

Drainage Area (mi²)	Number of Reservoirs	Peak Flow Reduction at Subwatershed Outlet (%)
10	1	90
100	6	40
500	25	30

watershed in that flood reduction evaluations are made for the effect of one reservoir on a 10-mi² subwatershed. Six reservoirs are located in the headwater areas of a 100-mi² subwatershed that is a tributary to the 500-mi² watershed. This is a large number of reservoirs. A flood control reservoir density of 1 for every 100 to 1000 mi² of watershed is more common in many parts of the U.S.

In regions of nearly level topography, floodwater detention reservoirs on the streams are not practical. Channel modification in these areas would be possible if not for the expense involved and public concern over the effect of such works on ecology. Nolte and Schwab (1971) reported that floodwater damage reduction in these regions may be possible by pumping excess storm runoff from streams into detention reservoirs located off stream and later releasing the stored water back to the stream in dry weather.

Small reservoirs and farm ponds have an unknown effect on flooding. There are at least 2.6 million of these (about 10 ha or 25 acre and smaller) in the U.S., mostly in the East. Such small reservoirs also affect sediment yields, geochemistry, and ecology (Smith et al., 2003).

12.5.4 CHANNEL MODIFICATION FOR FLOOD MITIGATION

An alternative, or sometimes a supplement, to flood mitigation techniques discussed is channel modification. For smaller streams, the process is usually to make the channel straighter, deeper, and smoother, which increases the slope and hydraulic efficiency. This increases the velocity, thus keeping the river at a lower stage for a given discharge and reducing the likelihood of overbank flow. Such channel excavations make an environmental mess, and the resulting earthen channels are usually unstable. We demonstrate such a project using a fictional location, River City, IA (**Figure 12.12**). Because of upstream urban development, floods have become more common, and the meandering stream that flows past River City can no longer convey the 1-year flood of 500 ft³/sec, so City Park is frequently flooded. Moreover, a flood study estimated that the 100-year flood would be about 4000 ft³/sec, and such a discharge would seriously endanger the city. Thus, River City

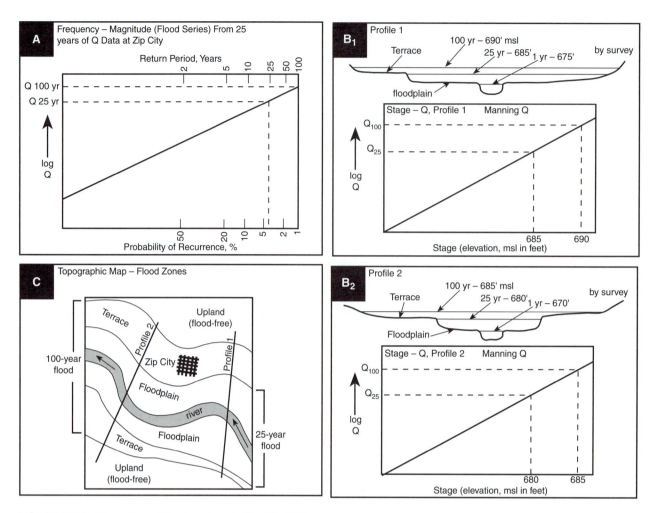

FIGURE 12.12 Illustration of flood zoning for Zip City, U.S.

is constructing a new channel that will convey all but the largest floods (**Figure 12.13**).

Inspection of **Figure 12.14** shows that, by virtue of shortening the channel to half of its former length, the slope is doubled. Moreover, straightening alone lowers Manning's *n* value (**Table 7.1**) because the stream no longer dissipates energy by flowing around bends. The new channel is a deep trapezoid that presents a better hydraulic cross section with a larger Manning *R* value. Initially, the channel was to remain earthen, but calculations of stream power showed that the new channel would probably erode, thus destabilizing the channel, creating an environmental and aesthetic mess, and sending a massive amount of sediment downstream, possibly impairing or even occluding downstream channel reaches or filling lakes or estuaries. For example, eroding "improved" earthen channels in urbanizing areas of Orange County, California, have furnished about two thirds of the sediment that has been filling Newport Bay, an important ecological reserve. Many channels there have eroded to several times their original size (see sediment budget section, Chapter 9). Control of erosion in such channels is difficult because

increased stream discharges produce so much force, and earthen channels, even when vegetated, offer so little resistance relative to the increased force. The frequent result is that channels have to be "hardened" with concrete stabilization structures or riprap to reduce or prevent erosion. In the worst cases, the channel bed and banks are built with solid reinforced concrete. While such channels may be unsightly, the old saying "beauty is as beauty does" applies here: These channels and banks do not erode (some urban property is worth millions of dollars per acre), the smooth concrete gives lower *n* values, and for intermittent channels, there are no stagnant pools where mosquitoes can breed during the dry season. Therefore, River City decided to install a concrete channel.

Note that the new channel can convey the estimated 100-year flood of 4000 ft³/sec. Any event greater than that would have to occupy the old floodplain, but that clearly would occur infrequently. The calculations of unit stream power in pounds per foot of width per second show that, at bankfull, the new channel develops over 40 times as much erosive and transporting power. Hence, the need for a concrete channel.

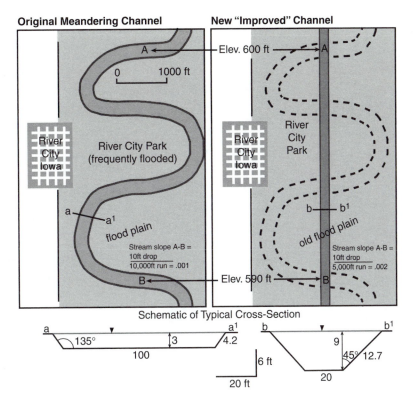

FIGURE 12.13 Construction of a new, "improved" channel for River City, IA.

Bankfull Discharge, Q$_{BF}$ = va (Equation 7.1)

Original Meandering Channel

a = 100(3) + 2(1/2)(3 × 3) ft^2
a = 309 ft^2

$v = \frac{1.5}{n} R^{2/3} S^{1/2}$ (Equation 7.2)

n = 0.045 (Table 7.1, stream on plain clean, winding)
S = 0.001 (see map above)

$R = \frac{a}{P} = \frac{309 \text{ ft}^2}{100 + 2(4.2) \text{ ft}} = \frac{309}{108.4} = 2.85$ ft

$v = \frac{1.5}{0.045} (2.85)^{2/3}(0.001)^{1/2}$ ft/sec

v = 33.3(2.0)(0.032) ft/sec
v = 2.1 ft/sec

Q$_{BF}$ = 309 ft^2 (2.1ft/sec) = 649 cfs

New "Improved" Channel

a = 9(20) + 2(1/2)(9 × 9) ft^2
a = 261 ft^2

$v = \frac{1.5}{n} R^{2/3} S^{1/2}$ (Equation 7.2)

n = 0.014 (Table 7.1, trowel finished concrete)
S = 0.002 (see map above)

$R = \frac{a}{P} = \frac{261}{20 + 2(12.7)} = \frac{261}{45.4} = 5.75$ ft

$v = \frac{1.5}{0.014} (5.75)^{2/3}(0.002)^{1/2}$ ft/sec

v = 107(3.2)(0.045) ft/sec
v = 15.4 ft/sec

Q$_{BF}$ = 261 ft^2 (15.4ft/sec) = 4020 cfs

Unit Stream Power at Q$_{BF}$ (in English Units)

Assume weight of water = 65 lbs/ft^3 (includes suspended sediment)

sp = zvDS

sp = (65 lbs/ft^3)(2.1 ft/sec)(3 ft)(0.001)

sp = 0.41 lbs/ft/sec

sp = zvDS

sp = (65 lbs/ft^3)(15.4 ft/sec)(9 ft)(0.002)

sp = 18 lbs/ft/sec

The improved channel can convey 6 times more discharge but the unit stream power is 45 times larger!

FIGURE 12.14 Unit stream power calculations that relate to the channels in **Figure 12.13**.

FIGURE 12.15 Large levees along the Yellow River in China.

Larger channels and rivers cannot be so easily channelized or "ditched" for flood control. In those cases, only moderate channel changes can be made, and the extra flood capacity must come from levees along either side that allow the stream to flow at higher stages without flooding the floodplains and low terraces (**Figure 12.15**). Such designs (in fact, all channel modifications) are expensive to build and maintain, but they do enjoy some degree of success in keeping out small-to-medium floods. However, they do have many drawbacks. The foremost of these is that, because the floodplain areas are lost for water storage during the floods (and thus flood diminution downstream), downstream floods will be larger than under natural conditions. Hence, downstream channel reaches must also be treated so the flooding and associated problems cascade downstream. Another problem is that, because flood discharges are restricted to the area between the levees, the flood stages must be higher than under natural conditions. So, when a levee is overtopped or breaks, the local flooding is worse than it would have been under natural conditions. This is what happened in several places during the great Midwest floods of 1993.

Yet another problem common to channelization of any kind, as well as to stream reaches below reservoirs, is that the ecological health of floodplains depends on periodic flooding (NRC, 1999), and this beneficial flooding is patently reduced. Likewise, the much more rapid flow in the restricted channels may have deleterious biophysical effects.

12.6 REDUCING RUNOFF FROM URBAN AREAS

The Ohio State University Extension (2000) provided the following guidelines on designing a stormwater management plan:

1. Consider the total environmental impact of the proposed system.
2. Consider water quality as well as water quantity.
3. Minimize the amount of impervious area to be created.
4. Be consistent with the local comprehensive land use plan and any existing watershed management plan.
5. Coordinate stormwater management practices with erosion control measures and aquifer protection.
6. Minimize disturbance of natural grades and vegetation and utilize existing topography for natural drainage systems if adequate.
7. Preserve natural vegetated buffers along water bodies and wetlands.
8. Maximize infiltration of cleansed runoff to appropriate soils.
9. Reduce peak flow to minimize soil erosion, stream channel instability, flooding, and habitat destruction.
10. Use wetlands and water bodies to receive or treat runoff only when it is ensured that these natural systems will not be overloaded or degraded.

TABLE 12.4
Urban Stormwater Management Best Management Practices (BMPs)

Strategy	BMP
Runoff Pollution Prevention	
Impervious surface reduction	Street, parking lot, and cul-de-sac design, green rooftops, turf pavers
Housekeeping	Pavement management, animal management, landscape design and maintenance, and BMP maintenance
Construction practices	Grading, sequencing, controlled traffic
Soil erosion control	Mulches, blankets, and mats; vegetative methods; structural methods; silt fences; sediment basins/traps; check dams; inlet protection
Stormwater Treatment BMPs	
Infiltration systems	On-lot infiltration, infiltration basins, infiltration trenches
Filtration systems	Bioretention systems, surface sand filters, underground filters, filter strips
Constructed wetlands	BMPs in series, stormwater wetlands, wet swales
Retention systems	Wet ponds, extended storage ponds, wet vaults
Detention systems	Dry ponds, oversize pipes, oil/grit separators, dry swales
Flow control structures	Permeable weirs, flow splitters, proprietary flow control devices

11. Provide a maintenance schedule for management practices, including designation of maintenance responsibilities.

To accomplish this 11-point plan, they described the main principles of stormwater management as "the four Cs" of stormwater management: control, collection, conveyance, and cleansing. Control measures are better described as prevention measures that reduce the amount of runoff and reduce the potential for runoff to become polluted. Collection and cleansing often go hand in hand. Retention basins are designed to hold the stormwater permanently until it evaporates or infiltrates. Detention basins temporarily store the water to facilitate treatment or peak flow reduction (**Figure 12.16A** and **12.16B**). Conveyance systems are designed to direct flow to collection and treatment systems and then to transfer the treated runoff to streams and other surface bodies. The term *best management practice* (BMP) is often used to describe a strategy (or set of strategies) to manage runoff; BMPs can be used to address any or all of the four Cs. A summary of BMPs is presented in **Table 12.4**.

A comparative assessment of several urban BMPs is presented in **Table 12.5** (based on Schueler, 1992). Typically, a combination of several strategies is needed to provide adequate management of stormwater. Most states have detailed guidelines and manuals that describe each of the strategies, where they might be used, and how they might perform. The Center for Watershed Protection (CWP) (www.cwp.org) has conducted extensive studies on BMPs and has assisted some states with the development of their stormwater management guidelines and manuals.

Detailed discussion of each of the strategies presented in **Table 12.4** and **Table 12.5** is beyond the scope of this text. We recommend that reference be made to state and

A

B

FIGURE 12.16 Urban detention basins: (A) small micropool extended detention pond; (B) multiple detention basins have improved water quality benefits. (Courtesy and copyright Center for Watershed Protection.)

federal guidelines and the studies documented by organizations such as the Center for Watershed Protection and the National NEMO (Nonpoint Education for Municipal Officials) Network. It is worthwhile to provide a brief

TABLE 12.5
Comparative Assessment of the Effectiveness of Urban Best Management Practices

Urban BMP	Reliability	Longevity	Application	Wildlife Habitat Potential	Environmental Concerns	Comparative Cost
Stormwater wetland	Moderate to high, depending on design	20+ years	Applicable to most sites if land is available	High	Stream warming, natural wetland alteration	Marginally higher than wet ponds
Extended detention ponds	Moderate, but not always reliable	20+ years, but frequent clogging and short detention common	Widely applicable, but requires at least 10 acres of drainage area	Moderate	Possible stream warming and habitat destruction	Lowest cost alternative in size range
Wet ponds	Moderate to high	20+ years	Widely applicable, but requires drainage area greater than 2 acres	Moderate to high	Possible stream warming, trophic shifts, habitat	Moderate to high compared to conventional
Multiple pond systems	Moderate to high; redundancy increases reliability	20+ years	Widely applicable	Moderate to high	Selection of appropriate pond option minimizes environmental impact	Most expensive pond option
Infiltration trenches	Presumed moderate	50% failure rate within 5 years	Highly restricted (soil, groundwater, slope area, sediment input)	Low	Slight risk of groundwater contamination	Cost-effective on smaller sites; rehab costs can be considerable
Infiltration basins	Presumed moderate if working	60–100% failure within 5 years	Highly restricted (see infiltration trench)	Low to moderate	Slight risk of groundwater contamination	Construction cost moderate, but rehab cost high
Porous pavement	High if working	75% failure within 5 years	Extremely restricted (traffic, soils, groundwater, slope, area, sediment input)	Low	Possible groundwater impacts; uncontrolled runoff	Cost-effective compared to conventional asphalt when working properly
Sand filters	Moderate to high	20+ years	Applicable (for smaller developments)	Low	Minor	Comparatively high construction costs and frequent maintenance
Grassed swales	Low to moderate, but unreliable	20+ years	Low-density development and roads	Low	Minor	Low compared to curb and gutter
Filter strips	Unreliable in urban settings	Unknown, but may be limited	Restricted to low-density areas	Moderate if forested	Minor	Low
Water quality inlets	Presumed low	20+ years	Small, highly impervious catchments (<2 acres)	Low	Resuspension of hydrocarbon loadings	High compared to trenches and sand filters

Source: From Schueler, T.R., *Design of Stormwater Wetland Systems: Guidelines for Creating Diverse and Effective Stormwater Wetland Systems in the Mid-Atlantic Region,* Anacostia Restoration Team, Department of Environmental Programs, Metropolitan Washington Council of Governments, Washington, DC, 1992.

FIGURE 12.17 Pervious pavement shortly after a storm event shows increased runoff storage and then infiltration. (Courtesy and copyright Center for Watershed Protection.)

FIGURE 12.18 Bioretention area. (Courtesy and copyright Center for Watershed Protection.)

discussion of prevention measures that may be considered. It will not always be possible to prevent the adverse impacts of development totally, but the smaller the problem that needs to be managed, the more likely the solution will be self-sustaining — and often more economic. These various concepts should be viewed as potential ideas, and their use should be weighed against site conditions, environmental benefit, costs, and other constraints that must be considered. Some of the BMPs listed in **Table 12.5** may not be appropriate for all areas; others may have limited application, but this should not dissuade their consideration in some circumstances. The cumulative effect of several management practices can have a major impact on obtaining both water quality and water quantity goals.

It is important to understand the concept of cumulative impact. No one structure or impervious parking area will cause a system to fail, but the cumulative impact of many parking areas and many structures, especially in the absence of a watershed or natural resource–based plan that regulates their number, design, and installation, can be very detrimental to the environment. The same concept holds true for the urban BMPs designed to mitigate the impact of stormwater runoff from impervious areas. No one pervious pavement, bioretention area, grassed swale, or filter strip (**Figure 12.17** to **Figure 12.19**) will provide all of the answers, but the cumulative effect of mitigation systems can have a meaningful impact on stream ecology and the environment, especially if done within a comprehensive land use plan. This is the underlying principle of the National NEMO Network now operating in over 50% of the states.

Programs like those established by the National NEMO Network provide frameworks with tiered approaches to stormwater management: natural resource-based land use planning on a watershed basis; low-impact site design that integrates the structural BMPs discussed in this section; and existing areas retrofitted with structural BMPs and maintenance practices. NEMO programs often use GIS tools to demonstrate the relationship between land use and water quality, and some programs (like Ohio

FIGURE 12.19 Bioswale and infiltration trench. (Courtesy and copyright Center for Watershed Protection.)

NEMO) are expanding that relationship to water quantity as well. One very useful tool that NEMO has developed along with the NOAA is the Impervious Surface Analysis Tool (ISAT; http://www.csc.noaa.gov/ crs/is/), which estimates the percentage of impervious surface area. The ISAT model also allows the user to conduct "what if" type of scenarios to determine the potential impact of additional impervious surface from new development on water quality using CWP criteria. Ohio NEMO is working with others to mirror the BDF with ISAT and integrating these tools into a GIS-based Natural Resource Evaluation and Site Assessment (NRESA) model to provide communities with a comprehensive land use planning tool.

When fully developed, NRESA will allow communities to determine their most important natural resources and to determine the areas that are best for development and minimize any adverse impact on those resources. In the future, it will be interesting to see how the results of these models can provide decision makers and stakeholders with information regarding mitigated and unmitigated impacts of stormwater runoff from impervious surface areas. For now, it is important to reiterate that urban BMPs will be far more effective if they are integrated into a comprehensive land use plan, and that it is the cumulative effect that these systems have that holds the greatest potential for meeting water quality and quantity goals.

12.7 DETENTION AND RETENTION PONDS

Stormwater detention ponds are designed to perform one or more of these functions: (1) reduce downstream flooding, (2) improve water quality, or (3) prevent or reduce downstream channel instability problems. Increases in the frequency and magnitude of discharges associated with urban development are discussed in Chapter 5 and Chapter 7, and procedures to calculate the temporary storage and spillway configuration associated with a flood reduction strategy in a reservoir are discussed in Chapter 8. Water quality problems associated with urbanization and management strategies to improve water quality are presented in this section.

Stream channel stability is often adversely impacted by urbanization due to increased bedload transport capability caused by higher peak discharges and larger volumes of runoff (Rhoads, 1995). A typical channel response is downcutting, entrenchment, and then widening. Straightening channels, filling floodplains, and reducing bedload supply exacerbate adverse impacts on the stream geomorphology. Degraded channels provide less ecological function, such as assimilative capacity of urban runoff, sediment pollution, and peak flow attenuation through valley storage and riparian and aquatic habitat. One common management practice has been to route stormwater through detention basins so that postdevelopment discharges for extreme events, typically 2-to 100-year return period peak discharges, do not exceed some assigned predevelopment peak discharge rate. However, extreme event peak discharge rates are only loosely correlated with fluvial processes that govern channel stability. More recent management strategies have better considered the discharges that influence channel stability. Mecklenburg and Ward (2002) hypothesized that effective postdevelopment discharge management strategies should produce bedload transport similar to the predevelopment flow regime in both total volume of bedload moved and recurrence interval of the dominant discharge. Procedures for sizing detention ponds based on channel stability goals are discussed in this section.

In designing a detention pond, it is first necessary to define the function of the pond (flood control, water quality, channel stability). Once the function is identified, the basic design considerations are deciding what pre- and postdevelopment events will be used to develop a design; selecting a target peak flow reduction strategy; determining what portion of the storm hydrograph will pass through the pond; and selecting a target mean detention/residence time of the flow in the pond. Historically, most strategies have focused on controlling postdevelopment peak discharges to some predevelopment level and trapping and settling solid materials carried by the stormwater.

TABLE 12.6
Types and Sources of Pollutants in Stormwater

Type of Pollutant	Examples and Sources
Biochemical oxygen demand (BOD)	Yard waste, animal and human waste, hydrocarbons
Metals	Arsenic, cadmium, copper, chromium, iron, lead, nickel, zinc; mainly from industrial activities, automobile emissions, and deicing agents
Nutrients	Nitrogen, phosphorus from fertilizers, sewerage overflows, and septic systems
Organic chemicals	Gasoline, grease, and oil from vehicles and industry; pesticides from golf courses, lawns, parks
Pathogens	Bacteria, protozoa, viruses due to sewer leaks, pets, rats, mice, and infected materials
Salts	Sodium chloride, calcium chloride mainly due to road, driveway, sidewalk, and parking lot deicing
Sediment	Sand, silt, and clay from construction activities

Source: From Schueler, 1992, *Design of Stormwater Wetland Systems,* Department of Environmental Programs, Washington, D.C. With permission.

There are numerous inlet and outlet designs and temporary vs. permanent storage sizes that might be used in developing a stormwater management strategy. Basically, the design revolves around the selection of a dry or wet pond, an in-line or offline structure, and whether the design is based on a single event or a range of events during the design life of the structure. Dry detention ponds hold water temporarily following a storm and do not contain water permanently between storms. Usually, they passively fill and empty due to gravity flow during and following each discharge event. Wet detention ponds maintain a permanent pool of water that enhances the removal of pollutants. These ponds fill with stormwater and passively release most of it over a period of a few days. Wet detention ponds remove pollutants by settling of suspended solids, biological uptake, decomposition, and consumption of pollutants by plants, algae, and bacteria.

With a passive in-line detention pond, all the stormwater passes through the pond, and the peak discharge for the outflow hydrograph occurs on the receding limb of the inflow hydrograph. Typically, for the design event the outflow peak discharge is some function of a predevelopment peak discharge. The inlet or bypass channel of a passive off-line detention pond can be designed either to fill the pond with flow that occurs at the beginning of an event or to start filling when some design threshold flow rate on the storm hydrograph is exceeded.

Numerous pond strategies together with their advantages and disadvantages are documented in the literature. Refer to the various reviews, manuals, and publications of the Center for Watershed Protection to ascertain the advantages and disadvantages of the more commonly used strategies and to develop a design to address a particular management concern (http://www.cwp.org; Schueler, 1992; T. Brown and Caraco, 2001).

12.7.1 USING DETENTION/RETENTION PONDS TO IMPROVE WATER QUALITY

Urban runoff often contains large quantities of nutrients, pesticides, heavy metals, oils, and grease. In addition,

sediment associated with urban development, salt use during winter storms, pathogens, and elevated biochemical oxygen demand (BOD) are common problems. A summary of the main pollutants in stormwater is presented in **Table 12.6**. The 1996 National Water Quality Inventory reported that construction and urban/suburban storm water runoff impaired the water quality of 19, 32, and 56% of the nation's rivers, lakes, and estuaries, respectively (http://www.epa.gov/305b/).

The prevention, control, and treatment of stormwater are primarily regulated through the USEPA National Pollution Discharge Elimination System (NPDES) program. Phase I of the EPA stormwater program was promulgated in 1990, and the Phase II Program was established in 1999. Phase I established permit requirements for (1) medium and large municipal separate storm sewer systems (MS4s) that serve populations of 100,000 or more people; (2) construction activities that disturbed 5 acres or more land; and (3) 10 categories of industrial activity. The Phase II Program essentially expanded the Phase I requirements to all urbanized areas with MS4s and to construction sites that disturbed 1 to 5 acres of land.

In addition to the NPDES program, which primarily relates to urban areas, all water bodies in the nation need to comply with a Total Maximum Daily Load (TMDL) requirement. A TMDL is the maximum amount of a pollutant that a water body can receive and still meet water quality standards for that water body. It is also an allocation of the amount of each pollutant that can occur from the pollutant's sources. The TMDL program and the establishment of water quality standards are specified in Section 303 of the Clean Water Act. Specific water quality standards are set by the USEPA or each state, territory, or tribe. These standards identify the uses (drinking water, fishing) for each water body and the scientific criteria to support that use. The development of TMDLs by each state was still in the process of development at the time this text was prepared.

In the following sections of this chapter, we consider retention/detention and construct wetland strategies, but we emphasize that, for all pollution problems, prevention strategies should be adopted as the first line of defense. Numerous studies have been conducted to evaluate the ability of BMPs to improve water quality. The performance of BMPs that have been studied is very variable and range from negative capture rates (more leaving a BMP than was received) to almost 100% capture (EPA, 1983; Schueler, 1992; ASCE, 1996; Schueler and Holland, 2000). Conclusions that can be drawn from studying the literature are as follows:

1. The performance of water quality BMPs is often lower than anticipated.
2. Designing BMPs to achieve a target water quality (such as a discharge concentration or TMDL) is difficult.
3. There is no single type of structure (dry or wet pond or wetland) that consistently performs better than another.
4. Optimal design based on space and cost considerations suggests that retention/detention/wetland structures should be sized based on capturing the runoff from 70 to 90% of the precipitation events at the site.
5. The captured runoff should then be discharged from the structure during a period of 24 h or longer.

The variable performance of pond-type structures is illustrated in **Figure 12.20** (Schueler, 1992). The results are summarized for more than 45 structures. Overall, the wet extended detention ponds and pond and wetlands in series performed slightly better than the other structures. However, in some individual structures, high removal of one pollutant was accompanied by low removal of the other pollutant. It should also be noted that only two pollutants are considered in this summary, and structures typically are designed to remove many more pollutants.

12.7.2 SIZING DETENTION PONDS TO IMPROVE WATER QUALITY

Pollutants on the surface of impervious areas can be washed away by small precipitation events. Even during large storm events, the highest loads are associated with the first part of the hydrograph — often called the first flush. Therefore, an effective water quality management strategy is to size pond-type structures to capture the initial flow from as many precipitation events as possible and then to provide sufficient detention/residence time of this flow in the pond. Much of the treatment is due to settling of particles, and detention times ranging from 12 to 24 h are adequate except for very fine silts and clays. Treatment of soluble constituents is more difficult and in some cases requires detention times that are longer than a few days. Sizing the pond therefore basically boils down to selecting

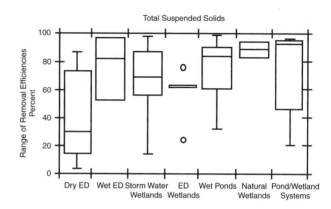

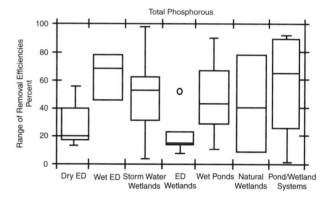

FIGURE 12.20 Pollutant removal performance for stormwater ponds and wetlands. (From Schueler, T.R., *Design of Stormwater Wetland Systems: Guidelines for Creating Diverse and Effective Stormwater Wetland Systems in the Mid-Atlantic Region*, Anacostia Restoration Team, Department of Environmental Programs, Metropolitan Washington Council of Governments, Washington, DC, 1992. With permission.)

a pond type, deciding which events to capture, deciding how long to temporarily detain these flows, and deciding what to do with the flows that are larger than those that will be captured.

Several studies have shown that there is a point at which capturing more of the annual rainfall events results in very small additional improvements in water quality (Urbonas et al., 1990; Roesner et al., 1991; Urbonas and Stahre, 1993). **Figure 12.21** shows the cumulative occurrence of precipitation events of different magnitudes for five cities. Precipitation events that are less than 0.05 in. were not considered as it has been assumed that they will produce little runoff and have a low potential to remove pollutants. Also, snow events were not separated from rainfall events. It can be seen that, for all the cities, there is a rapid increase in the number of events up until about 0.5 in. of precipitation and very little increase after about 1.5 in. Therefore, depending on the location, capturing the runoff from 0.5 to 1.5 in. of precipitation would result in treatment of 90% of the runoff events. For a 90% threshold, runoff from 0.5-, 0.7-, 0.8-, 1.25-, and 1.5-in. events would need

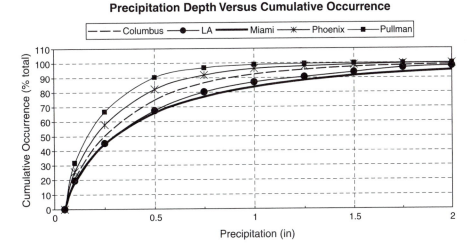

FIGURE 12.21 Cumulative occurrence of precipitation events for several U.S. cities.

to be captured in Pullman, WA; Phoenix, AZ; Columbus, OH; Los Angeles, CA; and Miami, FL, respectively.

Any of the runoff methods discussed in Chapter 5 can be used to determine the volume of runoff associated with the target precipitation event. However, the NRCS curve number method is the only true runoff volume/depth method presented. The rational C coefficient can be used as an indicator of the fraction of precipitation that will generate runoff. Many urban studies have related runoff to the amount of impervious area. Based on an analysis of data for more than 60 urban watersheds, Urbonas et al., (1990) developed the following regression equation:

$$C = 0.858i^3 - 0.78i^2 + 0.774i + 0.04 \qquad (12.1)$$

where C is a runoff coefficient, and i is the fraction of the watershed that is impervious. A procedure that relates the detention volume to the runoff coefficient and the mean storm precipitation volume was presented by Guo and Urbonas (1995) and ASCE (1998).

Schueler (1992) presented the following relationship between the runoff coefficient and impervious area:

$$C = 0.9i + 0.05 \qquad (12.2)$$

In **Table 12.7**, a summary is presented for different land uses of runoff coefficients derived from various methods, including Equation 12.1 and Equation 12.2. For the NRCS curve number, the runoff was calculated based on assuming Soil Type C and AMC III conditions and then calculating a runoff coefficient as the ratio of this runoff to the precipitation depth. For the basin development factor used in some of the USGS peak flow regression equations, we expanded the original BDF range of 0 to 12 to include values less than 0 and more than 12. The reason for doing this is because a BDF of 0 does not represent a truly rural condition, and a BDF of 12 does not represent the concrete-and-asphalt jungle associated with "downtown" in most large cities. While there is no perfect correlation between the various methods, they do provide reasonable agreement with each other and indicate that any of the methods might be used given the uncertainties associated with runoff estimate and runoff management. The clear message from the information in **Table 12.7** is the benefit obtained from incorporating pervious surfaces in land use development. The structure size needed for an industrial development will be six times larger than for an area that is mainly parks and twice as large as would be needed for a medium-density residential area.

EXAMPLE 12.1

Calculate the runoff depth and temporary detention volume for an urban development in Los Angeles with ¼-acre lots on a new 50-acre subdivision. Base the design on capturing runoff from at least 90% of the events and storing it for at least 24 h.

Solution

From **Figure 12.21**, it is determined that to capture at least 90% of the storms in Los Angeles, the design should be based on a 1.25-in. precipitation event.

From **Table 12.7**, a mean runoff coefficient of 0.35 is obtained for a medium-density residential development. Therefore, the runoff depth is 0.35 times 1.25 in., which equals approximately 0.44 in. The runoff volume is

$$V = (0.35)(1.25)\left(\frac{50}{12}\right) = 1.8 \text{ acre-ft}$$

TABLE 12.7

Alternative Approaches to Obtaining Runoff Coefficients for Several Land Uses

Runoff Approach[a]	Approach or Coefficient Method	Land Use and Runoff Attribute					
		Industrial Commercial	Residential Density[b]			Agriculture, Lawns, Parks	Woods, Forest, Prairie/Grassland
			High	Medium	Low		
Impervious Area	Fraction	0.9	0.6	0.4	0.2	0.05	0
NRCS CN	AMC 2	90	85	80	75	70	60
NRCS CN	AMC 3	96.3	93.5	91.2	87.8	84.7	78.0
Basin density function (BDF)		12+	9–12	6–9	3–6	0–3	<0
Runoff coefficient[c]							
Coefficient 1	Impervious area[d]	0.73	0.41	0.28	0.17	0.08	0.04
Coefficient 2	Impervious area[e]	0.86	0.59	0.41	0.23	0.095	0.05
Coefficient 3	CN, (P = 1.0 in)[f]	0.65	0.48	0.37	0.25	0.17	0.06
Coefficient 4	CN, (P = 0.75 in)[g]	0.57	0.38	0.27	0.16	0.09	0.02
Coefficient 5	Rational C soil[h]	0.8	0.5	0.35	0.25	0.15	0.01–0.05
Coefficient 6	Rational D soil[i]	0.8	0.55	0.4	0.3	0.2	0.01–0.05
Mean coefficient	Mean of 1 to 6	0.74	0.48	0.35	0.23	0.13	0.04

[a] Impervious area fractions based on authors' interpretation of various reports in the literature. Curve numbers from **Table 5.1** and assuming Soil Type C. BDF ranges conceptualized by the authors (note: original BDF range in the literature is from 0 to 12).

[b] High: about 1/8-acre lots; medium: 1/3- to 1/5-acre lots; low: 1-acre or more.

[c] Runoff coefficients based on runoff approach and land use as described in Footnotes d to i.

[d] $C = 0.85i^3 - 0.78i^2 + 0.04$ where i is the impervious area fraction (Urbonas et al., 1990), Equation 12.1.

[e] $C = 0.9i + 0.05$ where i is the impervious area fraction (Schueler, 1992), Equation 12.2.

[f] Coefficient based on NRCS CN method for a rainfall of 1.0 in. and assuming AMC III conditions (Coefficient = Runoff/rainfall).

[g] Coefficient based on NRCS CN method for a rainfall of 0.75 in. and assuming AMC III conditions (Coefficient = Runoff/rainfall).

[h] Rational runoff coefficients from **Table 5.5** for Hydrologic Soil Group C.

Answer

The temporary detention depth and volume should be 0.44 in. and 1.8 acre-ft, respectively.

EXAMPLE 12.2

It is decided to convert the detention pond from Example 12.1 into a wetland area. The average depth of water in the wetland will be less than 1 ft. For storm events, the maximum depth of water that is temporarily ponded in the wetland should not exceed 3 ft. Determine the minimum surface area of the wetland and the size pipe that should be used to drain the temporary storage volume.

Solution

From Example 12.1, it was determined that the storage volume should be 1.8 acre-ft. Therefore, the minimum surface area will be

$$A_s = \frac{1.8}{3} = 0.6 \text{ acres}$$

The average discharge rate during a 24-h release (dewatering) time would be

$$q_{ave} = \frac{(1.8)(43560)}{(24)(60)(60)} = 0.9 \text{ cfs}$$

From **Table 8.5** and **Table 8.6** in Chapter 8, it can be seen that the pipe size would probably need to be less than 6 in.

Answer

The minimum surface area would be 0.6 acres, and the spillway size would be less than 6 in.

In Example 12.2, it was found that a small pipe was needed to dewater the extended detention temporary storage volume slowly. An alternative to the use of small pipes is to use a perforated large-diameter riser. The crest of the riser is located at the bottom of the flood storage volume. This approach is often used in sediment ponds and some urban detention ponds. An approach that attempts to skim the cleanest water from near the water surface is to use a floating riser (sometimes called a skimmer). The riser moves up and down based on the depth of the flow. In 1985, Ward designed a floating riser for use in tailings dams and mine effluent treatment ponds in South Africa, but it has seen limited application. Recent research has demonstrated the merits of this approach (Federal Highway Administration, 2001), and it is beginning to see more application in highway construction and urban and mining situations.

12.7.3 DETENTION PONDS FOR CHANNEL STABILITY

Unfortunately, there is little evidence to suggest that any of the commonly used strategies will provide adequate protection of downstream channel stability. Useful insight on this problem was presented more than a decade ago by Booth (1991). Since then, there has been a shift away from designs based on 10- to 100-year recurrence interval events, but there is little evidence to suggest that designs based on 1- or 2-year recurrence interval events or providing 0.2 to 1.0 in. of storage at the start of an event or at some threshold flow rate will provide channel protection.

The most common strategy to provide channel protection is to hold postdevelopment peak discharge rates to 2-year predevelopment levels. However, this strategy often releases flows above the effective discharge for a longer period of time than occurred prior to development and theoretically results in greater transport of suspended sediment and bedload. An alternative overcontrol strategy is to control the 2-year postdevelopment peak discharge to a fraction of the predevelopment rate. A modification of this overcontrol strategy is to control the 2-year postdevelopment discharge to the 1-year predevelopment rate for a 24-h storm event.

Other strategies used frequently for water quality and channel stability are extended detention of a set volume of runoff. With the extended detention methods, a volume such as the 1-year, 24-h storm is released over a period of 12, 24, or more hours. The approach attempts to release flows at rates that do not exceed the critical erosive velocities in downstream channels.

A distributed runoff control strategy was proposed by MacRae (1993). The approach is based on fluvial geomorphology concepts and attempted to maintain the erosion potential of the channel boundary materials the same as predevelopment conditions over the range of available flows such that the channel is just able to move the dominant particle size of the bedload. Rhoads (1995) concluded that, "The concept of stream power provides a unifying theme for the broad range of issues emcompassed by urban fluvial geomorphology. Most importantly, this concept directly links channel instability with changes in fluvial energy expenditure and sediment transport capacity." Mecklenburg and Ward (2002) presented a new approach that is consistent with the philosophy that stormwater management strategies should be based on fluvial geomorphology concepts and the range of discharge events that might occur following development. The

approach is based on the premise that the bedload transport in alluvial streams is primarily responsible for maintaining the channel equilibrium.

The approach used by Mecklenburg and Ward (2002) provides useful insight into how a detention pond could be sized and how well, or poorly, various current practices might perform in maintaining channel stability as well as meeting a water quality objective. In their approach, total bedload transport over a period of 100 years was calculated based on assuming that all discharge events during this period could be approximated by a series of 0.098-, 0.195-, 0.391-, 0.781-, 1.56-, 3.13-, 6.25-, 12.5-, 25-, 50-, and 100-year RI events. On average, two 50-year or larger RI discharges would occur during the 100 years. It was assumed that one of these events was a 50-year event, and the other was a 100-year event. Similarly, there would be four 25-year or larger events, and two of these would be 25-year events, one the 50-year event and one the 100-year event. This approach was used to determine the number of events associated with each RI discharge that would occur on average during the 100-year period (**Table 12.8**). The RI doubling scale that was used (0.098 to 0.195 to 0.391 years, etc.) ensures an unbiased distribution of different magnitude events.

Determining bedload movement requires knowledge of the fluvial geomorphology of the stream system that the detention pond is designed to protect. In a real-world setting, this information would be obtained by conducting a survey of the dimensions, pattern, profile, and bed material of the stream system using procedures such as those described by Harrelson et al. (1994). However, in the method outlined by Mecklenburg and Ward (2002), bedload transport was estimated by using the Meyer–Peter–Muller method, and a theoretical channel geometry was estimated based on published regional curves for the eastern U.S.

The USGS peak discharge methods developed by Sherwood (1993) and presented in Chapter 5 were used to determine the peak flows. Sherwood also developed volume–duration–frequency relationships for urban areas in Ohio with drainage areas less than 6.5 mi^2. Inflow and outflow hydrographs were approximated as triangles using the procedure described by Equation 12.16 through Equation 12.21. For flows unregulated by detention, the USGS peak discharge was used to define the height of the triangle, and the runoff volume defined the duration of the flow and area of the triangular hydrograph. For flows regulated by detention, the outflow hydrographs were based on the inflow hydrographs, as they would be influenced by detention.

On average, the approach used by Mecklenburg and Ward resulted in 10.3 storms annually. However, 7.7 of these annual storms (5.1 plus 2.6) were associated with the 0.098- and 0.195-year RI storms and produced no bedload transport for the conditions that were considered.

TABLE 12.8
Calculated Runoff, Detention Storage, and Bedload Transports
for a 100-Year Period and Different Magnitude Recurrence Interval Events[a]

Recurrence Interval (years)	Occurrences in 100 years	Runoff (in.) for Different Basin Density Function (BDF) Values						
		0	2	4	6	8	10	12
0.10	513	0.01	0.01	0.01	0.01	0.01	0.01	0.02
0.20	257	0.06	0.06	0.06	0.07	0.08	0.09	0.12
0.39	128	0.15	0.15	0.16	0.17	0.18	0.20	0.25
0.78	64	0.29	0.29	0.30	0.32	0.34	0.37	0.44
1.6	32	0.50	0.52	0.53	0.55	0.57	0.61	0.70
3.1	16	0.64	0.66	0.68	0.71	0.74	0.81	0.96
6.3	8	0.76	0.79	0.82	0.87	0.94	1.04	1.32
12.5	4	1.01	1.05	1.09	1.14	1.22	1.35	1.68
25	2	1.28	1.32	1.37	1.43	1.53	1.68	2.08
50	1	1.50	1.55	1.61	1.69	1.80	1.98	2.44
100	1	1.70	1.75	1.82	1.91	2.04	2.25	2.77
Detention storage (in.)		0.00	0.07	0.12	0.17	0.22	0.30	0.47
Relative bedload		1.0	1.1	1.2	1.5	1.8	2.3	4.0

[a] Central Ohio conditions with annual precipitation of 39.4 in., a 0.19-mi^2 watershed, and a bed slope of 1%. Effective discharge channel cross-sectional area 6.5 ft^2, floodplain width 66 ft, and bed material d_{50} = 30 mm. Runoff was calculated using the USGS empirical methods for Ohio (see Chapter 5, **Table 5.8**).

Another 1.3 annual events were associated with the 0.391-year RI storms and only produced bedload transport for high-BDF conditions (9 to 12). Therefore, on average, significant bedload transport for the conditions considered only occurred 1 to 3 times annually. The use of bedload as an indicator of sediment transport and stream stability can be debated, and certainly there will be places where there will be no bedload transport or suspended loads will be the dominant load. However, of interest is how the results provided by this approach, which is based on fluvial concepts, compare to other methods. First, it can be seen from **Table 12.8** that high-BDF (urban) areas have the potential to transport two to four times as much bedload. Under some conditions, the amount transported might be more than six times larger than the predevelopment potential. Also, these high-BDF areas require extended detention of 0.22 to 0.47 in. of runoff for BDF values of 8 to 12. An evaluation of other conditions showed that these amounts might vary from 0.15 to 0.65 in. From **Figure 12.21**, we can see that the runoff from about 0.8 in. of rainfall will result in treatment of about 90% of the rainfall events in Columbus. From **Table 12.8**, the mean runoff coefficients from BDFs ranging from 6 to 12+ are 0.35 to 0.74. If we multiply the rainfall of 0.8 by these coefficients, we obtain runoff estimates of 0.28 to 0.59. These storage estimates to provide water treatment are very similar to those needed to manage runoff based on a channel stability objective.

Figure 12.23 shows the influence of different detention strategies on bedload transport for the watershed conditions summarized in **Table 12.8**. The approach proposed by

Mecklenburg and Ward (2002) provides fairly close agreement with the predevelopment conditions. Reducing postdevelopment 2-year peaks to 50% of the predevelopment 2-year peaks or to the predevelopment 1-year peaks provides fairly good agreement for events larger than the 1-year RI. All other strategies and several not shown, such as the Ohio critical storm method, greatly under- or over-controlled bedload transport. Surprisingly, the fluvial approach resulted in the smallest temporary storage volume (**Figure 12.23**). It would appear from these results that providing a water treatment volume that captures 90% of the annual rainfall events and then releases this volume slowly over a period of 12 h or longer might also be an effective strategy to protect downstream channel stability.

12.8 WETLANDS AS WATER TREATMENT SYSTEMS

The use of wetlands, particularly constructed wetlands, as water treatment systems has grown rapidly in the last two decades. Perhaps in part this is due to a societal response to attempting to reverse the loss of wetlands that has occurred worldwide, particularly in the U.S. Mitsch and Gosselink (2000) reported that 53% of the wetlands in the U.S. have been lost. In some parts of the Midwest, such as Ohio, more than 90% of the wetlands have been drained, and these areas are now primarily used for agricultural purposes. However, the increase in the use of wetland as treatment systems is also due to other factors, including that (1) they have a greater potential than ponds

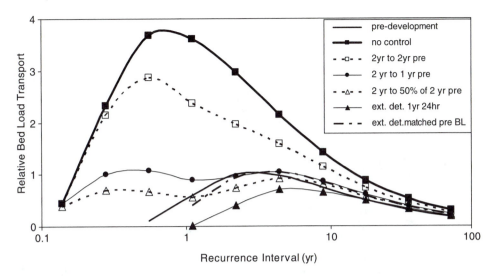

FIGURE 12.22 Relative bedload transport for different detention strategies (predevelopment 2-year transport = 1). (©Ward and Mecklenburg, 2002.)

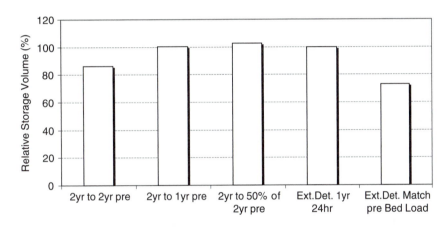

FIGURE 12.23 Relative storage volume for various detention strategies (extended detention 1-year 24-h storage = 100%).

to enhance water quality, (2) treatment ponds and lagoons are difficult to manage, and (3) wetlands often have better ecological function and provide better habitats than constructed treatment ponds. As Mitsch and Gosselink (2000) stated:

> Wetlands are sometimes described as "the kidneys of the landscape" because they function as the downstream receivers of water and waste from both natural and human sources. They stabilize water supplies, thus ameliorating both floods and droughts. They have been found to cleanse polluted water, protect shorelines, and recharge groundwater aquifers.

Wetlands are commonly used to treat municipal wastewater, stormwater, acid mine drainage, agricultural wastewater associated with livestock, and landfill leachate. Like ponds, wetlands are efficient at removing suspended solids,

but they have added treatment benefits associated with aquatic plants and richer and more diverse microorganism populations of fungi, bacteria, and actinomycetes. The microorganisms break down dissolved and particulate organic material into carbon dioxide and water. In addition to providing "homes" and an oxygen source for the microorganisms, aquatic plants play a role in the uptake of nutrients and other compounds. When plants die and decompose, some of the nutrients are released back into the water. Much of the nitrogen is then lost through denitrification processes. Simeral (1998) listed the primary processes in a wetland as (1) uptake and transformation of nutrients by microorganisms and plants; (2) breakdown and transformation of pollutants by microorganisms and plants; (3) filtration and chemical precipitation through contact with substrate and litter; (4) settling of suspended particulate matter; (5) chemical transformation of pollutants

FIGURE 12.24 Constructed wetland. (Courtesy and copyright Center for Watershed Protection.)

(i.e., ammonification of nitrogen); (6) absorption and ion exchange on the surfaces of plants, sediment, and litter; and (7) predation and natural die-off of pathogens.

There are two types of constructed wetlands: (1) surface-flow systems similar to natural wetlands and (2) subsurface flow systems in which water flows through a porous medium such as sand or gravel and there is little or no standing water in the wetland. In this text, we only discuss the first type of system (**Figure 12.24**).

12.8.1 HYDRAULICS AND TREATMENT CHARACTERISTICS

The detention time of water in a wetland can be calculated using the equations presented in Chapter 8. However, the optimal detention time in a wetland for most pollutants ranges from a few days to more than 2 weeks. Therefore, the theoretical detention time (sometimes called the *hydraulic residence time*) t_d is normally calculated as

$$t_d = \frac{Vn}{q_d} \qquad (12.3)$$

where V is the volume of the wetland (cubic feet or cubic meters), q_d is the daily inflow (cubic feet/day or cubic meters/day), and n is the porosity of the medium that water flows through for subsurface flow and is the ratio of water volume to the volume occupied by water and plants (0.85 for bulrushes, 0.95 for cattails, and 0.98 for reeds). The hydraulic loading rate q_{hlr} is defined as

TABLE 12.9
First-Order Areal Parameters for Use with Equation 12.27

Constituent	k_A (m/year)	C_b (g/m³)
Biochemical oxygen demand	34	$3.5 + 0.053\,C_i$
Suspended solids	1000	$5.1 + 0.16\,C_i$
Total phosphorus	12	0.02
Total nitrogen	22	1.5
Ammonia nitrogen	18	0
Nitrate nitrogen	35	0

Source: Kadlec, R.H., and R.L. Knight, *Treatment Wetlands*, Lewis, Boca Raton, FL, 1996.

$$q_{hlr} = \frac{q_d}{A} \qquad (12.4)$$

where A is the wetland surface area (square feet or square meters). Mitsch and Gosselink (2000) reported hydraulic loading rates that varied from less than 0.5 in./day to more than 9 in./day for wetlands that treated wastewater in Europe and North America. Fennessy and Mitsch (1989) recommended a hydraulic loading rate of about 2 in./day for acid mine drainage and a minimum theoretical detention time of 24 h.

The hydraulic loading rate can be related to changes in inflow and outflow chemical concentrations by the following first-order model:

$$\frac{C_o - C_b}{C_i - C_b} = \exp\left(\frac{-k_A}{q_{hlr}}\right) \qquad (12.5)$$

where C_i is the inflow concentration, C_o is the outflow concentration, C_b is the background concentration, and k_A is an areal removal rate constant. If Equation 12.4 is substituted into Equation 12.5 and the natural logarithm is taken of both sides of the equation, the following equation is obtained:

$$A = -\frac{q_d}{k_A}\ln\left[\frac{C_o - C_b}{C_i - C_b}\right] \qquad (12.6)$$

the units of q_d will depend on the units of the wetland area A and areal removal rate constant k_A. Estimates of k_A and C_b for several constituents that are often of concern are reported in **Table 12.9** (Kadlec and Knight, 1996).

Like ponds, some of the flow will short-circuit through the pond and will have a residence/detention time that is much less than the theoretical time, while some of the flow will be fairly stagnant and might stay in the wetland two or three times longer than the theoretical time. To reduce short-circuiting, the length-to-width ratio (called the *aspect ratio*) is made large. The NRCS recommends an overall aspect ratio of at least 3:1. However, they then divide the wetland ratio into cells with an aspect ratio of 10:1, which is consistent with the recommendations of Steiner and Freeman (1989).

An alternative approach to using areal rate constants is to use empirical equations, which are constituent specific. For example, Kadlec and Knight (1996) developed the following equation for nitrate nitrogen; the equation is based on data from 553 wetlands, but has an r^2 value of only 0.35:

$$C_0 = 0.093 C_i^{0.474} q_{hlr}^{0.745} \qquad (12.7)$$

where C_o and C_i are in grams/cubic meters, and q_{hlr} is in centimeters/day. When possible, avoid using empirical equations with r^2 values that are less than 0.7, and/or are based on only a few values, or relate to a narrow range of values or the range of values is unknown.

It should be recalled that Equation 12.7 assumes a first in, first out (plug flow) concept. An alternative approach is to assume there is complete mixing of the inflow with a portion of the water in the wetland, and that the wetland acts like a series of continuous stir reactors.

12.8.2 USING CONSTRUCTED WETLANDS TO TREAT LIVESTOCK WASTEWATER

This section is primarily based on fact sheets by Simeral (1998) and Tyson (1996).

Managing the waste produced by confined animal feeding operations is a major agricultural and environmental challenge. Confined animal feeding operations continually generate huge amounts of animal waste and must have waste management systems adequate to handle these large amounts of waste. In addition, waste management systems for different animal operations must process various types of waste. For dairy and swine farms, the animal waste system must process both liquid and solid waste; for poultry layer farms, it is primarily liquid waste; and for poultry broiler farms, it is mostly solid waste. Finally, most confined animal feeding operations apply both solid and liquid waste to nearby fields. Applying liquid animal waste to land has unique problems, including odor, high solids content, high nutrient concentrations, and limited pumping distances. In addition to these technical problems, other factors such as new regulations, more and closer residential neighbors, and increased animal numbers often cause existing land treatment sites to become inadequate rapidly. Properly applied to animal waste treatment, constructed wetlands can be a very important part of a total animal waste treatment plan. When combined with grass filter strips of cattails and bulrushes, constructed wetlands have demonstrated over 95% removal of nitrogen at a loading rate of slightly over 15 lbs of nitrogen per acre per day.

Most animal wastewater has such a high BOD that it either must be diluted with freshwater or must be pretreated in some manner to reduce the oxygen demand on the wetland system. During different seasons, weather conditions, and times of day, the oxygen concentration in wetlands will vary. The cooler the water temperature is, the lower the oxygen demand will be. These daily and seasonal changes affect the biological processes that remove carbon, nitrogen, and phosphorus along with reducing the high BOD in animal wastewater.

Constructed wetlands have limitations for treating animal waste. Some potential problems that can limit the success of this treatment method are:

- High nutrient levels in wastewater.
- High ammonia levels, which can kill aquatic plants on which the wetlands depend.
- Reduced treatment efficiency during the winter.
- High flows during heavy spring rainfalls.
- Plant residue buildup in the wetlands, which can contribute to the establishment of a potential nutrient sink.
- Zero discharge requirements for animal waste systems.
- Potential for muskrats and mosquitoes.

Several elements must be considered in designing constructed wetlands:

- Type of wastewater, which is influenced by the number and type of animals, lot runoff, rainfall collection, and stack pad drained liquid.
- Wastewater content, including BOD, total suspended solids, nitrogen, and phosphorus.
- Hydraulic flow through the system, which affects BOD reduction, fluid transport rate, and odor.
- Seepage, evaporation, and transportation losses.
- Suitable posttreatment of the outflow.
- Total land area needed for all components of the entire treatment system.

Wastewater "strength" is generally measured by how much oxygen is required to reduce wastewater contents to chemical compounds that are stable or nearly stable in the environment. Oxygen is necessary for most of the chemical and biological processes that "treat" or reduce animal wastewater BOD and nutrient content to more desirable levels.

EXAMPLE 12.3

Wastewater from a dairy farm is pretreated in an anaerobic lagoon and then discharged into a constructed wetland. The wastewater discharge is 500 ft³/day, and nitrate concentration when the wastewater enters the wetland is 100 mg/l. The target maximum effluent concentration of nitrate is 10 mg/l. Calculate the surface area of the wetland, the theoretical detention time, and the hydraulic loading rate. The average water depth in the wetland will be 9 in.

Solution

From **Table 12.9**, k_A is 35 m/year, and C_b is 0. First, convert k_A into feet/day by multiplying by 3.28 and dividing by 365. Therefore, k_A is

$$k_A = \frac{(35)(3.28)}{365} = 0.315 \text{ ft/day}$$

Now, solve Equation 12.6:

$$A = -\frac{q_d}{k_A} \ln\left[\frac{C_o - C_b}{C_i - C_b}\right]$$

and

$$A = -\frac{500}{0.315} \ln\left[\frac{10}{100}\right] = 3655 \text{ ft}^2$$

So, the wetland surface area is 3655 ft². The hydraulic loading rate is obtained from Equation 12.28:

$$q_{hlr} = \frac{q_d}{A} \quad \text{and} \quad q_{hlr} = \frac{500}{3655} = 0.137 \text{ ft/day}$$

or about 1.6 in./day

The theoretical detention time is obtained from Equation 12.3:

$$t_d = \frac{Vn}{q_d} \quad \text{so} \quad t_d = \frac{(3655)(9)(0.95)}{(500)(12)} = 5.2 \text{ days}$$

The theoretical detention time of 5.2 days was based on assuming that cattails grew in the wetland and by converting the depth of 9 in. to feet by dividing by 12.

Answer

The wetland surface area needs to be at least 3655 ft², the hydraulic loading rate is 1.6 in./day, and the theoretical detention time is 5.2 days.

12.9 LANDFILLS

12.9.1 INTRODUCTION

Solid waste (refuse, garbage, trash) is produced by almost every human activity. In the year 2000, each American produced 4.5 lb of solid waste each day. Despite the current emphasis placed on recycling and reuse, over half of this waste was landfilled. Given the current generation rates of solid waste in the U.S., the land required to accept the waste generated by a community of 10,000 people in a single year is a 1-acre area containing 8 ft of refuse.

Sanitary landfills were first constructed in the 1920s and 1930s to replace open dumps, which posed a significant threat to the environment and the community. Modern landfills are highly engineered containment systems designed to isolate the solid waste from the environment and therefore minimize the environmental impact of the solid waste. A schematic of a modern landfill is shown in **Figure 12.25**. Landfills are constructed to be the ultimate disposal sites for different types of wastes. *Construction and demolition debris landfills* are built for the disposal of asphalt, shingles, wood, bricks, and glass. *Sanitary landfills* are used for the disposal of municipal solid waste (residential, commercial, and some industrial wastes). *Secure landfills* are the final disposal option for hazardous wastes (including medical waste, flammable or toxic wastes).

The following sections provide an introduction to hydrology in relation to landfill design, operation, and management; more complete information may be found elsewhere (Tchobanoglous et al., 1993; Vesilind et al., 2002; Reinhart and Townsend, 1998; Bagchi, 1994).

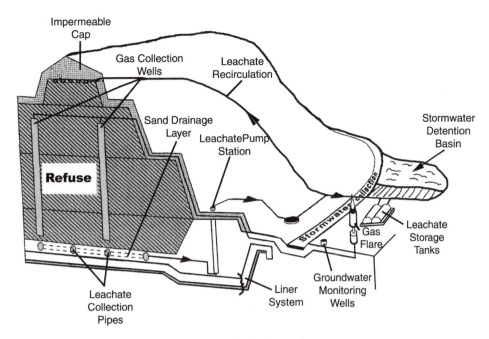

FIGURE 12.25 Schematic of a modern landfill showing hydrological control structures.

12.9.2 LANDFILL PROCESSES

A complex interplay of chemical, physical, and biological processes occurs within the landfill environment. Some of these processes are discussed below.

The organic components of the refuse may be degraded by microorganisms in the landfill. When oxygen is present, aerobic degradation is mediated by microorganisms in the refuse. As the oxygen is consumed, anaerobic microorganisms begin to degrade the refuse. The results of this degradation are gases (methane and carbon dioxide) and organic by-products. The extent and rate of the degradation are dependent on temperature, pH, composition of the refuse, availability of water and nutrients, and the presence of toxic substances in the refuse. In landfills, this degradation (or stabilization) usually occurs very slowly, taking place over several decades because conditions within the landfill are not optimal for biodegradation. Water that is present in the refuse and water that infiltrates the refuse combine with dissolved and particulate contaminants in the landfill to form leachate. Leachate may contain metals, ions (nitrate, sulfate, chloride), and organic compounds. Leachate may pose risks to human health and the environment if the management of leachate and water in the landfill are inadequate.

Chemical reactions that occur within the landfill include the dissolution of compounds, evaporation and vaporization of compounds into landfill gas, and sorption of compounds to landfill constituents. Physical processes of importance in the landfill are the diffusion of landfill gas out of the landfill, leachate movement, and settlement as a result of decomposition. Leachate moves through pores in the refuse (similar to water moving through soil);

however, the presence of many different size items in refuse (from soil particles to appliances and furniture) results in preferential flow paths (see Chapter 3).

All of these chemical, physical, and biological parameters interact in the landfill environment to create a complex system. In the soil system discussed in Chapter 2, soil was represented by soil particles, air, and water. The landfill system cannot be fully understood unless the biological component is included. Management of water within and around a landfill, therefore, must consider all aspects of the landfill system.

12.9.3 SURFACE WATER MANAGEMENT IN LANDFILLS

Intermediate covers are used at the end of each working day to reduce the amount of infiltration (by maximizing runoff from the landfill) and limit odors and windblown debris. Intermediate covers of compost and soil are commonly used.

Final covers (*caps*) are constructed once the landfill is filled and closed. The final cover minimizes infiltration; permits revegetation and reclamation; and reduces the uncontrolled migration of landfill gas. Final covers are constructed from a series of layers, including, but not limited to, subbase soil layer, barrier layer (impermeable geotextile), drainage layer, gas collection system, clay layer, protective soil layer, and topsoil layer to promote revegetation. These layers are constructed to maximize runoff and evapotranspiration while minimizing infiltration; this is achieved by grading the cap, using layers designed to have low permeability to water, and vegetating the cap. This limits the volume of water that enters the landfill and therefore minimizes the production of leachate and reduces the environmental threat posed by the landfill. Final covers require long-term maintenance to ensure that the integrity of the cover is maintained during settlement of the landfill.

The management of stormwater at landfills is mandated by federal regulations. Landfill operators must minimize run-on from surrounding land by contouring the landfill and land surrounding the active fill or by construction of ditches or culverts. Runoff from the landfill surface must be collected and directed to detention ponds by swales, ditches, berms, or culverts. Runoff must be stored and treated to prevent pollution of surface water and groundwater. Control and storage methods for run-on and runoff must be designed to handle peak volumes generated by a 24-h, 25-year return period storm event. This runoff and run-on must be controlled, stored, and treated to ensure that the landfill does not violate the Clean Water Act and pollute surface waters.

12.9.4 LEACHATE MANAGEMENT IN LANDFILLS

The production of leachate is affected by numerous factors, including the composition of the waste, precipitation characteristics of the site, site hydrology and drainage, cover design, waste age, site hydrogeology, and climate at the site. Design solutions for landfill leachate management (liners, leachate collection, storage, and treatment) must, therefore, consider these factors to minimize the adverse environmental impact of the leachate.

Landfill *liner systems* are the barrier between the landfill and the surrounding soil and groundwater. Liner components must be chosen to withstand the chemicals present within the leachate. Liners are designed to minimize the percolation of leachate from the base of the landfill into the surrounding soil and eventually into the groundwater. Frequently, a multilayer liner, consisting of a clay layer and geomembrane, is used to minimize leachate movement; there is a drainage layer of sand with leachate collection pipes and soil layers to protect the other liner components. An example of a liner system used in sanitary landfills is shown in **Figure 12.26**. The leak detection layer contains drainage pipes, and the presence of leachate in these pipes alerts landfill management to the fact that the primary liner has failed.

Liners may be characterized using *breakthrough time*, which is the time it takes for leachate to penetrate a liner. For a clay liner, the breakthrough time t may be calculated from the following equation:

$$t = d^2 \alpha / K(d + h) \qquad (12.8)$$

where d is the thickness of the clay liner (feet), α is the effective porosity, K is the coefficient of permeability (feet/year), and h is the hydraulic head (feet). The coefficient of permeability and effective porosity are dependent

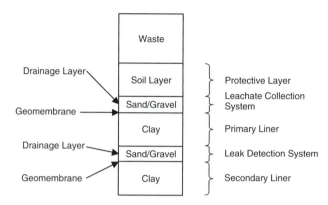

FIGURE 12.26 Example of a liner system used in a sanitary landfill.

on the type of clay used to construct the liner. This equation may be used to calculate the thickness of clay liner required if local or state regulations specify a minimum breakthrough time.

EXAMPLE 12.4

Determine the breakthrough time for a clay liner with the following characteristics: 0.2 effective porosity; 0.05 ft/year permeability coefficient; 3-ft thickness; 1-ft hydraulic head.

Solution

Substituting in Equation 12.8,

$$t = d^2 \alpha / K(d + h)$$

$$t = (3 \text{ ft})^2 (0.2)/(0.05 \text{ ft/year})(3 \text{ ft} + 1 \text{ ft})$$

$$t = 9 \text{ years}$$

Answer

The breakthrough time is 9 years.

Leachate Collection and Storage. The leachate collection system (a highly permeable drainage layer of sand or gravel and drainage channels and pipes) is built into the liner and is designed to maximize the drainage of leachate from the landfill. Leachate is directed to low points at the base of the landfill by grading the land before filling and using drainage layers of sand or gravel. The leachate is collected in perforated pipes that are sloped to move the leachate out of the landfill. Federal regulations restrict the depth of leachate on the liner to 30 cm, so leachate must be efficiently removed from the landfill. Gravity flow or pump systems may be used to remove the leachate. The leachate is then temporarily stored on site in surface impoundments or tanks until it can be treated (on site or off site) before disposal. These tanks or impoundments are sized, using the procedures presented in this chapter for detention ponds and methods for impoundments presented in Chapter 8, based on estimations of leachate generation rates.

Leachate Treatment. Leachate contains many organic compounds. Treatment of leachate is complicated by irregular production rates and composition. Some of the treatment methods used include stabilization basins, trickling filters, anaerobic lagoons, wetlands, neutralization, precipitation, oxidation, adsorption, reverse osmosis, and evaporation. Usually, a series of these processes is used to treat the leachate prior to discharge to a wastewater sewer or receiving body. The sequence and types of treatment options are determined by the chemical composition of the leachate at each landfill and the discharge standards that must be attained.

Leachate Recirculation. Leachate recirculation is a method of managing liquid within the landfill to create an active biological reactor that accelerates the decomposition of refuse by microorganisms in the landfill (a bioreactor landfill). The removal of leachate from landfills creates dry conditions that prohibit microbial growth. Leachate recirculation raises the moisture content in the landfill and promotes decomposition of the organic portion of the refuse. Gas production is also enhanced in landfills where leachate recirculation is practiced. A gas collection-and-recovery system, therefore, is required at these landfills. Stabilization of the refuse occurs more rapidly than in dry landfills, and as biodegradation of landfill components occurs, the leachate is partially remediated and poses less of a threat to the environment. A leachate recirculation system must include leachate collection, storage, and systems for the application of leachate to the refuse mass (trenches, wells, or irrigation systems). In some landfills, water is used to supplement leachate when insufficient leachate is produced to raise the moisture content of the refuse.

12.9.5 LANDFILL WATER BALANCE

Modern landfills have a liner system that prevents groundwater from entering the refuse mass. Once a landfill is closed, an impermeable cap is constructed to minimize the infiltration of rainwater. Despite these precautions, some water is present in the landfill (from infiltration or water present when the waste was placed in the landfill) or enters the landfill from precipitation. A diagram of components of the landfill water balance is shown in **Figure 12.27**.

If an impermeable layer was installed between the refuse and the vegetated cap, as is the recommended practice, this

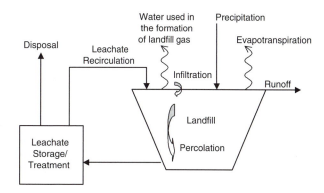

FIGURE 12.27 Water balance for a capped landfill.

permeate would be intercepted before it enters the refuse and drained to detention ponds. The presence of this impermeable layer therefore minimizes leachate generation by preventing percolation of water from the cap to the refuse. However, problems may arise if continuing maintenance is not performed on the cap. If the impermeable layer within the cap loses integrity, rainfall can permeate the refuse mass, and uncontrolled leachate generation may occur.

EXAMPLE 12.5

Estimate the leachate generation in the presence of a temporary cap of soil that is vegetated using a water balance method at a landfill in Columbus, OH. Assume no water is used in the generation of landfill gas, leachate is not recirculated, and the percentage runoff for the site is 10%.

Solution

$$\text{Precipitation } P \text{ at Columbus, OH} = 38 \text{ in./year (Chapter 2)}$$

$$\text{Runoff } R = (0.10)(38 \text{ in./year}) = 3.8 \text{ in./year (Chapter 5)}$$

$$\text{Evapotranspiration } E = 28 \text{ in./year (Chapter 4)}$$

$$\begin{aligned}\text{Water entering the refuse} &= P - R - E \\ &= (38 - 3.8 - 28) \text{ in./year} \\ &= 6.2 \text{ in./year}\end{aligned}$$

For every square foot of surface area, the volume of leachate generated will be

$$\begin{aligned}V &= (6.2 \text{ in./year})(12 \text{ in.}^2) \\ &= 74.4 \text{ in.}^3/\text{year} \\ &= 0.3 \text{ gal/year}\end{aligned}$$

Answer

The amount of leachate is 0.3 gal/year for every square foot of surface area.

12.9.6 HYDROLOGY AND LANDFILLS

Hydrology is an important parameter that must be considered when designing a landfill; these considerations include those mentioned in this section. When siting a new landfill, it is also vital to consider concepts discussed in this book. Some of these considerations are mentioned here.

Local weather conditions (e.g., precipitation type and duration; Chapter 2) determine the site runoff characteristics used to design detention ponds for runoff. Winter weather may restrict site access; wind patterns and strength must be examined to minimize windblown debris and odor migration off site. The surface water hydrology is important when determining the natural drainage and runoff (Chapter 5) patterns of the site. Existing channels and the watershed into which the landfill is constructed must be examined to minimize the disturbance and pollution of this water. The hydrogeology (Chapter 11) of the site is important when considering the type of liner required and the potential for groundwater contamination. Sites designated for landfills must not include wetlands, floodplains, or groundwater recharge areas. Other factors also influence siting a landfill, including site access, soil conditions, and land use in the proximity of the site.

Landfills are complex environments constructed by society to dispose of waste created by the community. The interplay of hydrology with the site of a landfill dictates the design, construction, and management of the landfill. Like more natural environments, landfills are influenced by physical, chemical, and biological processes that need to be understood to build and manage these facilities successfully.

13 Fundamentals of Remote Sensing and Geographic Information Systems for Hydrologic Applications

13.1 INTRODUCTION

13.1.1 WHAT ARE REMOTE SENSING AND GEOGRAPHIC INFORMATION SYSTEMS?

A variety of technologies offers great capabilities for evaluations of hydrology and related issues. The last 30 years have witnessed their application to experimental and operational efforts. Now, it is very appropriate to supply information on their capabilities and applications to facilitate the important work related to hydrological problems (Lyon and McCarthy, 1995; Lyon, 2001, 2003).

These modern technologies have historical antecedents. Remote sensing and geographic information systems (GIS) have melded a number of early technologies, and they are useful for many different applications. Remote sensing and GIS developed from earlier and still useful technologies, such as surveying, photogrammetry, photointerpretation, and the like.

Remote sensing is the practice of measuring an object from a distance or remote location using an instrument. In essence, it is instrumental sensing that is done without touching the object. This may be performed with an aircraft and a camera or sensor or with a spacecraft and sensor instrument system.

Geographic information systems are technologies with antecedents in surveying, mapping, cartography, and information management technologies. GIS technologies allow the storage and processing of data in a spatial or maplike reference system. The concept is over 30 years old, yet elements of GIS technologies have been applied for many more years. The myriad applications and the advent of lower cost computing capabilities have made the discipline grow in a rapid fashion.

13.1.2 MAPPING SCIENCE AND ENGINEERING TECHNOLOGIES

A very good way to conceptualize the common elements and heritage of these technologies is to view them broadly. It seems to be valuable to organize these technologies under the heading of the mapping sciences. This concept allows us to address the variety of technologies without excluding necessary examples. It also recognizes that the mapping sciences are like a "toolbox" of methods that can be applied when the need arises. And, like the old adage, one applies the "right tool for the job."

13.1.3 GEOGRAPHIC INFORMATION SYSTEMS

GISs are databases that usually have a spatial component in the storage and processing of data. Hence, they have the potential to both store and create maplike products. They offer the potential for performing multiple analyses or evaluations of scenarios resulting from model simulations (Kessler, 1995; Lyon, 2003).

Data are stored in multiple files. Each file contains data in a coordinate system that identifies a position for each data point or entry. Characteristics of the data point are stored as "attributes" of the point. A database of individual files is developed and may contain files with characteristics such as stream locations, topography, management practices, and more.

The strength of the GIS approach lies in the quality of the database and how it can be used to address the application of interest. Each variable or "layer" in the GIS can be used in the application to develop information on physical or chemical processes, characteristics of the features, or behavior of people or animals.

An important capability of GIS is the simulation of physical, chemical, and biological processes using models. GIS can be used with deterministic or complex models composed of algorithms that simulate processes, or they can be applied with statistical or stochastic models. The requirement is that the model to be applied has the capability to take spatial or multiple file or layer data as input to computations and yield some meaningful information.

GIS databases and products are also amenable to evaluations of quality. It is an important aspect of these efforts to document the quality of the data and determine the accuracy of the product. A number of methods can be used to assess accuracy (Congalton and Green, 1995; Lunetta and Lyon, 2004), and these approaches can be implemented in the experimental design (Bolstad and Smith, 1995).

13.1.4 Remote Sensing

Remote sensing is the science and engineering of collecting data about materials or features from a remotely located sensor. Remote sensor technologies can be used to acquire a variety of data for applications. A number of these technologies can supply a wealth of data and potentially can be accomplished at a lower cost than many other technologies (Lyon et al., 1995). These advantages have attracted great interest in the scientific and engineering communities.

To apply remote sensing technologies, it is necessary to identify the application and the characteristics that may be measured by the remote sensor. These characteristics or variables may be measured directly by the sensor or indirectly through measurement of a surrogate variable with behavior that is correlated to the variable of interest. Then, we can select the appropriate technology or tool from a suite of technologies. This selection is made with the knowledge of the characteristics or variables to be measured, the available budget and resources, and time frame of the project.

13.1.5 Photointerpretation

Photointerpretation is the science of deriving information from the characteristics of earth features recorded on photographs. Generally, the term refers to interpreting photographs taken from aircraft, although the same skills can be used to obtain information from any type of photograph or image. Photointerpretation utilizes several characteristics or elements of photographs to supply information. These elements include tone, shape, size, texture, pattern, shadow, and associations.

The *tone* refers to the absolute or relative gray tone of features within the photo. For example, clean or clear water may be dark toned, while soil is light toned on black-and-white aerial photographs or images. Tone can also be applied to interpretations of color photographs and the description of the relative or absolute reflectance or tone. We can describe a scale of, say, dark green approaching black and continuing onto bright green tones approaching green-white. This may also be done for the blue or red scales of the additive color primaries; that is, we can describe a black to bright red tone.

Shape refers to exterior configuration of materials or features. Many features or objects have distinctly regular or irregular shapes. Shape can be a unique clue as to the identity of the feature. For example, most people recognize the shape of certain buildings, such as the five-sided shape of the Pentagon, the horseshoe shape of the Ohio State University stadium, the shape of the arch in St. Louis, the pyramids of Egypt or the pyramid-shape Luxor hotel and casino in Las Vegas, and so on.

Size refers to the absolute or relative dimensions of the object or feature. Size can be very important as features are often indistinct on photographs, and knowledge of size can be a big clue as to the identity of the feature. Measurements can be made on the photo or image using a tool, and the measurements are converted to an absolute measure using scale. The value of size should not be underestimated as a clue to the identity of a feature.

Texture refers to the regular or irregular organization of features within the photo. Usually, the features are small relative to the resolution of the photo. For example, we may view a distinct texture resulting from small rocks breaking the surface of a lake, yet the rocks may be too small to view directly. Their presence results in a different texture from that of lake waters without the small rocks.

Pattern, conversely, is the regular or irregular distribution of large features on the earth. An example is the regular pattern of field boundaries in an agricultural area. When this regular field pattern is disturbed, say, by a stream course, we can infer certain characteristics of the stream channel, such as if it is too big for farm equipment to traverse.

The *shadow* of a feature can be an important clue to its size and shape and may help identify the feature as to type. Shadows cast by a feature can provide important shape and size information. A common example is that of interpreting the name of a business or building from the shadows of individual letters that form a sign. The individual letters "blend" into the building, but the shadows are distinct to the interpreter. Conversely, shadows can obscure detail. Because of the lower relative illumination in the shadow area compared to the overall illumination of the sunlit areas, it is difficult to view a feature hidden by a shadow.

The *association* of a feature is another characteristic or clue that is found together with the feature of interest. For example, the course of a stream is usually evident by shape, but often the tone of the water is hidden from view by trees. Stream courses are usually found in association with other clues, such as the meandering stream pattern, the branching drainage pattern, automobile and railroad bridges, pond or lakes, streamside vegetation, and lower relative elevation and a downhill course compared to the surrounding landscape (Argialas et al., 1988).

Figure 13.1 and **Figure 13.2** display many of the characteristics that can be used to derive information for an area using photointerpretation techniques. In **Figure 13.1**, we can see details of the near-shore topography and shape of coastal features due to the penetration of visible light into the shallow water column. Note the very light tone of bottom sediments, roads, and beach areas. In comparison, the vegetation of the swamp or upland forest areas is dark. **Figure 13.2** is a black-and-white infrared image of same location. Here, the areal extent of surface water can be seen by the very dark tone of water on the infrared image.

FIGURE 13.1 A typical black-and-white aerial photo of the central coast of Lake Michigan in Michigan (August 8, 1976). Note the stream entering the lake from the east and the trees that surround the stream; both help identify the stream and partially shield the water from view.

FIGURE 13.3 A black-and-white photo (July 29, 1952) from another date can supply additional details that may be unobservable in other photographs. The three images of **Figure 13.1** to **Figure 13.3** supply information over time, including changes in land cover and water resources.

(Lyon, 1987; D. Williams and Lyon, 1995). This is particularly true of hydrological characteristics in a given area, which change following rainstorms or flooding. It is always useful to obtain many aerial photos of a given study area in support of analyses, particularly hydrological analyses. This is because each photo supplies unique information, repetitive coverage adds to the value of multiple samples and statistical capabilities, and use of existing photos is very inexpensive.

13.1.6 PHOTOGRAMMETRY

Photogrammetry is a very valuable technology that has provided many years of service. Almost all maps showing horizontal positions, point elevations, elevation contours, or topography are made using photogrammetric technologies. *Photogrammetry* is the science of obtaining precise and accurate measurements from overlapping or stereoscopic photographs (USACE, 1992; Lyon et al., 1995; Falkner and Morgan, 2002).

Commonly, these photographs are taken from an aircraft platform, and individual exposures record the ground under the aircraft overflight. The photos are taken such that an individual photo includes ground coverage of the previous and next exposure in a series or flight line of overlapping photos. These photographs are called *stereoscopic pairs*. The difference in perspective of a common feature photographed from different positions is called *parallax*. The relative parallax or difference in position of a given feature on the pair of exposures creates the stereoscopic effect and allows determination of the elevational

FIGURE 13.2 The difference in this image (April 18,1981) of the same location in **Figure 13.1** is because the film is black-and-white infrared. Note the lack of detail within the water due to the absorbance of infrared light at the surface and near surface of the water. The infrared film helps identify the location and surface area of water bodies.

Figure 13.3 is an additional black-and-white image of the area. The use of a number of images and other, different film types can be a great help in photointerpretation

FIGURE 13.4 Lower altitude stereoscopic coverage (Ohio) acquired for engineering purposes, typical of the photographs available from state departments of transportation. Note the presence of two lakes, ditches, trees, and homes and how stereoscopic viewing supplies additional information about the area.

position of features in addition to their horizontal relative positions.

Photogrammetry is important because it is often the most cost-effective method to make topographic maps. Almost all USGS or NOAA topographic maps are made from aerial photographs using photogrammetric technologies. Most large-scale engineering maps are also made from aerial photographs. Hence, most of the products that are used in hydrological analyses are made from photogrammetric analysis of aerial photographs (USACE, 1992; Falkner and Morgan, 2002). Now, a number of watersheds are characterized using point elevations generated by laser LIDAR systems (Falkner and Morgan, 2002). Vertical, overlapping photographs can be valuable for photogrammetric measurements and for photointerpretation. **Figure 13.4** demonstrates a stereo triplet.

Photogrammetric measurements from aerial photographs can document historical conditions and change over time (Lyon, 1987; D. Williams and Lyon, 1995; Lyon, 2003). This is particularly true of hydrological events that cause change. **Figure 13.1** to **Figure 13.3** show the changes in land cover and water resources over a 29-year period. This information is valuable for scientific, engineering, and legal studies and is quantitative in quality.

Color and color infrared (CIR) photography is also very valuable for hydrology-related analyses. The CIR image can facilitate interpretation of water resource characteristics and general vegetation types and conditions. The stereoscopic coverage allows viewing of relative topography and the shape of tree canopies.

13.1.7 SURVEYING

Surveying is the science of obtaining precise and accurate measurements of the earth's surface or its waters using instruments. These measurements include horizontal locations as well as vertical or elevational locations. Commonly, these measurements are tied to some absolute reference to characterize the location of earth features in an absolute sense and to allow later relocation of these features using the original measurements or map products.

Surveying is important in hydrology as these measurements form the basis of many hydrological calculations. The surveyed positions of basins, subbasins, channels, control structures, and the like all form important inputs for calculations. These positions may also provide the basis for GIS databases and form GIS model calculations (Van Sickle, 2001).

Surveying is also important in the production of photogrammetric products (USACE, 1992). In general, to make photogrammetric calculations in absolute units and to reference measurements absolutely, it is necessary to collect ground-surveyed positions. These positions or ground control points (GCPs) allow the stereo model to be tied into absolute ground positions. Surveying is also used to mark the positions and follow the progress of field construction and ensure that the engineering design specifications are met.

An exciting arena is that of global positioning system (GPS) surveying. GPS can be used in measurement and reference control for photogrammetric measurements, among other applications (Falkner and Morgan, 2002). An example application of this work would be the staking of a river course to help modify its geomorphology to create more fish habitat and reduce erosion.

13.2 PRODUCTS

Products are the media that we use to handle the information. Often, the media are helpful in many aspects of the work. They may form original examples of products or may be derivative products optimized for a certain application.

13.2.1 PHOTOGRAPHS AND IMAGES

Photographic products have well-known characteristics and can be used in a variety of forms to collect and store original data or to produce derivatives. Photographs use silver-halide chemistry to capture black-and-white or color images on film (transparencies) or paper. As such, photography using cameras is a remote sensor technology. It has a number of advantages in that the products are usually available and are often relatively low cost compared to other sensor products.

Images can include photographs. Many images are digital, such as satellite images and those from digital cameras. The advent of relatively inexpensive software and very capable digital cameras has opened a completely new world for digital imaging.

13.2.2 MAPS

Maps are often the starting point in an analysis and often are used in the presentation of results in reports. USGS and other original source maps may be used to obtain a variety of information. They can be used to take point position information; topographic contour information on cultural or planimetric details such as roads, waterways, dwellings; or public land ownership boundaries. The maps can provide the starting point to form a variety of land cover themes or layers in a GIS.

Engineers often have large-scale maps available for site planning and design. The scale is commonly 1 in. equal to 100 ft or 200 ft. The contour interval is often 1 ft of elevation. These products are especially useful in hydrological analyses, and they are created through surveying or a combination of surveying, photogrammetric, or LIDAR technologies.

13.2.3 DIGITAL IMAGES

Images are defined as two-dimensional, pictorial representations of data. Images are often presented as nonphotographic data, and they are usually the result of computer systems. The word *image* helps to identify and separate images from photographic image products that are so common as the pictorial presentation of data.

Landsat data have been acquired over the years since 1972. These data are available in the form of computer image products or images, and much information can be developed from analyses of digital or image data. Later, Landsat satellite, SPOT Image, and other satellites have produced a variety of products. Landsat Multispectral Scanner (MSS) images demonstrate some of the general capabilities of these data. **Figure 13.5** and **Figure 13.6** are MSS images of the Brooks Range, Alaska, and the Arrigetch Peaks and Gates of the Arctic National Park and some of interior Alaska (Lyon, 2001). **Figure 13.5** is a Band 5 or red reflectance image of the area. The dark-tone areas are forest cover, and the whiter tone parts are bare areas. **Figure 13.6** holds some interesting general hydrological information. This image is in the infrared portion of the spectrum, and water resources appear dark black. We can see a variety of drainage patterns, including the

FIGURE 13.5 Digital satellite image from the MSS of Landsat (October 12, 1974) showing the red reflectance of the Brooks Range, Walker Lake, the Arrigetch Peaks, and some of interior Alaska. Note the small lake (Norutak) in the lower center and compare its size in **Figure 13.6**.

braided drainage of the Kobuk River and other meandering rivers in interior Alaska.

Figure 13.7 shows stereoscopic coverage of the Arrigetch Peaks in the Gates of the Arctic National Park, Brooks Range, Alaska. This area is also shown in **Figure 13.5** and **Figure 13.6**, although at much smaller scale. **Figure 13.8** to **Figure 13.10** also provide good information on water and terrestrial resources and provide some idea of scale. **Figure 13.8** was taken at the south end of Walker Lake looking to the northeast (Walker Lake is approximately 16 mi long in the north–south direction, and it is the largest lake shown on the Landsat images) and the headwaters of the Kobuk River and Arrigetch Peaks. **Figure 13.9** is a picture of the granite massif of the Arrigetch Peaks area partially shown on **Figure 13.7**. **Figure 13.10** is a small glacier and peak in the Awlinyak Creek drainage just north of the Arrigetch Peaks.

13.3 PARTS OF THE SPECTRUM AND RADIATION CHARACTERISTICS

Remote sensing inherently measures light. Light or electromagnetic radiation emanates from the sun, is propagated

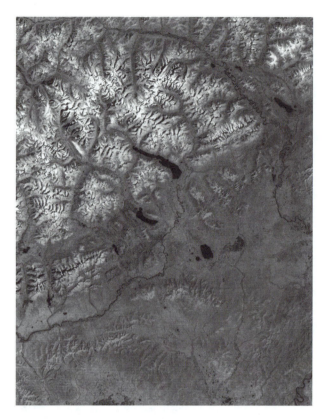

FIGURE 13.6 Image from the MSS of Landsat (June 9, 1974) showing the infrared reflectance of the Brooks Range and interior Alaska. The Arrigetch Peaks are in the top center of the image, north of Walker Lake. Note the Alatna River drainage from the very top center draining to the right and how the infrared highlights the river location.

through the atmosphere, strikes and interacts with the earth's surface, returns through the atmosphere, and is ultimately measured by the remote sensor or film above the earth. The study of these phenomena is *remote sensing*.

There are many valuable portions of the light spectrum that can be sensed. In particular, water characteristics are identified using visible, near, and thermal infrared (TIR); microwave or radar; and sonar sensors. The selection of the appropriate portion of the spectrum and sensor is based on the application and the cost of using the sensor.

There is a variety of light types, and these are usually measured based on the perceived information that can be obtained about a feature such as water. Hence, only a portion of the total spectrum of electromagnetic energy is measured; this is called a *spectral measurement*.

13.3.1 Visible, Infrared, and Thermal Infrared

The visible portion of the spectrum is found approximately from wavelengths between 0.3 µm and 0.7 µm. A micrometer (µm) is 1×10^{-6} m. This is the part of the spectrum to which the human eye is sensitive, and as no surprise, it is the same part of the spectrum that black-and-white and

color film or digital cameras sense to produce photographs that capture images in the way that humans see them. Good examples of multispectral images in the red (visible) portion of the spectrum are **Figure 13.5**, **Figure 13.11**, and **Figure 13.12**

The infrared portion of the spectrum is just beyond what a human may see. The *reflected infrared* is found approximately between 0.7 µm and 1.4 µm, and the *middle infrared* is found between 1.5 and 2.8 µm. **Figure 13.2**, **Figure 13.6**, **Figure 13.13**, and **Figure 13.14** are examples of infrared images that are sensitive to the reflected infrared portions of the spectrum.

The thermal infrared (TIR) is a measure of the heat emitted from features. TIR is found between 3.0 and 14.0 µm, and it provides very good information on features as a result of their emission of heat. All features emit thermal infrared radiation, and as such, TIR can be used for detection both night and day. Most TIR sensors that record the passive emission of light energy from a feature are called *passive sensors*.

13.3.2 Radar

Radar sensors work in the microwave region of the spectrum and have wavelengths that are relatively long compared to the visible and infrared. Commercial sensors generally operate between millimeter wavelengths and 24-cm and longer wavelengths. Radar sensors generate their own microwave wavelength electromagnetic radiation and can be used during night or day. Sensors that create their own radiation are called *active sensors*.

13.3.3 Radiation

One of the great capabilities of remote sensing is the fact that light can be measured or modeled in a quantitative manner. Photointerpretation is inherently qualitative, although photographic processes may be the subject of quantitative analyses. Remote sensor data are digital as they come from the sensor and can be processed in the same fashion as any digital product. With these characteristics, it is possible to model the characteristics of light and sensors and predict the behavior of measurement experiments beforehand.

Generally, electromagnetic radiation is radiated in three ways, and most materials exhibit a dominant way as well as a combination of other phenomena, depending on material characteristics. A *diffuse* or *Lambertian source* radiates light equally in all directions and is the most common or dominant phenomenon. A *specular source* has a particular angular orientation to the reradiation of light. *Bidirectional radiation* refers to the nondiffuse behavior typical of many materials. We can think of it as a combination of the other two types of radiation. Most objects that we measure on the earth exhibit a combination of these phenomena,

FIGURE 13.7 Stereo pair of aerial photographs of the Arrigetch Peaks, Alaska (July 1979). The Peaks are in stereo in the lower middle of the composite image. The Alatna River is at the top right, and the Awlinyak drainage runs from lower left to top center.

and the types and quantity of light encountered will be wavelength dependent.

13.3.4 LAMBERT'S LAW

A diffuse or Lambertian source will reflect light equally in all directions. This uniform radiation results from two effects. First, the area viewed by an observer or sensor will change with the cosine of the angle of a normal to the surface imaged. The Lambert cosine law states that the quantity of light will vary with the cosine of the viewing angle. Second, the area viewed by the observer, say, a plate of the material or feature, will change in geometry with the cosine of the angle. These two influences tend to cancel, resulting in a continuous level of illumination. The two phenomenon result in constant illumination when viewed from any angle (Lyon and Khumwaiter, 1987). Hence, the angle of illumination and viewing angle are important considerations. In measurements or in calculations, we measure the brightness; therefore, we must account for the projected angle of view and the area the sensor views on the ground or earth's surface. This makes it necessary to account for our geometry of measurement. The location viewed on the ground is usually expressed as an area, while in the atmosphere or in water, the light is propagated in three dimensions and is expressed as a volumetric quantity.

FIGURE 13.8 Terrestrial photo of Walker Lake, Brooks Range, Alaska, looking to the west at the headwaters of the Kobuk River and the Arrigetch Peaks area.

FIGURE 13.9 Photo of the granitic Arrigetch Peaks, gates of the Arctic National Park, Alaska.

FIGURE 13.10 Residual glacier and peak of the Awlinyak Creek area, Arrigetch Peaks, gates of the Arctic National Park.

13.3.5 PROJECTED SOURCE AREA

The area of the earth's surface measured by the sensor will change depending on the angle from which the sensor views the feature. This can be addressed by taking the

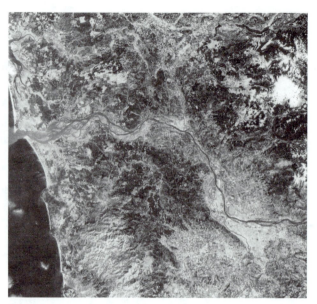

FIGURE 13.11 Landsat MSS image of Portland, OR, showing the red reflectance of water and terrestrial resources (April 7, 1973). Note the very white snow and glaciers of Mt. St. Helens in the top right before the explosion of 1980. Portland and the Columbia and Willamette Rivers are in the lower center of the image, with the Columbia flowing from lower right to the Columbia estuary in the center left of the image.

FIGURE 13.12 Landsat MSS image of Portland, OR, showing the red reflectance of water and terrestrial resources (September 11, 1974). Note the less extensive snow and glaciers on Mt. St. Helens late in the summer season.

cosine of the area of a plate of material or feature on the ground. This corrects for the geometry of the viewer or sensor at some other position than directly beneath the sensor or the "nadir" position.

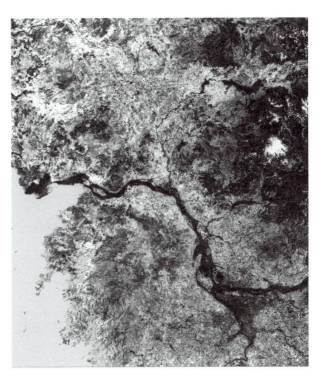

FIGURE 13.13 Landsat MSS image of Portland, OR, showing the infrared reflectance of water and terrestrial resources (July 24, 1973). Note the rivers are distinct, as are lakes and reservoirs, in the infrared.

FIGURE 13.14 Landsat MSS image of Portland, OR, showing the infrared reflectance of water and terrestrial resources. Note the large areal extent of black tones around Sauvie Island north of Portland and the Columbia River, indicative of higher water levels and potential flooding, as compared to the areal extent of water in **Figure 13.13** (June 13, 1974).

The area that is viewed by the remote sensor is equal to $A \cos \theta$.

$$L = \frac{\rho \, E(\text{on a feature})}{\pi \, A \, \cos\theta} \left(\frac{a}{R^2} \right) \qquad (13.1)$$

where L is the radiance, E is the irradiance, a is the area of the aperture, R is the distance between the feature and the sensor, ρ is the reflectance, θ is the angle from the perpendicular at which the sensor views the feature, and A is the area of the feature on the surface imaged.

As light is propagated through the atmosphere or water, the light is dissipated in the volume of gases or liquids. This propagation or *welling* can be thought of as a continuously spreading wave if viewed from one position or frame of reference. An *isotropic point source* can be thought of as a continuously expanding sphere for which any point on the sphere is the source of a continuously expanding volume of light later.

13.3.6 ISOTROPIC SOURCE

An *isotropic source* is a point source from which electromagnetic radiation radiates in all directions. It can be viewed as a spherical source of light or a point source at some distance. A very good example is a star or our sun. A correlate of this characteristic is that the further away

a sensor is from the observer, the less total light will be measured. The quantity of light received is the inverse square of the distance from the sensor to the source of the light. The inverse square law states that the fall off of light is proportional to the inverse square of the brightness of the feature. The light measured at the sensor is equal to $1/R^2$, where R is the distance or altitude between the source and sensor and is usually expressed in meters.

13.3.7 SOLID ANGLE

As mentioned, the geometry of measurements must be addressed. This is particularly important when light is radiated into and through a three-dimensional material such as water or the atmosphere. Here, the concept of solid angle is employed. For three-dimensional or volumetric cases, the concept of the steradian is most appropriate. The solid angle ω is subtended by an area on the surface of a sphere that is equal to the square of the radius of the sphere r_s, and a is the area of the entrance aperture of a sensor.

$$\omega = \frac{a}{r_s^2} \qquad (13.2)$$

The total solid angle about an isotrophic point feature is

$$\omega = 4\pi \frac{r_s^2}{r_s^2} = 4\pi \qquad (13.3)$$

It is common to cite the resolution of a sensor as the radians of the cone angle of the sensor. The sensor records information on the ground (e.g., 0.5 milliradians). This is the angle of the cone viewed by the sensor. An airborne sensor can fly at different altitudes and hence may have different ground resolutions depending on the height above ground.

To determine the solid angle, we use the equation

$$\omega = \pi \tan^2 \frac{\theta}{2} \qquad (13.4)$$

When provided with the cone angle in radians, we can calculate any other portion of the cone using the relationship

$$\tan \frac{\theta}{2} = \frac{r_s}{2} \qquad (13.5)$$

$\theta/2$ occurs because the entire cone angle is provided.

13.3.8 RADIANCE

Radiance L is equal to the irradiance E divided by the solid angle, E/ω, or E/π. The units are watts/square meter/steradian. The radiance L of a feature imaged or measured at a distance or altitude above the surface or earth R by the sensor is

$$L_{\text{at sensor}} = \frac{E(\text{on surface})}{\pi}(\cos\theta)\left(\frac{a}{R^2}\right) \qquad (13.6)$$

$$L_{\text{at sensor}}(\text{off surface}) \times (\text{viewing surface})$$
$$= \frac{M(\text{off surface})}{\pi}\left(\frac{a}{R^2}\right) \qquad (13.7)$$

where a is the area of the opening or aperture of the camera or sensor, R is the distance from the feature to the sensor or the altitude above the surface or above ground level (AGL), and M is the existence or irradiance off the surface or earth (watts/square meter/steradian, the same as E, but in the upwelling direction).

The energy received at the sensor is proportional to the distance from the sensor and the size of the entrance opening or aperture in the sensor. This simple, geometric approach utilizes the concept of the inverse square law, by which the quantity of light radiated into a volume will

decrease at the rate of 1 divided by the square of the distance or altitude R or distance from feature to sensor. The actual quantity of light measured by the sensor is therefore a function of distance and a function of the size area of the opening or aperture in the sensor a. Hence, a/R^2 accounts for the geometry of aerial or space measurements and the radiance at the sensor

$$L_{\text{at sensor}} = \frac{E(\text{off surface})}{\pi}\left(\frac{a}{R^2}\right) \qquad (13.8)$$

A correction can be made for measuring features away from the nadir of the sensor or from features that have a slope angle:

$$L_{\text{sensor}} = \frac{E(\text{off surface})}{\pi}(A\cos\theta)\left(\frac{a}{R^2}\right) \qquad (13.9)$$

where θ is the angle from which the sensor views the feature away from the normal vector view to the feature or ground surface.

Use of a/R^2 assumes that we are dealing with a point source. This is a good assumption if the distance or altitude R is at least 5 to 10 times the radius of the entrance aperture. For this case, any influence due to geometry is minimal and is ignored.

If this is true, we can ignore the difference in geometry of the flat base of a cone vs. the volume a circle would actually trace when taken at the outside of a sphere. For additional discussion of these geometric and electromagnetic radiation characteristics, please consult a detailed text (e.g., Suits, 1973; Slater, 1980; Elachi, 1987; Schott, 1997; Milman, 1999).

13.3.9 SPECTRAL RELATIVE UNITS

The measurement of light energy is usually separated into types based on the quantity and variety of light under study. Radiometry refers to measuring all available light or at least all the light that can be measured by the particular sensor. Spectral measurements refer to the fact that we measure discrete wavelengths or frequencies rather than the whole.

Spectral units are employed to record a quantity of energy per unit of wavelength. These units are encountered whenever per wavelength units are reported. Examples include graphs of reflectance, radiometer measurements, or engineering data on instrument response per unit wavelength in a certain portion of the spectrum. The dimensions naturally include a per wavelength unit. The units are generally micrometers (μm) (1×10^{-6} m) or nanometers (nm) (1×10^{-9} m). Typical dimensions are spectral radiance, λ = watts/(square meters $\cdot$ steradian $\cdot$ nanometer), or watts/(square meter $\cdot$ steradian $\cdot$ micrometer). For example, the flux or irradiance i in a narrow band is the value of the spectral flux at the interval center times the bandwidth interval. The sum of the fluxes in an adjacent narrow bandwidth provides the flux of the bandwidth.

$$\sum i = \sum \left[(\lambda)(\partial \lambda_i) \right] \qquad (13.10)$$

To obtain the spectral response of a given material in a certain bandwidth of operation, the area under a given spectral response curve must be integrated. This allows the modeling of spectral response of materials or combinations of material using existing spectral response curves from references. There are several methods to integrate the area under the curve within the wavelength region of operation. A simple approach is the trapezoidal rule:

$$A = \Delta\lambda \; (\text{nm}) \left(\frac{x_1}{2} + x_1 + x_3 + \dots\dots x_{n-1} + \frac{x_n}{2} \right) \qquad (13.11)$$

where A is the area under the curve.

For example, we can make a sample calculation using spectral flux measurements taken from a graph at the midpoints of the 20-nm intervals, and using the trapezoidal rule, with $\Delta\lambda = 20$ nm (flux of 400 to 600 nm):

Flux(400 to 600 nm)

$$= 20 \, \text{nm} \left(\frac{1.2}{2} + 1.1 + 1.0 + \dots x_{n-1} + \frac{x_n}{2} \frac{W}{\text{nm}} \right) \qquad (13.12)$$

$$= 20 \, \text{nm} \left(\begin{array}{l} \frac{1.2}{2} + 1.1 + 1.0 + 1.2 + 1.3 + 1.0 + \\[2mm] 1.1 + 1.4 + 1.1 + 1.2 + \frac{1.2}{2} \frac{W}{\text{nm}} \end{array} \right)$$

$$\text{Flux(400 to 600 nm)} = 20 \, \text{nm} \left(11.6 \frac{W}{\text{nm}} \right) = 232 \, W$$

In many applications using measurements or calculations, it is desirable to work in relative units. The relative units are usually simple ratios of some absolute measures of incoming radiation and some portion of the outgoing radiation. A common example of relative units is reflectance.

The incoming radiation is proportioned into three components. The sum of the components is equal to 1.00. The light components sum to 1.00 because energy is conserved per the first law of thermodynamics. The upwelling component is known as *reflectance*, or the light that upwells or is returned from the surface of the water body

into the sky, and is ultimately measured by the sensor. Radiation also travels through materials, and this concept is called *transmittance*; most people are familiar with light propagating through water. Radiation that is retained within the material is the *absorbance component*, such as light absorbed by plankton in water.

The general equations for these relative units are

Reflectance:

$$\rho = \frac{\text{radiation off of surface of materials}}{\text{radiation incident on materials}} \qquad (13.13)$$

Transmittance:

$$\tau = \frac{\text{radiation through the materials}}{\text{radiation incident on materials}} \qquad (13.14)$$

Absorbance:

$$\alpha = \frac{\text{radiation absorbed within the materials}}{\text{radiation incident on materials}} \qquad (13.15)$$

The relative units are calculated as ratios of the radiometric units measured or calculated. Commonly, we will ratio irradiances, or radiances, although any two equivalent units can be employed.

Use of the above measurement and modeling approaches allows the evaluation of remote sensor experiments. These elements can be combined to study the response of materials to light energy and to estimate the quantity of light that can be measured by remote sensors during experiments.

To learn more about these issues, please consult the following references, which explain the use of radiometric measurements and modeling in great detail: Suits (1973); Slater (1980); Elachi (1987); Schott (1997); and Milman (1999).

13.4 DATA TYPES AND DATABASES

13.4.1 DATA TYPES

Remote sensor and GIS technologies organize data according to two general protocols. Remote sensors collect their data in grid cell or raster format, and GIS technologies process data in raster or vector form or both as needed.

13.4.2 RASTER

GIS data are stored in files using one of two common methods, raster or vector. Raster storage is in grid cell form, and grid cells "build up" the image. Hence, a given area or polygon will be composed of a number of cells that have a certain area. The grid cells are referred to as *rasters* or as picture elements (pixels).

13.4.3 VECTOR

The other common way to store GIS data is in coordinate or "vector" form. A vector data set is stored as an *x,y* position for a point, two *x,y* positions and a line for an edge or curve vector, or a number of *x,y* points connected by line segments to form a polygon.

13.4.4 ATTRIBUTES

An *attribute* is a characteristic or value of a variable stored in a GIS file. Depending on the GIS requirements, one or several characteristics or attributes can be stored for a given point or area. Attributes can be almost anything. Examples for a given point or area could include elevation, ownership, size or concentration, grid or map position, or a whole series of characteristics, such as those pertaining to a given application.

13.4.5 DATABASES

A variety of data may be used in the assessment of hydrological features. This is truly a great capability of GIS and remote sensing. We often can access collections of data or databases and obtain needed information. These databases are usually less expensive than the fieldwork and data processing required to develop a custom database (Lyon, 2003). The use of existing data can expedite analyses, help focus field sampling activities, and generally ease the process and potentially lower costs of analyses.

An important thing to remember about data sources and databases is that they are inherently inexpensive once they have been created and are available to the public. On a relative basis, when compared to the costs of actual field sampling data, these sources can be acquired at lower relative costs and in a timely fashion. At a minimum, these sources can provide information for a reconnaissance-level analysis before the more expensive and more detailed field evaluations are conducted (Lyon, 1993, 2001). The data eases the planning process, allows the investigators to "come up to speed" quickly and at minimal relative cost. The implementation of existing databases is vital to most projects for the reasons above. Because of their importance, a number of sources for these data and databases are provided in Appendix E.

13.4.6 AERIAL PHOTO DATABASES

Aerial photos are acquired constantly to support science and engineering activities. These photographs are usually saved after the initial project and may be obtained later. Historical aerial photographs are particularly valuable as they are a record of a variety of conditions, including those relating to hydrology (Lyon, 2001). They can be obtained and used in a variety of analyses and applications (Lyon et al., 1986; Lyon, 1987, 2003; D. Williams and Lyon, 1995).

Aerial photographs can be obtained from a number of sources, including the USGS, the USDA, NOAA, state natural resource and transportation agencies, local aerial photography firms, and other valuable sources described in Appendix E.

13.4.7 DIGITAL ELEVATION MODEL

Digital elevation models (DEMs) are maplike products stored in digital computer files; they are composed of *x,y* grid locations and point elevation data or *z* variables. They are generated in a variety of ways and from a variety of map scales by the USGS and by commercial firms. The elevational data are very useful for GIS and landscape characterization activities (Lyon, 2001). The data can be further processed to yield important derivative products, including digital maps of slope or slope aspect. With more detailed image processing, stream channels and subbasin information can be derived. Stream channel or network information can be the data input for more complex analyses, including route of water flow (Argialas et al., 1988; Lyon, 2003).

13.4.8 DIGITAL TERRAIN MODEL, DIGITAL LINE GRAPH, AND OTHER DATA

Digital Terrain Model (DTM) data are an additional product created by a number of groups. DTM data provide digital topographic contours at a variety of scales. As with DEM products, these DTM data are available from commercial firms, the USGS, and local or regional sources and can be located from the addresses provided in Appendix E.

Digital line graph (DLG) data are available from the USGS and others, and they provide maplike presentations of the road network and stream courses for many USGS quadrangle areas. The data make very nice "overlays" of remote sensor or other map data and provide important details on the location of road and road types and streams.

TIGER (Topologically Integrated Geographic Encoding and Referencing) data are available from the Department of Commerce, Bureau of the Census. These data are by census tract and provide information on census-related issues. They are available in digital form and can be very helpful when human and census-related data can assist in a hydrological analysis. TIGER data are available from the Bureau of Census addresses provided in Appendix E. TIGER data in either original or enhanced forms may also be obtained through vendors.

13.4.9 North American Landscape Characterization

North American Landscape Characterization (NALC) is a multiple-year and multiple-agency project to develop current and historical Landsat MSS data sets for evaluation of continental land cover and change in land cover (Lunetta et al., 1993; Lunetta and Elvidge, 1998). The goals of NALC include developing MSS data for the North American continent; correcting and packaging the data and updating the images in the archive; and analyzing the data sets. The data sets are available as "triplicates" of MSS scenes from the USGS EROS Data Center (EDC). A given triplicate will contain three Landsat scenes from similar seasons during the epochs of 1970s (1973 ± 1 year), 1980s (1986 ± 1 year), and 1990s (1991 ± 1 year). The scenes are produced from the same ground area and with geometric corrections and ground registration of coordinates. The original radiometric data are provided in the above form using standardized methods (Lunetta et al., 1993). Additional files include DEM data for the scene area and "housekeeping" files.

Information about NALC data products may be obtained from User Services, EROS Data Center, Sioux Falls, SD 57198 (see Appendix E). Details concerning image products supplied by the EROS Data Center may be examined or browsed using Internet capabilities. Use of the EROS Data Center programs allows evaluation of data sets. The EROS Data Center metadata programs describe many data sets in general terms or variables. The "meta" level of detail in data sets allows the user to look at decompressed versions of the original images, read information on data characteristics, obtain ordering information, and collect appropriate documentation files from reading.

The historical Landsat data from NALC can be valuable for examining features over time. Examples are presented in **Figure 13.11** to **Figure 13.14** of the Portland, OR, area. **Figure 13.11** and **Figure 13.12** show the red reflectance of the earth. Note the change in tone from one image to the other. These tonal changes in multiple parts of the spectrum can be used to evaluate change in terrestrial and water resources. **Figure 13.13** and **Figure 13.14** show the near-infrared portion of the spectrum, which can be used to evaluate the areal extent of water bodies, including rivers and lakes. Note the great change in the dark-tone water area between the two images, which is indicative of a large surface extent of water or flooding.

13.5 REMOTE SENSING CHARACTERISTICS OF WATER

Water has a variety of known characteristics and reactions to light that facilitate its evaluation. These characteristics are based on the portion of the spectrum used to measure the water body and on the physical and chemical reactions of water to that portion of the spectrum.

Most features, including water, react to light in one to three ways. Water is well known for absorbance of light energy, and it is often much darker in the visible and near-infrared portion of the spectrum due to the absorbance.

Water also reflects light from the surface and from the water volume. Surface reflectance or sun glint can be seen near the water or on aerial photos as a very bright or white area (**Figure 13.1**). Reflectance from the water volume upwells to the surface from the water itself or the bottom sediments, and it supplies the blue and blue-green color of the water.

Water also transmits light from the surface to the bottom and back. Transmittance of light is the opposite of opaqueness, and it serves to illuminate the water volume or column. In particularly clean waters, one can observe the bottom of a lake or ocean at 60 ft or more in depth due to the transmittance of light from the surface to the bottom and back.

The first law of thermodynamics says that energy is conserved, and this is true for light energy. The incident quantity of light on a feature such as water is portioned into reflectance, absorbance, and transmittance, and the sum of the three is equal to the total of the incident light. The same relationships described above hold true for water and waterborne materials, as well as other familiar materials like snow, concrete, and wood.

13.6 APPLICATIONS

13.6.1 General Characteristics of Applications

The application of these technologies to hydrological and related problems is an important goal and a valuable effort. The landscape scale of hydrological problems requires methods to gather spatially distributed information. The problems also require repeated sampling of the variable of interest to acquire information over large areas.

The costs and logistics of these actions can be high and are always constrained. We need to develop more efficient technologies to gather spatially distributed information; hence, there is high interest in remote sensor and GIS technologies (Lyon, 1993, 2001).

13.6.2 Planning

The planning of scientific and engineered projects is a very appropriate application of remote sensor and GIS technologies. These technologies allow the collection of spatially distributed data, and these approaches facilitate the analysis of image or spatially distributed data.

These approaches also allow engaging in "scenarios" or in repetitive evaluations of project conditions. After the evaluations are made, the results can be produced in table or tabular form and in map or graphical forms. These capabilities can greatly facilitate any planning activity and the implementation of many of these technologies in business, government and university arenas. Their widespread application and utility is indicative of their value.

13.6.3 SITE DETERMINATION

The identification and selection of a site for a project, known as *siting*, is a type of planning with special value to engineers identifying the optimal locations for a given construction project. Examples could include siting a landfill, siting a raw materials processing plant in a good location between source materials and point of sale, optimizing the location of an office operation as compared to the adjacent markets, and establishing general site location and addressing the information requirements of zoning and environmental permitting activities.

An important advantage of using GIS in siting is the visualization capabilities (Lyon, 2001, 2003). Images can be developed that show the landscape, the engineered project, and related conditions. The computer manipulation of the image and model allow the user to judge from visualization the "look" of the final project. The repetition of different scenarios allows users to evaluate project and landscape conditions iteratively (Lyon, 2003).

13.6.4 MANAGEMENT

Management of a facility or a resource requires information from monitoring activities. Water is constantly changing in quality and quantity, and day-to-day information as to its characteristics is vital. Hence, a variety of problems and a variety of applications may be addressed.

13.6.5 BEST MANAGEMENT PRACTICES

Best management practices (BMPs) are the identification and implementation of management practices that minimize harm to the environment. BMP is a governmental term that encompasses a variety of activities that reduce and localize environmental problems. BMP have been applied to a number of commercial activities, including change in management practices that reduce erosion, pollution, and other nonpoint sources of pollution from agricultural, mining, and manufacturing activities.

Remote sensor and GIS technologies have a number of valuable applications in the BMP arena. They can help to identify the problems, monitor their influence and concentrations, map their locations, model the physical and chemical processes using GIS databases, and certify progress by determining the resulting impact of BMP as compared to past conditions.

13.6.6 WATER RESOURCE APPLICATIONS

There is a huge variety of water resource applications of remote sensor and GIS technologies (Lyon et al., 1988; Lyon and McCarthy, 1995; Lyon, 2001, 2003). The limit on the application of these technologies is that of the capabilities of scientists and engineers. The key to successful and lower relative cost applications is knowledge of the problem at hand and of the capabilities of remote sensor and GIS tools.

The hydrology applications of remote sensing and GIS technologies are many and are constantly growing (Maidment and Djokic, 2000; Lyon, 2003). This is not surprising because these are technologies with great applicability to spatial analysis and to the landscape scale of problems.

Many hydrological applications require land cover or water-type information. This is because land cover is an important variable that influences many hydrological processes. These issues and processes include erosion, storm runoff, groundwater recharge, storm hydrograph characteristics, permeability, and others.

Analyses of hydrological characteristics also benefit from data collected and summarized by watershed. A computer-based storage method allows data to be stored by an irregular boundary such as a watershed. It also allows the integration of hydrological characteristics for watershed or subwatershed areas, which is very useful in analyses.

13.6.7 QUANTITY

Remote sensor and GIS technologies have been used to great effect in the measurement or prediction of water quantities. Usually, these applications use land cover or topographic information. They may also measure water or snow/ice directly via remote sensing.

The prediction of snowmelt runoff volumes has been improved with the use of remote sensor data. Several experimental and operational programs have utilized weather satellite data, such as AVHRR (advanced very high resolution radiometer) and GOES (geostationary weather satellite) data and airborne measurement to quantify the snow moisture equivalent in target watershed snowpacks. An excellent body of work has been done by NOAA's National Weather Service National Operational Hydrologic Remote Sensing Center in Chanhassen, MN. They develop snow moisture equivalent data products for many areas in North America and make these data available for hydrologic forecasts and for studies of global change.

13.6.8 QUALITY AND NONPOINT SOURCES

Water quality characteristics can be measured from remote sensor technologies. The water is fairly transparent and absorbing, and as such it will exhibit characteristics of the material that are dissolved or suspended in the water. This makes the identification and measurement of the concentration of material fairly straightforward (Lyon et al., 1988).

A number of variables in water have been measured by remote sensor technologies. Some variables are measured directly and demonstrate a very good linear response of light reflectance or absorbance to concentration of the variable.

Surface-suspended sediment concentrations can be measured in a linear fashion over concentrations from 0 to 25 mg/l to as much as 600 mg/l or more. Chlorophyll a also shows good linear responses in concentrations of less than 1 mg/l to more than 25 mg/l and above, depending on water clarity.

GIS technologies can greatly facilitate the storage and analysis of measurements. Many water quality and sediment data sets have been developed over the years and in a number of data collection "campaigns." The differences in sampling can be characterized with attribute entries in GIS files. The sampling points maintain their unique characteristics, yet they can then be evaluated with GIS technologies.

GIS technologies allow a number of products to be created and models to be run in support of hydrological applications. Products include digital files and hardcopy image maps of single variables, variable combinations, thematic maps of drainage and watershed variables, and similar simple products.

Complex products can be generated to present results of statistical or deterministic models. These products would include visualizations of siting or management scenarios, land cover and land use change maps, tabular results of statistical analyses, and results of model simulations.

The addition of materials to water results in a reflectance change from the unaltered water body; it is a composite of the individual materials. Water that holds nonpoint materials such as suspended sediments or high concentrations of other pigments can be evaluated quantitatively using on-site sampling and remote sensor analyses (e.g., Lyon et al., 1988; Lyon, 2003).

The problems associated with nonpoint sources of pollution can be addressed with remote sensor and GIS technologies. The utility of the data and technologies is due to the size and distribution of the problem, the need for quantitative assessments or inventory of resources to manage, and the widespread sources of the problem.

13.6.9 EROSION STUDIES AND WETLANDS

Erosion is a widely distributed problem that has many sources (see Chapter 9 and Chapter 10). Identification of sources, inventory, and manage and improve problem areas requires a variety of measurement technologies. Remote sensor technologies can supply data on locations of sources and can be used to monitor conditions of sources and identify downstream impacts.

Sources of erosion products are usually bare earth areas. The location, extent, and temporal characteristics can be interpreted from aerial photographs or other appropriate scale remote sensor products. Erosion areas can be identified by the light tone or highly reflective bare soil conditions (Lyon, 2001). These light areas are in contrast to relative dark-tone areas of vegetation-covered areas and dark areas of clear water.

The downstream impact of erosional products can also be measured. Suspended sediment makes water highly reflective or lighter tone. High-level concentrations of suspended sediments are the result of erosion and are a nonpoint source of pollution. Suspended sediment conditions can be evaluated from a series of images, say, from the Landsat satellite systems. Quantitative assessments can be made if on-site measurements are taken in close temporal agreement with satellite data overpasses (Lyon et al., 1988; Lyon, 2003).

There is much interest in the identification of wetlands. Wetlands have taken on a vital position in any activity that changes the landscape or in activities that seek to manage landscapes for various purposes (Lyon, 1993, 2001). It is necessary to identify, map, and determine wetland type information in planning and management. Remote sensor technologies such as aerial photographs, topographic maps, and engineering style maps are vital for documenting size and location of wetlands (USACE, 1987; Lee and Lunetta, 1995; D. Williams and Lyon, 1995; Lee and Marsh, 1995; Lyon, 2001).

13.6.10 HAZARDOUS WASTE

A number of remote sensor and GIS technologies have been used in the identification, characterization, and management of hazardous waste sites. Often, the use of remote sensor technologies helps identify potential problems or educates the user as to the prevailing conditions of the feature or landscape (Lyon, 2001). These technological methods can also be generalized in their application to issues related to landfills and siting of these facilities. An example is the use of visualization technologies to help design the site (Lyon, 1987, 2003).

To implement a remote sensor technology for monitoring requires knowledge of the problem, the characteristics that can be measured remotely, and a plan for collecting the requisite data. The budget available to support the effort is very important. It can be influential in the selection of the methods and instruments to be used.

14 Practical Exercises on Conducting and Reporting Hydrologic Studies

14.1 INTRODUCTION

In the rest of this book, we provided technical information that should prove useful to scientists and engineers who will conduct studies relating to environmental hydrology. We provided information on individual principles and procedures and, in most cases, much of the basic data needed to understand the examples and problems presented in each chapter. However, when solving real-world problems, you will quickly discover that little or no information is initially available, each problem has unique aspects, and often the questions to be answered are not well defined. Also, many hydrologic studies require knowledge of several different topics to develop a comprehensive solution.

In this chapter, we present nine practical exercises and information on how to approach conducting a hydrologic study. Several of the exercises can be combined to develop comprehensive solutions to hydrologic issues on a watershed of interest. These exercises are designed to enhance and enrich your hydrologic experience by introducing some sources and procedures used by practicing hydrologists, especially those in the surface water and environmental fields. Most sources are available in university libraries. Keep in mind that potential employers seek people who have practical experience.

Like many scientists and engineers (including the authors), you have probably focused your academic interests on subjects relating to science, engineering, and mathematics. While this is very appropriate, you need to recognize that much of your life will be spent communicating. Effective oral and written communication skills are vital for a successful scientist or engineer. We have therefore included in this chapter a brief outline of how to prepare a technical report.

14.2 CONDUCTING A HYDROLOGIC STUDY

To conduct a hydrologic or environmental study, it will be necessary to (1) define the question, (2) conduct a preliminary investigation, (3) undertake detailed planning, (4) conduct the detailed study, and (5) report the results of the study. The first four aspects are discussed here and an outline of how to prepare a report is presented in Section 14.4.

14.2.1 DEFINE THE QUESTION

Before undertaking any study, the following questions need to be asked and answered:

- Who is the client?
- What is the purpose of the study?
- What is the motivation for the study?
- What issues are addressed?
- What resources are available?
- What are the deadlines?
- Are there specific methods that must be used?
- Are there any legal constraints?

It is important that each of these questions be addressed regardless of whether you are a junior scientist in an agency or a senior partner in a consulting company. Failure to ask appropriate questions might result in addressing the wrong issues, inappropriate (high or low) expenditure of resources, inadequate results, missing important deadlines, legal problems, the presentation of results at an inappropriate level of complexity, dissatisfaction of the client, and perhaps even failure to be paid for your work or the loss of your job.

As an outcome of asking these questions, you may need to prepare a proposal requesting resources to conduct the work, or you might simply prepare a statement that better defines the question. Regardless of the outcome, it is important that you carefully document all discussions, communications, and requests to conduct work. Make sure your document includes dates, participants in discussions, and communications of substance. Be professional and always adhere to high ethical standards.

14.2.2 CONDUCT A PRELIMINARY INVESTIGATION

Once the question has been defined and authorization has been provided to proceed, it will usually be necessary to conduct a preliminary investigation to provide sufficient information to (1) design the study; (2) select analytical methods, models, and procedures; and (3) prepare a budget. You will need to

- Become familiar with the system studied.
- Assemble and evaluate existing data.
- Obtain as much information as possible.

- Develop a scope of work.
- Identify data requirements, data deficiencies, and cost resource requirements.
- Identify system boundaries and boundary conditions.
- Prepare a preliminary diagram of the physical system.
- Identify issues that inhibit your ability to conduct the study or need further study.

A reconnaissance of the site should always be undertaken. Before visiting the site, it is important to define the purpose of the visit and prepare accordingly. Careful planning is critical to the success of any investigation. Obtain as much background information as possible. This should include reading reports on previous studies, obtaining topographic maps and aerial photographs (if available), and obtaining information on the climate, soils, geology, and land uses. Talk to people who are familiar with the location and communicate with local, state, and federal agencies to determine if they have any ongoing or recently completed studies of the region. Also, ascertain if there are climatic and stream gaging stations near the location.

Before leaving for a site visit (or any other fieldwork), prepare a checklist of equipment and materials needed. Obtain or prepare written descriptions of each task that will be performed. Make sure each of the people participating in the field reconnaissance is aware of the objectives and what they will be required to perform — this includes you. Make sure suitable data acquisition materials have been prepared to document all findings during the visit. Consider possible things that might go wrong and prepare contingency plans before going to the field. This is particularly important if electronic equipment is used.

A common error when performing fieldwork is to leave too much to memory. Take the time to organize before you leave. Check that all items work, electronic instruments are fully charged, and all items have been loaded into the vehicle used for the fieldwork. Make careful field notes and take photographs or videotapes whenever possible.

14.2.3 Undertaking Detailed Planning

After the preliminary investigation, select an approach based on the objectives, the required accuracy, how the data will be used, the time and resources needed to obtain the data, boundary conditions and assumptions, and data deficiencies and assumptions. Prepare a conceptual diagram of the system, identify data requirements, and identify sources and methods to obtain the data.

You then need to develop a detailed scope of work that specifies the administrative organization of the study, all resource requirements, operating procedures and methods,

a schedule of activities, and deadlines. Before proceeding with the detailed study, the detailed scope of work needs to be approved by your supervisor or the client. In some cases, it might serve as a proposal that will be submitted for funding consideration. Make sure you fully understand the status of the scope of work and have written confirmation of procedure before undertaking further work. Often, the scope of work will be modified by your supervisor or the client. Request to be part of any negotiations relating to a scope of work that you have prepared. You will be in the best position to explain to the participants what is proposed. Also, you will learn firsthand the rationale for any proposed changes.

14.2.4 Conducting the Detailed Study

Conducting the detailed study will require careful planning. Make sure you understand your role, the objectives, the resources that are available, the expectations of the client, and the deadlines. A job file should be prepared for each project. This will normally be divided into sections, such as communications, expenses, data files or databases, including raw data and field notes, files of calculations and work that was performed, drawings that were prepared, and reports. All materials need to be catalogued and archived carefully. Most organizations will have well-established procedures on how studies should be conducted and documented.

Most studies relating to environmental systems will either be used as the basis for a legal decision, such as the issuing of a permit, or have the potential to be used in some future litigation. We cannot overemphasize the importance of documenting all data sources, procedures, assumptions, and disagreements (if any) on procedures that should be used. Also, it is important that the documentation be securely archived for many years after the completion of a project. Legal action associated with design failures, loss of property, or loss of life could occur more than 50 years after a project was undertaken.

14.3 REPORTING A HYDROLOGIC STUDY

We strongly recommend taking additional courses on technical writing, English, word processing, and public speaking. Once you have completed your formal education, the opportunity to obtain expert knowledge and assistance in these areas will greatly diminish, although on-the-job experience will help you become a more accomplished communicator.

We both found that writing did not come easily. While we both still have much room for improvement, we are much better at writing than we used to be. The point is that you improve by doing.

In preparing this book, most students and reviewers commented on the clarity of the statements and how easy it was to read and understand the concepts presented. Reaching this level of acceptability of your writing style requires following a number of basic rules, feeding information to the reader in small doses, using simple words, and being as concise as possible.

In this section, we focus on the steps and rules that should be followed when preparing a technical report or paper. Modern word processing systems will help you overcome any spelling limitations. Most word processing systems also incorporate grammar checking capabilities. However, it has been our experience that these capabilities often recommend inappropriate or incorrect changes, require sophisticated knowledge of grammar by the user, and generally are not suitable for everyday use.

Many organizations have their own style guidelines. It is important that you become familiar with guidelines used by your organization. The guidelines presented below are based on our experience in preparing or reviewing many consulting reports, technical papers, theses, dissertations, proposals, book chapters, letters, and memos. It is also based in part on the work of Moore et al. (1990).

The following steps are critical to the timely completion of a well-written and technically correct report: (1) define the audience of the report; (2) find published reports that could serve as models; (3) prepare a topical outline and have it reviewed; and (4) prepare an annotated outline and have it reviewed. Also, send a copy to a cooperator for review, if appropriate.

You should begin writing parts of the introduction of the report during early stages of the project. Write, revise, and edit the first draft. It usually takes more than one draft to write a report. Try to prepare a draft as early as possible so you can give yourself the luxury of having plenty of time to think about it and rewrite. Some parts of this book have been rewritten more than six times. Have the report reviewed concurrently by within-office and out-of-office colleague reviewers.

The major objectives in having a report reviewed are to:

- Ensure that the report achieves the goals stated in the project description
- Ensure the readability of the report
- Ensure the technical quality of the report
- Evaluate the suitability of proposed publication media
- Evaluate the effectiveness of the presentation
- Correct errors and the other deficiencies that could embarrass the author or your organization

Respond, in writing and in an appropriate manner, to all reviewers. Have the report edited again, depending on the extent of revisions after colleague review.

Publish the report as quickly as possible after completing the study.

14.4 REPORT CONTENTS

Most reports and technical papers should be divided into distinct sections. This book has been divided into chapters, and then each chapter has been divided into several numbered sections and occasionally into subsections. It is not always necessary to number the sections, but if the report is long, numbering will facilitate cross-referencing and finding materials located in different parts of the reports. Generally, sections should have titles that are highlighted (are underlined or in bold and use some uppercase letters), centered on the page, and located on the left-hand side above the section. A brief description of the main sections that might appear in a report or technical paper is given here. Please note that sometimes several of these sections are combined, and often the results section will be divided into several separate sections.

Title: The title is a concise description of the subject of the report and, as such, has to convey to the reader the content of the report. Thus, a considerable amount of thought should be invested in the title.

Authors and Publication Date: List all authors and the publication date of the report. This information will usually be presented on a cover page that includes the name and address of the organization responsible for preparation of the report.

Table of Contents: The table of contents should be prepared before you begin writing the report as it will serve as a useful outline. It should include the main section headings and their page numbers. It might also include the headings and pages of subsections. The table of contents will often be followed by lists of the figures and tables. However, in short reports, this additional information is sometimes omitted.

Executive Summary: The executive summary is an optional section or independent document. Consideration should be given to preparing an executive summary if the main report is longer than 20 pages. The executive summary should not exceed 5 pages. It should include all the sections (but not the subsections) of the main report. An executive summary should be free of equations, references, and appendices. Summary tables and graphic materials can be used, but should not contain excessive detail.

Abstract or Summary: A well-prepared abstract tells the reader the basic content of the report. It should include the purpose, scope, results, and conclusions of the study. The use of citations, abbreviations, and acronyms should be avoided.

Introduction: The introduction should state why and where the study was conducted. A summary of the purpose and scope of the report should also be presented. The introduction might also provide information on the organization of the report.

Background or Justification: This section should include the problem addressed by the study; the objectives of the study; a statement regarding who commissioned the study and when; and a statement of cooperation, if applicable. (This section is sometimes called the *terms of reference*.) The introduction and background sections are often combined into a single section.

Purpose and Scope: This section describes the purpose and scope of the report, which may differ from that of the overall study. For example, the report might only address one aspect of a much larger study outlined in the background section. This section is often omitted, and the purpose and scope are presented in the introduction.

Description of Study Area: Briefly discuss the location and size of the study area; its climate; and its physiographic, geologic, hydrologic, or hydrogeologic setting. In short reports, this section is also often omitted, and the information is presented in the introduction; in other reports, separate sections might be devoted to attributes of the system, such as the climate, hydrology, or hydrogeology. It is sometimes a good idea to list the USGS topographic sheet for the study area. Sometimes, you also may need to list geologic, soils, or flood zone maps as well.

Methods or Procedures: Material under this heading pertains only to methods. No data should be included. Theory used in the study might be included in this section. However, be brief and only include key information that will allow the reader to understand better how the results were established. If an extensive literature review was conducted, it might be included in a separate section. If equations are presented, they should be numbered and cited in a sentence prior to the equations. Be sure that all terms in equations are defined.

Approach: The approach differs from the methods in that it presents the rationale behind the study and the manner in which the study was performed. However, the methods and approach sections are often combined into a single section. Avoid providing a "blow-by-blow" account of your actions and those of other participants in the study. For example, in describing a field reconnaissance, you might state that a team that included a hydrologist, soil scientist, geographer, environmental scientist, and so on participated in the reconnaissance. Their names might also be listed. However, avoid statements that say, "I did this, and then I did that." Instead, provide specific information on the observation and measurements made, how they were made, and why they were made.

Previous Studies: This section presents information concerning previous studies conducted at the site, or are reported in the literature, that provided information or knowledge used in the study. Sometimes, this information is presented in the background section. If no previous studies were conducted, a statement to this effect should be presented earlier in the report. All studies consulted for the report should be carefully cited in it. This would also include information on maps used.

Results: This section presents data developed to address the problems/objectives stated in the introduction. This is the most important part of the report. The results will often be presented in several sections. This section is a statement of facts resulting from the study. Do not include information from other studies, theory, or opinions. Each section should build on previous sections. Specific objectives of the study should have been presented in purpose and scope. When possible, organize the results in the same order that the objectives were presented and specifically tie the results to the objectives. Tables and figures should be used to summarize and illustrate important findings. Each table or figure (1) should have a number, (2) must have a caption, (3) should stand alone without the need to refer to other parts of the report to understand its meaning, and (4) must be mentioned in the text of the report. Figures and tables should be located in the body of the text shortly after they are first mentioned. Their use is not restricted to the results section. For example, a map or layout drawing of the site might be presented in the introduction or background section. Illustrations of procedures and concepts might be included in the methods and procedures sections.

Discussion of Results: This section presents an interpretation of the results. The discussion should be related to the objectives. It will often compare the reported results with previous findings reported in the literature. The results will also be related to the methods and approach. Uncertainties and assumptions that influenced the results should be discussed. Limitations of the application or usefulness of the results should also be presented. Many reports and papers combine a discussion of results with the results section. Try to avoid this practice as it is easy for facts based on the study to become cluttered with discussion on findings from other studies. Sometimes, additional work might be performed to evaluate a particular aspect of the results. If this work is not relevant to the specific objectives of the study, it should not be included in the report. However, if the work evaluates the validity of a result, it might be presented either in the results section or in the discussion of the results.

Summary or Conclusions: Conclusions summarize results and interpretations of a study. The section relates to the objectives presented in the introduction and focuses on significant findings. A summary is a restatement of all the main ideas presented in the report, beginning with the introduction. Unless required by your organization or the client, it is strongly recommended that only a single summary or abstract be placed at the beginning of the report.

Future Work: This is an optional section that identifies further work that should be performed.

Acknowledgments: This is an optional section that acknowledges extensive assistance by people other than the authors. It might be placed near either the beginning or the end of the report.

References: A list of literature cited in the report should be presented at the end of the body of the report. In a very long report or a book, the references for each section are sometimes located at the end of each section. References are usually listed alphabetically based on the surname of the first author. Sometimes, they are numbered in the order that they appear in the text and then are listed according to this number. A third approach in which the title is presented first is illustrated in the nine practical exercises that start in Section 14.6. A bibliography that includes literature in addition to that cited in the report might be presented. However, if a bibliography is prepared, it is important that the reader be informed that not all documents listed were cited in the report. The references in this book provide an example of how you might organize your references. In this section, we indicated that this outline of how to organize a report was based in part on Moore et al. (1990). This reference is: Moore, J.E., D.A. Aronson, J.H. Green, and C. Puente, *Report Planning, Preparation, and Review Guide*, Open-File Report 89-275, U.S. Geological Survey, Reston, VA, 1990.

Appendices: It is often desirable to include appendices that provide further detail on the scope of work, pertinent background information, details on methods and procedures, data, and secondary results. Each appendix must be cited in the report, and they are usually sequentially identified by the letters A, B, C, and so on. It is important that only materials that are not necessary for understanding the body of the report be included in the appendices.

14.5 GENERAL GUIDELINES FOR PREPARING EXERCISE REPORTS

In Section 14.6, we include nine practical exercises. Particular attention should be paid to specific directions provided by your instructor. These might include preparing a report based on the guidelines presented in this chapter. As a rule, we would recommend that in all class reports you:

1. Include exercise directions.
2. Give complete citations for all documents used, including maps. Maps should include scale, agency, and sheet name.
3. Include photocopies of all appropriate maps. Make tracings if necessary to avoid clutter. Do a professional job. Show all measurements that you have made or show how you made them.
4. Do a professional write-up in complete sentences. Tell what you did in such a way that others can understand and that you can understand years from now.
5. For locations you select, for example, give section number, township, and range, or you might use map coordinates for areas that are not included in the U.S. Rectilinear Land Survey.
6. Most exercises require graphing. For those with the competency, a computer spreadsheet of some type will aid matters. Not only will this be easier and neater, but also the computer can do many of the required calculations. Most environmental agencies and consulting firms would demand the use of this software, so now is a good time to start using it. However, do not force functions into an inappropriate computer program.
7. All reports are to be prepared as if you were doing them for a paying client. You must satisfy the client to be paid.

The only difference between the guidelines presented above and a report that you might prepare for your organization or a paying client is that some of these materials might form part of the job file or might only be presented in the appendices of the report.

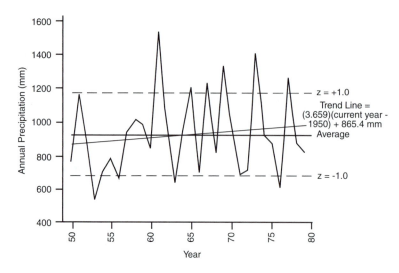

FIGURE 14.1 Annual precipitation 1950 to 1979, Kansas City, MO.

14.6 EXERCISE 14.1: PRECIPITATION (SEE THEORY IN CHAPTER 2)

14.6.1 INTRODUCTION

Precipitation is the most important hydrologic input. This exercise is designed to familiarize you with procuring, processing, and understanding precipitation data.

14.6.2 PART I. TIME TRENDS

A. Go to your library and find the U.S. Weather Bureau's *Climatic Summary for the United States,* QC/983/U58CL, Volumes 1 and 2 (Library of Congress subject headings), which cover up to 1930, and the 1931 to 1952 supplements. The 1951 to 1960 supplement and the annual supplements for individual years provide more recent data. Or, try www.wrcc.dri. edu or wrc.cornell.edu/other_rec.html. Alternatively, procure the *Hydrodata* CD-ROM from EarthInfo, Inc. This contains hydrologic data, including precipitation records, for stations throughout the U.S. and is updated frequently. For these exercises, the hardcopies are recommended because they allow you to see better the full range of materials available.

B. Select any station and extract at least 20 continuous years of annual precipitation values:
1. Plot the data against time (see Chapter 2).
2. Calculate the mean and standard deviation and indicate these as horizontal lines on the graph. Assuming an adequate sample and a normal distribution, it is probable that annual precipitation in the future should be within __ and __ in. 68% of the time, between __

and __ in. 95% of the time, and between __ and __ in. 99.7% of the time. The standard deviation *SD* is determined as follows:

$$SD = \left[\sum (x - \bar{x})^2 / (n-1) \right]^{1/2}$$

where *x* is each entry in the distribution, and *n* is the number of entries in the distribution.
3. Divide the standard deviation by the mean (*SD*/*x*). This is the coefficient of variation.
4. Calculate and plot a 5-year moving average (Chapter 2; **Figure 14.1**). Is there a time trend?
5. Calculate a trend line for the raw data. Does it show a time trend? How does it compare with the mean and the 5-year moving average (see **Figure 14.1**)?

14.6.3 PART II. FREQUENCY AND MAGNITUDE

A. **Annual Means.** Using data from the station in Part II and the procedure in Chapter 2, construct a frequency table of annual precipitation magnitudes. Use the graph matrix or make your own. Probability paper is available at www. weibull.com/GPaper.

B. **Individual Events.** This exercise introduces two of the most valuable documents available to hydrologists, geomorphologists, and planners of all stripes. They are: "Rainfall Frequency Atlas of the U.S.," U.S. Weather Bureau Tech Paper 40, 1961, QC/851/U58t; and "Two to Ten Day Precipitation ... 2–100 years ... U.S.," U.S. Weather Bureau Tech Paper 49,

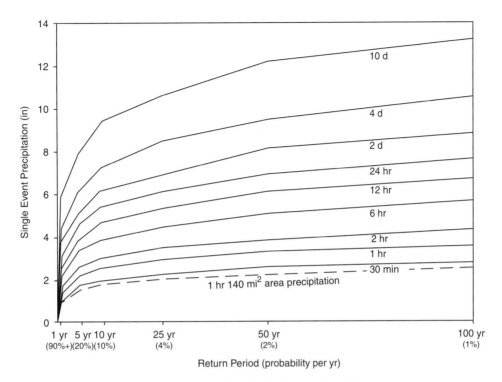

FIGURE 14.2 Precipitation depth, duration, frequency relationship for Kansas City, MO.

QC/851/U58t. These two extremely useful publications have been modified and enhanced over the years, and updated versions are available online for the West, Midwest, and Northeast at www.ncdc.noaa.gov/oa/documentlibrary/rainfall.html or www.nws.noaa.gov/oh/hdsc/studies/prcpfreq.html.

C. **Exercises:**

1. **Frequency Magnitude of Events for a Location.** Using 1-in. ruled graph paper, set up the ordinate as 2 in. of precipitation/inch and the abscissa as 10 years/in. Mark the 1-, 5-, 10-, 25-, 50-, and 100-year points. These points may be sublabeled as 90+, 20, 10, 4, 2, and 1% probability, respectively. From the maps of the contiguous U.S., select any point locality of interest to you and graph the probable 30-min, 1-h, 2-h, 6-h, 12-h, 24-h, 2-day, 4-day, and 10-day values with a continuous line (see **Figure 14.2**). Prepare the graph well enough to be presented in class and put your name on it. Write a short paragraph explaining the meteorological–climatological–landform conditions governing frequency and magnitude of precipitation events for your location.

2. **Seasonal Distribution.** Inspect and understand Charts 52 to 54 of Tech Paper 40. For example, compare Parts (Regions) 2 and 6. Why are they so different?

14.6.4 PART III. SPATIAL DISTRIBUTION

A. The values for your graph are point values. Assume that you are working with a watershed of 140 mi^2. Using **Figure 2.8** in Chapter 2, sketch a new 1-h duration event curve on your graph for this area. Make this curve dashed to distinguish it from the point values (**Figure 14.2**).

B. Find U.S. Weather Bureau (now National Climate Center) Climatological Data QC/983/U58c (the same data as for Part 1A). These reports give daily precipitation data for all U.S. stations. While looking at these, find a fairly recent report and look at all the other data included, such as Class A pan evaporation. Or, try www.wrcc.dri.edu.

1. Select a watershed with at least five weather stations (it might be a good idea to be armed with a 1:250,000 topographic map of your area of interest). For an excellent set of U.S. basin maps at 1:500,000, see *River Basin Maps Showing Hydrologic Stations,* Interagency Comm. or Water Resources, Washington, DC, Note on Hydrologic Activities 11,1961. Further, select some discrete storm event (i.e., one that is isolated in time and space). Compute the storm precipitation for the basin by (a) the arithmetic mean, (b) the Thiessen method, and (c) the isohyetal

method. For the last two methods, you will need to measure areas. This can be done by an overlay dot grid or using a polar planimeter. The dot grid may be easily constructed from a piece of clear plastic etched on one side so you can mark on it (such as Herculene). On a piece of graph paper, marked 1-in. squares subdivided into hundredths of a square inch, outline several square inches, and put a dot to represent each hundredth. Trace this on the plastic. Then, by placing the plastic grid over the area to be measured, you count squares and dots to obtain the total area of an irregularly shaped area. Of course, you can also do this in metric units.

2. Draw decent diagrams of your endeavors. Write a paragraph comparing the three methods (for your case) telling which you think best and why.

14.7 EXERCISE 14.2: EVAPORATION (SEE THEORY IN CHAPTER 4)

14.7.1 PART I. ANNUAL LAKE EVAPORATION (SURROGATE FOR POTENTIAL EVAPOTRANSPIRATION)

A. Find USGS Water Supply Paper 1838, "Reservoirs in the U.S." (www.cpc.ncep.noaa.gov/soilmst/e.html). Select any U.S. reservoir of interest to you. (The locational index is in a pocket in the back of the Water Supply Paper. Please treat it with care.) Find and record the surface area of the reservoir.

B. Find "Evaporation Maps for the U.S.," U.S. Weather Bureau Technical Paper 37, 1959. Read this paper; it is an important data source. From Plate 2, locate the average annual evaporation depth for your lake (or see **Figure 4.16**, this text). Using the area from A above, calculate the volume of average annual evaporation from your lake in acre-inches, acre-feet, and cubic meters.

C. From Plate 1, determine the average annual Class A pan evaporation at your reservoir location (or use **Figure 4.14**, this text).

D. Estimate the annual lake evaporation from annual pan evaporation using the pan coefficient (Plate 3 or **Figure 4.15**, this text). Multiply the pan evaporation by the pan coefficient to get an estimate of the lake evaporation. Your number should agree with the figure from Plate 2.

E. What proportion of your total annual lake evaporation (in inches) should occur during the stress (hot) period between May and October (Plate 4, or **Figure 4.18**, this text)? Assume that the same proportion for Class A pans applies to lakes. You should be able to explain the distribution shown on Plate 4.

F. What is the standard deviation of pan evaporation (in inches) at your location (Plate 5, or **Figure 4.17**)? Assuming this standard deviation is proportionately applicable to reservoirs, use the pan coefficient to estimate the standard deviation of annual evaporation from your reservoir in inches. Assuming you have a reasonably correct standard deviation of lake evaporation, future evaporation from your lake should be between __ and __ inches 68% of the time, between __ and __ inches 95% of the time, and between __ and __ inches 99.7% of the time.

G. If you were to map the coefficient of variation (*SD*/mean) for evaporation, how would the distribution vary over the U.S. (in very general terms)? Why?

14.7.2 PART II. DAILY AND MONTHLY LAKE EVAPORATION

A. The nomograph on page 2 of USWB Technical Paper 37 (**Figure 4.12**, this text) allows the estimation of daily pan (and thus lake) evaporation for specified climatic elements. If you have weather data for a certain day and want evaporation, just plug in your data. You need average data if you want the average for a given month.

B. For average data, obtain the *Climatic Atlas of the U.S.*, NOAA, 1974, ESSA, 1968 (QC/983/US7c). Data are on the following maps:
1. Mean daily air temperature (°F): pp. 1–24.
2. Solar radiation (Langleys/day): pp. 69–70.
3. Mean daily dewpoint temperature (°F): pp. 57–58.
4. Wind movement (miles/hour): pp. 73–74. You must multiply by 24 to get miles/day. Weather Service anemometers are located 22 ft above the ground, and you need wind speeds for 2 ft above the ground. Assume a roughness height of 0.1 ft.

C. The nomograph is entered at the upper left, and you proceed in both directions to obtain daily evaporation. This figure is an average day of an average month. Use any month for this exercise, but July should yield more interesting figures.

Note 1: Your estimates of lake evaporation will probably be somewhat low if the lake is shallow and high if the lake is very deep.

Note 2: Actual *ET* rates are often quite similar to the lake values. This will depend, in large part, on vegetation type and cover and available soil moisture.

14.7.3 PART III. ESTIMATING ACTUAL EVAPOTRANSPIRATION RATES

A. The water balance equation is

$$P = AET + Q \pm \Delta G \pm \Delta\theta$$

where *P* is the precipitation depth, *AET* is actual *ET*, *Q* is runoff depth, ΔG is groundwater inflow or outflow, and $\Delta\theta$ is soil water change. Transposing, we get:

$$AET = P - Q \pm \Delta G \pm \Delta\theta$$

Over long periods, ΔG and $\Delta\theta$ become negligible, so the equation reduces to

$$AET = P - Q$$

This equation should properly be assigned to watersheds, but a reasonable estimate can be obtained for areas, especially in the eastern U.S.

B. Long-term average precipitation can be handily obtained from *Climate and Man,* USDA Yearbook of Agriculture, 1941, from most state departments of natural resources. Or, try www.wrcc.dri.edu. These data are also available on the EarthInfo CD-ROM. Find the average precipitation for the station nearest your reservoir.

C. Streamflow (runoff) can be obtained handily from "Annual Runoff in the Conterminous U.S.," USGS *Hydrologic Atlas* 212, 1965; or USGS *Hydrologic Atlas* 710, 1987. Or, try www.cpc.ncep.noaa.gov/soilmst/r.html. Extrapolate a value with reference more to the precipitation station than to the lake. What is the estimated AET? How does this compare with the lake (freshwater surface evaporation) value that you found? Why is there a difference, and what is its significance?

D. Now, see Thornthwaite's map of PET values (e.g., in Strahler and Strahler, *Modern Physical Geography,* 4th ed., p. 169). This map has received wide, and often uncritical, acceptance and has been in use since the 1940s. It was compiled from the Thornthwaite PET equation,

which uses only temperature. Unlike the Blaney–Criddle equation, which you have used, the Thornthwaite equation does not consider vegetation; like the Blaney–Criddle equation, it does not consider humidity, radiation, and wind. Keeping all this in mind, compare and contrast your estimated AET, Thornthwaite's PET, and lake evaporation (surrogate PET). A short paragraph is sufficient.

14.8 EXERCISE 14.3: RUNOFF (SEE THEORY IN CHAPTER 5)

14.8.1 INTRODUCTION

A data set is furnished for this exercise (**Table 14.1** and **Table 14.2**), or you can procure your own from either USGS Water Supply Papers or the CD-ROM *Hydrodata* from EarthInfo, Inc. The supplied data set consists of 2 months of daily streamflow records for Coon Creek and Little La Crosse River, two streams in Wisconsin (**Table 14.1**) and rainfall records for the same period of time (**Table 14.2**). Select 1 month for analysis.

A. On the furnished annotated graph paper, graph daily flow including storm peaks (**Figure 14.3**). Use the sheet for one stream for 1 month. Sketch a line to separate baseflow from stormflow. A commonly used method is to begin at the upturn of the rising limb and increase the slope of the separation line at 0.05 ft³/sec/mi²/h (Hewlett, 1982), but there are other methods (Dunne and Leopold, 1978, p. 287). Note that baseflow varies day to day even when there is no precipitation. This is possibly attributable to (1) *ET* conditions, (2) barometric pressure fluctuations, (3) upstream water withdrawals, and (4) measurement errors. Despite these fluctuations, can you detect a baseflow recession curve (a line on this log paper) after precipitation events? Do you see that such a line would have to be a trend line because of the masking effect of the aforementioned fluctuations? According to Hewlett, baseflow accounts for about 70% of total streamflow (water yield) in the eastern U.S.

B. Make a hyetograph to accompany your hydrograph. From the precipitation data, take an arithmetic average for appropriate stations. The stations are well distributed, so Thiessen weighting is not needed. The hyetograph bars should extend downward from the top according to the scale. Since the 24-h precipitation period ends at 8 a.m. of the noted day, the bar (0.25-in. wide) should be centered over 8 p.m.

TABLE 14.1A
Data Set, Exercise 14.3, Storm Discharge and Suspended Sediment, July 1938

	Coon Creek (77.2 mi²)				Little LaCrosse River (77.1 mi²)				
Mean Daily Discharge		Maximum Rate of Discharge			Suspended Matter	Mean Daily Discharge	Maximum Rate of Discharge			Suspended Matter
Day	ft³/sec	ft³/sec	ft³/sec/mi²	Time	Tons per Day	ft³/sec	ft³/sec	ft³/sec/mi²	Time	Tons per Day
1	47				30.00	74	120	1.56	2:00	176.00
2	302	1,702	22.2	7:30	24,700.00	180	360	4.67	14:30	2,820.00
3	53				68.00	60				157.00
4	44				18.00	45				40.00
5	64	122	1.58	11:50	170.00	46				49.00
6	48				25.00	50	109	1.41	22:15	200.00
7	45				13.00	45				183.00
8	41				9.70	39				36.00
9	40				22.00	40				64.00
10	46	55	0.711	6:00	18.00	40				29.00
11	39				3.30	37				19.00
12	39				4.60	36				54.00
13	40	40	0.518	Constant	4.30	159	330	4.28	8:00	1,800.00
14	36				4.10	47				68.00
15	35				2.90	39				23.00
16	35				4.00	36				16.00
17	34				3.60	37				14.00
18	33				3.00	32				11.00
19	33				2.70	32				11.00
20	33				3.60	31				8.10
21	101	400	5.18	17:00	2,050.00	38	90	1.17	21:30	203.00
22	55				128.00	38				122.00
23	38				13.00	34				17.00
24	36				6.40	32				11.00
25	34				2.70	32				9.80
26	37				4.40	32				8.50
27	37				6.50	36				16.00
28	37				5.30	38	49	0.635	4:45	39.00
29	33				6.00	32				12.00
30	33				3.70	31				9.10
31	33				3.70	32				8.70

of the previous day. Note that this lack of synchronism can sometimes make the precipitation appear to occur after the runoff peak. If the average is less than 0.01 in., put "T" for trace.

C. Note stream response to each storm (response = stormflow/precipitation). Response varies within a basin depending on (1) precipitation amount, (2) precipitation intensity, (3) antecedent soil water (antecedent baseflow is used as a reasonable surrogate for soil water θ because drainage from unsaturated [often below FC] soil augments baseflow from deep groundwater), and (4) infiltration changes, especially from varying land use. Between basins, the list is longer; most important are (1) geology/soils, (2) slope, (3) drainage density (DD), and (4) infiltration, especially regarding land use.

D. In your graph, response varies primarily as a function of antecedent θ and precipitation intensity. Note how response increases after a rainy period of a few days. According to Hewlett, average response for all eastern streams is about 10%, but varies (2 to 24%) according to physiographic province. Season or storm attributes and land use cause added variation. Additional variables ground frost and snowmelt (effects of which often combine) must be considered for data sets in winter and spring. This is very important in March and April.

E. Graph one storm event at a large arithmetic scale of your choice. Draw the hydrograph from the information given using the technique shown in **Figure 14.4**. Separate baseflow from

TABLE 14.1B
Data Set, Exercise 14.3, Storm Discharge and Suspended Sediment, August 1938

	Coon Creek (77.2 mi²)					Little LaCrosse River (77.1 mi²)				
Mean Daily Discharge		Maximum Rate of Discharge			Suspended Matter	Mean Daily Discharge	Maximum Rate of Discharge			Suspended Matter
Day	ft³/sec	ft³/sec	ft³/sec/mi²	Time	Tons per day	ft³/sec	ft³/sec	ft³/sec/mi²	Time	Tons per Day
1	32				1.60	31				6.50
2	31				5.70	30				8.50
3	31				2.30	30				8.30
4	29				1.90	29				6.80
5	91	480	6.21	11:00	2,210.00	52	120	1.56	15:30	212.00
6	37				20.00	36				28.00
7	33				7.50	44	102	1.32	18:40	160.00
8	34				7.40	52	117	1.52	18:30	272.00
9	39	52	0.673	7:15	13.00	40				54.00
10	33				3.30	34				20.00
11	32				4.30	31				27.00
12	29				3.20	30				11.00
13	29				2.60	29				8.70
14	29				4.20	28				7.80
15	36				6.10	30				9.70
16	225	860	11.1	14:30	2,550.00	252				1,380.00
17	62				41.00	205	730	9.46	0:30	450.00
18	42				6.80	45				23.00
19	38				4.90	40				14.00
20	35				5.00	38				10.00
21	32				4.20	37				7.30
22	33				2.60	35				6.80
23	47	65	0.84	9:00	12.00	200	385	5	11:00	1,640.00
24	34				2.00	51				47.00
25	34				1.60	39				16.00
26	34				2.60	36				10.00
27	32				1.60	34				7.70
28	32				2.20	32				6.70
29	32				1.60	32				6.40
30	32				1.90	32				6.60
31	31				2.00	32				6.30

stormflow as described above. Baseflow should usually increase somewhat during the storm, depending on infiltration and duration.

F. Measure the area under the hydrograph above baseflow; this is total stormflow. This can be described by the expression

$$Q_T \text{ (stormflow)} = \int_{t=0}^{t=n} Q\,(dt)$$

G. Feel free to derive equations for the curves and solve them with integral calculus. Most will simply measure the area with a planimeter or dot grid and, using the graph scale, convert to total volume of stormflow (cubic feet or cubic meters of water).

H. Now, divide total volume by the watershed area to get average watershed depth. To do this, you have to obtain the same units. An easy way to do this is to convert the measured stormflow (cubic feet) to acre-feet, convert the drainage area of 77.2 mi² to acres, and then divide volume by area to obtain depth. This will give you a fraction of a foot, so convert to inches. Subtracting the stormflow depth from the precipitation depth, you will get an approximate idea of θ recharge.

I. Next, divide total stormflow (depth) by precipitation depth to get stream response in percent. Refer to your monthly hydrographs and see how the response of the stream to this storm compares with other storms. It is important to note that "stormflow" includes overland flow,

TABLE 14.2A
Exercise 14.3, Rainfall Data (in.), July 1938

| Date | Coon Creek | | | | | | Little LaCrosse River | | | | |
| | Raingage Number | | | | | | | | | | |
	1	2	7	8	9	13	14	5	6	11	12
1	0.46	0.38	0.46	0.58	0.60	0.42	0.43	0.63	0.53	1.32	0.72
2	1.25	1.02	1.57	0.49	1.22	1.38	0.70	0.37	0.51	0.22	0.38
3											
4					0.01						0.01
5	0.47	0.48	0.30	0.05	0.55	0.30	0.30	0.42	0.25	0.19	0.11
6		0.02	0.02	0.20	0.25	0.32	0.05	0.05	0.10	0.01	0.04
7	0.09	0.37		0.14	0.11	0.05	0.22	0.06	0.04	0.02	0.03
8				0.01	0.01						
9	0.16	0.06	0.07	0.03	0.12	0.25	0.05	0.10	0.04	0.40	0.42
10	0.45	0.45	0.29	0.41	0.54	0.40	0.55	0.32	0.26	0.32	0.27
11											
12											
13	0.12	0.12	0.28	0.12	0.17	0.18	0.20	0.31	0.70	0.95	1.62
14	T			0.06	0.02		0.07		0.02	0.02	0.02
15		0.12									
16		T									
17											
18											
19											
20											
21											
22	1.21	1.51	0.60	0.90	1.50	0.95	1.45	1.14	0.41	0.27	0.27
23	1.21	0.02					0.03	0.03			0.01
24											
25		0.01						0.01			0.01
26											
27	0.50	0.34	0.44	0.17	0.38	0.47	0.25	0.48	0.34	0.49	0.29
28		0.29						0.04	0.04		
29											
30	0.05	0.04	0.01	0.02	0.06	0.52	0.10	0.04	0.03	0.01	0.03
31	0.50		0.01		0.04	T	T	0.07	0.01	0.06	0.03
Total	4.78	4.82	4.52	3.23	5.65	4.97	4.40	4.41	3.19	4.31	4.27

T = trace.

throughflow, and channel precipitation, perhaps in different proportions for different storms. There is no reliable method of separating the hydrograph into overland flow, throughflow, channel precipitation, and baseflow.

J. **Sediment Yield** (optional). Note that sediment transported past the gaging station is an exponential function of stream discharge, especially peak discharge (properly, the values should be instantaneous). The ambitious may want to graph sediment discharge against stream discharge. If so, use 3 × 5 cycle log–log paper

(sediment discharge is the 5-cycle ordinate). Erosion and sedimentation are discussed in the text.

14.9 EXERCISE 14.4: FLOW DURATION (SEE THEORY IN CHAPTER 5)

14.9.1 INTRODUCTION

One of the most important characteristics of a stream is flow duration. For example, does the stream have a very equitable flow regime, or is it constantly up and down?

TABLE 14.2B
Exercise 14.3, Rainfall Data (in.), August 1938

| | Coon Creek | | | | | | | Little LaCrosse River | | | |
| | | | | Raingage Number | | | | | | | |
Date	1	2	7	8	9	13	14	5	6	11	12
1											
2											
3											
4											
5	1.47	0.64	0.76	0.64	0.49	1.28	0.85	0.91	0.79	0.38	0.61
6	0.10	0.06	0.20	0.28	0.18	0.07	0.05	0.10	0.19	0.10	0.29
7											0.21
8				0.34				0.05			0.82
9	0.70	0.57	0.26	0.72	0.16	0.45	0.30	0.55	0.20	0.49	0.56
10		0.02							0.49		
11											
12											
13											
14											
15	0.48	0.54	0.31	0.42	0.60	0.43	0.50	0.41	0.21	0.33	0.21
16	0.30	0.71	0.33	0.23	0.24	0.27	0.40	0.42	0.50	0.31	0.41
17	1.47	1.78	2.08	1.74	2.13	1.58	2.30	2.44	2.56	2.44	2.71
18											
19											
20											
21											
22											
23	0.65	0.63	0.88	0.68	0.54	0.80	0.55	1.04	1.32	1.22	1.13
24											
25	0.65		0.02								
26		0.04						0.01			
27											
28							T				
29											
30											
31											
Total	5.17	4.99	4.84	5.05	4.34	4.88	4.95	5.93	6.26	5.27	6.95

T = trace.

Does it have permanent flow? What are the low-flow and high-flow characteristics (is it "flashy")? If the stream is to furnish power, transport sewage effluent, or just dilute pollution, how much of the time can these jobs be accomplished? A flow duration curve can answer these questions, but you must know how to use it.

To extract enough flow data from USGS Water Supply Papers to construct a flow duration curve would take much time and effort. Instead, you are supplied with a data set for which the computer has done most of the work (**Table 14.3**). Daily flow for a stream has been tabulated and categorized and is ready for you to plot the flow duration curve. Those with the requisite computer skills might wish to obtain and process data from the *Hydrodata* CD-ROM from EarthInfo, Inc.

A. Refer to **Figure 14.5** and **Figure 14.6**. Plotting starts at the top left with the highest discharge Q values (lowest percentage of time equaled or exceeded) and progresses to the lower Q values. Your set includes data for one stream during two time periods. Land use and land treatment have been improved considerably from the first period to the second; thus, the stream regime might (but not necessarily) be more uniform (less flashy) for the second period.

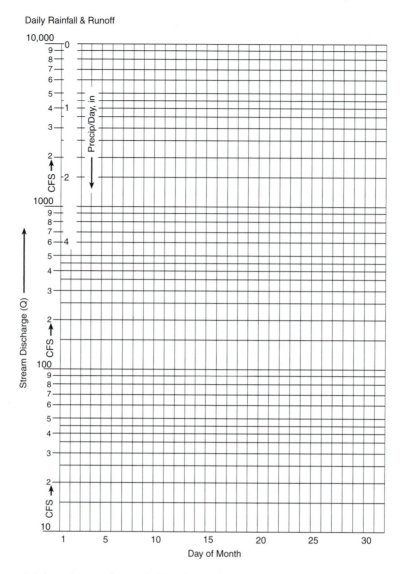

FIGURE 14.3 Typical annotated graph paper for use in Exercise 14.3.

B. After plotting the curve for each period on the same sheet (furnished), write a short paragraph pointing out differences in the two curves and giving the significance of these differences. Feel free to write on the graph. Has the stream been "tamed"? Keep in mind that the period is only a sample of the stream's behavior, and the samples may be poor to good in terms of representativeness. Generally, the quality of the sample, like precipitation, improves with number of years. Probably 15 to 20 years are required for a dependable curve, but this varies. Some factors other than land use that affect flow duration are precipitation and other climatic considerations, geology and landforms, vegetation and other infiltration factors, and reservoirs. The

latter may be the most important factor in some cases.

14.10 EXERCISE 14.5: STORM RUNOFF, TOTALS, AND PEAKS (SEE THEORY IN CHAPTER 5)

This exercise is designed to expose you to (1) a more sophisticated version of the SCS triangular hydrograph technique; (2) the USDA *Soil Surveys,* a source of much hydrologic information; and (3) the possible effects of changing land use and land management on total and peak stormflow (direct runoff). You first need to obtain:

A. A modern (post-1955) soil survey (often catalogued as S591/AZ). Find a watershed of interest to you with *at least* two different soil

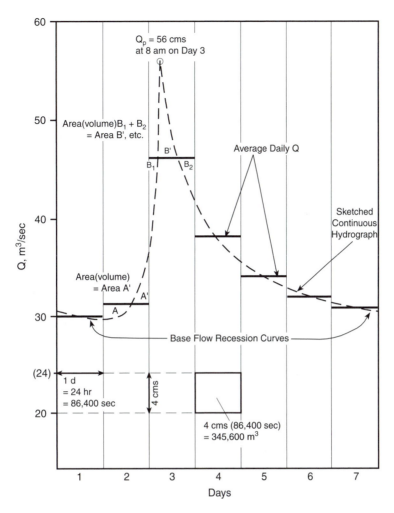

FIGURE 14.4 Mass rate budgeting to create a continuous hydrograph from average daily Q and peak Q values. (Procedure from S.W. Trimble, *Am. J. Sci.*, 277:876–887, 1983. With permission.)

TABLE 14.3A
Cumulative Flow Duration (CFS) for Zumbro River at Zumbro Falls, Minnesota, 1909 to 1945, Duration of Daily Values for Year Ending December 31

CFS	Σ%	CFS	Σ%	CFS	Σ%	CFS	Σ%
0	100.00	180	65.50	960	8.50	5,000	0.70
42	100.00	220	53.10	1,200	6.40	6,100	0.50
50	100.00	260	43.50	1,400	5.00	7,300	0.40
61	99.90	320	34.80	1,700	3.80	8,800	0.20
73	99.60	380	28.50	2,000	3.00	11,000	0.10
88	97.80	460	23.60	2,400	2.30	13,000	
110	92.00	550	18.70	2,900	1.80	15,000	
130	84.10	670	14.50	3,500	1.30	18,000	
150	77.70	800	11.20	4,200	1.00		

95th percentile = 1,400; 90th percentile = 870; 75th percentile = 440; 70th percentile = 370; 50th percentile = 230; 25th percentile = 160; 10th percentile = 120.

TABLE 14.3B
Cumulative Flow Duration for Zumbro River at Zumbro Falls, Minnesota, 1954 to 1974, Duration of Daily Values for Year Ending December 31

CFS	Σ%	CFS	Σ%	CFS	Σ%	CFS	Σ%
0	100.00	210	64.50	1,100	8.60	6,200	0.60
47	100.00	250	56.00	1,400	6.20	7,500	0.50
57	100.00	310	45.60	1,700	4.60	9,000	0.30
68	99.30	370	38.30	2,000	3.60	11,000	0.20
83	96.80	450	31.60	2,400	2.90	13,000	0.10
100	92.30	540	25.80	2,900	2.10	16,000	0.1
120	86.60	650	20.20	3,500	1.60	19,000	
150	78.50	790	17.30	4,300	1.10	23,000	
170	73.70	950	11.90	5,100	0.90		

95th percentile = 1,600; 90th percentile = 1,000; 75th percentile = 560; 70th percentile = 470; 50th percentile = 280; 25th percentile = 160.

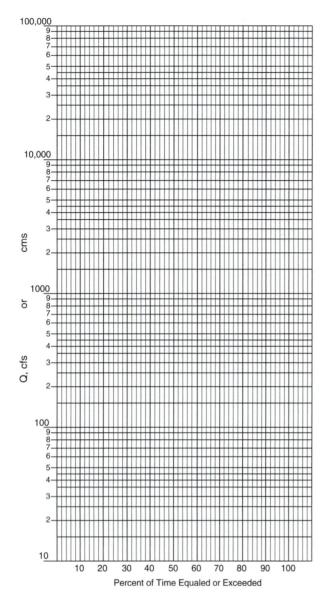

FIGURE 14.5 Typical graph paper for plotting flow duration exceedence relationships.

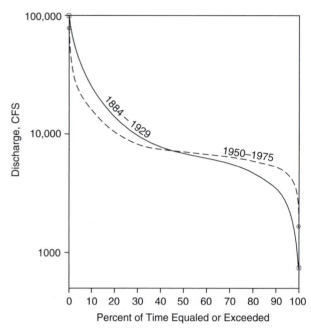

FIGURE 14.6 Flow duration exceedence relationships for the Savannah River, Augusta, GA. (Data from USGS; plotted by Linda O'Hirok.)

mapping categories. Note that each soil mapping category or "unit" usually includes (1) soil series, (2) slope, and (3) degree of previous erosion. These designations are explained in the description for each soil. Average weighted slope should be less than 20%. Select a small watershed to make things easier and faster. Trace the watershed (with the soil mapping units) onto a sheet of tracing paper. To determine the watershed boundaries properly, you will probably need a topographic map (**Figure 14.7**). Include a photocopy of the topographic map with the basin outlined. For slopes,

take the average for each category. For example, if a B slope is 2 to 6%, use 4%. The "length of water flow" is measured from the interfluve along the valley to the lower end of your basin. Soil Hydrologic Groups (A, B, C, D) for soil series are usually given in a table in your survey. While you are looking, consider all other good information (hydrologic and otherwise) given in the tables. You will need a planimeter to obtain the area of the watershed in each soil mapping unit, or you may use a dot-grid overlay.

B. Obtain "A Method for Estimating Volume and Rate of Runoff in Small Watersheds" (USDA-SCS TP 149) or see Chapter 5. The booklet gives complete details for the SCS triangular hydrography method. For land use and management, assume "before" and "after" conditions. That is, you will examine the effects of changing land use and management simply by assigning appropriate values from Table 2 for two time periods. Do whatever you wish, but keep it realistic. Assume any precipitation event that you wish, consistent with USWB Technical Publications 37 and 40, but hold it constant for both land use and conditions so that a comparison can be made. Calculate Q and Q_p for both land use conditions. Use Figure 3 and Table 2 in the SCS booklet, but use the techniques described in Chapter 5.

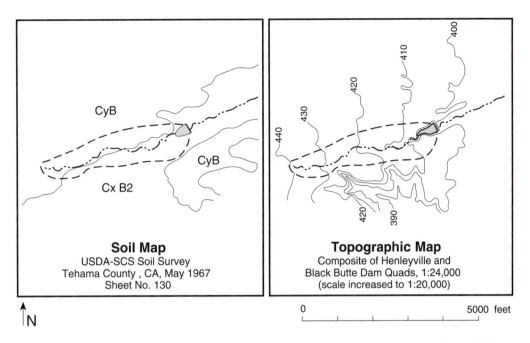

FIGURE 14.7 Same-scale comparison of stream basin on soil and topographic maps for Exercises 14.5 and 14.7.

14.11   EXERCISE 14.6: EROSION AND ELEMENTARY SEDIMENT ROUTING (SEE THEORY IN CHAPTER 6)

This exercise will familiarize you with the universal soil loss equation, sediment delivery ratios (SDRs), and reservoir sediment trap efficiencies. Review Chapter 9 or USDA-SCS Technical Release 51 (Rev.), "Procedure for Computing Sheet and Rill Erosion " and your watershed data from Exercise 14.5. Also review Chapter 8.6.3 concerning sediment trapping efficiency of reservoirs.

A.   Calculate the total average annual soil loss for your watershed for one set of land conditions assumed in Exercise 14.5. Perhaps you will want to find soil losses for both sets of land conditions in Exercise 14.5. Soil erodibility k ranges from about 0.1 to 0.8; assume 0.3 for this exercise if values are not available.

B.   **Sediment Delivery Ratios.** Soil loss is not the same as sediment yield. Eroded soil has many opportunities for deposition as colluvium or alluvium. Thus, not all eroded soil is delivered (at least in the short run) to some downstream point. That is, material goes into storage (see Chapter 9). This generalization has been embodied in a sediment delivery curve (**Figure 14.8**). It shows how sediment delivery decreases as an exponential function of area. This curve is an average value. Basins with a high relief ratio (RR) or fine-textured soil may have higher SDRs, and the converse is true for low RR and coarse texture. Find the appropriate SDR for your basin size; multiply this by the total soil loss from your basin, and you have estimated the sediment yield from your basin. By the way, this SDR cannot hold forever; eventually, the storage may become unsteady, causing further transportation. For details, see S.W. Trimble, *Science*, 188, 1207–1208, and *Science*, 191, 871.

C.   **Sediment Trap Efficiencies of Reservoirs.** Reservoirs make excellent sediment traps. Generally, the trap efficiency (TE) increases with the ratio of reservoir volume to average annual inflow. This is shown in **Figure 14.8B** and **14.8C**. Where the sediment texture is very fine or the stream regime is highly variable, the TE is decreased. TE increases with coarse texture and equitable flow. Assume that a reservoir of 100 acre-ft has been built at the mouth of your basin. What proportion of your sediment yield will be trapped, and what proportion passes over the dam? Average annual runoff is available from USGS HA-212 or HA-710 (see Exercise 14.2).

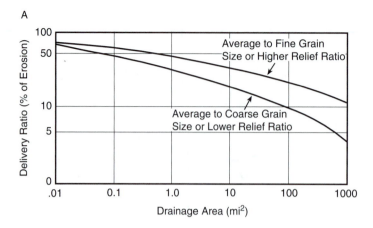

FIGURE 14.8A Sediment delivery ratio vs. size of drainage area (Roehl curve).

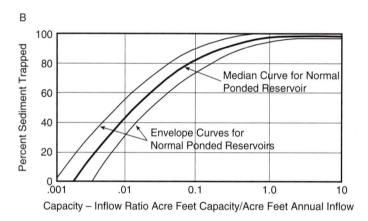

FIGURE 14.8B Trap efficiency of reservoirs as related to capacity–inflow ratio (Brune curve).

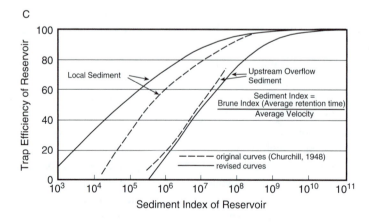

FIGURE 14.8C Trap efficiency vs. sedimentation index. (From Trimble and Bube, *Environ. Prof.,* 12, 255–272, 1990. With permission.)

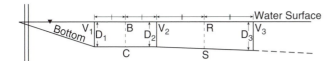

FIGURE 14.9 Geometric relationships for determining streamflow based on current meter measurements.

14.12 EXERCISE 14.7: STREAMFLOW MEASUREMENT IN THE FIELD (SEE THEORY IN CHAPTER 7)

14.12.1 PART I. GAGING STREAMFLOW BY INSTRUMENT

Using a current meter, measure the velocity of your stream in fairly short segments. Take the velocity about one third of the depth below the surface (two thirds from the bottom). The total discharge is determined by the procedure in **Figure 14.9**.

14.12.2 PART II. GAGING STREAMFLOW BY THE FLOAT METHOD

Measure along your stream and measure the time it takes some floating object to pass the distance. An apple or orange works best. Multiply the average velocity times the cross-sectional area to get q. How does this compare with your measured q? Usually, a diminished coefficient is necessary. For relatively deep streams, 0.8 is a common coefficient, but for shallow streams, it may fall below 0.7. What is your value?

14.12.3 PART III. STREAM GAGING EXERCISE

The purpose of this exercise is to construct a stage–discharge relationship for a stream as if you intended to install a gaging station. Check out an automatic level, tripod, rod, tape, compass (or transit), and chaining pins. If possible, select a reach of stream that is reasonably straight and uniform with minimum vegetative growth. Select a fixed local point for a benchmark. Survey a profile across the stream at above bankfull level so you can include the floodplain. Make sure you describe the profile well, including the azimuth. Then, carefully measure the slope of the floodplain. On a well-formed floodplain, you might do this by measuring the difference in elevation between two points about 500 to 1000 ft apart. However, streams in many areas tend to have poorly formed floodplains, so you may have to survey a long profile along the floodplain and then fit a regression line to the points, which will give you a slope for the stream at flood level. Look for evidence of "effective discharge" (Chapter 6), such as the top of recently formed point bars.

Use the Manning equation to construct a stage–discharge relationship from channel bottom (zero discharge)

to bankfull. Let each member of the team calculate discharge for at least one stage between zero and bankfull, but have at least three points on your curve. Manning's n factors are found in Chapter 7, **Table 7.1** and **Table 7.2**. Each stage will probably have a different n value. It is a good idea to assign an n value for each foresight taken in surveying the profile. Then, average the n values for each stage. Show your work for each stage and plot your stage–discharge relationship on semilog graph paper. For multiple channels and flows over bankfull (over the floodplain), contrast dummy channels as shown in the text. Give a short description of what you did so that you can do it later.

Your calculations should be done by hand to understand the procedure fully. However, most state highway departments have a computer program that will use your profile data to describe a stage–discharge relationship for whichever stage interval you specify. For instruction on surveying, see Harrelson et al., 1994.

14.13 EXERCISE 14.8: WATERSHED OR DRAINAGE BASIN MORPHOLOGY (SEE THEORY IN CHAPTER 7)

14.13.1 INTRODUCTION

The watershed is the focal point for surface water hydrology as well as fluvial geomorphology. Watershed morphology is greatly influenced by hydrologic processes, or in turn, watershed morphology greatly influences hydrologic processes. Over a long period of time, a generally efficient system of surface and subsurface drainage develops. A hydrologist should be familiar with the characteristics of watersheds and the interrelationships of these variables. This exercise is intended as a primer.

A. Using a 1:24,000 topographic map, select a watershed small enough to fit on an 8½ × 11 in. sheet of color copy paper. Select a fourth- or fifth-order watershed if possible. Topographic maps are available in many libraries and can be purchased from the U.S. Geological Survey. You may also download free topographic coverage from Topozone.com. Several scales are available, but the maps do not always print at the designated scale.

B. Trace the drainage pattern of your watershed (**Figure 14.10**). Extend the stream network up valleys to the highest contour line which has been crenelated to indicate the stream valley. Streams may be classified by the period of time during which flow occurs (but there is no way you can discriminate among these solely from the topographic map alone):

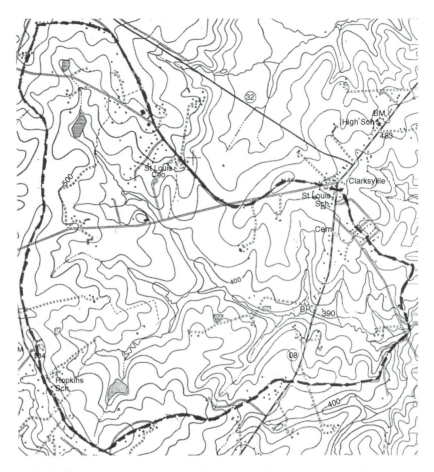

FIGURE 14.10 A 1:24,000 USGS topographic map showing Clarksville, MD, watershed. This figure has been reduced.

TABLE 14.4
Bifurcation Ratios and Stream Orders, Numbers, Lengths, and Slopes

Stream Order	Segments (*n*)	Total Length per Order (ft)	Average Length (ft)	Average Subwatershed Area (acres)	Average Slope (ft/ft)	Bifurcation Ratio
1						
2						
3						
4						

1. **Perennial** — shown on the map by solid blue lines; exist during the greater part of the year, usually in a well-formed channel.
2. **Intermittent** — shown by dashed blue line; flow only during the wet season, usually less than half the time.
3. **Ephemeral** — not shown on the map; flow occurs in (and formed) higher portions of valleys above intermittent streams; channels are usually poor or nonexistent.

If you can closely inspect your basin in the field, give a general description of your basin regarding the incidence of these three types.

C. **Stream Orders: Numbers, Lengths, Slopes (Horton Analysis).** On a separate sheet of paper, set up and complete **Table 14.4** based on the following instructions:

1. The Strahler stream order system is most commonly used, but Shreve's system appears to correlate better with hydrologic processes because it is additive, as is stream-flow (see **Figure 14.11**). For this exercise, use the Strahler system. Indicate stream order on your stream network outline. This is most easily done by color or by tick marks. Then, count the number of segments, or

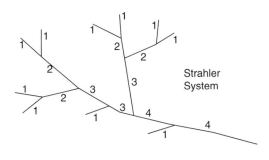

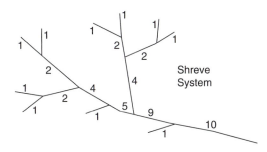

FIGURE 14.11 Comparison of the Strahler and Shreve stream-ordering systems.

links, for each order and enter them in the table.

2. **Length:** Just measure valley length instead of stream length. Stream length is always greater than valley length, the degree depending on sinuosity, but there will not be too much difference in lower-order streams such as your basin. To measure, lay a piece of paper alongside the valley, mark off the distance on the paper, and repeat the process with each stream valley segment. Compute the total length and average length per segment.

3. **Average Basin Size by Order:** Select a few typical basins for each order and planimeter them. Average the sizes and enter them in the table.

4. **Average Slope by Order:** Select a few typical streams for each order and measure the slopes (drop in elevation divided by valley length, expressed as a dimensionless decimal fraction, e.g., 0.0042).

5. **Bifurcation Ratio (BR):** Indicates the number of streams of one order feeding into streams of the next higher order (number of first-order streams divided by number of second-order streams). The BR is determined from data already calculated and tends to be constant for each order, so an average is usually given for the whole basin.

D. **Graph.** To see interrelationships among the above characteristics, graph each against stream

order. With stream order as the abscissa, plot number of segments, average length (basin size), and average slope. Fashion your graphs after those in **Figure 14.12**. In fact, you can just photocopy those matrices and use them, but you may have to revise the ordinate numbering. Be sure to use semilog paper.

E. **Profile of Main Channel (Valley) and Relief Ratio.** Draw a topographic profile of the main stream valley (the longest possible stream) from interfluve to the lowest end of the watershed (**Figure 14.13**). Use at least five points in this profile. Calculate the overall slope. This is the RR, which is significant in predicting storm discharge rapidity and magnitude as well as sediment yield.

F. **Drainage Density (DD).** DD is the total length of stream channels per unit area. DD equals total stream length divided by the total basin area. DD is indicative of the surface flow from an area and is thus highly significant. DD is related to precipitation (frequency and magnitude), other climatic conditions, and factors of infiltration and percolation (e.g., soils, geology, and land use). DD can be changed by varying land use and land management (**Figure 9.13**). DD is often called texture, with low DD coarse and high DD fine. Some examples of DD (in miles/square miles) are

Humid forested areas in eastern U.S.	8–16
Humid forested Rocky Mountains	8–16
California coast ranges, igneous rocks	20–30
California coast ranges, weak sediments	30–40
Drier areas of Rocky Mountains	50–100
Badlands, SD	200–400

G. **Constant of Channel Maintenance.** This constant is the reciprocal of DD. It is an interesting and significant variable because it tells how much area (e.g., square feet) is required to maintain 1 unit (e.g., foot) of drainage channel.

H. **Basin Shape.** Our interest here is the *elongation ratio* (diameter of a circle with the same area as the basin divided by basin length). For basin length, use the length of the main valley from (E.) above. The diameter of a circle is

$$2 \cdot (A/\pi)^{1/2}$$

where A is the area of your basin. Elongation of a basin is important in regulating the rapidity and magnitude of flood peaks. Elongation is often related to bifurcation ratio; an elongated basin may have a high BR.

Horton Analysis:

Stream Order	Number of Segments	Total Length Per Order, mi	Average Length, mi	Average Basin Size, mi²	Average Slope	Bifurcation Ratio
1	81	14	.17	.1	.044	5.4
2	15	4.76	.32	.25	.025	5.0
3	3	1.52	.51	.7	.017	3
4	1	1.90	1.9	3.05	.009	
						4.5 Average

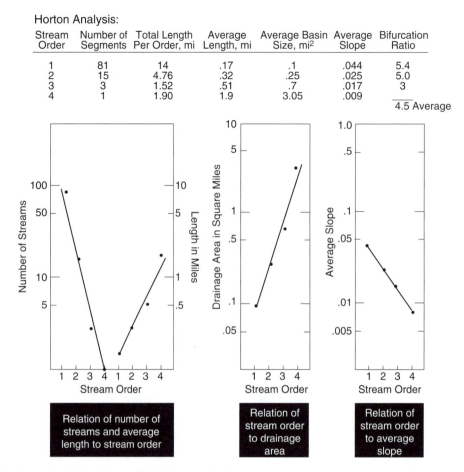

Relation of number of streams and average length to stream order

Relation of stream order to drainage area

Relation of stream order to average slope

FIGURE 14.12 Stream ordering for Clarksville, MD, watershed.

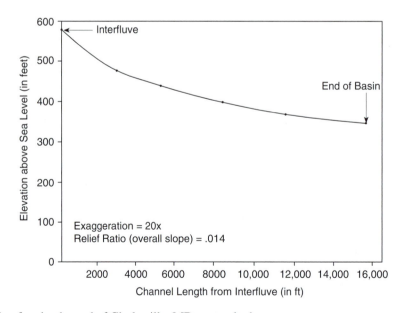

Exaggeration = 20x
Relief Ratio (overall slope) = .014

FIGURE 14.13 Profile of main channel of Clarksville, MD, watershed.

14.14 EXERCISE 14.9: THE PRACTICAL USE OF SOIL SURVEYS FOR ENVIRONMENTAL MANAGEMENT

14.14.1 INTRODUCTION

Soil surveys are useful for almost all aspects of environmental management. This exercise introduces some of the features of a modern soil survey.

> **Reference:** Any modern U.S. soil survey mapped over 1:24,000 orthophotos (orthophotos are exactly the same scale and projection as USGS 1:24,000 sheets, so the sheets cover the identical landscape). Most of these surveys are done by

USDA, usually with the cooperation of state agencies. If you have no particular choice, you may use *Soil Survey of Riverside Area, California* (USDA, 1971), which will fit well with this exercise (table numbers given in the survey accord with this survey only).

> **Instructions:** Select a parcel of real estate that interests you. You may have a place in mind, or you may randomly select a lot. The index map for the soils maps (superimposed over 1:24,000 air photos) immediately precedes the soils maps and must be folded out. For terminology, see the glossary in the survey.

14.14.2 THE PRACTICAL USE OF SOILS MAPS FOR MANAGEMENT

Name _____

Reference: *Soil Survey of Western Riverside Area, California* (USDA-SCS, 1971)

Soil surveys are found in most libraries, and any other modern soil survey is okay.

Instructions: Select a parcel of real estate that interests you. You may have a place in mind or you may randomly select a lot. The index map for the soils map (superimposed over large-scale aerial photographs) immediately precedes the soils maps and must be folded out. For terminology, see the glossary.

1. On which aerial photograph sheet is your parcel of land located? _____
2. What is the soil mapping unit for the soil on your parcel of land? _____
   What is the full name and nomenclature of this soil mapping unit? (See "Soil Legend," including thermal and moisture regimes and taxonomy after the aforementioned index map) _____
3. Read the general description of this soil ("Descriptions of the Soils").
   a. Draw and describe the profile (if one is given); indicate textures, thickness, and other pertinent data (use the back of this page).
   b. How eroded is this soil? _____
      Is this erosion a hazard? Why? _____
   c. Is this soil well drained? Why? _____
   d. Generally, what is the current usage of this soil? _____
   e. Into which capability unit does this soil fall? _____
      In a few words, what does this mean? (see later section in survey) _____
      _____
      _____
   f. Into which range site is this soil classified? _____
      What does this mean? (see later section) _____
   g. Is there any wildlife value associated with this soil? _____
4. Average crop yield (Table 2). For your soil, what is the productivity for the various crops shown under Common (A) Management and Optimum (B) Management?
   For one crop only, describe the two levels of management.
   For each crop, how does your soil compare with other soils of the area (high, medium, or low)?
5. What is the Storie Index Rating for your soil? What does this mean?

6. Give the following physical and chemical characteristics of your soil:
    a. Dry density _____
    b. Percentage clay (0.002 mm or 2 µm) _____
    c. Percentage fines (0.074 mm or #200 sieve, includes fine sand; these materials can be carried in suspension by a stream) _____
    d. Liquid limit (percentage) _____
    e. Plasticity index (liquid limit – plastic limit) _____
    f. Depth to bedrock or hardpan _____
    g. Permeability _____
    h. Available water capacity _____
    i. pH value _____
    j. Salinity _____
    k. Shrink–swell potential _____
    l. Hydrologic group _____
    m. Structure _____
7. How suitable is this soil for the following purposes (Table 6)?
    a. Topsoil _____
    b. Road fill _____
    c. Road location _____
    d. Embankments _____
    e. Reservoir area _____
    f. Septic tank field line _____
    g. Drainage _____
    h. Irrigation _____
8. How suitable is your soil for recreation? Consider play, camping, and picnic areas (Table 7).
9. If your soil is listed, give the following characteristics (Tables 9 and 10).
    a. Moisture held at 15 bars (WP) _____
    b. Cation exchange capacity _____
    c. Percentage organic carbon _____
10. Select one half of any map sheet that has some alluvial soil. Photocopy the sheet, color the alluvial soils (include low terraces), denote them as to limitations for a septic tank filter field (slight, moderate, severe). Denote any soils that have a seasonal water table depth less than 5 ft. Congratulations; you have just made a land use planning map!

References

Due to name changes through time, Soil Conservation Service, Natural Resources Conservation Service and USDA SCS are references to the same organization in this reference section.

Abrahams, A. and R. Marston, 1993. Drainage Basin Sediment Budgets and Change. *Phys. Geogr.* 14: 221–224.

Adler L.L., 1980. Adjustment of the Yuba River, California to the Influx of Hydraulic Mining Debris, 1849–1979, MA thesis, University of California, Los Angeles.

Alexander, R.B., R.A. Smith, and G.E. Schwarz. 2000. Effect of Stream Size on the Delivery of Nitrogen to the Gulf of Mexico. *Nature* 403, 758–761.

Allan, J.D., 1995. *Stream Ecology: Structure and Function of Running Waters.* Chapman & Hall, London.

Allen, R.G., M.E. Jensen, J.L. Wright, and R.D. Burman, 1989. Operational Estimates of Reference Evapotranspiration. *Agron. J.* 81:650–662.

Aller, L., T. Bennett, J. Lehr, R. Petty, G. Hackett, 1987. DRASTIC: A Standardized System for Evaluating Ground Water Pollution Using Hydrogeologic Settings. Washington (DC): U.S. Environmental Protection Agency. EPA/600/2-87/035. 455 p.

Amatya, D.M., R.W. Skaggs, and J.D. Gregory, 1992. Comparison of Methods for Estimating Potential Evapotranspiration. ASAE Winter Meeting Paper No. 92-2630.

American Society of Agricultural Engineers, 2001. Preferential Flow: Water Movement and Chemical Transport in the Environment. Proceedings of the 2nd International Symposium, Honolulu, Hawaii. American Society of Agricultural Engineers, St. Joseph, MI.

American Society of Civil Engineers and Water Environment Federation, 1998. Urban Runoff Quality Management. WEF Manual of Practice No. 23 and ASCE Manual and Report on Engineering Practice No. 87. Joint Task Force of the WEF and the ASCE, Alexandria and Reston, VA.

American Society for Testing and Materials, 1993. *Annual Book of ASTM Standards, Section 4: Construction - Soil and Rock, Building Stones; Geotextiles.* Vol. 04.08. ASTM , Philadelphia.

American Water Works Association, 1988. *New Dimensions in Safe Drinking Water.* AWWA.

Amoozegar, A. and A.W. Warrick, 1986. *Hydraulic Conductivity of Saturated Soils: Field Methods.* American Society of Agronomy, part 1, chap. 9:735–770.

Anderson, M.G., D.Walling, and P.Bates (Editors), 1996. *Floodplain Processes.* John Wiley & Sons, Chichester.

Andrews, E.D. and J.M. Nankervis, 1995. Effective Discharge and the Design of Channel Maintenance Flows for Gravel-Bed Rivers. Natural and Anthropogenic Influences in Fluvial Geomorphology. Geophysical Monograph 89: 151–164, American Geophysical Union.

Argialas, D.,J. Lyon, and O. Mintzer, 1988. Quantitative Description and Classification of Drainage Patterns. *Photogr. Eng. Remote Sensing*, 54:505–509.

Arms, K., 1990. *Environmental Science.* Saunders College Publishing Orlando, FL. 467 pp.

Arno, S.F. and S. Allison-Bunnell. 2002. *Flames in our Forests: Disaster or Renewal?* Island Press, Washington DC., 228 pp.

Arnold, J.G., J.R. Williams, and D.A. Maidment. 1995. A Continuous Time Water and Sediment Routing Model for Large Basins, *J. Hydraul. Div., ASCE* 121(2):171–183.

Auer, A.H. 1974. The Rain versus Snow Threshold Temperatures. *Weatherwise*, 27:67.

Bagchi, A., 1994. *Design, Construction, and Monitoring of Landfills*, 2nd ed. John Wiley & Sons, New York.

Bagnold, R.A., 1966. An Approach to the Sediment Transport Problem from General Physics, U.S. Geological Survey Professional Paper 422-1.

Balek, J., 1983. *Hydrology and Water Resources in Tropical Regions. (Developments in Water Science,* Vol. 18). Elsevier. New York. 271 pp.

Barfield, B.J., R.C. Warner, and C.T. Haan, 1981. *Applied Hydrology and Sedimentology for Disturbed Areas.* Oklahoma Technical Press, Stillwater, OK.

Barnes, H.H., 1967. *Roughness Characteristics of Natural Channels.* U.S. Geological Survey Water-Supply Paper 1749, U.S. Government Printing Office, Washington DC.

Bates, A.K., 1990. *Climate in Crisis: The Greenhouse Effect and What We Can Do.* The Book Publishing Company, Summertown, TN.

Baumer, O.W., R.D. Wenberg, and J.F. Rice, 1987. The Use of Soil Water Retention Curves in DRAINMOD. In: *Proceedings of the 3rd International Workshop on Land Drainage.* Department of Agricultural Engineering, Ohio State University, Columbus. pp. Al–A10.

Bedient, P.B. and W.C. Huber, 1988. *Hydrology and Floodplain Analysis.* Addison-Wesley Publishing Company, Reading, MA.

Beecher, J.A. and A.P. Laubach, 1989. *Compendium on Water Supply, Drought, and Conservation.* Report NRRI 89–15, National Regulatory Research Institute, Columbus, Ohio.

Bengtson, R.L. and G. Sabbagh, 1990. USLE P Factors for Subsurface Drainage on Low Slopes in a Hot, Humid Climate. *J. Soil Water Cons.* 45(4):480–482.

Berris, S.N. and R.D. Harr, 1987. Comparative Snow Accumulation and Melt During Rainfall in Forested and Clear-cut Plots in the Western Cascades of Oregon. *Water Resour. Res.*, 23(1):135–142.

Berthouex, P.M., and L.C. Brown, 2002. *Statistics for Environmental Engineers*, 2nd ed. Lewis Publishers, Boca Raton, FL. 489 pp.

Bertoni, J., and F. Lombardi Neto, 1993. *Conservação do solo,* 3rd ed. Ìcone Editora Ltda. São Paula, Brazil.

Betson, R., 1976. *Urban Hydrology: A Systems Study in Knoxville, Tennessee.* Tennessee Valley Authority, Knoxville. 138 pp

Bilger, B., 1992. *Global Warming: Earth at Risk.* Chelsea House Publishers, New York.

Biswas, A.K., 1969. A Short History of Hydrology. In*: The Progress of Hydrology.* Proceedings of the First International Seminar for Hydrology Professors, Volume II, Specialized Hydrologic Subjects. p. 914–935.

Blake, G.R. and K.H. Hartge, 1986a. Bulk Density. In: A. Klute (Ed.), *Methods of Soil Analysis, Part 1: Physical and Mineralogical Methods,* 2nd ed. American Society of Agronomy, Madison, WI.

Blake, G.R. and K.H. Hartge, 1986b. Particle density. In: A. Klute (Ed.), *Methods of Soil Analysis, Part 1: Physical and Mineralogical Methods,* 2nd ed. American Society of Agronomy, Madison, WI.

Bodman, G.B. and E.A. Coleman, 1943. Moisture and Energy Conditions During Downward Entry of Water Into Soils. *Soil Sci. Soc. Am. Proc.* 7:116–122.

Bolstad, P. and J. Smith, 1995. Errors in GIS: Assessing Spatial Data Accuracy. In: Lyon, J. and J. McCarthy (Eds.), *Wetland and Environmental Applications of GIS.* Lewis Publishers, Boca Raton. pp. 301–312.

Booth, D.B., 1991. Urbanization and the Natural Drainage System-Impacts, Solutions, and Prognoses. *The Northwest Environmental Journal* 7:93–118.

Booth, D.B., and C.R. Jackson. 1997. Urbanization of Aquatic Systems: Degradation Thresholds, Stormwater Detection, and the Limits of Mitigation. *J. Am. Water Resour. Assoc.* 35(5):1077–1090.

Borland, W.M., 1971. Reservoir Sedimentation, In: H.W. Shen (Ed.). *River Mechanics, Vol. 2.* Ft. Collins, CO.

Borland, W.M. 1951. Unpublished data in files of Bureau of Reclamation, Denver, CO.

Bos, M.G., J.A. Replogle, and A.J. Clemmens., 1991. *Flow Measuring Flumes for Open Channel Systems.* American Society of Agricultural Engineers, St. Joseph, MI.

Bosch, J.M. and J.D. Hewlett, 1982. A Review of Catchment Experiments to Determine the Effect of Vegetation Changes on Water Yield and Evapotranspiration. *Hydrology* 55: 2–23.

Bouma, J., 1986. Using Soil Survey Information to Characterize the Soil Water State. *J. Soil Sci.* 37:1–7.

Bouwer, H. and R.D. Jackson, 1974. Determining Soil Properties, In: J. van Schilfgaarde (Ed.), *Drainage for Agriculture.* ASA Monograph 17. American Society of Agronomy, Madison, WI.

Bouwer, H., 1986. Intake Rate: Cylinder Infiltrometer. In: A. Klute (ed.), *Methods of Soil Analysis, Part 1: Physical and Mineralogical Methods,* 2nd ed. American Society of Agronomy, Madison, WI.

Bradford, J.M., D.A. Farrell, and W.E. Larson, 1973. Mathematical Evaluation of Factors Affecting Gully Stability. *Soil Sci. Soc. Am. Proc.* 37:103–107.

Bradley, A. A. and K. W. Potter, 1992. Flood Frequency Analysis of Simulated Flows. *Water Resour. Res.,* 28(9): 2375–2385.

Bradley, G. E., Ed., 1984. *Land Use and Forest Resources in a Changing Environment. The Urban/Forest Interface.* University of Washington Press. Seattle, WA. 222 pp.

Brakensiek, D.L., H.B. Osborn, and W.J. Rawls, 1979. *Field Manual for Research in Agricultural Hydrology,* USDA Agriculture Handbook 224, 550 pp.

Branson, F.A., G. Gifford, K. Renard, and R. Hadley, 1981. *Rangeland Hydrology.* Kendall/Hunt, Dubuque, Iowa.

Brater, E.F. and H.W. King, 1976. *Handbook of Hydraulics* (6th Ed.). McGraw Hill, New York.

Breen, K.J., 1988. Geochemistry of the Stratified-drift Aquifer in Killbuck Creek Valley West of Wooster, Ohio. In: *Regional Aquifer Systems of the United States—Northeast Glacial Aquifers,* A.D. Randall and A.I. Johnson, Eds. American Water Resources Association Monograph Series No. 11, p. 105–131.

Brockman, C.S., and J.P. Szabo, 2000. Fractures and Their Distribution in the Tills of Ohio. *Ohio J. of Sci.* 100(3/4): 39–55.

Brookes, A. 1996. Floodplain Restoration and Rehabilitation. In: *Floodplain Processes.* M.G. Anderson, D.E Walling, and P.D. Bates, (Eds). John Wiley & Sons, Chichester, England pp. 554–576.

Brown, A.G. and T. Quine (Eds.), *Fluvial Processes and Environmental Change.* John Wiley & Sons, Chichester.

Brown, C.B. 1943. *The Control of Reservoir Silting.* U.S. Department of Agriculture Misc. Pub. 521, Washington, D.C.

Brown, L.R. (Project Director), 1992. *State of the World.* A Worldwatch Institute Report on Progress Toward a Sustainable Society, W.W. Norton & Company. New York.

Brown, L.R., 1995. *Who Will Feed China? Wake-Up Call for a Small Planet.* A Worldwatch Environmental Alert Series (Linda Starke, Series Ed.). Worldwatch Institute, W.W. Norton & Company. New York. 163 pp.

Brown, L.R., G. Gardner, and B. Halweil, 1999. *Beyond Malthus: Nineteen Dimensions of the Population Challenge.* The Worldwatch Environmental Alert Series. W.W. Norton and Co., New York. 167 pp.

Brown, M.L. and D.J. Austen, 1996. Data Management and Statistical Techniques. In: B.R. Murphy and D.W. Willis (Eds.), *Fisheries Techniques,* 2nd ed., American Fisheries Society, Bethesda, MD.

Brown, R.E., 1993. Runoff and Sediment Production from Forested Hillslope Segments in the Georgia Piedmont. MS thesis, University of Georgia, Athens.

Brown, T. and D. Caraco, 2001. Channel Protection. *Water Resour. Impact.* November: 16–19

Brune, G.M., 1953. Trap Efficiency of Reservoirs. *Trans. Am. Geophys. Union,* 34(3), 407–418.

Bube, K.P., and S.W. Trimble, 1986. Revision of the Churchill Reservoir Trap Efficiency Curves Using Smoothing Splines. *Water Resour. Bull.,* 22, 305–309.

Buchanan, T.J. and W.P. Somers, 1969. Discharge Measurements at Gaging Stations, In: *Techniques of Water-Resources Investigations of the United States Geological Survey, Book 3, Applications of Hydraulics.* U.S. Geological Survey, Alexandria, VA. chap. A7.

Bureau of Land Management, 1994. *Rangeland Reform '94: draft Environmental Impact Statement.* Department of the Interior. Washington, D.C.

Bureau of Reclamation, 1977. Ground Water Manual: U.S. Department of the Interior, 480 p.

Burn, D.H. and D.B. Boorman, 1993. Estimation of Hydrological Parameters at Ungauged Catchments. *J. Hydrol.*, 143:429–454.

Burroughs, E.R., Jr. and B.R. Thomas, 1977. Declining Root Strength in Douglas Fir After Felling as a Factor in Slope Stability. Research Paper INT-190. USDA Forest Service, Intermountain Forest and Range Experiment Station. Ogden, UT. 27 pp.

Calder, I.R., 1993. Hydrologic effects of land use. In: D. Maidment (Ed.), *Handbook of Hydrology*, McGraw-Hill, New York. pp. 13.1–13.50.

Carroll, S., T. Carroll, and R. Poston, 1999. Spatial Modeling and Prediction of Snow-water Equivalent Using Ground-based, Airborne, and Satellite Snow Data. *J. Geophys. Res.*, 104:19,623–19,629.

Cassel, D.K. and D.R. Nielsen, 1986. Field Capacity and Available Water Content. In: A. Klute (Ed.), *Methods of Soil Analysis, Part 1: Physical and Mineralogical Methods*, 2nd ed. American Society of Agronomy, Madison, WI.

Chang, H.H., 1998. *Fluvial Processes in River Engineering.* Krieger Publishing Company, Malabar, FL.

Chanson, H., 1999. *The Hydraulics of Open Channel Flow.* Arnold Publishers, London, England.

Chartrand, S.M. amd P.J. Whiting, 2000. Alluvial Architecture in *Headwater Streams with Special Emphasis on Step-Pool Topography Earth Surface Processes and Landforms*, 25:573–600, John Wiley & Sons.

Chen, C., 1975. Design of Sediment Retention Basins, Proceedings of the National Symposium on Urban Hydrology and Sediment Control, University of Kentucky, Lexington, July 28–31.

Chen, Y.H. 1976. *Sediment Transport and Flood Routing. A Short Course on Numerical Modeling for Engineers.* USAE Waterways Experiment Station Hydraulics Laboratory, Vicksburg, MS.

Chin A., 1999a. The Morphologic Structure of Step-Pools in Mountain Streams. *Geomorphology* 27:1919–204, Elsevier.

Chin, A., 1999b. On the Origin of Step-Pool Sequences in Mountain Streams. *Geophys. Res. Lett.*, 26(2): 231–234.

Chin, D.A., 1999. *Water-Resources Engineering.* Prentice-Hall, New Jersey. 750 pp.

Chow, V.T. (ed), 1964. *Handbook of Applied Hydrology.* McGraw-Hill, New York.

Chow, V.T., 1959. *Open-Channel Hydraulics.* McGraw-Hill, New York.

Church, M. and O. Slaymaker, 1989, Disequilibrium of Holocene Sediment Yield in Glaciated British Columbia. *Nature* 337:452–454.

Church, M., and D. Jones, 1982. Channel Bars in Gravel-bed Rivers, Hey, R.D., J.D. Bathurst, and C.R. Thorne, eds., *Gravel-bed Rivers: Fluvial Processes, Engineering and Management*, John Wiley & Sons, pp. 291–338.

Churchill, M.A., 1948. Discussion of Analysis and Use of Reservoir Sedimentation Data by L.C. Gottschalk, In: *Proceedings of the Federal Inter-Agency Sedimentation Conference,* Denver, 1947. Bureau of Reclamation, U.S. Department of the Interior, Washington, D. C. pp. 139–140.

Colebrook, C.F. and C.M. White, 1937. Experiments with Fluid Friction in Roughened Pipes. *Proc. R. Soc. London,* Ser. A 161:367–387.

Congalton, R. and K. Green, 1995. The ABCs of GIS: An Introduction to Geographic Information Systems. In Lyon, J. and J. McCarthy (Eds.), *Wetland and Environmental Applications of GIS.* Lewis Publishers, Boca Raton. pp. 9–24.

Congalton, R. and K. Green, 1998. *Accuracy Assessment of Remotely Sensed Data.* CRC/Lewis Publishers, Boca Raton, FL, 137 pp.

Cooke, R.U. and J. Doornkamp, 1990, *Geomorphology in Environmental Management.* Clarenden Press, Oxford.

Cooperative Extension Service and The Ohio State University, 1979. Ohio Erosion Control and Sediment Pollution Abatement Guide. Columbus, OH. 24 pp.

Costa, J. , 1978. *Hydrology and Hydraulics of Cherry Creek,* University of Denver, Department of Geography.

Cowan, W.L., 1956. Estimating Hydraulic Roughness Coefficients. *Agric. Eng.,* 37(7):473–475.

Cowardin, L., V. Carter, F. Golet, and E. LaRoe, 1979. Classification of Wetlands and Deepwater Habitats of the United States. U.S. Fish and Wildlife Service, Report FWS/OBS-79/31, Washington, D.C. 103 pp.

Cudworth, A.G., 1989. *Flood Hydrology Manual.* A Water Resources Technical Publication. U.S. Department of the Interior Bureau of Reclamation, Denver, CO.

Cummins, K.W., 1975. The Ecology of Running Waters; Theory and Practice. In: Baker, D.B., W.B. Jackson, and B.L. Prater (eds.) *Proceedings of the Sandusky River Basin Symposium, International Joint Commission, Great Lakes Pollution*, Environmental Protection Agency, Washington D.C. pp. 227–93.

Darby, S.E. and A. Simon (Eds.), 1999. *Incised River Channels.* John Wiley & Sons, Chichester, 442 pp.

de Roche, J.Z., 1993. An Examination of the Characteristics of Seasonal Recharge in Ohio Based on Well Hydrographs: Unpublished Senior Thesis, Department of Geological Sciences, The Ohio State University, 37 pp.

DeBano, L.F., 1981. Water Repellent Soils: A State-of-the-Art. USDA Forest Service General Technical Report PSW-46. Berkeley, CA.

DeBano, L.F., 2000. The Role of Fire and Soil Heating on Water Repellency in Wildland Environments: A Review. *J. Hydrology* 232: 195–206.

DeCoursey, D.G., 1966. *A Runoff Hydrograph Equation.* Publication ARS 41-116, U.S. Department of Agriculture, Washington, D.C.

Dendy, F.E., and W. Champion, 1978. Sediment Deposition in U.S. Reservoirs: Summary of Data Reported through 1975. U.S. Department of Agriculture Miscellaneous Publication 1362.

Derrick, D.L., T.J. Pokrefke, M.B. Boyd, J.P. Crutfield, and R.R. Henderson, 1994. Design and Development of Bendway Weirs for the Dogtooth Bend Reach, Mississippi River, Technical Report HL-84-10, U.S. Army Corps of Engineers Waterways Experiment Station, Vicksburg, MS.

DeShon, J.E., 1995. Development and Application of the Invertebrate Community Index (ICI). In: *Biological Assessment and Criteria: Tools for Water Resource Planning and Decision Making.* W.S. Davis, T. P. Simon (Eds.), Lewis Publishers, Boca Raton, FL

Dingman, S.L., 1994. *Physical Hydrology.* Macmillan. New York.

Dissmeyer, G.E. and G.R. Foster, 1981. Estimating the Management Factor (C) in the Universal Soil Loss Equation for Forest Conditions. *J. Soil Water Conserv.,* 36(4): 235–240.

Doerr, S.H., Shakesby, R., and Walsh, R., 2000. Soil Water Repellency: Its Cause, Characteristics and Hydro-geomorphological Significance. *Earth-Sci. Rev.* 51:33–65.

Doorenbos, J. and W.O. Pruitt, 1977. *Guidelines for Prediction of Crop Water Requirements.* FAO Irrigation and Drainage Paper No. 24, 2nd ed., FAO Rome, Italy. 156 pp.

Douglas, L.A. and M.L. Thompson (eds.), 1985. *Soil Micromorphology and Soil Classification.* Soil Science Society of America, Special Publication 15, Madison, WI. 216 pp.

Doyle M., J. Harbor, C. Rich, and A. Spacie, 2000. Examining the Effects of Urbanization on Streams Using Indicators of Geomorphic Stability. *Physic. Geogr.,* Volume: 21 , Number: 2 (MAR–APR) , Page: 155–181.

Dunne, T. and L.B. Leopold, 1978. *Water in Environmental Planning.* W. H. Freeman and Co., San Francisco.

Dunne, T. and R.D. Black, 1970. Partial Area Contributions to Storm Runoff in a Small New England Watershed. *Water Resour. Res.,* 6(5):1296–1311.

Egan, D. and E. Howell, 2001. *The Historical Ecology Handbook.* Island Press. Covelo, CA.

Einstein, H.A., 1950. The Bed-Load Function for Sediment Transportation in Open Channel Flow. USDA-SCS, Technical Bulletin 1026, Washington D.C., pp. 71.

Ekanayake, S.T. and J.F. Cruise, 1993. Comparisons of Weibull and Exponential-Based Partial Duration Stochastic Flood Models. *Stochastic Hydrology and Hydraulics,* 7(4):283–297.

Elachi, C., 1987. *Introduction to the Physics and Techniques of Remote Sensing.* J. Wiley & Sons, New York. 413 pp.

Elliot, W.J., J.M. Laflen, and G.R. Foster, 1993. Soil Erodibility Nomographs for the WEPP Model. Paper No. 932046, American Society of Agricultural Engineers, St. Joseph, MI.

Ellison, W.D., 1947. Soil Erosion Studies, Part I. *Agric. Eng.,* April 1947, pp. 145–146.

Elrick, D.E. and W.D. Reynolds, 1986. An Analysis of the Percolation Test Based on Three-dimensional, Saturated-unsaturated Flow from a Cylindrical Test Hole. *Soil Sci.* 142(5): 308–321.

Emmett, W.W. and M.G. Wolman, 2001. Effective Discharge and Gravel-bed Rivers. *Earth Surface Processes and Landforms* 26:1369–1380.

Erie and Niagara Counties Regional Planning Board, 1981. *Storm Drainage Design Manual.* Erie and Niagara Counties Regional Planning Board, Grand Island, New York.

Falkner, E. and D. Morgan, 2002. *Aerial Mapping, Methods and Applications.* CRC/Lewis Publishers, Boca Raton, FL, 192 pp.

Fausch K.D., J.R. Karr, and P.R. Yant. 1984. Regional Application of an Index of Biotic Integrity Based on Stream Fish Communities. Transactions of the American Fisheries Society 113(1):39–55.

Federal Council for Science and Technology, 1962. *Report by an Ad Hoc Panel on Hydrology and Scientific Hydrology,* Washington, D.C.

Federal Highway Adminstration. 2001. State Focus on Erosion Control in Construction and Maintenance Activities, *FOCUS,* April/May Issue, URL: www.tfhrc.gov/focus/apr01/erosion.htm (viewed 5-18-03)

Fennessy, M.S. and W.J. Mitsch, 1989. Treating Coal Mine Drainage with Artificial Wetland. Research *J. Water Pollut. Control Fed.,* 61:1691–1701.

Finkenbine, J.K., J.W. Atwater, and D.S. Mavinic, 2000. Stream Health After Urbanization. *J. Am. Water Resour. Assoc.* 36(5):1149–1160.

Finton, C.D., 1994. Simulation of Advective Flow and Attenuation of Two Agricultural Chemicals in an Alluvial-valley Aquifer, Piketon, Ohio: Unpublished M.S. Thesis, Department of Geological Sciences, The Ohio State University, 318 p.

Food and Agriculture Organization of the United Nations, 1993. *Forest Resources Assessment 1990: Tropical Countries.* FAO Forestry Paper 112. Rome, Italy.

Food and Agriculture Organization of the United Nations, 1999. *FAO Yearbook: Production – 1999.* Food and Agriculture Organization of the United Nations, Rome, Italy

Food and Agriculture Organization of the United Nations, 2002. *Unlocking the Water Potential of Agriculture. Fact Sheet.* Food and Agriculture Organization of the United Nations, www.fao.org.

Foster, D.R. and J. O'Keefe, 2000. *New England Forests Through Time.* Harvard University Press. Cambridge, Mass.

Foster, D., F. Swanson, J. Aber, I. Burke, N. Brokaw, D. Tilman and A. Knapp, 2003. The Importance of Land-use Legacies to Ecology and Conservation, *BioScience* 53:77–88.

Foster, G.R., 1988. *User Requirements, USDA-Water Erosion Prediction Project (WEPP).* USDA-ARS National Soil Erosion Lab. W. Lafayette, IN.

Foster, G.R., D.K. McCool, K.G. Renard, and W.C. Moldenhauer, 1981. Conversion of the Universal Soil Loss Equation to SI Metric Units. *J. Soil Water Conserv.* 36(6):355–359.

Foster, I., A. Gurnell, and B. Webb (Eds.), 1995. *Sediment and Water Quality in River Catchments.* John Wiley & Sons, Chichester.

Foster, P., 1992. *The World Food Problem: Tackling the Causes of Undernutrition in the Third World.* Lynne Rienner Publishers, Boulder, CO. 367 pp.

Franklin, J.F. and C.T. Dyrness, 1973. The Natural Vegetation of Oregon and Washington, USDA Forest Service, General Technical Report PNW-8. Portland, OR.

Freeze, R.A. and J.A. Cherry, 1979. *Groundwater*. Prentice-Hall, Englewood Cliffs, NJ.

Freeze, R.A. and P.A. Witherspoon, 1967. Theoretical Analysis of Regional Groundwater Flow: 1. Effect of Water-table Configuration and Subsurface Permeability Variation: *Water Resour. Res.*, 3(2):623–634.

Frissell, C.A, W.J. Wiss, C.E. Warren, and M.D. Huxley, 1986. A Hierachical Framework for Stream Classification: Viewing Streams in a Watershed Context. *Environ. Manage.* 10:199–214.

Gabler, R., 1987. *Is Your Water Safe to Drink?* Consumer Reports Books, Consumers Union, Mount Vernon, New York.

Gammon, J.R. 1976. The Fish Populations of the Middle 340 km of the Wabash River. Purdue University Water Resources Research Center Technical Report 86. 73pp.

Gammon, J.R., A. Spacie, J.L. Hamelink, and R.L. Kaesler, 1981. Role of Electrofishing in Assessing Environmental Quality of the Wabash River, In: Bates, J.M. and Weber, C.I. (Eds.), *Ecological Assessments of Effluent Impacts on Communities of Indigenous Aquatic Organisms*, ASTM STP 730, 307 pp.

Gardner, W.H., 1986. Water content. In: A. Klute (Ed.), *Methods of Soil Analysis, Part 1: Physical and Mineralogical Methods,* 2nd ed. American Society of Agronomy, Madison, WI.

Gebert, W. and W. Krug, 1996. Streamflow Trends in Wisconsin's Driftless area. *Water Resour. Bull.*, 32:733–744.

Gee, G.W. and J.W. Bauder, 1986. Particle-size Analysis. In: A. Klute (Ed.), *Methods of Soil Analysis, Part 1: Physical and Mineralogical Methods,* 2nd ed. American Society of Agronomy, Madison, WI.

GeoTrans Newsletter, 1987. Woburn Toxic Trial: GeoTrans, Inc. Herndon, VA, June, p. 1–3.

Gilbert, G.K., 1917, Hydraulic Mining Debris in the Sierra Nevada. USGS Professional Paper 105.

Giller, P.S. and B. Malmqvist, 1998. *The Biology of Streams and Rivers*. Oxford University Press, Oxford.

Glanz, J., 1999. Sharp Drop Seen in Erosion Rates. *Science,* 285, 1186–1187.

Goodwin, C.N., 1999. Fluvial Classification: Neanderthal Necessity or Needless Normalcy. In: D.S. Olson and J.P. Potyondy (Eds.), *Wildland Hydrology*. American Water Resources Association, TPS-99-3, Herndon, VA, pp. 229–236.

Goolsby, D.A. 2000. Mississippi Basin Nitrogen Flux Believed to Cause Gulf Hypoxia: EOS Transactions, *Am. Geophys. Union*, 81(29):321,326–327.

Goolsby, D.A., W.A. Battaglin, G.B. Lawrence, R.S. Artz, B.T. Aulenbach, R.P. Hooper, D.R. Keeney, and G.J. Stensland, 1999. Flux and Sources of Nutrients in the Mississippi-Atchafalaya River Basin–Topic 3, Report for the Integrated Assessment of Hypoxia in the Gulf of Mexico: NOAA Coastal Ocean Program Decision Analysis Series, 17, 129 pp.

Gore, A., *Earth in the Balance*. Houghton Mifflin, Boston.

Gottlieb, R., 1988. *A Life of its Own: The Politics and Power of Water*. Harcourt, Brace Jovanovich, San Diego.

Graf, W.G., 1988. *Fluvial Processes in Dryland Rivers*. Springer-Verlag, New York, 223 p.

Green, R.E., L.R. Ahuja, and S.K. Chong, 1986. Hydraulic Conductivity, Diffusivity, and Ssorptivity of Unsaturated Soils: Field Methods. In: A. Klute (Ed.), *Methods of Soil Analysis, Part 1: Physical and Mineralogical Methods,* 2nd ed. American Society of Agronomy, Madison, WI.

Gregory, S.V., F.J. Swanson, W.A. McKee, and K.W. Cummins, 1991. An Ecosystem Perspective of Riparian Zones. *Bioscience* 41:540–551.

Guo, C.Y. and B.R. Urbonas, 1995. Special Report to the Urban Drainage and Flood Control District on Stormwater BMP Capture Volume Probabilities in United States, Denver, CO.

Gurnell, A.G. and G. Petts (Eds.), 1995. *Changing River Channels*. John Wiley & Sons, Chichester.

Haan, C.T., 1970. A Dimensionless Hydrograph Equation. File Report, Agricultural Engineering Department, University of Kentucky, Lexington.

Haan, C.T., 1997. Statistical Methods in Hydrology. Iowa State University Press, Ames. 378 p.

Haan, C.T., B.J. Barfield, and J.C. Hayes, 1994. *Design Hydrology and Sedimentology for Small Catchments*. Academic Press, Orlando, FL. 588 p.

Haan, C.T., H.P. Johnson, and D.L. Brakensiek (Eds.), 1982. *Hydrologic Modeling of Small Watersheds*. ASAE Monograph 5, American Society of Agricultural Engineers, St. Joseph, MI.

Hack, J.T., 1957. Studies of Longitudinal Stream Profiles in Virginia and Maryland. Professional Paper 294B, U.S. Geological Survey, 97 pp.

Haefner, R.J., 2000. Characterization Methods for Fractured Glacial Tills. *Ohio J. Sci.* 100(3/4): 73–87.

Haktanir, T. and H.B. Horlacher, 1993. Evaluation of Various Distributions for Flood Frequency Analysis. *Hydrologic Sci. J.*, 38(1):15–32.

Hamblin, W.K. and E.H. Christiansen, 1995. *Earth's Dynamic Systems*. Prentice-Hall, Inc. Englewood Cliffs. 710 p.

Hambrook, J.A., G.F. Koltun, B.B. Palcsak, and J.S. Tertuliani, 1997. Hydrologic Disturbance and Response of Aquatic Biota in Big Darby Creek Basin, Ohio. U.S. Geological Survey Water-Resources Investigation Report 96-4315. USGS Branch of Information Services, Denver, CO.

Hamilton, C.L. and H.G. Jepson, 1940. Stock–water Developments; Wells, Springs, and Ponds, USDA Farmers Bulletin 1859.

Happ, S.C., G. Rittenhouse, and G. Dobson, 1940. Some Principles of Accelerated Stream and Valley Sedimentation. USDA Technical Bulletin 695.

Harman, W.A., G.D. Jennings, J.M. Patterson, D.R. Clinton, L.O Slate, A.G. Jessup, J.R. Everhart, and R.E. Smith, 1999. Bankfull Hydraulic Geometry Relationships for North Carolina Streams. In: D.S. Olsen and J.P. Potyondy (Eds.), *Wildland Hydrology*. AWRA Symposium Proceedings, American Water Resources Association. June 30-July 2, Bozeman, MT.

Harr, R.D., 1977. Water Flux in Soil and Subsoil on a Steep Forested Slope. *J. Hydrology*, 33:37–58.

Harr, R.D., 1982. Fog Drip in the Bull Run Municipal Watershed, Oregon. *Water Resour. Bull.*, 18(5):785–789.

Harrelson, C.C., C.L. Rawlins, and J.P. Potyondy, 1994. Stream Channel Reference Sites: An Illustrated Guide to Field Technique. USDA Forest Service, General Technical Report RM-245, Fort Collins, Colorado, 62 pp.

Harrison, L.P., 1963. Fundamental Concepts and Definitions Relating to Humidity. In: Wexler, A. (Ed.). *Humidity and Moisture*, Vol. 3, Reinhold Publishing, New York.

Harrold, L.L., 1957. Minimum Water Yield from Small Agricultural Watersheds. *Trans. Am. Geo. Union.* 38(2): 201–208.

Harrold, L.L., G.O. Schwab, and B.L. Bondurant, 1986. *Agricultural and Forest Hydrology.* Agricultural Engineering Department, Ohio State University, Columbus. 271 pp.

Harvey, J.G., 1994. Water Labyrinth, Policy Reform for Reallocation of California Water. unpublished PhD dissertation, University of California, Los Angeles.

Hathaway, G.A. et al., 1956. *Snow Hydrology.* North Pacific Division Corps of Engineers. U.S. Army. Portland, OR. 437 pp.

Hauer, F.R., and G.A. Lambert, 1996. *Methods in Stream Ecology.* Academic Press, San Diego.

Haupt, H.F. and B.L. Jeffers, 1967. A System for Automatically Recording Weight Changes in Sapling Trees. USDA Forest Service, Intermountain Forest and Range Experiment Station. Research Note INT-71. Ogden, UT. 4 pp.

Hazen, A. 1904. On Sedimentation. *Trans. Am. Soc. Civ. Eng.,* Paper 980:45–87.

Heath, R.C., 1983. *Basic Ground-water Hydrology,* U.S. Geological Survey Water-Supply Paper 2220, 84 p.

Heinemann, H.G., 1981. A New Sediment Trap Efficiency Curve for Small Reservoirs. *Water Resour. Bull.,* 17, 825–830.

Helvey, J.D., 1971. A Summary of Rainfall Interception by Certain Conifers of North America. In: E.J. Monke (Ed.), *Proceedings of the Third International Seminar for Hydrology Professors: Biological Effects in the Hydrological Cycle.* Purdue University, West Lafayette, IN. pp. 103–113.

Henderson, F.M. and R.A. Wooding, 1964. Overland Flow and Groundwater Flow from a Steady Rainfall of Finite Duration. *Geophys. Res.* 69:1531–1540.

Hershfield, D.N., 1961 Rainfall Frequency Atlas of the United States. U.S. Weather Bureau Technical Paper 40.

Heubner, G.L., 1985. Use of Radar for Precipitation Measurement. In: M.G. Anderson and T.P. Burt (Eds.) *Hydrological Forecasting.* John Wiley & Sons, Chichester.

Hewlett, J.D. and A.R. Hibbert, 1967. Factors Affecting the Response of Small Watersheds to Precipitation in Humid Areas, In: Sopper, W.E. and H.W. Lull (Eds.), *Forest Hydrology.* Pergamon Press, Oxford. pp. 275–290.

Hewlett, J.D., 1961. *Watershed Management in Report for 1961 Southeastern Forest Experiment Station,* U.S. Forest Service, Ashville, N.C.

Hewlett, J.D., 1969. Tracing Storm Base Flow to Variable Source Areas on Forested Headwaters. Technical Report 2, School of Forest Resources, University of Georgia, Athens.

Hewlett, J.D., 1971. Comments on the Catchment Experiment to Determine Vegetal Effects on Water Yield. *Water Resour. Bull.,* 7, 376–377.

Hewlett, J.D., 1982. *Principles of Forest Hydrology.* University of Georgia Press, Athens. 183 pp.

Hillel, D.J., 1992. *Out of the Earth: Civilization and the Life of the Soil.* The Free Press, MacMillan, Inc, New York.

Hillel, D., 1982. *Introduction to Soil Physics.* Academic Press, Orlando, FL.

Hoag, B.L., 1982. Urbanization and its Effects on the Los Angeles River, unpublished paper, Department of Geography, UCLA.

Hodge, S.A., and G.D. Tasker, 1995. Magnitude and Frequency of Floods in Arkansas: U.S. Geological Survey Water-Resources Investigations Report 95-4224, 52 p.

Hole, F.D., 1981. Effects of Animals on Soils. Geoderma 25: 75–112.

Holtan, H.N. and M. Kirkpatick, 1950. Rainfall Infiltration and Hydraulics of Flow in Runoff Computation. *Trans. Am. Geophys. Union* 31; 771–779.

Holtan, H.N., N.E. Minshall, and L.L. Harrold, 1962. Field Manual for Research in Agricultural Hydrology. USDA Handbook No. 224. 214 pp.

Horton, R.E., 1939. Analysis of Runoff Plot Experiments with Varying Infiltration-capacity. *Trans. Am. Geophys. Union, Hydrology Papers,* 693–711.

Horton, R.E., 1945. Erosional Development of Streams and their Drainage Basins: Hydrophysical Approach to Quantitative Morphology. *Bull. Geol. Soc. Am.,* 56:275–370.

Horton, T., and W.M. Eichbaum, 1991. *Saving the Chesapeake Bay,* The Chesapeake Bay Foundation, Island Press, Washington, D.C.

Hosking, J.R.M. and J.R. Wallis, 1993. Some Statistics Useful in Regional Frequency Analysis. *Water Resour. Res.,* 29(2):271–281.

Hoskins, W., 1979. Shoshone National Forest Stream Classification System. USDA Forest Service Shoshone National Forest.

Huber, W.C., J.P. Heaney, S.J. Nix, R.E. Dickinson, and D.J. Polmann, 1981. *Storm Water Management Model User's Manual Version III,* EPA-600/2-84-109a, Environmental Protection Agency, Athens, GA.

Huff, F.A., 1967. Time Distribution of Rainfall in Heavy Storms. *Water Resour. Res.* 3(4):1007–1019.

Hundley, N., Jr., 1992. *The Great Thirst: Californians and their Water.* University of California Press, Berkeley.

Hwang, N.H.C., and C.E. Hita, 1987. *Fundamentals of Hydraulic Engineering Systems,* 2nd ed. Prentice-Hall, Inc., Englewood Cliffs.

Hynes, H.B.N., 1970. *The Ecology of Running Waters.* Liverpool University Press.

Ikeda, S. and G. Parker, 1989. River Meandering. Water Resources Monograph 12, American Geophysical Union, Washington D. C.

Imeson, A.C., J. Verstraten, E. Van Mullingen, and J. Sevink, 1992. The Effects of Fire and Water Repellency on Infiltration and Runoff under Mediterranean Type Forests. *Catena,* 19: 1177–1182.

International Commission on Irrigation and Drainage, 1987. Editorial. *ICID Bull.* 38(2), ICID, New Delhi, India

International Institute for Applied Systems Analysis, 2002. Global-AEZ 2000 Website http://www.iiasa.ac.at/collections/IIASA_Research/Research/LUC/GAEZ/index.htm.

Iseri, K.T. and W.B. Langbein, 1974. Large Rivers of the United States. U.S. Geological Survey Circular 686, Washington, D.C.

James, L.A., 1989. Sustained Storage and Transport of Hydraulic Gold Mining Sediment in the Bear River, California. *Ann. Assoc. Am. Geographers.* 37: 570–592.

Jarvis, R.S., and M.J. Woldenberg (Eds.), 1984. *River Networks. Benchmark Papers in Geology. Vol. 80,* Hutchinson Ross, Stroudsburg, PA.

Jenkins, C.T., 1970. *Computation of Rate and Volume of Stream-flow Depletion: U.S. Geological Survey Techniques of Water Resources Investigations,* Book 4, Chapter Dl, 17 p.

Jennings, M.E., W.O. Thomas, Jr., and H.C. Riggs, 1994. Nationwide Summary of U.S. Geological Survey Regional Regression Equations for Estimating Magnitude and Frequency of Floods for Ungaged Sites, 1993: U.S. Geological Survey Water-Resources Investigations Report 94-4002, 196 p.

Jensen, J., 1997. *Introductory Digital Image Processing: A Remote Sensing Perspective.* Prentice Hall, Upper Saddle River, NJ.

Jensen, M.E. and H.R. Haise, 1963. Estimating Evapotranspiration from Solar Radiation. *J. Irrig. and Drainage Div., ASCE,* 89: 15–41.

Jensen, M.E., J.L. Wright, and B.J. Pratt, 1971. Estimating Soil Moisture Depletion from Climate, Crop and Soil Data. *Trans. ASAE,* 14:954–959.

Jensen, M.E., R.D. Burman, and R.G. Allen, 1990. *Evapotranspiration and Irrigation Water Requirements.* American Society of Civil Engineers, New York. 332 pp.

Johnson, A.I., 1967. *Specific Yield-compilation of Specific Yields for Various Materials,* U.S. Geological Survey Water-Supply Paper 1662-D, 74 p.

Johnson, C., 2003. Five Low-Cost Methods for Slowing Streambank Erosion. *J. Soil Water Conserv.* 58(1):12A–17A.

Johnson, P.A. R.D. Hey, E.R. Brown, and D.L. Rosgen, 2002. Stream Restoration in the Vicinity of Bridges, *J. Am. Water Resour. Assoc.,* 38(1):55–67.

Johnson, W.C., 1976. The Impact of Environmental Change on Fluvial Systems: Kickapoo River, Wisconsin, unpublished PhD dissertation. University of Wisconsin, Madison.

Johnston, R.S., 1970. Evapotranspiration from Bare, Herbaceous, and Aspen Plots: A Check on a Former Study. *Water Resour. Res.,* 6(1):324–327.

Jorgensen, E.P. (Ed.), 1989. *New Strategies for Groundwater Protection: The Poisoned Well. Sierra Club Legal Defense Fund,* Island Press, Washington, D.C.

Junk, W.J., P.B. Bayley, and R.E. Sparks, 1989. The Flood Pulse Concept in River-floodplain Systems. In: D.P. Dodge (Ed.), Proceedings of the International Large River Symposium. Canadian Special Publication of Fisheries and Aquatic Sciences 106:110–127.

Jurmu, M.C. and R. Andrle, 1997. Morphology of a Wetland Stream. Environmental Management 21, No. 6, pp. 921–941.

Jury, W.A., W.R. Gardner, and W.H. Gardner, 1991. *Soil Physics* (5th edition). John Wiley & Sons, New York.

Kadlec, R.H. and R.L. Knight, 1996. *Treatment Wetlands.* Lewis Publishers, Boca Raton, FL, 893 p.

Karr, J.R., 1981. Assessment of Biological Integrity Using Fish Communities. *Fisheries.* 6:21–27.

Karr, J.R., 1987. Biological Monitoring and Environmental Assessment: a Conceptual Framework. *Environ. Manage.* 11:249–256.

Karr, J.R., and I.J. Schlosser, 1978. Water Resources and the Land Water Interface. *Science.* 201:229–234.

Karr, J.R., K.D. Fausch, P.L. Angermeier, P.R. Yant, and I.J. Schlosser, 1986. Assessing Biological Integrity in Running Waters: a Method and its Rationale. Illinois Natural Hististory Surv. Special Publication 5.

Kennedy, E.J., 1984. Discharge Ratings at Gaging Stations. U.S. Geological Survey, Series Techniques of water-resources investigations of the United States Geological Survey. Book 3, *Applications of Hydraulics.* Chap. A10

Kent, C.A., 1999. The Influence of Changes in Land Cover and Agricultural Land Management Practices on Baseflow in Southwest Wisconsin, 1969–1998. unpublished PhD. dissertation. University of Wisconsin, Madison.

Kessler, B., 1995. Glossary of GIS Terms. In: Lyon, J. and J. McCarthy, *Wetland and Environmental Applications of GIS.* Lewis, Boca Raton. pp. 331–373.

Kincaid, D.C., 1986. Intake Rate: Border and Furrow. In: A. Klute (Ed.), *Methods of Soil Analysis, Part 1: Physical and Mineralogical Methods,* 2nd edition. American Society of Agronomy, Madison, WI.

Kinnell, P.I.A. and D. Cummings, 1993. Soil/slope Gradient Interactions in Erosion by Rain-impacted Flow. *Trans. ASAE,* 36(2):318–387.

Kirkby, M.J. (Ed.), 1978. *Hillslope Hydrology.* John Wiley & Sons, New York.

Kirkby, M.J., and R.J. Chorley, 1967. Throughflow, Overland flow, and Erosion, *Int. Assoc. Hydrol. Sci. Bull.* 12:5–21.

Kirpich, P.Z., 1940. Time of Concentration of Small Agricultural Watersheds. *Civil Eng.* 10:362.

Kittredge, J. 1948. *Forest Influences.* McGraw-Hill, New York. 360 pp.

Klint, K.E.S. and C.D. Tsakiroglou, 2000. A New Method of Fracture Aperture Characterisation. In: V.A. Tsihrintzis, G.P. Korfiatis, K.L. Katsifarakis, A.C. Demetracopoulos (Eds.), *Protection and Restoration of the Environment V.* Proceedings of an International Conference, Thassos, Greece, Vol 1. pp. 127–136.L

Klute, A. and C. Dirksen, 1986. Hydraulic Conductivity and Diffusivity: Laboratory Methods. Ch. 28 In: A. Klute (Ed.), *Methods of Soil Analysis, Part 1: Physical and Mineralogical Methods,* 2nd Edition. American Society of Agronomy, Madison, WI.

Knighton, D., 1984. *Fluvial Forms and Processes.* Edward Arnold Ltd., London.

Knisel, W.G. (Ed.), 1980. *CREAMS: A Field-Scale Model for Chemicals, Runoff, and Erosion from Agricultural Management Systems.* Conservation Research Report 26. USDA-Science and Education Administration, Washington, D.C.

Knox, J.C. and J.C. Hudson, 1995. Physical and Cultural Change in the Driftless Area of Soutwest Wisconsin, In: M. Conzen (Ed.), *Geographical Excursions in the Chicago Region.* Association of American Geographers, Washington, D.C. pp. 17–131.

Knox, J. C., 1977. Human Impacts on Wisconsin Stream Channels. *Ann. Assoc. Am. Geographers,* 67: 323–42.

Knox, J.C., 1999. Long-term Episodic Changes in Magnitudes and Frequencies of Floods in the Upper Mississippi River Valley. In: A.G. Brown and T. Quine, (Eds.). *Fluvial Processes and Environmental Change.* John Wiley & Sons, Chichester. pp. 255–282.

Kohler, M.A., T. Nordenson, and D. Baker, 1959. Evaporation Maps for the United States. U.S. Weather Bureau Technical Paper 37.

Kollar, K.L., and P. MacAuley, 1980. Water Requirements of Industrial Development, *J. Am. Water Works Assoc.,* (72)1.

Koltun, G.F. and J.W. Roberts, 1990. Techniques for Estimating Flood-Peak Discharges of Rural, Unregulated Streams in Ohio. U.S. Geological Survey, Water Resources Investigations Report 89-4126, USGS Denver, CO.

Kondolf, G.M., 1995. Geomorphological Stream Channel Classification in Aquatic Habitat Restoration: Uses and Limitations. *Aquatic Conservation: Marine and Freshwater Ecosystems,* 5:127–141.

Kraijenhoff, D.A. and J.R. Moll, 1986. *River Flow Modelling and Forecasting.* D. Reidel Publishing, Dordrecht.

Kramer, P.J., 1983. *Water Relations of Plants.* Academic Press, Inc. Orlando, FL. 489 pp.

Krug, W., 1996. Simulation of Temporal Changes in Rainfall-runoff Characteristics, Coon Creek Wisconsin. Water Resources Bulletin 32-745-752.

Ladle, M., 1991. Running Waters: a Conservationist's Nightmare. In: I.F. Spellberg, F.B. Goldsmith, and M.G. Morris (Eds.), *The Scientific Management of Temperate Communities for Conservation.* 31st Symposium of the British Ecological Society. Blackwell, Oxford.

Laflen, J.M., W.J. Elliot, J.R. Simanton, C.S. Holzhey and K.D. Kohl, 1991. WEPP Soil Erodibility Experiments for Rangeland and Cropland Soils. *J. Soil Water Conserv.* 46(1):39–44.

Lal, R., J.M. Kimble, R.F. Follett, and C.V. Cole. 1998. *The Potential of U.S. Cropland to Sequester Carbon and Mitigate the Greenhouse Effect.* Ann Arbor Press, MI, 128 pages.

Lane, E.W., 1955a. Design of Stable Channels. *Trans. Am. Soc. Civil Eng.* 120:1234–60.

Lane, E.W., 1955b. The Importance of Fluvial Morphology in Hydraulic Engineering, *Am. Soc..Civil Eng., Proc.,* 81, paper 745:1–17.

Lane, L.J., and M.A. Nearing, 1989. USDA-Water Erosion Prediction Project: Hill-slope Profile Model Documentation. NSERL Report No. 2. USDA-ARS National Soil Erosion Research Laboratory. West Lafayette, IN.

Langbein, W.B., 1959. Water Yields and Reservoir Storage in the United States. U.S. Geological Survey Circular 409.

Lara, J.M. and E.L. Pemberton, 1963. Initial Unit Weight of Deposited Sediments, *Proc. Federal Inter-Agency Sediment Conf.,* U.S. Department of Agriculture Misc. Publication 970.

Lee, C.T. and S. Marsh, 1995. The Use of Archival Landsat MSS and Ancillary Data in a GIS Environment to Map Historical Change in an Urban Riparian Habitat, In: Lyon, J. and J. McCarthy (Eds.), *Wetland and Environmental Applications of GIS.* Lewis Publishers, Boca Raton, FL. 373 pp.

Lee, K. and R. Lunetta, 1995. Wetland Detection Methods. In: Lyon, J. and J. McCarthy (Eds.), Wetland and Environmental Applications of GIS. Lewis Publishers, Boca Raton, FL. 373 pp.

Leonard, R.A., W.G. Knisel, and D.A. Still, 1987. GLEAMS: Groundwater Loading Effects of Agricultural Management Systems. Trans. ASAE. 30(5):1403–1418.

Leopold, L.B., 1994. *A View of the River.* Harvard University Press. Cambridge.

Leopold, L.B. and M.G. Wolman, 1957. River Channel Patterns — Braided, Meandering and Straight. Professional Paper 282-B, U.S. Geological Survey.

Leopold, L.B., M.G. Wolman, and J.P. Miller, 1964. *Fluvial Processes in Geomorphology.* W. H. Freeman, San Francisco.

Leopold, L.B., M.G. Wolman, and J.P. Miller, 1995. *Fluvial Processes in Geomorphology,* Dover, New York.

Liebenow, A.M., W.J. Elliot, J.M. Laflen, and K.D. Kohl, 1990. Interrill Erodibility: Collection and Analysis of Data from Cropland Soils. *Trans. ASAE* 33(6):1882–1888.

Lighthill, F.R.S. and C.B. Whitham, 1955. On Kinematic Waves, Flood Movement in Long Rivers. *Proc. R. Soc. London,* Vol. 229 (1178):281–316.

Lowdermilk, W.C., 1953. Conquest of the Land Through 7,000 Years. Agric. Information Bulletin No. 99. USDA SCS, Washington, D.C.

Luce, C.H. and T.W. Cundy, 1994. Parameter Identification for a Runoff Model for Forest Roads. *Water Resour. Res.,* 30(4):1057–1069.

Lull, H.W., 1964. Ecological and silvicultural aspects. In: Chow,V.T., *Handbook of Applied Hydrology,* McGraw-Hill, New York. Section 6. 30pp.

Lull, H.W. and K.G. Reinhart, 1972. Forests and Floods in the Eastern United States. USDA Forest Service Research Paper NE-226. Upper Darby, PA.

Lunetta, R. and C. Elvidge, 1998. *Remote Sensing Change Detection.* Taylor and Francis, UK, 318 pp.

Lunetta, R. and J. Lyon, 2004. Geospatial Data Accuracy Assessment, EPA Report EPA/600/R-03/064, Las Vegas.

Lunetta, R., J. Lyon, J. Sturdevant, J. Dwyer, C. Elvidge, L. Fenstermaker, D. Yuan, S. Hoffer and R. Weerackoon, 1993. North American Landscape Characterization (NALC), Technical Research Plan. USEPA EPA/600/R-93/135, Environmental Systems Monitoring Laboratory, Las Vegas.

Lyon, J., 1987. Maps, Aerial Photographs and Remote Sensor Data for Practical Evaluations of Hazardous Waste Sites. Photogrammetric Engineering and Remote Sensing, 53:515–519.

Lyon, J., 1993. *Practical Handbook for Wetland Identification and Delineation.* Lewis, Chelsea, MI, 157 pp.

Lyon, J., 2001. *Wetland Landscape Characterization: GIS, Remote Sensing, and Image Analysis.* Taylor and Francis, UK, 135 pp.

Lyon, J., 2003. *GIS for Water Resources and Watershed Management*. Taylor and Francis, UK, 266 pp.

Lyon, J. and I. Khumwaiter, 1989. Cropland Measurements Using Thematic Mapper Data and Radiometric Model. *ASCE J. Aerosp. Eng.*, 2:130–140.

Lyon, J. and J. McCarthy, 1995. *Wetland and Environmental Applications of GIS*. CRC/Lewis, Boca Raton, FL, 373 pp.

Lyon, J., E. Falkner, and Bergen, 1995. Cost Estimating Photogrammetric and Aerial Photography Services. *ASCE J. Surv. Eng.*, 121:63–86.

Lyon, J., K. Bedford, J. Chien-ching, D. Lee, and D. Mark, 1988. Suspended Sediment Concentrations as Measured from Multidate Landsat and AVHRR data. Remote Sensing of Environment, 25:107–115.

Lyon, J., R. Drobney, and C. Olson, 1986. Effects of Lake Michigan Water Levels on Wetland Soil Chemistry and Distribution of Plants in the Straits of Mackinac. *J. Great Lakes Res.*, 12:175–183.

Lyon, J.G., 1993. *Practical Handbook for Wetland Identification and Delineation*. Lewis, Boca Raton, FL. 157 pp.

MacDonald, G.M., 2002. *Biogegraphy*, John Wiley & Sons, New York.

Mackin, J.H., 1948. Concept of the Graded River. *Bull. Geol. Soc. Am.*, 59:463–512.

MacRae, C., 1993. An Alternative Design Approach for the Control of Instream Erosion Potential in Urbanizing Watersheds. In: Proceedings of the Sixth International Conference on Urban Storm Drainage. Marsalek and Torno (Eds). Niagara Falls, Ontario, pp. 1086–1091.

MacRae, C., 1996. Experience from Morphological Research on Canadian Streams: Is Control of the Two-year Frequency Runoff Event the Best Basis for Stream Channel Protection In: Larry Roesner (Ed.), *Effects of Watershed Development and Management on Aquatic Systems*. Engineering Foundation Conference Proceedings, Snowbird, Utah, August 4–9, pp. 144–160.

Madison, R.J. and J.O. Brunett, 1985. Overview of the Occurrence of Nitrate in Ground Water in the United States. National Water Summary 1984. Water Supply Paper 2275, U.S. Geological Survey, Washington, D.C., pp. 93–105.

Magalhaes, L. and T.S. Chau, 1983. Initiation of Motion Conditions for Shale Sediments. *Can. J Civil Eng.*, 10(3):549–554.

Magilligan, F.J., 1985. Historical floodplain sedimentation in the Galena River basin, Wisconsin and Illinois. *Ann. Assoc. Am. Geographers*, 75:583–594.

Maidment, D., 1993. *Handbook of Hydrology*. McGraw-Hill, New York.

Maidment, D. and D. Djokic, 2000. *Hydrologic and Hydraulic Modeling Support with Geographic Information Systems*. ESRI Press, Redlands, CA. 216 pp.

Magner, J. and L. Steffen, 2000. *Steam Morphological Response to Climate and Land-Use in the Minnesota River Basin*, ASCE Resources Engineering, Planning and Mangement, Reston, VA.

Manning, J.C., 1997. *Applied Principles of Hydrology*, 3rd ed. Prentice Hall, Upper Saddle River, New Jersey. 276 pp.

Marks, D. and J. Dozier, 1992. Climate and Energy Exchange at the Snow Surface in the Alpine Region of the Sierra Nevada. *Water Resour. Res.,* 28(11):3043–3054.

Mather, J.R., 1978. *The Climatic Water Budget in Environmental Analysis*. Lexington Books, Lexington, MA.

Mavis, F.T., 1942. The Hydraulics of Culverts, The Pennsylvania State College, Engineering Experiment Station, Bulletin 56, Oct. 1.

McBurnie, J.C., B.J. Barfield, M.L. Clar, and E. Shaver, 1990. Maryland Sediment Detention Pond Design Criteria and Performance. *Appl. Eng. Agric.*, 6(2):167–173.

McCuen, R.H., 1989. Hydrologic Analysis and Design. Prentice-Hall., Totowa, NJ, 867pp.

McCuen, R.H., S.L. Wong, and W.J. Rawls, 1983. Estimating the Time of Concentration of Urban Watersheds. Proceedings of the ASCE Conference on Frontiers in Hydraulic Engineering, Cambridge, MA, August 9–12, pp. 547–552.

McGraw-Hill, 1974. *Encyclopedia of Environmental Science*. McGraw-Hill, New York. 754pp.

McKay, L.D., J.A. Cheery, and R.W. Gillham, 1993. Field Experiments in a Fractured Clay Till: 1. Hydraulic conductivity and fracture aperture. *Water Resour. Res.*, 29(4): 1149–1162.

McKelvey, V.E., 1939. Stream and Valley Sedimentation in the Coon Creek Drainage Basin, Wisconsin. unpublished M.S. Thesis, University of Wisconsin, Madison.

Meade, R.H., 1982. Sources, Sinks, and Storage of River Sediment in the Atlantic Drainage of the United States. *J. Geol.*, 90, 235–252.

Meade, R.H., T. Yuzyk, and T. Day, 1990. Movement and Storage of Sediment in Rivers of the United States and Canada, In: M.G. Wolman and H.C. Riggs (Eds.), *Surface Water Hydrology, Geology of North America*. Geological Society of America., Boulder, pp. 255–280.

Mecklenburg, 2002. Personal Communication.

Mecklenburg, D., 1999. Typical Natural Channel Dimensions. Course Materials for the Rosgen Course, April 12–16, Lake County, Ohio. Ohio Department of Natural Resources.

Mecklenburg, D. and A. Ward, 2002. Quantifying and Managing the Impacts of Urbanization on the Effective Discharge and Stream Stability. Electronic Proceedings, Environmental Flows for River Systems, Cape Town, South Africa, March 3–8.

Mehaffey, M.H., M.S. Nash, T.G. Wade, C.M. Edmonds, D.W. Ebert, K.B. Jones, and A. Rager, 2001. A Landscape Assessment of the Catskill/Delaware Watersheds 1975–1998: New York City's Water Supply Watershed. EPA/600/R-01/075 (September). U. S. Environmental Protection Agency, Office of Research and Development. Las Vegas. 117 pp.

Mein, R.G. and C.L. Larson, 1973. Modeling Infiltration During a Steady Rain. *Water Resour. Res.* 9(2):384–394.

Melton, F., 1936, An Empirical Classification of Flood-plain Streams. *Geogr. Rev.,* 26:593–609.

Mendenhall, W. and T. Sincich, 1995. Statistics for Engineering and the Sciences, 4th ed. Prentice Hall, Upper Saddle River, NJ, 1182 pp.

Merva, G.E., 1979. Falling Head Permeameter for Field Investigation of Hydraulic Conductivity. ASAE paper No. 79-2515, American Society of Agricultural Engineers, St. Joseph, MI.

Meyer, A.F., 1915. Computing Run-Off from Rainfall and Other Physical Data. *Trans. ASCE*, 79:1056–1155.

Miller, I., and M. Miller, 1995. *Statistical Methods for Quality With Applications to Engineering and Management.* Prentice Hall, Englewood Cliffs, NJ. 368 pp.

Miller, J.F., 1964. Two-to Ten-Day Precipitation for Return Periods of 2 to 100 Years in the Contiguous United States. US Weather Bureau Technical Paper 49.

Milman, A., 1999. *Mathematical Principles of Remote Sensing.* Taylor and Francis, UK, 406 pp.

Mitsch, W.J. and J.G. Gosselink, 2000. *Wetlands*, 3rd ed. John Wiley & Sons Inc., New York.

Monteith, J.L., 1981. Evaporation and Surface Temperature. *Q. J. Roy. Meteorol. Soc.*, 107:1–27.

Montgomery, D. R. amd L. H. MacDonald, 2002. Diagnostic Approach to Stream Channel Assessment and monitoring. JWRA, 1:1–16

Montgomery, D.R. and J.M. Buffington, 1997. Channel-reach Morphology in Mountain Drainage Basins. GSA Bulletin, 109(5):596–611.

Montgomery, D.R. and K.B. Gran, 2001. Downstream Variations in the Width of Bedrock Channels. *Water Resour. Res.*, 37(6):1841–1846.

Montgomery, D.R., T.B. Abbe, J.M. Buffington, N.P. Peterson, K.M. Schmidt, and J.D. Stock, 1996. Distribution of Bedrock and Alluvial Channels in Forested Mountain Drainage Basins. *Nature* , 381:587–589.

Moody, W.T., 1966. Nonlinear Differential Equation of Drain Spacing. *Proc. ASCE*, 92(R2):1–9.

Morisawa, M., 1968. *Streams: Their Dynamics and Morphology.* McGraw-Hill, New York.

Mualem, Y., 1986. Hydraulic Conductivity of Unsaturated Soils: Prediction and Formulas. Ch. 31 In: A. Klute (ed.), *Methods of Soil Analysis, Part 1: Physical and Mineralogical Methods*, 2nd ed. American Society of Agronomy, Madison, WI.

Murray, R., 1952. Rain and Snow in Relation to the 100-700mb and 1000-500mb Thicknesses and the Freeezing Level. *The Meterolog. Mag.* 81, 5–8.

Myers, N. 1991. Trees by the Billions. International Wildlife Magazine. Sept/Oct, 12–15.

Myers, V.A. and R.M. Zehr, 1980. A Methodology for Point-to-Area Rainfall Frequency Ratios. NOAA Technical Report NWS 24. National Oceanic and Atmospheric Admin., U.S. Department of Commerce, Washington, D.C.

Nash, D.B., 1994. Effective Sediment-Transporting Discharge from Magnitude Frequency Analysis. *J. Geol.*, 102: 79–95.

National Research Council, 1992. Water Transfers in the West, Efficiency, Equity, and the Environment. National Academy Press, Washington, D. C.

National Research Council, 1999. New Strategies for America's Watersheds. National Academy Press, Washington, D.C.

Natural Resources Conservation Service, 1974. Engineering Field Manual (including Ohio Supplement). U. S. Department of Agriculture, Washington D. C.

Nature Conservancy, 1990. The Darby Book. A Guide for Residents of The Darby Creek Watershed. Nature Conservancy-Ohio Chapter, Columbus, OH, p 28.

NCCI, 1986. DRAINMOD: Documentation for the Water Management Simulation Model, *Software J.* 2(1), North Central Computer Institute, Madison, WI.

Nearing, M.A., G.R. Foster, L.J. Lane, and S.C. Finkner, 1989. A Process-based Soil Erosion Model for USDA-Water Erosion Prediction Project Technology. *Trans. ASAE*, 32(5):1587–1593.

Newbury, R.W. and M.N. Gaboury, 1993. *Stream Analysis and Fish Habitat Design.* Newbury Hydraulics Ltd. British Columbia, Canada. 262 pp

Nobel, P.S., 1983. *Biophysical Plant Physiology and Ecology.* W.H. Freeman and Company, New York.

Nolte, B.H. and G.G. Schwab, 1971. An Alternate to Channelization. *OH Agric. Res. and Devel. Center, Ohio Report* 56(5):70–71.

Norris, S.E. and R.E. Fidler, 1969. *Hydrogeology of the Scioto River Valley Near Piketon, South-Central Ohio.* U.S. Geological Survey Water-Supply Paper 1872. U.S. Government Printing Office, Washington D.C., 71.

Nortz, P.E., E.S. Bair, A.D. Ward, and D. White, 1994. Interactions Between an Alluvial-Aquifer Wellfield and the Scioto River, Ohio USA. *App. Hydrol.*, 4, 23–34.

OEPA, 1988a. *Biological Criteria for the Protection of Aquatic Life, Volume I: Role of Biological Data in Water Quality Assessment.* Ohio Environmental Protection Agency, Division of Water Quality Monitoring and Assessment, Surface Water Section, Columbus.

OEPA, 1988b. *Biological Criteria for the Protection of Aquatic Life, Volume II: Users Manual for Biological Field Assessment of Ohio Surface Waters.* Ohio Environmental Protection Agency, Division of Water Quality Monitoring and Assessment, Surface Water Section, Columbus.

OEPA, 1989. *Biological Criteria for the Protection of Aquatic Life. Volume III: Standardized Biological Field Sampling and Laboratory Methods for Assessing Fish and Macroinvertebrate Communities.* Ohio Environmental Protection Agency, Division of Water Quality Monitoring and Assessment, Columbus.

Ohio State University Extension, 2000. Stormwater and Your Community, Factsheet AEX-442. Columbus.

Osborne, L.L. and M.J. Wiley, 1988. Empirical Relationships Between Land Use/Cover and Stream Water Quality in an Agricultural Watershed. *J. Environ. Manage.*, 26: 9–27.

Ott, L., 1984. *An Introduction to Statistical Methods and Data Analysis,* 2nd ed., Prindle, Webber and Schmidt, Boston.

Overton, D.E. and E.C. Crosby, 1979. *Effects of Contour Coal Strip Mining on Stormwater Runoff and Quality.* Report to the U.S. Department of Energy. Civil Engineering Department, University of Tennessee, Knoxville.

Overton, D.E. and W.L. Troxler, 1978. Regionalization of Stormwater Response. Paper presented at the American Geophysical Union Meeting, Miami, April 17–21.

Palmer, L., 1976. River Management Criteria for Oregon and Washington, In: Coates, D.R. (Ed.), *Geomorphology and Engineering,* Allen and Unwin, London, 329–346.

Patrick, R., 1998. *Rivers of the United States* (Vols. I-IV). John Wiley & Sons, New York.

Patrick R., E. Ford and J. Quarles, 1987. *Groundwater Contamination in the United States*, 2nd ed., University of Pennsylvania Press, Philadelphia.

Penman, H.L., 1948. Natural Evaporation from Open Water, Bare Soil and Grass. *Proc. Roy. Soc. London.* A193: 120–146.

Penman, H.L., 1956. Evaporation: An Introductory Survey. *Neth. J. Agric. Sci.,* 1:9–29, 87–97, 151–153.

Penman, H.L., 1963. Vegetation and Hydrology. Technical Communication 53, Commonwealth Bureau of Soils, Harpenden, UK, 125 pp.

Peters, J.G., 1987. Description and Comparison of Selected Models for Hydrologic Analysis of Ground-water Flow, St. Joseph River Basin, Indiana: U.S. Geological Survey Water-Resources Investigations Report 86-4199, 125 p.

Peterson, A.E. and G.D. Bubenzer, 1986. Intake Rate: Sprinkler Infiltrometer. In: A. Klute (ed.), *Methods of Soil Analysis, Part 1: Physical and Mineralogical Methods,* 2nd ed. American Society of Agronomy, Madison, WI.

Petts, G. and P. Calow, 1996. *River Restoration.* Blackwell Science, Oxford, UK.

Pimentel, D, C. Harvey, P. Resosudarmo, K. Sinclair, D. Kurz, M. McNair, S. Crist, L. Shpritz, L. Fitton, R. Saffouri, and R. Blair, 1995. Environmental and Economic Costs of Soil Erosion and Conservation Benefits. *Science,* 267:1117–1123.

Plafkin, J.L., M.T. Barbour, K.D. Porter, S.K. Gross, and R.M. Hughes, 1989. *Rapid Bioassessment Protocols for Use in Streams and Rivers: Benthic Macroinvertebrates and Fish.* U.S. EPA, EPA/444/4-89-011, Washington D.C.

Posey, C.J., 1967. Computation of Discharge including Over-Bank Flow, *Civ. Eng.,* ASCE, April, 62–63.

Postel, S., 1984. *Water: Rethinking Management in a Age of Scarcity.* Worldwatch Paper 62, Worldwatch Institute, Washington, D.C.

Postel, S., 1992. *Last Oasis: Facing Water Scarcity.* Worldwatch Environmental Alert Series. W.W. Norton, New York, 239 pp.

Potter, K., 1991. Hydrological Impacts of Changing Land Management Practices in a Moderate-sized Agricultural Catchment. *Water Resour. Res.,* 27:845–855.

Price, A.P., 1998. The Effect of Climate and Land Use on the Hydrology of the Upper Oconee River Basin, Georgia. unpublished PhD dissertation, University of California, Los Angeles.

Price, M., 1985. *Introducing Groundwater,* George Allen and Unwin, Boston. 195 p.

Puckett, W. E., J. H. Dane, and B. F. Hajek, 1985. Physical and Mineralogical Data to Determine Soil Hydraulic Properties. Soil Sci. Soc. Am. J. 49: 831–836.

Pyne, S.J., 1982. *Fire in America. A Cultural History of Wild Land on Rural Fire.* Princeton University Press.

Pyne, S.J., 1997. *Fire: A Brief History,* University of Washington Press, reprint ed., 680 pp.

Quisenberry, V.L. and R.E. Phillips, 1976. Percolation of Surface-Applied Water in the Field. *Soil Sci. Soc. Am. J.,* 40:484–489.

Ramsey, R., 1998. Radar Remote Sensing of Wetlands. In: Lunetta, R. and C. Elvidge, (Eds.), *Remote Sensing Change Detection,* Ann Arbor Press, Chelsea, MI, pp. 211–243.

Rankin, E.T. 1989. *The Qualitative Habitat Evaluation Index (QHEI): Rationale, Methods, and Application.* Ohio Environmental Protection Agency, Division of Surface Water, Ecological Assessment Section, Columbus.

Rantz, S.E. et al., 1982. *Measurement and Computation of Streamflow.* U.S. Geological Survey Water-Supply Paper 2175, 2 volumes., 631 p.

Rawls, W.J., D.L. Brakensiek, and K.E. Saxton, 1982. Estimation of Soil Water Properties. *Trans. ASAE.* 25(5):1316–1320, 1328.

Reid, L.M. and T. Dunne, 1996. *Rapid Evaluation of Sediment Budgets.* Catena Verlag, Reiskirchen, Germany.

Reinhart, D.R. and T.G. Townsend, 1998. *Landfill Bioreactor Design and Operation.* Lewis, Boca Raton, FL.

Renard, K.G., G.R. Foster, F.A. Weesies, and J.P. Porter, 1991. RUSLE Revised Universal Soil Loss Equation. *J. of Soil Water Conserv.* 46(1):30–33.

Renfro, G.W. 1975. Use of Erosion Equations and Sediment-Delivery Ratios for Predicting Sediment Yield. In: *Present and Prospective Technology for Predicting Sediment Yields and Sources.* ARS-S-40, USDA-ARS.

Renwick, W.H., 1996, Continental-scale Reservoir Sedimentation Patterns in the United States, In: Walling,D. and B. Webb (Eds.) *Erosion and Sediment Yield: Global and Regional Perspectives.* IASH Pub. 236, pp. 513–522.

Reynolds, W.D. and D.E. Elrick, 1985. In-situ Measurement of Field-saturated Hydraulic Conductivity, Sorptivity, and the *A*-parameter Using the Guelph Parmeameter. *Soil Sci..* 140(4):292–302.

Rhoads, B.L. and E.E. Herricks, 1996. Naturalization of Headwater Agricultural Streams in Illinois: Challenges and Possibilities, In: Brookes, A. and D. Shields (Eds.), *River Channel Restoration,* John Wiley & Sons, Chichester. pp. 331–367

Rhoads, B.L., 1995. Stream Power: A Unifying Theme for Urban Fluvial Geomorphology. In: E.E. Herricks (Ed.). *Stormwater Runoff and Receiving Systems. Impact, Monitoring, and Assessment.* Lewis, Boca Raton, FL, pp. 65–73.

Richards, K., 1982. *Rivers, Form and Process in Alluvial Channels.* Methuen & Co., New York.

Ries, K.G., III, and M.Y. Crouse, 2002. The National Flood Frequency Program, Version 3: A Computer Program for Estimating Magnitude and Frequency of Floods for Ungaged Sites: U.S. Geological Survey Water-Resources Investigations Report 02-4168, 42 p.

Riley A.L., 1998. *Restoring Streams in Cities: A Guide for Planners, Policymakers, and Citizens.* Island Press, Washington, D.C. 423 pp.

Riley, J.P., E.K. Israelsen, and K.O. Eggleston, 1973. Some Approaches to Snowmelt Prediction, In: *The Role of Snowmelt and Ice in Hydrology.* International Association of Hydrological Sciences, 107(2):956–971.

Roberts, W.J., 1969. Significance of Evaporation in Hydrologic Education. In: *The Progress of Hydrology. Proceedings of The First International Seminar for Hydrology Professors, Volume II, Specialized Hydrologic Subjects.* p. 672.

Roehl, J.N., 1962. Sediment Source Areas, Delivery Ratios and Influencing Morphological Factors. International Association Scientific Hydrology, Commission of Land Erosion. Publication 59. pp. 202–213.

Roesner, L.A. et al., 1991. Hydrology of Urban Runoff Quality Management. Proceedings 18th Conference on Water Resources Planning and Management. ASCE, New Orleans, LA.

Rosegrant, M.W., X. Cai, and S.H. Cline. *World Water and Food to 2025: Dealing with Scarcity.* International Food Policy Research Institute, Washington D.C., 322 pp.

Rosgen, D., 1994. A Classification of Natural Rivers. *Catena,* 22:169–199.

Rosgen, D., 1996. *Applied River Morphology.* Wildland Hydrology. Pagosa Springs, CO.

Rosgen, D.L., 1998. The Reference Reach-A-Blueprint for Natural Channel Design. Proceedings ASCE Wetlands Engineering and River Restoration Conference. March, Denver, CO. http://www.wildlandhydrology.com/

Sanders, R.E. (Ed.), 2001. *A Guide to Ohio Streams.* Ohio Chapter of the American Fisheries Society, Columbus.

Sauer, V. B., W.O. Thomas, Jr., V.A. Stricker, and K. Wilson, 1983. *Flood Characteristics of Urban Watersheds in the United States.* USGS Water-Supply Paper 2207, Washington D.C.

Schertz, D.L., W.C. Moldenhauer, S.J. Livingston, G.A. Weesies, and E.A. Hintz, 1989. Effect of Past Soil Erosion on Crop Productivity in Indiana. *J. Soil Water Conserv. Soc.* 44(6):604–608.

Schott, J., 1997. *Remote Sensing, the Image Chain Approach.* Oxford University Press, New York, 394 pp.

Schueler, T. R., 1992. *Design of Stormwater Wetland Systems: Guidelines for Creating Diverse and Effective Stormwater Wetland Systems in the Mid-Atlantic Region. Anacostia Restoration Team,* Department of Environmental Programs, Metropolitan Washington Council of Governments, Washington D.C.

Schueler, T.R. and H.K. Holland (Eds.), 2000. *The Practice of Watershed Protection.* Center for Watershed Protection, Ellicott City, MD. 742 pp

Schumm, S.A., 1956. The Evolution of Drainage Systems and Slopes in Badlands at Perth Amboy, New Jersey. *Bull. Geolog. Soc. of Am.,* 67:597–646.

Schumm, S.A., 1977. The Fluvial System. Wiley-Interscience, New York.

Schumm, S.A., M.D. Harvey, and C.C. Watson, 1984. *Incised Channels: Morphology, Dynamics and Control.* Water Resources Publication, LLC. BookCrafters, Inc. Chelsea, MI, 200 pp.

Schwab, G.O., D.D. Fangmeier, W.J. Elliot, and R.K. Frevert. 1993. *Soil and Water Conservation Engineering,* 4th ed., John Wiley & Sons, Inc., New York. 507 pp.

Schwab, G.O., R.K. Frevert, T.W. Edminster, and K.K. Barnes, 1966. *Soil and Water Conservation Engineering,* 2nd ed., John Wiley & Sons, New York.

Shakesby, R.A., S. Doerr, R. Walsh, 2000. The Erosional Impact of Soil Hydrophobicity: Current Problems and Future Research Directions. *J. Hydrol.,* 232: 178–191.

Shaw, E.M. 1988. *Hydrology in Practice,* Van Nostrand-Reinhold, London.

Sherman, L.K., 1932. Stream Flow from Rainfall by Unit Graph Method, *Eng. News-Record,* 108:501–505.

Sherwood, J.M., 1986. Estimating Peak Discharge, Flood Volumes, and Hydrograph Shapes of Small Ungaged Urban Streams in Ohio. USGS Water Resources Investigation Report 86-4197, 52 pp.

Sherwood, J.M., 1993. Estimation of Peak-Frequency Relations, Flood Hydrographs, and Volume-Duration-Frequency Relations of Ungaged Small Urban Streams in Ohio. USGS Water Resources Open-File Report 93-135, 52 pp.

Sherwood, J.M., 1994. Estimation of Volume-Duration-Frequency Relations of Ungaged Small Urban Streams in Ohio. *AWRA Water Resour. Bull.* 30(2):261–269.

Shipitalo, M.J., W.A. Dick, and W.M. Edwards, 2002. Conservation Tillage and Macropore Factors that Affect Water Movement and the Fate of Chemicals. *Soil Tillage Res.,* 53:167–183.

Shreve, R.L., 1967. Infinite Topologically Random Channel Network. *J. Geol.,* 75:178–186.

Silver, C.S., and R.S. DeFries, 1990. *One Earth, One Future: Our Changing Global Environment.* National Academy of Sciences, National Academy Press, Washington, D.C.

Simeral, K.D., 1998. Using Constructed Wetlands for Removing Contaminants from Livestock Wastewater. The Ohio State University University Extension Factsheet A-5-98. Columbus.

Simon, A., 1989. A Model of Channel Response in Disturbed Alluvial Channels. *Earth Surface Processes and Landforms,* 14:11–26.

Simon, A., 1992. Energy, Time, and Channel Evolution in Catastrophically Disturbed Fluvial Systems, *Geomorphology,* 5:345–372.

Simon, A. and C.R. Hupp, 1986. Channel Evolution in Modified Tennessee Channels. In: *Proc. 4th Federal Interagency Sedimentation Conf.,* Las Vegas, U.S. Government Printing Office, Washington DC, 5.71–5.82.

Simon, A., and M. Rinaldi, 2000. Channel Instability in the Loess Area of the Midwestern United States. *J. AWRA,* 36(1):133–164.

Simon, A. and C.R. Thorne, 1996. Channel Adjustment of an Unstable Coarse-grained Stream: Opposing Trens of Boundary and Critical Shear Stress, and the Applicability of Extremal Hypotheses. *Earth Surface Processes and Landforms,* 21:155–180.

Singer, M.J. and D.N. Munns, 1999. *Soils: An Introduction,* 4th ed. Prentice-Hall, Upper Saddle River, NJ.

Singer, M.J. and D.N. Munns, 2002. *Soils: An Introduction,* 5th ed. Prentice-Hall, Upper Saddle River, NJ.

Skaggs, R.W., 1980. *Drainmod Reference Report. Methods for Design and Evaluation of Drainage-water Management Systems for Soils with High Water Tables.* North Carolina State University, Raleigh.

Slater, P., 1980. *Remote Sensing, Optics and Optical Systems.* Addison-Wesley, Reading, MA, 575 pp.

Slaymaker, O., 1993. The Sediment Budget of Lilliooet River Basin, British Columbia. *Phys. Geogr.*, 14: 304–320.

Smil, V., 2002. *Feeding the World: A Challenge for the Twenty-First Century.* The MIT Press, Cambridge, MA, 360 pp.

Smith, S.V., W. Renwick, J. Bartley, and R. Buddemeier, 2002. Distribution and Significance of Small, Artificial Water Bodies across the United States Landscape, *The Sci. Total Environ.*, 299:2–36.

Soil Conservation Service, 1972. *National Engineering Handbook, Section 4, Hydrology*, NRCS.

Soil Conservation Service, 1973. Computer Program for Project Formulation Hydrology.Technical Release No. 20, U.S. Department of Agriculture, Washington, D.C.

Soil Conservation Service, 1984. *Engineering Field Manual (including Ohio Supplement)*. U.S. Department of Agriculture, Washington, D.C.

Soil Conservation Service, 1986. *Urban Hydrology for Small Watersheds*, 2nd ed. Technical Release 55, U.S. Department of Agriculture, Washington, D.C.

Steiner, G.R. and R.J. Freeman, 1989. Configuration and Substrate Design Considerations for Constructed Wetlands Wastewater Treatment. In: *Constructed Wetlands for Wastewater Treatment: Municipal, Industrial and Agricultural.* Lewis, Chelsea, MI, pp. 363–377.

Stott, T., 1997. A Comparison of Stream Bank Erosion Processes on Forested and Moorland Streams in the Balquhidder Catchments, Central Scotland, *Earth Surface Processes and Landforms,* 22, 383–399.

Strahler A.H. and Strahler, A., 2002. *Physical Geography*, John Wiley & Sons, New York.

Strahler, A.N., 1952. Hypsometric (area-altitude) Analysis of Erosional Topography. *Bull. Geolog. Soc. Am.*, 63: 1117–42.

Subcommittee for Global Change Research, 2000. Our Changing Planet: The FY2000 U.S. Global Change Research Program. Implementation Plan and Budget Overview. Washington D.C. 100 pp.

Suits, G., 1973. *Manual of Remote Sensing.* American Society for Photogrammetry and Remote Sensing, Falls Church, VA.

Sullivan, K., T.E. Lisle, C.A. Dolloff, G.E. Grant, and L.M. Reid, 1987. Stream Channels: The Link Between Forests and Fishes. In: Salo, E.O. and T.W. Cundy, (Eds.), *Streamside Management: Forestry and Fishery Interactions.* University of Washington Institute of Forest Resources, Seattle. pp. 39–97.

Swanson, F.J., J. Swanson, R. Janda, T. Dunne, and D. Swanston, 1982. Sediment Budgets and Routing in Forested Drainage Basins. USFS General Technical Report PNW-141.

Swanston, D.N., G.W. Lienkamper, R.C. Mersereau, and A.B. Levno, 1988. Timber Harvest and Progressive Deformation of Slopes in Southwestern Oregon. *Bull. Assoc. Eng. Geol.*, 25(3):372–381.

Swartzendruber, D., 1969. The Flow of Water in Unsaturated Soils. In: R.M. DeWiest (Ed.) *Flow Through Porous Media.* Academic Press, New York, pp. 215–292.

Tabios, G. and J. Salas, 1985. A Comparative Analysis of Techniques for Spatial Interpolation of Precipitation. *Water Resour. Bull.*, 21, 365–380.

Tabler, R.D., 1994. Design Guidelines for Control of Blowing and Drifting Snow. SHRP-H-381, Strategic Highway Research Program, National Research Council, Washington, D.C.

Tandarich, J.P., R.G. Darmody, and L.R. Follmer, 1994. The Pedo-weathering Profile: A Paradigm for Whole-regolith Pedology from the Glaciated Midcontinental United States of America. p. 97 –117. In: D.L. Cremeens, R.B. Brown, and J.H. Huddleston (Eds.) *Whole Regolith Pedology*, SSSA Spec. Publ. 34, Soil Science Society of America, Madison, WI.

Tarrant J. (Ed.), 1991. *Farming and Food.* Oxford University Press, New York.

Taylor, S.A. and G.L. Ashcroft, 1972. *Physical Edaphology.* W.H. Freeman, San Francisco.

Tchobanoglous, G., H. Theisen, and S. Vigil, 1993. *Integrated Solid Waste Management: Engineering Principles and Management Issues.* McGraw Hill, Boston.

Teigen, L.D. and F. Singer, 1992. Weather in U.S. Agriculture: Monthly Temperature and Precipitation by State and Farm Region, 1950–1990. USDA Statistical Bulletin 834. 129 pp.

Tennessee Valley Authority, 1973. Storm Hydrographs Using a Double-triangle Model. Research Paper 9, Knoxville.

Terstriep, M.L., and J.B. Stall, 1974. The Illinois Urban Drainage Area Simulator, ILLUDAS. Bulletin 85, Illinois State Water Survey, Urbana.

Thomas, G.W. and R.E. Phillips, 1979. Consequences of Water Movement in Macropores. *J. Environ. Qual.*, 8:149–152.

Thorn, W.C., C. Anderson, W. Lorenzen, D. Hendrickson, and J. Wagner, 1997. A Review of Trout Management in Southeast Minnesota, *North Am. J. Fish. Manage.*, 17: 860–872.

Thornthwaite, C.W., 1948. An Approach Toward a Rational Classification of Climate. *Geogr. Rev.*, 38:55–94.

Tornes, L.A., K.E. Miller, J.C. Gerken, and N.E. Smeck, 2000. Distribution of Soils in Ohio that are Described with Fractured Substratums in Unconsolidated Materials. *OH J. Sci.*, 100(3/4): 56–62.

Toth, J.A., 1962. A Theory of Ground-water Motion in Small Drainage Basins in Central Alberta, Canada. *J. Geophys. Res.*, 67, no. 11, 4375–4387.

Toy, T.J., G.R. Foster and K.G. Renard, 2002. *Soil Erosion: Processes, Prediction, Measurement, and Control.* John Wiley & Sons, New York.

Trautman, M.B., 1981. *The Fishes of Ohio.* Ohio State University Press, Columbus.

Trimble, S.W., 1970. *Culturally Accelerated Sedimentation on the Middle Georgia Piedmont.* USDA-SCS, Fort Worth.

Trimble, S.W., 1974. *Man-Induced Soil Erosion on the Southern Piedmont, 1700–1970.* Soil Conservation Society of America, Ankeny, IA.

Trimble, S.W., 1975a. Denudation Studies: Can We Assume Stream Steady State? *Science,* 188: 1207–1208.

Trimble, S.W., 1975b. Response of Coon Creek, Wisconsin, to Soil Conservation Measures, In: B. Zakrewska-Borowiecki (Ed.), In: *Landscapes of Wisconsin*, Association of American Geographers, Washington, D. C., pp. 24–29.

Trimble, S.W., 1976. Sedimentation in Coon Creek, Wisconsin, In: *Proc. 3rd Federal Inter-Agency Sedimentation Conf.*, Denver, Water Resources Council, Washington, D.C. Part 5, pp. 110–112.

Trimble. S.W., 1983. *Am J. Sci.*, 277:876–887.

Trimble, S.W., 1983. A sediment budget for Coon Creek in the Driftless Area, Wisconsin, 1853–1977. *Am. J. Sci.,* 283: 454–474.

Trimble, S.W., 1988. The Impact of Organisms on Overall Erosion Rates within Catchments in Temperate Regions. In: Viles, H., (Ed.), *Biogeomorphology*, Blackwell, Oxford, UK, pp. 83–142.

Trimble, S.W., 1990. Geomorphic Effects of Vegetation Cover and Management: Some Time and Space Considerations in Prediction of Erosion and Sediment Yield. In; J. Thornes (Ed.). Vegetation and Erosion. John Wiley & Sons, Chichester. pp. 55–65.

Trimble, S.W., 1993. The Distributed Sediment Budget Model and Watershed Management in the Paleozoic Plateau of the Upper Midwestern United States, *Phys. Geogr.,* 14: 285–303.

Trimble, S.W., 1994. Erosional Effects of Cattle on Streambanks in Tennessee, U.S.A. *Earth Surface Processes and Landforms,* 19, 451–464.

Trimble, S.W., 1995. Catchment Sediment Budgets and Change. In: A Gurnell and G. Petts, (Eds.), *Changing River Channels*. John Wiley & Sons, Chichester. pp. 201–215.

Trimble, S.W., 1997a. Contribution of Stream Channel Erosion to Sediment Yield from an Urbanizing Watershed. *Science,* 278:1442–1444.

Trimble, S.W., 1997b. Steam Channel Erosion and Change Resulting from Riparian Forests. *Geology,* 25, 467–469.

Trimble, S.W., 1999. Decreased Rates of Alluvial Sediment Storage in the Coon Creek Basin, Wisconsin, 1975–93. *Science,* 285:1244–46.

Trimble, S.W. and K. Bube, 1990. Improved Reservoir Trap Efficiency Prediction. *The Environ. Prof.,* 12, 255–272.

Trimble, S.W. and W.P. Carey, 1992. A Comparison of the Brune and Churchill Methods for Computing Sediment Yields Applied to a Reservoir System. In: S. Subitsky (Ed.), *Selected Papers in the Hydrologic Sciences,* 1988–1992. USGS Water Supply Paper 2340, pp. 195–202.

Trimble, S.W. and R.U. Cooke, 1991. Historical Sources for Geomorphological Research in the United States. *Prof. Geogr.,* 43:212–228.

Trimble, S.W. and P. Crosson, 2000. U.S. Soil Erosion Rates-Myth and Reality. *Science,* 289:248–250.

Trimble, S.W. and S.W. Lund, 1982. Soil Conservation and the Reduction of Erosion and Sedimentation in the Coon Creek Basin, Wisconsin. USGS Professional Paper 1234. 35pp.

Trimble, S.W. and A.C. Mendel, 1995. The Cow as a Geomorphic Agent-A Critical Review. *Geomorphology,* 13:233–253.

Trimble, S.W., F. Weirich, and B.L. Hoag, 1987. Reforestation and the Reduction of Water Yield on the Southern Piedmont Since C. 1940. *Water Resour. Res.,* 23:425–437.

Troeh, F.R. and L.M. Thompson, 1993. *Soils and Fertility,* 5th ed., Oxford University Press, New York.

Tyson, T.W., 1996. Constructed Wetlands For Animal Waste Treatment. ANR–965, Alabama Cooperative Extension System, Auburn University.

University of Illinois Department of Atmospheric Sciences, 2000. World Weather Project 2010 Website http://ww2010.atmos.uiuc.edu/(Gh)/home.rxml.

Urbonas, B.R. and P. Stahre, 1993. *Stormwater-Best Management Practices and Detention*. Prentice Hall, Englewood Cliffs, NJ.

Urbonas, B.R., C.Y. Guo, and L.S. Tucker, 1990. *Optimization of Stormwater Quality Capture Volume. Urban Stormwater Quality Enhancement*. American Society of Civil Engineers, New York.

U.S. Army Corps of Engineers, 1960. *Runoff from Snowmelt,* (EM 1110 2 1406):68 U. S. Government Printing Office, Washington, D.C.

U.S. Army Corps of Engineers, 1987. *Corps of Engineers Wetlands Delineation Manual*. Technical Report Y-87-1, Department of the Army, Washington, DC.

U.S. Army Corps of Engineers, 1992. *Photogrammetric mapping. Engineering Manual,* Washington, D.C.

U.S. Census 2000 (http://www.census.gov/)

U.S. Department of Agriculture, 1941. *Climate and Man,* U.S. Government Printing Office, Washington, D.C.

U.S. Department of Agriculture, 1975. *Soil Taxonomy: A Basic System of Soil Classification for Making and Interpreting Soil Surveys*, NRCS. USDA Handbook 436, U.S. Government Printing Office, Washington D.C., pp. 754.

U.S. Department of Agriculture, 1987. *Farm Drainage in the United States: History, Status, and Prospects*. Miscellaneous Publication 1455, Economic Research Service, U.S. Department of Agriculture, Washington, D.C.

U.S. Department of Agriculture, 1993. *Soil Survey Manual*. U.S. Department of Agriculture, Handbook 18, U.S. Governemnt Printing Office, Washington DC. 438 pages.

U.S. Department of Agriculture–Soil Conservation Service, 1972. *SCS National Engineering Handbook on Hydrology,* Washington, D.C.

U.S. Department of Agriculture–Soil Conservation Service, 1982. *Soil Survey of Clayton County, Iowa.* Washington, D.C. 356 pp.

U.S. Department of Agriculture–Soil Conservation Service, 1984. Snow Survey Sampling Guide. Agriculture Handbook 169.

U.S. Department of Agriculture Soil Conservation Service, 1994. *National Resources Inventory*. Washington, D.C.

U.S. Department of Commerce. 1987. *Statistical Abstract of the United States*. U.S. Department of Commerce, Washington, D.C.

U.S. Environmental Protection Agency, 2000. National Water Quality Inventory: 2000 Report. U.S. Environmental Protection Agency, Office of Water. EPA-841-F-02-001. www.epa.gov/305b.

U.S. Geological Survey, 1965. Annual Runoff from the United States. *Hydrolog. Atlas,* U.S. Government Printing Office, Washington, D.C., p. 212.

U.S. Geological Survey, 1973. *Water Resources Data for Ohio, Part 1, Surface Water Records,* 1972. U.S. Department of Interior, p 223.

U.S. Geological Survey, 1984. *National Water Summary 1983-Hydrologic Events and Issues*. U.S. Geological Survey Water-Supply Paper 2250, U.S. Government Printing Office, Washington, D.C.

U.S. Geological Survey, 1986. *National Water Summary. The Association of State and Interstate Water Pollution Control Administrators, in cooperation with the EPA. America's Clean Water: The States' Nonpoint Source Assessment, 1985*. U.S. Geological Survey, Washington, D.C.

U.S. Geological Survey. 1987. Annual Runoff from the United States, *Hydrolog. Atlas*, U.S. Government Printing Office, Washington, D.C. p. 710.

U.S. Weather Bureau, 1947. General Estimate of Probable Maximum Precipitation for the United States East of the 105th Meridian for Areas to 400 Square Miles and Durations to 24 Hours. Hydrometeorological Report 23.

van der Leeden, F., F.L. Troise, and D.K. Todd, 1991. *The Water Encyclopedia*, 2nd ed. Lewis, Chelsea, MI.

Van Schilfgaarde, J., 1963. Tile Drainage Design Procedure for Falling Water Tables. *Proc. ASCE,* 89(1R2).

Van Sickle, J., 2001. *GPS for Land Surveyors*. Taylor and Francis, London. 284 pp.

Vannote, R.L., G.W. Minshall, K.W. Cummins, J.R. Sedell, and C.E. Cushing, 1980. The River Continuum Concept. *Can. J. Fish. Aquatic Sci.*, 37:130–137.

Varnes, D.J., 1978. Slope Movement Types and Processes, In: *Landslides: Analysis and Control*, Special Report 176. Transport. Res. Board, National Academy of Science. Natl. Res. Counc., Washington, D.C. pp. 11–33.

Veimeyer, F.J., 1964. Evaporation In: Chow, V.T. (Ed), *Handbook of Applied Hydrology*. McGraw-Hill, New York, Section 11, 38pp.

Ver Steeg, K., 1946. The Teays River. *OH J. Sci.*, 46 (6):297–307.

Vesilind, P.A., W. Worrell, and D.R. Reinhart, 2002. *Solid Waste Engineering*. Brooks/Cole, Pacific Grove, CA.

Viessman, W. Jr., J.W. Knapp, G.L. Lewis, and T.E. Harbaugh, 1977. *Introduction to Hydrology*, 2nd edition, Harper & Row, New York. 704pp.

Vieux, B.E., 2001. *Distributed Hydrologic Modeling Using GIS*. Kluwer Academic, London. 293 pp.

Vita-Finzi, C., 1969. *The Mediterranean Valleys*. Cambridge University Press, Cambridge.

Wahl, K.L., W.O. Thomas Jr., and R.M. Hirsch, 1995. Stream-Gaging Program of the U.S. Geological Survey. U.S. Geological Survey Circular 1123. USGS, Reston, VA.

Wallace, J.B., S.L. Eggert, J.L. Meyer, and J.R. Webster, 1997. Multiple Trophic Linkages of a Forest Stream Linked to Terrestrial Litter Inputs. *Science*, 277:102–104.

Walling, D.E., 1983. The Sediment Delivery Problem. *J. Hydrol.*, 65: 209–237.

Wallworth, J.A., 1970. *Ecology of Soil Animals*. McGraw-Hill, London.

Wang, C., J.A. McKeague, and G.C. Topp, 1985. Comparison of Estimated and Measured Horizontal Ksat Values. *Can. J. Soil Sci.*, 65:707–715.

Ward, A.D. and W.J. Elliot, 1995. *Environmental Hydrology*. CRC/Lewis, Boca Raton, FL.

Ward, A.D., C.T. Haan, and B.J. Barfield, 1977. *Simulation of the Sedimentology of Sediment Detention Basins*. Research Report 103. Water Resources Research Institute, University of Kentucky, Lexington.

Ward, A.D., C.T. Haan, and J. Tapp, 1979. *The DEPOSITS Sedimentation Pond Design Manual*. Institute for Mining and Minerals Research, University of Kentucky, Lexington.

Ward, A.D., J.L. Hatfield, J.A. Lamb, E.E. Alberts, T.J. Logan, and J.L. Anderson, 1994. The Management Systems Evaluation Areas Program: Tillage and Water Quality Research. *Soil Tillage Res.*, 30:49–74.

Ward, A., D. Mecklenburg, and L. Brown, 2002. Using Knowledge of Fluvial Processes to Design Self-Maintaining Agricultural Ditches in the Midwestern Region of the USA, Proceedings of the International Conference on Environmental Flows for River Systems, Cape Town, South Africa, March 3–8.

Ward, A.D., D. Mecklenburg, D. Farver, J. Witter, and A. Jayakaran, 2003. Insight on the Variable Nature of Channel and Bank-Forming Discharges. ASAE Paper 032279. American Society of Agricultural Engineers, St. Joseph, MI.

Ward, A.D., B.J. Middleton, C.T. Haan, and G.V. Campbell, 1989. An Evaluation of Small Catchment Flood Estimation Techniques. *Trans. ASAE*, 32(2):6114–619.

Ward, A.D., L.G. Wells and R.E. Phillips, 1983. Infiltration Through Reconstructed Surface Mine Spoils and Soils. *Trans. ASAE*, 26(3):821–832, St Joseph, MI.

Ward, A.D., B.N. Wilson, T. Bridges, and B.J. Barfield, 1980. An Evaluation of Hydrologic Modeling Techniques for Determining A Design Storm Hydrograph, In: *Proc., International Symposium on Urban Storm Runoff*, University of Kentucky, Lexington, July 28–31.

Ward, R.C. and M. Robinson, 2000. *Principles of Hydrology*, 3rd ed, McGraw-Hill, London.

Ward, R.C., 1967. Principles of Hydrology, McGraw-Hill, Maidenhead, UK.

Waring, R.H. and W.H. Schlesinger, 1985. *Forest Ecosystems, Concepts and Management*. Academic Press. Orlando, FL. 340 pp.

Warner, R.C. and P.J. Schwab, 1992. *SEDCAD + Version 3 Training Manual*. Civil Software Design, Ames, Iowa.

Waters, T.F., 1995. *Sediment in Streams: Sources, Biological Effects and Control*. American Fisheries Society Monograph 7, American Fisheries Society, Bethesda, MD.

Watson, D.A., and J.M. Laflen, 1986. Soil Strength, Slope, and Rainfall Intensity Effects on Interrill Erosion. *Am. Soc. Agric. Eng. Trans.* 29(1):98–102.

Weiss, L.L., 1962. A General Relation Between Frequency and Duration of Precipitation. *Monthly Weather Rev.*, 90: 87–88.

Wells, L.G., A.D. Ward, I.D. Moore, and R.E. Phillips, 1986. Comparison of Four Infiltration Models in Characterizing Infiltration through Surface Mine Profiles. *Trans. ASAE*, (29)3:785–793.

Whiting, P.J., 1998. Floodplain Maintenance Flows. *Rivers*, 6(3): 160–170.

Whiting, P.J., 2002. Streamflow Necessary for Environmental Maintenance. *Ann. Rev. Earth Planet Sci.*, 30:181–206.

Whiting, P.J., and J.B. Bradley, 1993. A Process-Based Classification System for Headwater Streams. *Earth Surface Processes and Landforms*, 18:603–612.

Whiting, P.J., J.F. Stamm, D.B. Moog, and R.L. Orndorff, 1999. Sediment Transporting Flows of Headwater Streams. *GSA Bull.* 111(3): 450–466.

Whiting, P.J., W.E. Dietrich, L.B. Leopold, T.G. Drake, and R.L. Shreve, 1988. Bedload Sheets in Heterogeneous Sediment. *Geology*, 16:105–108.

Whittaker, R.H., 1975. *Communities and Ecosystems*, 2nd ed., MacMillan, New York.

Williams, D. and J. Lyon, 1995. Use of a Geographic Information System Data base to Measure and Evaluate Wetland Changes in the St. Marys River, Michigan. In Lyon, J. and J. McCarthy (eds.), *Wetland and Environmental Applications of GIS*. Lewis, Boca Raton, FL. pp. 125–139.

Williams, G.P., 1978. Bankfull Discharge of Rivers, *Water Resour. Res.,* 14:1141–1154.

Williams, G.P., 1986. River Meanders and Channel Size. *J. Hydrol.*, 88:147–164.

Williams, G.P. and M.G. Wolman, 1984. Downstream Effects of Dams on Alluvial Rivers. USGS Professional Paper 1286.

Williams, J.L., 1975. Sediment-yield Prediction with Universal Equation Using Runoff Factor. In: *Present and Prospective Technology for Predicting Sediment Yields and Sources*. USDA-ARS, West Lafayette, IN, Pub. 540. pp. 244–251.

Williams, M., 1989. *Americans and their Forests*. Cambridge University Press, Cambridge.

Wilson, B.N. et al., 1983. *A Hydrology and Sedimentology Watershed Model. Special Publication: SEDIMOT II Design Manual*. Agricultural Engineering Department, University of Kentucky, Lexington.

Wilson, B.N., B.J. Barfield, and R.C. Warner, 1982. *A Simulation Model of the Hydrology and Sedimentology of Surface Mined Lands. I. Modeling Techniques*. Special Publication. University of Kentucky Agricultural Engineering Department, Lexington.

Wischmeier, W.H., 1959. A Rainfall Erosion Index for a Universal Soil Loss Equation. *Soil Sci. Soc. Am. Proc.*, 23: 246–249.

Wischmeier, W.H. and D.D. Smith, 1978. *Predicting Rainfall Erosion Losses - A Guide to Conservation Planning*. U.S.D.A. Handbook 537. USDA-Science and Education Administration, Washington, D.C. 58pp.

Wisler, C.O. and E.F. Brater, 1959. *Hydrology*. John Wiley & Sons, Inc. New York.

Wolman, M.G. and J.P. Miller, 1960. Magnitude and Frequency of Forces in Geomorphic Processes, *J. Geol.* 68:54–74.

Wolman, M.G., 1954. A Method of Sampling Coarse Bed Material. *Trans. Am. Geophys. Union*, 35:951–956.

Wolman, M.G., 1967. A Cycle of Sedimentation and Erosion in Urban River Channels. *Geografiska Annaler* 49A: 209–237.

Woolhiser, D.A., 1982. Hydrologic System Synthesis. In: Haan, C.T., H.P. Johnson, and D.L. Brakensiek (Eds.). *Hydrologic Modeling of Small Watersheds*. ASAE, St. Joseph, MI. pp. 3–16.

World Resources Institute, 1986. *World Resources 1986*. Basic Books, New York.

Wright, J.L., 1982. New Evapotranspiration Crop Coefficients. *J. Irrig. Drain. Div., ASCE,* 108(IR2): 57–74.

Wu, Y.S., 2002. Numerical Simulation of Single-Phase and Multiphase Non-Darcy Flow in Porous and Fractured Reservoirs. *Transp. Porous Media,* 49(2): 209–240.

Yoder, C.O. and E.T. Rankin, 1997. Assessing the Condition and Status of Aquatic Life Designated Uses in Urban and Suburban Watersheds. In: Roesner, L.A. (Ed.). *Effects of Watershed Development and Management on Aquatic Ecosystems*, American Society of Civil Engineers, New York, NY. 201–227.

Yoder, C.O. and E.T. Rankin, 1998. The Role of Biological Indicators in a State Water Quality Management Process. *Environ. Monitoring Assess.*, 51(1–2): 61–88.

Zrinji, Z. and D.H. Burn, 1994. Flood Frequency Analysis for Ungauged Sites Using a Region of Influence Approach. *J. Hydrol.*, 153:1–21.

Appendix A: Unit Conversion Factors

Multiply the U.S. Customary Unit		By	To Obtain the SI Unit	
Name	**Symbol**		**Symbol**	**Name**
Acceleration				
feet per second squared	ft/sec^2	0.3048	m/sec^2	meter per second squared
inch per second squared	in./sec^2	0.0254	m/sec^2	meter per second squared
Area				
acre	acre	0.4047	ha	hectare
acre	acre	4.0469×10^{-3}	km^2	square kilometer
square foot	ft^2	9.2903×10^{-2}	m^2	square meter
square inch	in.^{2}	6.4516	cm^2	square centimeter
square mile	mi^2	2.5900	km^2	square kilometer
square yard	yd^2	0.8361	m^2	square meter
Energy				
British thermal unit	Btu	1.0551	kJ	joule
foot pound-force	ft·lb$_f$	1.3558	J	joule
horsepower-hour	hp·h	2.6845	MJ	megajoule
kilowatt-hour	kWh	3600	kJ	kilojoule
kilowatt-hour	kWh	3.600×10^6	J	joule
watthour	W·h	3.600	kJ	kilojoule
watt-second	W·sec	1.000	J	joule
Force				
pound-force	lb$_f$	4.4482	N	newton
Flow Rate				
cubic foot per second	ft^3/sec	2.8317×10^{-2}	m^3/sec	cubic meter per second
gallon per day	gal/d	4.3813×10^{-2}	L/sec	liter per second
gallon per day	gal/d	3.7854×10^{-3}	m^3/d	cubic meter per day
gallon per minute	gal/min	6.3090×10^{-5}	m^3/sec	cubic meter per second
gallon per minute	gal/min	6.3090×10^{-2}	L/sec	liter per second
million gallon per day	Mgal/d	43.8126	L/sec	liter per second
million gallon per day	Mgal/d	3.7854×10^3	m^3/d	cubic meter per day
million gallon per day	Mgal/d	4.3813×10^{-2}	m^3/sec	cubic meter per second
Length				
foot	ft	0.3048	m	meter
inch	in.	2.54	cm	centimeter
inch	in.	0.0254	m	meter
inch	in.	25.4	mm	millimeter
mile	mi	1.6093	km	kilometer
yard	yd	0.9144	m	meter
Mass				
ounce	oz	28.3495	g	gram
pound	lb	4.5359×10^2	g	gram
pound	lb	0.4536	kg	kilogram
ton (short: 2000 lb)	ton	0.9072	Mg (metric ton)	megagram (10^3 kilogram)
ton (long: 2240 lb)	ton	1.0160	Mg (metric ton)	megagram (10^3 kilogram)

Multiply the U.S. Customary Unit		By	To Obtain the SI Unit	
Name	Symbol		Symbol	Name
Power				
British thermal units per second	Btu/sec	1.0551	kW	kilowatt
foot-pound (force) per second	ft·lb$_f$/sec	1.3558	W	watt
horsepower	hp	0.7457	kW	kilowatt
Pressure (force/area)				
atmosphere (standard)	atm	1.0133×10^2	kPa (kN/m^2)	kilopascal (kilonewton per square meter)
inches of mercury (60°F)	in. Hg (60°F)	3.3768×10^3	Pa (N/m^2)	pascal (newton per square meter)
inches of water (60°F)	in. H$_2$O (60°F)	2.4884×10^2	Pa (N/m^2)	pascal (newton per square meter)
pound-force per square foot	lb$_f$·ft^2	47.8803	Pa (N/m^2)	pascal (newton per square meter)
pound-force per square inch	lb$_f$·in.^{2}	6.8948×10^3	Pa (N/m^2)	pascal (newton per square meter)
pound-force per square inch	lb$_f$·in.^{2}	6.8948	kPa (kN/m^2)	kilopascal (kilonewton per square meter)
Temperature				
degree Fahrenheit	°F	0.555(°F − 32)	°C	degree Celsius (centigrade)
degree Fahrenheit	°F	0.555 (°F + 459.67)	°K	degree Kelvin
Velocity				
feet per second	ft/s	0.3048	m/sec	meters per second
mile per hour	mi/h	4.4704×10^{-1}	km/sec	kilometer per second
Volume				
acre-foot	acre-ft	1.2335×10^3	m^3	cubic meter
cubic foot	ft^3	28.3168	L	liter
cubic foot	ft^3	2.8317×10^{-2}	m^3	cubic meter
cubic inch	in.^{3}	16.3871	cm^3	cubic centimeter
cubic yard	yd^3	0.7646	m^3	cubic meter
gallon	gal	3.7854×10^{-3}	m^3	cubic meter
gallon	gal	3.7854	L	liter
ounce (U.S. fluid)	oz (U.S. fluid)	2.9573×10^{-2}	L	liter

Appendix B: Glossary

Most of the definitions in Appendix B are based on *ASAE Standard: ASAE 5256 Soil and Water Engineering Terminology*, American Society of Agricultural Engineers, St. Joseph, MI.

Terms not contained in Appendix B can possibly be found at these glossary Web sites:

- EPA Terms of Environment: http://www.epa.gov/docs/OCEPAterms/
- USGS Unofficial Glossary: http://wwwga.usgs.gov/edu/dictionary.html
- Water Quality Association Glossary of Terms: http://www.wqa.org/

Accelerated erosion Erosion much more rapid than normal, natural, or geological erosion, primarily as a result of the influence of the activities of humans or, in some cases, of animals.

Accuracy The proximity of the measured value to the actual value.

Acid mine drainage Water draining from areas that have been mined for coal or other mineral ores. The drainage water is acidic, sometimes with a pH less than 2.0 because of its contact with sulfur-bearing material.

Acid rain Precipitation that has a low pH (less than 5.6, which is normal for "natural" precipitation). The precipitation becomes acidic when moisture in the air reacts with sulfur and nitrogen pollutants in the atmosphere. Acid rain has a harmful effect on some plants, aquatic organisms, soils, and buildings.

Actual evaporation Evaporation depends on climatic conditions and conditions at the surface from which the water will evaporate. Actual evaporation is a measure of the amount of water actually evaporated from a surface and accounts for the surface conditions as well as climatic conditions. Potential evaporation only accounts for climatic conditions.

Actual evapotranspiration Evapotranspiration depends on climatic conditions and conditions of the plants and soil surface. Actual evapotranspiration is a measure of the amount of water evapotranspired from the plants and soil system and takes into account the conditions of the plant/soil system.

Actual vapor pressure The pressure a gas exerts on the liquid it is contacting is its *vapor pressure*. In the context of this book, the gas is water vapor, and the liquid is water. The actual vapor pressure is the amount of pressure the water vapor in the air exerts on the surface it contacts.

Adhesion The attraction of water to the sides of the pore.

Advection The process by which solutes are transported by the motion of flowing groundwater.

Aerodynamic roughness A measure of the roughness of a surface regarding the disturbance that surface would cause for wind moving over it. A measure of the ability of the surface and its roughness elements to create a sufficiently large turbulent boundary layer as to reach the surface.

Albedo A measure of the light reflectance properties of a soil and crop surface; the ratio of shortwave electromagnetic radiation reflected from a soil and crop surface to the amount incident on that surface.

Alkali soil Soil containing sufficient exchangeable sodium to interfere with water penetration and the growth of most crops. The exchangeable sodium content is greater than 15% (preferred term is *saline–sodic soil*).

Anaerobic decomposition The decay of organic matter by microorganisms in the absence of oxygen.

Anisotropic soils Soils not having the same physical properties when the direction of measurement is changed. Commonly used in reference to permeability changes with direction of measurement.

Application rate Rate that water is applied to a given area. Usually expressed in units of depth per time.

Appreciable meandering Sinuosities of 1.2 to 1.5.

Aquiclude Underground geologic formation that neither yields nor allows the passage of an appreciable quantity of water, although it may be saturated with water itself.

Aquifer A geologic formation that holds and yields usable amounts of water. Aquifers can be classified as confined or unconfined.

Aquitard Underground geologic formation that is slightly permeable, but yields inappreciable amounts of water compared to an aquifer.

Area (of precipitation) The area covered by a given event, usually given in acres or hectares or in square miles or square kilometers.

Arid climate Climate characterized by low rainfall and high evaporation potential. A region is usually considered arid when precipitation averages less than 250 mm (10 in.) per year.

Artesian aquifer Aquifer that contains water under pressure as a result of hydrostatic head. For artesian conditions to exist, an aquifer must be overlain by a confining material or aquiclude and receive a supply of water. The free water surface stands at a higher elevation than the top confining layer.

Atmospheric instability Atmospheric vertical temperature gradients are high, and the added heat energy from the *latent heat of condensation* helps drive convection.

Atmospheric pressure The force per unit area exerted on a surface by the weight of the air above that surface.

Available plant water The portion of water in a soil that can be readily absorbed by plant roots. It is the amount of water released between *in situ* field capacity and the permanent wilting point.

Average rate of precipitation Derived by dividing the amount of precipitation that occurs during a given time period by the length of that period. Common units of intensity are inches, millimeters, or centimeters per hour. Average rate is also known as intensity.

Bank storage Water leaving a stream channel during rising stages of streamflow, during falling stages. *See* Floodplain storage.

Bankfull discharge The streamflow that fills the main channel and begins to spill onto the active floodplain (Wolman and Leopold, 1956; Wolman and Miller, 1960). It is a range of flows that is most effective in forming a channel, benches (floodplains), banks, and bars (G.P. Williams, 1978). The bankfull discharge is "considered to be the channel-forming or effective discharge" (Leopold, 1994).

Bankfull indicators Physical characteristics of a channel that denote bankfull.

Bankfull width, depth, or cross-sectional area The channel width, depth, or cross-sectional area, respectively, when the streamflow is at bankfull discharge.

Barometer An instrument that measures atmospheric or "barometric" pressure.

Baseflow Sustained low flow of a stream often due to groundwater inflow to the stream channel. Often written as a single word.

Baseflow depletion curve The declining rate of baseflow between storms.

Bed features The sequence of bed forms found in streams, such as riffle–pools, step–pools, cascades, and convergence/divergence. The particular form a stream achieves is dependent on channel plan form and gradient.

Bedload Coarse sediment or material moving on or near the bottom of a flowing channel by rolling, sliding, or bouncing.

Berm Strip or area of land, usually level, between the upper edge of a spoil bank and the edge of a ditch or canal.

Best management practice (BMP) Structural, nonstructural, and managerial techniques recognized as the most effective and practical means to reduce surface and groundwater contamination while still allowing the productive use of resources.

Bias Taking measurements with an uncalibrated or incorrectly calibrated piece of equipment, also known as systematic error.

Biodegradation Breaking down of natural or synthetic organic materials by microorganisms in soils, natural bodies of water, or wastewater treatment systems.

Biome types Unique terrestrial biological communities related to climatic zones and vegetation structure.

Biopores Root holes and earthworm burrows in soil.

Bivariate data Occurs when two variables are related systematically such that one is a fairly constant multiple of the other.

Broad-crested weir Weir of water measurement having a rounded or wide crest in the direction of the stream.

Bulk density (Soil) The mass of dry soil per unit bulk volume. The bulk volume is determined before drying to constant weight at 105°C (220°F).

Calculated risk Calculated likelihood of an unacceptable event occurring. In hydrology, the probability of a certain magnitude rain event occurring.

Canopy Vegetative cover over the land surface of a catchment area.

Capacity (of streams) Streams have a finite sediment transport rate.

Capillary fringe A zone in the soil just above the water table that remains saturated or almost saturated. The extent depends on the size distribution of pores.

Capillary pressure head Height water will rise by surface tension above a free water surface in the soil; expressed as length unit of water. Sometimes called *capillary rise*.

Capillary soil moisture Preferred term is *soil water potential*.

Capture zone, steady state The region surrounding a well that contributes flow to the well and extends upgradient to the groundwater divide of the drainage basin.

Capture zone, travel time related The region surrounding a well that contributes flow to the well within a specified period of time.

Catchment *See* Watershed.

Cation exchange capacity A measure of the quantity of cations a given mass of soil can hold. It is related to clay content and type and organic matter content.

Channel The bed and banks of a stream or river.

Channel capacity Flow rate in a ditch, canal, or natural channel when flowing full or at design flow.

Channel evolution Channels will change shape over time to seek equilibrium of the multiple factors, such as sediment supply, valley geology, water surface slope, vegetation, and the like, which influence their form. The stages of the evolution process are typically predictable.

Channel improvement Increasing the cross section, straightening, or clearing vegetation from a channel to change its hydraulic characteristics, increase its flow capacity, and reduce flooding. Often used perjoratively.

Channel stabilization Erosion prevention and stabilization of a channel using vegetation, jetties, drops, revetments, or other measures.

Channel storage (1) (Hydrology) Water temporarily stored in channels while en route to an outlet. (2) (Drainage) The volume of water that can be stored above the start-pumping level in ditches or floodways without flooding cropland.

Chlorinated hydrocarbon Synthetic compound that contains chlorine, hydrogen, and carbon; a main ingredient in some pesticides.

Clay A soil separate consisting of particles less than 2 μm in equivalent diameter.

Climatology The study of the day-to-day movement of moist air masses from their sources (lakes, oceans, transpiration from land areas) and their associated effects on precipitation and temperature over time.

Coefficient of variation The standard deviation expressed as a percentage of the mean.

Cohesion The attraction of water to water.

Concentration gradient A concentration gradient exists when there is more of a substance in one place than in another and the two places are in contact, either directly or by a material through which this substance can flow.

Concrete frost A condition that occurs when very wet, frozen soil becomes practically impervious.

Cone of depression or influence The water table or piezometric surface, roughly conical in shape, produced by the extraction of water from a well.

Confined aquifer An aquifer with an upper, and perhaps lower, boundary that is defined by a layer of natural material that does not transmit water readily.

Confining layer A body of material of low hydraulic conductivity that overlies or underlies an aquifer.

Conservation tillage A tillage practice that leaves plant residues on the soil surface for erosion control and water conservation.

Consumptive use The total amount of water taken up by vegetation for transpiration or building of plant tissue plus the unavoidable evaporation of soil moisture, snow, and intercepted precipitation associated with vegetal growth.

Continuous random variable Exists when a random variable can take on any numerical value along a continuum, such as a real line. In this way, the variable takes on an infinite number of values along that line.

Convective Air that expands when heated by solar energy and becomes lighter than the air around it. The lighter air rises by convection, potentially causing convective precipitation. Convection can also occur in most fluids including water.

Conventional tillage The traditional tillage practice that involves inverting the tillage layer, burying most of the plant residues, and leaving the soil bare.

Correlation The intensity or level of association between the two variables.

Cover crop Close-growing crop that provides soil protection, seeding protection, and soil improvement between periods of normal crop production or between trees in orchards and vines in vineyards. When plowed under and incorporated into the soil, cover crops may be referred to as *green manure crops.*

Creek A small stream (smaller than a river), often a shallow or intermittent tributary to a river. Also called regional branch, brook, kill, or run.

Crop residue Portion of a plant or crop left in the field after harvest.

Crop rotation A system of farming in which a succession of different crops is planted on the same land area, as opposed to growing the same crop time after time (monoculture).

Crusting The dry condition when water will not enter the coarse-textured subsurface layer until the suction forces at the interface between the two layers is equal; it affects many soils, especially those with low organic matter and unstable structure.

Curve number An index of the runoff potential that is related to the soil and vegetation conditions of the site. Used in SCS runoff equations.

Darcy's law A concept formulated by Henry Darcy in 1856 to describe the rate of flow of water through porous media. The rate of flow of water in porous media is proportional to, and in the direction of, the hydraulic gradient and inversely proportional to the thickness of the bed.

Deep percolation Water that moves downward through the soil profile below the root zone and cannot be used by plants.

Dendritic A treelike system of channels.

Dendrochronology The study of estimating annual precipitation rates using tree growth records. Rings are thin for dry years and thick for wet years, and the record can be extended back for centuries.

Descriptive statistics Involves the organization, summarization, and description of data sets.

Design runoff rate Maximum runoff rate expected for a given design return period storm.

Detention storage Water in excess of depression storage that is temporarily stored in the watershed while en route to streams. Most eventually becomes surface runoff, but some may infiltrate or evaporate.

Dewpoint When moist air has cooled below the temperature of saturation.

Dewpoint temperature Air typically contains water vapor. The amount of water vapor a given parcel of air can hold depends on the temperature of the air. Warmer air can hold more water than cooler air. The dewpoint temperature is the temperature to which a given parcel of air must be cooled (at constant pressure and water vapor content) for the air to be saturated with water.

Diffuse The process of a substance moving from an area of higher concentration to an area of lower concentration of that substance.

Diffusion coefficient A measure of the ease with which a particular substance can diffuse in a given system.

Discharge Rate of water movement.

Discharge curve Rating curve that shows the relation between stage and flow rate of a stream, channel, or conduit.

Discrete random variable Occurs when occurrences of a certain event can be counted, as opposed to something measured, as in the case of a continuous random variable.

Dominant channel materials (D_{50}) The median size particle determined by a channel material size distribution analysis, typically a pebble count.

Drain Any closed conduit (perforated tubing or tile) or open channel used for removal of surplus ground or surface water.

Drain tile Short length of pipe made of burned clay, concrete, or similar material. Usually laid with open joints to collect and remove subsurface water.

Drainage Process of removing surface or subsurface water from a soil or area.

Drainage basin *See* Watershed.

Effective discharge The streamflow that transports the most sediment over the long term (Wolman and Miller, 1960).

Emissivity The ratio of the emittance of a given surface at a specified wavelength and temperature to the emittance of an ideal black body at the same wavelength and temperature.

Entrenchment The vertical containment of a stream and the degree to which it is incised in the valley floor.

Entrenchment ratio A computed index used to describe the degree of vertical containment of a river channel. It is calculated as the flood-prone width divided by the bankfull width. In other words, the width of the water surface in a stream or river at a water elevation of two times the maximum bankfull depth divided by the bankfull width.

Ephemeral gully Small channels eroded by runoff that can be easily filled and removed by normal tillage, only to re-form again in the same location.

Ephemeral stream A stream that is dry most of the year and only contains water during and immediately after a rainfall event.

Equipotential line A contour line of a potentiometric surface along which the hydraulic head of the groundwater flow system is the same for all points on the line.

Erosion The wearing away of the land surface by running water, wind, ice, or other geological agents, including such processes as gravitational creep.

Erosivity (EI) The ability of rainfall to detach and transport soil particles.

Evaporation The physical process by which a liquid is transformed to a gaseous state.

Evaporation pan A pan, typically of specific materials and dimensions, that is filled with water and left open to the environment. Evaporation from the pan is measured, and this evaporation can be related to evapotranspiration from a nearby crop.

Evapotranspiration The combination of water transpired from vegetation and evaporated from the soil and plant surfaces.

Exceedence probability The probability that an event with a specified magnitude and duration will be exceeded in one time period, which is most often assumed to be 1 year.

Exchangeable cation A positively charged ion held on or near the surface of a solid particle by a negative surface charge of a colloid and that may be replaced by other positively charged ions in the soil solution.

Exchangeable sodium percentage The fraction of the cation exchange capacity of a soil occupied by sodium ions.

Experimental error The fluctuation or discrepancy in replicate observations from one experiment to another.

Field capacity Amount of water remaining in a soil when the downward water flow due to gravity becomes negligible.

Final covers (caps) Covering of landfill to limit infiltration and odors once the landfill is filled and closed.

Flood routing Process of determining stage height, storage volume, and outflow from a reservoir or reach of a stream for a given hydrograph of inflow.

Flood spillway An auxiliary channel to carry a flood flow that exceeds a given design rate to the channel downstream (the preferred term is *emergency spillway*).

Floodplain The area on either side of the bankfull channel that carries the flow greater than the bankfull flow, that is, all storms greater than the 1- to 2-year storm.

Floodplain storage Volume of water that spreads out and is temporarily stored in a floodplain.

Flood-prone width Width of the water surface in a stream or river at a water elevation of two times the maximum bankfull depth.

Flow duration curve The amount of time a flow event of a certain magnitude occurs.

Flow rate Rate of water movement. Often written as a single word and expressed in cubic feet per second or cubic meters per second.

Flowline A line indicating the instantaneous direction of groundwater flow throughout a flow system, at all times in a steady-state flow system, or at a specific time in a transient flow system. In an isotropic medium, flowlines are drawn perpendicular to equipotential lines.

Flume (1) Open conduit for conveying water across obstructions. (2) An entire canal elevated above natural ground; an aqueduct. (3) A specifically calibrated structure for measuring open channel flows.

Flux density The rate of flow of any quantity, such as water vapor, through a unit area of specified surface.

Freeboard Vertical distance between the maximum water surface elevation anticipated in design and the top of retaining banks, pipeline vents, or other structures, provided to prevent overtopping because of unforeseen conditions.

Frequency–magnitude (of storm events) The relationship between frequency of occurrence and the amount of rainfall or runoff.

Friction slope Friction head loss per unit length of conduit.

Front The zone where two unlike (e.g., warm moist air vs. cooler heavier air) air masses meet.

Gabion Rectangular or cylindrical wire mesh cage filled with rock for protecting aprons, stream banks, shorelines, and the like, against erosion.

Gage height (1) (Surveying) The vertical distance from the sight bar, batter board, or receiver to the bottom of the finished cut. (2) (Hydraulics) Elevation of a water surface measured by a gage.

Gaging station Section in a stream channel equipped with a gage or facilities for obtaining streamflow data.

Gaining stream Stream or part of a stream that has an increase in flow because of inflow from groundwater.

Geographic Information System (GIS) Computer database management system for spatially distributed attributes.

Geological erosion The normal or natural erosion caused by geological processes acting over long geological periods (synonymous with natural erosion).

Glaciotectonic The structural deformation of sediment or bedrock as a direct result of glacier ice movement or loading.

Grade (noun) Degree of slope to a road, channel, or ground surface usually expressed as a dimensionless fraction. (verb) To finish the surface of a canal bed, roadbed, top of embankment, or bottom of excavation.

Gradually varied flow Steady, nonuniform open channel flow in which the changes in depth and velocity from section to section are gradual enough so that accelerative forces are negligible.

Grassed waterway Natural or constructed channel covered with an erosion-resistant grass that transports surface runoff to a suitable discharge point at a nonerosive rate.

Gravimetric soil water content Soil water content is determined as a function of mass or weight.

Gravitational The second force (along with tension) that causes the downward movement of water through a soil profile.

Gravitational water Soil water that moves into, through, or out of the soil under the influence of gravity (the preferred term is *soil water potential*).

Groundwater Water occurring in the zone of saturation in an aquifer or soil.

Groundwater divide A ridge in the water table or potentiometric surface from which groundwater moves away in both directions.

Groundwater flow Flow of water in an aquifer or soil. The portion of the discharge of a stream derived from groundwater.

Growing season The period, often the frost-free period, during which the climate is such that crops can be produced.

Guard cells Specialized cells in a leaf that surround the stomates. The guard cells can expand and contract to control the loss of water vapor through stomates.

Gully Eroded channel where runoff concentrates, usually so large that it cannot be obliterated by normal tillage operations.

Gully erosion The erosion process by which water accumulates in narrow channels and, over short periods, removes the soil from this narrow area to considerable depths, ranging from 0.5 m (1.6 ft) to as much as 30 m (97 ft.)

Gully head advance Upstream migration of the upper end of a gully. Sometimes referred to as a head cut.

Gypsum block An electrical resistance block in which the absorbent material is gypsum.

Hazen method A method of determining the statistical distribution of rainfall amounts for the duration of interest, plotting that distribution on probability paper, and interpolating or extrapolating from the graph to determine the storm associated with the return period of interest.

Head The height to which water can raise itself above a known datum (commonly sea level), exerting a pressure on a given area, at a given point. It is synonymous with hydraulic head.

Head loss Energy loss in fluid flow.

Heterogeneous Pertaining to a nonuniform geologic material having different characteristics and hydraulic properties at different locations.

Homogeneous Pertaining to a uniform geologic material having identical characteristics and hydraulic properties everywhere.

Humid climate Climate characterized by high rainfall and low evaporation potential. A region is usually considered humid when precipitation averages more than 500 mm (20 in.) per year.

Hydraulic conductivity The ability of a porous medium to transmit a specific fluid under a unit hydraulic gradient; a function of both the characteristics of the medium and the properties of the fluid transmitted. Usually a laboratory measurement corrected to a standard temperature and expressed in units of length/time. Although the term hydraulic conductivity is sometimes used interchangeably with the term *permeability*, the user should be aware of the differences.

Hydraulic gradient Change in the hydraulic head per unit distance.

Hydraulic head The total mechanical energy per unit weight of water that is equal to the sum of the elevation head, pressure head, and the velocity head at a given point in the flow system.

Hydraulic length Longest flow path on a watershed.

Hydraulic resistance Friction along the wetted boundary of a channel or conduit that causes a loss in head.

Hydrograph Graphical or tabular representation of the flow rate of a stream with respect to time.

Hydrograph separation Separation of a hydrograph into direct runoff and baseflow portions.

Hydrologic cycle Term used to describe the movement of water in and on the earth and atmosphere. Numerous processes, such as precipitation, evaporation, condensation, and runoff, comprise the hydrologic cycle.

Hydrologist A person who considers such phenomena as precipitation and temperature in relation to water supplies and movement to, on, and beneath the earth's surface.

Hyetograph A plot of rainfall intensity with respect to time.

Hygroscopic Water that is tightly held by molecular attraction and is not removed under normal climatic conditions.

Hysteresis The condition that is caused during wetting when the small pores fill first, while during drainage and drying, the large pores empty first.

Impeller meter A rotating mechanical device for measuring flow rate in a pipe or open channel.

Impermeable layer (Soil) Layer of soil resistant to penetration by water, air, or roots.

In situ In position; in its physical place.

Incised stream A stream that has, through degradation, cut its channel into the bed of the valley.

Independence When the simple multiplicative laws of probability are true. This means that the product of the probabilities of each individual occurrence gives the probability of the joint occurrence of two events.

Inferential statistics The process of drawing conclusions about an entire population based only on the results obtained from a small sample of that population.

Infiltration The downward entry of water through the soil surface into the soil.

Infiltration capacity The point at which additional precipitation is not infiltrated.

Infiltration rate The quantity of water that enters the soil surface in a specified time interval. Often expressed in volume of water per unit of soil surface area per unit of time.

Influent stream Stream or portion of stream that contributes water to the groundwater supply. *See* Losing stream.

Initial storage The portion of precipitation required to satisfy canopy interception, the wetting of the soil surface, and depression storage. Sometimes called *initial abstraction.*

Intensity–duration–frequency (IDF) The relationship that develops from the Hazen method.

Interception The portion of precipitation caught by vegetation and prevented from reaching the soil surface.

Interflow Water that infiltrates into the soil and moves laterally through the upper soil horizons until it returns to the surface, often in a stream channel.

Interfluve The boundary between streams.

Intermediate covers Compost, soil, or other material used at the end of each working day to reduce the amount of infiltration (by maximizing runoff from the landfill) and limit odors and wind-blown debris at a landfill.

Intermittent stream Natural channel in which water does not flow continuously.

Interrill erosion The removal of a fairly uniform layer of soil on a multitude of relatively small areas due to raindrop impact and by shallow surface flow.

Intrinsic permeability The property of a porous material that expresses the ease with which gases or liquids flow through it. *See* Permeability.

Isolation Received solar radiation that evaporation requires to supply energy for the latent heat of vaporization.

Isostatic Denotes a condition in which there is equal pressure on every side.

Isotropic (Soil) The condition of a soil or other porous media for which physical properties, particularly hydraulic conductivity, are equal in all directions.

Kinematic wave A method of mathematical analysis of unsteady open channel flow in which the dynamic terms are omitted because they are small and assumed negligible.

Lag time (1) (Hydrology) The interval between the time when one half of the equivalent uniform excess rain (runoff) has fallen and the time when the peak of the runoff hydrograph occurs.

(2) (Irrigation) The interval after water is turned off at the upper end of a field until it recedes (disappears from that point).

Laminar flow Flow in which there are no cross currents or eddies and where the fluid elements move in approximately parallel directions. Flow through granular materials is usually laminar. Sometimes called *streamline* or *viscous flow.*

Langley A unit of energy per unit area commonly used in radiation measurements. One Langley is equal to 1 gram calorie/cm^2.

Latent heat of condensation Heat released when water vapor changes to water.

Latent heat of vaporization The heat released or absorbed per unit mass of water when evaporation occurs (1 kg of water at 20°C requires 2.45 MJ of heat to vaporize the water).

Lateral point bar accretion The process by which point bars are created when scour on the inside of the bend where the flow and velocity are lower deposits material, thereby forming a point bar.

Leachate Mixture of water and contaminants flowing from a landfill. Leachate may pose risks to human health and the environment if not managed adequately.

Leachate Water that moves downward through some porous media and contains dissolved substances removed from the media.

Leaching Removal of soluble material from soil or other permeable material by the passage of water through it.

Losing stream A channel that loses water into the bed or banks. *See* Influent stream.

Lysimeter An isolated block of soil, usually undisturbed and *in situ*, for measuring the quantity, quality, or rate of water movement through or from the soil.

Macropores Places (pores) in the soil profile where the dominant flow process is gravity flow, and soil tension forces are negligible.

Mass flow Movement of a substance that occurs when force is exerted on the substance by some outside influence, such as pressure or gravity, such that all the molecules of the substance tend to move in the same direction.

Meandering The propensity of natural channels to vary from a straight line and follow a winding and turning course.

Median (M_d) The midpoint or middle observation in an ordered distribution.

Meniscus The curved surface created when water is at a higher level on the sides of a pore due to adhesion.

Meterologist A person who is generally concerned with the day-to-day movement of moist air masses from their sources (lakes, oceans, transpiration from land areas) and their associated effects on precipitation and temperature.

Minor (meandering) Sinuosities of 1.0 to 1.2.

Mode (M_o) The value that occurs most frequently in a distribution.

Natural erosion Wearing away of the earth's surface by water, ice, or other natural agents under natural environmental conditions of climate, vegetation, and the like undisturbed by humans. *See* Geological erosion.

No tillage or no till A tillage system in which the soil is not tilled except during planting, when a small slit is made in the soil for seed and agrichemical placement. Pest control is achieved using pesticides, crop rotation, and biological control rather than tillage. Sometimes called *zero tillage*.

Nonpoint source pollution (NPS) Pollution originating from diffuse areas (land surface or atmosphere) with no well-defined source.

Nonsaline–alkali soil Soil containing sufficient exchangeable sodium to interfere with the growth of most crops (the preferred term is *sodic soil*).

Normal depth Depth of flow in an open channel during uniform flow for the given conditions.

Normal distribution Distribution characterized completely by its mean and variance.

Normal erosion The gradual erosion of land used by society; does not greatly exceed natural erosion. *See* Natural erosion.

Normality Describes the fact that the error term in a measurement e_i is assumed to come from a normal probability distribution, although many statistical procedures tend to yield correct conclusions even when applied to data that are not normally distributed.

Null hypothesis The hypothesis to be tested.

Observation well Hole bored to a desired depth below the ground surface for observing the water table level.

Orographic storm A weather pattern in which precipitation is caused by the rising and cooling of air masses as they are forced upward by topography.

Overbank vertical accretion The process by which materials that are finer than those deposited during lateral point bar accretion are deposited on the banks and floodplain when flows exceed the effective discharge and overtop the banks (except in deeply incised channels).

Overland flow Surface runoff occurring at relatively shallow depths across the land surface prior to

concentration in drainage ways. May cause sheet and still erosion.

Particle size analysis Determination of the various amounts of different separates in a soil sample, usually by sedimentation, sieving, or micrometry.

Pattern, profile, and dimension These terms are used by geomorphologists to describe completely the features of a stream channel. *Pattern* refers to the plan view of a stream. The meander is the pattern of concern. *Profile* refers to a longitudinal cross section of the channel. Slope is observed in the profile. *Dimension* refers to a cross section. The flow cross section is observed here.

Pebble count A systematic method of random selection of channel particles to perform a particle size distribution analysis.

Perched water table A water table, usually of limited area, maintained above large groundwater bodies by the presence of an intervening, relatively impervious, confining stratum.

Percolating water Subsurface water that flows through the soil or rocks. *See* Seepage.

Percolation (1) Downward movement of water through porous media such as soil. (2) Intake rate used for designing wastewater absorption systems.

Perennial stream A stream that flows throughout the year.

Permanent wilting point Soil water content below which plants cannot readily obtain water and permanently wilt. Sometimes called *permanent wilting percentage*.

Permeability (1) (Qualitative) The ease with which gases, liquids, or plant roots penetrate or pass through a layer of soil or porous media. (2) (Quantitative) The specific soil property designating the rate at which gases and liquids can flow through the soil or porous media.

Permeameter Device for containing a soil sample and subjecting it to fluid flow to measure permeability or hydraulic conductivity.

Permissible velocity Highest water velocity that does not cause erosion in a channel or conduit.

Pheatophytes Plants with root systems that can reach down into groundwater, or the phreatic zone.

Pipe drain Any circular subsurface drain, including corrugated plastic tubing and concrete or clay tile.

Pipe spillway A pipe drain for transporting water through an embankment. Sometimes called a *culvert*.

Piston flow Water moving down as a front with no mixing.

Plant available water *See* Available plant water.

Population The data set that is the focus of interest and involves all elements under investigation (e.g., all of the streams in your state).

Porosity (1) (Soil) The volume of pores in a soil sample divided by the combined volume of the pores and the soil of the sample. (2) (Aquifer) The sum of the specific yield and the specific retention.

Porosity, effective The volume of the void spaces through which water or other fluids can travel in a rock or sediment divided by the total volume of the rock or sediment.

Potential evaporation Evaporation from a surface when all surface–atmosphere interfaces are wet so there is no restriction on the rate of evaporation from the surface.

Potential evapotranspiration Rate at which water, if available, would be removed from soil and plan surfaces.

Potentiometric surface An imaginary surface representing the static head of groundwater and defined by the level to which water will rise in a well.

Precipitation depth The total amount of precipitation, usually in inches, centimeters, or millimeters.

Precipitation duration The time from the beginning of the storm until the end of the storm.

Precipitation intensity Rate of precipitation, generally expressed in units of depth per time. *See* Rainfall intensity.

Precipitation mechanisms The mechanism by which air is cooled that results in frontal, convective, or orographic precipitation.

Precipitation types Hail, rain, snow, sleet.

Precision The scatter resulting between repeated measurements.

Preferential flow Flow into and through porous media or soil by way of cracks, root holes, and other paths of low resistance rather than uniformly through the entire media.

Pressure head Pressure expressed as a height of water. The density of mercury is 13.6 times that of water, so 29.92 inHg is equivalent to 406.6 in. (33.9 ft or 1034 cm) of water. Therefore, a pressure of 1 atm is the same as that exerted by a 1-ft^2 column of water that is 33.9 ft high. Also known as *head of water* or just *head*.

Probability distribution A function with values of a random variable and with functional values that are probabilities. In defining a probability distribution (**Figure 1.13**), the following are always true: Each value of $f(x)$ must lie between 0 and 1, inclusive, and the sum of all values of $f(x)$ must equal 1.

Probability histogram A pictorial representation of a probability distribution for which the areas of each bar represent the probabilities.

Probability values (p values) The odds that we are wrong when we state that the null hypothesis is correct.

Raindrop erosion Soil detachment resulting from the impact of raindrops on the soil. *See* Erosion.

Rainfall erosivity A measure of rainfall's ability to detach and transport soil particles.

Rainfall frequency Frequency of occurrence of a rainfall event with an intensity and duration that can be expected to be equaled or exceeded (the preferred term is *return period*).

Rainfall intensity Rate of rainfall for any given time interval, usually expressed in units of depth per time.

Random sampling When samples are drawn from a population, they are taken in a manner that ensures that every member or element of a population has an equal chance to be drawn.

Range Equal to the difference between the smallest and largest measurement in a data set.

Rating curve Graphic or tabular presentation of the discharge of or flow through a structure or channel section as a function of water stage or depth of flow. Sometimes called a *rating table*.

Reach A length of a stream or channel with relatively constant characteristics.

Receiving waters Distinct bodies of water, such as streams, lakes, or estuaries, that receive runoff or wastewater discharges.

Recession curve Descending portion of a streamflow or hydrograph.

Recharge Process by which water is added to the zone of saturation to replenish an aquifer.

Recharge area Land area over which water infiltrates and percolates downward to replenish an aquifer. For unconfined aquifers, the area is essentially the entire land surface overlaying the aquifer; for confined aquifers, the recharge area may be a part of or unrelated to the overlaying area.

Recurrence interval *See* Return period.

Reference crop evapotranspiration The evapotranspiration predicted from a specific crop, arbitrarily called a reference crop, under a given climatic condition assuming water is available to the crop.

Regime Condition of a stream with respect to its rate of flow.

Relative humidity Ratio of the amount of water present in the air to the amount required for saturation of the air at the same dry bulb temperature and barometric pressure, expressed as a percentage.

Reservoir Body of water, such as a natural or constructed lake, in which water is collected and stored for use.

Retention Precipitation on an area that does not escape as runoff; the difference between total precipitation and total runoff.

Return period The frequency of occurrence of a hydrologic event with an intensity and duration that can be expected to be equaled or exceeded, usually expressed in years.

Riffle The shallow area of a stream where water accelerates, the water surface becomes rippled, and in more turbulent waters, contains a hydraulic jump.

Riffle–pool sequence The pattern of consecutive riffle and pools that naturally form in many streams and rivers. In steeper valleys, a step–pool pattern will form.

Rill Small channel eroded into the soil surface by runoff that can be filled easily and removed by normal tillage.

Rill erosion An erosion process by concentrated overland flow in which numerous small channels only several centimeters deep are formed; occurs mainly on recently cultivated soils. *See* Rill.

Riparian (1) Pertaining to the banks of a body of water, a riparian owner is one who owns the banks. (2) A riparian water right is the right to use and control water by virtue of ownership of the banks.

River A large natural stream of water emptying into an ocean, lake, or other body of water and usually fed along its course by converging tributaries.

River flow Flow of water in a channel.

Root zone Depth of soil that plant roots readily penetrate and in which the predominant root activity occurs.

Row grade The slope in the direction of crop rows.

Runoff The portion of precipitation, snowmelt, or irrigation that flows over and through the soil, eventually making its way to surface water supplies.

Runoff coefficient Ratio of peak runoff rate to rainfall intensity.

Runoff duration Elapsed time between the beginning and end of a runoff event.

Safe well yield Amount of groundwater that can be withdrawn from an aquifer without degrading quality or reducing pumping level.

Saline–sodic soil Soil containing sufficient exchangeable sodium to interfere with the growth of most crops and containing appreciable quantities of soluble salts. The exchangeable sodium percentage is greater than 15, and the electrical conductivity of the saturation extract is greater than 4 mS/cm (0.01 mho/in.).

Sample A subset of data collected from the population or the collection of observations actually available for analysis.

Sample mean ($\bar{y}$) The arithmetic average of all samples collected.

Sample standard deviation A statistical measure of the distance a quantity is likely to lie from its average value. Square root of the sample variance.

Sample variance Square of the standard deviation.

Sand Soil particles ranging from 50 to 200 μm in diameter. Soil material containing 85% or more particles in this size range.

Saturated air When air's capacity to hold water is reached.

Saturated flow Flow of water through a porous material under saturated conditions.

Saturation vapor pressure The vapor pressure at which a liquid–vapor system is in a state of dynamic equilibrium, with the number of molecules escaping from the liquid equal to the number of molecules leaving the vapor and recaptured into the liquid. The saturation vapor pressure increases exponentially with temperature.

Sealing The wet condition by which water will not enter the coarse-textured subsurface layer until the suction forces at the interface between the two layers is equal; it affects many soils, especially those with low organic matter and unstable structure.

Sediment basin Pond at the upper end of a conveyance or reservoir for detaining particle-laden water for a sufficient length of time for deposition to occur.

Sediment load Amount of sediment carried by running water or wind.

Sediment rating curve The relationship between sediment load and stream discharge.

Sediment transport Amount of sediment transported by a stream; calculated from the sediment rating curve and the flow duration curve, usually given in tons/year or, more commonly, tons/unit area/year.

Sedimentation Deposition of waterborne or windborne particles resulting from a decrease in transport capacity.

Seepage The movement of water into and through the soil from unlined canals, ditches, and water storage facilities.

Semiarid climate Climate characterized as neither entirely arid nor humid, but intermediate between the two conditions. A region is usually considered semiarid when precipitation averages between 250 mm (10 in.) and 500 mm (20 in.) per year.

Severe (meandering) Sinuosities of 1.5 and greater.

Sheet erosion The removal of soil from the land surface by rainfall and surface runoff. Often interpreted to include rill and interrill erosion.

Sheet flow Water, usually storm runoff, flowing in a thin layer over the soil or other smooth surface.

Silt (1) A soil separate consisting of particles between 2 and 50 μm in diameter. (2) (Colloquial) Deposits of sediment that may contain soil particles of all sizes.

Silt bar A deposition of sediment in a channel.

Sinuosity A bending or curving shape or movement. Defined as the length of the channel to the valley length.

Sinuosity Ratio of the stream length to the valley length.

Sludge The solids removed from raw water or wastewater during water treatment.

Snow course A designated line along which the snow is sampled at appropriate times to determine its depth and density (water content) for forecasting water supplies.

Snow density Water content of snow expressed as a percentage by volume.

Sodic soil A nonsaline soil containing sufficient exchangeable sodium to affect crop production and soil structure adversely. The exchangeable sodium percentage is greater than 15, and the electrical conductivity of the saturation extract is less than 4 mS/cm (0.01 mbo/in.).

Sodium adsorption ratio (SAR) The proportion of soluble sodium ions in relation to the soluble calcium and magnesium ions in the soil water extract. (Can be used to predict the exchangeable sodium percentage.)

Sodium percentage Percentage of total cations that is sodium in water or soil solution.

Soil The unconsolidated minerals and material on the immediate surface of the earth that serves as a natural medium for the growth of plants.

Soil aeration Process by which air and other gases enter the soil or are exchanged.

Soil compaction Consolidation, reduction in porosity, and collapse of the structure of soil when subjected to surface loads.

Soil conservation Protection of soil against physical loss by erosion and chemical deterioration by the application of management and land use methods that safeguard the soil against all natural and human-induced factors.

Soil erodibility A measure of the soil's susceptibility to erosional processes.

Soil erosion Detachment and movement of soil from the land surface by wind or water. *See* Erosion.

Soil map unit A delineation on a map of an area dominated by one major kind of soil or several soils with similar properties.

Soil series A group of soils that have similar characteristics and a similar arrangement of soil layers (horizons).

Soil structure The combination or arrangement of primary soil particles into secondary particles, units, or peds that make up the soil mass. These secondary units may be, but usually are not, arranged in the profile in such a manner as to give a distinctive characteristic pattern. The principal types of soil structure are platy, prismatic, columnar, blocky, and granular.

Soil texture Classification of soil by the relative proportions of sand, silt, and clay present.

Soil water All water stored in the soil.

Soil water characteristic curve Soil-specific relationship between the soil water matric potential and soil water content.

Soil and water conservation district (SWCD) A local governmental entity within a defined water or soil protection area that provides assistance to residents in conserving natural resources, especially soil and water.

Soil water deficit or depletion Amount of water required to raise the soil water content of root zone to field capacity.

Spatial Area or space.

Specific heat of water The amount of heat required to raise the temperature of 1 g of water 1°C.

Specific retention Amount of water that a unit volume of porous media or soil, after saturation, will retain against the force of gravity (compare to specific yield).

Specific yield Amount of water that a unit volume of porous media or soil, after saturation, will yield when drained by gravity (compare to specific retention).

Spillway Conduit through or around a dam for the passage of excess water. May have controls.

Splash erosion The detachment and airborne movement, caused by the impact of raindrops on soils, of small soil particles.

Staff gage Graduated scale, generally vertical, from which the water surface elevation may be read.

Stage Elevation of a water surface above or below an established datum; gage height.

Standard error The estimate of the variation of a statistic.

Static lift Vertical distance between source and discharge water levels in a pump installation.

Statistical hypothesis A statement that something is true.

Steady flow Open channel flow in which the rate and cross-sectional area remain constant with time at a given station.

Steady-state flow A condition of groundwater flow for which there is no change in head with time; occurs when, at any point in a flow field, the magnitude and direction of the flow velocity are constant with time.

Stefan–Boltzmann constant A universal constant used in the equation relating the rate of emission of radiant energy from the surface of a body to the emissivity of the surface and the temperature of the body.

Stem flow (1) Vegetation-intercepted precipitation that reaches the ground by flowing down the stems or trunks of vegetation. (2) Flow in the xylem of plants.

Stomata Pores on the leaf surface that lead to the intracellular spaces within the leaves. It is through the stomata that the water vapor transpired exits the leaf.

Storage, specific The volume of water released from or taken into storage per unit volume of a porous medium per unit change in head.

Storativity The volume of water an aquifer releases from or takes into storage per unit surface area of the aquifer per unit change in head. It is equal to the product of specified storage and aquifer thickness. In an unconfined aquifer, the storativity is equivalent to the specific yield.

Stratified soils Soils that are composed of layers, usually varying in permeability and texture.

Stratosphere Lies above the troposphere and contains very little moisture, but is the location of some of the fast-moving upper-level winds, called *jet streams*, that circulate around the world and are one of the steering mechanisms for low- and high-pressure systems.

Stream A current or flow of water running along the surface of the earth; specifically, a creek or small river.

Stream A flow of running water that runs along a channel and has a surface open to the atmosphere. If it flows under the ground, it is called a subterranean stream.

Stream bank stabilization Vegetative or mechanical control of erodible stream banks, including measures to prevent stream banks from caving or sloughing, such as lining banks with riprap or matting and constructing jetties or revetments, as necessary, for permanent protection.

Stream instability A stream that is not self-sustaining, changes geometry, and is not balanced between import and export of sediment.

Stream stability Requires a stream to be self-sustaining, retain the same general geometry over time (decades), and be balanced between the import and export of sediment.

Stream–channel erosion Scouring of soil and the cutting of channel banks or beds by running water. Sometimes called *streambed erosion* or *stream bank erosion*.

Streamflow The rate of water movement in a stream. Often written as two words.

Subcritical (flow) Hydraulic jump.

Subgrade Earth material beneath a subsurface drain or foundation.

Subsoiling Tillage operation to loosen the soil below the tillage zone without inversion and with a minimum of mixing with the tilled zone.

Subsurface drain Subsurface conduits used primarily to remove subsurface water from soil. Classifications of subsurface drains include pipe drains, tile drains, and blind drains.

Surface albedo The proportion of solar radiation that is reflected from a soil and crop surface.

Surface drainage The diversion or orderly removal of excess water from the surface of land by improved natural or constructed channels, supplemented when necessary by shaping and grading of land surfaces to such channels.

Surface inlet Structure for diverting surface water into an open ditch, subsurface drain, or pipeline.

Surface irrigation Broad class of irrigation methods in which water is distributed over the soil surface by gravity flow.

Surface retention The portion of precipitation required to satisfy interception, the wetting of the soil surface, and depression storage. *See* Initial storage.

Surface roughness A measure of the impact of surface vegetation on wind speed. Surface roughness is equal to 0.123 times the height of the vegetation.

Surface runoff Precipitation, snowmelt, or irrigation in excess of what can infiltrate or be stored in small surface depressions.

Surface sealing Reorienting and packing of dispersed soil particles in the immediate surface layer of soil and clogging of surface pores, resulting in reduced infiltration.

Surface storage Sum of detention and channel storage excluding depression storage; represents at any given moment the total water en route to an outlet from an area or watershed.

Surface water Water flowing or stored on the earth's surface.

Suspended load Fine materials such as clay, silt, and fine sand that remain suspended in the water column, but can settle in locations where the travel velocity is low or the settling depth is small. The suspended load can be more than 90% of the transported material.

Suspended sediment Material moving in suspension in a fluid; caused by the upward components of the turbulent currents or by colloidal suspension. Sometimes called *suspended load.*

T **distribution** Also known as the *Student t distribution;* a way of creating a distribution for samples collected from a population when actual population values are unknown.

t **statistic** Value calculated to determine whether two samples are from the same population.

Taxonomic class A set of soil characteristics with precisely defined limits. In the U.S., these classes are mainly based on the kind and character of soil properties and the arrangement of the soil horizons in a profile. The classification is based on these six categories: order, suborder, great group, subgroup, family, and series.

Temporal Lasting only for a time, transitory; temporary.

Tensiometer Instrument, consisting of a porous cup filled with water and connected to a manometer or vacuum gage, used for measuring the soil water matric potential.

Tension One of the forces that helps cause the downward movement of water through a soil profile.

Terrace (1) A broad channel, bench, or embankment constructed across the slope to intercept runoff and detain or channel it to protected outlets. (2) A level plain, usually with a steep front, bordering a river, lake, or sea. (3) A relief floodplain.

Threshold value (level of significance) Risk of falsely rejecting the null hypothesis.

Time of concentration The time it takes water to travel along the hydraulic length.

Time domain reflectometry An apparatus that sends a step pulse of electromagnetic radiation along dual probes inserted into the soil. The pulse is "reflected" and returned to the source with a velocity characteristic to a specific dielectric constant. This instrument can be calibrated to give soil water content based on the measured dielectric constant.

Transient flow Unsteady flow that occurs when, at any point in a flow field, the magnitude or direction of the flow velocity changes with time.

Transmissivity The rate at which water of a prevailing density and viscosity is transmitted through a unit width of an aquifer or confining bed under a unit hydraulic gradient. It is a function of properties of the liquid, the porous media, and the thickness of the porous media.

Transpiration The process by which water in plants is transferred to the atmosphere as water vapor.

Transpiration ratio The ratio of the weight of water transpired to the weight of dry matter contained in the plant.

Trapezoidal weir A sharp-crested weir of trapezoidal shape.

Trend line or regression line (of precipitation) A statistical technique to determine the annual average of precipitation for a given period of time.

Troposphere The layer of the atmosphere closest to the earth; it varies in thickness from 11 mi (17.6 km) at the equator to 4 mi (6.4 km) at the poles. This layer produces precipitation.

Unconfined aquifer An aquifer with an upper boundary that consists of relatively porous natural material that transmits water readily and does not confine water. The water level in the aquifer is the water table.

Uniform flow Flow in which the velocity and depth are the same at each cross section.

Unsaturated flow Movement of water in soil in which the pores are not completely filled with water.

Unsaturated zone The part of the soil profile in which the voids are not completely filled with water.

Vapor pressure The pressure a gas exerts on the liquid with which it is in contact. In the context of this book, the gas is water vapor, and the liquid is water.

Vapor pressure deficit Difference between the existing vapor pressure and that of a saturated atmospheric vapor pressure at the same temperature.

Vegetation Plant material. It affects resistance to flow and therefore affects flow rates. The effect of vegetation on flow resistance is also a function of flow depth relative to vegetation height and density. The effect can be qualified as low, medium, high, or very high.

Wash load That part of the sediment load of a stream composed of suspended clay and silt particles.

Wastewater Water of reduced quality that has been used for some purpose and discarded.

Water rights Legal rights, derived from common law, court decisions, or statutory enactments, to use water supplies.

Water table The upper surface of saturated zone below the soil surface where the water is at atmospheric pressure.

Water-holding capacity Amount of soil water available to plants. *See* Available plant water.

Watershed Land area that contributes runoff (drains) to a given point in a stream or river. Synonymous with catchment and drainage or river basin.

Watershed gradient The average slope in a watershed; measured along a path of water flow from a given point in the stream channel to the most remote point in the watershed.

Weir (1) Structure across a stream to control or divert the flow. (2) Device for measuring the flow of water. Classification includes sharp crested or broad crested with rectangular, trapezoidal, or triangular cross sections.

Well casing Pipe installed within a borehole to prevent collapse of sidewall material, to receive and protect pump and pump column, and to allow water flow from the aquifer to pump intake.

Well screen The part of the well casing that has openings through which water enters.

Well test Determination of the well yield vs. drawdown relationship with time.

Well yield Discharge that can be sustained from a well through some specified period of time. *See* Safe well yield.

Wetlands Area of wet soil that is inundated or saturated under normal circumstances and would support a prevalence of hydrophytic plants.

Wetted perimeter Length of the wetted contact between a conveyed liquid and the open channel or closed conduit conveying it; measured in a plane at right angles to the direction of flow.

Wind erosion Detachment, transportation, and deposition of soil by the action of wind. The removal and redeposition may be in more or less uniform layers or as localized blowouts and dunes.

Xylem The woody vascular tissue of a plant that conducts water and mineral salts in the stems, roots, and leaves and gives support to the softer tissues.

Appendix C: Precipitation Frequency–Magnitude Information for the United States

(Provided by the U.S. Department of Agriculture Natural Resources Conservation Service)
For additional information, see Chapter 14, Section 14.6.3.B

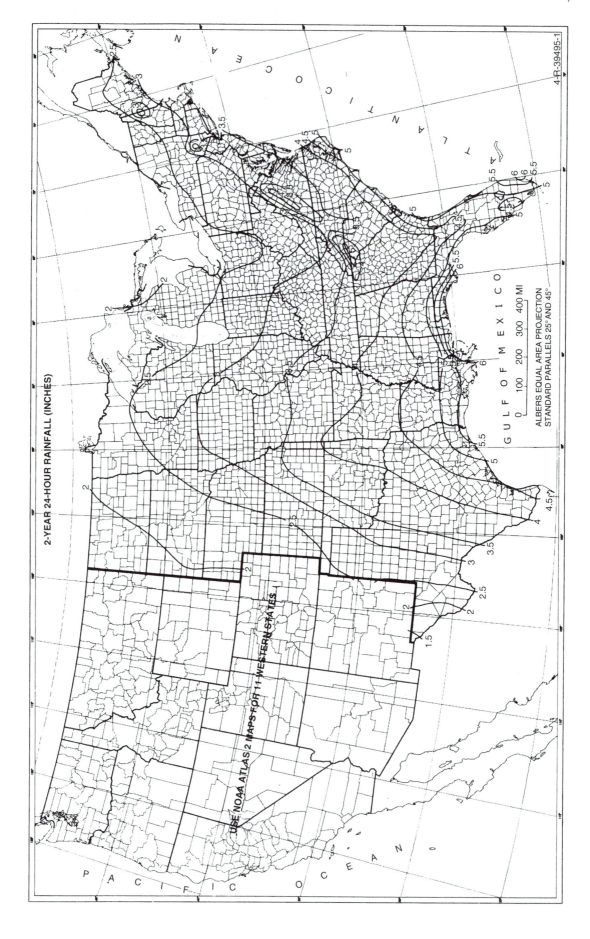

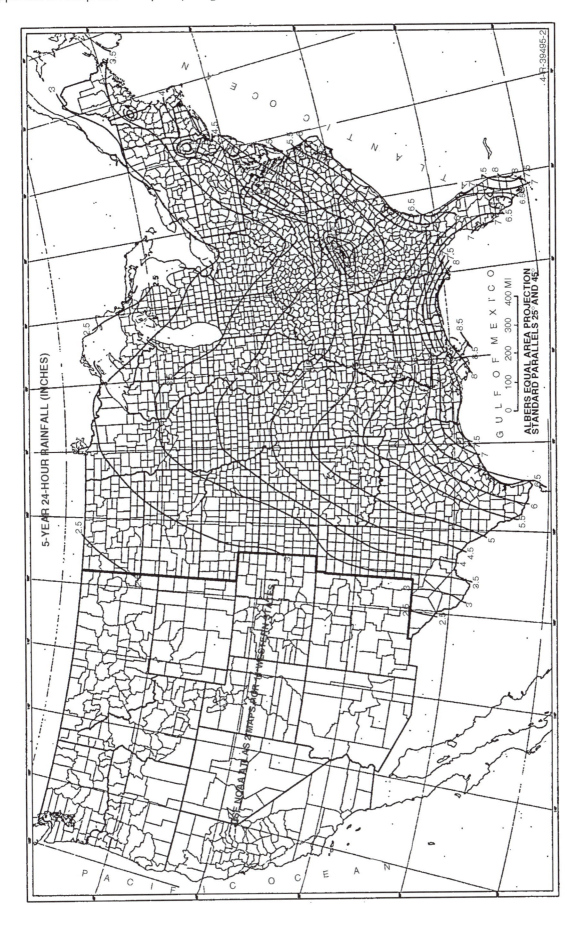

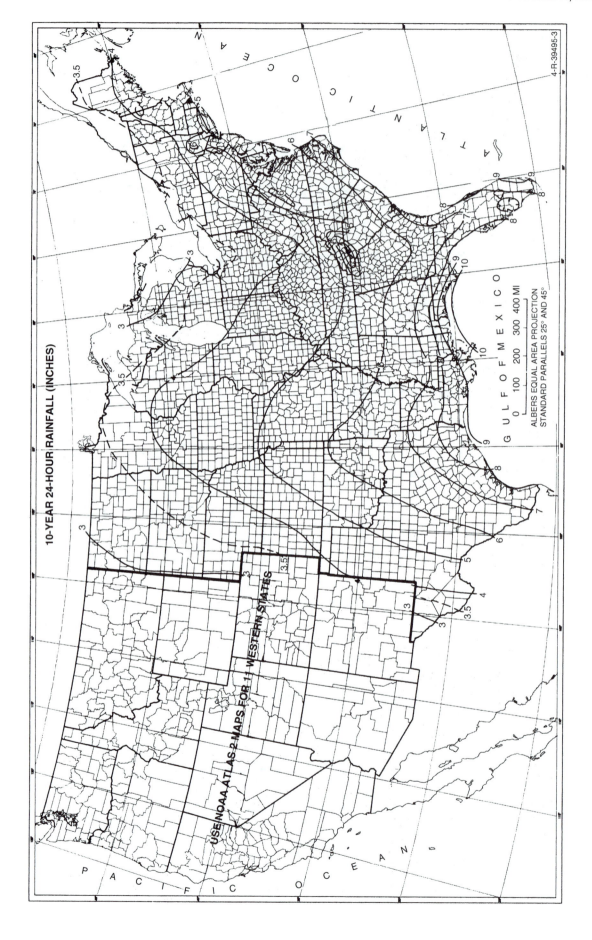

10-YEAR 24-HOUR RAINFALL (INCHES)

USE NOAA ATLAS 2 MAPS FOR 11 WESTERN STATES

GULF OF MEXICO

0   100   200   300   400 MI

ALBERS EQUAL AREA PROJECTION
STANDARD PARALLELS 25° AND 45°

4-R-39495-3

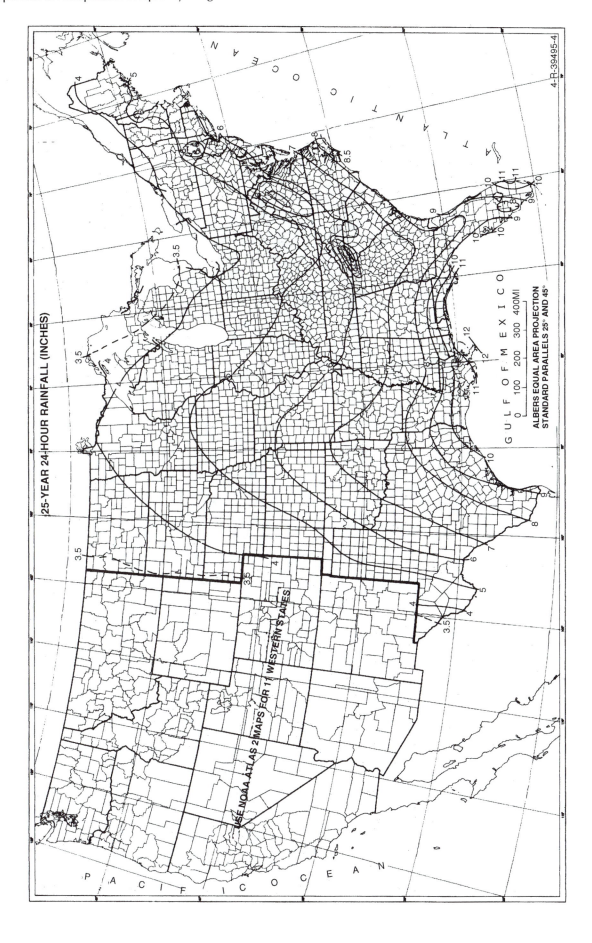

25-YEAR 24-HOUR RAINFALL (INCHES)

USE NOAA ATLAS 2 MAPS FOR 11 WESTERN STATES

GULF OF MEXICO

ALBERS EQUAL AREA PROJECTION
STANDARD PARALLELS 25° AND 45°

0   100   200   300   400MI

4-R-39495-4

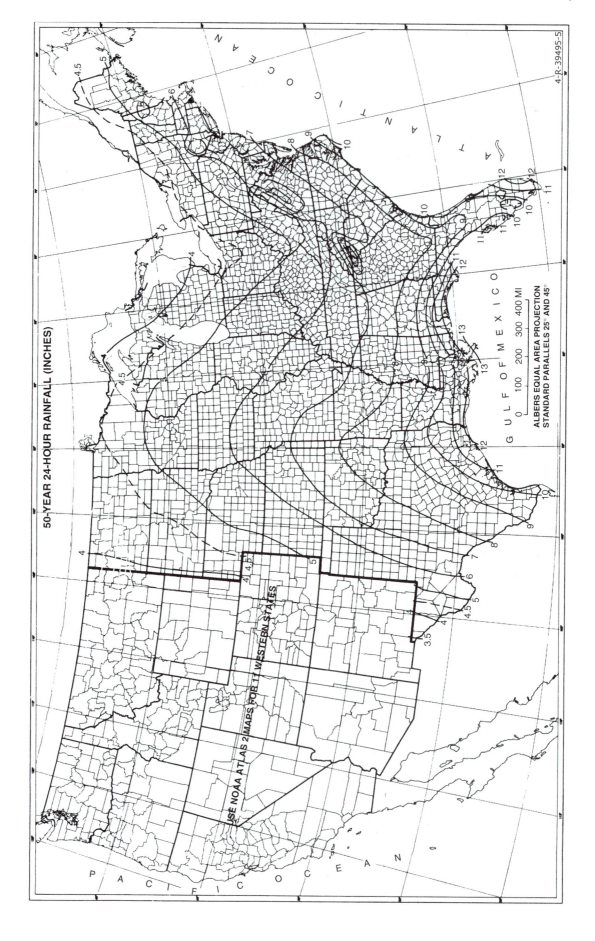

50-YEAR 24-HOUR RAINFALL (INCHES)

USE NOAA ATLAS 2 MAPS FOR 11 WESTERN STATES

ALBERS EQUAL AREA PROJECTION
STANDARD PARALLELS 25° AND 45°

0   100 200 300 400 MI

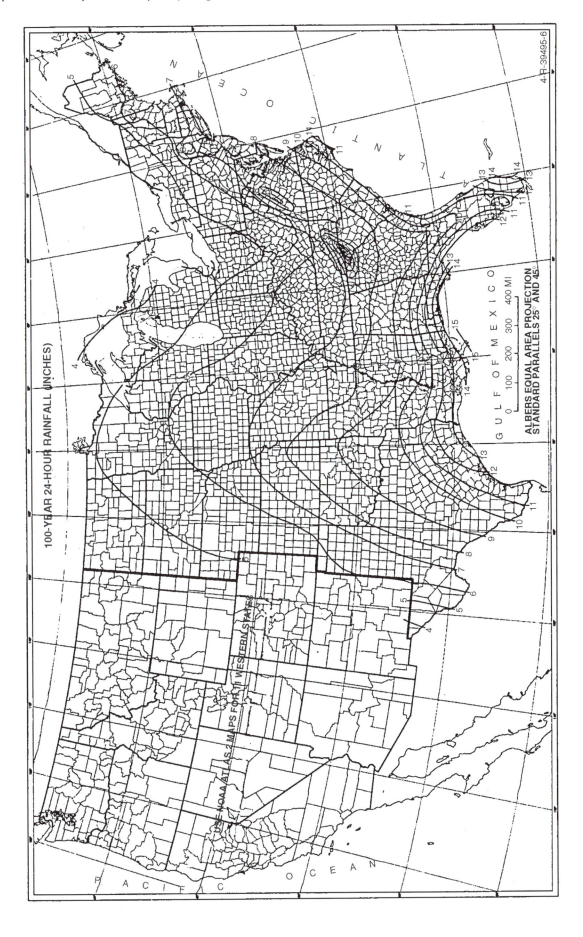

Appendix D: Hydrologic Soil Group and Erodibility Factors for the Most Common Soils in Each State

Provided by the U.S. Department of Agriculture Natural Resources Conservation Service

Soil Series	Group Type	K Values		Soil Series	Group Type	K Values
			Alabama			
Troup	A	0.10–0.10		Almontevallo	D	0.28–0.28
Luverne	C	0.24–0.24		Algorgas	D	0.17–0.20
Smithdale	B	0.17–0.17		Townley	C	0.28–0.37
Nauvoo	B	0.28–0.28		Bibb	D	0.20–0.20
Lorangeburg	B	0.10–0.10		Dothan	B	0.15–0.15
			Alaska			
Tolstoi	D	0.37–0.37		Akina	D	0.05–0.05
Salamatof	D	0.05–0.05		Mosman	D	0.15–0.15
Maybeso	D	0.05–0.05		Godstream	D	0.05–0.05
Kupreanof	B C	0.24–0.24		Strandline	B	0.37–0.37
McGilvery	D	0.05–0.05		Kushneahin	D	0.05–0.05
			Arizona			
Winona	D	0.15–0.32		Sheppard	A	0.10–0.20
Thunderbird	D	0.05–0.24		Mohall	B	0.20–0.43
Barkerville	C	0.20–0.20		Gilman	B D	0.28–0.55
Springerville	D	0.05–0.28		Denure	B	0.10–0.55
Clovis	B	0.17–0.28		Gunsight	B	0.02–0.15
			Arkansas			
Enders	C	0.32–0.32		Nella	B	0.15–0.15
Mountainburg	D	0.17–0.24		Leadvale	C	0.43–0.43
Linker	B	0.24–0.28		Sacul	C	0.28–0.28
Sharkey	D	0.43–0.43		Guyton	D	0.43–0.43
Carnasaw	C	0.24–0.32		Perry	D	0.37–0.37
			California			
Cajon	A/B	0.10–0.24		Gaviota	D	0.24–0.24
Cieneba	C	0.20–0.20		Auburn	D	0.15–0.32
Hanford	B	0.20–0.32		Millsholm	D	0.20–0.37
San Joaquin	D	0.32–0.32		Rositas	A C	0.10–0.43
Maymen	D	0.20–0.24		Los Osos	C	0.32–0.32
			Colorado			
Valent	A	0.17–0.24		Olney	B	0.15–0.28
Wiley	B	0.32–0.32		Platner	C	0.20–0.28
Ascalon	B	0.15–0.24		Baca	B C	0.20–0.37
Vona	B	0.20–0.28		Manvel	BC	0.28–0.43
Weld	C	0.20–0.37		Stoneham	B	0.37–0.37

Soil Series	Group Type	K Values	Soil Series	Group Type	K Values
Connecticut					
Charlton	B	0.20–0.24	Canton	B	0.20–0.24
Hollis	C/D	0.20–0.20	Sutton	B	0.20–0.24
Paxton	C	0.20–0.24	Ridgebury	C	0.20–0.24
Woodbridge	C	0.20–0.24	Leicester	C	0.24–0.28
Hinckley	A	0.20–0.20	Wethersfield	C	0.24–0.28
Delaware					
Fallsington	B/D	0.24–0.24	Matapeake	B	0.49–0.49
Sassafras	B	0.28–0.28	Rumford	B	0.17–0.17
Evesboro	A	0.17–0.17	Chincoteague	D	0.32–0.32
Pocomoke	B/D	0.20–0.20	Johnston	D	0.17–0.17
Woodstown	C	0.24–0.24	Keyport	C	0.43–0.43
Florida					
Myakka	D B/D	0.10–0.10	Blanton	A B	0.10–0.10
Candler	A	0.10–0.10	Riviera	D B/D C/D	0.10–0.10
Immokalee	D B/D	0.10–0.10	Basinger	D B/D	0.10–0.10
Lakeland	A	0.10–0.10	Beaugallie	D B/D	0.10–0.10
Smyrna	D B/D	0.10–0.10	Pineda	D B/D	0.10–0.15
Georgia					
Tifton	B	0.10–0.10	Pelham	B/D	0.10–0.10
Cecil	B	0.28–0.28	Orangeburg	B	0.10–0.10
Madison	B	0.24–0.24	Dothan	B	0.15–0.15
Fuquay	B	0.15–0.15	Leefield	C	0.10–0.10
Pacolet	B	0.15–0.20	Cowarts	C	0.15–0.15
Hawaii					
Akaka	A	0.05–0.05	Kahaluu	D	0.02–0.02
Kekake	D	0.02–0.02	Mawae	A	0.02–0.02
Kiloa	A	0.02–0.02	Guam	D	0.05–0.05
Keei	D	0.02–0.02	Hanipoe	B C	0.10–0.17
Puu Pa	A	0.10–0.20	Honokaa	A	0.05–0.05
Idaho					
Portneuf	B	0.49–0.49	Power	B	0.43–0.43
Purdam	C	0.32–0.43	Rexburg	B	0.32–0.49
Pancheri	B	0.43–0.49	Arbidge	C	0.20–0.24
Ririe	B	0.43–0.49	Chilcott	C D	0.24–0.49
Wickahoney	D	0.20–0.20	Declo	B	0.37–0.49
Illinois					
Drummer	B/D	0.28–0.28	Rozetta	B	0.37–0.43
Tama	B	0.28–0.43	Sable	B/D	0.28–0.28
Ipava	B	0.28–0.28	Bluford	C	0.43–0.43
Hickory	C	0.37–0.37	Flanagan	B	0.28–0.28
Fayette	B	0.32–0.43	Hosmer	C	0.43–0.43
Indiana					
Crosby	C	0.37–0.43	Wellston	B	0.37–0.37
Blount	D	0.43–0.43	Glynwood	C	0.43–0.43
Pewamo	C/D	0.24–0.24	Fincastle	C	0.37–0.37
Miami	B	0.37–0.37	Crider	B	0.32–0.32
Brookston	B/D	0.28–0.28	Cincinnati	C	0.43–0.43
Iowa					
Clarion	B	0.24–0.28	Canisteo	B/D	0.24–0.24
Fayette	B	0.32–0.43	Webster	B/D	0.24–0.24
Colo	B B/D	0.28–0.37	Marshall	B	0.28–0.43
Galva	B	0.28–0.32	Tama	B	0.28–0.43
Nicollet	B	0.24–0.24	Monona	B	0.28–0.43

Soil Series	Group Type	K Values		Soil Series	Group Type	K Values
			Kansas			
Harney	B	0.32–0.32		Holdrege	B	0.32–0.32
Ulysses	B	0.32–0.32		Uly	B	0.32–0.32
Richfield	B	0.32–0.32		Irwin	D	0.32–0.37
Keith	B	0.32–0.32		Colby	B	0.43–0.43
Crete	C	0.37–0.37		Clime	C	0.20–0.28
			Kentucky			
Shelocta	B	0.32–0.32		Zanesville	C	0.43–0.43
Eden	C	0.17–0.43		Faywood	C	0.32–0.37
Latham	D	0.43–0.43		Crider	B	0.32–0.32
Lowell	C	0.37–0.37		Caneyville	C	0.28–0.43
Loring	C	0.49–0.49		Baxter	B	0.28–0.28
			Louisiana			
Sharkey	D	0.20–0.43		Allemands	D	0.32–0.32
Guyton	D	0.43–0.43		Fausse	D	0.20–0.20
Sacul	C	0.28–0.28		Barbary	D	
Commerce	C	0.37–0.37		Moreland	D	0.43–0.43
Ruston	B	0.15–0.15		Clovelly	D	
			Maine			
Lyman	C/D	0.20–0.28		Plaisted	C	0.20–0.20
Monarda	D	0.15–0.28		Burnham	D	0.24–0.24
Marlow	C	0.20–0.24		Colonel	C	0.17–0.20
Thorndike	C/D	0.17–0.20		Hermon	A	0.10–0.17
Telos	C	0.15–0.28		Dixfield	C	0.17–0.20
			Maryland			
Sassafras	B	0.20–0.28		Beltsville	C	0.43–0.43
Manor	B	0.32–0.37		Mattapex	C	0.37–0.37
Glenelg	B	0.24–0.32		Matapeake	B	0.43–0.49
Othello	D C/D	0.24–0.37		Elkton	D C/D	0.24–0.43
Fallsington	B/D	0.24–0.24		Woodstown	C	0.24–0.28
			Massachusetts			
Paxton	C	0.20–0.24		Woodbridge	C	0.20–0.24
Hinckley	A	0.17–0.20		Freetown	D	
Lyman	C/D	0.20–0.28		Canton	B	0.20–0.24
Carver	A	0.10–0.10		Hollis	C/D	0.20–0.24
Merrimac	A	0.24–0.24		Charlton	B	0.20–0.24
			Michigan			
Kalkaska	A	0.15–0.15		Capac	C	0.32–0.32
Marlette	B	0.32–0.32		Parkhill	B/C	0.17–0.24
Spinks	A	0.15–0.15		Graycalm	A	0.10–0.10
Rubicon	A	0.10–0.15		Houghton	D A/D	
Oshtemo	B	0.24–0.24		Boyer	B	0.17–0.24
			Minnesota			
Canisteo	D B/D	0.16–0.24		Hamerly	C	0.28–0.28
Clarion	B	0.24–0.28		Glencoe	D B/D	0.28–0.28
Webster	B/D	0.24–0.24		Ves	B	0.17–0.24
Barnes	B	0.20–0.28		Nicollet	B	0.24–0.24
Lester	B	0.28–0.28		Seelyeville	D A/D	0.10–0.10
			Mississippi			
Smithdale	B	0.17–0.17		Sweatman	C	0.37–0.37
Sharkey	D	0.20–0.43		Alligator	D	0.37–0.37
Providence	C	0.49–0.49		Ruston	B	0.15–0.15
Memphis	B	0.49–0.49		Ora	C	0.28–0.28
Loring	C	0.49–0.49		Savannah	C	0.37–0.37

Soil Series	Group Type	K Values		Soil Series	Group Type	K Values
Missouri						
Clarksville	B	0.28–0.28		Keswick	C	0.32–0.37
Goss	B	0.24–0.24		Shelby	B	0.28–0.37
Armstrong	C	0.32–0.37		Lamoni	C	0.37–0.37
Mexico	D	0.43–0.43		Gara	C	0.28–0.37
Menfro	B	0.37–0.37		Lagonda	C	0.37–0.37
Montana						
Cabbart`	D	0.37–0.37		Phillips	C	0.43–0.43
Yawdim	D	0.32–0.32		Delpoint	C	0.20–0.37
Williams	B	0.20–0.43		Neldore	D	0.32–0.32
Cabba	D	0.17–0.24		Scobey	C	0.20–0.43
Zahill	C	0.28–0.37		Cambert	C	0.37–0.37
Nebraska						
Valentine	A	0.15–0.17		Uly	B	0.32–0.32
Valent	A	0.17–0.24		Nora	B	0.32–0.32
Coly	B	0.43–0.43		Crete	C	0.37–0.37
Holdrege	B	0.17–0.32		Moody	B	0.32–0.32
Hastings	B	0.32–0.32		Hobbs	B	0.32–0.32
Nevada						
Stewval	D	0.10–0.10		Theon	D	0.10–0.20
Palinor	D	0.24–0.24		Mazuma	B C	0.28–0.55
Cleavage	D	0.10–0.20		Orovada	B	0.15–0.49
Unsel	B	0.10–0.24		Downeyville	D	0.05–0.24
Chiara	D	0.20–0.55		Sumine	C	0.17–0.24
New Hampshire						
Marlow	C	0.20–0.24		Canton	B	0.20–0.24
Becket	C	0.17–0.20		Tunbridge	C	0.20–0.24
Monadnock	B	0.17–0.28		Berkshire	B	0.20–0.24
Lyman	C/D	0.20–0.28		Peru	C	0.20–0.24
Hermon	A	0.10–0.17		Colton	A	0.15–0.15
New Jersey						
Downer	B	0.20–0.20		Rockaway	C	0.17–0.24
Sassafras	B	0.20–0.28		Lakewood	A	0.10–0.10
Atsion	D C/D	0.17–0.17		Manahawkin	D	0.05–0.05
Evesboro	A	0.17–0.17		Aura	B	0.43–0.43
Lakehurst	A	0.17–0.17		Freehold	B	0.28–0.28
New Mexico						
Deama	D	0.05–0.20		Pastura	D	0.37–0.37
Ector	D	0.10–0.15		Travessilla	D	0.10–0.55
Kimbrough	D	0.37–0.37		Upton	C	0.15–0.15
Clovis	B	0.28–0.28		Lozier	D	0.15–0.15
Amarillo	B	0.24–0.24		Berino	B	0.17–0.17
New York						
Volusia	C	0.24–0.37		Bath	C	0.24–0.24
Mardin	C	0.24–0.32		Honeoye	B	0.24–0.32
Lordstown	C	0.20–0.20		Ontario	B	0.24–0.32
Arnot	C/D	0.24–0.24		Oquaga	C	0.20–0.28
Nassau	C	0.32–0.32		Howard	A	0.32–0.32
North Carolina						
Pacolet	B	0.15–0.20		Georgeville	B	0.24–0.43
Cecil	B	0.24–0.28		Goldsboro	B	0.20–0.20
Norfolk	B	0.20–0.20		Appling	B	0.24–0.24
Rains	B/D	0.15–0.15		Chewacla	C	0.24–0.24
Evard	B	0.15–0.24		Badin	B	0.15–0.32

Soil Series	Group Type	K Values	Soil Series	Group Type	K Values
North Dakota					
Barnes	B	0.20–0.28	Buse	B	0.20–0.28
Williams	B	0.15–0.28	Fargo	D	0.32–0.32
Svea	B	0.28–0.28	Cabba	D	0.20–0.24
Hamerly	C	0.28–0.28	Parnell	C/D	0.28–0.28
Zahl	B	0.28–0.28	Tonka	C/D	0.32–0.32
Ohio					
Blount	C	0.43–0.43	Westmoreland	B	0.28–0.37
Hoytville	C/D	0.24–0.28	Bennington	C	0.43–0.43
Pewamo	C/D	0.24–0.24	Miamian	C	0.37–0.37
Gilpin	C	0.24–0.32	Upshur	D	0.37–0.43
Crosby	C	0.43–0.43	Mahoning	D	0.43–0.43
Oklahoma					
Dennis	C	0.43–0.43	Hector	D	0.10–0.15
Stephenville	B	0.17–0.20	Port	B	0.37–0.37
Richfield	B	0.32–0.32	Darnell	C	0.20–0.20
Quinlan	C	0.37–0.37	Clarksville	B	0.28–0.28
Carnasaw	C	0.32–0.32	Woodward	B	0.37–0.37
Oregon					
Lickskillet	C D	0.05–0.24	Condon	C	0.43–0.43
Bohannon	C	0.10–0.15	Bakeoven	D	0.05–0.10
Ritzville	B	0.49–0.49	Klickitat	B	0.20–0.20
Walla Walla	B	0.43–0.43	Peavine	C	0.28–0.28
Preacher	B	0.17–0.17	Simas	C	0.17–0.37
Pennsylvania					
Hazleton	B	0.15–0.17	Berks	C	0.17–0.17
Gilpin	C	0.24–0.32	Cookport	C	0.24–0.32
Dekalb	A C	0.17–0.17	Ernest	C	0.32–0.43
Weikert	B/D	0.20–0.28	Wellsboro	C	0.24–0.32
Oquaga	C	0.20–0.28	Buchanan	C	0.24–0.24
Rhode Island					
Canton	B	0.20–0.24	Paxton	C	0.20–0.24
Charlton	B	0.20–0.24	Newport	C	0.24–0.28
Hinckley	A	0.20–0.20	Bridgehampton	B	0.43–0.49
Woodbridge	C	0.20–0.24	Ridgebury	C	0.20–0.24
Merrimac	A	0.24–0.24	Sutton	B	0.20–0.24
South Carolina					
Cecil	B	0.28–0.28	Coxville	D	0.24–0.24
Pacolet	B	0.20–0.20	Rains	B/D	0.15–0.15
Lakeland	A	0.10–0.10	Wilkes	C	0.24–0.24
Lynchburg	C	0.15–0.15	Madison	B	0.24–0.24
Goldsboro	B	0.20–0.20	Johnston	D	0.17–0.17
South Dakota					
Sansarc	D	0.37–0.37	Highmore	B	0.32–0.32
Opal	D	0.37–0.37	Williams	B	0.15–0.28
Clarno	B	0.20–0.20	Houdek	B	0.20–0.28
Pierre	D	0.37–0.37	Promise	D	0.37–0.37
Samsil	D	0.37–0.37	Lakoma	D	0.37–0.37
Tennessee					
Bodine	B	0.28–0.28	Lexington	B	0.49–0.49
Memphis	B	0.49–0.49	Talbott	C	0.32–0.37
Smithdale	B	0.17–0.17	Grenada	C	0.49–0.49
Loring	C	0.49–0.49	Ramsey	D	0.17–0.20
Baxter	B	0.28–0.37	Mimosa	C	0.28–0.37

Soil Series	Group Type	K Values	Soil Series	Group Type	K Values
Texas					
Tarrant	D	0.10–0.20	Miles	B	0.24–0.24
Pullman	D	0.32–0.32	Houston Black	D	0.32–0.32
Ector	D	0.15–0.15	Olton	C	0.32–0.32
Amarillo	B	0.24–0.24	Crockett	D	0.43–0.43
Reagan	B	0.32–0.37	Sherman	D	0.32–0.32
Utah					
Rizno	D	0.28–0.32	Chipeta	D	0.43–0.43
Skumpah	B D	0.43–0.55	Tooele	B	0.17–0.37
Saltair	D	0.49–0.49	Hiko Peak	B	0.10–0.20
Moenkopie	D	0.10–0.15	Sheppard	A	0.20–0.20
Amtoft	D	0.10–0.24	Begay	B	0.49–0.49
Vermont					
Tunbridge	C	0.20–0.24	Vergennes	C	0.49–0.49
Berkshire	B	0.20–0.24	Rawsonville	C	0.43–0.49
Lyman	C/D	0.20–0.28	Cabot	D	0.28–0.32
Peru	C	0.20–0.24	Houghtonville	C	0.43–0.49
Marlow	C	0.20–0.24	Woodstock	D	0.24–0.24
Virginia					
Cecil	B	0.28–0.28	Madison	B	0.24–0.24
Appling	B	0.24–0.24	Hayesville	B C	0.15–0.24
Frederick	B	0.28–0.32	Nason	B C	0.24–0.43
Berks	C	0.17–0.17	Weikert	B/D	0.20–0.28
Emporia	C	0.28–0.28	Tatum	B	0.20–0.37
Washington					
Ritzville	B	0.49–0.49	Walla Walla	B	0.43–0.43
Alderwood	C	0.15–0.15	Palouses	B	0.32–0.32
Shano	B	0.55–0.55	Warden	B	0.55–0.55
Athena	B	0.37–0.37	Newbell	B	0.24–0.28
Quincy	A	0.15–0.32	Aits	B	0.24–0.37
West Virginia					
Gilpin	C	0.24–0.32	Weikert	B/D	0.20–0.28
Dekalb	A C	0.17–0.17	Pineville	B	0.20–0.24
Berks	C	0.17–0.17	Westmoreland	B	0.37–0.37
Upshur	D	0.37–0.43	Calvin	C	0.15–0.37
Muskingum	C	0.24–0.37	Cateache	C	0.28–0.32
Wisconsin					
Pence	B	0.24–0.24	Kewaunee	C	0.17–0.37
Plainfield	A	0.15–0.17	Magnor	C	0.37–0.37
Menahga	A	0.15–0.15	Fayette	B	0.32–0.43
Padus	B	0.24–0.24	Seaton	B	0.37–0.37
Newglarus	B C	0.37–0.37	Valton	B C	0.32–0.32
Wyoming					
Shingle	D	0.02–0.37			
Hiland	B	0.20–0.37			
Forkwood	B C	0.32–0.43			
Kishona	B C	0.28–0.43			
Theedle	C	0.32–0.32			
American Samoa					
Aua	B	0.17–0.17	Puapua	D	0.10–0.10
Pavaiai	C	0.10–0.10	Sogi	C	0.10–0.10
Ofu	B	0.10–0.10	Leafu	C	0.17–0.17
Oloava	B	0.10–0.10	Tafuna	A	0.02–0.02
Iliili	D	0.05–0.05	Fagasa	C	0.10–0.10

Soil Series	Group Type	K Values	Soil Series	Group Type	K Values
Fed. Sts. Micronesia					
Tolonier	B	0.05–0.05	Umpump	B	0.15–0.15
Dolen	B	0.05–0.05	Rumung	C	0.10–0.10
Fomseng	C	0.10–0.10	Weloy	C	0.10–0.10
Naniak	D	0.05–0.05	Yap	B	0.10–0.10
Dolekei	B	0.10–0.10	Ilachetomel	D	0.05–0.05
Guam					
Guam	D	0.05–0.05	Inarajan	C	0.17–0.24
Akina	B	0.20–0.20	Ylig	C	0.24–0.24
Pulantat	C	0.24–0.24	Togcha	B	0.15–0.15
Agfayan	D	0.20–0.20	Atate	B	0.15–0.15
Ritidian	D	0.02–0.02	Shioya	A	0.15–0.15
Marshall Islands					
Ngedebus	A	0.05–0.10	Majuro	A	0.02–0.02
North Mariana Islands					
Chinen	D	0.10–0.15	Laolao	B	0.15–0.15
Luta	D	0.10–0.10	Kagman	C	0.05–0.15
Takpochao	D	0.10–0.10	Shioya	A	0.15–0.15
Dandan	C	0.15–0.15	Banaderu	D	0.20–0.20
Saipan	B	0.02–0.15	Akina	B	0.20–0.20
Palau					
Aimeliik	B	0.10–0.10	Ngardmau	B	0.05–0.05
Palau	B	0.10–0.10	Dechel	D	0.15–0.15
Ilachetomel	D	0.05–0.05	Wollei	D	0.10–0.10
Ngardok	B	0.15–0.15	Peleliu	D	0.05–0.05
Babelthuap	B	0.05–0.05	Tabecheding	C	0.17–0.17
Puerto Rico					
Mucara	D	0.10–0.10	Descalabrado	D	0.24–0.24
Caguabo	D	0.24–0.24	Pandura	D	0.17–0.17
Humatas	C	0.02–0.02	Soller	D	0.17–0.17
Consumo	B	0.10–0.10	Naranjito	C	0.10–0.10
Los Guineos	C	0.10–0.10	Callabo	C	0.10–0.10

Appendix E:
Sources of Data and Images

SOURCES OF AERIAL PHOTOGRAPHS AND SENSOR IMAGES

U.S. DEPARTMENT OF THE INTERIOR

The U.S. Geological Survey (USGS) EROS Data Center (EDC) archives and produces copies of aerial photographs and other imagery acquired by Department of the Interior agencies. These acquisitions include USGS mapping photographs, products from high-altitude aerial photography programs such as NAPP and NHAP, and products of different emulsion types, such as color infrared, color, and black-and-white infrared and aerial radar images. EDC also archives Landsat satellite data and NASA (National Aeronautic and Space Administration) space and aerial images and U.S. Bureau of Land Management, U.S. National Park Service, and other agency photos and images. A computerized database of products can be accessed by latitude and longitude locators or by USGS quadrangle or other reference location. Computer printouts and microfiche of high-altitude aerial photographs are available at no charge. Copies of products can be ordered from EDC.

For assistance, please contact User Services, EROS Data Center, Sioux Falls, SD 57198 or www.usgs.gov.

U.S. DEPARTMENT OF AGRICULTURE

The former U.S. Department of Agriculture–Agriculture Stabilization and Conservation Service (USDA-ASCS) holds photographs acquired by agencies of the Department of Agriculture. These groups include the U.S. Forest Service, U.S. Natural Resources Conservation Service (NRCS; formerly the Soil Conservation Service, SCS), Farm Services Agency (the former ASCS), and other agencies. Listings of holdings are supplied by computer printout and are available at no cost. Information about their holdings can be accessed by latitude and longitude of the site or by county name.

For information, contact the Aerial Photography Field Office, USDA, Sales Branch, 2222 West 2300 South, Salt Lake City, UT 84125 or www.usda.gov.

NATIONAL ARCHIVES AND RECORD ADMINISTRATION

The National Archives and Record Administration (NARA) archives a variety of photo and map data, particularly government archive aerial photographs taken prior to 1945. Generally, one or two dates of aerial photography coverage are available for counties in the United States.

For more information, please contact the National Archives and Records Administration, 700 Pennsylvania Avenue NW, Washington, DC 20408 or www.nara.gov.

NATIONAL OCEANIC AND ATMOSPHERIC ADMINISTRATION

The National Environmental Satellite, Data, and Information Service (NESDIS) provides a variety of image and digital data products from weather and environmental satellites. These products are from the geostationary weather satellites (GOES), the moderate-resolution satellite data used to make biomass maps of regional and continental areas (advanced very high resolution radiometer, AVHRR), and other atmospheric and weather satellite data. Technical and ordering information are available from NOAA-NEDIS, Room 100, Princeton Executive Square, Washington, DC 20233 or www.noaa.gov.

The NOAA Coast and Geodetic Survey (CGS) has acquired aerial photography and other image data for coastal and offshore areas in support of its mapping mandates. For more information, contact NOAA-CGS, 6001 Executive Boulevard, Rockville, MD 20852, or www.noaa.gov.

OTHER SOURCES OF DATA OR INFORMATION

Research and engineering efforts often involve federal lands or lands in-holding found within federal management and ownership boundaries. In these cases, appropriate aerial photographs may be available locally or regionally from the agencies discussed next.

Sources of Maps

Small-scale topographic and other maps made by the U.S. Geological Survey are available. These maps may be purchased locally from vendors such as map stores, climbing and outdoor shops, and hunting and fishing stores. These maps also can be obtained from the USGS. One should be aware of the time frame necessary for delivery. It may be faster to obtain maps locally or to copy maps available in an archive such as a library or state natural resource agency, or try topozone.com.

Map orders may be placed through USGS Map Sales, Box 25286, Denver, CO 80225 or www.usgs.gov.

The U.S. Geological Survey also vends a variety of digital cartographic and geographic data. These computer-compatible files allow the user to access data digitally and to conduct computer processing and graphical display exercises. The most interesting data are the digital elevation model (DEM) products, which are files of point elevations, and the digital line graph (DLG) products, which display cultural or planimetric details such as roads and resource data such as drainage systems.

Should an application require this sort of information, DEM and DLG data sets can be of great assistance in regional analyses of hydrological variables. Other sources of digital data may be identified.

Conventional navigation maps and bathymetric maps are available from Mapping and Charting, NOAA, 6001 Executive Boulevard, Rockville, MD 20852 or, for orders, contact the Distribution Branch, National Ocean Service, NOAA, Riverdale, MD 20737 or www.noaa.gov.

The following are USGS geographic information system (GIS) data Web sites:

- USGS Water Resources:
  http://water.usgs.gov/GIS/
- USGS TerraServer:
  http://mapping.usgs.gov/digitalbackyard/
- USGS EROS:
  http://edcwww.cr.usgs.gov/doc/edchome/
  ndcdb/ndcdb.html

Other sites for topographic maps, orthophotos, and most GIS layers are

- http://gisdatadepot.com/
- http://www.delorme.com/
- http://www.topozone.com/find.asp
- http://www.esri.com/
- http://www.geocomm.com/

National Wetland Inventory Maps

National Wetland Inventory (NWI) Maps and information are available from the USGS, 507 National Center, Reston, VA 22092 or www.usgs.gov.

Floodplain Maps

Floodplain maps are available from the Federal Emergency Management Agency (FEMA) through its local, state, and regional offices. These offices can be located through numbers listed in the government section of the telephone directory. These maps and information can also be obtained from FEMA, 500 C Street, Washington, DC 20472 or www.fema.gov.

Watershed Boundaries

For watershed boundaries, contact NRCS Hydrologic Units at http://www.ftw.nrcs.usda.gov/huc_data.html.

OTHER DATA SOURCES

The National Weather Service (NWS) of the Department of Commerce can be contacted for a variety of weather records. Products that record the weather conditions at a neighboring station are available on a daily basis. Summary statistics by location, region, and state are also available. Many times, these publications and general records are deposited in libraries and at universities, where they may be accessed quickly and at no charge.

For weather records, contact the NOAA National Climatic Data Center, Federal Building, 37 Battery Park Avenue, Ashville, NC 28801-2733 or www.noaa.gov.

The Hydrometeorological Design Studies Center is part of the National Weather Service's Office of Hydrologic Development, Hydrology Laboratory. Their home page is for those interested in probable maximum precipitation (PMP) and precipitation frequency (PF); see the NWS Hydrometeorological Design Studies Center at http://www.nws.noaa.gov/oh/hdsc/.

Additional climatic data can be found at these Web sites:

- National Climatic Data Center:
  http://lwf.ncdc.noaa.gov/oa/ncdc.html
- Intellicast Realtime Weather and Weather Forecasts: http://www.intellicast.com/

For airborne measurements of hydrological variables such as airborne-measured snow moisture equivalents, maps of snow cover during the winter, and other synoptic hydrological data, contact the NOAA-NWS, National Operational Hydrologic Remote Sensing Center, 1735 Lake Drive West, Chanhassen, MN 55317-8582 or www.nohrsc.nws.gov.

REGIONAL HYDROLOGICAL INFORMATION

The U.S. Geological Survey maintains offices in most states, and many of these offices are responsible for collecting hydrological information. The information may include water gaging station data for major and minor rivers and streams and water quality sampling information for selected locations.

The local USGS offices concerned with hydrology can be accessed locally through numbers listed in the government pages of the telephone directory. One can also contact the USGS directly at the larger centers listed in the Sources of Maps section and through the USGS national clearinghouses of information. The Hydrologic Information Unit, U.S. Geological Survey, 419 National Center, Reston, VA 22092 or www.usgs.gov can also be contacted. Discharge data are available at USGS Water Resources at http://water.usgs.gov/. Additional hydrologic data are available for USGS environmental impact analysis data at http://water.usgs.gov/eap/env_data.html#HDR3.

WATER QUALITY INFORMATION

For EPA water quality criteria and standards, information is available at http://www.epa.gov/wqsdatabase.

WATERSHED INFORMATION AND RESEARCH

- EPA watershed:
  http://www.epa.gov/OWOW/watershed/
- Center for Watershed Protection:
  http://www.cwp.org/
- National Wild and Scenic Rivers System:
  http://www.nps.gov/rivers/
- Stroud Water Research Center:
  http://www.stroudcenter.org/
- European Center for River Restoration:
  http://www.ecrr.org/
- Dave Rosgen's Wildland Hydrology home page: http://www.wildlandhydrology.com/

REGIONAL SOILS INFORMATION

The USDA and the NRCS (formerly SCS) have created several databases of soil boundaries and soil attributes for the United States. These include programs on local soils (SSURGO), regional soils (SSURGO, STATSGO), and national soils (STATSGO, NATSGO). These databases are very useful for GIS studies. Information can be obtained from National Cartographic and Geographic Information Systems Center, USDA, P.O. Box 6567, Fort Worth, TX 76115 or www.usda.gov.

BUREAU OF THE CENSUS

The Bureau of the Census of the Department of Commerce creates a number of useful products. In particular, the Topologically Integrated Geographic Encoding and Referencing (TIGER) system databases are used to form GIS variables for analyses. TIGER line files and other extract-type products provide administrative boundaries, water

and coastal boundaries, and other geographic information, including attribute characteristics. Information is available from Customer Services, Bureau of the Census, Washington Plaza, Room 315, Washington, DC 20233 or www.census.gov.

U.S. ARMY CORPS OF ENGINEERS COMMANDS

When corresponding with U.S. Army Corps of Engineers (USACE) commands, it is best to locate the district-level office that has regional jurisdiction in the area of interest. The district jurisdictions are often based on the boundaries of watersheds and may not be easy to ascertain. Use the local telephone directory and the government pages to locate telephone numbers for the appropriate command. The USACE districts and divisions are also listed in Lyon (1993).

The field activities of USACE are organized under divisions, and the activities are conducted by the districts. The initial point of contact for hydrological or wetland-related issues is the Hydraulics and Hydrology Branch or the Regulatory Functions Branch of the district. Information for USACE is available online at http://www.usace.army.mil/.

FEDERAL AGENCIES

- U.S. Bureau of Reclamation:
  http://www.usbr.gov/main/index.html
- Coastal and Hydraulics Laboratory:
  http://chl.wes.army.mil/
- U.S. Environmental Protection Agency:
  http://www.epa.gov/
- EPA River Corridor and Wetland Restoration:
  http://www.epa.gov/owow/wetlands/restore/
- EPA Rapid Bioassessment Protocol:
  http://www.epa.gov/owow/monitoring/rbp/
- U.S. Department of Agriculture Agricultural Research Service: http://www.ars.usda.gov/
- Natural Resources Conservation Service:
  http://www.nrcs.usda.gov/
- NRCS National Water and Climate Center:
  http://www.wcc.nrcs.usda.gov/
- U.S. Fish and Wildlife Service:
  http://www.fws.gov/

RIVER AND STREAM MANUALS AND GUIDELINES

- U.S. Fish and Wildlife Service:
  http://www.r6.fws.gov/pfw/r6pfw2h.htm
- Federal Agency Stream Corridor Restoration document:
  www.usda.gov/stream_restoration/newgra.html

- Habitat Technical Assistance:
  http://www.wa.gov/wdfw/hab/ahg/strmbank.
  htm
- Management Recommendations for
  Washington's Priority Habitats: Riparian:
  http://www.wa.gov/wdfw/hab/ripxsum.htm

WATER REGULATIONS

- Anderson's Ohio Revised Code (all state statutes): http://onlinedocs.andersonpublishing.
  com/revisedcode/
- Clean Water Act (Cornell law site):
  http://www4.law.cornell.edu/uscode/
  unframed/33/ch26.html
- National Pollution Discharge Elimination
  System Permit Program:
  http://cfpub.epa.gov/npdes/
- Total Maximum Daily Load (TMDL) program:
  http://www.epa.gov/owow/tmdl/

Index

Note: Italicized pages refer to figures and tables